D0463904

COMMUNICATION SYSTEMS
An Introduction to Signals and Noise
in Electrical Communication

McGraw-Hill Series in Electrical Engineering

Consulting Editor
Stephen W. Director, Carnegie-Mellon University

Circuits and Systems
Communications and Signal Processing
Control Theory
Electronics and Electronic Circuits
Power and Energy
Electromagnetics
Computer Engineering
Introductory
Radar and Antennas
VLSI

Previous Consulting Editors

Ronald N. Bracewell, Colin Cherry, James F. Gibbons, Willis H. Harman,
Hubert Heffner, Edward W. Herold, John G. Linvill, Simon Ramo, Ronald A. Rohrer,
Anthony E. Siegman, Charles Susskind, Frederick E. Terman, John G. Truxal,
Ernst Weber, and John R. Whinnery

Communications and Signal Processing

Consulting Editor
Stephen W. Director, Carnegie-Mellon University

COMMUNICATION SYSTEMS
An Introduction to Signals and Noise in Electrical Communication

Third Edition

A. Bruce Carlson

Rensselaer Polytechnic Institute

McGraw-Hill, Inc.
New York St. Louis San Francisco Auckland Bogotá
Caracas Lisbon London Madrid Mexico City Milan
Montreal New Delhi San Juan Singapore
Sydney Tokyo Toronto

This book was set in Times Roman.
The editor was Sanjeev Rao;
the production supervisor was Phil Galea;
the cover was designed by Rafael Hernandez;
Project supervision was done by Albert Harrison, Harley Editorial Services.

COMMUNICATION SYSTEMS
An Introduction to Signals and Noise in Electrical Communication

10 11 12 13 14 BKMBKM 9 9 8 7 6 5 4

ISBN 0-07-009960-X

Library of Congress Cataloging in Publication Data

Carlson, A. Bruce, date
    Communication systems.

    (McGraw-Hill series in electrical engineering.
Communications and signal processing)
    Bibliography: p.
    Includes index.
    1. Signal theory (Telecommunication)   2. Modulation
(Electronics)   3. Digital communications.   I. Title.
II. Series.
TK5102.5.C3   1986   621.38′043   85-4314
ISBN 0-07-009960-X
ISBN 0-07-009961-8 (solutions manual)

To the memory of my father and mother
Albin John Carlson
Mildred Elizabeth Carlson

CONTENTS

The numbers in parentheses after each section title identify previous sections that contain the minimum prerequisite material. The symbol ★ identifies optional material within a section.

Part 2   Analog Communication

Part 3 Digital Communication

PREFACE

This text, like its previous editions, is an introduction to communication systems written at a level appropriate for advanced undergraduates and first-year graduate students in electrical engineering. An initial study of signal transmission and the inherent limitations of physical systems establishes unifying concepts of communication. Equal attention is then given to the analysis and design of analog and digital communication systems.

Mathematical techniques and models necessarily play an important role throughout the book, but always in the engineering context as means to an end. Numerous applications have been incorporated for their practical significance and as illustrations of concepts and design strategies. Hardware considerations are also included to stimulate interest and to bring out connections with other branches of electrical engineering.

PREREQUISITE BACKGROUND

The assumed background is equivalent to the first two or three years of an electrical engineering curriculum. Essential prerequisites are differential equations, steady state and transient circuit analysis, and a first course in electronics. Students should also have some familiarity with operational amplifiers, digital logic, and matrix notation. Prior exposure to linear systems analysis, Fourier transforms, and probability theory is helpful but not required.

CONTENTS AND ORGANIZATION

A distinctive feature of this edition is the division of material into three major parts, each part consisting of several short chapters. Part 1 introduces signal and system analysis methods. These tools are applied to the study of analog communication in Part 2, and to digital communication in Part 3. The specific topics are listed in the table of contents and discussed in Sect. 1.4. Optional topics within chapters are marked by the symbol ★. Three appendixes contain supplemental material.

The coverage of digital communication has been substantially increased compared to the second edition, especially in the areas of spectral analysis, error-control coding, and carrier modulation. Other topics given expanded treatment are: filtering, random signals, multiplexing, synchronization, coded pulse modulation, optimum detection, and information theory. Added topics include: FM distortion, phase-lock loops, spread spectrum, predictive coding, discrete Fourier transforms, and hardware implementation using digital circuits.

All three parts can be presented in a year-long undergraduate course with minimum prerequisites. If linear systems and probability theory are covered in prerequisite courses, then Chap. 5 and most of Parts 2 and 3 can be included in a one-term senior/graduate course. As a guide to topic selection, the table of contents indicates the minimum prerequisites for each section.

The modular chapter structure also allows considerable latitude for other formats. For instance, a one-term introduction to signal transmission and analog modulation could consist of Chaps. 2, 3, 6, 7, 8, and 10 (omitting Sect. 10.4). A subsequent course emphasizing random signals and digital communication might include Chaps. 4, 5, 11, 12, and 13 and Chaps. 9, 14, or 15.

INSTRUCTIONAL AIDS

Each chapter contains several examples and exercises. The examples have been chosen to illuminate concepts and techniques that students often find troublesome. Answers to the exercises are supplied so that students can test their grasp of new material.

Problems at the end of each chapter are numbered by text section. They range from basic manipulations and computations to more advanced analysis and design tasks identified by the symbol ‡. A manual of solutions for the problems and exercises is available to instructors from the publisher.

Tables at the back of the book include transform pairs, probability functions, and mathematical relations for convenient reference. An annotated bibliography is also provided at the back in the form of a supplementary reading list.

ACKNOWLEDGMENTS

I am indebted to many people for advice and assistance. Professors John B. Anderson, David G. Gisser, James W. Modestino, William A. Pearlman, and Henry Stark of Rensselaer Polytechnic Institute freely shared their wisdom and their libraries. Professor John C. Lindenlaub of Purdue University contributed to the previous edition, from which this third edition evolved. Pedagogical improvements were stimulated by discussions with Rensselaer students and teaching assistants, particularly Gordon and Kathy Bechtel. Suggestions from reviewers for McGraw-Hill helped shape the final manuscript.

Special thanks for support and encouragement go to Lester A. Gerhardt, Chairman of the Electrical, Computer, and Systems Engineering Department, to Charles M. Close, Curriculum Chairman, and to the administration of Rensselaer Polytechnic Institute. Thanks also to my wife and family for patience and understanding.

A. Bruce Carlson

ONE

INTRODUCTION

"Attention, the Universe! By kingdoms, right wheel!" This prophetic phrase is the first telegraph message on record. It was sent over a 16-km line by Samuel F. B. Morse in 1838. Thus was born a new era, the era of electrical communication.

Over the next century and a half, communication engineering had advanced to the point that earthbound TV viewers could watch astronauts working in space. Telephone, radio, and television have become integral parts of modern life. Long-distance circuits span the globe carrying text, data, voice, and images. Computers talk to computers via intercontinental networks. Certainly great strides have been made since the days of Morse. Equally certain, coming decades will bring many new achievements of communication engineering.

This textbook presents an introduction to electrical communication systems, including analysis methods, design principles, and hardware considerations. We begin here with a descriptive overview that establishes a perspective for the chapters that follow.

1.1 ELEMENTS AND LIMITATIONS OF COMMUNICATION SYSTEMS

A communication system conveys information from its source to a destination some distance away. There are so many different applications of communication systems that we cannot attempt to cover every type. Nor can we discuss in detail all the individual parts that make up a specific system. A typical system involves numerous components that run the gamut of electrical engineering—circuits,

1

electronics, electromagnetics, and computers, to name a few of the relevant fields. Moreover, a piece-by-piece treatment would obscure the essential point that a communication system is an integrated whole that really does exceed the sum of its parts.

We therefore approach the subject from a more general viewpoint. Recognizing that all communication systems have the same basic function of *information transfer*, we'll seek out and isolate the principles and problems of conveying information in electrical form. These will be examined in sufficient depth to develop analysis and design methods suited to a wide range of applications. In short, this text is concerned with communication systems as *systems*.

Information, Messages, and Signals

Clearly, the concept of *information* is central to communication. But information is a loaded word, implying semantic and philosophical notions that defy precise definition. We avoid these difficulties by dealing instead with the *message*, defined as the physical manifestation of information as produced by the source. Whatever form the message takes, the goal of a communication system is to reproduce at the destination an acceptable replica of the source message.

There are many kinds of information sources, including machines as well as people, and messages appear in various forms. Nonetheless, we can identify two distinct message categories, *analog* and *digital*. This distinction, in turn, determines the criterion for successful communication.

An *analog* message is a physical quantity that varies with time, usually in a smooth and continuous fashion. Examples of analog messages are the acoustic pressure produced when you speak, the angular position of an aircraft gyro, or the light intensity at some point in a television image. Since the information resides in a time-varying waveform, an analog communication system should deliver this waveform with a specified degree of *fidelity*.

A *digital* message is an ordered sequence of symbols selected from a finite set of discrete elements. Examples of digital messages are the letters printed on this page, a listing of hourly temperature readings, or the keys you press at a computer terminal. Since the information resides in discrete symbols, a digital communication system should deliver these symbols with a specified degree of *accuracy* in a specified amount of time.

Whether analog or digital, few message sources are inherently electrical. Consequently, most communication systems have input and output *transducers* as shown in Fig. 1.1-1. The input transducer converts the message to an electrical *signal*, say a voltage or current, and another transducer at the destination converts the output signal to the desired message form. For instance, the transducers in a voice communication system could be a microphone at the input and a loudspeaker at the output. We'll assume hereafter that suitable transducers exist, and we'll concentrate primarily on the task of *signal transmission*. In this context the terms *signal* and *message* will be used interchangeably since the signal, like the message, is a physical embodiment of information.

Figure 1.1-1 Communication system with input and output transducers.

Elements of a Communication System

Figure 1.1-2 depicts the elements of a communication system, omitting transducers but including unwanted contaminations. There are three essential parts of any communication system, the transmitter, transmission channel, and receiver. Each part plays a particular role in signal transmission, as follows.

The *transmitter* processes the input signal to produce a transmitted signal suited to the characteristics of the transmission channel. Signal processing for transmission almost always involves *modulation* and may also include *coding*.

The *transmission channel* is the electrical medium that bridges the distance from source to destination. It may be a pair of wires, a coaxial cable, or a radio wave or laser beam. Every channel introduces some amount of transmission *loss* or *attenuation*, so the signal power progressively decreases with increasing distance.

The *receiver* operates on the output signal from the channel in preparation for delivery to the transducer at the destination. Receiver operations include *amplification* to compensate for transmission loss, and *demodulation* and *decoding* to reverse the signal-processing performed at the transmitter. *Filtering* is another important function at the receiver, for reasons discussed next.

Various unwanted undesirable effects crop up in the course of signal transmission. Attenuation is undesirable since it reduces signal *strength* at the receiver. More serious, however, are distortion, interference, and noise, which appear as alterations of the signal *shape*. Although such contaminations may occur at any point, the standard convention is to blame them entirely on the channel, treating the transmitter and receiver as being ideal. Figure 1.1-2 reflects this convention.

Figure 1.1-2 Elements of a communication system.

Distortion is waveform perturbation caused by imperfect response of the system to the desired signal itself. Unlike noise and interference, distortion disappears when the signal is turned off. If the channel has a linear but distorting response, then distortion may be corrected, or at least reduced, with the help of special filters called *equalizers*.

Interference is contamination by extraneous signals from human sources — other transmitters, power lines and machinery, switching circuits, etc. Interference occurs most often in radio systems whose receiving antennas usually intercept several signals at the same time. Radio-frequency interference (RFI) also appears in cable systems if the transmission wires or receiver circuitry pick up signals radiated from nearby sources. Appropriate filtering removes interference to the extent that the interfering signals occupy different frequency bands than the desired signal.

Noise refers to random and unpredictable electrical signals produced by natural processes both internal and external to the system. When such random variations are superimposed on an information-bearing signal, the message may be partially corrupted or totally obliterated. Filtering reduces noise contamination, but there inevitably remains some amount of noise that cannot be eliminated. This noise constitutes one of the fundamental system limitations.

Finally, it should be noted that Fig. 1.1-2 represents one-way or *simplex* (SX) transmission. Two-way communication of course requires a transmitter and receiver at each end. A *full-duplex* (FDX) system has a channel that allows simultaneous transmission in both directions. A *half-duplex* (HDX) system allows transmission in either direction but not at the same time.

Fundamental Limitations

An engineer faces two general kinds of constraints when designing a communication system. On the one hand are the *technological problems*, including such diverse considerations as hardware availability, economic factors, federal regulations, etc. These are problems of feasibility that can be solved in theory, even though perfect solutions may not be practical. On the other hand are the *fundamental physical limitations*, the laws of nature as they pertain to the task in question. These limitations ultimately dictate what can or cannot be accomplished, irrespective of the technological problems. The fundamental limitations of information transmission by electrical means are *bandwidth* and *noise*.

The concept of bandwidth applies to both signals and systems as a measure of *speed*. When a signal changes rapidly with time, its frequency content or *spectrum* extends over a wide range and we say that the signal has a large bandwidth. Similarly, the ability of a system to follow signal variations is reflected in its usable frequency response or *transmission bandwidth*. Now all electrical systems contain energy-storage elements, and stored energy cannot be changed instantaneously. Consequently, every communication system has a finite bandwidth B that limits the rate of signal variations.

Communication under real-time conditions requires sufficient transmission bandwidth to accommodate the signal spectrum; otherwise, severe distortion will result. Thus, for example, a bandwidth of several megahertz is needed for a TV video signal, while the much slower variations of a voice signal fit into $B \approx 3$ kHz. For a digital signal with r symbols per second, the bandwidth must be $B \geq r/2$. In the case of information transmission without a real-time constraint, the available bandwidth determines the maximum signal speed. The time required to transmit a given amount of information is therefore inversely proportional to B.

Noise imposes a second limitation on information transmission. Why is noise unavoidable? Rather curiously, the answer comes from kinetic theory. At any temperature above absolute zero, thermal energy causes microscopic particles to exhibit random motion. The random motion of charged particles such as electrons generates random currents or voltages called *thermal noise*. There are also other types of noise, but thermal noise appears in every communication system.

We measure noise relative to an information signal in terms of the *signal-to-noise power ratio* S/N. Thermal noise power is ordinarily quite small, and S/N can be so large that the noise goes unnoticed. At lower values of S/N, however, noise degrades fidelity in analog communication and produces errors in digital communication. These problems become most severe on long-distance links when the transmission loss reduces the received signal power down to the noise level. Amplification at the receiver is then to no avail, because the noise will be amplified along with the signal.

Taking both limitations into account, Shannon (1948)† stated that the rate of information transmission cannot exceed the *channel capacity*

$$C = B \log (1 + S/N)$$

This relationship, known as the *Hartley-Shannon law*, sets an upper limit on the performance of a communication system with a given bandwidth and signal-to-noise ratio.

1.2 MODULATION AND CODING

Modulation and coding are operations performed at the transmitter to achieve efficient and reliable information transmission. So important are these operations that they deserve further consideration here. Subsequently, we'll devote several chapters to modulating and coding techniques.

---

† References are indicated in this fashion throughout the text. Complete citations are listed alphabetically by author in the References at the end of the book.

Modulation Methods

Modulation involves two waveforms: a *modulating signal* that represents the message, and a *carrier wave* that suits the particular application. A modulator systematically alters the carrier wave in correspondence with the variations of the modulating signal. The resulting modulated wave thereby "carries" the message information. We generally require that modulation be a *reversible* operation, so the message can be retrieved by the complementary process of *demodulation*.

Figure 1.2-1 depicts a portion of an analog modulating signal (part *a*) and the corresponding modulated waveform obtained by varying the amplitude of a *sinusoidal* carrier wave (part *b*). This is the familiar amplitude modulation (AM) used for radio broadcasting and other applications. A message may also be impressed on a sinusoidal carrier by frequency modulation (FM) or phase modulation (PM). All methods for sinusoidal carrier modulation are grouped under the heading of *continuous-wave* (CW) modulation.

Incidentally, you act as a CW modulator whenever you speak. The transmission of voice through air is accomplished by generating carrier tones in the vocal cords and modulating these tones with muscular actions of the oral cavity. Thus, what the ear hears as speech is a modulated acoustic wave similar to an AM signal.

Most long-distance transmission systems employ CW modulation with a carrier frequency much higher than the highest frequency component of the

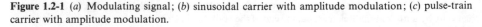

Figure 1.2-1 (*a*) Modulating signal; (*b*) sinusoidal carrier with amplitude modulation; (*c*) pulse-train carrier with amplitude modulation.

modulating signal. The spectrum of the modulated signal then consists of a band of frequency components clustered around the carrier frequency. Under these conditions, we say that CW modulation produces *frequency translation*. In AM broadcasting, for example, the message spectrum typically runs from 100 Hz to 5 kHz; if the carrier frequency is 600 kHz, then the spectrum of the modulated carrier covers 595–605 kHz.

Another modulation method, called *pulse modulation*, has a periodic train of short pulses as the carrier wave. Figure 1.2-1c shows a waveform with pulse amplitude modulation (PAM). Notice that this PAM wave consists of short samples extracted from the analog signal at the top of the figure. *Sampling* is an important signal-processing technique and, subject to certain conditions, it's possible to *reconstruct* an entire waveform from periodic samples.

But pulse modulation by itself does not produce the frequency translation needed for efficient signal transmission. Some transmitters therefore combine pulse and CW modulation. Other modulation techniques, described shortly, combine pulse modulation with *coding*.

Modulation Benefits and Applications

The primary purpose of modulation in a communication system is to generate a modulated signal suited to the characteristics of the transmission channel. Actually, there are several practical benefits and applications of modulation briefly discussed below.

Modulation for efficient transmission Signal transmission over appreciable distance always involves a traveling electromagnetic wave, with or without a guiding medium. The efficiency of any particular transmission method depends upon the frequency of the signal being transmitted. By exploiting the frequency-translation property of CW modulation, message information can be impressed on a carrier whose frequency has been selected for the desired transmission method.

As a case in point, efficient line-of-sight ratio propagation requires antennas whose physical dimensions are at least 1/10 of the signal's wavelength. Unmodulated transmission of an audio signal containing frequency components down to 100 Hz would thus call for antennas some 300 km long. Modulated transmission at 100 MHz, as in FM broadcasting, allows a practical antenna size of about one meter. At frequencies below 100 MHz, other propagation modes have better efficiency with reasonable antenna sizes. DeFrance (1972, Chap. 11) gives a compact treatment of radio propagation and antennas.

For reference purposes, Fig. 1.2-2 shows those portions of the electromagnetic spectrum suited to signal transmission. The figure includes the free-space wavelength, frequency-band designations, and typical transmission media and propagation modes. Also indicated are representative applications authorized by the U.S. Federal Communications Commission.

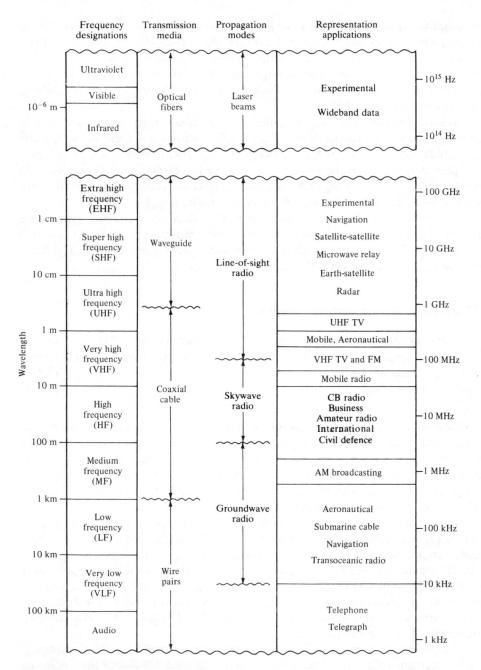

Figure 1.2-2 The electromagnetic spectrum.

Modulation to overcome hardware limitations The design of a communication system may be constrained by the cost and availability of hardware, hardware whose performance often depends upon the frequencies involved. Modulation permits the designer to place a signal in some frequency range that avoids hardware limitations. A particular concern along this line is the question of *fractional bandwidth*, defined as absolute bandwidth divided by the center frequency. Hardware costs and complications are minimized if the fractional bandwidth is kept within 1–10%. Fractional-bandwidth considerations account for the fact that modulation units are found in receivers as well as in transmitters.

It likewise follows that signals with large bandwidth should be modulated on high-frequency carriers. Since information rate is proportional to bandwidth, according to the Hartley-Shannon law, we conclude that a high information rate requires a high carrier frequency. For instance, a 5-GHz microwave system can accommodate 10,000 times as much information in a given time interval as a 500-kHz radio channel. Going even higher in the electromagnetic spectrum, one optical laser beam has a bandwidth potential equivalent to 10 million TV channels.

Modulation to reduce noise and interference A brute-force method for combating noise and interference is to increase the signal power until it overwhelms the contaminations. But increasing power is costly and may damage equipment. (One of the early transatlantic cables was apparently destroyed by high-voltage rupture in an effort to obtain a usable received signal.) Fortunately, FM and certain other types of modulation have the valuable property of suppressing both noise and interference.

This property is called *wideband noise reduction* because it requires the transmission bandwidth to be much greater than the bandwidth of the modulating signal. Wideband modulation thus allows the designer to exchange increased bandwidth for decreased signal power, a trade-off implied by the Hartley-Shannon law. Note that a higher carrier frequency may be needed to accommodate wideband modulation.

Modulation for frequency assignment When you tune a radio or television set to a particular station, you are selecting one of the many signals being received at that time. Since each station has a different assigned carrier frequency, the desired signal can be separated from the others by filtering. Were it not for modulation, only one station could broadcast in a given area; otherwise, two or more broadcasting stations would create a hopeless jumble of interference.

Modulation for multiplexing Multiplexing is the process of combining several signals for simultaneous transmission on one channel. Frequency-division multiplexing (FDM) uses CW modulation to put each signal on a different carrier frequency, and a bank of filters separates the signals at the destination. Time-division multiplexing (TDM) uses pulse modulation to put samples of different signals in nonoverlapping time slots. Back in Fig. 1.2-1c, for instance, the gaps between pulses could be filled with samples from other signals. A switching circuit at the destination then separates the samples for signal reconstruction.

Applications of multiplexing include data telemetry, FM stereophonic broadcasting, and long-distance telephone. As many as 1800 voice signals can be multiplexed on a coaxial cable less than one centimeter in diameter. Multiplexing thereby provides another way of increasing communication efficiency.

Coding Methods and Benefits

We've described modulation as a *signal*-processing operation for effective transmission. *Coding* is a *symbol*-processing operation for improved communication when the information is digital or can be approximated in the form of discrete symbols. Both coding and modulation may be necessary for reliable long-distance digital transmission.

The operation of *encoding* transforms a digital message into a new sequence of symbols. *Decoding* converts an encoded sequence back to the original message with, perhaps, a few errors caused by transmission contaminations. Most coding methods involve digital logic circuits and *binary* symbols, corresponding to the binary digits 0 and 1.

Consider a computer terminal or some other digital source having $M \gg 2$ symbols. Uncoded transmission of a message from this source would require M different waveforms, one for each symbol. Alternatively, each symbol could be represented by a *binary codeword* consisting of K binary digits. Since there are 2^K possible codewords made up of K binary digits, we need $K \geq \log_2 M$ digits per codeword to encode M source symbols. If the source produces r symbols per second, the binary code will have Kr digits per second and the transmission bandwidth requirement is K times the bandwidth of an uncoded signal.

In exchange for increased bandwidth, binary encoding of M-ary source symbols offers two advantages. First, less complicated hardware is needed to handle a binary signal composed of just two different waveforms. Second, contaminating noise has less effect on a binary signal than it does on a signal composed of M different waveforms, so there will be fewer errors caused by the noise. Hence, this coding method is essentially a digital technique for wideband noise reduction.

Error-control coding goes further in the direction of wideband noise reduction. By appending extra *check digits* to each binary codeword, we can detect or even correct most of the errors that do occur. Error-control coding increases both bandwidth and hardware complexity, but it pays off in terms of nearly error-free digital communication despite a low signal-to-noise ratio.

Now, going in the opposite direction, suppose we have a binary data source and a communication system with adequate signal-to-noise ratio but limited bandwidth. This situation is typically encountered in computer communication networks using voice-telephone channels. Encoding blocks of K binary digits as

M-ary symbols *reduces* the signal bandwidth by a factor of $K = \log_2 M$, thereby allowing an increased data rate on a limited-bandwidth channel. A more sophisticated statistical technique known as *source coding* may yield additional bandwidth compression.

Finally, the benefits of digital coding can be incorporated in *analog* communication with the help of an analog-to-digital conversion method such as *pulse-code modulation* (PCM). A PCM signal is generated by sampling the analog message, digitizing (quantizing) the sample values, and encoding the sequence of digitized samples. In view of the reliability, versatility, and efficiency of digital transmission, PCM has become an important method for analog communication. Furthermore, when coupled with high-speed microprocessors, PCM makes it possible to substitute *digital signal processing* for analog operations.

1.3 HISTORICAL PERSPECTIVE

The organization of this text is dictated by pedagogical considerations and does not necessarily reflect the evolution of communication systems. To provide at least some historical perspective, a chronological outline of electrical communication is presented in Table 1.3-1. The table lists key inventions, scientific discoveries, important papers, and the names associated with these events.

Table 1.3-1 A chronology of electrical communication

Year	Event
1800–1837	*Preliminary developments* Volta discovers the primary battery; the mathematical treatises by Fourier, Cauchy, and Laplace; experiments on electricity and magnesium by Oersted, Ampere, Faraday, and Henry; Ohm's law (1826); early telegraph systems by Gauss, Weber, and Wheatstone.
1838–1866	*Telegraphy* Morse prefects his system; Steinheil finds that the earth can be used for a current path; commercial service initiated (1844); multiplexing techniques devised; William Thomson (Lord Kelvin) calculates the pulse response of a telegraph line (1855); transatlantic cables installed by Cyrus Field and associates.
1845	Kirchhoff's circuit laws enunciated.
1864	Maxwell's equations predict electromagnetic radiation.
1876–1899	*Telephony* Acoustic transducer perfected by Alexander Graham Bell, after earlier attempts by Reis; first telephone exchange, in New Haven, with eight lines (1878); Edison's carbon-button transducer; cable circuits introduced; Strowger devises automatic step-by-step switching (1887); the theory of cable loading by Heaviside, Pupin, and Campbell.
1887–1907	*Wireless telegraphy* Heinrich Hertz verifies Maxwell's theory; demonstrations by Marconi and Popov; Marconi patents a complete wireless telegraph system (1897); the theory of tuning circuits developed by Sir Oliver Lodge; commercial service begins, including ship-to-shore and transatlantic systems.
1892–1899	Oliver Heaviside's publications on operational calculus, circuits, and electromagnetics.

(continued)

Table 1.3-1 A chronology of electrical communication (*continued*)

Year	Event
1904–1920	*Communication electronics* Lee De Forest invents the Audion (triode) based on Fleming's diode; basic filter types devised by G. A. Campbell and others; experiments with AM radio broadcasting; transcontinental telephone line with electronic repeaters completed by the Bell System (1915); multiplexed carrier telephony introduced; E. H. Armstrong perfects the superheterodyne radio receiver (1918); first commercial broadcasting station, KDKA, Pittsburgh.
1920–1928	*Transmission theory* Landmark papers on the theory of signal transmission and noise by J. R. Carson, H. Nyquist, J. B. Johnson, and R. V. L. Hartley.
1923–1938	*Television* Mechanical image-formation system demonstrated by Baird and Jenkins; theoretical analysis of bandwidth requirements; Farnsworth and Zworykin propose electronic systems; vacuum cathode-ray tubes perfected by DuMont and others; field tests and experimental broadcasting begin.
1927	Federal Communications Commission established.
1931	Teletypewriter service initiated.
1934	H. S. Black develops the negative-feedback amplifier.
1936	Armstrong's paper states the case for FM radio.
1937	Alec Reeves conceives pulse code modulation.
1938–1945	*World War II* Radar and microwave systems developed; FM used extensively for military communications; improved electronics, hardware, and theory in all areas.
1944–1947	*Statistical communication theory* Rice develops a mathematical representation of noise; Weiner, Kolmogoroff, and Kotel'nikov apply statistical methods to signal detection.
1948–1950	*Information theory and coding* C. E. Shannon publishes the founding papers of information theory; Hamming and Golay devise error-correcting codes.
1948–1951	Transistor devices invented by Bardeen, Brattain, and Shockley.
1950	Time-division multiplexing applied to telephony.
1953	Color TV standards established in the United States.
1955	J. R. Pierce proposes satellite communication systems.
1956	First transoceanic telephone cable (36 voice channels).
1958	Long-distance data transmission system developed for military purposes.
1960	Maiman demonstrates the first laser.
1961	Integrated circuits go into commercial production.
1962	Satellite communication begins with Telstar I.
1962–1966	*High-speed digital communication* Data transmission service offered commercially; wideband channels designed for digital signaling; pulse code modulation proves feasible for voice and TV transmission; major breakthroughs in the theory and implementation of digital transmission, including error-control coding methods by Viterbi and others, and the development of adaptive equalization by Lucky and coworkers.
1963	Solid-state microwave oscillators perfected by Gunn.
1964	Fully electronic telephone switching system (No. 1 ESS) goes into service.
1965	Mariner IV transmits pictures from Mars to earth.
1966–1975	*Wideband communication systems* Cable TV systems; commercial satellite relay service becomes available; optical links using lasers and fiber optics; ARPANET created (1969) followed by other intercontinental computer networks.
1975–1985	*State of the art* Integrated-circuit communication modules; high-frequency power MOS devices; digital signal processing with microprocessors; filter circuits using switched capacitors and surface acoustic waves; rate distortion theory and predictive coding applied to data compression.

Several of the terms in the chronology have been mentioned already, while others will be described in later chapters when we discuss the impact and inter-relationships of particular events. You may therefore find it helpful to refer back to this table from time to time.

1.4 PROSPECTUS

This text is divided into three major parts. Part I, *Signal and System Analysis*, develops the concepts and mathematical tools needed for the study of communication systems. Chapters 2 and 3 deal with deterministic signals, emphasizing time-domain and frequency-domain analysis of signal transmission, distortion, and filtering. Chapters 4 and 5 apply probability theory and statistics to the representation of random signals and noise. Throughout these chapters we rely heavily on mathematical models to cut through to the heart of complex problems. Keep in mind, however, that such models must be combined with physical reasoning and engineering judgment.

Part II, *Analog Communication*, begins with four chapters (6 through 9) devoted to the *how* and the *why* of various types of CW modulation. Particular topics include modulated waveforms and transmission bandwidth, implementation of transmitters and receivers, frequency-division multiplexing, and system performance in the presence of noise. Chapter 10 introduces sampling theory, analog pulse modulation, and time-division multiplexing.

Part III, *Digital Communication*, starts in Chap. 11 with baseband (unmodulated) transmission so we can focus on the important concepts of digital signals and spectra, noise and errors, and synchronization. Chapter 12 then draws upon the previous two chapters for the study of coded pulse modulation, including PCM and digital multiplexing systems. A short survey of error-control coding is presented in Chap. 13. Chapter 14 analyzes digital transmission systems with CW modulation, culminating in a performance comparison of various methods. An introduction to information theory in Chap. 15 provides a retrospective view of digital communication and returns us to the Hartley-Shannon law.

Each chapter contains several exercises designed to clarify and reinforce the concepts and analytic techniques. You should work these exercises as you come to them, checking your results with the answers provided at the back of the book. Also at the back you'll find tables containing handy summaries of important text material and mathematical relations pertinent to the exercises and to the problems at the end of each chapter.

Although we mostly describe communication systems in terms of "black boxes" with specified properties, we'll occasionally lift the lid to look at electronic circuits that carry out particular operations. Such digressions are intended to be illustrative rather than a comprehensive treatment of communication electronics. Besides discussions of electronics, certain optional or more advanced topics are interspersed in various chapters and identified by the symbol ★. These topics may be omitted without loss of continuity. Other optional material of a supplementary nature is contained in three appendices.

Two types of references have been included. Books and papers cited within chapters provide further information about specific items. Additional references are collected in a supplementary reading list that serves as an annotated bibliography for those who wish to pursue subjects in greater depth.

Finally, as you have probably observed, communications engineers use many abbreviations and acronyms. Most abbreviations defined in this book are also listed in the index, to which you can refer if you happen to forget a definition.

PART
ONE

SIGNAL AND SYSTEM ANALYSIS

TWO

SIGNALS AND SPECTRA

Electrical communication signals are time-varying quantities such as voltage or current. Although a signal physically exists in the time domain, we can also represent it in the *frequency domain* where we view the signal as consisting of sinusoidal components at various frequencies. This frequency-domain description is called the *spectrum*.

Spectral analysis, using the Fourier series and transform, is one of the fundamental methods of communication engineering. It allows us to treat entire *classes* of signals that have similar properties in the frequency domain, rather than getting bogged down in detailed time-domain analysis of individual signals. Furthermore, when coupled with the frequency-response characteristics of filters and other system components, the spectral approach provides valuable insight for design work.

This chapter therefore is devoted to signals and spectral analysis, giving special attention to the frequency-domain interpretation of signal properties. We'll examine line spectra based on the Fourier series expansion of periodic signals, and continuous spectra based on the Fourier transform of nonperiodic signals. These two types of spectra will ultimately be merged with the help of the impulse concept.

As the first step in spectral analysis we must write equations representing signals as functions of time. But such equations are only mathematical *models* of the real world, and imperfect models at that. In fact, a completely faithful description of the simplest physical signal would be quite complicated and impractical for engineering purposes. Hence, we try to devise models that represent with minimum complexity the significant properties of physical signals. The study of many different signal models provides us with the background needed to choose appropriate models for specific applications.

2.1 LINE SPECTRA AND FOURIER SERIES

This section introduces and interprets the frequency domain in terms of rotating phasors. We'll begin with the line spectrum of a sinusoidal signal. Then we'll invoke the Fourier series expansion to obtain the line spectrum of any periodic signal that has finite average power.

Phasors and Line Spectra

Consider the familiar sinusoidal or AC (alternating-current) waveform $v(t)$ plotted in Fig. 2.1-1. By convention, we express sinusoids in terms of the cosine function and write

$$v(t) = A \cos (\omega_0 t + \phi) \tag{1}$$

where A is the peak value or *amplitude* and ω_0 is the *radian frequency*. The *phase angle* ϕ represents the fact that the peak has been shifted away from the time origin and occurs at $t = -\phi/\omega_0$. Equation (1) implies that $v(t)$ repeats itself for all time, with repetition *period* $T_0 = 2\pi/\omega_0$. The reciprocal of the period equals the *cyclical frequency*

$$f_0 \triangleq \frac{1}{T_0} = \frac{\omega_0}{2\pi} \tag{2}$$

measured in cycles per second or hertz.

Obviously, no real signal goes on forever, but Eq. (1) could be a reasonable model for a sinusoidal waveform that lasts a long time compared to the period. In particular, AC steady-state circuit analysis depends upon the assumption of an eternal sinusoid — usually represented by a complex exponential or *phasor*. Phasors also play a major role in spectral analysis.

The phasor representation of a sinusoidal signal comes from *Euler's theorem*

$$e^{\pm j\theta} = \cos \theta \pm j \sin \theta \tag{3}$$

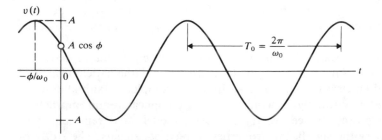

Figure 2.1-1 A sinusoidal waveform $v(t) = A \cos (\omega_0 t + \phi)$.

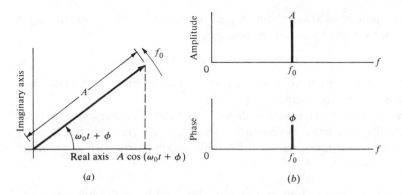

Figure 2.1-2 Representations of $A \cos(\omega_0 t + \phi)$. (a) Phasor diagram; (b) line spectrum.

where $j \triangleq \sqrt{-1}$ and θ is an arbitrary angle. If we let $\theta = \omega_0 t + \phi$, we can write any sinusoid as the real part of a complex exponential, namely

$$A \cos(\omega_0 t + \phi) = A \, \text{Re} \, [e^{j(\omega_0 t + \phi)}]$$

$$= \text{Re} \, [Ae^{j\phi} e^{j\omega_0 t}] \qquad (4)$$

This is called a phasor representation because the term inside the brackets may be viewed as a rotating vector in a complex plane whose axes are the real and imaginary parts, as Fig. 2.1-2a illustrates. The phasor has length A, rotates counterclockwise at a rate f_0 revolutions per second, and at time $t = 0$ makes an angle ϕ with respect to the positive real axis. The projection of the phasor on the real axis equals the sinusoid in Eq. (4).

Now observe that only three parameters completely specify a phasor: amplitude, phase angle, and rotational frequency. To describe the same phasor in the *frequency domain*, we must associate the corresponding amplitude and phase with the particular frequency f_0. Hence, a suitable frequency-domain description would be the *line spectrum* in Fig. 2.1-2b, which consists of two plots: amplitude versus frequency, and phase versus frequency. While this figure appears simple to the point of being trivial, it does have great conceptual value when extended to more complicated signals. But before taking that step, four conventions regarding line spectra should be stated.

1. In all our spectral drawings the independent variable will be *cyclical frequency* f in hertz, rather than radian frequency ω, and any specific frequency such as f_0 will be identified by a subscript. (We'll still use ω with or without subscripts as a shorthand notation for $2\pi f$ since that combination occurs so often.)
2. Phase angles will be measured with respect to *cosine* waves or, equivalently, with respect to the positive real axis of the phasor diagram. Hence, sine waves need to be converted to cosines via the identity

$$\sin \omega t = \cos(\omega t - 90°) \qquad (5)$$

3. We regard amplitude as always being a *positive* quantity. When negative signs appear, they must be absorbed in the phase using

$$-A \cos \omega t = A \cos (\omega t \pm 180°) \qquad (6)$$

It does not matter whether you take $+180°$ or $-180°$ since the phasor ends up in the same place either way.

4. Phase angles usually are expressed in degrees even though other angles such as ωt are inherently in radians. No confusion should result from this mixed notation since angles expressed in degrees will always carry the appropriate symbol.

To illustrate these conventions and to carry further the idea of line spectrum, consider the signal

$$w(t) = 7 - 10 \cos (40\pi t - 60°) + 4 \sin 120\pi t$$

which is sketched in Fig. 2.1-3a. Converting the constant term to a zero frequency or DC (direct-current) component and applying Eqs. (5) and (6) gives the sum of cosines

$$w(t) = 7 \cos 2\pi 0t + 10 \cos (2\pi 20t + 120°) + 4 \cos (2\pi 60t - 90°)$$

whose spectrum is shown in Fig. 2.1-3b.

Drawings like Fig. 2.1-3b, called *one-sided* or *positive*-frequency line spectra, can be constructed for any linear combination of sinusoids. But another spectral representation turns out to be more valuable, even though it involves *negative* frequencies. We obtain this representation from Eq. (4) by recalling that Re $[z] = \frac{1}{2}(z + z^*)$, where z is any complex quantity with complex conjugate z^*. Hence, if $z = Ae^{j\phi}e^{j\omega_0 t}$ then $z^* = Ae^{-j\phi}e^{-j\omega_0 t}$ and Eq. (4) becomes

$$A \cos (\omega_0 t + \phi) = \frac{A}{2} e^{j\phi}e^{j\omega_0 t} + \frac{A}{2} e^{-j\phi}e^{-j\omega_0 t} \qquad (7)$$

so we now have a *pair of conjugate phasors*.

The corresponding phasor diagram and line spectrum are shown in Fig. 2.1-4. The phasor diagram consists of two phasors with equal lengths but opposite angles and directions of rotation. The phasor sum always falls along the real axis to yield $A \cos (\omega_0 t + \phi)$. The line spectrum is *two-sided* since it must include negative frequencies to allow for the opposite rotational directions, and one-half of the original amplitude is associated with each of the two frequencies $\pm f_0$. The amplitude spectrum has *even symmetry* while the phase spectrum has *odd symmetry* because we are dealing with conjugate phasors. This symmetry appears more vividly in Fig. 2.1-5 which is the two-sided version of Fig. 2.1-3b.

It should be emphasized that these line spectra, one-sided or two-sided, are just pictorial ways of representing sinusoidal or phasor time functions. A single line in the one-sided spectrum represents a *real cosine wave*, whereas a single line in the two-sided spectrum represents a *complex exponential* and the conjugate term must be added to get a real cosine wave. Thus, whenever we speak of some

(a)

(b)

Figure 2.1-3

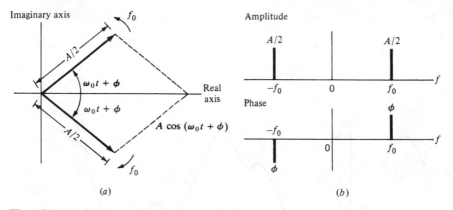

Figure 2.1-4 (a) Conjugate phasors; (b) two-sided line spectrum.

Figure 2.1-5

frequency interval such as f_1 to f_2 in a two-sided spectrum, we should also include the corresponding negative-frequency interval $-f_1$ to $-f_2$. A simple notation for specifying both intervals is $f_1 \leq |f| \leq f_2$.

Finally, note that the amplitude spectrum in either version conveys more information than the phase spectrum. Both parts are required to define the time-domain function, but the amplitude spectrum by itself tells us what frequencies are present and in what proportion. Putting this another way, the amplitude spectrum displays the signal's *frequency content*.

Exercise 2.1-1 Construct the one-sided and two-sided spectrum of $v(t) = -3 - 4 \sin 30\pi t$.

Periodic Signals and Average Power

Sinusoids and phasors are members of the general class of *periodic signals*. These signals obey the relationship

$$v(t \pm mT_0) = v(t) \qquad -\infty < t < \infty \tag{8}$$

where m is any integer. This equation simply says that shifting the signal by an integer number of periods to the left or right leaves the waveform unchanged. Consequently, a periodic signal is fully described by specifying its behavior over any one period.

The frequency-domain representation of a periodic signal is a line spectrum obtained by Fourier series expansion. The expansion requires that the signal have finite average power. Because average power and other time averages are important signal properties, we'll formalize these concepts here.

Given any time function $v(t)$, its *average value* over all time is defined as

$$\langle v(t) \rangle \triangleq \lim_{T \to \infty} \frac{1}{T} \int_{-T/2}^{T/2} v(t) \, dt \tag{9}$$

The notation $\langle v(t) \rangle$ represents the averaging operation on the right-hand side, which comprises three steps: integrate $v(t)$ to get the net area under the curve from $-T/2 \le t \le T/2$; divide that area by the duration T of the time interval; then let $T \to \infty$ to encompass all time. In the case of a *periodic* signal, Eq. (9) reduces to the average over any interval of duration T_0. Thus

$$\langle v(t) \rangle = \frac{1}{T_0} \int_{t_1}^{t_1 + T_0} v(t) \, dt = \frac{1}{T_0} \int_{T_0} v(t) \, dt \tag{10}$$

where the shorthand symbol $\int_{T_0}$ stands for an integration from any time t_1 to $t_1 + T_0$.

If $v(t)$ happens to be the voltage across a resistance R, it produces the current $i(t) = v(t)/R$ and we could compute the resulting average power by averaging the instantaneous power $v(t)i(t) = v^2(t)/R = Ri^2(t)$. But we don't necessarily know whether a given signal is a voltage or current, so let's *normalize* power by assuming henceforth that $R = 1 \, \Omega$. Our definition of the *average power* associated with an arbitrary periodic signal then becomes

$$P \triangleq \langle |v(t)|^2 \rangle = \frac{1}{T_0} \int_{T_0} |v(t)|^2 \, dt \tag{11}$$

where we have written $|v(t)|^2$ instead of $v^2(t)$ to allow for the possibility of *complex* signal models. In any case, the value of P will be *real* and *nonnegative*.

When the integral in Eq. (11) exists and yields $0 < P < \infty$, the signal $v(t)$ is said to have well-defined average power, and will be called a *periodic power signal*. Almost all periodic signals of practical interest fall in this category. The average value of a power signal may be positive, negative, or zero, but is bounded by

$$|\langle v(t) \rangle| \le \sqrt{PT_0}$$

which follows from the integral relation $|\int v(t) \, dt|^2 \le \int |v(t)|^2 \, dt$.

Some signal averages can be found by inspection, using the physical interpretation of averaging. As a specific example take the sinusoid

$$v(t) = A \cos(\omega_0 t + \phi)$$

which has

$$\langle v(t) \rangle = 0 \qquad P = \frac{A^2}{2} \tag{12}$$

You should have no trouble confirming these results if you sketch one period of $v(t)$ and $|v(t)|^2$.

Fourier Series

The signal $w(t)$ back in Fig. 2.1-3 was generated by summing a DC term and two sinusoids. Now we'll go the other way and decompose periodic signals into sums of sinusoids or, equivalently, rotating phasors. We invoke the *exponential Fourier series* for this purpose.

Let $v(t)$ be a power signal with period $T_0 = 1/f_0$. Its exponential Fourier series expansion is

$$v(t) = \sum_{n=-\infty}^{\infty} c_n e^{j2\pi n f_0 t} \qquad n = 0, 1, 2, \ldots \tag{13}$$

The series coefficients are related to $v(t)$ by

$$c_n = \frac{1}{T_0} \int_{T_0} v(t) e^{-j2\pi n f_0 t} \, dt \tag{14}$$

so c_n equals the average of the product $v(t)e^{-j2\pi n f_0 t}$. Since the coefficients are complex quantities in general, they can be expressed in the polar form

$$c_n = |c_n| e^{j \arg c_n}$$

where $\arg c_n$ stands for the *angle* of c_n. Equation (13) thus expands a periodic power signal as an infinite sum of phasors, the nth term being

$$c_n e^{j2\pi n f_0 t} = |c_n| e^{j \arg c_n} e^{j2\pi n f_0 t}$$

The series convergence properties will be discussed after considering its spectral implications.

Observe that $v(t)$ in Eq. (13) consists of phasors with amplitude $|c_n|$ and angle $\arg c_n$ at the frequencies $nf_0 = 0, \pm f_0, \pm 2f_0, \ldots$. Hence, the corresponding frequency-domain picture is a two-sided line spectrum defined by the series coefficients. We emphasize the spectral interpretation by writing

$$c(nf_0) \triangleq c_n$$

so that $|c(nf_0)|$ represents the *amplitude spectrum* as a function of f, and $\arg c(nf_0)$ represents the *phase spectrum*. Three important spectral properties of periodic power signals are listed below.

1. All frequencies are integer multiples or *harmonics* of the *fundamental frequency* $f_0 = 1/T_0$. Thus the spectral lines have uniform spacing f_0.

2. The DC component equals the *average value* of the signal, since setting $n = 0$ in Eq. (14) yields

$$c(0) = \frac{1}{T_0} \int_{T_0} v(t) \, dt = \langle v(t) \rangle \tag{15}$$

Calculated values of $c(0)$ may be checked by inspecting $v(t)$ — a wise practice when the integration gives an ambiguous result.

3. If $v(t)$ is a *real* (noncomplex) function of time, then

$$c_{-n} = c_n^* = |c_n| e^{-j \arg c_n} \tag{16a}$$

which follows from Eq. (14) with n replaced by $-n$. Hence

$$|c(-nf_0)| = |c(nf_0)| \qquad \arg c(-nf_0) = -\arg c(nf_0) \tag{16b}$$

which means that the amplitude spectrum has even symmetry and the phase spectrum has odd symmetry.

When dealing with real signals, the property in Eq. (16) allows us to regroup the exponential series into *complex-conjugate pairs*, except for c_0. Equation (13) then becomes

$$v(t) = c_0 + \sum_{n=1}^{\infty} |2c_n| \cos (2\pi n f_0 t + \arg c_n) \tag{17}$$

which is the *trigonometric Fourier series* and suggests a one-sided spectrum. Most of the time, however, we'll use the exponential series and two-sided spectra.

One final comment should be made before taking up an example. The integration for c_n often involves a phasor average in the form

$$\frac{1}{T} \int_{-T/2}^{T/2} e^{j2\pi f t} \, dt = \frac{1}{j2\pi f T} (e^{j\pi f T} - e^{-j\pi f T})$$

$$= \frac{1}{\pi f T} \sin \pi f T \tag{18}$$

Since this expression occurs time and again in spectral analysis, we'll now introduce the *sinc function* defined by

$$\operatorname{sinc} \lambda \triangleq \frac{\sin \pi \lambda}{\pi \lambda} \tag{19}$$

where λ represents the independent variable.† Figure 2.1-6 shows that sinc λ is an even function of λ having its peak at $\lambda = 0$ and zero crossings at all other integer values of λ, so

$$\operatorname{sinc} \lambda = \begin{cases} 1 & \lambda = 0 \\ 0 & \lambda = \pm 1, \pm 2, \ldots \end{cases}$$

---

† Some authors use the related *sampling function* defined as Sa $(x) \triangleq (\sin x)/x$ so that sinc $\lambda = $ Sa $(\pi \lambda)$.

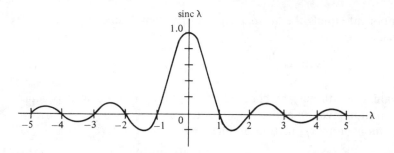

Figure 2.1-6 The function sinc $\lambda = (\sin \pi\lambda)/\pi\lambda$.

Numerical values of sinc λ and sinc^2 λ are given in Table T.4 at the back of the book, while Table T.3 includes several mathematical relations that you'll find helpful for Fourier analysis.

> **Example 2.1-1 Rectangular pulse train** Consider the periodic train of rectangular pulses in Fig. 2.1-7. Each pulse has height or amplitude A and width or duration τ. There are stepwise *discontinuities* at each pulse-edge location, $t = \pm\tau/2$, etc., so the values of $v(t)$ are *undefined* at these points of discontinuity. This brings out another possible difference between a physical signal and its mathematical model, for a physical signal never makes a perfect stepwise transition. However, the model may still be reasonable if the actual transition times are quite small compared to the pulse duration.
>
> To calculate the Fourier coefficients, we'll take the range of integration in Eq. (14) over the central period $-T_0/2 \le t \le T_0/2$, where
>
> $$v(t) = \begin{cases} A & |t| < \tau/2 \\ 0 & |t| > \tau/2 \end{cases}$$
>
> Thus
>
> $$c_n = \frac{1}{T_0} \int_{-T_0/2}^{T_0/2} v(t)e^{-j2\pi nf_0 t}\, dt = \frac{1}{T_0} \int_{-\tau/2}^{\tau/2} Ae^{-j2\pi nf_0 t}\, dt$$
>
> $$= \frac{A}{-j2\pi nf_0 T_0} (e^{-j\pi nf_0\tau} - e^{+j\pi nf_0\tau})$$
>
> $$= \frac{A}{T_0} \frac{\sin \pi nf_0\tau}{\pi nf_0}$$
>
> Multiplying and dividing by τ finally gives
>
> $$c_n = \frac{A\tau}{T_0} \text{ sinc } nf_0\tau \tag{20}$$
>
> which follows from Eq. (19) with $\lambda = nf_0\tau$.

Figure 2.1-7 Rectangular pulse train.

The amplitude spectrum obtained from $|c(nf_0)| = |c_n| = Af_0\tau|\text{sinc } nf_0\tau|$ is shown in Fig. 2.1-8a for the case of $\tau/T_0 = f_0\tau = 1/4$. We construct this plot by drawing the continuous function $Af_0\tau|\text{sinc } f\tau|$ as a dashed curve, which becomes the *envelope* of the lines. The spectral lines at $\pm 4f_0$, $\pm 8f_0$, etc., are "missing" since they fall precisely at multiples of $1/\tau$ where the envelope equals zero. The DC component has amplitude $c(0) = A\tau/T_0$ which should be recognized as the average value of $v(t)$ by inspection of Fig. 2.1-7. Incidentally, τ/T_0 equals the ratio of "on" time to period, frequently designated as the *duty cycle* in pulse electronics work.

The phase spectrum in Fig. 2.1-8b is obtained by observing that c_n is always real but sometimes negative. Hence, arg $c(nf_0)$ takes on the values $0°$ and $\pm 180°$, depending on the sign of sinc $nf_0\tau$. Both $+180°$ and $-180°$ were used here to bring out the odd symmetry of the phase.

Having decomposed the pulse train into its frequency components, let's build it back up again. For that purpose, we'll write out the trigonometric

(a)

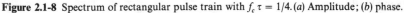

(b)

Figure 2.1-8 Spectrum of rectangular pulse train with $f_c\tau = 1/4$. (a) Amplitude; (b) phase.

Figure 2.1-9 Fourier-series reconstruction of a rectangular pulse train.

series in Eq. (17), still taking $\tau/T_0 = f_0 \tau = 1/4$ so $c_0 = A/4$ and $|2c_n| = (2A/4)|\text{sinc } n/4| = (2A/\pi n)|\sin \pi n/4|$. Thus

$$v(t) = \frac{A}{4} + \frac{\sqrt{2}\,A}{\pi} \cos \omega_0 t + \frac{A}{\pi} \cos 2\omega_0 t + \frac{\sqrt{2}\,A}{3\pi} \cos 3\omega_0 t + \cdots$$

Summing terms through the third harmonic gives the approximation of $v(t)$ sketched in Fig. 2.1-9a. This approximation contains the gross features of the pulse train but lacks sharp corners. A more accurate approximation shown in Fig. 2.1-9b comprises all components through the seventh harmonic. Note that the small-amplitude higher harmonics serve primarily to square up the corners. Also note that the series is converging toward the midpoint value $A/2$ at $t = \pm\tau/2$ where $v(t)$ has discontinuities.

Exercise 2.1-2 Sketch the amplitude spectrum of a rectangular pulse train for each of the following cases: $\tau = T_0/5$, $\tau = T_0/2$, $\tau = T_0$. In the last case the pulse train degenerates into a constant for all time; how does this show up in the spectrum?

Convergence Conditions and Gibbs Phenomenon

We've seen that a periodic signal can be approximated with a finite number of terms of its Fourier series. But does the infinite series *converge* to $v(t)$? The study of convergence involves subtle mathematical considerations that we'll not go into here. Instead, we'll state without proof some of the important results. Further details are given by Stuart (1966, Chap. 1) or Stark and Tuteur (1979, Chap. 2).

The *Dirichlet conditions* for Fourier series expansion are as follows: If a periodic function $v(t)$ has a finite number of maxima, minima, and discontinuities per period, and if $v(t)$ is *absolutely integrable*, so that $v(t)$ has a finite area per period, then the Fourier series exists and converges uniformly wherever $v(t)$ is continuous. These conditions are sufficient but not strictly necessary.

An alternative condition is that $v(t)$ be *square integrable*, so that $|v(t)|^2$ has finite area per period—equivalent to a power signal. Under this condition, the series *converges in the mean* such that if

$$v_N(t) = \sum_{n=-N}^{N} c_n e^{j2\pi n f_0 t}$$

then

$$\lim_{N \to \infty} \int_{T_0} |v(t) - v_N(t)|^2 \, dt = 0$$

In other words, the mean square difference between $v(t)$ and the partial sum $v_N(t)$ vanishes as more terms are included.

Regardless of whether $v(t)$ is absolutely integrable or square integrable, the series exhibits a behavior known as *Gibbs phenomenon* at points of discontinuity. Figure 2.1-10 illustrates this behavior for a stepwise discontinuity at $t = t_0$. The partial sum $v_N(t)$ converges to the *midpoint* at the discontinuity, which seems quite reasonable. However, on each side of the discontinuity, $v_N(t)$ has oscillatory over-shoot with period $T_0/2N$ and peak value of about 18% of the step height, independent of N. Thus, as $N \to \infty$, the oscillations collapse into nonvanishing spikes called "Gibbs ears" above and below the discontinuity.

Figure 2.1-10 Gibbs phenomenon at a step discontinuity.

Since a real signal must be continuous, Gibbs phenomenon does not occur and we're justified in treating the Fourier series as being *identical* to $v(t)$. But idealized signal models like the rectangular pulse train often do have discontinuities. You therefore need to pay attention to convergence when working with such models.

Parseval's Power Theorem

Parseval's theorem relates the average power P of a periodic signal to its Fourier coefficients. To derive the theorem, we start with

$$P = \frac{1}{T_0} \int_{T_0} |v(t)|^2 \, dt = \frac{1}{T_0} \int_{T_0} v(t) v^*(t) \, dt$$

Now replace $v^*(t)$ by its exponential series

$$v^*(t) = \left[\sum_{n=-\infty}^{\infty} c_n e^{j2\pi nf_0 t} \right]^* = \sum_{n=-\infty}^{\infty} c_n^* e^{-j2\pi nf_0 t}$$

so that

$$P = \frac{1}{T_0} \int_{T_0} v(t) \left[\sum_{n=-\infty}^{\infty} c_n^* e^{-j2\pi nf_0 t} \right] dt$$

$$= \sum_{n=-\infty}^{\infty} \left[\frac{1}{T_0} \int_{T_0} v(t) e^{-j2\pi nf_0 t} \, dt \right] c_n^*$$

and the integral inside the sum equals c_n. Thus

$$P = \sum_{n=-\infty}^{\infty} c_n c_n^* = \sum_{n=-\infty}^{\infty} |c_n|^2 \tag{21}$$

which is Parseval's theorem.

The spectral interpretation of this result is extraordinarily simple: the average power can be found by squaring and adding the heights $|c_n| = |c(nf_0)|$ of the amplitude lines. Observe that Eq. (21) does not involve the phase spectrum, underscoring our prior comment about the dominant role of the amplitude spectrum relative to a signal's frequency content. For further interpretation of Eq. (21) recall that the exponential Fourier series expands $v(t)$ as a sum of phasors of the form $c_n e^{j2\pi nf_0 t}$. You can easily show that the average power of each phasor is

$$\langle |c_n e^{j2\pi nf_0 t}|^2 \rangle = |c_n|^2 \tag{22}$$

Therefore, Parseval's theorem implies *superposition of average power*, since the total average power of $v(t)$ is the sum of the average powers of its phasor components.

Several other theorems pertaining to Fourier series could be stated here. However, they are more conveniently treated as special cases of Fourier transform theorems covered in Sect. 2.3. Table T.2 lists some of the results, along with the Fourier coefficients for various periodic waveforms encountered in communication systems.

Exercise 2.1-3 Use Eq. (21) to calculate P from Fig. 2.1-5.

2.2 FOURIER TRANSFORMS AND CONTINUOUS SPECTRA

Now let's turn from periodic signals that last forever (in theory) to nonperiodic signals concentrated over relatively short-time durations. If a nonperiodic signal has finite total energy, its frequency-domain representation will be a continuous spectrum obtained from the Fourier transform.

Fourier Transforms

Figure 2.2-1 shows two typical nonperiodic signals. The single rectangular pulse (Fig. 2.2-1*a*) is *strictly timelimited* since $v(t)$ is identically zero outside the pulse duration. The other signal is *asymptotically timelimited* in the sense that $v(t) \rightarrow 0$ as $t \rightarrow \pm \infty$. Such signals may also be described loosely as "pulses." In either case, if you attempt to average $v(t)$ or $|v(t)|^2$ over all time you'll find that these averages equal zero. Consequently, instead of talking about average power, a more meaningful property of a nonperiodic signal is its energy.

If $v(t)$ is the voltage across a resistance, the total delivered energy would be

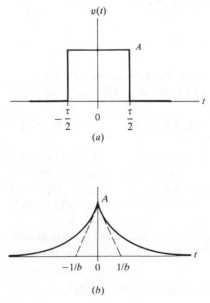

(a)

(b)

Figure 2.2-1

found by integrating the instantaneous power $v^2(t)/R$. We therefore define normalized *signal energy* as

$$E \triangleq \int_{-\infty}^{\infty} |v(t)|^2 \, dt \tag{1}$$

Some energy calculations can be done by inspection, since E is just the total area under the curve of $|v(t)|^2$. For instance, the energy of a rectangular pulse with amplitude A is simply $E = A^2\tau$.

When the integral in Eq. (1) exists and yields $0 < E < \infty$, the signal $v(t)$ is said to have well-defined energy and will be called a *nonperiodic energy signal*. Almost all timelimited signals of practical interest fall in this category, which is the essential condition for spectral analysis using the Fourier transform.

To introduce the Fourier transform, we'll start with the Fourier series representation of a periodic power signal

$$v(t) = \sum_{n=-\infty}^{\infty} c(nf_0)e^{j2\pi nf_0 t}$$

$$= \sum_{n=-\infty}^{\infty} \left[\frac{1}{T_0} \int_{T_0} v(t)e^{-j2\pi nf_0 t} \, dt \right] e^{j2\pi nf_0 t} \tag{2}$$

where the integral expression for $c(nf_0)$ has been written out in full. According to the *Fourier integral theorem* there's a similar representation for a nonperiodic energy signal, namely

$$v(t) = \int_{-\infty}^{\infty} \left[\int_{-\infty}^{\infty} v(t)e^{-j2\pi ft} \, dt \right] e^{j2\pi ft} \, df \tag{3}$$

The bracketed term is the *Fourier transform* of $v(t)$, symbolized by $V(f)$ or $\mathcal{F}[v(t)]$ and defined as

$$V(f) = \mathcal{F}[v(t)] \triangleq \int_{-\infty}^{\infty} v(t)e^{-j2\pi ft} \, dt \tag{4}$$

an integration over all time that yields a function of the continuous variable f.

The time function $v(t)$ is recovered from $V(f)$ by the *inverse Fourier transform*

$$v(t) = \mathcal{F}^{-1}[V(f)] \triangleq \int_{-\infty}^{\infty} V(f)e^{j2\pi ft} \, df \tag{5}$$

an integration over all frequency f. To be more precise, it should be stated that $\mathcal{F}^{-1}[V(f)]$ *converges in the mean* to $v(t)$, similar to Fourier series convergence, with Gibbs phenomenon occurring at discontinuities. But we'll regard Eq. (5) as being an equality for most purposes. A proof that $\mathcal{F}^{-1}[V(f)] = v(t)$ will be outlined in Sect. 2.4.

Equations (4) and (5) constitute the pair of *Fourier integrals*†. At first glance, these integrals seem to be a closed circle of operations. In a given problem,

---

† Other definitions take ω for the frequency variable and therefore include $1/2\pi$ or $1/\sqrt{2\pi}$ as multiplying terms.

however, you usually know either $V(f)$ or $v(t)$. If you know $V(f)$, you can find $v(t)$ from Eq. (5); if you know $v(t)$, you can find $V(f)$ from Eq. (4).

Turning to the frequency-domain picture, a comparison of Eqs. (2) and (5) indicates that $V(f)$ plays the same role for nonperiodic signals that $c(nf_0)$ plays for periodic signals. Thus, $V(f)$ is the *spectrum* of the nonperiodic signal $v(t)$. But $V(f)$ is a continuous function defined for all values of f whereas $c(nf_0)$ is defined only for discrete frequencies. Therefore, a nonperiodic signal will have a *continuous spectrum* rather than a line spectrum. Again, comparing Eqs. (2) and (5) helps explain this difference: in the periodic case we return to the time domain by *summing* discrete-frequency phasors, while in the nonperiodic case we *integrate* a continuous frequency function. Three major properties of $V(f)$ are listed below.

1. The Fourier transform is a complex function, so $|V(f)|$ is the amplitude spectrum of $v(t)$ and arg $V(f)$ is the phase spectrum.
2. The value of $V(f)$ at $f = 0$ equals the *net area* of $v(t)$, since

$$V(0) = \int_{-\infty}^{\infty} v(t)\, dt \qquad (6)$$

which compares with the periodic case where $c(0)$ equals the average value of $v(t)$.
3. If $v(t)$ is *real*, then

$$V(-f) = V^*(f) \qquad (7a)$$

and

$$|V(-f)| = |V(f)| \qquad \text{arg } V(-f) = -\text{arg } V(f) \qquad (7b)$$

so again we have even amplitude symmetry and odd phase symmetry. The term *hermitian symmetry* describes complex functions that obey Eq. (7).

Example 2.2-1 Rectangular pulse In the last section we found the line spectrum of a rectangular pulse train. Now consider the single rectangular pulse in Fig. 2.2-1a. This is so common a signal model that it deserves a symbol of its own. Let's adopt the pictorial notation

$$\Pi(t/\tau) \triangleq \begin{cases} 1 & |t| < \tau/2 \\ 0 & |t| > \tau/2 \end{cases} \qquad (8)$$

which stands for a rectangular function with unit amplitude and duration τ centered at $t = 0$. The pulse in the figure is then written

$$v(t) = A\Pi(t/\tau) \qquad (9a)$$

Inserting $v(t)$ in Eq. (4) yields

$$V(f) = \int_{-\tau/2}^{\tau/2} A e^{-j2\pi f t}\, dt = \frac{A}{\pi f} \sin \pi f \tau$$

$$= A\tau \text{ sinc } f\tau \qquad (9b)$$

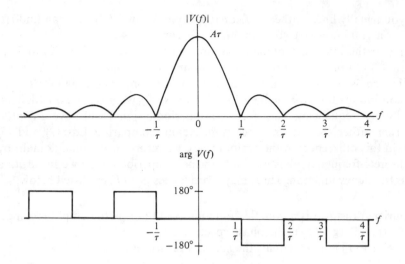

Figure 2.2-2 Rectangular pulse spectrum $V(f) = A\tau$ sinc $f\tau$.

so $V(0) = A\tau$, which clearly equals the pulse's area. The corresponding spectrum, plotted in Fig. 2.2-2, should be compared with Fig. 2.1-8 to illustrate the similarities and differences between line spectra and continuous spectra.

Further inspection of Fig. 2.2-2 reveals that the significant portion of the spectrum is in the range $|f| < 1/\tau$ since $|V(f)| \ll |V(0)|$ for $|f| > 1/\tau$. We therefore may take $1/\tau$ as a measure of the spectral "width." Now if the pulse duration is reduced (small τ), the frequency width is increased, whereas increasing the duration reduces the spectral width. Thus, short pulses have broad spectra, and long pulses have narrow spectra. This phenomenon, called *reciprocal spreading*, is a general property of all signals, pulses or not, because high-frequency components are demanded by rapid time variations while smoother and slower time variations require relatively little high-frequency content.

Symmetric and Causal Signals

When a signal possesses symmetry with respect to the time axis, its transform integral can be simplified. Of course any signal symmetry depends upon both the waveshape and the location of the time origin. But we're usually free to choose the time origin since it's not physically unique — as contrasted with the frequency-domain origin which has a definite physical meaning.

To develop the time-symmetry properties, we'll write ω in place of $2\pi f$ for notational convenience and expand Eq. (4) using $e^{-j2\pi ft} = \cos \omega t - j \sin \omega t$. Thus, in general,

$$V(f) = V_e(f) + jV_o(f) \tag{10a}$$

where

$$V_e(f) \triangleq \int_{-\infty}^{\infty} v(t) \cos \omega t \, dt$$

$$V_o(f) \triangleq - \int_{-\infty}^{\infty} v(t) \sin \omega t \, dt$$

(10b)

which are the *even* and *odd* parts of $V(f)$, regardless of $v(t)$. Incidentally, note that if $v(t)$ is *real*, then

$$\text{Re } [V(f)] = V_e(f) \qquad \text{Im } [V(f)] = V_o(f)$$

so $V^*(f) = V_e(f) - jV_o(f) = V(-f)$, as previously asserted in Eq. (7).

When $v(t)$ has time symmetry, we simplify the integrals in Eq. (10b) by applying the general relationship

$$\int_{-\infty}^{\infty} w(t) \, dt = \int_{0}^{\infty} w(t) \, dt + \int_{-\infty}^{0} w(t) \, dt$$

$$= \begin{cases} 2 \displaystyle\int_{0}^{\infty} w(t) \, dt & w(t) \text{ even} \\[2mm] 0 & w(t) \text{ odd} \end{cases}$$

(11)

where $w(t)$ stands for either $v(t) \cos \omega t$ or $v(t) \sin \omega t$. If $v(t)$ has *even symmetry* so that

$$v(-t) = v(t)$$

(12a)

then $v(t) \cos \omega t$ is even whereas $v(t) \sin \omega t$ is odd. Hence, $V_o(f) = 0$ and

$$V(f) = V_e(f) = 2 \int_{0}^{\infty} v(t) \cos \omega t \, dt$$

(12b)

Conversely, if $v(t)$ has *odd symmetry* so that

$$v(-t) = -v(t)$$

(13a)

then

$$V(f) = jV_o(f) = -j2 \int_{0}^{\infty} v(t) \sin \omega t \, dt$$

(13b)

and $V_e(f) = 0$.

Equations (12) and (13) further show that the spectrum of a *real symmetrical* signal will be either purely real and even or purely imaginary and odd. For instance, the rectangular pulse in Example 2.2-1 is a real and even time function and its spectrum was found to be a real and even frequency function.

Now consider the case of a *causal* signal, defined by the property that

$$v(t) = 0 \qquad t < 0$$

(14a)

This simply means that the signal "starts" at or after $t = 0$. Since causality precludes any time symmetry, the spectrum consists of both real and imaginary parts computed from

$$V(f) = \int_0^\infty v(t)e^{-j2\pi ft}\, dt \qquad (14b)$$

This integral bears a resemblance to the *Laplace transform* commonly used for the study of transients in linear circuits and systems. Therefore, we should briefly consider the similarities and differences between these two types of transforms.

The unilateral or one-sided Laplace transform is a function of the complex variable $s = \sigma + j\omega$ defined by

$$\mathscr{L}[v(t)] \triangleq \int_0^\infty v(t)e^{-st}\, dt$$

which implies that $v(t) = 0$ for $t < 0$. Comparing $\mathscr{L}[v(t)]$ with Eq. (14b) shows that if $v(t)$ is a *causal energy* signal, you can get $V(f)$ from the Laplace transform by letting $s = j2\pi f$. But a typical table of Laplace transforms includes many *non-energy* signals whose Laplace transforms exist only with $\sigma > 0$ so that $|v(t)e^{-st}| = |v(t)e^{-\sigma t}| \rightarrow 0$ as $t \rightarrow \infty$. Such signals do not have a Fourier transform because $s = \sigma + j\omega$ falls outside the frequency domain when $\sigma \neq 0$. On the other hand, the Fourier transform exists for noncausal energy signals that do not have a Laplace transform. See McGillem and Cooper (1974, Chap. 6) for further discussion.

Example 2.2-2 Causal exponential pulse Figure 2.2-3a shows a causal waveform that decays exponentially with time constant $1/b$, so

$$v(t) = \begin{cases} Ae^{-bt} & t > 0 \\ 0 & t < 0 \end{cases} \qquad (15a)$$

The spectrum can be obtained from Eq. (14b) or from the Laplace transform $\mathscr{L}[v(t)] = A/(s + b)$, with the result that

$$V(f) = \frac{A}{b + j2\pi f} \qquad (15b)$$

which is a complex function in unrationalized form. Multiplying numerator and denominator of Eq. (15b) by $b - j2\pi f$ yields the rationalized expression

$$V(f) = \frac{b - j2\pi f}{b^2 + (2\pi f)^2}\, A$$

and we see that

$$V_e(f) = \text{Re }[V(f)] = \frac{bA}{b^2 + (2\pi f)^2}$$

$$V_o(f) = \text{Im }[V(f)] = -\frac{2\pi f A}{b^2 + (2\pi f)^2}$$

Figure 2.2-3 Causal exponential pulse. (a) Waveform; (b) spectrum.

Conversion to polar form then gives the amplitude and phase spectrum

$$|V(f)| = \sqrt{V_e^2(f) + V_o^2(f)} = \frac{A}{\sqrt{b^2 + (2\pi f)^2}}$$

$$\arg V(f) = \arctan \frac{V_o(f)}{V_e(f)} = -\arctan \frac{2\pi f}{b}$$

which are plotted in Fig. 2.2-3b.

The phase spectrum in this case is a smooth curve that includes all angles from $-90°$ to $+90°$. This is due to the signal's lack of time symmetry. But $V(f)$ still has hermitian symmetry since $v(t)$ is a real function. Also note that the spectral width is proportional to b, whereas the time "width" is proportional to the time constant $1/b$—another illustration of reciprocal spreading.

Exercise 2.2-1 Find and sketch $V(f)$ for the symmetrical decaying exponential $v(t) = Ae^{-b|t|}$ in Fig. 2.2-1b. (You may use a definite integral from Table T.3.) Compare your result with $V_e(f)$ in Example 2.2-2. Confirm the reciprocal-spreading effect by calculating the frequency range such that $|V(f)| \geq (1/2)|V(0)|$.

Rayleigh's Energy Theorem

Rayleigh's energy theorem is analogous to Parseval's power theorem. It states that the energy E of a signal $v(t)$ is related to the spectrum $V(f)$ by

$$E = \int_{-\infty}^{\infty} V(f)V^*(f) \, df = \int_{-\infty}^{\infty} |V(f)|^2 \, df \tag{16}$$

Therefore, integrating the square of the amplitude spectrum over all frequency yields the total energy.

The value of Eq. (16) lies not so much in computing E, since the time-domain integration of $|v(t)|^2$ often is easier. Rather, it implies that $|V(f)|^2$ gives the distribution of energy in the frequency domain, and therefore may be termed the *energy spectral density*. By this we mean that the energy in any differential frequency band df equals $|V(f)|^2 \, df$, an interpretation we'll further justify in Sect. 3.5. That interpretation, in turn, lends quantitative support to the notion of spectral width in the sense that most of the energy of a given signal should be contained in the range of frequencies taken to be the spectral width.

By way of illustration, Fig. 2.2-4 is the energy spectral density of a rectangular pulse, whose spectral width was claimed to be $|f| < 1/\tau$. The energy in that band is the shaded area in the figure, namely

$$\int_{-1/\tau}^{1/\tau} |V(f)|^2 \, df = \int_{-1/\tau}^{1/\tau} (A\tau)^2 \operatorname{sinc}^2 f\tau \, df = 0.92A^2\tau$$

a calculation that requires numerical methods. But the total pulse energy is $E = A^2\tau$, so the asserted spectral width encompasses more than 90% of the total energy.

Figure 2.2-4 Energy spectral density of a rectangular pulse.

Rayleigh's theorem is actually a special case of the more general integral relationship

$$\int_{-\infty}^{\infty} v(t)w^*(t)\, dt = \int_{-\infty}^{\infty} V(f)W^*(f)\, df \tag{17}$$

where $v(t)$ and $w(t)$ are arbitrary energy signals with transforms $V(f)$ and $W(f)$. Equation (17) yields Eq. (16) if you let $w(t) = v(t)$ and note that $\int_{-\infty}^{\infty} v(t)v^*(t)\, dt = E$. Other applications of Eq. (17) will emerge subsequently.

The proof of Eq. (17) follows the same lines as our derivation of Parseval's theorem. We substitute for $w^*(t)$ the inverse transform

$$w^*(t) = \left[\int_{-\infty}^{\infty} W(f)e^{j\omega t}\, df \right]^* = \int_{-\infty}^{\infty} W^*(f)e^{-j\omega t}\, df$$

Interchanging the order of time and frequency integrations then gives

$$\int_{-\infty}^{\infty} v(t)w^*(t)\, dt = \int_{-\infty}^{\infty} v(t)\left[\int_{-\infty}^{\infty} W^*(f)e^{-j\omega t}\, df \right] dt$$

$$= \int_{-\infty}^{\infty} \left[\int_{-\infty}^{\infty} v(t)e^{-j\omega t}\, dt \right] W^*(f)\, df$$

which completes the proof since the bracketed term equals $V(f)$.

The *interchange of integral operations* illustrated here is a valuable technique in signal analysis, leading to many useful results. However, you should not apply the technique willy-nilly without giving some thought to the validity of the interchange. As a pragmatic guideline, you can assume that the interchange is valid if the results make sense. If in doubt, test the results with some simple cases having known answers.

Exercise 2.2-2 Calculate the energy of a causal exponential pulse by applying Rayleigh's theorem to $V(f)$ in Eq. (15b). Then check the result by integrating $|v(t)|^2$.

Duality Theorem

If you reexamine the pair of Fourier integrals, you'll see that they differ only by the variable of integration and the sign in the exponent. A fascinating consequence of this similarity is the *duality theorem*. The theorem states that if $v(t)$ and $V(f)$ constitute a known transform pair, and if there exists a *time function* $z(t)$ related to the function $V(f)$ by

$$z(t) = V(t) \tag{18a}$$

then

$$\mathscr{F}[z(t)] = v(-f) \tag{18b}$$

where $v(-f)$ equals $v(t)$ with $t = -f$.

Proving the duality theorem hinges upon recognizing that Fourier transforms are definite integrals whose variables of integration are *dummy* variables. Therefore, we may replace f in Eq. (5) with the dummy variable λ and write

$$v(t) = \int_{-\infty}^{\infty} V(\lambda)e^{j2\pi\lambda t}\, d\lambda$$

Furthermore, since t is a dummy variable in Eq. (4) and since $z(t) = V(t)$ in the theorem,

$$\mathscr{F}[z(t)] = \int_{-\infty}^{\infty} z(\lambda)e^{-j2\pi f\lambda}\, d\lambda = \int_{-\infty}^{\infty} V(\lambda)e^{j2\pi\lambda(-f)}\, d\lambda$$

Comparing these integrals then confirms that $\mathscr{F}[z(t)] = v(-f)$.

Although the statement of duality in Eq. (18) seems somewhat abstract, it turns out to be a handy way of generating new transform pairs without the labor of integration. The theorem works best when $v(t)$ is real and even so $z(t)$ will also be real and even, and $Z(f) = \mathscr{F}[z(t)] = v(-f) = v(f)$. The following example should clarify the procedure.

Example 2.2-3 Sinc pulse A rather strange but important time function in communication theory is the *sinc pulse* plotted in Fig. 2.2-5a and defined by

$$z(t) = A \,\text{sinc}\, 2Wt \tag{19a}$$

We'll obtain $Z(f)$ by applying duality to the transform pair

$$v(t) = B\Pi(t/\tau) \qquad V(f) = B\tau \,\text{sinc}\, f\tau$$

Rewriting Eq. (19a) as

$$z(t) = \left(\frac{A}{2W}\right)(2W) \,\text{sinc}\, t(2W)$$

brings out the fact that $z(t) = V(t)$ with $\tau = 2W$ and $B = A/2W$. Duality then says that $\mathscr{F}[z(t)] = v(-f) = B\Pi(-f/\tau) = (A/2W)\Pi(-f/2W)$ or

$$Z(f) = \frac{A}{2W}\, \Pi\!\left(\frac{f}{2W}\right) \tag{19b}$$

since the rectangle function has even symmetry.

Figure 2.2-5 A sinc pulse and its bandlimited spectrum.

The plot of $Z(f)$, given in Fig. 2.2-5b, shows that the spectrum of a sinc pulse equals zero for $|f| > W$. Thus, the spectrum has clearly defined width W, measured in terms of positive frequency, and we say that $Z(f)$ is *bandlimited*. Note, however, that the signal $z(t)$ goes on forever and is only asymptotically timelimited.

Exercise 2.2-3. Find the transform of $z(t) = B/[1 + (2\pi t)^2]$ by applying duality to the result of Exercise 2.2-1.

Transform Calculations

Except in the case of a very simple waveform, brute-force integration should be viewed as the method of last resort for transform calculations. Other, more practical methods are discussed here.

When the signal in question is defined mathematically, you should first consult a table of Fourier transforms to see if the calculation has been done before. Both columns of the table may be useful, in view of the duality theorem. A table of Laplace transforms also has some value, as mentioned in conjunction with Eq. (14).

Besides duality, there are several additional transform theorems covered in the next section. These theorems often help you decompose a complicated waveform into simpler parts whose transforms are known. Along this same line, you may find it expedient to *approximate* a waveform in terms of idealized signal models. Suppose $\tilde{z}(t)$ approximates $z(t)$ and the magnitude-squared error $|z(t) - \tilde{z}(t)|^2$ is a small quantity. If $Z(f) = \mathscr{F}[z(t)]$ and $\tilde{Z}(f) = \mathscr{F}[\tilde{z}(t)]$ then

$$\int_{-\infty}^{\infty} |Z(f) - \tilde{Z}(f)|^2 \, df = \int_{-\infty}^{\infty} |z(t) - \tilde{z}(t)|^2 \, dt \tag{20}$$

which follows from Rayleigh's theorem with $v(t) = z(t) - \tilde{z}(t)$. Thus, the integrated approximation error has the same value in the time and frequency domains.

The above methods are easily modified for the calculation of Fourier series coefficients. Specifically, let $v(t)$ be a periodic signal and let $z(t) = v(t)\Pi(t/T_0)$, a nonperiodic signal consisting of one period of $v(t)$. If you can obtain

$$Z(f) = \mathscr{F}[v(t)\Pi(t/T_0)] \tag{21a}$$

then, from Eq. (14), Sect. 2.1, the coefficients of $v(t)$ are given by

$$c_n = \frac{1}{T_0} Z(nf_0) \tag{21b}$$

This relationship facilitates the application of transform theorems to Fourier series calculations.

Finally, if the signal is defined in numerical form, its transform can be found via numerical calculations. For this purpose, the FFT computer algorithm is especially well suited. The algorithm and the supporting theory of discrete Fourier transforms is the subject of App. B.

2.3 TIME AND FREQUENCY RELATIONS

Rayleigh's theorem and the duality theorem in the previous section helped us draw useful conclusions about the frequency-domain representation of energy signals. Now we'll look at some of the many other theorems associated with Fourier transforms. They are included not just as manipulation exercises but for two very practical reasons. First, the theorems are invaluable when interpreting spectra, for they express relationships between time-domain and frequency-domain operations. Second, we can build up an extensive catalog of transform pairs by applying the theorems to known pairs—and such a catalog will be useful as we seek new signal models.

In stating the theorems, we indicate a signal and its transform (or spectrum) by lowercase and uppercase letters, as in $V(f) = \mathscr{F}[v(t)]$ and $v(t) = \mathscr{F}^{-1}[V(f)]$. This is also denoted more compactly by $v(t) \leftrightarrow V(f)$. Table T.1 at the back lists the theorems and transform pairs covered here, plus a few others.

Superposition

Superposition applies to the Fourier transform in the following sense. If a_1 and a_2 are constants and

$$v(t) = a_1 v_1(t) + a_2 v_2(t)$$

then

$$\mathscr{F}[v(t)] = a_1 \mathscr{F}[v_1(t)] + a_2 \mathscr{F}[v_2(t)]$$

Generalizing to sums with an arbitrary number of terms, we write the superposition (or linearity) theorem as

$$\sum_k a_k v_k(t) \leftrightarrow \sum_k a_k V_k(f) \tag{1}$$

This theorem simply states that linear combinations in the time domain become linear combinations in the frequency domain.

Although proof of the theorem is trivial, its importance cannot be overemphasized. From a practical viewpoint Eq. (1) greatly facilitates spectral analysis when the signal in question is a linear combination of functions whose individual spectra are known. From a theoretical viewpoint it underscores the applicability of the Fourier transform for the study of linear systems.

Time Delay and Scale Change

Given a time function $v(t)$, various other waveforms can be generated from it by modifying the argument of the function. Specifically, replacing t by $t - t_d$ produces the *time-delayed* signal $v(t - t_d)$. The delayed signal has the same shape as $v(t)$ but shifted t_d units to the right along the time axis. In the frequency domain, time delay causes an added linear phase with slope $-2\pi t_d$, so that

$$v(t - t_d) \leftrightarrow V(f)e^{-j2\pi f t_d} \tag{2}$$

If t_d is a negative quantity, the signal is *advanced* in time and the added phase has positive slope. The amplitude spectrum remains unchanged in either case, since $|V(f)e^{-j2\pi f t_d}| = |V(f)||e^{-j2\pi f t_d}| = |V(f)|$.

Proof of the time-delay theorem is accomplished by making the change of variable $\lambda = t - t_d$ in the transform integral. Thus, using $\omega = 2\pi f$ for compactness, we have

$$\mathscr{F}[v(t - t_d)] = \int_{-\infty}^{\infty} v(t - t_d)e^{-j\omega t}\, dt$$

$$= \int_{-\infty}^{\infty} v(\lambda)e^{-j\omega(\lambda + t_d)}\, d\lambda$$

$$= \left[\int_{-\infty}^{\infty} v(\lambda)e^{-j\omega\lambda}\, d\lambda\right]e^{-j\omega t_d}$$

The integral in brackets is just $V(f)$, so $\mathscr{F}[v(t - t_d)] = V(f)e^{-j\omega t_d}$.

Another time-axis operation is *scale change*, which produces a horizontally scaled image of $v(t)$ by replacing t with αt. The scaled signal $v(\alpha t)$ will be *expanded* if $|\alpha| < 1$ or *compressed* if $|\alpha| > 1$; a negative value of α yields time *reversal* as well as expansion or compression. These effects may occur during playback of recorded signals, for instance.

Scale change in the time domain becomes *reciprocal* scale change in the frequency domain, since

$$v(\alpha t) \leftrightarrow \frac{1}{|\alpha|} V\left(\frac{f}{\alpha}\right) \qquad \alpha \neq 0 \tag{3}$$

Hence, compressing a signal expands its spectrum, and vice versa. If $\alpha = -1$, then $v(-t) \leftrightarrow V(-f)$ so both the signal and spectrum are reversed.

We'll prove Eq. (3) for the case $\alpha < 0$ by writing $\alpha = -|\alpha|$ and making the change of variable $\lambda = -|\alpha|t$. Therefore, $t = \lambda/\alpha$, $dt = -d\lambda/|\alpha|$, and

$$\mathscr{F}[v(-|\alpha|t)] = \int_{-\infty}^{+\infty} v(-|\alpha|t)e^{-j\omega t}\, dt$$

$$= \frac{-1}{|\alpha|} \int_{+\infty}^{-\infty} v(\lambda)e^{-j\omega\lambda/\alpha}\, d\lambda$$

$$= \frac{1}{|\alpha|} \int_{-\infty}^{+\infty} v(\lambda)e^{-j2\pi(f/\alpha)\lambda}\, d\lambda$$

$$= \frac{1}{|\alpha|} V\left(\frac{f}{\alpha}\right)$$

Observe how this proof uses the general relationship

$$\int_a^b x(\lambda)\, d(-\lambda) = -\int_{-a}^{-b} x(\lambda)\, d\lambda = \int_{-b}^{-a} x(\lambda)\, d\lambda$$

Hereafter, the intermediate step will be omitted when this type of manipulation occurs.

Example 2.3-1 The signal in Fig. 2.3-1a has been constructed using two rectangular pulses $v(t) = A\Pi(t/\tau)$ such that

$$z_a(t) = v(t - t_d) + (-1)v[t - (t_d + T)]$$

Application of the superposition and time-delay theorems yields

$$Z_a(f) = V(f)e^{-j2\pi f t_d} + (-1)V(f)e^{-j2\pi f(t_d + T)}$$
$$= V(f)[e^{-j2\pi f t_d} - e^{-j2\pi f(t_d + T)}]$$

where $V(f) = A\tau \operatorname{sinc} f\tau$.

The bracketed term in $Z_a(f)$ is a particular case of the expression $e^{j2\theta_1} \pm e^{j2\theta_2}$ which often turns up in Fourier analysis. A more informative version of this expression is obtained by factoring and using Euler's theorem, as follows:

$$e^{j2\theta_1} \pm e^{j2\theta_2} = [e^{j(\theta_1 - \theta_2)} \pm e^{-j(\theta_1 - \theta_2)}]e^{j(\theta_1 + \theta_2)}$$
$$= \begin{cases} 2\cos(\theta_1 - \theta_2)\, e^{j(\theta_1 + \theta_2)} \\ j2\sin(\theta_1 - \theta_2)\, e^{j(\theta_1 + \theta_2)} \end{cases} \tag{4}$$

The upper result in Eq. (4) corresponds to the upper $(+)$ sign and the lower result to the lower $(-)$ sign.

In the problem at hand we have $\theta_1 = -\pi f t_d$ and $\theta_2 = -\pi f(t_d + T)$, so $\theta_1 - \theta_2 = -\pi f T$ and $\theta_1 + \theta_2 = -2\pi f t_0$ where $t_0 = t_d + T/2$ as marked in Fig. 2.3-1a. Therefore, after substituting for $V(f)$, we obtain

$$Z_a(f) = (A\tau \operatorname{sinc} f\tau)(j2\sin \pi f T\, e^{-j2\pi f t_0})$$

Note that $Z_a(0) = 0$, agreeing with the fact that $z_a(t)$ has zero net area.

Figure 2.3-1 Signals in Example 2.3-1.

If $t_0 = 0$ and $T = \tau$, $z_a(t)$ degenerates to the waveform in Fig. 2.3-1b where

$$z_b(t) = A\Pi\left(\frac{t + \tau/2}{\tau}\right) - A\Pi\left(\frac{t - \tau/2}{\tau}\right)$$

The spectrum then becomes

$$Z_b(f) = (A\tau \text{ sinc } f\tau)(j2 \sin \pi f\tau)$$
$$= (j2\pi f\tau)A\tau \text{ sinc}^2 \pi f\tau$$

This spectrum is purely imaginary because $z_b(t)$ has odd symmetry.

Exercise 2.3-1 Let $v(t)$ be a *real* but otherwise arbitrary energy signal. Show that if

$$z(t) = a_1 v(t) + a_2 v(-t) \tag{5a}$$

then

$$Z(f) = (a_1 + a_2)V_e(f) + j(a_1 - a_2)V_o(f) \tag{5b}$$

where $V_e(f)$ and $V_o(f)$ are the real and imaginary parts of $V(f)$.

Frequency Translation and Modulation

Besides generating new transform pairs, duality can be used to generate transform theorems. In particular, a dual of the time-delay theorem is

$$v(t)e^{j\omega_c t} \leftrightarrow V(f - f_c) \qquad \omega_c = 2\pi f_c \tag{6}$$

We designate this as *frequency translation* or *complex modulation*, since multiplying a time function by $e^{j\omega_c t}$ causes its spectrum to be translated in frequency by $+f_c$.

To see the effects of frequency translation, let $v(t)$ have the bandlimited spectrum of Fig. 2.3-2a, where the amplitude and phase are plotted on the same axes

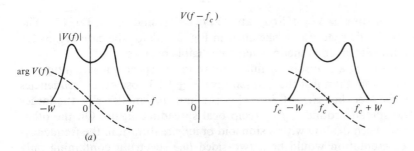

(a)

Figure 2.3-2 Frequency translation of a bandlimited spectrum.

using solid and broken lines, respectively. Also let $f_c > W$. Inspection of the translated spectrum $V(f - f_c)$ in Fig. 2.3-2b reveals the following:

1. The significant components are concentrated around the frequency f_c.
2. Though $V(f)$ was bandlimited in W, $V(f - f_c)$ has a spectral width of $2W$. Translation has therefore doubled spectral width. Stated another way, the negative-frequency portion of $V(f)$ now appears at positive frequencies.
3. $V(f - f_c)$ is not hermitian but does have symmetry with respect to translated origin at $f = f_c$.

These considerations may appear somewhat academic in view of the fact that $v(t)e^{j\omega_c t}$ is not a real time function and cannot occur as a communication signal. However, signals of the form $v(t) \cos(\omega_c t + \phi)$ are common—in fact, they are the basis of carrier modulation—and by direct extension of Eq. (6) we have the following *modulation theorem*:

$$v(t) \cos(\omega_c t + \phi) \leftrightarrow \frac{e^{j\phi}}{2} V(f - f_c) + \frac{e^{-j\phi}}{2} V(f + f_c) \tag{7}$$

In words, multiplying a signal by a sinusoid translates its spectrum *up and down* in frequency by f_c. All the comments about complex modulation also apply here. In addition, the resulting spectrum is hermitian, which it must be if $v(t) \cos(\omega_c t + \phi)$ is a real function of time. The theorem is easily proved with the aid of Euler's theorem and Eq. (6).

Example 2.3-2 RF pulse Consider the finite-duration sinusoid of Fig. 2.3-3a, sometimes referred to as an RF pulse when f_c falls in the radio-frequency band. Since

$$z(t) = A\Pi\left(\frac{t}{\tau}\right) \cos \omega_c t$$

we have immediately

$$Z(f) = \frac{A\tau}{2} \text{ sinc } (f - f_c)\tau + \frac{A\tau}{2} \text{ sinc } (f + f_c)\tau.$$

obtained by setting $v(t) = A\Pi(t/\tau)$ and $V(f) = A\tau$ sinc $f\tau$ in Eq. (7). The resulting amplitude spectrum is sketched in Fig. 2.3-3b for the case of $f_c \gg 1/\tau$ so the two translated sinc functions have negligible overlap.

Because this is a sinusoid of finite duration, its spectrum is continuous and contains more than just the frequencies $f = \pm f_c$. Those other frequencies stem from the fact that $z(t) = 0$ for $|t| > \tau/2$, and the smaller τ is, the larger the spectral spread around $\pm f_c$—reciprocal spreading, again. On the other hand, had we been dealing with a sinusoid of *infinite* duration, the frequency-domain representation would be a two-sided *line* spectrum containing only the discrete frequencies $\pm f_c$.

Figure 2.3-3 (a) RF pulse; (b) amplitude spectrum.

Differentiation and Integration

Certain processing techniques involve differentiating or integrating a signal. The frequency-domain effects of these operations are indicated in the theorems below. A word of caution, however: The theorems should not be applied before checking to make sure that the differentiated or integrated signal is Fourier-transformable; the fact that $v(t)$ has finite energy is not a guarantee that the same holds true for its derivative or integral.

To derive the differentiation theorem, we replace $v(t)$ by the inverse transform integral and interchange the order of operations, as follows:

$$\frac{d}{dt} v(t) = \frac{d}{dt} \left[\int_{-\infty}^{\infty} V(f) e^{j2\pi ft} \, df \right]$$

$$= \int_{-\infty}^{\infty} V(f) \left(\frac{d}{dt} e^{j2\pi ft} \right) df$$

$$= \int_{-\infty}^{\infty} [j2\pi f V(f)] e^{j2\pi ft} \, df$$

Referring back to the definition of the inverse transform reveals that the bracketed term must be $\mathscr{F}[dv(t)/dt]$, so

$$\frac{d}{dt} v(t) \leftrightarrow j2\pi f V(f)$$

and by iteration we get

$$\frac{d^n}{dt^n} v(t) \leftrightarrow (j2\pi f)^n V(f) \qquad (8)$$

which is the *differentiation theorem.*

Now suppose we generate another function from $v(t)$ by integrating it over all past time. We write this operation as $\int_{-\infty}^{t} v(\lambda)\, d\lambda$, where the dummy variable λ is needed to avoid confusion with the independent variable t in the upper limit. The *integration theorem* says that if

$$V(0) = \int_{-\infty}^{\infty} v(\lambda)\, d\lambda = 0 \qquad (9a)$$

then

$$\int_{-\infty}^{t} v(\lambda)\, d\lambda \leftrightarrow \frac{1}{j2\pi f}\, V(f) \qquad (9b)$$

The zero net area condition in Eq. (9a) ensures that the integrated signal goes to zero as $t \to \infty$. (We'll relax this condition in Sect. 2.4.)

To interpret these theorems, we see that differentiation enhances the high-frequency components of a signal, since $|j2\pi f V(f)| > |V(f)|$ for $|f| > 1/2\pi$. Conversely, integration suppresses the high-frequency components. Spectral interpretation thus agrees with the time-domain observation that differentiation accentuates time variations while integration smoothes them out.

Example 2.3-3 Triangular pulse The waveform $z_b(t)$ in Fig. 2.3-1*b* has zero net area, and integration produces a *triangular* pulse shape. Specifically, let

WIDTH $= 2\tau$

$$w(t) = \frac{1}{\tau} \int_{-\infty}^{t} z_b(\lambda)\, d\lambda = \begin{cases} A\left(1 - \dfrac{|t|}{\tau}\right) & |t| < \tau \\ 0 & |t| > \tau \end{cases}$$

which is sketched in Fig. 2.3-4*a*. Applying the integration theorem to $Z_b(f)$ from Example 2.3-1, we obtain

$$W(f) = \frac{1}{\tau}\frac{1}{j2\pi f}\, Z_b(f) = A\tau \operatorname{sinc}^2 f\tau$$

as shown in Fig. 2.3-4*b*. A comparison of this spectrum with Fig. 2.2-2 reveals that the triangular pulse has less high-frequency content than a rectangular pulse with amplitude A and duration τ, although they both have area $A\tau$.

Figure 2.3-4 A triangular pulse and its spectrum.

The difference is traced to the fact that the triangular pulse is spread over 2τ seconds and does not have the sharp, stepwise time variations of the rectangular shape.

This transform pair can be written more compactly by defining the *triangular function*

$$\Lambda\left(\frac{t}{\tau}\right) \triangleq \begin{cases} 1 - \dfrac{|t|}{\tau} & |t| < \tau \\ 0 & |t| > \tau \end{cases} \tag{10}$$

Then $w(t) = A\Lambda(t/\tau)$ and

$$A\Lambda\left(\frac{t}{\tau}\right) \leftrightarrow A\tau \, \text{sinc}^2 \, f\tau \tag{11}$$

It so happens that triangular functions can be generated from rectangular functions by another mathematical operation, namely, convolution. And convolution happens to be the next item on our agenda.

Exercise 2.3-2 A dual of the differentiation theorem is

$$t^n v(t) \leftrightarrow \frac{1}{(-j2\pi)^n} \frac{d^n}{df^n} V(f) \tag{12}$$

Derive this relationship for $n = 1$ by differentiating the transform integral $\mathscr{F}[v(t)]$ with respect to f.

Convolution and Multiplication

The mathematical operation known as convolution ranks high among the tools used by communication engineers. Its applications include system analysis and probability theory as well as transform calculations. Here we're concerned with convolution in the time and frequency domains.

The convolution of two functions of the same variable, say $v(t)$ and $w(t)$, is defined by

$$v * w(t) \triangleq \int_{-\infty}^{\infty} v(\lambda) w(t - \lambda) \, d\lambda \tag{13}$$

The notation $v * w(t)$ merely stands for the operation on the right-hand side of Eq. (13) and the asterisk ($*$) has nothing to do with complex conjugation. Equation (13) is the *convolution integral*, often denoted $v * w$ when the independent variable is unambiguous. At other times the notation $[v(t)] * [w(t)]$ is necessary for clarity. Note carefully that the independent variable here is t, the same as the independent variable of the functions being convolved; the integration is always performed with respect to a dummy variable (such as λ) and t is a constant insofar as the integration is concerned.

Calculating $v * w(t)$ is no more difficult than ordinary integration when the two functions are continuous for all t. Often, however, one or both of the functions is defined in a piecewise fashion, and the graphical interpretation of convolution becomes especially helpful.

By way of illustration, take the functions in Fig. 2.3-5a where

$$v(t) = Ae^{-t} \qquad 0 < t < \infty$$

$$w(t) = t/T \qquad 0 < t < T$$

For the integrand in Eq. (13), $v(\lambda)$ has the same shape as $v(t)$ and

$$w(t - \lambda) = \frac{t - \lambda}{T} \qquad 0 < t - \lambda < T$$

But obtaining the picture of $w(t - \lambda)$ as a function of λ requires two steps: first, we reverse $w(t)$ in time and replace t with λ to get $w(-\lambda)$; second, we shift $w(-\lambda)$

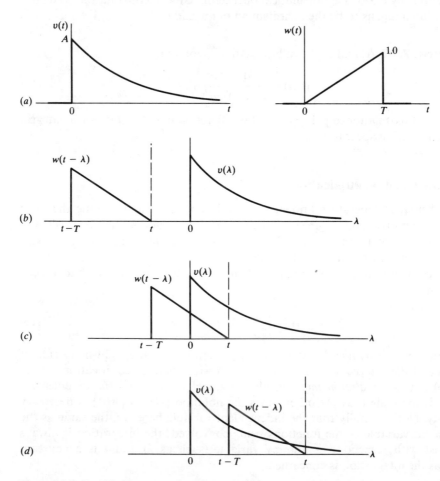

Figure 2.3-5 Graphical interpretation of convolution.

to the right by t units to get $w[-(\lambda - t)] = w(t - \lambda)$ for a given value of t. Figure 2.3-5b shows $v(\lambda)$ and $w(t - \lambda)$ with $t < 0$. The value of t always equals the distance from the origin of $v(\lambda)$ to the shifted origin of $w(-\lambda)$ indicated by the dashed line.

As $v * w(t)$ is evaluated for $-\infty < t < \infty$, $w(t - \lambda)$ slides from left to right with respect to $v(\lambda)$, so the convolution integrand changes with t. Specifically, we see in Fig. 2.3-5b that the functions don't overlap when $t < 0$; hence, the integrand equals zero and

$$v * w(t) = 0 \qquad t < 0$$

When $0 < t < T$ as in Fig. 2.3-5c, the functions overlap for $0 < \lambda < t$, so t becomes the upper limit of integration and

$$v * w(t) = \int_0^t A e^{-\lambda} \left(\frac{t - \lambda}{T} \right) d\lambda$$

$$= \frac{A}{T} (t - 1 + e^{-t}) \qquad 0 < t < T$$

Finally, when $t > T$ as in Fig. 2.3-5d, the functions overlap for $t - T < \lambda < t$ and

$$v * w(t) = \int_{t-T}^t A e^{-\lambda} \left(\frac{t - \lambda}{T} \right) d\lambda$$

$$= \frac{A}{T} (T - 1 + e^{-T}) e^{-(t-T)} \qquad t > T$$

The complete result plotted in Fig. 2.3-6 shows that convolution is a *smoothing* operation in the sense that $v * w(t)$ is "smoother" than either of the original functions.

Further study of Fig. 2.3-5 should reveal that you get the same result by reversing v and sliding it past w, so convolution is *commutative*. This property is listed below along with the *associative* and *distributive* properties.

$$v * w = w * v \qquad (14a)$$

$$v * (w * z) = (v * w) * z \qquad (14b)$$

$$v * (w + z) = (v * w) + (v * z) \qquad (14c)$$

All of these can be derived from Eq. (13).

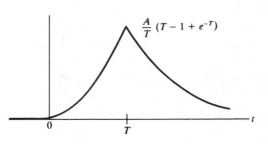

$$\frac{A}{T} (T - 1 + e^{-T})$$

0   T t

Figure 2.3-6 Result of the convolution in Fig. 2.3-5.

Having defined and examined the convolution operation, we now list the two *convolution theorems*:

$$v * w(t) \leftrightarrow V(f)W(f) \tag{15}$$

$$v(t)w(t) \leftrightarrow V * W(f) \tag{16}$$

These theorems state that convolution in the time domain becomes multiplication in the frequency domain, while multiplication in the time domain becomes convolution in the frequency domain. Both of these relationships are important for future work.

The proof of Eq. (15) uses the time-delay theorem, as follows:

$$
\begin{aligned}
\mathscr{F}[v * w(t)] &= \int_{-\infty}^{\infty} \left[\int_{-\infty}^{\infty} v(\lambda)w(t - \lambda)\, d\lambda \right] e^{-j\omega t}\, dt \\
&= \int_{-\infty}^{\infty} v(\lambda) \left[\int_{-\infty}^{\infty} w(t - \lambda)e^{-j\omega t}\, dt \right] d\lambda \\
&= \int_{-\infty}^{\infty} v(\lambda)[W(f)e^{-j\omega\lambda}]\, d\lambda \\
&= \left[\int_{-\infty}^{\infty} v(\lambda)e^{-j\omega\lambda}\, d\lambda \right] W(f) = V(f)W(f)
\end{aligned}
$$

Equation (16) can be proved by writing out the transform of $v(t)w(t)$ and replacing $w(t)$ by the inversion integral $\mathscr{F}^{-1}[W(f)]$.

Example 2.3-4 Trapezoidal pulse To illustrate the convolution theorem — and obtain yet another transform pair — let's convolve the rectangular pulses in Fig. 2.3-7a. This is a relatively simple task using the graphical interpretation and symmetry considerations. If $\tau_1 > \tau_2$, the problem breaks up into three cases: no overlap, partial overlap, and full overlap. The result is the *trapezoidal* pulse shown in Fig. 2.3-7b, whose transform will be the product $V(f)W(f) = (A_1\tau_1 \ \text{sinc}\ f\tau_1)(A_2\tau_2 \ \text{sinc}\ f\tau_2)$.

Figure 2.3-7 Convolution of rectangular pulses.

Now let $\tau_1 = \tau_2 = \tau$ so the trapezoidal shape reduces to the *triangular* pulse back in Fig. 2.3-4a with $A = A_1 A_2 \tau$. Correspondingly, the spectrum becomes $(A_1\tau \operatorname{sinc} f\tau)(A_2\tau \operatorname{sinc} f\tau) = A\tau \operatorname{sinc}^2 f\tau$, which agrees with our prior result.

Exercise 2.3-3 Let $v(t) = A \operatorname{sinc} 2Wt$, whose spectrum is bandlimited in W. Use Eq. (16) with $w(t) = v(t)$ to show that the spectrum of $v^2(t)$ will be bandlimited in $2W$.

2.4 IMPULSES AND TRANSFORMS IN THE LIMIT

So far we've maintained a distinction between two spectral classifications: line spectra that represent periodic power signals, and continuous spectra that represent nonperiodic energy signals. But the distinction poses something of a quandary when you encounter a signal consisting of periodic *and* nonperiodic terms. We'll resolve this quandary here by allowing *impulses* in the frequency domain for the representation of discrete frequency components. The underlying notion of *transforms in the limit* also permits the spectral representation of time-domain impulses and other signals whose transforms don't exist in the usual sense.

Properties of the Unit Impulse

The *unit impulse* or *Dirac delta function* $\delta(t)$ is not a function in the strict mathematical sense. Rather, it belongs to a special class known as *generalized functions* or *distributions* whose definitions are stated by assignment rules. In particular, the properties of $\delta(t)$ will be derived from the defining relationship

$$\int_{t_1}^{t_2} v(t)\, \delta(t)\, dt = \begin{cases} v(0) & t_1 < 0 < t_2 \\ 0 & \text{otherwise} \end{cases} \tag{1}$$

where $v(t)$ is any ordinary function that's continuous at $t = 0$. This rule assigns a number — either $v(0)$ or 0 — to the expression on the left-hand side. Equation (1) and all subsequent expressions will also apply to the frequency-domain impulse $\delta(f)$ by replacing t with f.

If $v(t) = 1$ in Eq. (1), it then follows that

$$\int_{-\infty}^{\infty} \delta(t)\, dt = \int_{-\epsilon}^{\epsilon} \delta(t)\, dt = 1 \tag{2}$$

with ϵ being arbitrarily small. We interpret Eq. (2) by saying that $\delta(t)$ has *unit area* concentrated at the discrete point $t = 0$ and no net area elsewhere. Carrying this argument further suggests that

$$\delta(t) = 0 \qquad t \neq 0 \tag{3}$$

Equations (2) and (3) are the more familiar definitions of the impulse, and lead to the common graphical representation. For instance, the picture of $A\, \delta(t - t_d)$ is

Figure 2.4-1 Graphical representation of $A\delta(t - t_d)$.

shown in Fig. 2.4-1, where the letter A next to the arrowhead means that $A\, \delta(t - t_d)$ has unit area or weight A located at $t = t_d$.

Although an impulse does not exist physically, there are numerous conventional functions that have all the properties of $\delta(t)$ in the limit as some parameter ϵ goes to zero. In particular, if the function $\delta_\epsilon(t)$ is such that

$$\lim_{\epsilon \to 0} \int_{-\infty}^{\infty} v(t)\, \delta_\epsilon(t)\, dt = v(0) \tag{4a}$$

then we say that

$$\lim_{\epsilon \to 0} \delta_\epsilon(t) = \delta(t) \tag{4b}$$

Two functions satisfying Eq. (4a) are

$$\delta_\epsilon(t) = \frac{1}{\epsilon}\, \Pi\!\left(\frac{t}{\epsilon}\right) \tag{5}$$

$$\delta_\epsilon(t) = \frac{1}{\epsilon}\, \text{sinc}\, \frac{t}{\epsilon} \tag{6}$$

which are protted in Fig. 2.4-2. You can easily show that Eq. (5) satisfies Eq. (4a) by expanding $v(t)$ in a Maclaurin series prior to integrating. An argument for Eq. (6) will be given shortly when we consider impulses and transforms.

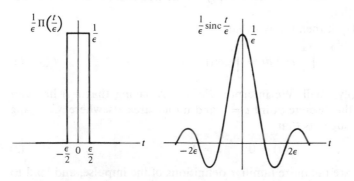

Figure 2.4-2 Two functions that become impulses as $\epsilon \to 0$.

By definition, the impulse has no mathematical or physical meaning unless it appears under the operation of integration. Two of the most significant integration properties are

$$v(t) * \delta(t - t_d) = v(t - t_d) \tag{7}$$

$$\int_{-\infty}^{\infty} v(t)\, \delta(t - t_d)\, dt = v(t_d) \tag{8}$$

both of which can be derived from Eq. (1). Equation (7) is a *replication* operation, since convolving $v(t)$ with $\delta(t - t_d)$ reproduces the entire function $v(t)$ delayed by t_d. In contrast, Eq. (8) is a *sampling* operation that picks out or samples the value of $v(t)$ at $t = t_d$—the point where $\delta(t - t_d)$ is "located."

Given the stipulation that any impulse expression must eventually be integrated, you can use certain nonintegral relations to simplify expressions *before* integrating. Two such relations are

$$v(t)\, \delta(t - t_d) = v(t_d)\, \delta(t - t_d) \tag{9a}$$

$$\delta(\alpha t) = \frac{1}{|\alpha|}\, \delta(t) \qquad \alpha \neq 0 \tag{9b}$$

which are justified by integrating both sides over $-\infty < t < \infty$. The product relation in Eq. (9a) simply restates the sampling property. The scale-change relation in Eq. (9b) says that, relative to the independent variable t, $\delta(\alpha t)$ acts like $\delta(t)/|\alpha|$. Setting $\alpha = -1$ then brings out the even-symmetry property $\delta(-t) = \delta(t)$.

Exercise 2.4-1 Evaluate or simplify each of the following expressions with $v(t) = (t - 3)^2$:

(a) $\displaystyle\int_{-\infty}^{\infty} v(t)\, \delta(t + 4)\, dt$;   (b) $v(t) * \delta(t + 4)$

(c) $v(t)\, \delta(t + 4)$;   (d) $v(t) * \delta(-t/4)$.

Impulses in Frequency

Impulses in frequency represent phasors or constants. In particular, let $v(t) = A$ be a constant for all time. Although this signal has infinite energy, we can obtain its transform in a limiting sense by considering that

$$v(t) = \lim_{W \to 0} A \operatorname{sinc} 2Wt = A \tag{10a}$$

Now we already have the transform pair $A \operatorname{sinc} 2Wt \leftrightarrow (A/2W)\Pi(f/2W)$, so

$$\mathscr{F}[v(t)] = \lim_{W \to 0} \frac{A}{2W} \Pi\left(\frac{f}{2W}\right) = A\, \delta(f) \tag{10b}$$

which follows from Eq. (5) with $\epsilon = 2W$ and $t = f$. Therefore,

$$A \leftrightarrow A \, \delta(f) \tag{11}$$

and the spectrum of a constant in the time domain is an impulse in the frequency domain at $f = 0$.

This result agrees with intuition in that a constant signal has no time variation and its spectral content ought to be confined to $f = 0$. The impulsive form results simply because we use integration to return to the time domain, via the inverse transform, and an impulse is required to concentrate the nonzero area at a discrete point in frequency. Checking this argument mathematically using Eq. (1) gives

$$\mathscr{F}^{-1}[A \, \delta(f)] = \int_{-\infty}^{\infty} A \, \delta(f) e^{j2\pi ft} \, dt = Ae^{j2\pi ft}\bigg|_{f=0} = A$$

which justifies Eq. (11) for our purposes. Note that the impulse has been integrated to obtain a physical quantity, namely the signal $v(t) = A$.

As an alternative to the above procedure, we could have begun with a rectangular pulse, $A\Pi(t/\tau)$, and let $\tau \to \infty$ to get a constant for all time. Then, since $\mathscr{F}[A\Pi(t/\tau)] = A\tau \, \mathrm{sinc} \, f\tau$, agreement with Eq. (11) requires that

$$\lim_{\tau \to \infty} A\tau \, \mathrm{sinc} \, f\tau = A \, \delta(f)$$

And this supports the earlier assertion in Eq. (6) that a sinc function becomes an impulse under appropriate limiting conditions.

To generalize Eq. (11), direct application of the frequency-translation and modulation theorems yields

$$Ae^{j\omega_c t} \leftrightarrow A \, \delta(f - f_c) \tag{12}$$

$$A \cos \left(\omega_c t + \phi\right) \leftrightarrow \frac{Ae^{j\phi}}{2} \delta(f - f_c) + \frac{Ae^{-j\phi}}{2} \delta(f + f_c) \tag{13}$$

Thus, the spectrum of a single phasor is an impulse at $f = f_c$ while the spectrum of a sinusoid has two impulses, shown in Fig. 2.4-3. Going even further in this direction, if $v(t)$ is an arbitrary periodic signal whose exponential Fourier series is

$$v(t) = \sum_{n=-\infty}^{\infty} c(nf_0)e^{j2\pi nf_0 t} \tag{14a}$$

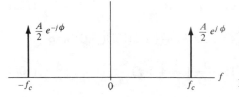

Figure 2.4-3 Spectrum of $A \cos \left(\omega_c t + \phi\right)$.

then its Fourier transform is

$$V(f) = \sum_{n=-\infty}^{\infty} c(nf_0)\, \delta(f - nf_0) \tag{14b}$$

where superposition allows us to transform the sum term by term.

By now it should be obvious from Eqs. (11)–(14) that any two-sided line spectrum can be converted to a "continuous" spectrum using this rule: convert the spectral lines to impulses whose weights equal the line heights. The phase portion of the line spectrum is absorbed by letting the impulse weights be complex numbers. Hence, with the aid of transforms in the limit, we can represent both periodic and nonperiodic signals by continuous spectra. That strange beast the impulse function thereby emerges as a key to unifying spectral analysis.

But you may well ask: What's the difference between the line spectrum and the "continuous" spectrum of a period signal? Obviously there can be no physical difference; the difference lies in the mathematical conventions. To return to the time domain from the line spectrum, we sum the phasors which the lines represent. To return to the time domain from the continuous spectrum, we integrate the impulses to get phasors.

Example 2.4-1 The sinusoidal waveform in Fig. 2.4-4a has constant frequency f_c except for the interval $-1/f_c < t < 1/f_c$ where the frequency jumps to $2f_c$. Such a signal might be produced by the process of *frequency modulation*, to be discussed in Chap. 7. Our interest here is the spectrum, which consists of both impulsive and nonimpulsive components.

For analysis purposes, we'll let $\tau = 2/f_c$ and decompose $v(t)$ into a sum of three terms as follows:

$$v(t) = A\cos\omega_c t - A\Pi(t/\tau)\cos\omega_c t + A\Pi(t/\tau)\cos 2\omega_c t$$

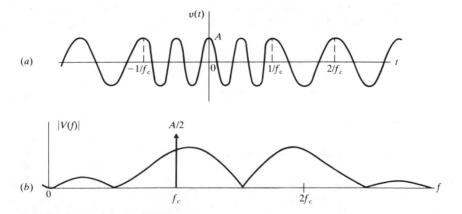

Figure 2.4-4 Waveform and amplitude spectrum in Example 2.4-1.

The first two terms represent a cosine wave with a "hole" to make room for an RF pulse at frequency $2f_c$ represented by the third term. Transforming $v(t)$ term by term then yields

$$V(f) = \frac{A}{2} \left[\delta(f - f_c) + \delta(f + f_c) \right]$$

$$- \frac{A\tau}{2} \left[\text{sinc } (f - f_c)\tau + \text{sinc } (f + f_c)\tau \right]$$

$$+ \frac{A\tau}{2} \left[\text{sinc } (f - 2f_c)\tau + \text{sinc } (f + 2f_c)\tau \right]$$

where we have drawn upon Eq. (13) and the results of Example 2.3-2. The amplitude spectrum is sketched in Fig. 2.4-4b, omitting the negative-frequency portion. Note that $|V(f)|$ is not symmetric about $f = f_c$ because the nonimpulsive component must include the term at $2f_c$.

Step and Signum Functions

We've seen that a constant for all time becomes a DC impulse in the frequency domain. Now consider the *unit step function* in Fig. 2.4-5a which steps from "off" to "on" at $t = 0$ and is defined as

$$u(t) \triangleq \begin{cases} 1 & t > 0 \\ 0 & t < 0 \end{cases} \tag{15}$$

This function has several uses in Fourier theory, especially with regard to *causal* signals since any time function multiplied by $u(t)$ will equal zero for $t < 0$. However, the lack of symmetry creates a problem when we seek the transform in the limit, because limiting operations are equivalent to *contour integrations* and must be performed in a symmetrical fashion — as we did in Eq. (10). To get

Figure 2.4-5 (*a*) Unit step function; (*b*) signum function.

around this problem, we'll start with the *signum function* (also called the *sign function*) plotted in Fig. 2.4-5b and defined as

$$\text{sgn } t \triangleq \begin{cases} +1 & t > 0 \\ -1 & t < 0 \end{cases} \tag{16}$$

which clearly has odd symmetry.

The signum function is a limiting case of the energy signal $z(t)$ in Fig. 2.4-6 where $v(t) = e^{-bt}u(t)$ and

$$z(t) = v(t) - v(-t) = \begin{cases} +e^{-bt} & t > 0 \\ -e^{+bt} & t < 0 \end{cases}$$

so that $z(t) \to \text{sgn } t$ if $b \to 0$. Combining the results of Example 2.2-3 and Exercise 2.3-1 yields

$$\mathscr{F}[z(t)] = Z(f) = j2V_o(f) = \frac{-j4\pi f}{b^2 + (2\pi f)^2}$$

Therefore,

$$\mathscr{F}[\text{sgn } t] = \lim_{b \to 0} Z(f) = \frac{-j}{\pi f}$$

and we have the transform pair

$$\text{sgn } t \leftrightarrow \frac{1}{j\pi f} \tag{17}$$

We then observe from Fig. 2.4-5 that the step and signum functions are related by

$$u(t) = \tfrac{1}{2}(\text{sgn } t + 1) = \tfrac{1}{2} \text{sgn } t + \tfrac{1}{2}$$

Hence,

$$u(t) \leftrightarrow \frac{1}{j2\pi f} + \frac{1}{2}\,\delta(f) \tag{18}$$

since $\mathscr{F}[1/2] = \tfrac{1}{2}\delta(f)$

Note that the spectrum of the signum function does not include a DC impulse. This agrees with the fact that sgn t is an odd function with *zero average*

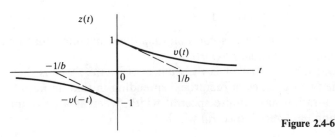

Figure 2.4-6

value when averaged over all time, as in Eq. (9), Sect. 2.1. In contrast, the average value of the unit step is $\langle u(t) \rangle = 1/2$ so its spectrum includes $\frac{1}{2}\delta(f)$—just as the transform of a periodic signal with average value $c(0)$ would include the DC term $c(0)\, \delta(f)$.

An impulsive DC term also appears in the *integration theorem* when the signal being integrated has nonzero net area. We derive this property by convolving $u(t)$ with an arbitrary energy signal $v(t)$ to get

$$v * u(t) = \int_{-\infty}^{\infty} v(\lambda)u(t - \lambda)\, d\lambda$$

$$= \int_{-\infty}^{t} v(\lambda)\, d\lambda \tag{19}$$

since $u(t - \lambda) = 0$ for $\lambda > t$. But from the convolution theorem and Eq. (18)

$$\mathscr{F}[v * u(t)] = V(f)\left[\frac{1}{j2\pi f} + \frac{1}{2}\,\delta(f)\right]$$

so

$$\int_{-\infty}^{t} v(\lambda)\, d\lambda \leftrightarrow \frac{1}{j2\pi f}\, V(f) + \frac{1}{2}\, V(0)\, \delta(f) \tag{20}$$

where we have used $V(f)\,\delta(f) = V(0)\,\delta(f)$. Equation (20) reduces to our previous statement of the integration theorem when $V(0) = 0$.

Exercise 2.4-2 Apply the modulation theorem to obtain the spectrum of the *causal sinusoid* $v(t) = Au(t) \cos \omega_c t$.

Impulses in Time

Although the time-domain impulse $\delta(t)$ seems a trifle farfetched as a signal model, we'll run into meaningful practical applications in subsequent chapters. Equally important is the value of $\delta(t)$ as an analytic device. To derive its transform, we let $\tau \to 0$ in the known pair

$$\frac{A}{\tau}\,\Pi\left(\frac{t}{\tau}\right) \leftrightarrow A \text{ sinc } f\tau$$

which becomes

$$A\,\delta(t) \leftrightarrow A \tag{21}$$

Hence, the transform of a time impulse has *constant amplitude*, meaning that its spectrum contains *all frequencies in equal proportion*.

You may have noticed that $A\,\delta(t) \leftrightarrow A$ is the dual of $A \leftrightarrow A\,\delta(f)$. This dual relationship embraces the two extremes of reciprocal spreading in that an impulsive signal with "zero" duration has infinite spectral width, whereas a constant signal with infinite duration has "zero" spectral width.

Applying the time-delay theorem to Eq. (21) yields the more general pair

$$A\,\delta(t - t_d) \leftrightarrow Ae^{-j2\pi f t_d} \tag{22}$$

It's a simple matter to confirm the direct transform relationship $\mathscr{F}[A\,\delta(t - t_d)] = Ae^{-j2\pi d t_d}$; consistency therefore requires that $\mathscr{F}^{-1}[Ae^{-j2\pi f t_d}] = A\,\delta(t - t_d)$, which leads to a significant integral expression for the unit impulse. Specifically, since

$$\mathscr{F}^{-1}[e^{-j2\pi f t_d}] = \int_{-\infty}^{\infty} e^{-j2\pi f t_d} e^{j2\pi f t}\,df$$

we conclude that

$$\int_{-\infty}^{\infty} e^{j2\pi f(t - t_d)}\,df = \delta(t - t_d) \tag{23}$$

Thus, the integral on the left side may be evaluated in the limiting form of the unit impulse — a result we'll put immediately to work in a proof of the Fourier integral theorem.

Let $v(t)$ be a continuous time function with a well-defined transform $V(f) = \mathscr{F}[v(t)]$. Our task is to show that the inverse transform does, indeed, equal $v(t)$. From the definitions of the direct and inverse transforms we can write

$$\mathscr{F}^{-1}[V(f)] = \int_{-\infty}^{\infty} \left[\int_{-\infty}^{\infty} v(\lambda)e^{-j2\pi f \lambda}\,d\lambda \right] e^{j2\pi f t}\,df$$

$$= \int_{-\infty}^{\infty} v(\lambda) \left[\int_{-\infty}^{\infty} e^{j2\pi(t - \lambda)f}\,df \right] d\lambda$$

But the bracketed integral equals $\delta(t - \lambda)$, from Eq. (23), so

$$\mathscr{F}^{-1}[V(f)] = \int_{-\infty}^{\infty} v(\lambda)\,\delta(t - \lambda)\,d\lambda = v(t) * \delta(t) \tag{24}$$

Therefore $\mathscr{F}^{-1}[V(f)]$ equals $v(t)$ in the same sense that $v(t) * \delta(t) = v(t)$. A more rigorous proof, including Gibbs' phenomenon at points of discontinuity, is given by Papoulis (1962, Chap. 2).

Lastly, we relate the unit impulse to the *unit step* by means of the integral

$$\int_{-\infty}^{t} \delta(\lambda - t_d)\,d\lambda = \begin{cases} 1 & t > t_d \\ 0 & t < t_d \end{cases}$$

$$= u(t - t_d) \tag{25}$$

Differentiating both sides then yields

$$\delta(t - t_d) = \frac{d}{dt}\,u(t - t_d) \tag{26}$$

which provides another interpretation of the impulse in terms of the derivative of a step discontinuity.

Equations (26) and (22), coupled with the differentiation theorem, expedite certain transform calculations and help us predict a signal's high-frequency spectral rolloff. The method is as follows. Repeatedly differentiate the signal in question until one or more stepwise discontinuities first appear. The next derivative, say the nth, then includes an impulse $A_k \, \delta(t - t_k)$ for each discontinuity of height A_k at $t = t_k$, so

$$\frac{d^n}{dt^n} v(t) = w(t) + \sum_k A_k \, \delta(t - t_k) \tag{27a}$$

where $w(t)$ is a nonimpulsive function. Transforming Eq. (27a) gives

$$(j2\pi f)^n V(f) = W(f) + \sum_k A_k e^{-j2\pi f t_k} \tag{27b}$$

which can be solved for $V(f)$ if we know $W(f) = \mathscr{F}[w(t)]$.

Furthermore, if $|W(f)| \rightarrow 0$ as $f \rightarrow \infty$, the high-frequency behavior of $|V(f)|$ will be proportional to $|f|^{-n}$ and we say that the spectrum has an *nth-order rolloff*. A large value of n thus implies that the signal has very little high-frequency content—an important consideration in the design of many communication systems.

Example 2.4-2 Raised cosine pulse Figure 2.4-7a shows a waveform called the *raised cosine pulse* because

$$v(t) = \frac{A}{2} \left(1 + \cos \frac{\pi t}{\tau} \right) \Pi \left(\frac{t}{2\tau} \right)$$

We'll use the differentiation method to find the spectrum $V(f)$ and the high-frequency rolloff. The first three derivatives of $v(t)$ are sketched in Fig. 2.4-7b, and we see that

$$\frac{dv(t)}{dt} = -\left(\frac{\pi}{\tau} \right) \frac{A}{2} \sin \frac{\pi t}{\tau} \, \Pi \left(\frac{t}{2\tau} \right)$$

which has no discontinuities. However, $d^2 v(t)/dt^2$ is discontinuous at $t = \pm\tau$ so

$$\frac{d^3}{dt^3} v(t) = \left(\frac{\pi}{\tau} \right)^3 \frac{A}{2} \sin \frac{\pi t}{\tau} \, \Pi \left(\frac{t}{2\tau} \right) + \left(\frac{\pi}{\tau} \right)^2 \frac{A}{2} \, \delta(t + \tau) - \left(\frac{\pi}{\tau} \right)^2 \frac{A}{2} \, \delta(t - \tau)$$

This expression has the same form as Eq. (27a), but we do not immediately know the transform of the first term.

Fortunately, a comparison of the first and third derivatives reveals that the first term of $d^3 v(t)/dt^3$ can be written as $w(t) = -(\pi/\tau)^2 \, dv(t)/dt$. Therefore, $W(f) = -(\pi/\tau)^2 (j2\pi f) V(f)$ and Eq. (27b) gives

$$(j2\pi f)^3 V(f) = -\left(\frac{\pi}{\tau} \right)^2 (j2\pi f) V(f) + \left(\frac{\pi}{\tau} \right)^2 \frac{A}{2} (e^{j2\pi f \tau} - e^{-j2\pi f \tau})$$

(a)

(b)

(c)

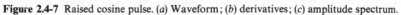

Figure 2.4-7 Raised cosine pulse. (a) Waveform; (b) derivatives; (c) amplitude spectrum.

Routine manipulations finally produce the result

$$V(f) = \frac{jA \sin 2\pi f \tau}{j2\pi f + (\tau/\pi)^2 (j2\pi f)^3} = \frac{A\tau \operatorname{sinc} 2f\tau}{1 - (2f\tau)^2}$$

whose amplitude spectrum is sketched in Fig. 2.4-7c for $f \geq 0$. Note that $|V(f)|$ has a third-order rolloff ($n = 3$), whereas a rectangular pulse with $|V(f)| = |\operatorname{sinc} f\tau| = |(\sin \pi f\tau)/(\pi f\tau)|$ would have only a first-order rolloff.

Exercise 2.4-3 Let $v(t) = (2At/\tau)\Pi(t/\tau)$. Sketch $dv(t)/dt$ and use Eq. (27) to find $V(f)$.

2.5 PROBLEMS

2.1-1 Consider the phasor signal $v(t) = Ae^{j\phi}e^{j2\pi m f_0 t}$. Confirm that Eq. (14) yields just one nonzero coefficient c_m having the appropriate amplitude and phase.

2.1-2 If a periodic signal has the *even-symmetry* property $v(-t) = v(t)$, then Eq. (14) may be written as

$$c_n = \frac{2}{T_0} \int_0^{T_0} v(t) \cos (2\pi nt/T_0) \, dt$$

Use this expression to find c_n when $v(t) = A$ for $|t| < T_0/4$ and $v(t) = -A$ for $T_0/4 < |t| < T_0/2$. As a preliminary step you should sketch the waveform and determine c_0 directly from $\langle v(t) \rangle$. Then sketch and label the spectrum after finding c_n.

2.1-3 Do Prob. 2.1-2 with $v(t) = 2A|t|/T_0$ for $|t| < T_0/2$.

2.1-4 Do Prob. 2.1-2 with $v(t) = A|\sin (2\pi t/T_0)|$ for $|t| < T_0/2$.

2.1-5 If a periodic signal has the *odd-symmetry* property $v(-t) = -v(t)$, then Eq. (14) may be written as

$$c_n = -j \frac{2}{T_0} \int_0^{T_0} v(t) \sin (2\pi nt/T_0) \, dt$$

Use this expression to find c_n when $v(t) = A$ for $0 < t < T_0/2$ and $v(t) = -A$ for $-T_0/2 < t < 0$. As a preliminary step you should sketch the waveform and determine c_0 directly from $\langle v(t) \rangle$. Then sketch and label the spectrum after finding c_n.

2.1-6 Do Prob. 2.1-5 with $v(t) = 2At/T_0$ for $-T_0/2 < t < T_0/2$.

2.1-7‡ Consider a periodic signal with the *half-wave symmetry* property $v(t \pm T_0/2) = -v(t)$, so the second half of any period looks like the first half inverted. Show that $c_n = 0$ for all *even* harmonics.

2.1-8 Suppose all frequency components at $|f| > 1/\tau$ are removed from the spectrum of a rectangular pulse train. Use Parseval's theorem to calculate the percentage of the signal power that remains when $\tau = T_0/2$ and $\tau = T_0/5$.

2.1-9 Let $v(t)$ be the sawtooth wave listed in Table T.2, and let $v'(t)$ be the approximation obtained with the first three nonzero terms of the trigonometric Fourier series. (*a*) What percentage of the total signal power is contained in $v'(t)$? (*b*) Sketch $v'(t)$ for $0 < t < T_0$.

2.1-10 Do Prob. 2.1-9 for the half-rectified sine wave in Table T.2.

2.1-11‡ Calculate P for the sawtooth wave listed in Table T.2. Then apply Parseval's power theorem show that the infinite sum $1/1^2 + 1/2^2 + 1/3^2 + \cdots$ equals $\pi^2/6$.

2.1-12‡ Calculate P for the triangular wave listed in Table T.2. Then apply Parseval's power theorem to show that the infinite sum $1/1^4 + 1/3^4 + 1/5^4 + \cdots$ equals $\pi^4/96$.

2.2-1 Consider the *cosine pulse* $v(t) = A \cos (\pi t/\tau)\Pi(t/\tau)$. Show that $V(f) = (A\tau/2) [\mathrm{sinc} (f\tau - 1/2) + \mathrm{sinc} (f\tau + 1/2)]$. Then sketch and label $| V(f)|$ for $f \geq 0$ to verify reciprocal spreading.

2.2-2 Consider the *sine pulse* $v(t) = A \sin (2\pi t/\tau)\Pi(t/\tau)$. Show that $V(f) = -j(A\tau/2) [\mathrm{sinc} (f\tau - 1) + \mathrm{sinc} (f\tau + 1)]$. Then sketch and label $| V(f)|$ for $f \geq 0$ to verify reciprocal spreading.

2.2-3 Find $V(f)$ when $v(t) = (A|t|/\tau)\Pi(t/2\tau)$. Express your result in terms of the sinc function.

2.2-4 Find $V(f)$ when $v(t) = (At/\tau)\Pi(t/2\tau)$. Express your result in terms of the sinc function.

2.2-5 Use a table of Laplace transforms to obtain the even and odd parts of $V(f)$ for the causal signal $v(t) = Ate^{-2\pi at}, t \geq 0$.

2.2-6 Do Prob. 2.2-5 with $v(t) = Ae^{-2\pi t} \cos 2\pi bt$.

2.2-7 Let $v(t)$ be the causal exponential pulse in Example 2.2-2. Use Rayleigh's theorem to calculate the percentage of the total energy contained in $| f | < W$ when $W = b/2\pi$ and $W = 2h/\pi$

2.2-8 Suppose the left-hand side of Eq. (17) had been written as

$$\int_{-\infty}^{\infty} v(t)w(t) \, dt$$

Find the resulting right-hand side and simplify for the case when $v(t)$ is real and $w(t) = v(t)$.

2.2-9 Show that $\mathscr{F}[w^*(t)] = W^*(-f)$. Then use Eq. (17) to obtain a frequency-domain expression for $\int_{-\infty}^{\infty} v(t)z(t) \, dt$.

2.2-10 Apply duality to the result of Prob. 2.2-1 to find $z(t)$ when $Z(f) = (A/2W) \cos (\pi f/2W) \times \Pi(f/2W)$.

2.2-11 Apply duality to the result of Prob. 2.2-2 to find $z(t)$ when $Z(f) = -j(A/2W) \sin (\pi f/W) \times \Pi(f/2W)$.

2.2-12‡ Use Eq. (16) and a known transform pair to show that

$$\int_{0}^{\infty} (a^2 + x^2)^{-2} \, dx = \pi/4a^3$$

2.2-13‡ Use Eq. (17) and known transform pairs to show that

$$\int_{0}^{\infty} (\mathrm{sinc} \; x)/(a^2 + x^2) \, dx = (1 - e^{-\pi a})/2a^2$$

2.3-1 Find $V(f)$ for $v(t)$ in Fig. P2.3-1. Then sketch $| V(f)|$ taking $\tau \ll T$.

Figure P2.3-1

2.3-2 Find $V(f)$ for $v(t)$ in Fig. P2.3-2. Then sketch $| V(f)|$ taking $\tau \ll T$.

Figure P2.3-2

2.3-3 Find $V(f)$ for $v(t)$ in Fig. P2.3-3. Then sketch $|V(f)|$ taking $\tau \ll T$.

Figure P2.3-3

2.3-4 Find $V(f)$ for $v(t)$ in Fig. P2.3-4.

Figure P2.3-4

2.3-5 Find $V(f)$ for $v(t)$ in Fig. P2.3-5.

Figure P2.3-5

2.3-6 Find $Z(f)$ in terms of $V(f)$ when $z(t) = v(at - t_d)$.

2.3-7 Prove Eq. (7) from Eq. (6).

2.3-8 Prove Eq. (6).

2.3-9 Use Eq. (7) to obtain the transform pair in Prob. 2.2-1.

2.3-10 Use Eq. (7) to obtain the transform pair in Prob. 2.2-2.

2.3-11 Use Eq. (7) to find $Z(f)$ when $z(t) = Ae^{-|t|} \sin \omega_c t$.

2.3-12 Use Eq. (7) to find $Z(f)$ when $z(t) = Ae^{-t} \cos \omega_c t$ for $t \geq 0$ and $z(t) = 0$ for $t < 0$.

2.3-13 Prove Eq. (9b) by letting $w(t) = \int_{-\infty}^{t} v(\lambda) \, d\lambda$ so $v(t) = dw(t)/dt$.

2.3-14 Use Eq. (12) to do Prob. 2.2-4.

2.3-15 Use Eq. (12) to find $Z(f)$ when $z(t) = Ate^{-bt}$ for $t \geq 0$ and $z(t) = 0$ for $t < 0$.

2.3-16 Use Eq. (12) to find $Z(f)$ when $z(t) = At^2 e^{-b|t|}$.

2.3-17 Consider the gaussian pulse listed in Table T.1. Generate a new transform pair by: (a) applying Eq. (8) with $n = 1$; (b) applying Eq. (12) with $n = 1$.

2.3-18 Find and sketch $y(t) = v * w(t)$ when $v(t) = t$ for $0 < t < 2$ and $w(t) = A$ for $t > 0$. Both signals equal zero outside the specifed ranges.

2.3-19 Do Prob. 2.3-18 with $w(t) = A$ for $0 < t < 3$.

2.3-20 Do Prob. 2.3-18 with $w(t) = A$ for $0 < t < 1$.

2.3-21 Prove Eq. (14a) from Eq. (13).

2.3-22 Let $v(t)$ and $w(t)$ have even symmetry. Show from Eq. (13) that $v * w(t)$ will have even symmetry.

2.3-23 Let $v(t)$ and $w(t)$ have odd symmetry. Show from Eq. (13) that $v * w(t)$ will have even symmetry.

2.3-24 Find and sketch $v * v * v$ when $v(t) = \Pi(t/\tau)$. You may use the symmetry property stated in Prob. 2.3-22.

2.3-25 Use Eq. (15) to prove Eq. (14b).

2.3-26 Prove Eq. (16).

2.4-1 Show that Eq. (5) satisfies Eq. (4a) by expanding $v(t)$ in a Maclaurin series before integrating.

2.4-2 Let $v(t)$ be a periodic signal whose Fourier series coefficients are denoted by $c_v(nf_0)$. Use Eq. (14) and an appropriate transform theorem to express $c_w(nf_0)$ in terms of $c_v(nf_0)$ when $w(t) = v(t - t_d)$.

2.4-3 Do Prob. 2.4-2 with $w(t) = dv(t)/dt$.

2.4-4‡ Do Prob. 2.4-2 with $w(t) = v(t) \cos m\omega_0 t$.

2.4-5 Let $v(t) = A$ for $0 < t < \tau$ and $v(t) = 0$ otherwise. Use Eq. (18) to find $V(f)$. Check your result by writing $v(t)$ in terms of the rectangle function.

2.4-6 Let $v(t) = A$ for $|t| > \tau/2$ and $v(t) = 0$ otherwise. Use Eq. (18) to find $V(f)$. Check your result by writing $v(t)$ in terms of the rectangle function.

2.4-7 Let $v(t) = -A$ for $t < -T$, $v(t) = A$ for $t > T$, and $v(t) = 0$ otherwise. Use Eq. (18) to find $V(f)$. Check your result by letting $T \to 0$.

2.4-8 Let

$$w(t) = \int_{-\infty}^{t} v(\lambda)\, d\lambda$$

with $v(t) = (1/\epsilon)\Pi(t/\epsilon)$. Sketch $w(t)$ and use Eq. (20) to find $W(f)$. Then let $\epsilon \to 0$ and compare your results with Eq. (18).

2.4-9 Do Prob. 2.4-8 with $v(t) = (1/\epsilon)\Lambda(t/\epsilon)$.

2.4-10 Obtain the transform of the signal in Fig. P2.3-1 by expressing $v(t)$ as the convolution of a rectangle function with impulses.

2.4-11 Do Prob. 2.4-10 for the signal in Fig. P2.3-2.

2.4-12 Do Prob. 2.4-10 for the signal in Fig. P2.3-3.

2.4-13 Use Eq. (27) to obtain the transform of $v(t) = A[1 - (2t/\tau)^2]\Pi(t/\tau)$.

2.4-14 Use Eq. (27) to obtain the transform of $v(t) = (A|t|/\tau)\Pi(t/2\tau)$.

2.4-15 Use Eq. (27) to obtain the transform of $v(t) = A \cos (\pi t/\tau)\Pi(t/\tau)$.

2.4-16 Use Eq. (27) to obtain the transform of $v(t) = A \sin (2\pi t/\tau)\Pi(t/\tau)$.

2.4-17 Suppose a continuous spectrum $V(f)$ is differentiated $n - 1$ times until one or more stepwise discontinuities first appear. Write expressions analogous to Eq. (27). What information about the behavior of $v(t)$ can be gained from this process?

THREE

SIGNAL TRANSMISSION AND FILTERING

Signal transmission is the process whereby an electrical waveform gets from one location to another and, ideally, arrives without distortion. In contrast, signal filtering is an operation that purposefully distorts a waveform by altering its spectral content. Nonetheless, most transmission systems and filters have in common the properties of linearity and time invariance. These properties allow us to model both transmission and filtering in the time domain in terms of the impulse response, or in the frequency domain in terms of the frequency response.

This chapter begins with a general consideration of system response in both domains. Then we'll apply our results to the analysis of signal transmission, distortion, and filtering. Some related topics — notably transmission loss, Hilbert transforms, and correlation — are included as starting points for subsequent development.

3.1 RESPONSE OF LTI SYSTEMS

Figure 3.1-1 depicts a system inside a "black box" with an external *input signal* $x(t)$ and an *output signal* $y(t)$. In the context of electrical communication, the system usually would be a two-port network driven by an applied voltage or current at the input port, producing another voltage or current at the output port. Energy storage elements and other internal effects may cause the output waveform to look quite different from the input. But regardless of what's in the box, the system is characterized by an *excitation-and-response* relationship between input and output.

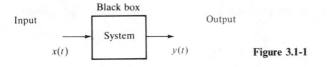

Figure 3.1-1

Here we're concerned with the special but important class of *linear time-invariant* systems — or LTI systems for short. We'll develop the input–output relationship in the time domain using the superposition integral and the system's impulse response. Then we'll turn to frequency-domain analysis expressed in terms of the system's transfer function.

Impulse Response and the Superposition Integral

Let Fig. 3.1-1 be an LTI system having no internal stored energy at the time the input $x(t)$ is applied. The output $y(t)$ is then the *forced response* due entirely to $x(t)$, as represented by

$$y(t) = F[x(t)] \tag{1}$$

where $F[x(t)]$ stands for the functional relationship between input and output. The *linear* property means that Eq. (1) obeys the *principle of superposition*. Thus, if

$$x(t) = \sum_k a_k x_k(t) \tag{2a}$$

where a_k are constants, then

$$y(t) = \sum_k a_k F[x_k(t)] \tag{2b}$$

The *time-invariance* property means that the system's characteristics remain *fixed with time*. Thus, a time-shifted input $x(t - t_d)$ produces

$$F[x(t - t_d)] = y(t - t_d) \tag{3}$$

so the output is time-shifted but otherwise unchanged.

Most LTI systems consist entirely of *lumped-parameter* elements (such as resistors capacitors, and inductors), as distinguished from elements with spatially distributed phenomena (such as transmission lines). Direct analysis of a lumped-parameter system starting with the element equations leads to the input–output relation as a linear differential equation in the form

$$a_n \frac{d^n y(t)}{dt^n} + \cdots + a_1 \frac{dy(t)}{dt} + a_0 y(t) = b_m \frac{d^m x(t)}{dt^m} + \cdots + b_1 \frac{dx(t)}{dt} + b_0 x(t) \tag{4}$$

where the a's and b's are constant coefficients involving the element values. The number of independent energy-storage elements determines the value of n, known as the *order* of the system. Unfortunately, Eq. (4) doesn't provide us with a direct expression for $y(t)$.

To obtain an explicit input–output equation, we must first define the system's *impulse response*

$$h(t) \triangleq F[\delta(t)] \tag{5}$$

which equals the forced response when $x(t) = \delta(t)$. But any continuous input signal can be written as the convolution $x(t) = x(t) * \delta(t)$, so

$$y(t) = F\left[\int_{-\infty}^{\infty} x(\lambda)\,\delta(t - \lambda)\,d\lambda\right]$$

$$= \int_{-\infty}^{\infty} x(\lambda)F[\delta(t - \lambda)]\,d\lambda$$

in which the interchange of operations is allowed by virtue of the system's linearity. Now, from the time-invariance property, $F[\delta(t - \lambda)] = h(t - \lambda)$ and hence

$$y(t) = \int_{-\infty}^{\infty} x(\lambda)h(t - \lambda)\,d\lambda \tag{6a}$$

$$= \int_{-\infty}^{\infty} h(\lambda)x(t - \lambda)\,d\lambda \tag{6b}$$

where we have drawn upon the commutivity of convolution.

Either form of Eq. (6) is called the *superposition integral*. It expresses the forced response as a convolution of the input $x(t)$ with the impulse response $h(t)$. System analysis in the time domain therefore requires knowledge of the impulse response along with the ability to carry out the convolution.

Various techniques exist for determining $h(t)$ from a differential equation or some other system model; see, for instance, Frederick and Carlson (1971, Chap. 4). But you may be more comfortable taking $x(t) = u(t)$ and calculating the system's *step response*

$$g(t) \triangleq F[u(t)] \tag{7a}$$

from which

$$h(t) = \frac{dg(t)}{dt} \tag{7b}$$

This derivative relation between the impulse and step response follows from the general convolution property

$$\frac{d}{dt}[v * w(t)] = v(t) * \left[\frac{dw(t)}{dt}\right]$$

Thus, since $g(t) = h * u(t)$ by definition, $dg(t)/dt = h(t) * [du(t)/dt] = h(t) * \delta(t) = h(t)$.

Example 3.1-1 Time response of a first-order system The simple RC circuit in Fig. 3.1-2 has been arranged as a two-port network with input voltage $x(t)$

Figure 3.1-2 RC lowpass filter.

and output voltage $y(t)$. The reference voltage polarities are indicated by arrows whose tips point to the assumed higher potential, like the $+$ sign in the $+/-$ notation. This circuit is a first-order system governed by the differential equation

$$RC \frac{dy(t)}{dt} + y(t) = x(t)$$

Similar expressions describe certain transmission lines and cables, so we're particularly interested in the system response.

From either the differential equation or the circuit diagram, the step response is readily found to be

$$g(t) = (1 - e^{-t/RC})u(t) \qquad (8a)$$

Interpreted physically, the capacitor starts at zero initial voltage and charges toward $y(\infty) = 1$ with time constant RC when $x(t) = u(t)$. Figure 3.1-3a plots this behavior, while Fig. 3.1-3b shows the corresponding impulse response

$$h(t) = \frac{1}{RC} e^{-t/RC} u(t) \qquad (8b)$$

obtained by differentiating $g(t)$. Note that $g(t)$ and $h(t)$ are *causal* waveforms since the input equals zero for $t < 0$.

The response to an arbitrary input $x(t)$ can now be found by putting Eq. (8b) in the superposition integral. For instance, take the case of a rectangular pulse applied at $t = 0$, so $x(t) = A$ for $0 < t < \tau$. The convolution

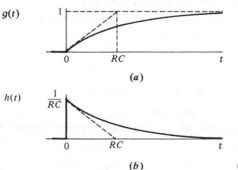

Figure 3.1-3 Output of an RC lowpass filter. (a) Step response; (b) impulse response.

Figure 3.1-4 Rectangular pulse response of an RC lowpass filter. (a) $\tau \gg RC$; $(b)\tau \approx RC$; (c) $\tau \ll RC$.

$y(t) = h * x(t)$ divides into three parts, like the example back in Fig. 2.3-5, with the result that

$$y(t) = \begin{cases} 0 & t < 0 \\ A(1 - e^{-t/RC}) & 0 < t < \tau \\ A(1 - e^{-\tau/RC})e^{-(t-\tau)/RC} & t > \tau \end{cases} \tag{9}$$

as sketched in Fig. 3.1-4 for three values of τ/RC.

Exercise 3.1-1 Let the resistor and the capacitor be interchanged in Fig. 3.1-2. Find the step and impulse response.

Transfer Functions and Frequency Response

Time-domain analysis becomes increasingly difficult for higher-order systems, and the mathematical complications tend to obscure significant points. We'll gain a different and often clearer view of system response by going to the frequency domain. As a first step in this direction, we define the system *transfer function* to be the Fourier transform of the impulse response, namely,

$$H(f) \triangleq \mathscr{F}[h(t)] = \int_{-\infty}^{\infty} h(t)e^{-j2\pi ft} \, dt \tag{10}$$

This definition requires that $H(f)$ exists, at least in a limiting sense. In the case of an *unstable* system, $h(t)$ *grows with time* and $H(f)$ does not exist.

When $h(t)$ is a *real* time function, $H(f)$ has the hermitian symmetry

$$H(-f) = H^*(f) \tag{11a}$$

so that

$$|H(-f)| = |H(f)| \qquad \arg H(-f) = -\arg H(f) \tag{11b}$$

We'll assume this property holds unless otherwise stated.

The frequency-domain interpretation of the transfer function comes from $y(t) = h * x(t)$ with a *phasor* input, say

$$x(t) = A_x e^{j\phi_x} e^{j2\pi f_0 t} \qquad -\infty < t < \infty \tag{12a}$$

The stipulation that $x(t)$ persists for all time means that we're dealing with *steady state* conditions, like the familiar case of AC steady state circuit analysis. The steady state forced response is

$$y(t) = \int_{-\infty}^{\infty} h(\lambda) A_x e^{j\phi_x} e^{j2\pi f_0(t-\lambda)} \, d\lambda$$

$$= \left[\int_{-\infty}^{\infty} h(\lambda) e^{-j2\pi f_0\lambda} \, d\lambda \right] A_x e^{j\phi_x} e^{j2\pi f_0 t}$$

$$= H(f_0) A_x e^{j\phi_x} e^{j2\pi f_0 t}$$

where, from Eq. (10), $H(f_0)$ equals $H(f)$ with $f = f_0$. Converting $H(f_0)$ to polar form then yields

$$y(t) = A_y e^{j\phi_y} e^{j2\pi f_0 t} \qquad -\infty < t < \infty \qquad (12b)$$

in which we have identified the output phasor's amplitude and angle

$$A_y = |H(f_0)| A_x \qquad \phi_y = \arg H(f_0) + \phi_x \qquad (13)$$

Using conjugate phasors and superposition, you can similarly show that if

$$x(t) = A_x \cos (2\pi f_0 t + \phi_x)$$

then

$$y(t) = A_y \cos (2\pi f_0 t + \phi_y)$$

with A_y and ϕ_y as in Eq. (13).

Since $A_y/A_x = |H(f_0)|$ at any frequency f_0, we conclude that $|H(f)|$ represents the system's *amplitude ratio* as a function of frequency (sometimes called the *amplitude response* or *gain*). By the same token, $\arg H(f)$ represents the *phase shift*, since $\phi_y - \phi_x = \arg H(f_0)$. Plots of $|H(f)|$ and $\arg H(f)$ versus frequency give us the frequency-domain representation of the system or, equivalently, the system's *frequency response*. Henceforth, we'll refer to $H(f)$ as either the transfer function or frequency-response function.

Now let $x(t)$ be any signal with spectrum $X(f)$. Calling upon the convolution theorem, we take the transform of $y(t) = h * x(t)$ to obtain

$$Y(f) = H(f)X(f) \qquad (14)$$

This elegantly simple result constitutes the basis of frequency-domain system analysis. It says that the output spectrum $Y(f)$ equals the input spectrum $X(f)$ multiplied by the transfer function $H(f)$. The corresponding amplitude and phase spectra are

$$|Y(f)| = |H(f)||X(f)|$$

$$\arg Y(f) = \arg H(f) + \arg X(f)$$

which compare with the single-frequency expressions in Eq. (13). If $x(t)$ is an energy signal, then $y(t)$ will be an energy signal whose spectral density and total energy are given by

$$|Y(f)|^2 = |H(f)|^2|X(f)|^2 \tag{15a}$$

$$E_y = \int_{-\infty}^{\infty} |H(f)|^2|X(f)|^2 \, df \tag{15b}$$

as follows from Rayleigh's energy theorem.

Equation (14) sheds new light on the meaning of the system transfer function and the transform pair $h(t) \leftrightarrow H(f)$. For if we let $x(t)$ be a unit impulse, then $X(f) = 1$ and $Y(f) = H(f)$ — in agreement with the definition $y(t) = h(t)$ when $x(t) = \delta(t)$. From the frequency-domain viewpoint, the "flat" input spectrum $X(f) = 1$ contains all frequencies in equal proportion and, consequently, the output spectrum takes on the shape of the transfer function $H(f)$.

Figure 3.1-5 summarizes our input–output relations in both domains. Clearly, when $H(f)$ and $X(f)$ are given, the output spectrum $Y(f)$ is much easier to find than the output signal $y(t)$. In principle, we could compute $y(t)$ from the inverse transform

$$y(t) = \mathcal{F}^{-1}[H(f)X(f)] = \int_{-\infty}^{\infty} H(f)X(f)e^{j2\pi ft} \, df$$

But this integration does not necessarily offer any advantages over time-domain convolution. Indeed, the power of frequency-domain system analysis largely depends on staying in that domain and using our knowledge of spectral properties to draw inferences about the output signal.

Finally, we point out two ways of determining $H(f)$ that don't involve $h(t)$. If you know the *differential equation* for a lumped-parameter system, you can immediately write down its transfer function as the ratio of polynomials

$$H(f) = \frac{b_m(j2\pi f)^m + \cdots + b_1(j2\pi f) + b_0}{a_n(j2\pi f)^n + \cdots + a_1(j2\pi f) + a_0} \tag{16}$$

whose coefficients are the same as those in Eq. (4). Equation (16) follows from Fourier transformation of Eq. (4).

Alternatively, if you can calculate a system's *steady state phasor response*, Eqs. (12) and (13) show that

$$H(f) = \frac{y(t)}{x(t)} \quad \text{when} \quad x(t) = e^{j2\pi ft} \tag{17}$$

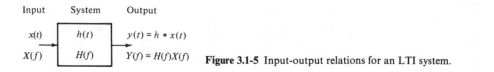

Figure 3.1-5 Input-output relations for an LTI system.

This method corresponds to impedance analysis of electrical circuits, but is equally valid for any LTI system. Furthermore, Eq. (17) may be viewed as a special case of the s domain transfer function $H(s)$ used in conjunction with Laplace transforms. Since $s = \sigma + j\omega$ in general, $H(f)$ is obtained from $H(s)$ simply by letting $s = j2\pi f$. These methods assume, of course, that the system is stable.

Example 3.1-2 Frequency response of a first-order system The *RC* circuit from Example 3.1-1 has been redrawn in Fig. 3.1-6a with the impedances $Z_R = R$ and $Z_C = 1/j\omega C$ replacing the elements. Since $y(t)/x(t) = Z_C/(Z_C + Z_R)$ when $x(t) = e^{j\omega t}$, Eq. (17) gives

$$H(f) = \frac{(1/j2\pi f C)}{(1/j2\pi f C) + R} = \frac{1}{1 + j2\pi f RC}$$

$$= \frac{1}{1 + j(f/B)} \tag{18a}$$

where we have introduced the system parameter

$$B \triangleq \frac{1}{2\pi RC} \tag{18b}$$

Identical results would have been obtained from Eq. (16), or from $H(f) = \mathscr{F}[h(t)]$. (In fact, the system's impulse response has the same form as

Figure 3.1-6 RC lowpass filter. (a) circuit; (b) transfer function.

the causal exponential pulse discussed in Example 2.2-2.) The amplitude ratio and phase shift are

$$|H(f)| = \frac{1}{\sqrt{1 + (f/B)^2}} \qquad \arg H(f) = -\arctan \frac{f}{B} \qquad (18c)$$

as plotted in Fig. 3.1-6b for $f \geq 0$. The hermitian symmetry allows us to omit $f < 0$ without loss of information.

The amplitude ratio $|H(f)|$ has special significance relative to any *frequency-selective* properties of the system. We call this particular system a *lowpass filter* because it has almost no effect on the amplitude of low-frequency components, say $|f| \ll B$, while it drastically reduces the amplitude of high-frequency components, say $|f| \gg B$. The parameter B serves as a measure of the filter's *passband* or *bandwidth*.

To illustrate how far you can go with frequency-domain analysis, let the input $x(t)$ be an arbitrary signal whose spectrum has negligible content for $|f| > W$. There are three possible cases to consider, depending on the relative values of B and W:

1. If $W \ll B$, as shown in Fig. 3.1-7a, then $|H(f)| \approx 1$ and $\arg H(f) \approx 0$ over the signal's frequency range $|f| < W$. Thus, $Y(f) = H(f)X(f) \approx X(f)$ and $y(t) \approx x(t)$ so we have *undistorted transmission* through the filter.

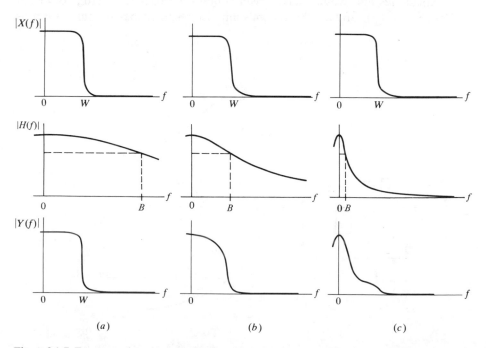

(a)   (b)   (c)

Figure 3.1-7 Frequency-domain analysis of a first-order lowpass filter.   (a) $B \gg W$; (b) $B \approx W$; (c) $B \ll W$.

2. If $W \approx B$, as shown in Fig. 3.1-7b, then $Y(f)$ depends on both $H(f)$ and $X(f)$. We can say that the output is *distorted*, since $y(t)$ will differ significantly from $x(t)$, but time-domain calculations would be required to find the actual waveform.

3. If $W \gg B$, as shown in Fig. 3.1-7c, the input spectrum has a nearly constant value $X(0)$ for $|f| < B$ so $Y(f) \approx X(0)H(f)$. Thus, $y(t) \approx X(0)h(t)$, and the output signal now looks like the filter's *impulse response*. Under this condition, we can reasonably model the input signal as an impulse.

Our previous time-domain analysis with a rectangular input pulse confirms these conclusions since the nominal spectral width of the pulse is $W = 1/\tau$. The case $W \ll B$ thus corresponds to $1/\tau \ll 1/2\pi RC$ or $\tau/RC \gg 1$, and we see in Fig. 3.1-3a that $y(t) \approx x(t)$. Conversely, $W \gg B$ corresponds to $\tau/RC \ll 1$ as in Fig. 3.1-3c where $y(t)$ looks more like $h(t)$.

Exercise 3.1-2 Find $H(f)$ when $Z_L = j\omega L$ replaces Z_C in Fig. 3.1-6a. Express your result in terms of the system parameter $f_\ell = 2\pi L/R$, and justify the name "highpass filter" by sketching $|H(f)|$ versus f.

Block-Diagram Analysis

More often than not, a communication system comprises many interconnected building blocks or subsystems. Some blocks might be two-port networks with known transfer functions, while other blocks might be given in terms of their time-domain operations. Any LTI operation, of course, has an equivalent transfer function. For reference purposes, Table 3.1-1 lists the transfer functions obtained by applying transform theorems to four primitive time-domain operations.

When the subsystems in question are described by individual transfer functions, it is possible and desirable to lump them together and speak of the overall system transfer function. The corresponding relations are given below for two blocks connected in parallel, cascade, and feedback. More complicated configurations can be analyzed by successive application of these basic rules. One essential assumption must be made, however, namely, that any interaction or *loading* effects have been accounted for in the individual transfer functions so that they

Table 3.1-1

Time-domain operation		Transfer function
Scalar multiplication	$y(t) = \pm K x(t)$	$H(f) = \pm K$
Differentiation	$y(t) = \dfrac{dx(t)}{dt}$	$H(f) = j2\pi f$
Integration	$y(t) = \displaystyle\int_{-\infty}^{t} x(\lambda)\, d\lambda$	$H(f) = \dfrac{1}{j2\pi f}$
Time delay	$y(t) = x(t - t_d)$	$H(f) = e^{-j2\pi f t_d}$

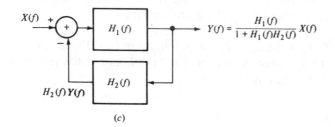

Figure 3.1-8 (*a*) Parallel connection; (*b*) cascade connection; (*c*) feedback connection.

represent the actual response of the subsystems in the context of the overall system. (A simple op-amp voltage follower might be used to provide isolation between blocks and prevent loading.)

Figure 3.1-8*a* diagrams two blocks in *parallel*: both units have the same input and their outputs are summed to get the system's output. From superposition it follows that $Y(f) = [H_1(f) + H_2(f)]X(f)$ so the overall transfer function is

$$H(f) = H_1(f) + H_2(f) \qquad \text{Parallel connection} \qquad (19a)$$

In the *cascade* connection, Fig. 3.1-8*b*, the output of the first unit is the input to the second, so $Y(f) = H_2(f)[H_1(f)X(f)]$ and

$$H(f) = H_1(f)H_2(f) \qquad \text{Cascade connection} \qquad (19b)$$

The *feedback* connection, Fig. 3.1-8*c*, differs from the other two in that the output is sent back through $H_2(f)$ and subtracted from the input. Thus,

$$Y(f) = H_1(f)[X(f) - H_2(f)Y(f)]$$

and rearranging yields $Y(f) = \{H_1(f)/[1 + H_1(f)H_2(f)]\}X(f)$ so

$$H(f) = \frac{H_1(f)}{1 + H_1(f)H_2(f)} \qquad \text{Feedback connection} \qquad (19c)$$

This case is more properly termed the *negative* feedback connection as distinguished from positive feedback, where the returned signal is added to the input
instead of subtracted.

Example 3.1-3 Zero-order hold The zero-order hold system in Fig. 3.1-9a has
several applications in electrical communication. Here we take it as an
instructive exercise of the parallel and cascade relations. But first we need the
individual transfer functions, determined as follows: the upper branch of the
parallel section is a straight-through path so, trivially, $H_1(f) = 1$; the lower
branch produces pure time delay of T seconds followed by sign inversion,
and lumping them together gives $H_2(f) = -e^{-j2\pi fT}$; the integrator in the
final block has $H_3(f) = 1/j2\pi f$. Figure 3.1-9b is the equivalent block diagram
in terms of these transfer functions.

Having gotten this far, the rest of the work is easy. We combine the
parallel branches in $H_{12}(f) = H_1(f) + H_2(f)$ and use the cascade rule to
obtain

$$H(f) = H_{12}(f)H_3(f) = [H_1(f) + H_2(f)]H_3(f)$$

$$= [1 - e^{-j2\pi fT}] \frac{1}{j2\pi f}$$

$$= \frac{e^{j\pi fT} - e^{-j\pi fT}}{j2\pi f} e^{-j\pi fT} = \frac{\sin \pi fT}{\pi f} e^{-j\pi fT}$$

$$= T \operatorname{sinc} fT e^{-j\pi fT}$$

Hence we have the unusual result that the amplitude ratio of this sytem is a
sinc function in frequency!

To confirm this result by another route, let's calculate the impulse
response $h(t)$ drawing upon the definition that $y(t) = h(t)$ when $x(t) = \delta(t)$.

(a)

(b)

Figure 3.1-9 Block diagrams of a zero-order
hold. (a) Time domain; (b) frequency domain.

Inspection of Fig. 3.1-9a shows that the input to the integrator then is $x(t) - x(t - T) = \delta(t) - \delta(t - T)$, so

$$h(t) = \int_{-\infty}^{t} [\delta(\lambda) - \delta(\lambda - T)] \, d\lambda = u(t) - u(t - T)$$

which represents a rectangular pulse starting at $t = 0$. Rewriting the impulse response as $h(t) = \Pi[(t - T/2)/T]$ helps verify the transform relation $h(t) \leftrightarrow H(f)$.

Exercise 3.1-3 Let $x(t) = A\Pi(t/\tau)$ be applied to the zero-order hold. Use frequency-domain analysis to find $y(t)$ when $\tau \ll T$, $\tau = T$, and $\tau \gg T$.

3.2 SIGNAL DISTORTION IN TRANSMISSION

A signal transmission system is the electrical channel between an information source and destination. These systems range in complexity from a simple pair of wires to a sophisticated laser-optics link. But all transmission systems have two physical attributes of particular concern in communication: internal power dissipation that reduces the size of the output signal, and energy storage that alters the shape of the output.

Our purpose here is to formulate the conditions for distortionless signal transmission, assuming an LTI system so we can work with its transfer function. Then we'll define various types of distortion and address possible techniques for minimizing their effects.

Distortionless Transmission

Distortionless transmission means that the output signal has the same "shape" as the input. More precisely, given an input signal $x(t)$, we say that the output is undistorted if it differs from the input only by a multiplying constant and a finite time delay. Analytically, we have distortionless transmission if

$$y(t) = Kx(t - t_d) \tag{1}$$

where K and t_d are constants.

The properties of a distortionless system are easily found by examining the output spectrum

$$Y(f) = \mathscr{F}[y(t)] = Ke^{-j\omega t_d}X(f)$$

Now by definition of transfer function, $Y(f) = H(f)X(f)$, so

$$H(f) = Ke^{-j\omega t_d} \tag{2a}$$

In words, a system giving distortionless transmission must have *constant amplitude response* and negative *linear phase shift*, so

$$|H(f)| = |K| \qquad \arg H(f) = -2\pi t_d f \pm m180° \tag{2b}$$

$|X(f)|^2$

0   200   3200

f

Figure 3.2-1 Energy spectral density of an average voice signal.

Note that arg $H(f)$ must pass through the origin or intersect at an integer multiple of $\pm 180°$. We have added the term $\pm m180°$ to the phase to account for K being positive or negative. In the case of zero time delay, the phase is constant at 0 or $\pm 180°$.

An important and rather obvious qualification to Eq. (2) should be stated immediately. The conditions on $H(f)$ are required only over those frequencies where the input signal has significant spectral content. To underscore this point, Fig. 3.2-1 shows the energy spectral density of an average voice signal obtained from laboratory measurements. Since the spectral density is quite small for $f < 200$ Hz and $f > 3200$ Hz, we conclude that a system satisfying Eq. (2) over $200 \le |f| \le 3200$ Hz would yield nearly distortion-free voice transmission.

However, the stringent demands of distortionless transmission can only be satisfied approximately in practice, so transmission systems always produce some amount of signal distortion. For the purpose of studying distortion effects on various signals, we'll define three major types of distortion:

1. Amplitude distortion, which occurs when

$$|H(f)| \neq |K|$$

2. Delay distortion, which occurs when

$$\arg H(f) \neq -2\pi t_d f \pm m180°$$

3. Nonlinear distortion, which occurs when the system includes nonlinear elements

The first two types can be grouped under the general designation of *linear* distortion, described in terms of the transfer function of a linear system. For the third type, the nonlinearity precludes the existence of a transfer function.

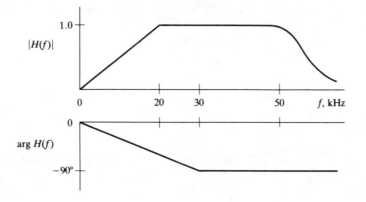

Figure 3.2-2

Example 3.2-1 Suppose a transmission system has the frequency response plotted in Fig. 3.2-2. This system satisfies Eq. (2) for $20 \leq |f| \leq 30$ kHz. Otherwise, there's amplitude distortion for $|f| < 20$ kHz and $|f| > 50$ kHz, and delay distortion for $|f| > 30$ kHz.

Linear Distortion

Linear distortion includes any amplitude or delay distortion associated with a linear transmission system. Amplitude distortion is easily described in the frequency domain; it means simply that the output frequency components are not in correct proportion. Since this is caused by $|H(f)|$ not being constant with frequency, amplitude distortion is sometimes called *frequency distortion*.

The most common forms of amplitude distortion are excess attenuation or enhancement of extreme high or low frequencies in the signal spectrum. Less common, but equally bothersome, is disproportionate response to a band of frequencies within the spectrum. While the frequency-domain description is easy, the effects in the time domain are far less obvious, save for very simple signals. For illustration, a suitably simple test signal is $x(t) = \cos \omega_0 t - \frac{1}{3} \cos 3\omega_0 t + \frac{1}{5} \cos 5\omega_0 t$, a rough approximation to a square wave sketched in Fig. 3.2-3. If the low-frequency or high-frequency component is attenuated by one-half, the resulting outputs are as shown in Fig. 3.2-4. As expected, loss of the high-frequency term reduces the "sharpness" of the waveform.

Figure 3.2-3 Test signal $x(t) = \cos \omega_0 t - \frac{1}{3} \cos 3\omega_0 t + \frac{1}{5} \cos 5\omega_0 t$.

Figure 3.2-4 Test signal with amplitude distortion. (*a*) Low frequency attenuated; (*b*) high frequency attenuated.

Beyond qualitative observations, there's little more we can say about amplitude distortion without experimental study of specific signal types. Results of such studies are usually couched in terms of required "flat" frequency response — meaning the frequency range over which $|H(f)|$ must be constant to within a certain tolerance so that the amplitude distortion is sufficiently small.

We now turn our attention to phase shift and time delay. If the phase shift is not linear, the various frequency components suffer different amounts of time delay, and the resulting distortion is termed *phase* or *delay distortion*. For an arbitrary phase shift, the time delay is a function of frequency and can be found by writing arg $H(f) = -2\pi f t_d(f)$ with all angles expressed in *radians*. Thus

$$t_d(f) = -\frac{\arg H(f)}{2\pi f} \tag{3}$$

which is independent of frequency only if arg $H(f)$ is linear with frequency.

A common area of confusion is *constant time delay* versus *constant phase shift*. The former is desirable and is required for distortionless transmission. The latter, in general, causes distortion. Suppose a system has the constant phase shift θ not equal to $0°$ or $\pm m180°$. Then each signal frequency component will be delayed by $\theta/2\pi$ *cycles* of its own frequency; this is the meaning of constant phase shift. But the time delays will be different, the frequency components will be scrambled in time, and distortion will result.

That constant phase shift does give distortion is simply illustrated by returning to the test signal of Fig. 3.2-3 and shifting each component by one-fourth cycle, $\theta = -90°$. Whereas the input was roughly a square wave, the output will look like the triangular wave in Fig. 3.2-5. With an arbitrary nonlinear phase shift, the deterioration of waveshape can be even more severe.

You should also note from Fig. 3.2-5 that the *peak* excursions of the phase-shifted signal are substantially greater (by about 50%) than those of the input test signal. This is not due to amplitude response, since the output amplitudes of the three frequency components are, in fact, unchanged; rather, it is because the components of the distorted signal all attain maximum or minimum values at the same time, which was not true of the input. Conversely, had we started with Fig. 3.2-5 as the test signal, a constant phase shift of $+90°$ would yield Fig. 3.2-3 for the output waveform. Thus we see that *delay distortion alone* can result in an increase or decrease of peak values as well as other waveshape alterations.

Figure 3.2-5 Test signal with constant phase shift $\theta = -90°$.

Clearly, delay distortion can be critical in pulse transmission, and much labor is spent *equalizing* transmission delay for digital data systems and the like. On the other hand, the human ear is curiously insensitive to delay distortion; the waveforms of Figs. 3.2-3 and 3.2-5 would sound just about the same when driving a loudspeaker. Thus, delay distortion is seldom of concern in voice and music transmission.

Exercise 3.2-1 Use Eq. (3) to plot $t_d(f)$ from arg $H(f)$ given in Fig. 3.2-2.

Equalization

Linear distortion — both amplitude and delay — is theoretically curable through the use of *equalization* networks. Figure 3.2-6 shows an equalizer $H_{eq}(f)$ in cascade with a distorting transmission channel $H_C(f)$. Since the overall transfer function is $H(f) = H_C(f)H_{eq}(f)$ the final output will be distortionless if $H_C(f)H_{eq}(f) = Ke^{-j\omega t_d}$, where K and t_d are more or less arbitrary constants. Therefore, we require that

$$H_{eq}(f) = \frac{Ke^{-j\omega t_d}}{H_C(f)} \qquad (4)$$

wherever $X(f) \neq 0$.

Rare is the case when an equalizer can be designed to satisfy Eq. (4) exactly — which is why we say that equalization is a *theoretical* cure. But excellent approximations often are possible so that linear distortion can be reduced to a tolerable level. Probably the oldest equalization technique involves the use of *loading coils* on twisted-pair telephone lines. These coils are lumped inductors placed in shunt across the line every kilometer or so, giving the improved amplitude ratio typically illustrated in Fig. 3.2-7. Other lumped-element circuits have been designed for specific equalization tasks.

Figure 3.2-6 Channel with equalizer for linear distortion

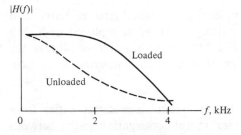

Figure 3.2-7 Amplitude ratio of a typical telephone line with and without loading coils for equalization.

More recently, the *tapped-delay-line equalizer* or *transversal filter* has emerged as a convenient and flexible device. To illustrate the principle, Fig. 3.2-8 shows a delay line with total time delay 2Δ having taps at each end and the middle. The tap outputs are passed through adjustable gains, c_{-1}, c_0, and c_1, and summed to form the final output. Thus

$$y(t) = c_{-1}x(t) + c_0 x(t - \Delta) + c_1 x(t - 2\Delta) \tag{5a}$$

and

$$H_{eq}(f) = c_{-1} + c_0 e^{-j\omega\Delta} + c_1 e^{-j\omega2\Delta}$$

$$= (c_{-1}e^{+j\omega\Delta} + c_0 + c_1 e^{-j\omega\Delta})e^{-j\omega\Delta} \tag{5b}$$

Generalizing Eq. (5b) to the case of a $2M\Delta$ delay line with $2M + 1$ taps yields

$$H_{eq}(f) = \left(\sum_{m=-M}^{M} c_m e^{-j\omega m\Delta} \right) e^{-j\omega M\Delta} \tag{6}$$

which has the form of an *exponential Fourier series* with frequency periodicity $1/\Delta$. Therefore, given a channel $H_c(f)$ to be equalized over $|f| < W$, you can approximate the right-hand side of Eq. (4) by a Fourier series with frequency periodicity $1/\Delta \geq W$ (thereby determining Δ), estimate the number of significant terms (which determines M), and match the tap gains to the series coefficients.

In many applications, the tap gains must be readjusted from time to time to compensate for changing channel characteristics. Adjustable equalization is especially important in switched communication networks, such as a telephone system, since the route between source and destination cannot be determined in advance. Sophisticated *adaptive equalizers* have therefore been designed with provision for automatic readjustment.

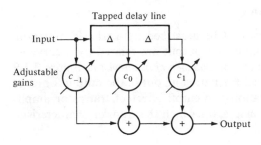

Figure 3.2-8 Transversal filter with three taps.

Adaptive equalization is usually implemented with digital circuitry and microprocessor control, in which case the delay line may be replaced by a shift register or charge-coupled device (CCD). For fixed (nonadjustable) equalizers, the transversal filter can be fabricated in an integrated circuit using a surface-acoustic-wave (SAW) device.

Example 3.2-2 Multipath distortion Radio systems sometimes suffer from multipath distortion caused by two (or more) propagation paths between transmitter and receiver. Reflections due to mismatched impedance on a cable system produce the same effect. As a simple example, suppose the channel output is

$$y(t) = K_1 x(t - t_1) + K_2 x(t - t_2)$$

whose second term corresponds to an *echo* of the first if $t_2 > t_1$. Then

$$H_C(f) = K_1 e^{-j\omega t_1} + K_2 e^{-j\omega t_2}$$

$$= K_1 e^{-j\omega t_1}(1 + ke^{-j\omega t_0}) \tag{7}$$

where $k = K_2/K_1$ and $t_0 = t_2 - t_1$.

If we take $K = K_1$ and $t_d = t_1$ for simplicity in Eq. (4), the required equalizer characteristic becomes

$$H_{eq}(f) = \frac{1}{1 + ke^{-j\omega t_0}} = 1 - ke^{-j\omega t_0} + k^2 e^{-j2\omega t_0} + \cdots$$

The binomial expansion has been used here because, in this case, it leads to the form of Eq. (6) without any Fourier-series calculations. Asssuming a small echo, so that $k^2 \ll 1$, we drop the higher-power terms and rewrite $H_{eq}(f)$ as

$$H_{eq}(f) \approx (e^{+j\omega t_0} - k + k^2 e^{-j\omega t_0})e^{-j\omega t_0}$$

Comparison with Eqs. (5b) or (6) now reveals that a three-tap transversal filter will do the job if $c_{-1} = 1$, $c_0 = -k$, $c_1 = k^2$, and $\Delta = t_0$.

Exercise 3.2-2 Sketch $|H_{eq}(f)|$ and arg $H_{eq}(f)$ needed to equalize the frequency response in Fig. 3.2-2 over $5 \le |f| \le 50$ kHz. Take $K = \frac{1}{4}$ and $t_d = 1/120$ ms in Eq. (4).

Nonlinear Distortion and Companding

A system having nonlinear elements cannot be described by a transfer function. Instead, the instantaneous values of input and output are related by a curve or function $y(t) = T[x(t)]$, commonly called the *transfer characteristic*. Figure 3.2-9 is a representative transfer characteristic; the flattening out of the output for large input excursions is the familiar saturation-and-cutoff effect of transistor amplifiers. We'll consider only *memoryless* devices, for which the transfer characteristic is a complete description.

$y = T[x]$

Figure 3.2-9 Transfer characteristic of a nonlinear device.

Under small-signal input conditions, it may be possible to linearize the transfer characteristic in a piecewise fashion, as shown by the thin lines in the figure. The more general approach is a polynomial approximation to the curve, of the form

$$y(t) = a_1 x(t) + a_2 x^2(t) + a_3 x^3(t) + \cdots \tag{8a}$$

and the higher powers of $x(t)$ in this equation give rise to the nonlinear distortion.

Even though we have no transfer function, the output spectrum can be found, at least in a formal way, by transforming Eq. (8a). Specifically, invoking the convolution theorem,

$$Y(f) = a_1 X(f) + a_2 X * X(f) + a_3 X * X * X(f) + \cdots \tag{8b}$$

Now if $x(t)$ is bandlimited in W, the output of a linear network will contain no frequencies beyond $|f| < W$. But in the nonlinear case, we see that the output includes $X * X(f)$, which is bandlimited in $2W$, $X * X * X(f)$, which is bandlimited in $3W$, etc. The nonlinearities have therefore created output frequency components that were not present in the input. Furthermore, since $X * X(f)$ may contain components for $|f| < W$, this portion of the spectrum overlaps that of $X(f)$. Using filtering techniques, the added components at $|f| > W$ can be removed, but there is no convenient way to get rid of the added components at $|f| < W$. These, in fact, constitute the nonlinear distortion.

A quantitative measure of nonlinear distortion is provided by taking a simple cosine wave, $x(t) = \cos \omega_0 t$, as the input. Inserting in Eq. (8a) and expanding yields

$$y(t) = \left(\frac{a_2}{2} + \frac{3a_4}{8} + \cdots \right) + \left(a_1 + \frac{3a_3}{4} + \cdots \right) \cos \omega_0 t$$

$$+ \left(\frac{a_2}{2} + \frac{a_4}{4} + \cdots \right) \cos 2\omega_0 t + \cdots$$

Therefore, the nonlinear distortion appears as *harmonics* of the input wave. The amount of second-harmonic distortion is the ratio of the amplitude of this term to that of the fundamental, or in percent:

$$\text{Second-harmonic distortion} = \left| \frac{a_2/2 + a_4/4 + \cdots}{a_1 + 3a_3/4 + \cdots} \right| \times 100\%$$

Higher-order harmonics are treated similarly. However, their effect is usually much less, and many can be removed entirely by filtering.

If the input is a sum of two cosine waves, say $\cos \omega_1 t + \cos \omega_2 t$, the output will include all the harmonics of f_1 and f_2, plus crossproduct terms which yield $f_2 - f_1, f_2 + f_1, f_2 - 2f_1$, etc. These sum and difference frequencies are designated as *intermodulation distortion*. Generalizing the intermodulation effect, if $x(t) = x_1(t) + x_2(t)$, then $y(t)$ contain the *cross product* $x_1(t)x_2(t)$ (and higher-order products, which we ignore here). In the frequency domain $x_1(t)x_2(t)$ becomes $X_1 * X_2(f)$; and even though $X_1(f)$ and $X_2(f)$ may be separated in frequency, $X_1 * X_2(f)$ can overlap both of them, producing one form of *crosstalk*. This aspect of nonlinear distortion is of particular concern in telephone transmission systems. On the other hand the crossproduct term is the desired result when nonlinear devices are used for modulation purposes.

Although nonlinear distortion has no perfect cure, it too can be minimized by careful design. The basic idea is to make sure that the signal does not exceed the linear operating range of the channel's transfer characteristic. Ironically, one strategy along this line utilizes two nonlinear signal processors, a *compressor* at the input and an *expander* at the output, as shown in Fig. 3.2-10.

A compressor has greater amplification at low signal levels than at high signal levels, similar to Fig. 3.2-9, and thereby compresses the range of the input signal. If the compressed signal falls within the linear range of the channel, the signal at the channel output is proportional to $T_{\text{comp}}[x(t)]$ which is distorted by the compressor but not the channel. Ideally, then, the expander has a characteristic that perfectly complements the compressor so the expanded output is proportional to $T_{\text{exp}}\{T_{\text{comp}}[x(t)]\} = x(t)$, as desired.

The joint use of compressing and expanding is called *companding* (surprise?) and is of particular value in telephone systems. Besides reducing nonlinear distortion, companding tends to compensate for the signal-level difference between loud and soft talkers. Indeed, the latter is the key advantage of companding compared to the simpler technique of linearly attenuating the signal at the input (to keep it in the linear range of the channel) and linearly amplifying it at the output.

Figure 3.2-10 Companding system.

3.3 TRANSMISSION LOSS AND DECIBELS

In addition to any signal distortion, a transmission system also reduces the power level or "strength" of the output signal. This signal-strength reduction is expressed in terms of transmission power loss. Although transmission loss can be compensated by power amplification, the ever-present electrical noise may prevent successful signal recovery in the face of large transmission loss.

This section describes transmission loss encountered on cable and radio communication systems. We'll start with a brief review of the more familiar concept of power gain, and we'll introduce decibels as a handy measure of power ratios used by communication engineers.

Power Gain

Let Fig. 3.3-1 represent an LTI system whose input signal has average power P_{in}. If the system is distortionless, the average signal power at the output will be proportional to P_{in}. Thus, the system's *power gain* is

$$g \triangleq P_{out}/P_{in} \tag{1}$$

a constant parameter not to be confused with our step-response notation $g(t)$. Systems that include amplification may have very large values of g, so we'll find it convenient to express power gain in *decibels* (dB) defined as

$$g_{dB} \triangleq 10 \log_{10} g \tag{2}$$

The "B" in dB is capitalized in honor of Alexander Graham Bell who first used logarithmic power measurements.

Since the decibel is a logarithmic unit, it converts powers of 10 to products of 10. For instance, $g = 10^m$ becomes $g_{dB} = m \times 10$ dB. Power gain is always positive, of course, but negative dB values occur when $g \leq 1.0 = 10^0$ and hence $g_{dB} \leq 0$ dB. Note carefully that 0 dB corresponds to *unity* gain ($g = 1$). Given a value in dB, the ratio value is

$$g = 10^{(g_{dB}/10)} \tag{3}$$

obtained by inversion of Eq. (2).

While decibels always represent *power ratios*, signal power itself may be expressed in dB if you divide P by one watt or one milliwatt, as follows:

$$P_{dBW} = 10 \log_{10} \frac{P}{1\text{ W}} \qquad P_{dBm} = 10 \log_{10} \frac{P}{1\text{ mW}} \tag{4}$$

$P_{in} \longrightarrow \boxed{ g } \longrightarrow P_{out} = gP_{in}$

Figure 3.3-1 LTI system with power gain g.

Rewriting Eq. (1) as $(P_{out}/1 \text{ mW}) = g(P_{in}/1 \text{ mW})$ and taking the logarithm of both sides then yields the dB equation

$$P_{out_{dBm}} = g_{dB} + P_{in_{dBm}}$$

Such manipulations have particular advantages for the more complicated relations encountered subsequently, where multiplication and division become addition and subtraction of known dB quantities. Communication engineers usually work with dBm because the signal powers are quite small at the output of a transmission system.

Now consider a system described by its transfer function $H(f)$. A sinusoidal input with amplitude A_x produces the output amplitude $A_y = |H(f)|A_x$, and the *normalized* signal powers are $P_x = A_x^2/2$ and $P_y = A_y^2/2 = |H(f)|^2 P_x$. These normalized powers do not necessarily equal the actual powers in Eq. (1). However, when the system has the same impedance level at input and output, the ratio P_y/P_x does equal P_{out}/P_{in}. Therefore, if $H(f) = Ke^{-j\omega t_d}$, then

$$g = |H(f)|^2 = K^2 \tag{5}$$

In this case, the power gain also applies to *energy* signals in the sense that $E_y = gE_x$. When the system has unequal input and output impedances, the power (and energy) gain is proportional to K^2.

If the system is frequency-selective, Eq. (5) does not hold but $|H(f)|^2$ still tells us how the gain varies as a function of frequency. For a useful measure of frequency dependence in terms of signal power we take

$$|H(f)|_{dB} \triangleq 10 \log_{10} |H(f)|^2 \tag{6}$$

which represents the *relative gain* in dB.

Exercise 3.3-1 (a) Verify that $P_{dBm} = P_{dBW} + 30$ dB. (b) Show that if $|H(f)|_{dB} = -3$ dB then $|H(f)| \approx 1/\sqrt{2}$.

Transmission Loss and Repeaters

Any passive transmission medium has power loss rather than gain, since $P_{out} < P_{in}$. We therefore prefer to work with the transmission *loss* or *attenuation*

$$L \triangleq 1/g = P_{in}/P_{out}$$

$$L_{dB} = -g_{dB} = 10 \log_{10} P_{in}/P_{out} \tag{7}$$

Hence, $P_{out} = P_{in}/L$ and $P_{out_{dBm}} = P_{in_{dBm}} - L_{dB}$.

In the case of transmission lines, coaxial cables, and waveguides, the output power decreases *exponentially* with distance. We'll write this relation in the form

$$P_{out} = 10^{-(\alpha\ell/10)} P_{in}$$

where ℓ is the *path length* between source and destination and α is the *attenuation coefficient* in dB per unit length. Equation (7) then becomes

$$L = 10^{(\alpha\ell/10)} \qquad L_{dB} = \alpha\ell \tag{8}$$

Table 3.3-1 Typical values of transmission loss

Transmission medium	Frequency	Loss, dB/km
Open-wire pair (0.3 cm diameter)	1 kHz	0.05
Twisted-wire pair (16 gauge)	10 kHz	2
	100 kHz	3
	300 kHz	6
Coaxial cable (1 cm diameter)	100 kHz	1
	1 MHz	2
	3 MHz	4
Coaxial cable (15 cm diameter)	100 MHz	1.5
Rectangular waveguide (5 × 2.5 cm)	10 GHz	5
Helical waveguide (5 cm diameter)	100 GHz	1.5
Fiber-optic cable	4×10^{14} Hz	10

showing that the dB loss is proportional to the length. Table 3.3-1 lists some typical values of α for various transmission media and signal frequencies.

Attenuation values in dB somewhat obscure the dramatic decrease of signal power with distance. To bring out the implications of Eq. (8) more clearly, suppose you transmit a signal on a 30-km length of cable having $\alpha = 3$ dB/km. Then $L_{dB} = 3 \times 30 = 90$ dB, $L = 10^9$, and $P_{out} = 10^{-9} P_{in}$. Doubling the path length doubles the attenuation to 180 dB, so that $L = 10^{18}$ and $P_{out} = 10^{-18} P_{in}$. This loss is so great that you'd need an input power of one megawatt (10^6 W) to get an output power of one picowatt (10^{-12} W)!

Large attenuation certainly calls for *amplification* to boost the output signal. As an example, Fig. 3.3-2 represents a cable transmission system with an output amplifier and a *repeater amplifier* inserted near the middle of the path. (Any *pre-amplification* at the input would be absorbed in the value of P_{in}.) Since power gains multiply in a cascade connection like this,

$$P_{out} = (g_1 g_2 g_3 g_4)P_{in} = \frac{g_2 g_4}{L_1 L_3} P_{in} \tag{9a}$$

which becomes the dB equation

$$P_{out} = (g_2 + g_4) - (L_1 + L_3) + P_{in} \tag{9b}$$

We've dropped the dB subscripts here for simplicity, but the addition and subtraction in Eq. (9b) unambiguously identifies it as a dB equation. Of course, the units of P_{out}(dBW or dBm) will be the same as those of P_{in}.

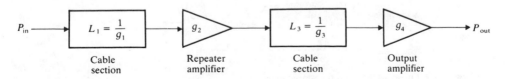

Cable   Repeater   Cable   Output
section   amplifier   section   amplifier

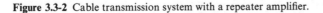

Figure 3.3-2 Cable transmission system with a repeater amplifier.

The repeater in Fig. 3.3-2 has been placed near the middle of the path to prevent the signal power from dropping down into the noise level of the amplifier. Long-haul cable systems have repeaters spaced every few kilometers for this reason, and a transcontinental telephone link might include more than 2000 repeaters. The signal-power analysis of such systems follows the same lines as Eq. (9). The noise analysis is presented in App. A.

Radio Transmission★

Signal transmission by radiowave propagation can reduce the required number of repeaters, and has the additional advantage of eliminating long cables. Although radio involves modulation processes described in later chapters, it seems appropriate here to examine the transmission loss for *line-of-sight propagation* illustrated in Fig. 3.3-3 where the radio wave travels a direct path from transmitting to receiving antenna. This propagation mode is commonly employed for long-distance communication at frequencies above about 100 MHz.

The *free-space loss* on a line-of-sight path is due to spherical dispersion of the radio wave. This loss is given by

$$L = \left(\frac{4\pi\ell}{\lambda}\right)^2 = \left(\frac{4\pi f \ell}{c}\right)^2 \tag{10a}$$

in which λ is the wavelength, f the signal frequency, and c the speed of light. If we express ℓ in kilometers and f in gigahertz (10^9 Hz), Eq. (10a) becomes

$$L_{dB} = 92.4 + 20 \log_{10} f_{GHz} + 20 \log_{10} \ell_{km} \tag{10b}$$

We see that L_{dB} increases as the logarithm of ℓ, rather than in direct proportion to path length. Thus, for instance, doubling the path length increases the loss by only 6 dB.

Furthermore, directional antennas have a focusing effect that acts like amplification in the sense that

$$P_{out} = \frac{g_T g_R}{L} P_{in} \tag{11}$$

Figure 3.3-3 Line-of-sight radio transmission.

where g_T and g_R represent the antenna gains at the transmitter and receiver. The maximum transmitting or receiving gain of an antenna with effective *aperture area A_e* is

$$g = \frac{4\pi A_e}{\lambda^2} = \frac{4\pi A_e f^2}{c^2} \qquad (12)$$

where $c \approx 3 \times 10^5$ km/s. The value of A_e for a horn or dish antenna approximately equals its physical area, and large parabolic dishes may provide gains in excess of 60 dB.

Example 3.3-1 Satellite relay system Figure 3.3-4 shows a transoceanic radio system with a satellite relay serving as a repeater. The satellite is about 40,000 km from either ground station, and the signal frequency is 6 GHz. Equation (10*b*) gives the uplink and downlink path loss

$$L_u = L_d = 92.4 + 20 \log_{10} 6 + 20 \log_{10} 4 \times 10^4 = 200 \text{ dB}$$

The antenna gains in dB are given on the drawing with subscripts identifying the various functions—for example, g_{Ru} stands for the receiving antenna gain on the uplink from ground to satellite. The satellite has a repeater amplifier with $g_{amp} = 80$ dB, and the input power at the ground station is $P_{in} = 100 \text{ W} = 50 \text{ dBm}$.

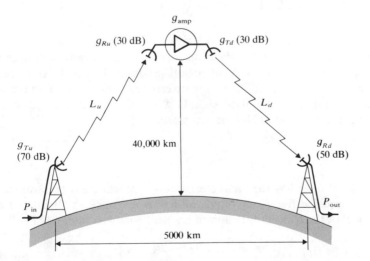

Figure 3.3-4 Satellite relay system.

Since we have all the data in dB, we can compute P_{out} by adding gains and subtracting losses in the following tabulation:

P_{in}	+50 dBm
g_{Tu}	+70 dB
L_u	−200 dB
g_{Ru}	+30 dB
g_{amp}	+80 dB
g_{Td}	+30 dB
L_d	−200 dB
g_{Rd}	+50 dB
P_{out}	−90 dBm

This negative result means that P_{out} is less than one milliwatt. Indeed, by inverting Eq. (4) we obtain

$$P_{out} = 10^{(-90 \text{ dBm}/10)} \times 1 \text{ mW} = 10^{-9} \text{ mW} = 10^{-12} \text{ W}$$

Such minute power levels are typical for satellite systems.

Exercise 3.3-2 A 40-km cable system has $P_{in} = 2$ W and a repeater with 64-dB gain is inserted 24 km from the input. The cable sections have $\alpha = 2.5$ dB/km. Use dB equations to find the signal power at: (a) the repeater's input; (b) the final output.

3.4 FILTERS AND FILTERING

Virtually every communication system includes one or more filters for the purpose of separating an information-bearing signal from unwanted contaminations such as interference, noise, and distortion products. In this section we'll define ideal filters, describe the differences between real and ideal filters, and examine the effect of filtering on pulsed signals. We also introduce quadrature phase-shift filtering and the related Hilbert transform.

Ideal Filters

By definition, an ideal filter has the characteristics of distortionless transmission over one or more specified frequency bands and zero response at all other frequencies. In particular, the transfer function of an ideal *bandpass* filter (BPF) is

$$H(f) = \begin{cases} Ke^{-j\omega t_d} & f_\ell \le |f| \le f_u \\ 0 & \text{otherwise} \end{cases} \tag{1}$$

as plotted in Fig. 3.4-1. The parameters f_ℓ and f_u are the lower and upper *cutoff*

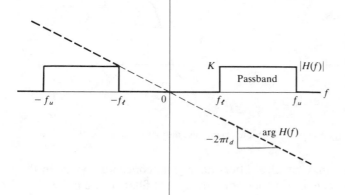

Figure 3.4-1 Transfer function of an ideal bandpass filter.

frequencies, respectively, since they mark the end points of the *passband*. The filter's *bandwidth* is

$$B = f_u - f_\ell$$

which we measure in terms of the positive-frequency portion of the passband.

In similar fashion, an ideal *lowpass* filter (LPF) is defined by Eq. (1) with $f_\ell = 0$, so $B = f_u$, while an ideal *highpass* filter (HPF) has $f_\ell > 0$ and $f_u = \infty$. Ideal *band-rejection* or *notch* filters provide distortionless transmission over all frequencies except some *stopband*, say $f_\ell \leq |f| \leq f_u$, where $H(f) = 0$.

But all such filters are physically *unrealizable* in the sense that their characteristics cannot be achieved with a finite number of elements. We'll skip the general proof of this assertion. Instead, we'll give an instructive plausibility argument based on the impulse response.

Consider an ideal LPF whose transfer function, shown in Fig. 3.4-2a, can be written as

$$H(f) = Ke^{-j\omega t_d}\Pi\left(\frac{f}{2B}\right) \tag{2a}$$

Its impulse response will be

$$h(t) = \mathscr{F}^{-1}[H(f)] = 2BK \operatorname{sinc} 2B(t - t_d) \tag{2b}$$

which is sketched in Fig. 3.4-2b. Since $h(t)$ is the response to $\delta(t)$ and $h(t)$ has nonzero values for $t < 0$, the *output appears before the input is applied*. Such a filter is said to be *anticipatory* or *noncausal*, and the portion of the output appearing before the input is called a *precursor*. Without doubt, such behavior is physically impossible, and hence the filter must be unrealizable. Like results hold for the ideal BPF and HPF.

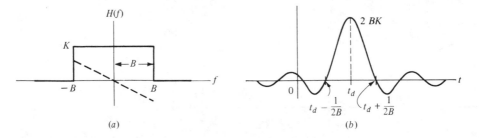

Figure 3.4-2 Ideal lowpass filter. (*a*) Transfer function; (*b*) impulse response.

Fictitious though they may be, ideal filters have great conceptual value in the study of communication systems. Furthermore, many real filters come quite close to ideal behavior.

> **Exercise 3.4-1** Show that the impulse response of an ideal BPF is
> $$h(t) = 2BK \text{ sinc } B(t - t_d) \cos \omega_c(t - t_d)$$
> where $\omega_c = \pi(f_\ell + f_u)$.

Real Filters

The design of realizable filters that approach ideal behavior is an advanced topic outside the scope of this book. But we should at least look at the major differences between real and ideal filters to gain some understanding of the approximations implied by the assumption of an ideal filter. Further information on filter design and implementation can be found in texts such as Van Valkenburg (1982).

To begin our discussion, Fig. 3.4-3 shows the amplitude ratio of a typical real bandpass filter. Compared with the ideal BPF in Fig. 3.4-1, we see a passband where $|H(f)|$ is relatively large (but not constant) and stopbands where $|H(f)|$ is quite small (but not zero). The end points of the passband are usually defined by

$$|H(f)| = \frac{1}{\sqrt{2}} |H(f)|_{\max} = \frac{K}{\sqrt{2}} \qquad f = f_\ell, f_u \qquad (3)$$

so that $|H(f)|^2$ falls no lower than $K^2/2$ for $f_\ell \le |f| \le f_u$. The bandwidth $B = f_u - f_\ell$ is then called the *half-power* or 3-dB bandwidth. Similarly, the end points of

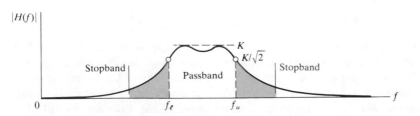

Figure 3.4-3 Typical amplitude ratio of a real bandpass filter.

the stopbands can be taken where $|H(f)|$ drops to a suitably small value such as $K/10$ or $K/100$.

Between the passband and stopbands are *transition* regions, shown shaded, where the filter neither "passes" nor "rejects" frequency components. Therefore, effective signal filtering often depends on having a filter with very narrow transition regions. We'll pursue this aspect by examining one particular class of filters in some detail. Then we'll describe other popular designs.

The simplest of the standard filter types is the nth-order *Butterworth* LPF, whose circuit contains n reactive elements (capacitors and inductors). The transfer function with $K = 1$ has the form

$$H(f) = \frac{1}{P_n(jf/B)} \tag{4a}$$

where B equals the 3-dB bandwidth and $P_n(jf/B)$ is a complex polynomial. The family of Butterworth polynomials is defined by the property

$$|P_n(jf/B)|^2 = 1 + (f/B)^{2n}$$

so that

$$|H(f)| = \frac{1}{\sqrt{1 + (f/B)^{2n}}} \tag{4b}$$

Consequently, the first n derivatives of $|H(f)|$ equal zero at $f = 0$ and we say that $|H(f)|$ is *maximally flat*. Table 3.4-1 lists the Butterworth polynomials for $n = 1$ through 4, using the normalized variable $p = jf/B$.

A first-order Butterworth filter has the same characteristics as an RC lowpass filter and would be a poor approximation of an ideal LPF. But the approximation improves as you increase n by adding more elements to the circuit. For instance, the impulse response of a third-order filter sketched in Fig. 3.4-4a bears obvious resemblance to that of an ideal LPF — without the precursors, of course. The frequency-response curves of this filter are plotted in Fig. 3.4-4b. Note that the phase shift has a reasonably linear slope over the passband, implying time delay plus some delay distortion.

A clearer picture of the amplitude ratio in the transition region is obtained from a *Bode diagram*, constructed by plotting $|H(f)|$ in dB versus f on a logarithmic scale. Figure 3.4-5 shows the Bode diagram for Butterworth lowpass filters with various values of n. If we define the edge of the stopband at $|H(f)| = -20$ dB, the width of the transition region when $n = 1$ is $10B - B = 9B$ but only

Table 3.4-1 Butterworth polynomials

n	$P_n(p)$
1	$1 + p$
2	$1 + \sqrt{2}\,p + p^2$
3	$(1 + p)(1 + p + p^2)$
4	$(1 + 0.765p + p^2)(1 + 1.848p + p^2)$

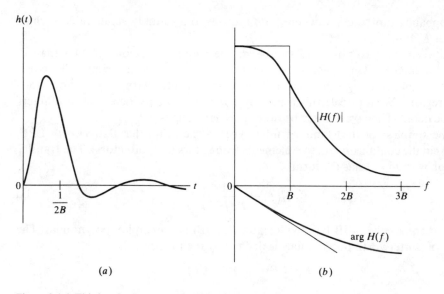

Figure 3.4-4 Third-order Butterworth LPF. (*a*) Impulse response; (*b*) transfer function.

Figure 3.4-5 Bode diagram for Butterworth LPFs.

$1.25B - B = 0.25B$ when $n = 10$. Clearly, $|H(f)|$ approaches the ideal square characteristic in the limit as $n \rightarrow \infty$. At the same time, however, the slope of the phase shift (not shown) increases with n and the delay distortion may become intolerably large.

In situations where potential delay distortion is a major concern, a *Bessel-Thomson* filter would be the preferred choice. This class of filters is characterized by *maximally linear phase shift* for a given value of n, but has a wider transition region. At the other extreme, the class of *equiripple* filters (including Chebyshev and elliptic filters) provides the sharpest transition for a given value of n; but these filters have small amplitude ripples in the passband and significantly non-linear phase shift. Equiripple filters would be satisfactory in audio applications, for instance, whereas pulse applications might call for the superior transient performance of Bessel-Thomson filters.

All three filter classes can be implemented with active devices (such as operational amplifiers) that eliminate the need for bulky inductors. *Switched-capacitor filter* designs go even further and eliminate resistors that would take up too much space in a large-scale integrated circuit. All three classes can also be modified to obtain highpass or bandpass filters. However, some practical implementation problems do arise when you want a bandpass filter with a narrow but reasonably square passband. Special designs that employ electromechanical phenomena have been developed for such applications. For example, Fig. 3.4-6 shows the amplitude ratio of a seventh-order monolithic crystal BPF intended for use in an AM radio.

Figure 3.4-6 Amplitude ratio of a mechanical filter.

Exercise 3.4-2 Show that a Butterworth LPF has $|H(f)|_{dB} \approx -20n \log_{10} (f/B)$ when $f > B$. The find the minimum value of n needed so that $|H(f)| \leq 1/100$ for $f \geq 2B$.

Pulse Response and Risetime

A rectangular pulse, or any other signal with an abrupt transition, contains significant high-frequency components that will be attenuated or eliminated by a lowpass filter. Pulse filtering therefore produces a smoothing or smearing effect that must be studied in the time domain. The study of pulse response undertaken here leads to useful information about pulse transmission systems.

Let's begin with the unit step input signal $x(t) = u(t)$, which could represent the leading edge of a rectangular pulse. In terms of the filter's impulse response $h(t)$, the step response will be

$$g(t) \triangleq \int_{-\infty}^{\infty} h(\lambda)u(t - \lambda) \, d\lambda = \int_{-\infty}^{t} h(\lambda) \, d\lambda \tag{5}$$

since $u(t - \lambda) = 0$ for $\lambda > t$. We saw in Example 3.1-1, for instance, that a first-order lowpass filter has

$$g(t) = (1 - e^{-2\pi Bt})u(t)$$

where B is the 3-dB bandwidth.

Of course a first-order LPF doesn't severely restrict high-frequency transmission. So let's go to the extreme case of an ideal LPF, taking unit gain and zero time delay for simplicity. From Eq. (2b) we have $h(t) = 2B \sin 2Bt$ and Eq. (5) becomes

$$g(t) = \int_{-\infty}^{t} 2B \text{ sinc } 2B\lambda \, d\lambda$$

$$= \int_{-\infty}^{0} \text{sinc } \mu \, d\mu + \int_{0}^{2Bt} \text{sinc } \mu \, d\mu$$

where $\mu = 2B\lambda$. The first integral is known to equal $1/2$, but the second requires numerical evaluation. Fortunately, the result can be expressed in terms of the tabulated *sine integral function*

$$\text{Si } (\theta) \triangleq \int_{0}^{\theta} \frac{\sin \alpha}{\alpha} \, d\alpha = \pi \int_{0}^{\theta/\pi} \text{sinc } \mu \, d\mu \tag{6}$$

which is plotted in Fig. 3.4-7 for $\theta > 0$ and approaches the value $\pi/2$ as $\theta \to \infty$. The function is also defined for $\theta < 0$ by virtue of the odd-symmetry property $\text{Si } (-\theta) = -\text{Si } (\theta)$. Using Eq. (6) in the problem at hand we get

$$g(t) = \frac{1}{2} + \frac{1}{\pi} \text{ Si } (2\pi Bt) \tag{7}$$

obtained by setting $\theta/\pi = 2Bt$.

Figure 3.4-7 The sine integral function.

For comparison purposes, Fig. 3.4-8 shows the step response of an ideal LPF along with that of a first-order LPF. The ideal LPF completely removes all high frequencies $|f| > B$, producing *precursors*, *overshoot*, and *oscillations* in the step response. (This behavior is the same as Gibbs' phenomenon illustrated in Fig. 2.1-10.) None of these effects appears in the response of the first-order LPF, which gradually attenuates but does not eliminate high frequencies. The step response of a more selective filter — a third-order Butterworth LPF, for example — would more nearly resemble a time-delayed version of the ideal LPF response.

Before moving on to pulse response per se, there's an important conclusion to be drawn from Fig. 3.4-8 regarding *risetime*. Risetime is a measure of the "speed" of a step response, usually defined as the time interval t_r between $g(t) = 0.1$ and $g(t) = 0.9$ and known as the 10–90% risetime. The risetime of a first-order lowpass filter can be computed from $g(t)$ as $t_r \approx 0.35/B$, while the ideal filter has $t_r \approx 0.44/B$. Both values are reasonably close to $0.5/B$ so we'll use the approximation

$$t_r \approx \frac{1}{2B} \qquad (8)$$

for the risetime of an arbitrary lowpass filter with bandwidth B.

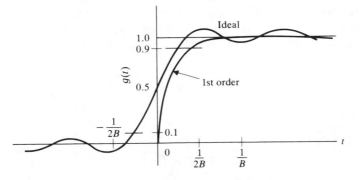

Figure 3.4-8 Step response of ideal and first-order LPFs.

Our work with step response pays off immediately in the calculation of pulse response if we take the input signal to be a unit-height rectangular pulse with duration τ starting at $t = 0$. Then we can write

$$x(t) = u(t) - u(t - \tau)$$

and hence

$$y(t) = g(t) - g(t - \tau)$$

which follows from superposition.

Using $g(t)$ from Eq. (7), we obtain the pulse response of an ideal LPF as

$$y(t) = \frac{1}{\pi} \{ \text{Si} \, (2\pi Bt) - \text{Si} \, [2\pi B(t - \tau)] \} \tag{9}$$

which is plotted in Fig. 3.4-9 for three values of the product $B\tau$. The response has a more-or-less rectangular shape when $B\tau \geq 2$, whereas it becomes badly smeared and spread out if $B\tau \leq 1/4$. The intermediate case $B\tau = 1/2$ gives a recognizable but not rectangular output pulse. The same conclusions can be drawn from the pulse response of a first-order lowpass filter previously sketched in Fig. 3.1-3, and similar results would hold for other input pulse shapes and other lowpass filter characteristics.

Now we're in a position to make some general statements about *bandwidth requirements for pulse transmission.* Reproducing the actual pulse shape requires a large bandwidth, say

$$B \gg \frac{1}{\tau_{\min}}$$

where $\tau_{\min}$ represents the smallest output pulse duration. But if we only need to *detect* that a pulse has been sent, or perhaps measure the pulse amplitude, we can get by with the smaller bandwidth

$$B \geq \frac{1}{2\tau_{\min}} \tag{10}$$

an important and handy rule of thumb.

Equation (10) also gives the condition for distinguishing between, or *resolving*, output pulses spaced by $\tau_{\min}$ or more. Figure 3.4-10 shows the resolution condition for an ideal lowpass channel with $B = 1/2\tau$. A smaller bandwidth or

Figure 3.4-9 Pulse response of an ideal LPF.

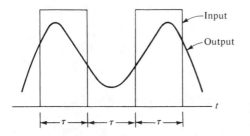

Figure 3.4-10 Pulse resolution of an ideal LPF, $B = 1/2\tau$.

smaller spacing would result in considerable overlap, making it difficult to identify separate pulses.

Besides pulse detection and resolution, we'll occasionally be concerned with *pulse position* measured relative to some reference time. Such measurements have inherent ambiguity due to the rounded output pulse shape and nonzero risetime of leading and trailing edges. For a specified minimum risetime, Eq. (8) yields the bandwidth requirement

$$B \geq \frac{1}{2t_{r_{min}}} \tag{11}$$

another handy rule of thumb.

Throughout the foregoing discussion we've tacitly assumed that the transmission channel has satisfactory phase-shift characteristics. If not, the resulting delay distortion could render the channel useless for pulse transmission, regardless of the bandwidth. Therefore, our bandwidth requirements in Eqs. (10) and (11) imply the additional stipulation of nearly linear phase shift over $|f| \leq B$. A phase equalization network may be needed to achieve this condition.

Exercise 3.4-3 A certain signal consists of pulses whose durations range from 10 to 25 μs; the pulses occur at random times, but a given pulse always starts at least 30 μs after the starting time of the previous pulse. Find the minimum transmission bandwidth required for pulse detection and resolution, and estimate the resulting risetime at the output.

Quadrature Filters and Hilbert Transforms

While the Fourier transform serves most of our needs in the study of filtered signals, the Hilbert transform provides a more convenient approach to certain particular topics. We'll introduce this specialized transform in conjunction with quadrature filtering.

A *quadrature filter* is an allpass network that merely shifts the phase of positive frequency components by $-90°$ and negative frequency components by $+90°$. Since a $\pm 90°$ phase shift is equivalent to multiplying by $e^{\pm j90°} = \pm j$, the transfer function can be written in terms of the signum function as

$$H_Q(f) = -j \operatorname{sgn} f = \begin{cases} -j & f > 0 \\ +j & f < 0 \end{cases} \tag{12a}$$

Figure 3.4-11 Transfer function of a quadrature phase shifter.

which is plotted in Fig. 3.4-11. The corresponding impulse response is

$$h_Q(t) = \frac{1}{\pi t} \tag{12b}$$

We obtain this result by applying duality to $\mathscr{F}[\text{sgn } t] = 1/j\pi f$ which yields $\mathscr{F}[1/j\pi t] = \text{sgn } (-f) = -\text{sgn } f$, so $\mathscr{F}^{-1}[-j \text{ sgn } f] = j/j\pi t = 1/\pi t$.

Now let an arbitrary signal $x(t)$ be the input to a quadrature filter. The output signal $y(t) = x(t) * h_Q(t)$ will be defined as the *Hilbert transform* of $x(t)$, denoted by $\hat{x}(t)$. Thus

$$\hat{x}(t) \triangleq x(t) * \frac{1}{\pi t} = \frac{1}{\pi} \int_{-\infty}^{\infty} \frac{x(\lambda)}{t - \lambda} \, d\lambda \tag{13}$$

Note that Hilbert transformation is a *convolution* and does not change the domain, so both $x(t)$ and $\hat{x}(t)$ are functions of time. Even so, we can easily write the spectrum of $\hat{x}(t)$, namely

$$\mathscr{F}[\hat{x}(t)] = (-j \text{ sgn } f)X(f) \tag{14}$$

since phase shifting produces the output spectrum $H_Q(f)X(f)$.

The catalog of Hilbert transform pairs is quite short compared to our Fourier transform catalog, and the Hilbert transform does not even exist for many common signal models. Mathematically, the trouble comes from potential singularities in Eq. (13) when $\lambda = t$ and the integrand becomes undefined. Physically, we see from Eq. (12b) that $h_Q(t)$ is *noncausal*, which means that the quadrature filter is *unrealizable*—although its behavior can be approximated over a finite frequency band using a real network.

The simplest and most obvious Hilbert transform pair follows directly from the phase shift property of the quadrature filter. Specifically, if

$$x(t) = A \cos (\omega_0 t + \phi) \tag{15a}$$

then a phase shift of $-90°$ produces

$$\hat{x}(t) = A \sin (\omega_0 t + \phi) \tag{15b}$$

Furthermore, Eq. (15) readily generalizes to any signal that consists of a sum of sinusoids. However, most other Hilbert transforms involve performing the convolution operation in Eq. (13), as illustrated by the following example.

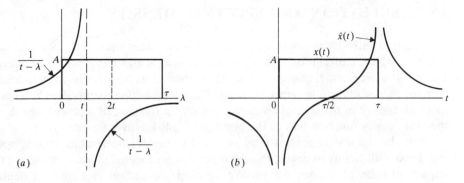

Figure 3.4-12 Hilbert transform of a rectangular pulse. (*a*) Convolution; (*b*) result.

Example 3.4-1 Hilbert transform of a rectangular pulse Consider the delayed rectangular pulse $x(t) = A[u(t) - u(t - \tau)]$. The Hilbert transform is

$$\hat{x}(t) = \frac{A}{\pi} \int_0^\tau \frac{1}{t - \lambda}\, d\lambda$$

whose evaluation requires graphical interpretation. Figure 3.4-12*a* shows the case $0 < t < \tau/2$ and we see that the areas cancel out between $\lambda = 0$ and $\lambda = 2t$, leaving

$$\hat{x}(t) = \frac{A}{\pi} \int_{2t}^\tau \frac{d\lambda}{t - \lambda} = \frac{A}{\pi} \left[\ln(-t) - \ln(t - \tau) \right]$$

$$= \frac{A}{\pi} \ln\left(\frac{-t}{t - \tau}\right) = \frac{A}{\pi} \ln\left(\frac{t}{\tau - t}\right)$$

This result also holds for $\tau/2 < t < \tau$, when the areas cancel out between $\lambda = 2t - \tau$ and $\lambda = \tau$. There is no area cancellation for $t < 0$ or $t > \tau$, and

$$\hat{x}(t) = \frac{A}{\pi} \int_0^\tau \frac{d\lambda}{t - \lambda} = \frac{A}{\pi} \ln\left(\frac{t}{t - \tau}\right)$$

These separate cases can be combined in one expression

$$\hat{x}(t) = \frac{A}{\pi} \ln \left| \frac{t}{t - \tau} \right| \tag{16}$$

which is plotted in Fig. 3.4-12*b* along with $x(t)$.

The infinite spikes in $\hat{x}(t)$ at $t = 0$ and $t = \tau$ can be viewed as an extreme manifestation of *delay distortion*. See Fig. 3.3-3 for comparison.

Exercise 3.4-4 The inverse Hilbert transform recovers $x(t)$ from $\hat{x}(t)$. Use spectral analysis to show that $\hat{x}(t) * (-1/\pi t) = x(t)$.

3.5 CORRELATION AND SPECTRAL DENSITY

This section introduces *correlation functions* as another approach to signal and system analysis. Correlation focuses on time averages and signal power or energy. Taking the Fourier transform of a correlation function leads to frequency-domain representation in terms of *spectral density functions*, equivalent to energy spectral density in the case of an energy signal. In the case of a power signal, the spectral density function tells us the power distribution over frequency.

But the signals themselves need not be Fourier transformable. Hence, spectral density allows us to deal with a broader range of signal models, including the important class of *random* signals. We develop correlation and spectral density here as analytic tools for nonrandom signals. You should then feel more comfortable with them when we get to random signals in Chap. 5.

Correlation of Power Signals

Let $v(t)$ be a power signal, but not necessarily real nor periodic. Our only stipulation is that it must have well-defined *average power*

$$P_v \triangleq \langle |v(t)|^2 \rangle = \langle v(t)v^*(t) \rangle \geq 0 \tag{1}$$

The time-averaging operation here is interpreted in the general form

$$\langle z(t) \rangle = \lim_{T \to \infty} \frac{1}{T} \int_{-T/2}^{T/2} z(t) \, dt$$

where $z(t)$ is an arbitrary time function. For reference purposes, we note that this operation has the following properties:

$$\langle z^*(t) \rangle = \langle z(t) \rangle^* \tag{2a}$$

$$\langle z(t - t_d) \rangle = \langle z(t) \rangle \qquad \text{any } t_d \tag{2b}$$

$$\langle a_1 z_1(t) + a_2 z_2(t) \rangle = a_1 \langle z_1(t) \rangle + a_2 \langle z_2(t) \rangle \tag{2c}$$

We'll have frequent use for these properties in conjunction with correlation.

If $v(t)$ and $w(t)$ are power signals, the average $\langle v(t)w^*(t) \rangle$ is called the *scalar product* of $v(t)$ and $w(t)$. The scalar product is a number, possibly complex, that serves as a measure of *similarity* between the two signals. *Schwarz's inequality* relates the scalar product to the signal powers P_v and P_w, in that

$$|\langle v(t)w^*(t) \rangle|^2 \leq P_v P_w \tag{3}$$

You can easily confirm that the equality holds when $v(t) = aw(t)$, with a being an arbitrary constant. Hence, $|\langle v(t)w^*(t) \rangle|$ is maximum when the signals are *proportional*. We'll soon define correlation in terms of the scalar product.

First, however, let's further interpret $\langle v(t)w^*(t) \rangle$ and prove Schwarz's inequality by considering

$$z(t) = v(t) - aw(t) \tag{4a}$$

The average power of $z(t)$ is

$$P_z = \langle z(t)z^*(t) \rangle = \langle [v(t) - aw(t)][v^*(t) - a^*w(t)] \rangle$$
$$= \langle v(t)v^*(t) \rangle + aa^*\langle w(t)w^*(t) \rangle - a^*\langle v(t)w^*(t) \rangle - a\langle v^*(t)w(t) \rangle$$
$$= P_v + aa^*P_w - 2 \text{ Re } [a^*\langle v(t)w^*(t) \rangle] \tag{4b}$$

where Eqs. (2a) and (2c) have been used to expand and combine terms. If $a = 1$, then $z(t) = v(t) - w(t)$ and

$$P_z = P_v + P_w - 2 \text{ Re } \langle v(t)w^*(t) \rangle$$

A large value of the scalar product thus implies similar signals, in the sense that the difference signal $v(t) - w(t)$ has small average power. Conversely, a small scalar product implies dissimilar signals and $P_z \approx P_v + P_w$.

To prove Schwarz's inequality from Eq. (4b), let $a = \langle v(t)w^*(t) \rangle / P_w$ so

$$aa^*P_w = a^*\langle v(t)w^*(t) \rangle = |\langle v(t)w^*(t) \rangle|^2 / P_w$$

Then $P_z = P_v - |\langle v(t)w^*(t) \rangle|^2 / P_w \geq 0$, which reduces to Eq. (3) and completes the preliminary work.

Now we define the *crosscorrelation* of two power signals as†

$$R_{vw}(\tau) \triangleq \langle v(t)w^*(t - \tau) \rangle = \langle v(t + \tau)w^*(t) \rangle \tag{5}$$

This is a scalar product with the second signal delayed by τ relative to the first or, equivalently, the first signal advanced by τ relative to the second. The relative displacement τ is the *independent variable* in Eq. (5), the variable t having been washed out in the time average. General properties of $R_{vw}(\tau)$ are

$$|R_{vw}(\tau)|^2 \leq P_v P_w \tag{6a}$$

$$R_{wv}(\tau) = R_{vw}^*(-\tau) \tag{6b}$$

Equation (6a) simply restates Schwarz's inequality, while Eq. (6b) points out that $R_{wv}(\tau) \neq R_{vw}(\tau)$.

We conclude from our previous observations that $R_{vw}(\tau)$ measures the similarity between $v(t)$ and $w(t - \tau)$ as a function of τ. Crosscorrelation is thus a more sophisticated measure than the ordinary scalar product since it detects time-shifted similarities or differences that would be ignored in $\langle v(t)w^*(t) \rangle$.

But suppose we correlate a signal with itself, generating the *autocorrelation function*

$$R_v(\tau) \triangleq R_{vv}(\tau) = \langle v(t)v^*(t - \tau) \rangle = \langle v(t + \tau)v^*(t) \rangle \tag{7}$$

This autocorrelation tells us something about the *time variation* of $v(t)$, at least in an averaged sense. If $|R_v(\tau)|$ is large, we infer that $v(t - \tau)$ is very similar to $v(t)$ for that particular value of τ; whereas if $|R_v(\tau)|$ is small, then $v(t)$ and $v(t - \tau)$ must look quite different.

---

† Another definition used by some authors is $\langle v^*(t)w(t + \tau) \rangle$, equivalent to interchanging the subscripts on $R_{vw}(\tau)$ in Eq. (5).

Properties of the autocorrelation function include

$$R_v(0) = P_v \tag{8a}$$

$$|R_v(\tau)| \le R_v(0) \tag{8b}$$

$$R_v(-\tau) = R_v^*(\tau) \tag{8c}$$

Hence, $R_v(\tau)$ has hermitian symmetry and a maximum value at the origin equal to the signal power. If $v(t)$ is real, then $R_v(\tau)$ will be real and even. If $v(t)$ happens to be periodic, $R_v(\tau)$ will have the same periodicity.

Lastly, consider the sum or difference signal

$$z(t) = v(t) \pm w(t) \tag{9a}$$

Upon forming its autocorrelation, we find that

$$R_z(\tau) = R_v(\tau) + R_w(\tau) \pm [R_{vw}(\tau) + R_{wv}(\tau)] \tag{9b}$$

If $v(t)$ and $w(t)$ are *uncorrelated* for all τ, so

$$R_{vw}(\tau) = R_{wv}(\tau) = 0$$

then $R_z(\tau) = R_v(\tau) + R_w(\tau)$ and setting $\tau = 0$ yields

$$P_z = P_v + P_w$$

Superposition of average power therefore holds for uncorrelated signals.

Example 3.5-1 Correlation of phasors and sinusoids The calculation of correlation functions for phasors and sinusoidal signals is expedited by calling upon Eq. (18), Sect. 2.1, written as

$$\langle e^{j\omega_1 t} e^{-j\omega_2 t} \rangle = \lim_{T \to \infty} \frac{1}{T} \int_{-T/2}^{T/2} e^{j(\omega_1 - \omega_2)t}\, dt$$

$$= \lim_{T \to \infty} \text{sinc} \frac{(\omega_1 - \omega_2)T}{2\pi} = \begin{cases} 0 & \omega_2 \ne \omega_1 \\ 1 & \omega_2 = \omega_1 \end{cases} \tag{10}$$

We'll apply this result to the phasor signals

$$v(t) = C_v e^{j\omega_v t} \qquad w(t) = C_w e^{j\omega_w t} \tag{11a}$$

where C_v and C_w are complex constants incorporating the amplitude and phase angle. The crosscorrelation is

$$R_{vw}(\tau) = \langle [C_v e^{j\omega_v t}][C_w e^{j\omega_w (t-\tau)}]^* \rangle$$

$$= C_v C_w^* e^{j\omega_w \tau} \langle e^{j\omega_v t} e^{-j\omega_w t} \rangle$$

$$= \begin{cases} 0 & \omega_w \ne \omega_v \\ C_v C_w^* e^{j\omega_v \tau} & \omega_w = \omega_v \end{cases} \tag{11b}$$

Hence, the phasors are uncorrelated unless they have identical frequencies.

The autocorrelation function is

$$R_v(\tau) = |C_v|^2 e^{j\omega_v \tau} \tag{11c}$$

which drops out of Eq. (11b) when $w(t) = v(t)$.

Now it becomes a simple task to show that the sinusoidal signal

$$z(t) = A \cos(\omega_0 t + \phi) \tag{12a}$$

has

$$R_z(\tau) = \frac{A^2}{2} \cos \omega_0 \tau \tag{12b}$$

Clearly, $R_z(\tau)$ is real, even, and periodic, and has the maximum value $R_z(0) = A^2/2 = P_z$. This maximum also occurs whenever $\omega_0 \tau$ equals a multiple of 2π radians, so $z(t \pm \tau) = z(t)$. On the other hand, $R_z(\tau) = 0$ when $z(t \pm \tau)$ and $z(t)$ are in phase quadrature.

But notice that the phase angle ϕ does not appear in $R_z(\tau)$, owing to the averaging effect of correlation. This emphasizes the fact that the autocorrelation function does not uniquely define a signal.

Exercise 3.5-1 Derive Eq. (12b) by writing $z(t)$ as a sum of conjugate phasors and applying Eqs. (9) and (11).

Correlation of Energy Signals

Averaging products of energy signals over all time yields zero. But we can meaningfully speak of the *total energy*

$$E_v \triangleq \int_{-\infty}^{\infty} v(t)v^*(t) \, dt \geq 0 \tag{13}$$

Similarly, the correlation functions for energy signals can be defined as

$$R_{vw}(\tau) \triangleq \int_{-\infty}^{\infty} v(t)w^*(t - \tau) \, dt \tag{14a}$$

$$R_v(\tau) \triangleq R_{vv}(\tau) \tag{14b}$$

Since the integration operation $\int_{-\infty}^{\infty} z(t) \, dt$ has the same mathematical properties as the time-average operation $\langle z(t) \rangle$, all of our previous corelation relations hold for the case of energy signals if we replace average power P_v with total energy E_v. Thus, for instance, we have the property

$$|R_{vw}(\tau)|^2 \leq E_v E_w \tag{15}$$

as the energy-signal version of Eq. (6a).

Closer examination of Eq. (14) reveals that energy-signal correlation is a type of *convolution*. For with $z(t) = w^*(-t)$ and $t = \lambda$, the right-hand side of Eq. (14a) becomes

$$\int_{-\infty}^{\infty} v(\lambda)z(\tau - \lambda) \, d\lambda = v(\tau) * z(\tau)$$

and therefore

$$R_{vw}(\tau) = v(\tau) * w^*(-\tau) \qquad (16)$$

Likewise, $R_v(\tau) = v(\tau) * v^*(-\tau)$.

Some additional relations are obtained in terms of the Fourier transforms $V(f) = \mathcal{F}[v(t)]$, etc. Specifically, from Eqs. (16) and (17), Sect. 2.2,

$$R_v(0) = E_v = \int_{-\infty}^{\infty} |V(f)|^2 \, df$$

$$R_{vw}(0) = \int_{-\infty}^{\infty} v(t)w^*(t) \, dt = \int_{-\infty}^{\infty} V(f)W^*(f) \, df$$

Combining these integrals with $|R_{vw}(0)|^2 \le E_v E_w = R_v(0)R_w(0)$ yields

$$\left| \int_{-\infty}^{\infty} V(f)W^*(f) \, df \right|^2 \le \int_{-\infty}^{\infty} |V(f)|^2 \, df \int_{-\infty}^{\infty} |W(f)|^2 \, df \qquad (17)$$

Equation (17) is a frequency-domain statement of Schwarz's inequality. The equality holds when $V(f)$ and $W(f)$ are proportional.

Exercise 3.5-2 Let $v(t) = A[u(t) - u(t - D)]$ and $w(t) = v(t - t_d)$. Use Eq. (16) with $z(\tau) = w^*(-\tau)$ to sketch $R_{vw}(\tau)$. Confirm from your sketch that $|R_{vw}(\tau)|^2 \le E_v E_w$ and that $|R_{vw}(\tau)|^2_{\max} = E_v E_w$ at $\tau = -t_d$.

Input-Output Correlations

We next investigate system analysis in the "τ domain," as represented by Fig. 3.5-1. A signal $x(t)$ having known autocorrelation $R_x(\tau)$ is applied to an LTI system with impulse response $h(t)$, producing the output signal

$$y(t) = h(t) * x(t) = \int_{-\infty}^{\infty} h(\lambda)x(t - \lambda) \, d\lambda$$

We'll show that the input-output crosscorrelation function is

$$R_{yx}(\tau) = h(\tau) * R_x(\tau) = \int_{-\infty}^{\infty} h(\lambda)R_x(\tau - \lambda) \, d\lambda \qquad (18)$$

Figure 3.5-1

and that the output autocorrelation function is

$$R_y(\tau) = h^*(-\tau) * R_{yx}(\tau) = \int_{-\infty}^{\infty} h^*(-\mu)R_{yx}(\tau - \mu) \, d\mu \qquad (19a)$$

Substituting Eq. (18) into (19a) then gives

$$R_y(\tau) = h^*(-\tau) * h(\tau) * R_x(\tau) \qquad (19b)$$

Note that these τ-domain relations are convolutions, similar to the time-domain relation.

For derivation purposes, let's assume that $x(t)$ and $y(t)$ are power signals so we can use the compact time-averaged notation. Obviously, the same results will hold when $x(t)$ and $y(t)$ are both energy signals. The assumption of a stable system ensures that $y(t)$ will be the same type of signal as $x(t)$.

Starting with the crosscorrelation $R_{yx}(\tau) = \langle y(t)x^*(t - \tau) \rangle$, we insert the superposition integral $h(t) * x(t)$ for $y(t)$ and interchange the order of operations to get

$$R_{yx}(\tau) = \int_{-\infty}^{\infty} h(\lambda)\langle x(t - \lambda)x^*(t - \tau) \rangle \, d\lambda$$

But since $\langle z(t) \rangle = \langle z(t + \lambda) \rangle$ for any λ,

$$\langle x(t - \lambda)x^*(t - \tau) \rangle = \langle x(t + \lambda - \lambda)x^*(t + \lambda - \tau) \rangle$$
$$= \langle x(t)x^*[t - (\tau - \lambda)] \rangle$$
$$= R_x(\tau - \lambda)$$

Hence,

$$R_{yx}(\tau) = \int_{-\infty}^{\infty} h(\lambda)R_x(\tau - \lambda) \, d\lambda$$

Proceeding in the same fashion for $R_y(\tau) = \langle y(t)y^*(t - \tau) \rangle$ we arrive at

$$R_y(\tau) = \int_{-\infty}^{\infty} h^*(\lambda)\langle y(t)x^*(t - \tau - \lambda) \rangle \, d\lambda$$

in which $\langle y(t)x^*(t - \tau - \lambda) \rangle = R_{yx}(\tau + \lambda)$. Equation (19a) follows from the change of variable $\mu = -\lambda$.

Spectral Density Functions

At last we're prepared to discuss spectral density functions. Given a power or energy signal $v(t)$, its spectral density function $G_v(f)$ represents the distribution of power or energy in the frequency domain and has two essential properties. First, the area under $G_v(f)$ equals the average power or total energy, so

$$\int_{-\infty}^{\infty} G_v(f) \, df = R_v(0) \qquad (20)$$

Second, if $x(t)$ is the input to an LTI system with $H(f) = \mathscr{F}[h(t)]$, then the input and output spectral density functions are related by

$$G_y(f) = |H(f)|^2 G_x(f) \tag{21}$$

since $|H(f)|^2$ is the power or energy gain at any f. These two properties are combined in

$$R_y(0) = \int_{-\infty}^{\infty} |H(f)|^2 G_x(f) \, df \tag{22}$$

which expresses the output power or energy $R_y(0)$ in terms of the input spectral density.

Equation (22) leads to a physical interpretation of spectral density with the help of Fig. 3.5-2. Here, $G_x(f)$ is arbitrary and $|H(f)|^2$ acts like a narrowband filter with unit gain, so

$$G_y(f) = \begin{cases} G_x(f) & f_c - \dfrac{\Delta f}{2} < f < f_c + \dfrac{\Delta f}{2} \\ \\ 0 & \text{otherwise} \end{cases}$$

If Δf is sufficiently small, the area under $G_y(f)$ will be $R_y(0) \approx G_x(f_c)\,\Delta f$ and

$$G_x(f_c) \approx R_y(0)/\Delta f$$

We conclude that at any frequency $f = f_c$, $G_x(f_c)$ equals the signal *power or energy per unit frequency*. We further conclude that any spectral density function must be *real* and *nonnegative* for all values of f.

But how do you determine $G_v(f)$ from $v(t)$? The *Wiener–Kinchine theorem* states that you first calculate the autocorrelation function and then take its

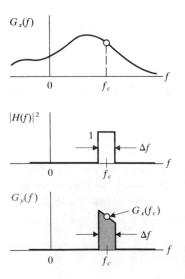

Figure 3.5-2 Interpretation of spectral density functions.

Fourier transform. Thus,

$$G_v(f) = \mathcal{F}_\tau[R_v(\tau)] \triangleq \int_{-\infty}^{\infty} R_v(\tau) e^{-j2\pi f \tau} \, d\tau \tag{23a}$$

where $\mathcal{F}_\tau$ stands for the Fourier transform operation with τ in place of t. The inverse transform is

$$R_v(\tau) = \mathcal{F}_\tau^{-1}[G_v(f)] \triangleq \int_{-\infty}^{\infty} G_v(f) e^{j2\pi f \tau} \, df \tag{23b}$$

so we have the Fourier transform pair

$$R_v(\tau) \leftrightarrow G_v(f)$$

All of our prior transform theorems therefore may be invoked to develop relationships between autocorrelation and spectral density.

If $v(t)$ is an *energy* signal with $V(f) = \mathcal{F}[v(t)]$, application of Eqs. (16) and (23a) shows that

$$G_v(f) = |V(f)|^2 \tag{24}$$

and we have the *energy spectral density*. If $v(t)$ is a *periodic power* signal with the Fourier series expansion

$$v(t) = \sum_{n=-\infty}^{\infty} c(nf_0) e^{j2\pi n f_0 t} \tag{25a}$$

the Wiener-Kinchine theorem gives the *power spectral density*, or *power spectrum*, as

$$G_v(f) = \sum_{n=-\infty}^{\infty} |c(nf_0)|^2 \, \delta(f - nf_0) \tag{25b}$$

This power spectrum consists of *impulses* representing the average phasor power $|c(nf_0)|^2$ concentrated at each harmonic frequency $f = nf_0$. Substituting Eq. (25b) into Eq. (20) then yields a restatement of Parseval's power theorem. In the special case of a sinusoidal signal

$$z(t) = A \cos(\omega_0 t + \phi)$$

we use $R_z(\tau)$ from Eq. (12b) to get

$$G_v(f) = \mathcal{F}_\tau[(A^2/2) \cos 2\pi f_0 \tau]$$

$$= \frac{A^2}{4} \delta(f - f_0) + \frac{A^2}{4} \delta(f + f_0)$$

which is plotted in Fig. 3.5-3.

All of the foregoing cases lend support to the Wiener-Kinchine theorem but do not constitute a general proof. To prove the theorem, we must confirm that taking $G_v(f) = \mathcal{F}_\tau[R_v(\tau)]$ satisfies the properties in Eqs. (20) and (21). The former

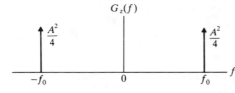

Figure 3.5-3 Power spectrum of $z(t) = A \cos(\omega_0 t + \phi)$.

immediately follows from the inverse transform in Eq. (23b) with $\tau = 0$. Now recall the output autocorrelation expression

$$R_y(\tau) = h^*(-\tau) * h(\tau) * R_x(\tau)$$

Since

$$\mathscr{F}_\tau[h(\tau)] = H(f) \qquad \mathscr{F}_\tau[h^*(-\tau)] = H^*(f)$$

the convolution theorem yields

$$\mathscr{F}_\tau[R_y(\tau)] = H^*(f)H(f)\mathscr{F}_\tau[R_x(\tau)]$$

and thus $G_y(f) = |H(f)|^2 G_x(f)$ if we take $\mathscr{F}_\tau[R_y(\tau)] = G_y(f)$, etc.

Example 3.5-2 Comb Filter Consider the *comb filter* in Fig. 3.5-4a. The impulse response is

$$h(t) = \delta(t) - \delta(t - T)$$

so

$$H(f) = 1 - e^{-j2\pi fT}$$

and

$$|H(f)|^2 = 2 - e^{-j2\pi fT} - e^{j2\pi fT}$$

$$= 4 \sin^2 2\pi(f/f_c) \qquad f_c = 1/2T$$

The sketch of $|H(f)|^2$ in Fig. 3.5-4b explains the name of this filter.

If we know the input spectral density, the output density and autocorrelation can be found from

$$G_y(f) = 4 \sin^2 2\pi(f/f_c) \, G_x(f)$$

$$R_y(\tau) = \mathscr{F}_\tau^{-1}[G_y(f)]$$

If we also know the input autocorrelation, we can write

$$R_y(\tau) = \mathscr{F}_\tau^{-1}[|H(f)|^2] * R_x(f)$$

(a)

(b)

Figure 3.5-4 Comb filter.

where, using the exponential expression for $|H(f)|^2$,

$$\mathcal{F}_\tau^{-1}[|H(f)|^2] = 2\,\delta(\tau) - \delta(\tau - T) - \delta(\tau + T)$$

Therefore,

$$R_y(\tau) = 2R_x(\tau) - R_x(\tau - T) - R_x(\tau + T)$$

and the output power or energy is $R_y(0) = 2R_x(0) - R_x(-T) - R_x(T)$.

Exercise 3.5-3 Let $v(t)$ be an energy signal, Show that $\mathcal{F}_\tau[v^*(-\tau)] = V^*(f)$. Then derive $G_v(f) = |V(f)|^2$ by applying Eq. (23a) to Eq. (16).

3.6 PROBLEMS

3.1-1 A given system has impulse response $h(t)$ and transfer function $H(f)$. Obtain expressions for $y(t)$ and $Y(f)$ when $x(t) = A[\delta(t + t_d) - \delta(t - t_d)]$.

3.1-2 Do Prob. 3.1-1 with $x(t) = A[\delta(t) + \delta(t - t_d)]$.

3.1-3 Do Prob. 3.1-1 with $x(t) = Ah(t - t_d)$.

3.1-4 Do Prob. 3.1-1 with $x(t) = Au(t - t_d)$.

3.1-5 Justify Eq. (7b) from Eq. (14) with $x(t) = u(t)$.

3.1-6 Find and sketch $|H(f)|$ and arg $H(f)$ for a system described by the differential equation $dy(t)/dt + 2\pi y(t) = dx(t)/dt + 20\pi x(t)$.

3.1-7 Do Prob. 3.1-6 with $dy(t)/dt + 20\pi y(t) = dx(t)/dt + 2\pi x(t)$.

3.1-8 Do Prob. 3.1-6 with $dy(t)/dt + 2\pi y(t) = -dx(t)/dt + 2\pi x(t)$.

3.1-9 Use frequency-domain analysis to obtain an approximate expression for $y(t)$ when $H(f) = B/(B + jf)$ and $x(t)$ is such that $X(f) \approx 0$ for $|f| < W$ with $W \gg B$.

3.1-10 Use frequency-domain analysis to obtain an approximate expression for $y(t)$ when $H(f) = jf/(B + jf)$ and $x(t)$ is such that $X(f) \approx 0$ for $|f| > W$ with $W \ll B$.

3.1-11 The input to an RC lowpass filter is $x(t) = \mathrm{sinc}\ 2Wt$. Plot the energy ratio E_y/E_x versus B/W.

3.1-12 Sketch and label the impulse response of the cascade system in Fig. 3.1-8b when the blocks represent zero-order holds with time delays $T_1 > T_2$.

3.1-13 Sketch and label the impulse response of the cascade system in Fig. 3.1-8b when $H_1(f) = [1 + j(f/B)]^{-1}$ and the second block represents a zero-order hold with time delay $T \gg 1/B$.

3.1-14 Find the step and impulse response of the feedback system in Fig. 3.1-8c when $H_1(f)$ is an integrator and $H_2(f)$ is a gain K.

3.1-15 Find the step and impulse response of the feedback system in Fig. 3.1-8c when $H_1(f)$ is a gain K and $H_2(f)$ is an integrator.

3.1-16‡ If $H(f)$ is the transfer function of a physically realizable system, then $h(t)$ must be real and causal. As a consequence, for $t \geq 0$ show that

$$h(t) = 4 \int_0^\infty H_r(f) \cos \omega t\ df = 4 \int_0^\infty H_i(f) \cos \omega t\ df$$

where $H_r(f) = \mathrm{Re}\ [H(f)]$ and $H_i(f) = \mathrm{Im}\ [H(f)]$.

3.2-1 Show that a first-order lowpass system yields essentially distortionless transmission if $x(t)$ is bandlimited to $W \ll B$.

3.2-2 Find and sketch $y(t)$ when the test signal in Fig. 3.2-3 is applied to a first-order lowpass system with $B = 3f_0$.

3.2-3 Find and sketch $y(t)$ when the test signal in Fig. 3.2-3 is applied to a first-order highpass with $H(f) = jf/(B + jf)$ and $B = 3f_0$.

3.2-4 Evaluate $t_d(f)$ at $f = 0, 0.25, 1,$ and 4 kHz for a first-order lowpass system with $B = 1$ kHz.

3.2-5 Consider a transmission channel with $H_c(f) = (1 + 2\alpha \cos \omega T)e^{-j\omega T}$, which has *amplitude ripples*. (a) Show that $y(t) = \alpha x(t) + x(t - T) + \alpha x(t - 2T)$, so the output includes a leading and trailing echo. (b) Let $x(t) = \Pi(t/\tau)$ and $\alpha = 1/2$. Sketch $y(t)$ for $\tau = 2T/3$ and $4T/3$.

3.2-6 Consider a transmission channel with $H_c(f) = \exp\left[-j(\omega T - \alpha \sin \omega T)\right]$, which has *phase ripples*. Assume $|\alpha| \ll \pi/2$ and use a series expansion to show that the output includes a leading and trailing echo.

3.2-7 Design a tapped-delay-line equalizer for $H_c(f)$ in Prob. 3.2-6 with $\alpha = 0.2$.

3.2-8 Design a tapped-delay-line equalizer for $H_c(f)$ in Prob. 3.2-5 with $\alpha = 0.2$.

3.2-9 Suppose $x(t) = A \cos \omega_0 t$ is applied to a nonlinear system with $y(t) = 3x(t) - x^3(t)$. Write $y(t)$ as a sum of cosines. Then evaluate the second-harmonic and third-harmonic distortion when $A = 1$ and $A = 1/2$.

3.2-10 Do Prob. 3.2-9 with $y(t) = 3x(t) + x^2(t) - x^3(t)$.

3.3-1 Let the repeater system in Fig. 3.3-2 have $P_{in} = 0.5$ W, $\alpha = 2$ dB/km, and a total path length of 40 km. Find the amplifier gains and the location of the repeater so that $P_{out} = 100$ mW and the signal power at the input to each amplifier equals 20 μW.

3.3-2 Do Prob. 3.3-1 with $P_{in} = 50$ mW and $P_{out} = 0.2$ W.

3.3-3 A 400-km repeater system consists of m identical cable sections with $\alpha = 0.4$ dB/km and m identical amplifiers with 30 dB maximum gain. Find the required number of sections and the gain per amplifier so that $P_{out} = 50$ mW when $P_{in} = 2$ W.

3.3-4 A 3000-km repeater system consists of m identical cable sections with $\alpha = 0.6$ dB/km and m identical amplifiers. Find the required number of sections and the gain per amplifier so that $P_{out} = P_{in} = 100$ mW and the input power to each amplifier is at least 25 μW.

3.3-5 Suppose the radio link in Fig. 3.3-3 has $f = 3$ GHz, $\ell = 40$ km, and $P_{in} = 5$ W. If both antennas are circular dishes with the same radius r, find the value of r that yields $P_{out} = 2 \mu$W.

3.3-6 Do Prob. 3.3-5 with $f = 500$ MHz and $\ell = 20$ km.

3.3-7 Suppose a radio repeater is inserted in the middle of the path in Fig. 3.3-3. Let the total gain of the repeater be g_{rpt}, including antennas and amplifier. Obtain the condition on the value of g_{rpt} so that P_{out} is increased.

3.4-1 Find and sketch the impulse response of the ideal HPF defined by Eq. (1) with $f_u = \infty$.

3.4-2 Find and sketch the impulse response of an ideal band-rejection filter having $H(f) = 0$ for $f_c - B/2 < |f| < f_c + B/2$ and distortionless transmission for all other frequencies.

3.4-3 Find the minimum value of n such that a Butterworth filter has $|H(f)| \geq -1$ dB for $|f| < 0.8B$. Then calculate $|H(2B)|$ in dB.

3.4-4 Find the minimum value of n such that a Butterworth filter has $|H(f)| \geq -1$ dB for $|f| < 0.9B$. Then calculate $|H(2B)|$ in dB.

3.4-5 The circuit in Fig. P3.4-5 is one implementation of a second-order Butterworth LPF with $B = 1/2\pi\sqrt{LC}$. Obtain the expression for $H(f)$ and find the required relationship between R, L, and C.

Figure P3.4-5

3.4-6 The impulse response of a second-order Butterworth LPF is $h(t) = 2be^{-bt} \sin bt \, u(t)$ with $b = 2\pi B/\sqrt{2}$. Derive this result using a table of Laplace transforms by taking $p = s/2\pi B$ in Table 3.4-1.

3.4-7 Let $R = \sqrt{L/C}$ in Fig. P3.4-5. (a) Show that $|H(f)|^2 = [1 - (f/f_0)^2 + (f/f_0)^4]^{-1}$ with $f_0 = 1/2\pi\sqrt{LC}$. (b) Find the 3-dB bandwidth in terms of f_0. Then sketch $|H(f)|$ and compare with a second-order Butterworth response.

3.4-8 Show that the 10–90% risetime of a first-order LPF equals $1/2.87B$.

3.4-9 Use $h(t)$ given in Prob. 3.4-6 to find the step response of a second-order Butterworth LPF. Then plot $g(t)$ and estimate the risetime in terms of B.

3.4-10 Let $x(t) = A$ sinc $2Wt$ be applied to an ideal LPF with bandwidth B. Taking the duration of sinc at to be $\tau = 2/a$, plot the ratio of output to input pulse duration as a function of B/W.

3.4-11‡ The *effective bandwidth* of an LPF and the *effective duration* of its impulse response are defined by

$$B_{\text{eff}} \triangleq \frac{\displaystyle\int_{-\infty}^{\infty} |H(f)|\, df}{2H(0)} \qquad \tau_{\text{eff}} \triangleq \frac{\displaystyle\int_{-\infty}^{\infty} h(t)\, dt}{|h(t)|_{\max}}$$

Obtain expressions for $H(0)$ and $|h(t)|$ from $\mathscr{F}[h(t)]$ and $\mathscr{F}^{-1}[H(f)]$, respectively. Then show that $\tau_{\text{eff}} \geq 1/2B_{\text{eff}}$.

3.4-12 Let $x(t) = \delta(t)$. (a) Find $\hat{x}(t)$ from Eq. (13) and use your result to confirm that $\mathscr{F}^{-1}[-j\operatorname{sgn} f] = 1/\pi t$. (b) Then derive another Hilbert transform pair from the property $\hat{x}(t) * (-1/\pi t) = x(t)$.

3.4-13 Use Eq. (3), Sect. 3.1, and the results in Example 3.4-1 to obtain the Hilbert transform of $A\Pi(t/\tau)$. Now show that if $v(t) = A$ for all time, then $\hat{v}(t) = 0$.

3.4-14 Use Eq. (14) to show that if $x(t) =$ sinc $2Wt$ then $\hat{x}(t) = \pi Wt$ sinc$^2 Wt$.

3.4-15‡ Let the transfer function of a filter be written in the form $H(f) = H_e(f) + jH_o(f)$, as in Eq. (10), Sect. 2.2. If the filter is physically realizable, then its impulse response must have the causal property $h(t) = 0$ for $t < 0$. Hence, we can write $h(t) = (1 + \operatorname{sgn} t)h_e(t)$ where $h_e(t) = \frac{1}{2}h(|t|)$ for $-\infty < t < \infty$. Show that $\mathscr{F}[h_e(t)] = H_e(f)$ and thus causality requires that $H_o(f) = -\hat{H}_e(f)$.

3.4-16‡ An ideal LPF with zero time delay has $H(f) = K\Pi(f/2B)$ so the even and odd parts of $H(f)$ are $H_e(f) = H(f)$ and $H_o(f) = 0$—which violates the causality condition given in Prob. 3.4-15. Consider, therefore, a filter with a causal impulse response and $H_e(f) = K\Pi(f/2B)$. Obtain the corresponding expression for $H_o(f)$ and sketch $|H(f)|^2$.

3.4-17‡ Let the impulse response of an ideal LPF be truncated to obtain the causal function

$$h(t) = 2KB \text{ sinc } 2B(t - t_d) \qquad 0 < t < 2t_d$$

and $h(t) = 0$ elsewhere. (a) Show by Fourier transformation that

$$H(f) = \frac{K}{\pi} e^{-j\omega t_d}\{\text{Si }[2\pi(f + B)t_d] - \text{Si }[2\pi(f - B)t_d]\}$$

(b) Sketch $h(t)$ and $|H(f)|$ for $t_d \gg 1/B$ and $t_d = 1/2B$.

3.5-1 Prove Eq. (6b).

3.5-2 Let $v(t)$ be periodic with period T_0. Show from Eq. (7) that $R_v(\tau)$ has the same periodicity.

3.5-3 Derive Eq. (8b) by taking $w(t) = v(t - \tau)$ in Eq. (3).

3.5-4 Use Eq. (24) to obtain the spectral density, autocorrelation, and signal energy when $v(t) = A\Pi[(t - t_d)/D]$.

3.5-5 Do Prob. 3.5-4 with $v(t) = A$ sinc $2W(t - t_d)$.

3.5-6 Do Prob. 3.5-4 with $v(t) = Ae^{-bt}u(t)$.

3.5-7 Do Prob. 3.5-4 with $v(t) = Ae^{-\pi(t/D)^2}$.

3.5-8 Use Eq. (25) to obtain the spectral density, autocorrelation, and signal power when $v(t) = A_0 + A_1 \cos(\omega_0 t + \phi)$.

3.5-9 Do Prob. 3.5-8 with $v(t) = A_1 \cos(\omega_0 t + \phi_1) + A_2 \cos(2\omega_0 t + \phi_2)$.

3.5-10 Obtain the autocorrelation of $v(t) = Au(t)$ from Eq. (7). Use your result to find the signal power and spectral density.

3.5-11 Do Prob. 3.5-10 with $v(t) = A$ sgn t.

3.5-12‡ Use Eqs. (7) and (10) to derive Eq. (25b).

CHAPTER

FOUR

PROBABILITY AND RANDOM VARIABLES

Chapters 2 and 3 dealt with *deterministic signals*, for when we write an explicit time function $v(t)$ we presume that the behavior of the signal is known or determined for all time. In the next chapter we'll deal with *random signals* whose exact behavior cannot be predicted in advance. Random signals occur in communication both as unwanted noise and as desired information-bearing signals.

Lacking detailed knowledge of the time variation of a random signal, we must speak instead of its statistical averages and various probabilities. This chapter therefore presents the groundwork for the description of random signals. The major topics will be probabilities and events, discrete and continuous random variables, probability functions, statistical averages, and important probability models. Our coverage is directed specifically toward those aspects used in later chapters, and we'll rely heavily on intuitive reasoning rather than mathematical rigor.

If you've previously studied probability and statistics, you can skim over this chapter and go on to the next. If you want to pursue the subject in more detail, you'll find a wealth of material in texts such as Papoulis (1984), Beckmann (1967), Drake (1967), or Peebles (1980).

4.1 PROBABILITIES AND EVENTS

Probability theory establishes a mathematical framework for the study of *random phenomena*. The theory does not deal with the nature of random processes per se, but rather with their experimentally observable manifestations. Accordingly, we'll introduce probability here in terms of events associated with the outcomes of experiments. Then we'll develop the sample space concept to obtain relationships for the probabilities of various types of events.

Probability as Frequency of Occurrence

Consider an experiment involving some element of chance, so the outcome is unknown in advance and varies from trial to trial. Tossing a coin is such an experiment, for a trial toss could result in the coin landing heads up or tails up or even on its edge, and we would be hard-pressed to predict the result of a single trial. However, we may be able to draw useful conclusions about the results of a large number of trials.

For this purpose, let's identify a specific *event A* as a possible outcome that either does or does not occur on any trial. We repeat the experiment N times and record the number of times N_A that A occurs. The ratio N_A/N equals the *relative frequency of occurrence* of the event A for that set of trials.

If N_A/N approaches a definite limit as N becomes very large, and if every set of trials yields the same limiting value, then we say that the experiment obeys the *empirical law of large numbers*. Under this condition we take the *probability* of A to be

$$P(A) = \frac{N_A}{N} \qquad N \to \infty \tag{1}$$

The functional notation emphasizes that the value of $P(A)$ depends upon the event in question. Nonetheless, every probability is a positive number bounded by

$$0 \le P(A) \le 1 \tag{2}$$

since $0 \le N_A \le N$ for any event A.

Our interpretation of probability as frequency of occurrence agrees with intuition and common experience in the following sense: We can't predict the exact result of a single trial of a chance experiment, but we expect that the number of times A occurs out of $N \gg 1$ trials will be

$$N_A \approx NP(A)$$

Probability therefore has meaning only in relation to a large number of trials.

By the same token, Eq. (1) implies the need for an *infinite* number of trials in order to measure *exact* probability values. Thus, in absence of additional information about the underlying physical process, we must be content with probability estimates based on a large but finite number of trials. Fortunately, many experiments of practical interest possess inherent symmetry that allows us to deduce probabilities by logical reasoning, without resorting to actual experimentation. We feel certain, for instance, that an honest coin would come up heads half the time in a large number of trials, so the probability of heads must be 1/2.

Formal probability theory avoids the problems associated with Eq. (1) by defining probability in an indirect axiomatic fashion discussed next. The axioms then provide the basis for all other probability relationships.

Sample Space and Axiomatic Probability

A typical experiment may have several possible outcomes, and there may be various ways of characterizing the associated events. To construct a systematic model of a chance experiment, we identify the *elementary nondecomposable outcomes* and collect them in a set S known as the *sample space*. The elements of S, called *sample points*, are subject to two requirements:

1. The set of sample points must be *exhaustive*, so that S consists of all possible outcomes of the experiment in question.
2. The outcomes must be *disjoint* or *mutually exclusive*, so that one and only one of them occurs on a given trial.

Consequently, any events of interest can be described by *subsets* of S containing one or more sample points.

As an illustration of these concepts, let the experiment be tossing a coin twice and observing the sequence of heads (H) and tails (T). The sample space then contains $2 \times 2 = 4$ distinct sequences represented in the set notation

$$S = \{TT, TH, HT, HH\}$$

where the order of the listing is unimportant. Clearly, the elements of S are mutually exclusive and qualify as sample points. If we're concerned with the event "tails on the first toss," we can describe it by the subset

$$A = \{TT, TH\}$$

since A occurs when the outcome is either TT, or TH. Similarly, the events "matching tosses" and "heads on both tosses" are described by the subsets

$$B = \{TT, HH\} \qquad C = \{HH\}$$

Note that A and C are mutually exclusive events having no common elements, whereas B has an element in common with A and with C. Figure 4.1-1 brings out the relationships between A, B, and C in the form of a *Venn diagram*.

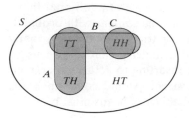

Figure 4.1-1 Sample space and Venn diagram of three events.

We may also describe events in terms of the union or intersection of event subsets.† If an event D corresponds to the *union* of A and B, then every element of D is an element of A or B or possibly both. We'll represent this union event by writing $D = A + B$, which stands for the occurrence of *either A or B*. If another event E corresponds to the *intersection* of A and B, then every element of E is an element of both A and B. We'll represent this intersection event by writing $E = AB$, which stands for the joint occurrence of *both A and B*. For instance, in Fig. 4.1-1 we see that

$$D = A + B = \{TT, TH, HH\} \qquad E = AB = \{TT\}$$

so D is the event "tails on the first toss *or* matching tosses" whereas E is the event "tails on the first toss *and* matching tosses."

Now we're prepared to state the axioms of probability theory. Let S be the sample space of a chance experiment, and let A be any subset of S. Probability is defined to be a function that assigns to every event a number $P(A)$ such that

$$P(A) \geq 0 \tag{3a}$$

$$P(S) = 1 \tag{3b}$$

Additionally, if $A_1 + A_2$ is the union of mutually exclusive events, then

$$P(A_1 + A_2) = P(A_1) + P(A_2) \tag{3c}$$

These three axioms constitute an *implicit* rather than *explicit* definition of probability, and they say nothing about frequency of occurrence.

However, Eqs. (3a) and (3b) are consistent with Eqs. (1) and (2), and Eq. (3b) can be interpreted by noting that one of the outcomes in S must occur on every trial so the *event S* should have $P(S) = 1$. To interpret Eq. (3c) we note that if A_1 occurs N_1 times in N trials and A_2 occurs N_2 times, then the event "A_1 or A_2" occurs $N_1 + N_2$ times since A_1 and A_2 are mutually exclusive; hence, as N becomes very large, $P(A_1 + A_2) = (N_1 + N_2)/N = (N_1/N) + (N_2/N) = P(A_1) + P(A_2)$. The axiomatic definition thus agrees with our frequency-of-occurrence interpretation which, in turn, links probability theory to the real world.

The specific advantage of the axiomatic approach is that it leads directly to various probability relationships. We'll state here several relationships needed for later use. The omitted derivations are exercises in elementary set theory.

Equation (3c) immediately generalizes to the case of any three or more mutually exclusive events, say $A_1, A_2, A_3, \ldots$, so that

$$P(A_1 + A_2 + A_3 + \cdots) = P(A_1) + P(A_2) + P(A_3) + \cdots \tag{4}$$

Furthermore, if M mutually exclusive events have the exhaustive property

$$A_1 + A_2 + \cdots + A_M = S$$

† Set theory notation uses $A \cup B$ for the union and $A \cap B$ for the intersection.

then Eq. (3b) gives

$$P(A_1 + A_2 + \cdots + A_M) = \sum_{i=1}^{M} P(A_i) = 1 \tag{5}$$

Incidentally, Eq. (5) also applies to the set of sample points. Sometimes we'll be concerned with the *nonoccurrence* of an event. The event "not A" is called the *complement* of A, symbolized† by A^c. It then follows that

$$P(A^c) = 1 - P(A) \tag{6}$$

since A and A^c are mutually exclusive and exhaustive.

Equation (5) takes on special importance when the M events happen to be *equally likely*, meaning that they have *equal probabilities*. The sum of the probabilities in this case reduces to $MP(A_i) = 1$, and hence

$$P(A_i) = \frac{1}{M} \qquad i = 1, 2, \ldots, M \tag{7}$$

This result allows you to calculate probabilities if you can identify all possible outcomes of an experiment in terms of mutually exclusive, equally likely events. The hypothesis of equal likelihood might be based on experimental data or on symmetry considerations — as in coin-tossing and other honest games of chance.

Finally, suppose A and B are *not* mutually exclusive so Eq. (3c) does not apply. The probability of the union event "A or B" is then given by

$$P(A + B) = P(A) + P(B) - P(AB) \tag{8}$$

in which $P(AB)$ is the probability of the joint event "A and B." We call $P(AB)$ the *joint probability* and interpret it as

$$P(AB) = N_{AB}/N \qquad N \to \infty$$

where N_{AB} stands for the number of times A and B occur together in N trials. Equation (8) reduces to Eq. (3c) when A and B *are* mutually exclusive and cannot occur together — so $P(AB) = 0$. Putting this another way, the intersection AB would be an empty subset and therefore an impossible event.

Example 4.1-1 To clarify the use of probability relationships, let's mentally toss three honest coins and observe the resulting sequence of heads and tails. The sample space for this experiment contains eight sample points listed in Fig. 4.1-2. These sequences are mutually exclusive and can be assumed to be equally likely. Hence, we apply Eq. (7) with $M = 8$ to get $P(A_i) = 1/8$ for each sequence.

Now let A be the event "first two tosses match," which corresponds to the subset $\{TTT, TTH, HHT, HHH\}$. Equation (4) then yields

$$P(A) = P(TTT + TTH + HHT + HHH) = 4P(A_i) = \tfrac{1}{2}$$

---

† Another symbol used for the complement is $\bar{A}$.

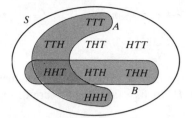

Figure 4.1-2 Sample space for Example 4.1-1.

Similarly, the event "exactly two heads" has probability

$$P(B) = P(HHT + HTH + THH) = 3P(A_i) = \tfrac{3}{8}$$

while the event "not two heads" has

$$P(B^c) = 1 - \tfrac{3}{8} = \tfrac{5}{8}$$

as obtained from Eq. (6).

Figure 4.1-2 shows that the intersection of A and B contains one element, the sequence HHT. Hence, the joint probability is

$$P(AB) = P(HHT) = \tfrac{1}{8}$$

From Eq. (8) we then obtain

$$P(A + B) = \tfrac{1}{2} + \tfrac{3}{8} - \tfrac{1}{8} = \tfrac{3}{4}$$

which agrees with the fact that $A + B = \{TTT, TTH, HHT, HTH, THH, HHH\}$ so $P(A + B) = 6P(A_i)$.

Exercise 4.1-1 Suppose an honest wheel of chance is divided into three equal segments colored green (G), red (R), and yellow (Y). You spin the wheel twice and take the outcome to be the resulting color sequence—GR, RG, etc. Let A be the event "neither color is yellow" and let B be "matching colors." Draw the Venn diagram and calculate $P(A)$, $P(B)$, $P(AB)$, and $P(A + B)$.

Conditional Probabilities and Statistical Independence

Suppose an event B depends in some way on another event A. We measure such dependence in terms of the *conditional probability*

$$P(B \mid A) \triangleq \frac{P(AB)}{P(A)} \tag{9a}$$

provided that $P(A) \neq 0$. The notation $P(B \mid A)$ represents the probability of B conditioned by the fact that A has occurred. Take the case of mutually exclusive events as an extreme example; if you know that A has occurred, you can be sure that B did not occur on the same trial. This information is reflected in the conditional probability $P(B \mid A) = 0$ when $P(AB) = 0$. For an interpretation of Eq. (9a) with $P(AB) \neq 0$, recall that $P(A) = N_A/N$ and $P(AB) = N_{AB}/N$ as $N \to \infty$. Thus,

$$P(B \mid A) = \frac{N_{AB}/N}{N_A/N} = \frac{N_{AB}}{N_A}$$

which says that $P(B \mid A)$ equals the relative frequency of A and B together in the N_A trials where A occurred with or without B. In like fashion, the conditional probability of A given B is

$$P(A \mid B) = \frac{P(AB)}{P(B)} \tag{9b}$$

provided that $P(B) \neq 0$.

Combining Eqs. (9a) and (9b) yields two relations for the joint probability, namely,

$$P(AB) = P(B \mid A)P(A) = P(A \mid B)P(B) \tag{10}$$

Or we could eliminate $P(AB)$ to obtain

$$P(B \mid A) = \frac{P(B)P(A \mid B)}{P(A)} \qquad P(A \mid B) = \frac{P(A)P(B \mid A)}{P(B)} \tag{11}$$

Equation (11), known as *Bayes' theorem*, plays an important role in statistical decision theory because it allows you to reverse the conditioning event. Another useful expression is the *total probability*

$$P(B) = \sum_{i=1}^{M} P(B \mid A_i)P(A_i) \tag{12}$$

where the conditioning events $A_1, A_2, \ldots, A_M$ must be mutually exclusive and exhaustive.

If events A and B do *not* depend on each other, the conditional probabilities simply equal the unconditional probabilities, i.e.,

$$P(B \mid A) = P(B) \qquad P(A \mid B) = P(A) \tag{13}$$

We then say that A and B are *statistically independent* events. Inserting Eq. (13) into Eq. (10) then gives

$$P(AB) = P(A)P(B)$$

so the joint probability equals the *product* of the individual event probabilities. Furthermore, if three or more events are all independent of each other, then

$$P(ABC \cdots) = P(A)P(B)P(C) \cdots \tag{14}$$

in addition to pairwise independence.

As a general rule, *physical* independence is a sufficient indication of statistical independence. We can thus apply Eq. (14) to any situations where the events have no physical connection. For instance, the three coin tosses in Example 4.1-1 are physically independent, and a sequence such as TTH may be viewed as a joint event. Invoking the equally likely argument for each toss alone gives $P(T) = 1/2$ and $P(H) = 1/2$, so $P(TTH) = P(T)P(T)P(H) = 1/8$ which agrees with our previous conclusion that $P(A_i) = 1/8$.

Example 4.1-2 In Example 4.1-1 we calculated the probabilities $P(A) = 1/2$, $P(B) = 3/8$, and $P(AB) = 1/8$. Since $P(AB) \neq P(A)P(B)$, the events A and B are not statistically independent. The dependence is reflected in the conditional probabilities

$$P(B \mid A) = \frac{1/8}{1/2} = \frac{1}{4} \qquad P(A \mid B) = \frac{1/8}{3/8} = \frac{1}{3}$$

obtained from Eq. (9).

Closer examination of Fig. 4.1-2 reveals why $P(B \mid A) = 1/4 < P(B)$. Specifically, the event A corresponds to any one of four equally likely outcomes, and only one of those outcomes corresponds to the event B. Hence, B occurs with frequency $N_{AB}/N_A = 1/4$ in the N_A trials in which A occurs — as contrasted with $N_B/N = P(B) = 3/8$ for the entire set of N trials. Similar reasoning justifies the value of $P(A \mid B)$.

Exercise 4.1-2 Let the event B be as defined in Exercise 4.1-1, and let $A_1 = \{GG, GY, YG, YY\}$ and $A_2 = A_1^c$. Calculate $P(B \mid A_1)$ and $P(B \mid A_2)$ using Eq. (9a). Then confirm that Eq. (12) yields the correct value for $P(B)$.

4.2 RANDOM VARIABLES AND PROBABILITY FUNCTIONS

The various games of chance are natural and fascinating subjects for probability calculations. But communication engineers are more concerned (professionally) with random processes that produce *numerical* outcomes — the instantaneous size of a noise voltage, the number of errors in a digital message, etc. The related concepts of random variables and probability functions will be presented in this section for the description and analysis of numerical-valued random events.

Discrete Random Variables and CDFs

Despite the name *random variable* — often abbreviated as r.v. — a random variable is neither random nor a variable. Instead, it's a *function* that generates numbers from the outcomes in the sample space S of a chance experiment. To formulate this definition, let s stand for an arbitrary sample point in S. An r.v. associated with the experiment is a rule or relationship $X(s)$ that assigns a *real number* X to every sample point s. If S contains a countable number of sample points, then $X(s)$ will be a *discrete* r.v. having a *countable number of distinct values*.

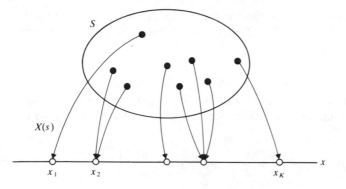

Figure 4.2-1 Sample points mapped by the discrete random variable $X(s)$ into numbers on the real line.

Virtually any reasonable function may serve as a random variable, subject to the conditions that it be real and single-valued. The essential property is that $X(s)$ maps the experimental outcomes in S into numbers along the real line $-\infty < x < \infty$. Figure 4.2-1 depicts this mapping process and introduces the notation $x_1 < x_2 < \cdots < x_K$ for the values of $X(s)$ in ascending order. Each outcome produces a single number, but two or more outcomes may map into the same number. Consequently, a discrete r.v. has $K \leq M$ possible values when S consists of M sample points.

Although a mapping relationship underlies every r.v., we're usually concerned only with the numbers resulting from $X(s)$. We'll therefore adopt a more direct viewpoint that treats X itself as the general symbol for the outcome of a chance experiment. This viewpoint allows us to deal with *numerical-valued events* such as $X = a$ or $X \leq a$, where a is some point on the real line. Furthermore, the event probabilities $P(X = a)$ and $P(X \leq a)$ become *probability functions* when we replace the constant a with the independent variable x. These functions provide a convenient way of calculating probabilities for numerical-valued events.

The probability function $P(X \leq x)$ is known as the *cumulative distribution function* (or CDF), symbolized by

$$F_X(x) \triangleq P(X \leq x) \qquad -\infty < x < \infty \tag{1}$$

Pay careful attention to the notation here: the subscript X identifies the r.v. whose characteristics determine the nature of the function $F_X(x)$; the argument x defines an event, namely $X \leq x$, and is *not* a random variable.

Since the CDF represents a probability, it must be bounded by

$$0 \leq F_X(x) \leq 1 \tag{2a}$$

with the extreme values

$$F_X(-\infty) = 0 \qquad F_X(\infty) = 1 \tag{2b}$$

$X \leq a$　　　　　$a < X \leq b$　　　　$X > b$

Figure 4.2-2 Numerical-valued events.

The lower limit assumes that $P(X = -\infty) = 0$, while the upper limit simply says that X always falls somewhere on the real line. The *complementary* events $X \leq x$ and $X > x$ encompass the entire real line, so

$$P(X > x) = 1 - F_X(x) \tag{3}$$

Other CDF properties emerge from the following consideration.

Suppose we know $F_X(x)$ and we want to find the probability of observing $a < X \leq b$. Figure 4.2-2 illustrates the relationship of this event to the events $X \leq a$ and $X > b$. The figure also brings out the difference between *open* and *closed* inequalities for specifying numerical-valued events. Clearly, the three events in question are *mutually exclusive* if $b > a$, and

$$P(X \leq a) + P(a < X \leq b) + P(X > b) = P(X \leq \infty) = 1$$

Substituting $P(X \leq a) = F_X(a)$ and $P(X > b) = 1 - F_X(b)$ finally yields

$$P(a < X \leq b) = F_X(b) - F_X(a) \qquad b > a \tag{4}$$

Besides being an important result in its own right, Eq. (4) shows that $F_X(x)$ has the *nondecreasing* property $F_X(b) \geq F_X(a)$ for any $b > a$. Furthermore, $F_X(x)$ is continuous from the right in the sense that $F_X(x + \epsilon^2) \to F_X(x)$ as $\epsilon^2 \to 0$.

Now let's take account of the fact that a discrete random variable is restricted to the *distinct* values $x_1, x_2, \ldots, x_K$. This restriction means that the possible outcomes $X = x_i$ constitute a set of mutually exclusive events. The corresponding set of probabilities will be written as

$$P_X(x_i) \triangleq P(X = x_i) \qquad i = 1, 2, \ldots, K \tag{5}$$

which we call the *frequency function* of a discrete r.v. If you can determine $P_X(x_i)$ from the description of an experiment, you can then calculate any probability of interest. By way of illustration, consider the event $a < X \leq b$ with $a < x_1$ and $x_k \leq b < x_{k+1}$. This event corresponds to any of the first k possible values, so

$$P(a < X \leq b) = P_X(x_1) + P_X(x_2) + \cdots + P_X(x_k)$$

$$= \sum_{i=1}^{k} P_X(x_i)$$

whereas $P(X \leq a) = 0$.

Similar reasoning gives the CDF of a discrete r.v. in terms of its frequency function in the form

$$F_X(x) = \begin{cases} 0 & x < x_1 \\ \sum_{i=1}^{k} P_X(x_i) & x_k \leq x < x_{k+1} \\ 1 & x \geq x_K \end{cases} \tag{6}$$

Therefore, as x goes from $-\infty$ to ∞, $F_X(x)$ looks like a *staircase* with upward steps of height $P_X(x_i)$ at each $x = x_i$. The CDF remains constant between steps, and

$$F_X(x_i) - F_X(x_{i-1}) = P_X(x_i)$$

which provides another relationship between these two probability functions.

Often, but not always, a discrete r.v. takes on only integer values. The following example discusses a significant application of integer-valued random variables. We'll generalize this example later on in conjunction with the binomial distribution.

Example 4.2-1 Consider transmitting a three-digit message over a noisy channel having *error probability* $P(E) = 2/5$ per digit, so the probability of a correct digit is $P(C) = 1 - 2/5 = 3/5$. The sample space for this experiment consists of $M = 2 \times 2 \times 2 = 8$ distinct error patterns, just like the head-tail sequences back in Fig. 4.1-2. But these patterns are not equally likely since the error-free pattern has $P(CCC) = P(C)P(C)P(C) = (3/5)^3$ whereas the all-error pattern has $P(EEE) = (2/5)^3$. Similarly, the three patterns with one error each have probability $(2/5)(3/5)^2$ and the three patterns with two errors each have probability $(2/5)^2(3/5)$. Note that these calculations assume statistical independence.

If we take $X(s)$ to be the *number of errors* in the received message, then X is a discrete random variable with $K = 4$ possible values, namely $x_i = 0, 1, 2,$ and 3. From the error-pattern probabilities we obtain the values of $P_X(x_i)$ tabulated in Fig. 4.2-3a along with $F_X(x_i)$. Note that $F_X(0) = P_X(0)$, $F_X(1) = P_X(0) + P_X(1)$, and so forth in accordance with Eq. (6). Figure 4.2-3b displays

(a)

x_i	$P_X(x_i)$	$F_X(x_i)$
0	$(3/5)^3 = 27/125$	27/125
1	$3(2/5)(3/5)^2 = 54/125$	81/125
2	$3(2/5)^2(3/5) = 36/125$	117/125
3	$(2/5)^3 = 8/125$	125/125

Figure 4.2-3 The discrete r.v. and CDF in Example 4.2-1.

the frequency function and CDF as functions of x. The probability of more than one error is then found from $1 - F_X(1) = 44/125$ or $P_X(2) + P_X(3) = 44/125$.

Exercise 4.2-1 Let $X(s)$ be defined for the experiment in Exercise 4.1-1 according to the following rule: The colors are assigned the numerical weights $G = 2$, $R = -1$, and $Y = 0$, and X is taken as the *average* of the weights observed on a given trial. For instance, the outcome RY maps into $X = (-1 + 0)/2 = -0.5$. Find and plot $P_X(x_i)$ and $F_X(x_i)$. Then calculate $P(-1.0 < X \le 1.0)$.

Continuous Random Variables and PDFs

A *continuous* random variable is not restricted to a finite number of distinct values; instead, it can have *any* value within a certain range of the real line. For instance, you might spin a pointer and measure the final angle θ, as shown in Fig. 4.2-4. If you take $X = \tan^2 \theta$, then every value in the range $0 \le x < \infty$ is a possible outcome of this experiment. Or you could take $X = \cos \theta$ whose values fall in the range $-1.0 \le x \le 1.0$.

Since a continuous r.v. has an *uncountable* number of possible values, the chance of observing $X = a$ must be vanishingly small in the sense that $P(X = a) = 0$ for any specific a. Consequently, we cannot define a meaningful frequency function. However, events such as $X \le a$ and $a < X \le b$ may have nonzero probabilities, so $F_X(x)$ still provides useful information. Indeed, the properties stated before in Eqs. (1)–(4) also apply to the CDF of a continuous random variable.

But a more convenient description of a continuous r.v. turns out to be its *probability density function* (or PDF), defined as the derivative

$$p_X(x) \triangleq \frac{dF_X(x)}{dx} \tag{7}$$

provided that this derivative exists. We don't lose information when differentiating $F_X(x)$ because we know that $F_X(-\infty) = 0$. Therefore, using the dummy variable λ for clarity,

$$P(X \le x) = F_X(x) = \int_{-\infty}^{x} p_X(\lambda) \, d\lambda \tag{8}$$

Figure 4.2-4 Mapping by a continuous random variable.

Figure 4.2-5 A typical PDF and the area interpretation of probabilities.

Other PDF properties include

$$p_X(x) \geq 0 \qquad \int_{-\infty}^{\infty} p_X(x) \, dx = 1 \tag{9}$$

$$P(a < X \leq b) = F_X(b) - F_X(a) = \int_a^b p_X(x) \, dx \tag{10}$$

Thus, $p_X(x)$ is a nonnegative function whose total area equals unity and whose area between a and b equals the probability of observing X in that range.

As a special case of Eq. (10), let $a = x - dx$ and $b = x$. The integral then reduces to the *differential area* $p_X(x) \, dx$ and we see that

$$p_X(x) \, dx = P(x - dx < X \leq x) \tag{11}$$

This relation serves as another interpretation of the PDF, emphasizing its nature as a probability *density*. Figure 4.2-5 shows a typical PDF and the areas involved in Eqs. (10) and (11).

Occasionally we'll encounter *mixed* random variables having both continuous and discrete values. We treat such cases using *impulses* in a manner analogous to the spectrum of a signal having both periodic and nonperiodic components. Specifically, for each discrete value x_0 with probability $P(X = x_0) = P_X(x_0)$, the PDF must include the impulsive term $P_X(x_0) \, \delta(x - x_0)$ so that $F_X(x)$ has an appropriate jump at $x = x_0$. If desired, the frequency function of a discrete r.v. can likewise be converted into a PDF consisting entirely of impulses.

When a PDF includes impulses, you need to be particularly careful with events specified by open and closed inequalities. For instance, the probability of $X \geq x_0$ should be written out as $P(X \geq x_0) = P(X > x_0) + P(X = x_0)$ if $p_X(x)$ has an impulse at x_0. In contrast, there's no difference between $P(X \geq x_0)$ and $P(X > x_0)$ for a strictly continuous r.v. since $P(X = x_0) = 0$.

Example 4.2-2 Uniform PDF Suppose you take $X = \theta$ in Fig. 4.2-4, with θ measured in radians. Presumably all angles between 0 and 2π are equally likely, so $p_X(x)$ has some constant value C for $0 \leq x < 2\pi$ and equals zero everywhere else. The unit-area property requires that

$$\int_{-\infty}^{\infty} p_X(x) \, dx = \int_0^{2\pi} C \, dx = 2\pi C = 1$$

Figure 4.2-6 PDF and CDF of a uniformly distributed continuous r.v.

Therefore $C = 1/2\pi$ and

$$p_X(x) = \frac{1}{2\pi} [u(x) - u(x - 2\pi)] = \begin{cases} \dfrac{1}{2\pi} & 0 \le x < 2\pi \\ 0 & \text{otherwise} \end{cases} \tag{12}$$

which is plotted in Fig. 4.2-6a. Equation (8) then yields the corresponding PDF

$$F_X(x) = \begin{cases} 0 & x < 0 \\ \dfrac{x}{2\pi} & 0 \le x < 2\pi \\ 1 & x \ge 2\pi \end{cases}$$

as shown in Fig. 4.2-6b. These functions describe a continuous r.v. that's *uniformly distributed* over the range $0 \le x < 2\pi$.

You could also define another random variable Z such that

$$Z = \begin{cases} \pi & X \le \pi \\ X & X > \pi \end{cases}$$

Then $P(Z < \pi) = 0$, $P(Z = \pi) = P(X \le \pi) = 1/2$, and $P(Z \le z) = P(X \le z)$ for $z > \pi$. Hence the PDF of Z is

$$p_Z(z) = \frac{1}{2} \delta(z - \pi) + \frac{1}{2\pi} [u(z - \pi) - u(z - 2\pi)]$$

in which the impulsive term accounts for the discrete value $Z = \pi$.

Exercise 4.2-2 Use $p_X(x)$ and $p_Z(z)$ from Example 4.2-2 to calculate the probabilities of the following events: $\pi < X \le 3\pi/2$, $X > 3\pi/2$, $\pi < Z \le 3\pi/2$, $\pi \le Z \le 3\pi/2$.

Transformations of Random Variables

The preceding example touched upon the *transformation* problem, wherein we seek the PDF that results when one random variable is defined as a function of another. We'll consider here the relatively simple but important class of problems that involve a continuous transformation of a continuous r.v.

Let X have a known PDF, and let the transformation function $Z = g(X)$ be such that the inverse function $X = g^{-1}(Z)$ exists. If $g(X)$ is *monotonic*, so that each value $Z = z$ corresponds to a unique value $X = x$, then the probability of observing Z in some differential range dz equals the probability that X occurs in the corresponding range dx. Figure 4.2-7 illustrates this point. From Eq. (11) we have $p_Z(z)|dz| = p_X(x)|dx|$ or

$$p_Z(z) = p_X(x)\left|\frac{dx}{dz}\right|$$

The absolute values of the differentials ensure that $p_Z(z)$ will be nonnegative when $g(X)$ is a decreasing function and $dx/dz < 0$ (as is the case in Fig. 4.2-7). Finally, noting that x transforms to $z = g(x)$, we insert $x = g^{-1}(z)$ and obtain

$$p_Z(z) = p_X[g^{-1}(z)]\left|\frac{dg^{-1}(z)}{dz}\right| \tag{13}$$

a result that's actually less complicated than it looks.

In the case of a *linear* transformation

$$Z = \alpha X + \beta \tag{14a}$$

where α and β are constants, Eq. (13) yields

$$p_Z(z) = \frac{1}{|\alpha|} p_X\left(\frac{z - \beta}{\alpha}\right) \tag{14b}$$

Hence $p_Z(z)$ has the same shape as $p_X(x)$ shifted by β and expanded or contracted by α.

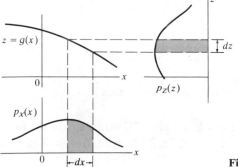

Figure 4.2-7 Transformation of a random variable.

If $g(X)$ is not monotonic, then two or more values of X produce the same value of Z. We handle this case by subdividing $g(X)$ into a set of monotonic functions, $g_1(X), g_2(X), \ldots$, defined over different ranges of X. These ranges constitute mutually exclusive events insofar as X is concerned. Hence, we can compute the right-hand side of Eq. (13) for each monotonic function and add the results to get $p_Z(z)$. The following example illustrates this method.

Example 4.2-3 Consider the transformation $Z = \cos X$, with X being the uniformly distributed angle from Example 4.2-2. The plot of Z versus X in Fig. 4.2-8a shows the range of each r.v., namely $-1 \leq Z \leq 1$ and $0 \leq X < 2\pi$, and brings out the fact that the transformation is not monotonic. We therefore subdivide $g(X)$ into the two monotonic functions

$$g_1(X) = \cos X \qquad 0 \leq X < \pi$$

$$g_2(X) = \cos X \qquad \pi \leq X < 2\pi$$

which happen to be identical except for the defining range.

Proceeding with the calculations, we note that $g_1^{-1}(Z) = \arccos Z$ so $dg_1^{-1}(z)/dz = -1/\sqrt{1 - z^2}$. For the interval $0 \leq X < \pi$, $p_X(x)$ equals $1/2\pi$ so $p_X[g_1^{-1}(z)] = 1/2\pi$ over the corresponding interval $-1 \leq Z \leq 1$ seen in Fig. 4.2-8a. The results for $g_2(Z)$ turn out to be the same because $g_2(z)$ equals $g_1(z)$ and $p_X(x)$ still equals $1/2\pi$ over $\pi \leq X < 2\pi$. Therefore

$$p_Z(z) = p_X[g_1^{-1}(z)] \left| \frac{dg_1^{-1}(z)}{dz} \right| + p_X[g_2^{-1}(z)] \left| \frac{dg_2^{-1}(z)}{dz} \right|$$

$$= 2 \times \frac{1}{2\pi} \left[u(z + 1) - u(z - 1) \right] \left| \frac{-1}{\sqrt{1 - z^2}} \right|$$

$$= \begin{cases} \dfrac{1}{\pi\sqrt{1 - z^2}} & -1 \leq z \leq 1 \\ 0 & \text{otherwise} \end{cases} \qquad (15)$$

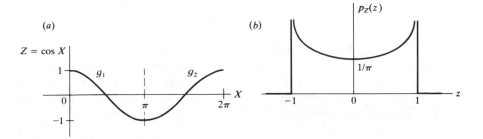

(a)

$Z = \cos X$

(b)

$p_Z(z)$

$1/\pi$

Figure 4.2-8

This PDF, sketched in Fig. 4.2-8b, has peaks at $z = \pm 1$ because $\cos X$ occurs more often near ± 1 than any other value when X is uniformly distributed.

Exercise 4.2-3 Carry out the steps leading to Eq. (14b). Then use this expression to sketch the PDF of $Z = \frac{1}{2}X - \pi$ when X has the uniform PDF in Eq. (12).

Joint and Conditional PDFs

Some chance experiments are characterized by several random variables, which may or may not be independent. We'll develop the PDF description in terms of joint and conditional density functions for the case of two random variables. Extension to three or more follows the same line of reasoning.

Suppose the continuous random variables X and Y can be observed simultaneously. Their *joint probability density function* is written $p_{XY}(x,y)$, where the comma stands for *and*. The meaning of the joint PDF follows from a generalization of Eq. (11) in the two-dimensional form

$$p_{XY}(x,y)\, dx\, dy = P(x - dx < X \le x, \, y - dy < Y \le y) \tag{16}$$

so the *differential volume* $p_{XY}(x,y)\, dx\, dy$ equals the probability of the joint event $x - dx < X \le x$ and $y - dy < Y \le y$. Hence, the probability of $a < X \le b$ and $c < Y \le d$ is

$$P(a < X \le b, c < Y \le d) = \int_c^d \int_a^b p_{XY}(x,y)\, dx\, dy \tag{17}$$

Note that the integral defines the volume between the x-y plane and the surface $p_{XY}(x,y)$.

If X and Y happen to be *statistically independent*, the joint PDF reduces to the product

$$p_{XY}(x,y) = p_X(x)p_Y(y) \tag{18}$$

If not independent, the dependence of X on Y is expressed by the *conditional* PDF

$$p_X(x \mid y) = \frac{p_{XY}(x,y)}{p_Y(y)} \tag{19}$$

which equals the PDF of X given that $Y = y$. We could also write PDF versions of the joint and conditional probability relations in Eqs. (10) and (11), Sect. 4.1.

Finally, the PDF for X alone is obtained from the joint PDF by the following argument. Since we're only concerned with X, the value of Y doesn't matter and $P(X \le x)$ is equivalent to $P(X \le x, -\infty < Y \le \infty)$. Thus

$$p_X(x) = \int_{-\infty}^{\infty} p_{XY}(x,y)\, dy \tag{20a}$$

and likewise

$$p_Y(y) = \int_{-\infty}^{\infty} p_{XY}(x,y) \, dx \tag{20b}$$

When $p_X(x)$ and $p_Y(y)$ are derived from $p_{XY}(x,y)$ they're said to be *marginal* PDFs.

4.3 STATISTICAL AVERAGES

For many of our purposes the PDF provides more information about a random variable than needed. In fact, such a complete description may prove to be an embarrassment of riches, more confusing than illumination. We'll therefore often find it simpler and more convenient to describe an r.v. by a few characteristic *numbers*. These numbers are the various statistical averages or mean values defined in this section.

Means, Moments, and Expectation

The *mean* m_x of a random variable X equals the average of the values of X weighted by their probabilities. This statistical average corresponds to an ordinary experimental average in the sense that the sum of the values observed over $N \gg 1$ trials is expected to be about Nm_x. For this reason, we call m_x the *expected value* of X and we use $E[X]$ to denote the *expectation operation* that yields m_x. Another common symbol for the mean of X is $\bar{x}$, so $E[X] = m_x = \bar{x}$. The lowercase x helps distinguish the mean, which is a constant, from the random variable X.

Let's begin our discussion of statistical averages by considering N independent observations of a *discrete* r.v. If the event $X = x_i$ occurs N_i times, the sum of the observed values is

$$N_1 x_1 + N_2 x_2 + \cdots + N_K x_K = \sum_{i=1}^{K} N_i x_i$$

Upon dividing the sum by N and letting $N \to \infty$, the relative frequency N_i/N becomes $P(X = x_i) = P_X(x_i)$. Thus

$$m_x \triangleq \sum_{i=1}^{K} x_i P_X(x_i) \tag{1}$$

which defines the mean of a discrete r.v. in terms of its frequency function.

To define the mean of a *continuous* r.v. we replace $P_X(x_i)$ with $P(x - dx < X \leq x) = p_X(x) \, dx$ and pass from summation to integration so that

$$m_x \triangleq \int_{-\infty}^{\infty} x p_X(x) \, dx \tag{2}$$

This expression actually includes Eq. (1) as a special case obtained by taking

$$p_X(x) = \sum_i P_X(x_i) \, \delta(x - x_i)$$

Hence, Eq. (2) applies to *any* type of r.v.— continuous, discrete, or mixed — when we allow impulses in the PDF. Hereafter, all statistical averages and expectation operations will be expressed in integral form with PDFs.

If the function $g(X)$ transforms X into another r.v., the expected value $\bar{z} = E[g(X)]$ could be found by integrating $z p_Z(z)$. But it's given more directly by

$$E[g(X)] = \int_{-\infty}^{\infty} g(x) p_X(x) \, dx \tag{3}$$

which follows from Fig. 4.2-7 plus the fact that the event $X = x$ transforms to $Z = g(x)$. When $g(X) = X^n$, the expectation $E[X^n] = \overline{x^n}$ is known as the *nth moment* of X. The first moment, of course, is just the mean value $\bar{x} = m_X$. The second moment $\overline{x^2}$, called the *mean square* value of X, plays a major role in statistical analysis and will be especially significant when we get to random signals and noise.

The expectation operation in Eq. (3) has many similarities to time averaging as defined in Sect. 2.1, and both are *linear* operations. Thus, if $g(X) = \alpha X + \beta$, where α and β are constants, then

$$E[\alpha X + \beta] = \alpha \bar{x} + \beta \tag{4}$$

While this result seems rather trivial, it leads to the not-so-obvious relation

$$E[\bar{x}X] = \bar{x}E[X] = \bar{x}^2$$

since $\bar{x}$ is a constant inside $E[\bar{x}X]$. Incidentally, watch for the difference between the mean value squared, $m_x^2 = \bar{x}^2 = E^2[X]$, and the mean square value, $\overline{x^2} = E[X^2]$. Interchanging the order of expectation and squaring generally produces different values.

Standard Deviation and Chebyshev's Inequality

The *standard deviation* of X, denoted σ_x, provides a measure of the *spread* of observed values of X relative to m_x. The square of the standard deviation is called the *variance* or *second central moment*, defined by

$$\sigma_x^2 \triangleq E[(X - m_x)^2] \tag{5a}$$

To obtain a more convenient expression for σ_x, we expand $(X - m_x)^2$ and invoke Eq. (4) as follows:

$$E[X^2 - 2m_x X + m_x^2] = \overline{x^2} - 2m_x \bar{x} + m_x^2 = \overline{x^2} - m_x^2$$

Hence

$$\sigma_x^2 = \overline{x^2} - m_x^2 \tag{5b}$$

and the standard deviation equals the square root of the mean square minus the mean squared.

For the interpretation of σ_x, let k be any positive quantity and consider the event $m_x - k\sigma_x < X < m_x + k\sigma_x$ and its complement $|X - m_x| \geq k\sigma_x$. *Chebyshev's inequality* (also spelled Tchebycheff) states that, regardless of $p_X(x)$,

$$P(|X - m_x| \geq k\sigma_x) \leq \frac{1}{k^2} \tag{6a}$$

and

$$P(|X - m_x| < k\sigma_x) > 1 - \frac{1}{k^2} \tag{6b}$$

Thus, the probability of observing any r.v. outside $\pm k$ standard deviations of its mean is no larger than $1/k^2$. With $k = 2$ for instance, the probability of $|X - m_x| > 2\sigma_x$ does not exceed $1/4$ or 25%. By the same token, we expect X to occur within the range $m_x \pm 2\sigma_x$ for more than 75% of the observations. A *small standard deviation* therefore implies a *small spread* of likely values, and vice versa.

The proof of Eq. (6a) starts by taking $Z = X - m_x$ and $a = k\sigma_x > 0$. Then let ϵ be a small positive quantity and note that

$$\overline{z^2} = \int_{-\infty}^{\infty} z^2 p_Z(z)\, dz \geq \int_{-\infty}^{-a} z^2 p_Z(z)\, dz + \int_{a-\epsilon}^{\infty} z^2 p_Z(z)\, dz$$

$$\geq a^2 \left[\int_{-\infty}^{-a} p_Z(z)\, dz + \int_{a-\epsilon}^{\infty} p_Z(z)\, dz \right]$$

since $z^2 \geq a^2$ over $a \leq |z| < \infty$. The first integral inside the brackets represents $P(Z \leq -a)$, while the second represents $P(Z \geq a)$. Therefore

$$P(|Z| \geq a) = P(Z \leq -a) + P(Z \geq a) \leq \overline{z^2}/a^2$$

and Eq. (6a) comes out of the substitutions $Z = X - m_x$, $\overline{z^2} = \sigma_x^2$, and $a = k\sigma_x$.

Example 4.3-1 A continuous r.v. is said to have a *Laplace* distribution when $p_X(x) = \frac{1}{2}e^{-|x|}$ for $-\infty < x < \infty$. Drawing upon the even symmetry of $p_X(x)$ we obtain

$$m_x = \int_{-\infty}^{\infty} x\, \frac{1}{2}\, e^{-|x|}\, dx = 0$$

$$\overline{x^2} = 2 \int_0^{\infty} x^2\, \frac{1}{2}\, e^{-x}\, dx = 2$$

and

$$\sigma_x = \sqrt{2 - 0} = \sqrt{2}$$

The probability that an observed value falls within $\pm 2\sigma_x$ of the mean is

$$P(|X - 0| < 2\sqrt{2}) = 2 \int_0^{2\sqrt{2}} \frac{1}{2}e^{-x}\, dx = 0.94$$

as compared with the lower bound 0.75 from Chebyshev's inequality.

Exercise 4.3-1 Let X have the uniform distribution in Example 4.2-2. Calculate m_x, $\overline{x^2}$, and σ_x. What's the probability of $|X - m_x| < 2\sigma_x$?

Multivariate Expectations

The term *multivariate* refers to any situation involving more than one r.v. Multivariate expections are calculated using multiple integration over joint PDFs. Specifically, if $g(X,Y)$ defines a function of the random variables X and Y, then

$$E[g(X,Y)] \triangleq \int\!\!\int_{-\infty}^{\infty} g(x,y)p_{XY}(x,y)\,dx\,dy \tag{7}$$

However, we'll restrict our attention to those cases in which the multiple integration reduces to separate integrals.

First suppose that X and Y are *independent*, so $p_{XY}(x,y) = p_X(x)p_Y(y)$. If we can write $g(X,Y)$ as a *product* — say $g(X,Y) = g_1(X)g_2(Y)$ — then Eq. (7) becomes

$$E[g(X,Y)] = \int\!\!\int_{-\infty}^{\infty} g_1(x)g_2(y)p_X(x)p_Y(y)\,dx\,dy$$

$$= \int_{-\infty}^{\infty} g_1(x)p_X(x)\,dx \int_{-\infty}^{\infty} g_2(y)p_Y(y)\,dy$$

$$= E[g_1(X)]E[g_2(Y)] \tag{8}$$

Taking $g(X,Y) = XY$ yields $E[XY] = E[X]E[Y]$ or

$$\overline{xy} = \bar{x}\bar{y} = m_x m_y \tag{9}$$

which says that the mean of the product equals the product of the means of independent random variables.

Next consider the sum $g(X,Y) = X + Y$, where X and Y are not necessarily independent. Routine manipulation of Eq. (7) now leads to $E[X+Y] = E[X] + E[Y]$ or

$$\overline{x+y} = \bar{x} + \bar{y} = m_x + m_y \tag{10}$$

Hence, the mean of the sum equals the sum of the means, irrespective of statistical dependence.

Finally, let $Z = X + Y$. We know that $m_z = m_x + m_y$, but what's the variance σ_z^2? To answer this question we calculate the mean square $\overline{z^2}$ via

$$\overline{z^2} = E[X^2 + 2XY + Y^2] = \overline{x^2} + \overline{2xy} + \overline{y^2}$$

Therefore

$$\sigma_z^2 = \overline{z^2} - (m_x + m_y)^2$$
$$= (\overline{x^2} - m_x^2) + (\overline{y^2} - m_y^2) + 2(\overline{xy} - m_x m_y)$$

whose last term disappears when Eq. (9) applies. Under this condition we get

$$\sigma_z^2 = \sigma_x^2 + \sigma_y^2 \tag{11}$$

for the variance of the sum of independent random variables.

Equation (11) readily extends to include three or more random variables, and so do Eqs. (8)–(10). But keep in mind that only Eq. (10) holds for *dependent* random variables.

Example 4.3-2 Sample mean and frequency of occurrence Let $X_1, X_2, \ldots,$ X_N be *sample values* obtained from N independent observations of an r.v. X having mean m_x and variance σ_x^2. Each sample value is a random variable, and so is the sum $Z = X_1 + X_2 + \cdots + X_N$ and the *sample mean* $\mu = Z/N$.

From Eqs. (10) and (11) we have $\bar{z} = \bar{x}_1 + \bar{x}_2 + \cdots + \bar{x}_N = Nm_x$ and $\sigma_z^2 = N\sigma_x^2$. Thus, $\bar{\mu} = \bar{z}/N = m_x$ while

$$\sigma_\mu^2 = E[(\mu - \bar{\mu})^2] = \frac{1}{N^2} E[(Z - \bar{z})^2]$$

$$= \frac{1}{N^2} \sigma_z^2 = \frac{1}{N} \sigma_x^2 \tag{12a}$$

Since the spread $\sigma_\mu = \sigma_x/\sqrt{N}$ decreases with increasing N, the sample mean μ approaches m_x as $N \to \infty$. Furthermore, the probability that μ differs from m_x by more than some amount ϵ is upper-bounded by

$$P(|\mu - m_x| \geq \epsilon) \leq \sigma_x^2/N\epsilon^2 \tag{12b}$$

obtained from Eq. (6a) with $k\sigma_\mu = \epsilon$ so $k = \sqrt{N}\,\epsilon/\sigma_x$.

Although it's not immediately obvious, Eq. (12b) provides further justification for the relative-frequency interpretation of probability. To develop this point, let A be an event associated with a chance experiment and let X be a discrete r.v. defined such that $X = 1$ when A occurs and $X = 0$ when A does not occur on a given trial. If A occurs N_A times in N independent trials of the experiment, then $Z = N_A$ and $\mu = N_A/N$. Thus, our definition of X makes the sample mean equal to the *frequency of occurrence* of A for that set of trials.

Now we compute the statistical averages of X by observing that $P_X(1) = P(A)$, where $P(A)$ is the probability of the event A. From Eq. (1) we obtain $m_x = 0 \times P_X(0) + 1 \times P_X(1) = P(A)$. Likewise, $\overline{x^2} = 0^2 \times P_X(0) + 1^2 \times P_X(1) = P(A)$, so $\sigma_x^2 = P(A) - P^2(A)$. Substituting these results into Eq. (12b) yields

$$P\left[\left|\frac{N_A}{N} - P(A)\right| \geq \epsilon\right] \leq \frac{P(A) - P^2(A)}{N\epsilon^2}$$

We therefore conclude that as $N \to \infty$, N_A/N must approach $P(A)$ in the sense that the probability of a significant difference becomes negligibly small.

Exercise 4.3-2 Prove Eq. (10) using the marginal PDFs defined by Eq. (20), Sect. 4.2.

Characteristic Functions★

Having found the mean and variance of a sum $Z = X + Y$, it seems appropriate to investigate $p_Z(z)$ and its relation to $p_X(x)$ and $p_Y(y)$. We'll assume that X and Y are independent, and we'll proceed in an indirect fashion using characteristic functions.

The *characteristic function* of the r.v. X is defined by

$$\Phi_X(v) \triangleq E[e^{jvX}] = \int_{-\infty}^{\infty} e^{jvx} p_X(x) \, dx \tag{13}$$

where v is an auxiliary independent variable and $j = \sqrt{-1}$. The presence of the complex exponential e^{jvx} reminds us of a *Fourier* integral. Indeed, letting $v = 2\pi t$ and $x = f$ in Eq. (13) shows that

$$\Phi_X(2\pi t) = \int_{-\infty}^{\infty} p_X(f) e^{j2\pi ft} \, df = \mathscr{F}^{-1}[p_X(f)]$$

and consequently

$$p_X(f) = \mathscr{F}[\Phi_X(2\pi t)] = \int_{-\infty}^{\infty} \Phi_X(2\pi t) e^{-j2\pi ft} \, dt \tag{14}$$

In other words, the characteristic function and PDF of a random variable constitute the *Fourier-transform pair* $\Phi_X(2\pi t) \leftrightarrow p_X(f)$.

The connection between characteristic functions and the sum of independent random variables now begins to emerge. For if $Z = X + Y$ then, from Eqs. (13) and (8),

$$\Phi_Z(v) = E[e^{jv(X+Y)}] = E[e^{jvX} e^{jvY}] = \Phi_X(v)\Phi_Y(v)$$

so

$$p_Z(f) = \mathscr{F}[\Phi_X(2\pi t)\Phi_Y(2\pi t)] = [p_X(f)] * [p_Y(f)]$$

where we have used the convolution theorem for the Fourier transform of a product. Appropriate change of variables yields the final result

$$p_Z(z) = \int_{-\infty}^{\infty} p_X(z - \lambda) p_Y(\lambda) \, d\lambda$$

$$= \int_{-\infty}^{\infty} p_X(\lambda) p_Y(z - \lambda) \, d\lambda \tag{15}$$

which says that the PDF of $X + Y$ equals the *convolution* of $p_X(x)$ with $p_Y(y)$. Similarly, the PDF of the sum of three or more independent random variables equals the convolution of the individual PDFs.

Other applications of characteristic functions are considered in Probs. 4.3-14 and 4.3-16

4.4 PROBABILITY MODELS

Numerous probability functions have been devised as models for a variety of random phenomena. Here we discuss the properties of two discrete functions (binomial and Poisson) and two continuous functions (gaussian and Rayleigh). These models cover most of the cases encountered in our later work. Table T.5 summarizes the results and includes a few other probability functions for reference purposes.

Binomial Distribution

The binomial model describes an integer-valued discrete r.v. associated with repeated trials of an experiment. Specifically, if you repeat an experiment n times and count the number of times a certain event A occurs, you have implicitly defined the *binomial random variable I* whose possible values are the integers $i = 0, 1, \ldots, n$. This model therefore applies to repeated coin tossings when I stands for the number of heads in n tosses. But more significantly in the study of communications, the binomial model also applies to *digital transmission* when I stands for the number of *errors* in a message of n digits.

To formulate the probability $P(I = i) = P_I(i)$, let the event A under consideration have $P(A) = \alpha$ so $P(A^c) = 1 - \alpha$. Any sequence of n independent trials in which A occurs i times then has the probability $\alpha^i (1 - \alpha)^{n-i}$. The number of such sequences is given by the *binomial coefficient*, denoted as $\binom{n}{i}$. Therefore, the *binomial frequency function* is

$$P_I(i) = \binom{n}{i} \alpha^i (1 - \alpha)^{n-i} \qquad i = 0, 1, \ldots, n \tag{1}$$

and the corresponding CDF can be written as

$$F_I(k) = \sum_{i=0}^{k} P_I(i) \qquad k = 0, 1, \ldots, n \tag{2}$$

These probability functions were previously evaluated in Fig. 4.2-3 for the case of $n = 3$ and $\alpha = 2/5$.

The binomial coefficient in Eq. (1) is defined by the factorial expression

$$\binom{n}{i} \triangleq \frac{n!}{i!(n-i)!} \tag{3}$$

where it's understood that $0! = 1$ when $i = 0$ or $i = n$. This quantity equals the $(i + 1)$th coefficient in the expansion of $(a + b)^n$, and has the symmetry property

$$\binom{n}{i} = \binom{n}{n-i}$$

Thus, we see that

$$\binom{n}{0} = \binom{n}{n} = 1 \qquad \binom{n}{1} = \binom{n}{n-1} = n$$

$$\binom{n}{2} = \binom{n}{n-2} = \frac{1}{2} n(n-1)$$

and so on. Other values can be found using Pascal's triangle, tables, or a calculator with a factorial key.

The statistical averages of a bonomial r.v. are obtained by inserting Eq. (1) into the appropriate expectation formulas. Some rather laborious algebra leads to simple results for the mean and variance, namely

$$m = n\alpha \qquad \sigma^2 = n\alpha(1 - \alpha) = m(1 - \alpha) \tag{4}$$

where we've omitted the subscripts for simplicity. Note that the *relative spread* σ/m decreases as $1/\sqrt{n}$, meaning that the likely values cluster around m when n is large.

Example 4.4-1 Suppose 10,000 digits are transmitted over a noisy channel with per-digit error probability $\alpha = 0.01$. Equation (4) gives $m = 100$ and $\sigma^2 = 99$ so the likely range $m \pm 2\sigma$ tells us to expect about 80 to 120 errors.

Poisson Distribution

The binomial distribution becomes awkward to handle when n is very large and α very small. But if the product $n\alpha$ remains finite, Eq. (1) can be approximated by the *Poisson* frequency function

$$P_I(i) = e^{-m} \frac{m^i}{i!} \tag{5}$$

where, as before,

$$m = n\alpha \qquad \sigma^2 = m(1 - \alpha) \approx m$$

Neither n nor α appears in Eq. (5) directly, since they're absorbed in the mean m.

Besides being an approximation for the binomial model, the Poisson model applies to any event A having probability $P(A) = \mu \, \Delta T \ll 1$ in a small time interval ΔT. If multiple occurrences are statistically independent, the probability that A occurs i times in time T is

$$P_I(i) = e^{-\mu T} \frac{(\mu T)^i}{i!}$$

and $m = \mu T$. This expression describes random phenomena such as *radioactive decay* and *shot noise* in electronic devices.

Exercise 4.4-1 Suppose 10,000 digits are transmitted over a noisy channel having $\alpha = 5 \times 10^{-5}$. Use Eq. (2) with $P_I(i)$ approximated by Eq. (5) to estimate the probability of $I \leq 2$ errors.

Gaussian PDF

A *gaussian random variable* is a continuous r.v. whose PDF is

$$p_X(x) = \frac{1}{\sqrt{2\pi\sigma^2}} \, e^{-(x-m)^2/2\sigma^2} \qquad -\infty < x < \infty \qquad (6)$$

where m and σ^2 are the mean and variance, respectively. This PDF defines the bell-shaped curve plotted in Fig. 4.4-1. The curve has even symmetry about the peak at $x = m$. Consequently

$$P(X \le m) = P(X > m) = \tfrac{1}{2} \qquad (7)$$

and observed values of X are just as likely to fall above or below the mean.

The gaussian distribution is also called the *normal* distribution because it pertains to so many different problems in engineering, physics, and statistics. The remarkable versatility of the gaussian model stems from the famous *central limit theorem* which essentially says that:

> If X represents the sum of N independent random components, and if each component makes only a small contribution to the sum, then the CDF of X approaches a gaussian CDF as N becomes large, regardless of the distribution of the individual components.

A more complete statement of the theorem and its proof involves sophisticated mathematical arguments and won't be pursued further. Instead, we'll draw upon the implication that the gaussian model often applies to situations in which the quantity of interest results from the summation of many irregular and fluctuating components. Thus, for instance, random errors in experimental measurements may cause measured values to have a gaussian distribution about the true value. Similarly, the random motion of thermally agitated electrons produces the gaussian random variable known as *thermal noise*—a random process we'll study in the next chapter.

Now assume that you know the mean and variance of a gaussian r.v. and you want to find the probability of the event $X > m + k\sigma$. The integral in question is

$$P(X > m + k\sigma) = \frac{1}{\sqrt{2\pi\sigma^2}} \int_{m+k\sigma}^{\infty} e^{-(x-m)^2/2\sigma^2} \, dx$$

Figure 4.4-1 Gaussian PDF.

Figure 4.4-2 Area interpretation of $Q(k)$.

which, unfortunately, cannot be solved in closed form. But numerical methods have been used to generate extensive tables of the normalized integral

$$Q(k) \triangleq \frac{1}{\sqrt{2\pi}} \int_k^\infty e^{-\lambda^2/2} \, d\lambda \tag{8}$$

The change of variable $\lambda = (x - m)/\sigma$ then shows that $Q(k) = P(X > m + k\sigma)$ so we'll call $Q(k)$ the *area under the gaussian tail*, illustrated in Fig. 4.4-2. The figure also brings out the relationship $P(m < X \leq m + k\sigma) = \frac{1}{2} - Q(k)$, which follows from Eq. (7) and the unit-area property of $p_X(x)$.

You can calculate any desired gaussian probability in terms of $Q(k)$ using Fig. 4.4-2 and the symmetry of $p_X(x)$. In particular, note that

$$P(X > m + k\sigma) = P(X \leq m - k\sigma) = Q(k) \tag{9a}$$

$$P(|X - m| > k\sigma) = 2Q(k) \tag{9b}$$

$$P(m < X \leq m + k\sigma) = P(m - k\sigma < X \leq m) = \frac{1}{2} - Q(k) \tag{9c}$$

$$P(|X - m| \leq k\sigma) = 1 - 2Q(k) \tag{9d}$$

Table 4.4-1 compares some values of $P(|X - m| \leq k\sigma) = 1 - 2Q(k)$ with the lower bound $1 - 1/k^2$ from Chebyshev's inequality. The lower bound is clearly conservative, and the likely range of observed values is somewhat less than $m \pm 2\sigma$. We usually take $m \pm \sigma$ as the likely range of a gaussian r.v. since $P(|X - m| \leq \sigma) \approx 0.68$.

Table T.6 at the back of the book gives a detailed plot of $Q(k)$ for $0 \leq k \leq 7$. The table also relates $Q(k)$ to other gaussian probability functions found in the literature.

Table 4.4-1

k	$1 - 2Q(k)$	$1 - 1/k^2$
0.5	0.38	
1.0	0.68	0.00
1.5	0.87	0.56
2.0	0.95	0.75
2.5	0.99	0.84

For large values of k, the area under the gaussian tail becomes too small for numerical integration. In this case, you can use an analytic approximation of $Q(k)$ obtained from Eq. (8) by integration by parts. The approximation is

$$Q(k) \approx \frac{1}{\sqrt{2\pi k^2}} e^{-k^2/2} \qquad (10)$$

which is quite accurate for $k > 3$.

Exercise 4.4-2 If X is gaussian with $m = 5$ and $\sigma = 8$, show that $P(9 < X \leq 25) = Q(0.5) - Q(2.5) \approx 0.30$. *Hint:* Sketch the PDF and mark the boundaries of the area in question in terms of $m + k\sigma$.

Rayleigh PDF

The Rayleigh model describes a continuous r.v. produced from two gaussian random variables by the transformation shown in Fig. 4.4-3a. Specifically, let X and Y be *independent* gaussians with

$$m_x = m_y = 0 \qquad \sigma_x = \sigma_y = \sigma$$

and let

$$R = +\sqrt{X^2 + Y^2}$$

Then R has the *Rayleigh* PDF

$$p_R(r) = \frac{r}{\sigma^2} e^{-r^2/2\sigma^2} u(r) \qquad (11)$$

which is sketched in Fig. 4.4-3b. Since R cannot be negative, by definition, it must have a nonzero mean value — even though X and Y have zero means. The mean and second moment of R are

$$\bar{r} = \sqrt{\frac{\pi}{2}} \sigma \qquad \overline{r^2} = 2\sigma^2 \qquad (12)$$

where σ^2 equals the gaussian variance.

Figure 4.4-3 (a) Rectangular to polar conversion; (b) Rayleigh PDF.

A happy feature of the Rayleigh PDF is its ease of integration for determining probabilities. In fact, the Rayleigh CDF takes the simple form

$$F_R(r) = P(R \le r) = (1 - e^{-r^2/2\sigma^2})u(r) \tag{13}$$

obtained by integrating $p_R(\lambda)$ over $0 \le \lambda \le r$.

As represented in Fig. 4.4-3a, the Rayleigh model applies to any *rectangular-to-polar conversion* when the rectangular coordinates are identical but independent gaussians with zero mean. Accordingly, we derive Eq. (11) by starting with the two-dimensional joint PDF relationship

$$p_{R\Phi}(r,\phi)|\, dr\, d\phi| = p_{XY}(x,y)|\, dx\, dy|$$

where

$$r^2 = x^2 + y^2 \qquad \phi = \arctan y/x$$

Now $dx\, dy = r\, dr\, d\phi$ and

$$p_{XY}(x,y) = p_X(x)p_Y(y) = \frac{1}{2\pi\sigma^2} e^{-(x^2+y^2)/2\sigma^2}$$

so

$$p_{R\Phi}(r,\phi) = \frac{r}{2\pi\sigma^2} e^{-r^2/2\sigma^2} \tag{14}$$

for $r \ge 0$. The angle ϕ does not appear explicitly in Eq. (14), but its range is clearly limited to 2π radians. We could take either $0 \le \phi < 2\pi$ or $-\pi \le \phi < \pi$, depending upon how we want to measure ϕ.

Taking $0 \le \phi < 2\pi$, we obtain the marginal PDF for R alone by integrating Eq. (14) with respect to ϕ. Thus

$$p_R(r) = \int_0^{2\pi} p_{R\Phi}(r,\phi)\, d\phi$$

which yields Eq. (11). Similarly, the marginal PDF for Φ alone is

$$p_\Phi(\phi) = \int_0^{\infty} p_{R\Phi}(r,\phi)\, dr = \frac{1}{2\pi}$$

so Φ has a *uniform* distribution over $0 \le \phi < 2\pi$. From these marginal PDFs we see that $p_{R\Phi}(r,\phi) = p_R(r)p_\Phi(\phi)$, and we conclude that the polar coordinates R and Φ are *statistically independent*. These results will be used in Chap. 9 for the representation of *bandpass noise*.

Example 4.4-2 Suppose you throw darts at a target whose bull's-eye has a radius of 3 cm. If the horizontal and vertical coordinates of the impact points are gaussianly distributed about the origin with spread $\sigma = 4$ cm in either direction, then your probability of hitting the bull's-eye is given by Eq. (13) as $P(R \le 3) = 1 - e^{-9/32} \approx 25\%$.

Exercise 4.4-3 Derive Eq. (13) from Eq. (11).

Bivariate Gaussian Distribution★

Lastly, we consider the joint PDF of two gaussian random variables that are neither identically distributed nor independent. To measure the interdependence between X and Y, we'll introduce the *correlation coefficient*

$$\rho \triangleq \frac{1}{\sigma_x \sigma_y} E[(X - m_x)(Y - m_y)] \tag{15}$$

where $m_x = \bar{x}$, $\sigma_x^2 = \overline{x^2} - m_x^2$, etc. The expectation in Eq. (15) is called the *covariance* of X and Y, and it equals zero if they're statistically independent. At the other extreme, if Y depends entirely on X in the sense that $Y = \pm \alpha X$, then $E[(X - m_x)(Y - m_y)] = \pm \alpha \sigma_x^2$ and $\sigma_y^2 = (\alpha \sigma_x)^2$ so $\rho = \pm 1$. Thus, the correlation coefficient ranges over $-1 \le \rho \le 1$ and $|\rho|$ reflects the degree of dependence. These general properties of ρ hold regardless of the distributions of X and Y.

The *bivariate gaussian* model describes two jointly gaussian random variables having the joint PDF

$$p_{XY}(x,y) = \frac{1}{2\pi \sigma_x \sigma_y \sqrt{1 - \rho^2}}$$

$$\exp \left[-\frac{(x - m_x)^2}{2\sigma_x^2(1 - \rho^2)} - \frac{(y - m_y)^2}{2\sigma_y^2(1 - \rho^2)} + \frac{\rho(x - m_x)(y - m_y)}{\sigma_x \sigma_y(1 - \rho^2)} \right] \tag{16}$$

This formidable-looking expression represents a bell-shaped surface over the x-y plane, with the peak at $x = m_x$ and $y = m_y$. If $\rho = 0$, the last term in the exponential disappears and $p_{XY}(x,y)$ reduces to a product of two gaussian PDFs. Thus, *uncorrelated gaussians are statistically independent.* (This property does not necessarily hold for other distributions, even though independence always guarantees that $\rho = 0$.)

For any value of ρ, it can be shown from Eq. (16) that all marginal PDFs and all conditional PDFs are gaussian functions. Furthermore, the *linear combination* $Z = \alpha X + \beta Y$ will also be a gaussian random variable. Again, these properties are unique characteristics of the bivariate gaussian case that do not necessarily hold for other distributions.

All of the foregoing analysis can be extended to three or more jointly gaussian random variables. However, the notation becomes quite messy unless you use matrix symbols. See, for instance, Sakrison (1968, chap. 3).

4.5 PROBLEMS

4.1-1 The outcome of an experiment is an integer whose value is equally likely to be any integer in the range $1 \le I \le 12$. Let A be the event that I is odd, let B be the event that I is divisible by 3, and let C be the event that I is divisible by 4. Draw the Venn diagram and find the probabilities of the events A, B, C, AB, AC, BC, and A^cB.

4.1-2 The outcome of an experiment is an integer whose value is equally likely to be any integer in the range $1 \le I \le 4$. The experiment is performed twice, yielding the outcomes I_1 and I_2. Let A be the event that $I_1 = I_2$, let B be the event that $I_1 > I_2$, and let C be the event that $I_1 + I_2 \geq 6$. Draw the Venn diagram and find the probabilities of the events A, B, C, AB, AC, BC, and $A^c B$.

4.1-3 If A and B are not mutually exclusive events, then the number of times A occurs in N trials can be written as $N_A = N_{AB} + N_{AB^c}$ where N_{AB^c} stands for the number of times A occurs without B. Use this notation to show that $P(AB^c) = P(A) - P(AB)$.

4.1-4 Use the notation in Prob. 4.1-3 to justify Eq. (8).

4.1-5 Let C stand for the event that either A or B occurs but not both. Use the notation in Prob. 4.1-3 to express $P(C)$ in terms of $P(A)$, $P(B)$, and $P(AB)$.

4.1-6 A biased coin is loaded such that $P(H) = (1 + \epsilon)/2$ with $0 < |\epsilon| < 1$. Show that the probability of a match in two independent tosses will be greater than 1/2.

4.1-7 A certain computer becomes inoperable if two components A and B both fail. The probability that A fails is 0.01 and the probability that B fails is 0.005. However, the probability that B fails increases by a factor of 4 if A has failed. Calculate the probability that the computer becomes inoperable. Also find the probability that A will fail if B has failed.

4.1-8 An honest coin is tossed twice and you are given partial information about the outcome. (a) Use Eq. (9) to find the probability of a match when you are told that the first toss came up heads. (b) Use Eq. (9) to find the probability of a match when you are told that heads came up on at least one toss. (c) Use Eq. (11) to find the probability of heads on at least one toss when you are told that a match occurred.

4.1-9 Do Prob. 4.1-8 with a loaded coin having $P(H) = 1/4$.

4.1-10 Derive from Eq. (9) the *chain rule*

$$P(XYZ) = P(X)P(Y|X)P(Z|XY)$$

4.1-11 A box contains 10 fair coins and 20 loaded coins with $P(H) = 1/4$. A coin is drawn at random from the box and tossed twice. (a) Use Eq. (12) to find the probability of the event "all tails." Let the conditioning event be the honesty of the coin. (b) If the event "all tails" occurs, what's the probability that the coin was loaded?

4.1-12 Do Prob. 4.1-11 for the case when the coin is tossed three times.

4.1-13 Two marbles are randomly withdrawn from a bag initially containing 5 red marbles, 3 white marbles, and 2 green marbles. (a) Use Eq. (12) to find the probability that the withdrawn marbles have matching colors. Let the conditioning event be the color of the first marble withdrawn. (b) If the withdrawn marbles have matching colors, what's the probability that they are white?

4.1-14 Do Prob. 4.1-3 for the case when three marbles are withdrawn from the bag.

4.2-1 Let $X = \frac{1}{2}N^2$, where N is a random integer whose value is equally likely to be any integer in the range $-1 \leq N \leq 3$. Plot the CDF of X and use it to evaluate the probabilities of the following events: $X \leq 0$, $2 < X \leq 3$, $X < 2$, and $X \geq 2$.

4.2-2 Do Prob. 4.2-1 with $X = 4 \cos \pi N/3$.

4.2-3 Let $p_X(x) = xe^{-x}u(x)$. Find $F_X(x)$ and use it to evaluate $P(X \leq 1)$, $P(1 < X \leq 2)$, and $P(X > 2)$.

4.2-4 Let $p_X(x) = \frac{1}{2}e^{-|x|}$. Find $F_X(x)$ and use it to evaluate $P(X \leq 0)$, $P(0 < X \leq 1)$, and $P(X > 1)$.

4.2-5 Suppose a certain random variable has the CDF

$$F_X(x) = \begin{cases} 0 & x \leq 0 \\ Kx^2 & 0 < x \leq 10 \\ 100K & x > 10 \end{cases}$$

Evaluate K, find the values of $P(X \leq 5)$ and $P(5 < X \leq 7)$, and plot the corresponding PDF.

4.2-6 Do Prob. 4.2-5 with

$$F_X(x) = \begin{cases} 0 & x \leq 0 \\ K \sin \pi x/40 & 0 < x \leq 10 \\ K \sin \pi/4 & x > 10 \end{cases}$$

4.2-7 Given that $F_X(x) = (\pi + 2 \arctan x)/2\pi$, find and sketch the CDF and PDF of the random variable Z defined by

$$Z = \begin{cases} 0 & X \le 0 \\ X & X > 0 \end{cases}$$

4.2-8‡ Do Prob. 4.2-7 for the random variable Z defined by

$$Z = \begin{cases} -1 & X \le 0 \\ X & X > 0 \end{cases}$$

4.2-9 Let $p_X(x) = 2e^{-2x}u(x)$. Find the PDF of the random variable Z defined by the transformation $Z = 2X - 5$. Then sketch both PDFs on the same set of axes.

4.2-10 Do Prob. 4.2-9 with $Z = -2X + 1$.

4.2-11 Let X have a uniform PDF over $-1 \le x \le 3$. Find and sketch the PDF of Z defined by the transformation $Z = \sqrt{X + 1}$.

4.2-12 Do Prob. 4.2-11 with $Z = |X|$.

4.2-13‡ Do Prob. 4.2-11 with $Z = \sqrt{|X|}$.

4.2-14 Consider the *square-law transformation* $Z = X^2$. Show that

$$p_Z(z) = \frac{1}{2\sqrt{z}} [p_X(\sqrt{z}) + p_X(-\sqrt{z})]u(z)$$

4.2-15 Find $p_Y(y)$ when $p_{XY}(x,y) = ye^{-y(x+1)}u(x)u(y)$. Then show that X and Y are not statistically independent, and find $p_X(x|y)$.

4.2-16 Do Prob. 4.2-15 with $p_{XY}(x,y) = [(x + y)^2/40]\Pi(x/2)\Pi(y/6)$.

4.2-17 Show that

$$\int_{-\infty}^{\infty} p_X(x|y)\, dx = 1$$

Explain why this must be true.

4.2-18 Obtain an expression for $p_Y(y|x)$ in terms of $p_X(x|y)$ and $p_Y(y)$.

4.3-1 Find the mean, second moment, and standard deviation of X when $p_X(x) = ae^{-ax}u(x)$ with $a > 0$.

4.3-2 Find the mean, second moment, and standard deviation of X when $p_X(x) = a^2xe^{-ax}u(x)$ with $a > 0$.

4.3-3 Find the mean, second moment, and standard deviation of X when

$$p_X(x) = \sqrt{2/\pi}[1 + (x - a)^4]$$

4.3-4 A discrete r.v. has two possible values, a and b. Find the mean, second moment, and standard deviation in terms of $p = P(X = a)$.

4.3-5 A discrete r.v. has K equally likely possible values, $0, a, 2a, \ldots, (K - 1)a$. Find the mean, second moment, and standard deviation.

4.3-6 Find the mean, second moment, and standard deviation of $Y = A \cos X$, where A is a constant and X has a uniform PDF over $\theta \le x \le \theta + 2\pi$.

4.3-7 Do Prob. 4.3-6 for the case when X has a uniform PDF over $\theta \le x \le \theta + \pi$.

4.3-8 Let $Y = \alpha X + \beta$. Show that $\sigma_y = |\alpha|\sigma_x$.

4.3-9 Let $Y = X + \beta$. What value of β minimizes $E[Y^2]$?

4.3-10‡ Let X be a nonnegative continuous r.v. and let a be any positive constant. By considering $E[X]$, derive *Markov's inequality* $P(X \ge a) \le m_x/a$.

4.3-11 Use $E[(X \pm Y)^2]$ to obtain upper and lower bounds on $E[XY]$ when X and Y are not statistically independent.

4.3-12 The *covariance* of X and Y is defined as

$$C_{xy} = E[(X - m_x)(Y - m_y)]$$

Expand this joint expectation and simplify it for the case when: (a) X and Y are statistically independent; (b) Y is related to X by $Y = \alpha X + \beta$.

4.3-13 In *linear estimation* we estimate Y from X by writing $\hat{Y} = \alpha X + \beta$. Obtain expressions for α and β to minimize the mean square error $\epsilon^2 = E[(Y - \hat{Y})^2]$.

4.3-14 Show that the nth moment of X can be found from its characteristic function via

$$E[X^n] = j^{-n} \frac{d^n \Phi_X(v)}{dv^n}\bigg|_{v=0}$$

4.3-15 Obtain the characteristic function of X when $p_X(x) = ae^{-ax}u(x)$ with $a > 0$. Then use the relation in Prob. 4.3-14 to find the first three moments.

4.3-16 Let X have a known PDF and let $Y = g(X)$, so

$$\Phi_Y(v) = E[e^{jvg(X)}] = \int_{-\infty}^{\infty} e^{jvg(x)} p_X(x)\, dx$$

If this integral can be rewritten in the form

$$\Phi_Y(v) = \int_{-\infty}^{\infty} e^{jv\lambda} h(\lambda)\, d\lambda$$

then $p_Y(y) = h(y)$. Use this method to obtain the PDF of $Y = X^2$ when $p_X(x) = 2axe^{-ax^2}u(x)$.

4.3-17‡ Use the method in Prob. 4.3-16 to obtain the PDF of $Y = \sin X$ when X has a uniform PDF over $|x| \leq \pi/2$.

4.4-1 Ten honest coins are tossed. What's the likely range of the number of heads? What's the probability that there will be fewer than three heads.

4.4-2 Do Prob. 4.4-1 with biased coins having $P(H) = 3/5$.

4.4-3 The *one-dimensional random walk* is described as follows. An inebriated man walking in a narrow hallway takes steps of equal length L. He steps forward with probability $\alpha = 3/4$ or backward with probability $1 - \alpha = 1/4$. Let X be his distance from the starting point after 100 steps. Find the mean and standard deviation of X.

4.4-4 A noisy transmission channel has per-digit error probability $\alpha = 0.01$. Calculate the probability of more than one error in 10 received digits. Repeat this calculation using the Poisson approximation.

4.4-5 A radioactive source emits particles at the average rate of 0.5 particles per second. Use the Poisson model to find the probability that: (a) exactly one particle is emitted in two seconds; (b) more than one particle is emitted in two seconds.

4.4-6‡ Show that the Poisson distribution in Eq. (5) yields $E[I] = m$ and $E[I^2] = m^2 + m$. The summations can be evaluated by writing the series expansion for e^m and differentiating it twice.

4.4-7 A gaussian r.v. has $E[X] = 10$ and $E[X^2] = 500$. Find $P(X > 20)$, $P(10 < X \leq 20)$, $P(0 < X \leq 20)$, and $P(X > 0)$.

4.4-8 When a binomial CDF has $n \gg 1$, it can be approximated by a gaussian CDF with the same mean and variance. Suppose an honest coin is tossed 100 times. Use the gaussian approximation to find the probability that: (a) heads occurs more than 70 times; (b) the number of heads is between 40 and 60.

4.4-9 Let X be a gaussian r.v. with mean m and variance σ^2. Write an expression in terms of the Q function for $P(a < X \leq b)$ with $a < m < b$.

4.4-10 A random noise voltage X is known to be gaussian with $E[X] = 0$ and $E[X^2] = 9$. Find the value of c such that $|X| < c$ for: (a) 90% of the time; (b) 99% of the time.

4.4-11 Write $e^{-\lambda^2/2}\, d\lambda = -(1/\lambda)d(e^{-\lambda^2/2})$ to show that the approximation in Eq. (10) is an upper bound on $Q(k)$. Then justify the approximation for $k \gg 1$.

4.4-12‡ Let X be a gaussian r.v. with mean m and variance σ^2. (a) Show that $E[(X - m)^n] = 0$ for odd n. (b) For even n, use integration by parts to obtain the recursion relation

$$E[(X - m)^n] = (n - 1)\sigma^2 E[(X - m)^{n-2}]$$

Then show that

$$E[(X - m)^n] = 1 \cdot 3 \cdot 5 \cdots (n - 1)\sigma^n$$

for $n = 2, 4, 6, \ldots$.

4.4-13 Let X be a gaussian r.v. with mean m and variance σ^2. Show that its characteristic function is

$$\Phi_X(v) = e^{-\sigma^2 v^2 / 2} e^{jmv}$$

4.4-14 Let $Z = X + Y$, where X and Y are independent gaussian r.v.'s with different means and variances. Use the characteristic function in Prob. 4.4-13 to show that Z has a gaussian PDF. Then extrapolate your results for

$$Z = \frac{1}{n} \sum_{i=1}^{n} X_i$$

where the X_i are mutually independent gaussian r.v.'s.

4.4-15 A random variable Y is said to be *log-normal* if the transformation $X = \ln Y$ yields a gaussian r.v. Use the gaussian characteristic function in Prob. 4.4-13 to obtain $E[Y]$ and $E[Y^2]$ in terms of m_x and σ_x^2. Do not find the PDF or characteristic function of Y.

4.4-16 Let $Z = X^2$, where X is a gaussian r.v. with zero mean and variance σ^2. (a) Use Eqs. (3) and (13), Sect. 4.3, to show that

$$\Phi_Z(v) = (1 - j2\sigma^2 v)^{-1/2}$$

(b) Apply the method in Prob. 4.3-14 to find the first three moments of Z. What statistical properties of X are obtained from these results?

4.4-17 Certain radio channels suffer from *Rayleigh fading* such that the received signal power is a random variable $Z = X^2$ and X has a Rayleigh distribution. Use Eq. (13), Sect. 4.2, to obtain the PDF

$$p_Z(z) = \frac{1}{m} e^{-z/m} u(z)$$

where $m = \bar{z}$. Evaluate the probability $P(Z \le km)$ for $k = 1$ and $k = 0.1$.

4.4-18‡ Let R_1 and R_2 be independent Rayleigh r.v.'s with $E[R_1^2] = E[R_2^2] = 2\sigma^2$. (a) Use the characteristic function from Prob. 4.4-16 to obtain the PDF of $A = R_1^2$. (b) Now apply Eq. (15), Sect. 4.3, to find the PDF of $W = R_1^2 + R_2^2$.

4.4-19 Let the bivariate gaussian PDF in Eq. (16) have $m_x = m_y = 0$ and $\sigma_x = \sigma_y = \sigma$. Show that $p_Y(y)$ and $p_X(x \mid y)$ are gaussian functions.

4.4-20 Find the PDF of $Z = X + 3Y$ when X and Y are gaussian r.v.'s with $m_x = 6$, $m_y = -2$, $\sigma_x = \sigma_y = 4$, and $E[XY] = -22$.

4.4-21 Let $Y = X^2$ so, clearly, X and Y are not independent. Nevertheless, show that they are uncorrelated if the PDF of X has even symmetry.

CHAPTER
FIVE

RANDOM SIGNALS AND NOISE

All meaningful communication signals are unpredictable or random, as viewed from the receiving end. Otherwise, there would be little value in transmitting a signal whose behavior was completely known beforehand. Furthermore, all communication systems suffer to some degree from the adverse effects of electrical noise. The study of random signals and noise is therefore essential for evaluating performance of communication systems.

The first two sections of this chapter combine concepts of signal analysis and probability to construct mathematical models of random electrical processes. Don't be discouraged if the material seems rather theoretical and abstract, for we'll put our models to practical use in the remainder of the chapter. Specifically, the third section is devoted to descriptions of noise per se, while the last section examines signal transmission in the presence of noise. Most of the topics introduced here will be further developed and extended in the latter half of the text.

5.1 RANDOM PROCESSES

A random signal is the manifestation of a *random electrical process* that takes place in time. When time enters the picture the complete description of random phenomena becomes quite complicated — especially if the statistical properties change with time. The assumption of an *ergodic* process, however, leads to rather simple and intuitively meaningful relationships between statistical properties, time averages, and spectral analysis. Moreover, many of the random processes encountered in a communication system fit the ergodic model to a reasonable degree.

Throughout the rest of this book we'll assume that random signals come from ergodic sources, unless otherwise indicated. The present section discusses the ensemble description of a random process, introduces ensemble averages, and sets forth the conditions implied by ergodicity.

Description of a Random Process

Previously we said that a random variable $X(s)$ maps the outcomes s of a chance experiment into numbers along the real line. A *random process* (also known as *stochastic process*) maps outcomes into real (noncomplex) *functions of time*. The collection of time functions is called an *ensemble*, represented formally by $v(t,s)$, and each member is called a *sample function*.

To clarify the concept of a random process, consider the set of voltage waveforms generated by thermal electron motion in a large number of identical resistors. The underlying experiment might be: pick a resistor at random and observe the waveform across its terminals. Figure 5.1-1 depicts some of the waveforms from the ensemble $v(t,s)$ associated with this experiment. A particular outcome (or choice of resistor) corresponds to the sample function $v_i(t) = v(t,s_i)$ having the value $v_i(t_1) = v(t_1,s_i)$ at time t_1. If you know the experimental outcome then, in principle, you know the entire behavior of the sample function and all randomness disappears.

But the basic premise regarding random processes is that you don't know which sample function you're observing. So at time t_1 you could expect *any* value from the ensemble of possible values $v(t_1,s)$. In other words, $v(t_1,s)$ constitutes a *random variable*, say V_1, defined by a vertical slice through the ensemble of sample functions at $t = t_1$. Likewise, the vertical slice at t_2 defines another r.v., V_2. Viewed in this light, a random process boils down to a *family of random variables*.

Now let's omit s and represent the random process by $v(t)$, just as we did when we wrote X instead of $X(s)$ for a single random variable. The context will

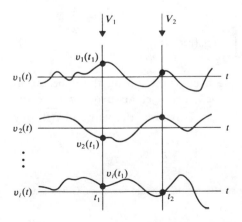

Figure 5.1-1 Waveforms in an ensemble $v(t,s)$.

make it clear when we're talking about a random process rather than a deterministic signal, so we'll not need a more formal notation.† The streamlined symbol $v(t)$ also agrees with the fact that we hardly ever know or care about the details of the underlying experiment. We do care about the random variables $V_1 = v(t_1)$, $V_2 = v(t_2)$, etc., whose probability functions describe the random process at the respective instants of time.

Suppose, for example, that we know the *probability density function* of V_1. We'll write its PDF as $p_{V_1}(v_1)$ where the dummy variable v_1 shouldn't be confused with a sample function. The probability of an event such as $v(t_1) \leq a$ can then be found by integrating $p_{V_1}(v_1)$ over $-\infty < v_1 \leq a$. Or we can calculate the mean value $\overline{v(t_1)}$ from the usual expection formula. Similarly, given the PDF of V_2, we could calculate the probability $P[v(t_2) \leq a]$ and the mean $\overline{v(t_2)}$, whose values will generally differ from those of $v(t_1)$. The set of PDFs $p_{V_1}(v_1)$, $p_{V_2}(v_2)$, ... provides a partial description of the random process referred to collectively as the *first-order density function*.

Consider next the *joint* probability $P[v(t_1) \leq a, v(t_2) \leq b]$. Since V_1 and V_2 are different and possibly dependent random variables, the joint probability calculation requires the joint PDF $p_{V_1 V_2}(v_1, v_2)$ — called a *second-order* density function because it involves two instants of time, t_1 and t_2. Upon extending this line of reasoning to three or more instants of time, we infer that the *complete* description of $v(t)$ requires *all higher-order joints* PDFs $p_{V_1 V_2, \ldots}(v_1, v_2, \ldots)$. Such a complete description demands a lot of work with a lot of data — data that's seldom available in practice. Fortunately, most engineering problems require only the first-order description plus one second-order function discussed below.

Ensemble Averages and Correlation Functions

For notational convenience, let the set of first-order PDFs be rolled into a single expression, denoted $p_V(v;t)$ and having the property that $p_V(v_1;t_1) = p_{V_1}(v_1)$, etc. Given $p_V(v;t)$, the statistical average or mean value of $v(t)$ at any time t is written as

$$\overline{v(t)} = E[v(t)] \triangleq \int_{-\infty}^{\infty} v p_V(v;t) \, dv \tag{1}$$

with the understanding that you treat t like a constant during the expectation operation. Equation (1) represents an *ensemble average*, an average over the ensemble of waveforms keeping time held fixed. However, the value of $\overline{v(t)}$ may turn out to depend on time.

---

† Some authors use $V(t)$ for the random process and V_{t_i} for the random variables. Others employ boldface notation to make this distinction.

Another important ensemble average is the *autocorrelation function* defined by

$$R_v(t_1, t_2) \triangleq E[v(t_1)v(t_2)]$$

$$= \int\!\!\int_{-\infty}^{\infty} v_1 v_2 \, p_{V_1 V_2}(v_1, v_2) \, dv_1 \, dv_2 \tag{2}$$

This joint expectation differs in three respects from the autocorrelation of deterministic power signals in Eq. (7), Sect. 3.5. First, we now deal with ensemble averaging instead of time averaging.† Second, $R_v(t_1, t_2)$ depends on t_1 and t_2 rather than a single variable τ. Third, we have limited our consideration to real processes so the complex conjugate does not appear in Eq. (2). Differences notwithstanding, $R_v(t_1, t_2)$ has an interpretation similar to the deterministic autocorrelation in the sense that it measures the relatedness or dependence between V_1 and V_2. If they happen to be statistically independent then $R_v(t_1, t_2) = \overline{v(t_1)v(t_2)}$. At the other extreme, if $t_1 = t_2 = t$ then

$$R_v(t,t) = E[v^2(t)] = \overline{v^2(t)}$$

which is the mean square value of $v(t)$ as a function of time. We'll say more about autocorrelation later on.

Sometimes we need to examine the joint statistics of two random processes, say $v(t)$ and $w(t)$. A useful measure of their relatedness is the *crosscorrelation function*

$$R_{vw}(t_1, t_2) \triangleq E[v(t_1)w(t_2)] \tag{3}$$

whose calculation involves the corresponding joint PDF. The processes are termed *uncorrelated* if, for all t_1 and t_2,

$$R_{vw}(t_1, t_2) = \overline{v(t_1)} \, \overline{w(t_2)} \tag{4}$$

which further reduces to $R_{vw}(t_1, t_2) = 0$ when either process has zero mean value. Physically independent processes are statistically independent and, hence, uncorrelated. However, uncorrelated processes are not necessarily independent.

Equations (1) and (2) become easier to handle when the random process has the functional form

$$v(t) = g(X,t)$$

with X being an ordinary random variable. Thus, at any time t_1, the relation $V_1 = g(X, t_1)$ is simply a random-variable *transformation*. Knowledge of the PDF of X then allows you to calculate $\overline{v(t)}$ at an arbitrary time t via

$$\overline{v(t)} = E[g(X,t)] = \int_{-\infty}^{\infty} g(x,t) p_X(x) \, dx \tag{5a}$$

---

† Some authors emphasize this difference by using different symbols for the two types of autocorrelation functions.

Likewise, the autocorrelation function becomes

$$R_v(t_1,t_2) = E[g(X,t_1)g(X,t_2)]$$

$$= \int_{-\infty}^{\infty} g(x,t_1)g(x,t_2)p_X(x) \, dx \tag{5b}$$

If desired, you could also obtain the first-order PDF of $v(t)$ using the transformation method in Sect. 4.2.

The foregoing expressions readily generalize to the case of a random process defined in terms of two or more r.v.'s. When $v(t) = g(X,Y,t)$, for instance, its autocorrelation function can be calculated from the joint expectation

$$R_v(t_1,t_2) = \iint_{-\infty}^{\infty} g(x,y,t_1)g(x,y,t_2)p_{XY}(x,y) \, dx \, dy \tag{6}$$

if you know the joint PDF of X and Y.

Example 5.1-1 Randomly phased sinusoid The randomly phased sinusoid is a random process defined by

$$v(t) = A \cos (\omega_0 t + \Phi)$$

in which A and ω_0 are constants while Φ is a random angle whose sample values are uniformly distributed over 2π radians. The underlying experiment might consist of picking an oscillator at random from a large collection of oscillators having the same amplitude and frequency but no phase synchronization. A particular oscillator with phase angle $\Phi = \phi_i$ generates the sample function $v_i(t) = A \cos (\omega_0 t + \phi_i)$.

Since $v(t)$ is defined by a transformation, we'll obtain its mean value and autocorrelation function using Eq. (5) with $g(\Phi,t) = A \cos (\omega_0 t + \Phi)$ and $p_\Phi(\phi) = 1/2\pi$ for $0 \le \phi \le 2\pi$. As a preliminary step for both calculations, consider the expected value of $\cos (\alpha + n\Phi)$ where n is an integer and α is treated as a constant with respect to the integration over ϕ. Thus, when $n = 0$,

$$E[\cos \alpha] = \int_0^{2\pi} \cos \alpha \, \frac{1}{2\pi} \, d\phi = \cos \alpha$$

while for $n \neq 0$,

$$E[\cos (\alpha + n\Phi)] = \frac{1}{2\pi} \int_0^{2\pi} \cos (\alpha + n\phi) \, d\phi$$

$$= [\sin (\alpha + 2\pi n) - \sin \alpha]/2\pi n = 0$$

Therefore, with $\alpha = \omega_0 t$ and $n = 1$, Eq. (5a) yields

$$\overline{v(t)} = E[g(\Phi, t)] = AE[\cos (\omega_0 t + \Phi)] = 0$$

and we see that the mean value equals zero for all time.

Now we insert $g(\Phi,t)$ into Eq. (5b) to find the atuocorrelation function. Letting $\alpha_1 = \omega_0 t_1$ and $\alpha_2 = \omega_0 t_2$ we have

$$
\begin{aligned}
R_v(t_1,t_2) &= A^2 E[\cos(\alpha_1 + \Phi)\cos(\alpha_2 + \Phi)] \\
&= \tfrac{1}{2}A^2 E[\cos(\alpha_1 - \alpha_2) + \cos(\alpha_1 + \alpha_2 + 2\Phi)] \\
&= \tfrac{1}{2}A^2\{E[\cos(\alpha_1 - \alpha_2)] + E[\cos(\alpha_1 + \alpha_2 + 2\Phi)]\} \\
&= \tfrac{1}{2}A^2 \cos(\alpha_1 - \alpha_2)
\end{aligned}
$$

and thus

$$
R_v(t_1,t_2) = \frac{A^2}{2}\cos\omega_0(t_1 - t_2)
$$

Finally, setting $t_1 = t_2 = t$ gives

$$
\overline{v^2(t)} = R_v(t,t) = A^2/2
$$

so the mean square value stays constant over time.

Exercise 5.1-1 Let $v(t) = X + 3t$ where the r.v. X has $\bar{x} = 0$ and $\overline{x^2} = 5$. Find $\overline{v(t)}$, $R_v(t_1,t_2)$, and $\overline{v^2(t)}$.

Stationary and Ergodic Processes

A *stationary* random process is one whose characteristics remain invariant over all time. Or, stated another way, any translation of the time origin along the ensemble does not affect the values of the ensemble averages. Among other consequences of stationarity, it follows that:

1. The mean value must be independent of time, so we can write $E[v(t)] = \bar{v} = m_v$.
2. The autocorrelation function must depend only on the time difference $t_1 - t_2$, so we can write $E[v(t_1)v(t_2)] = R_v(t_1 - t_2)$.

If these conditions hold, we say that the process is at least stationary in the *wide sense*.

When dealing with a wide-sense or strictly stationary process, we let $\tau = t_1 - t_2$ and rewrite the autocorrelation function as

$$
R_v(\tau) = E[v(t)v(t - \tau)] = E[v(t + \tau)v(t)] \tag{7}
$$

The form of this expression is now directly analogous to the autocorrelation of a real deterministic power signal. Setting $\tau = 0$ immediately shows that $E[v^2(t)] = R_v(0)$; hence, the mean square value and variance of a stationary process will be constants.

Properties of $R_v(\tau)$ that follow from Eq. (7) include

$$R_v(0) = \overline{v^2} \tag{8a}$$

$$|R_v(\tau)| \leq R_v(0) \tag{8b}$$

$$R_v(-\tau) = R_v(\tau) \tag{8c}$$

which tell us that $R_v(\tau)$ has even symmetry about its maximum value at $\tau = 0$. Furthermore, if $v(t)$ and $v(t \pm \tau)$ become *independent* as $\tau \to \infty$, then

$$\lim_{\tau \to \infty} R_v(\pm\tau) = \bar{v}^2 = m_v^2 \tag{8d}$$

However, if the sample functions are *periodic* with period T_0, then

$$R_v(\tau \pm nT_0) = R_v(\tau) \qquad n = 1, 2, \ldots \tag{8e}$$

and Eq. (8d) does not apply.

The randomly phased sinusoid in Example 5.1-1 illustrates a stationary process having $\bar{v} = 0$, $\overline{v^2} = A^2/2$, and $R_v(t_1 - t_2) = R_v(\tau) = (A^2/2) \cos \omega_0 \tau$ — in agreement with Eqs. (8a)–(8c). The autocorrelation function also exhibits the same periodicity as the sample functions, as stated in Eq. (8e). Hence, $R_v(\tau)$ does not have a unique limiting value as $|\tau| \to \infty$.

The randomly phased sinusoid further illustrates that some ensemble averages may equal the corresponding *time* averages of an arbitrary sample function. Specifically, if you take the sample function $v_i(t) = A \cos(\omega_0 t + \phi_i)$ and apply the time-average operation

$$\langle v_i(t) \rangle = \lim_{T \to \infty} \frac{1}{T} \int_{-T/2}^{T/2} v_i(t) \, dt$$

you'll find that $\langle v_i(t) \rangle = E[v(t)]$, $\langle v_i^2(t) \rangle = E[v^2(t)]$, and $\langle v_i(t)v_i(t - \tau) \rangle = E[v(t)v(t - \tau)]$.

When *all* ensemble averages equal the corresponding time averages the process is said to be *ergodic*. An ergodic process must be strictly stationary, but stationarity does not guarantee ergodicity. Ergodicity imposes the additional condition that a single sample function is representative of the entire process.

The randomly phased sinusoid is one of the processes of interest in communication that's known to be ergodic. Others include stationary gaussian processes and certain Poisson processes such as shot noise. But there exists no simple ergodic test for arbitrary processes. Consequently, we often take a more pragmatic approach and assume ergodicity if a process satisfies the conditions for wide-sense stationarity and if we can reasonably argue that a typical sample function exhibits all statistical variations of the process.

Example 5.1-2 Random Digital Wave The *random digital wave* is an ensemble of rectangular pulse trains like the sample function in Fig. 5.1-2a. All pulses have fixed duration D, but the delay parameter T_d is a continuous r.v. uniformly distributed over $0 \leq t_d < D$ — meaning that the ensemble consists

Figure 5.1-2 Random digital wave. (a) Sample function; (b) a single pulse.

of unsynchronized waveforms. The amplitude† a_k in the kth pulse interval $(k - 1)D < t - T_d < kD$ is a *discrete* r.v. with zero mean and variance σ^2. The amplitudes in different intervals are statistically independent, so $\overline{a_j a_k} = \bar{a}_j \bar{a}_k = 0$ for $j \neq k$.

Since $v(t) = a_k$ when t falls in the kth interval, we see that

$$E[v(t)] = E[a_k] = 0$$

$$E[v^2(t)] = E[a_k^2] = \sigma^2$$

Being independent of time, these results suggest a stationary process. To complete the test for wide-sense stationarity, we must find $R_v(t_1, t_2)$ from $E[v(t_1)v(t_2)]$. This calculation will be carried out by considering t_1 and t_2 to be in the same or different intervals.

Clearly, t_1 and t_2 must fall in different intervals when $|t_2 - t_1| > D$. Therefore

$$E[v(t_1)v(t_2)] = E[a_j a_k] = 0 \qquad |t_2 - t_1| > D$$

which also holds when $|t_2 - t_1| < D$ but t_1 and t_2 are in *adjacent* intervals. On the other hand, $E[v(t_1)v(t_2)] = E[a_k^2] = \sigma^2$ when t_1 and t_2 fall in the same interval. If we let A stand for the event "adjacent intervals," we can write

$$E[v(t_1)v(t_2)] = E[a_j a_k]P(A) + E[a_k^2][1 - P(A)]$$

$$= \sigma^2[1 - P(A)]$$

when $|t_2 - t_1| < D$.

To calculate the probability $P(A)$, it suffices to consider $0 < t < D$ shown in Fig. 5.1-2b since $P(A)$ will have the same value when t_1 and t_2 fall in any time range of length D. For $t_1 < t_2$ as shown, t_1 and t_2 are in adjacent pulse intervals if $t_1 < T_d < t_2$, and

$$P(A) = P(t_1 < T_d < t_2) = \int_{t_1}^{t_2} \frac{1}{D} \, dt_d = \frac{t_2 - t_1}{D}$$

† In the random *binary* wave the amplitude is restricted to two values, such as $a_k = \pm A$, and the duration is usually represented by T_b rather than D.

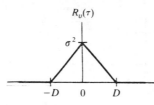

$R_v(\tau)$

σ^2

$-D$ 0 D

Figure 5.1-3 Autocorrelation of the random digital wave.

Had we taken $t_1 > t_2$, the term $t_2 - t_1$ would have become $t_1 - t_2$. Therefore, when $|t_1 - t_2| < D$

$$E[v(t_1)v(t_2)] = \sigma^2\left[1 - \frac{|t_1 - t_2|}{D}\right]$$

in which $|t_1 - t_2|/D$ represents $P(A)$ for both $t_1 < t_2$ and $t_1 > t_2$.

We've now found that $R_v(t_1,t_2) = E[v(t_1)v(t_2)]$ depends only on the time difference $t_1 - t_2$ so the random digital wave is a wide-sense stationary process. Accordingly, we let $\tau = t_1 - t_2$ and combine our results in the compact expression

$$R_v(\tau) = \sigma^2\Lambda\left(\frac{\tau}{D}\right) \tag{9}$$

where $\Lambda(\tau/D)$ is the triangle function. The plot of $R_v(\tau)$ in Fig. 5.1-3 illustrates the autocorrelation properties listed in Eq. (8) and deserves careful study.

Exercise 5.1-2 Let $v(t)$ be a stationary process and let $z(t_1,t_2) = v(t_1) \pm v(t_2)$. Use the fact that $E[z^2(t_1,t_2)] \geq 0$ to prove Eq. (8b).

Gaussian Processes

A random process $v(t)$ is called a *gaussian process* if $p_V(v;t)$ is a gaussian PDF for all values of t, and if $p_{V_1V_2}(v_1,v_2;t_1,t_2)$ is a bivariate gaussian PDF for all t_1 and t_2, and likewise for every higher-order joint PDF. As an alternative to the difficult task of finding all these PDFs, the central limit theorem may be invoked to determine if a given process is gaussian. Gaussian processes play a major role in the study of communications because the gaussian model applies to so many random electrical phenomena — at least as a first approximation.

Having determined that $v(t)$ is a gaussian process, several convenient properties flow therefrom. Specifically, it can be shown that:

1. The process is completely described by $\overline{v(t)}$ and $R_v(t_1,t_2)$.
2. If $R_v(t_1,t_2) = \overline{v(t_1)}\,\overline{v(t_2)}$ then $v(t_1)$ and $v(t_2)$ are uncorrelated and statistically independent.
3. If $v(t)$ satisfies the conditions for wide-sense stationarity then the process is strictly stationary and ergodic.
4. Any linear operation on $v(t)$ produces another gaussian process.

These properties greatly simplify the analysis of random signals. But bear in mind that they apply, in general, only to gaussian random signals.

5.2 RANDOM SIGNALS

In this section we consider random signals from ergodic or at least stationary sources. We'll invoke the Wiener–Kinchine theorem to obtain the power spectrum from the autocorrelation function. Then we'll use correlation and spectral analysis to study filtering and other operations on random signals.

Signal Power and Time Averages

Let $v(t)$ be a *stationary random signal*, meaning that $v(t)$ stands for the ensemble of sample functions $v(t,s)$ from a stationary random process. An immediate consequence of stationarity is that each sample function $v_i(t) = v(t,s_i)$ must be a *power signal*. To clarify this assertion, consider

$$P_T(s_i) = \frac{1}{T} \int_{-T/2}^{T/2} v_i^2(t)\, dt$$

which equals the signal power averaged over $-T/2 < t < T/2$. The total energy in $v_i(t)$ is

$$\lim_{T \to \infty} TP_T(s_i) = \int_{-\infty}^{\infty} v_i^2(t)\, dt$$

If the energy is finite, $v_i(t)$ must be timelimited such that $|v_i(t)| \to 0$ as $|t| \to \infty$. But then the ensemble averages at, say, $t = 0$ would differ from those at large values of $|t|$, which contradicts the definition of stationarity. Hence, we work with the power of a stationary random signal, rather than energy.

Now, as a generalization of $P_T(s_i)$, consider the random variable

$$P_T(s) = \frac{1}{T} \int_{-T/2}^{T/2} v^2(t)\, dt$$

which embraces the entire ensemble $v(t,s) = v(t)$. Since we don't know in advance the sample function the source produces, we define the *average power* of the random signal $v(t)$ to be

$$\bar{P} = \lim_{T \to \infty} E[P_T(s)] \tag{1a}$$

The notation $\bar{P}$ reflects the expectation operation $E[P_T(s)]$, which must be performed before taking the limit $T \to \infty$ to ensure that $\bar{P}$ exists. However, the order of time integration and expectation can be interchanged, so that

$$\bar{P} = \lim_{T \to \infty} \frac{1}{T} \int_{-T/2}^{T/2} E[v^2(t)]\, dt = \langle E[v^2(t)] \rangle \tag{1b}$$

This expression more clearly shows that $\bar{P}$ involves time averaging after ensemble averaging.

Drawing upon the assumed stationarity of $v(t)$, we write $E[v^2(t)] = \overline{v^2(t)} = \overline{v^2}$. Since $\overline{v^2}$ is a constant, $\langle \overline{v^2} \rangle = \overline{v^2}$ and Eq. (1b) becomes

$$\bar{P} = \overline{v^2} = \sigma_v^2 + m_v^2 = R_v(0) \tag{2}$$

Therefore, if $0 < \overline{v^2} < \infty$, we say that a stationary random signal is a *power signal* with average power $\overline{v^2}$. Similarly, the average value of a stationary random signal is taken to be $\langle E[v(t)] \rangle = \bar{v} = m_v$.

If the source is *ergodic* as well as stationary, then every sample function has the property that $\langle v_i(t) \rangle = \bar{v}$, $\langle v_i^2(t) \rangle = \overline{v^2}$, and so forth. Accordingly, we state the following relations between time averages and ensemble averages of an ergodic random signal:

1. The mean value m_v equals the *DC component* $\langle v(t) \rangle$.
2. The mean squared m_v^2 equals the *DC power* $\langle v(t) \rangle^2$.
3. The mean square value $\overline{v^2}$ equals the *total average power* $\langle v^2(t) \rangle$.
4. The variance σ_v^2 equals the *AC power*, or the power in the time-varying component of $v(t)$.
5. The standard deviation σ_v equals the *rms value* of the time-varying component.

These relations help make an electrical engineer feel at home in the world of random signals and they account, in part, for the intuitive appeal of ergodic processes.

If the process is stationary but not ergodic, we can still work with $\bar{P} = \overline{v^2}$ and other ensemble averages. However, the average power of a particular sample function does not necessarily equal $\bar{P}$.

Power Spectrum

The power spectral density or *power spectrum* $G_v(f)$ of a stationary random signal $v(t)$ tells us how the power $\bar{P}$ is distributed over the frequency domain. According to the *Wiener-Kinchine theorem*, $G_v(f)$ can be calculated from $R_v(\tau)$ by the Fourier transform

$$G_v(f) = \mathscr{F}_\tau[R_v(\tau)] \triangleq \int_{-\infty}^{\infty} R_v(\tau)e^{-j2\pi f\tau}\, d\tau \tag{3a}$$

Conversely

$$R_v(\tau) = \mathscr{F}_\tau^{-1}[G_v(f)] \triangleq \int_{-\infty}^{\infty} G_v(f)e^{j2\pi f\tau}\, df \tag{3b}$$

Thus, the autocorrelation and spectral density constitute a Fourier transform pair, just as in the case of deterministic signals. Properties of $G_v(f)$ include

$$\int_{-\infty}^{\infty} G_v(f)\, df = R_v(0) = \overline{v^2} = \bar{P} \tag{4}$$

$$G_v(f) \geq 0 \qquad G_v(-f) = G_v(f) \tag{5}$$

The even-symmetry property comes from the fact that $R_v(\tau)$ is real and even since $v(t)$ is real.

The power spectrum of a random signal may be continuous, impulsive, or mixed, depending upon the nature of the source. By way of illustration, the randomly phased sinusoid in Example 5.1-1 has

$$R_v(\tau) = \frac{A^2}{2} \cos 2\pi f_0 \tau \qquad (6a)$$

so

$$G_v(f) = \frac{A^2}{4} \delta(f - f_0) + \frac{A^2}{4} \delta(f + f_0) \qquad (6b)$$

This impulsive spectrum, shown in Fig. 5.2-1a, is identical to that of a deterministic sinusoid (Fig. 3.5-3) because the randomly phased sinusoid is an ergodic process whose sinusoidal sample functions differ only in the phase angle. As contrast, the random digital wave in Example 5.1-2 has

$$R_v(\tau) = \sigma^2 \Lambda(\tau/D) \qquad (7a)$$

and

$$G_v(f) = \sigma^2 D \operatorname{sinc}^2 fD \qquad (7b)$$

a continuous power spectrum shown in Fig. 5.2-1b.

Justification of the Wiener-Kinchine theorem for random signals could rest on our proof in Sect. 3.5 plus the fact that the autocorrelation of a stationary random signal has the same mathematical properties as those of a deterministic power signal. However, an independent derivation based on physical reasoning provides additional insight and a useful alternative definition of power spectrum.

Consider the finite-duration or *truncated* random signal

$$v_T(t) = \begin{cases} v(t) & |t| < T/2 \\ 0 & |t| > T/2 \end{cases}$$

(a)   (b)

Figure 5.2-1 Power spectra. (a) Randomly phased sinusoid; (b) random digital wave.

Since each truncated sample function has finite energy, we can write the Fourier transform

$$V_T(f,s) = \int_{-\infty}^{\infty} v_T(t)e^{-j\omega t}\, dt$$

$$= \int_{-T/2}^{T/2} v(t)e^{-j\omega t}\, dt \tag{8}$$

so $|V_T(f,s_i)|^2$ is the *energy spectral density* of the truncated sample function $v_T(t,s_i)$. Furthermore, using Rayleigh's energy theorem in the form

$$\int_{-T/2}^{T/2} v^2(t)\, dt = \int_{-\infty}^{\infty} v_T^2(t)\, dt = \int_{-\infty}^{\infty} |V_T(f,s)|^2\, df$$

the average power $\bar{P}$ in Eq. (1a) becomes

$$\bar{P} = \lim_{T\to\infty} E\left[\frac{1}{T}\int_{-\infty}^{\infty} |V_T(f,s)|^2\, df\right]$$

$$= \int_{-\infty}^{\infty} \lim_{T\to\infty} \frac{1}{T} E[|V_T(f,s)|^2]\, df$$

whose integrand may be interpreted as *power spectral density*.

Therefore, we now *define* the power spectrum of a random signal $v(t)$ to be

$$G_v(f) \triangleq \lim_{T\to\infty} \frac{1}{T} E[|V_T(f,s)|^2] \tag{9}$$

which agrees with the properties in Eqs. (4) and (5). Conceptually, this definition corresponds to the following steps: calculate the energy spectral density of the truncated sample functions, average over the ensemble, divide by T to convert energy to power, and take the limit $T\to\infty$.

Equation (9) also serves as the basis for experimental *spectral estimation*. After observing a sample function $v(t,s_i)$ over a long time interval T, you can estimate $G_v(f)$ from

$$\tilde{G}_v(f) = \frac{1}{T}|V_T(f,s_i)|^2$$

The spectral estimate $\tilde{G}_v(f)$ is called a *periodogram* because it was originally used to search for periodicities in seismic records and similar experimental data. For treatments of spectral estimation, see Blackman and Tukey (1958) and Schwartz and Shaw (1975).

Now we outline the proof that $G_v(f)$ in Eq. (9) equals $\mathcal{F}_\tau[R_v(\tau)]$. First, substitute Eq. (8) into $E[|V_T(f,s)|^2] = E[V_T(f,s)V_T^*(f,s)]$ and interchange integration and expectation to get

$$E[|V_T(f,s)|^2] = \iint_{-T/2}^{T/2} E[v(t)v(\lambda)]e^{-j\omega(t-\lambda)}\, dt\, d\lambda$$

Figure 5.2-2

in which

$$E[v(t)v(\lambda)] = R_v(t,\lambda) = R_v(t - \lambda)$$

Next, let $\tau = t - \lambda$ and $\mu = t$ so the double integration is performed over the region of the τ-μ plane shown in Fig. 5.2-2. Integrating with respect to μ for the two cases $\tau < 0$ and $\tau > 0$ then yields

$$E[|V_T(f,s)|^2] = \int_{-T}^{0} R_v(\tau)e^{-j\omega\tau}\left(\int_{-T/2}^{\tau+T/2} d\mu\right) d\tau + \int_{0}^{T} R_v(\tau)e^{-j\omega\tau}\left(\int_{\tau-T/2}^{T/2} d\mu\right) d\tau$$

$$= \int_{-T}^{0} R_v(\tau)e^{-j\omega\tau}(T + \tau) \, d\tau + \int_{0}^{T} R_v(\tau)e^{-j\omega\tau}(T - \tau) \, d\tau$$

Finally, since $\tau = -|\tau|$ for $\tau < 0$, we have

$$E[|V_T(f,s)|^2] = T \int_{-T}^{T} \left(1 - \frac{|\tau|}{T}\right) R_v(\tau)e^{-j\omega\tau} \, d\tau \qquad (10)$$

Therefore,

$$\lim_{T\to\infty} \frac{1}{T} E[|V_T(f,s)|^2] = \int_{-\infty}^{\infty} R_v(\tau)e^{-j\omega\tau} \, d\tau$$

which confirms that $G_v(f) = \mathcal{F}_\tau[R_v(\tau)]$.

Example 5.2-1 Random Telegraph Wave Figure 5.2-3a represents a sample function of a *random telegraph wave*. This signal makes independent random shifts between two equally likely values, A and 0. The number of shifts per unit time is governed by the Poisson distribution, with μ being the average shift rate. The autocorrelation function is

$$R_v(\tau) = \frac{A^2}{4} (e^{-2\mu|\tau|} + 1)$$

sketched in Fig. 5.2-3b. From $R_v(\tau)$ we see that

$$\bar{P} = \overline{v^2} = R_v(0) = \frac{A^2}{2} \qquad m_v^2 = \bar{v}^2 = R_v(\pm\infty) = \frac{A^2}{4}$$

so the rms value of $v(t)$ is $\sigma_v = (\overline{v^2} - \bar{v}^2)^{1/2} = A/2$.

(a)

(b)

(c)

Figure 5.2-3 Random telegraph wave. (a) Sample function; (b) autocorrelation; (c) power spectrum.

Taking the Fourier transform of $R_v(\tau)$ gives the power spectrum

$$G_v(f) = \frac{A^2}{4\mu[1 + (\pi f/\mu)^2]} + \frac{A^2}{4} \delta(f)$$

This mixed spectrum is plotted in Fig. 5.2-3c. Although μ equals the average shift rate, about 20% of the AC power (measured in terms of σ_v^2) is contained in the higher frequencies $|f| > \mu$.

Exercise 5.2-1 Physically, we expect that if $m_v \neq 0$, $G_v(f)$ will include an impulse representing the DC power — as illustrated by the previous example. To confirm this in general, let $z(t) = v(t) - m_v$, so $\overline{z(t)} = 0$. Then show from $R_z(\tau)$ that $G_v(f) = G_z(f) + m_v^2 \, \delta(f)$.

Superposition and Modulation

Some random signals may be viewed as a combination of other random signals. In particular, suppose that

$$z(t) = v(t) \pm w(t)$$

where $v(t)$ and $w(t)$ are *jointly stationary*, so $R_{vw}(t_1, t_2) = R_{vw}(t_1 - t_2)$, as well as individually stationary. Then

$$R_z(\tau) = R_v(\tau) + R_w(\tau) \pm [R_{vw}(\tau) + R_{wv}(\tau)]$$

and

$$G_z(f) = G_v(f) + G_w(f) \pm [G_{vw}(f) + G_{wv}(f)]$$

in which we have introduced the *cross-spectral density*

$$G_{vw}(f) \triangleq \mathscr{F}_\tau[R_{vw}(\tau)] \tag{11}$$

The cross-spectral density vanishes when $v(t)$ and $w(t)$ are uncorrelated and $m_v m_w = 0$, so

$$R_{vw}(\tau) = R_{wv}(\tau) = 0 \tag{12}$$

Under these conditions

$$R_z(\tau) = R_v(\tau) + R_w(\tau)$$
$$G_z(f) = G_v(f) + G_w(f) \tag{13}$$
$$\overline{z^2} = \overline{v^2} + \overline{w^2}$$

and we have *superposition* of autocorrelation, power spectra, and average power.

When Eqs. (12) and (13) hold, the random signals $v(t)$ and $w(t)$ are said to be *incoherent*. Signals from independent sources are usually incoherent, and superposition of average power is a common physical effect. For example, if two musicians play in unison but without perfect synchronization, the total acoustical power simply equals the sum of the individual powers.

Now consider the *modulation* operation defined by the product

$$z(t) = v(t) \cos (\omega_c t + \Phi) \tag{14}$$

where $v(t)$ is a stationary random signal and Φ is a random angle independent of $v(t)$ and uniformly distributed over 2π radians. If we didn't include Φ, $z(t)$ would be unstationary; including Φ merely recognizes the arbitrary choice of the time origin when $v(t)$ and $\cos \omega_c t$ come from independent sources.

Since modulation is a time-varying process, we must determine $R_z(\tau)$ from $R_z(t_1,t_2) = E[z(t_1)z(t_2)]$ by taking the joint expectation with respect to both v and Φ. After trigonometric expansion we get

$$R_z(t_1,t_2) = \tfrac{1}{2}E[v(t_1)v(t_2)]\{\cos \omega_c(t_1 - t_2) + E[\cos (\omega_c t_1 + \omega_c t_2 + 2\Phi)]\}$$
$$= \tfrac{1}{2}R_v(t_1,t_2) \cos \omega_c(t_1 - t_2)$$

Therefore, with $\tau = t_1 - t_2$,

$$R_z(\tau) = \tfrac{1}{2}R_v(\tau) \cos 2\pi f_c \tau \tag{15a}$$

and Fourier transformation yields

$$G_z(f) = \tfrac{1}{4}[G_v(f - f_c) + G_v(f + f_c)] \tag{15b}$$

Not surprisingly, modulation translates the power spectrum of $v(t)$ up and down by f_c units.

Modulation is a special case of the product operation

$$z(t) = v(t)w(t)$$

If $v(t)$ and $w(t)$ are independent and jointly stationary, then

$$R_z(\tau) = R_v(\tau)R_w(\tau) \qquad (16a)$$

and

$$G_z(f) = G_v(f) * G_w(f) \qquad (16b)$$

which follows from the convolution theorem.

Exercise 5.2-2 Derive Eq. (15b) from Eq. (16b) by making a judicious choice for $w(t)$.

Filtered Random Signals

Figure 5.2-4 represents a random signal $x(t)$ applied to the input of a filter (or any LTI system) having transfer function $H(f)$ and impulse response $h(t)$. The resulting output signal is then given by the convolution

$$y(t) = \int_{-\infty}^{\infty} h(\lambda)x(t - \lambda)\, d\lambda \qquad (17)$$

Since convolution is a linear operation, a *gaussian input* produces a *gaussian output* whose properties will be completely described by m_y and $R_y(\tau)$. These output statistics can be found from $H(f)$ or $h(t)$ and will be useful even in the nongaussian case.

We assume that $h(t)$ is *real*, so both $x(t)$ and $y(t)$ will be real. We also assume that $x(t)$ is a stationary power signal and that the system is *stable*. Under these assumptions, $y(t)$ will be a stationary power signal.

The input-output correlation relations are convolutions with the impulse response, namely

$$R_{yx}(\tau) = h(\tau) * R_x(\tau)$$
$$R_y(\tau) = h(-\tau) * R_{yx}(\tau) = h(-\tau) * h(\tau) * R_x(\tau) \qquad (18)$$

The spectra are related by

$$G_y(f) = |H(f)|^2 G_x(f) \qquad (19)$$

from which

$$R_y(\tau) = \mathscr{F}_\tau^{-1}[|H(f)|^2 G_x(f)]$$
$$\overline{y^2} = R_y(0) = \int_{-\infty}^{\infty} |H(f)|^2 G_x(f)\, df \qquad (20)$$

$$x(t) \qquad \qquad y(t)$$
$$R_x(\tau) \longrightarrow \boxed{\begin{array}{c} h(t) \\ H(f) \end{array}} \longrightarrow R_y(\tau)$$
$$G_x(f) \qquad \qquad G_y(f)$$

Figure 5.2-4

The mean value of the output is

$$m_y = \left[\int_{-\infty}^{\infty} h(\lambda) \, d\lambda \right] m_x = H(0)m_x \tag{21}$$

where $H(0)$ equals the system's DC gain.

Except for Eq. (21), all of these relations are identical to those derived in Sect. 3.5 for deterministic signals when $h(t)$ is real. The mean-value relation is easy to establish, and Eqs. (19) and (20) follow from Eq. (18) and the Wiener-Kinchine theorem. Therefore, we only need to confirm that the convolution relations in Eq. (18) are valid for stationary random signals.

For this task we start with $R_{yx}(t_1,t_2) = E[y(t_1)x(t_2)]$, insert Eq. (17) for $y(t_1)$, and again exchange the order of operations, with the result that

$$R_{yx}(t_1,t_2) = \int_{-\infty}^{\infty} h(\lambda)E[x(t_1 - \lambda)x(t_2)] \, d\lambda$$

Since $x(t)$ is stationary,

$$E[x(t_1 - \lambda)x(t_2)] = R_x(t_1 - \lambda, t_2) = R_x(t_1 - \lambda - t_2)$$

Thus letting $t_2 = t_1 - \tau$,

$$R_{yx}(t_1,t_1 - \tau) = R_{yx}(\tau) = \int_{-\infty}^{\infty} h(\lambda)R_x(\tau - \lambda) \, d\lambda$$

which proves that $R_{yx}(\tau) = h(\tau) * R_x(\tau)$. The proof of $R_y(\tau) = h(-\tau) * R_{yx}(\tau)$ proceeds in the same fashion and establishes the fact that $y(t)$ is at least wide-sense stationary.

The power spectrum relation in Eq. (19) has additional value for the analysis of *operations* on random signals, whether or not filtering is actually involved. In particular, suppose you know $G_x(f)$ and you want to find $G_y(f)$ when $y(t) = dx(t)/dt$. Conceptually, $y(t)$ could be obtained by applying $x(t)$ to an ideal differentiator. Since time-domain differentiation corresponds to $H(f) = j2\pi f$ in the frequency domain, we see that if

$$y(t) = \frac{dx(t)}{dt} \tag{22a}$$

then

$$G_y(f) = (2\pi f)^2 G_x(f) \tag{22b}$$

Conversely, if $m_x = 0$ and

$$y(t) = \int_{-\infty}^{t} x(\lambda) \, d\lambda \tag{23a}$$

then

$$G_y(f) = (2\pi f)^{-2} G_x(f) \tag{23b}$$

Equations (22) and (23) obviously parallel the differentiation and integration theorems for Fourier transforms.

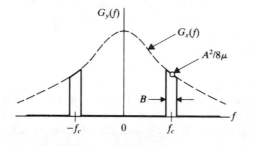

Figure 5.2-5 Filtered power spectrum in Example 5.2-2.

Example 5.2-2 Suppose the random telegraph wave in Example 5.2-1 is the input to an ideal bandpass filter with unit gain and narrow bandwidth $B \ll f_c$ centered at $f_c = \mu/\pi$. Figure 5.2-5 shows the resulting output power spectrum $G_y(f) = |H(f)|^2 G_x(f)$.

Since $H(0) = 0$ we know from Eq. (21) that $m_y = 0$, even though $m_x = A/2$. Furthermore, since $G_x(\pm f_c) = A^2/8\mu$ and $B \ll f_c$

$$G_y(f) \approx \begin{cases} \dfrac{A^2}{8\mu} & -\dfrac{B}{2} < |f - f_c| < \dfrac{B}{2} \\ 0 & \text{otherwise} \end{cases}$$

The output power then equals the area under $G_y(f)$, namely

$$\overline{y^2} \approx 2B \times (A^2/8\mu) = A^2 B/4\mu \ll A^2/4\pi$$

whereas $\overline{x^2} = A^2/2$. Note that we obtained this information without the added labor of finding $R_y(\tau)$.

Example 5.2-3 Hilbert Transform of a Random Signal The Hilbert transform $\hat{x}(t)$ was defined in Sect. 3.4 as the output of a quadrature phase-shifting filter with

$$h_Q(t) = \frac{1}{\pi t} \qquad H_Q(f) = -j \operatorname{sgn} f$$

Since $|H_Q(f)|^2 = 1$, we conclude that if $x(t)$ is a random signal and $y(t) = \hat{x}(t)$, then

$$G_{\hat{x}}(f) = G_x(f) \qquad R_{\hat{x}}(\tau) = \mathscr{F}_\tau^{-1}[G_{\hat{x}}(f)] = R_x(\tau)$$

which means that Hilbert transformation does not alter the values of m_x or $\overline{x^2}$. However, from Eq. (18),

$$R_{\hat{x}x}(\tau) = h_Q(\tau) * R_x(\tau) = \hat{R}_x(\tau)$$

where $\hat{R}_x(\tau)$ stands for the Hilbert transform of $R_x(\tau)$. It can also be shown that

$$R_{x\hat{x}}(\tau) = -\hat{R}_x(\tau)$$

We'll apply these results in Chap. 9.

Exercise 5.2-3 Let the random digital wave described by Eq. (7) be applied to a first-order LPF with $H(f) = [1 + j(f/B)]^{-1}$ and $B \ll 1/D$. Sketch $G_y(f)$ and obtain an approximate expression for $R_y(\tau)$.

5.3 NOISE

Unwanted electric signals come from a variety of sources, generally classified as human interference or naturally occurring noise. Human interference comes from other communication systems, ignition and commutator sparking, 60 cycle hum, and so forth; natural noise-producing phenomena include atmospheric disturbances, extraterrestrial radiation, and circuit noise. By careful engineering, the effects of many unwanted signals can be reduced or eliminated completely. But there always remain certain inescapable random signals, that present a fundamental limit to systems performance.

One unavoidable cause of electrical noise is the thermal motion of electrons in conducting media — wires, resistors, etc. As long as communication systems are constructed from such material, this *thermal noise* will be with us. Accordingly, this section begins with a brief discussion of thermal noise which, in turn, leads to the abstraction of white noise. We then consider filtered white noise and input–output relations. Other aspects of noise analysis are developed in subsequent chapters and App. A.

Thermal Noise and Available Power

Thermal noise, for our purposes, is the noise due to the random motion of charged particles (usually electrons) in conducting media. From kinetic theory, the average energy of a particle at absolute temperature $\mathcal{T}$ is proportional to $k\mathcal{T}$, k being the Boltzmann constant. We thus expect thermal-noise values to involve the product $k\mathcal{T}$; in fact we'll develop a measure of noise power in terms of temperature. Historically, Johnson (1928) and Nyquist (1928) first studied thermal noise in metallic resistors, hence the designation *Johnson noise* or *resistance noise*. There now exists an extensive body of theoretical and experimental studies pertaining to noise, from which we'll freely draw.

When a metallic resistance R is at temperature $\mathcal{T}$, random electron motion produces a noise voltage $v(t)$ at the open-circuited terminals. Consistent with the central limit theorem, $v(t)$ has a *gaussian distribution* with zero mean and variance

$$\overline{v^2} = \sigma_v^2 = \frac{2(\pi k \mathcal{T})^2}{3h} R \qquad \text{volts squared (V}^2) \tag{1}$$

where

k = Boltzmann constant = 1.37×10^{-23} joules per degree (J/deg)

h = Planck constant = 6.62×10^{-34} joule second (J·s)

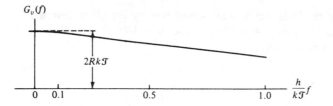

Figure 5.3-1 Thermal noise spectral density, V^2/Hz.

and $\mathcal{T}$ is measured in kelvins. The presence of the Planck constant in Eq. (1) indicates a result from quantum mechanics. Quantum mechanics further shows that the spectral density of thermal noise is[†]

$$G_v(f) = \frac{2Rh|f|}{e^{h|f|/k\mathcal{T}} - 1} \qquad \text{volts squared per hertz (V}^2/\text{Hz)} \qquad (2a)$$

as plotted in Fig. 5.3-1 for $f \geq 0$. This expression simplifies at "low" frequencies to

$$G_v(f) \approx 2Rk\mathcal{T}\left(1 - \frac{h|f|}{2k\mathcal{T}}\right) \qquad |f| \ll \frac{k\mathcal{T}}{h} \qquad (2b)$$

obtained with the help of series approximations.

Actually, communication engineers almost never have need for these expressions. To see why, let room temperature or the *standard temperature* be

$$\mathcal{T}_0 \triangleq 290 \text{ K (63°F)} \qquad (3a)$$

which is rather on the chilly side but does simplify numerical work since

$$k\mathcal{T}_0 \approx 4 \times 10^{-21} \qquad \text{watt second (W·s)} \qquad (3b)$$

Then if the resistance is at $\mathcal{T}_0$, $G_v(f)$ is essentially *constant* for $|f| < 0.1 \, k\mathcal{T}_0/h \approx 10^{12}$ Hz. But this upper limit falls in the infrared portion of the electromagnetic spectrum, far above the point where conventional electrical components have ceased to respond. Even at cryogenic temperatures ($\mathcal{T} \approx 0.01\mathcal{T}_0$) this conclusion holds.

Therefore, to all intents and purposes, we can say that the spectral density of thermal noise is constant at

$$G_v(f) = 2Rk\mathcal{T} \qquad V^2/Hz \qquad (4)$$

as obtained from Eq. (2) with $h|f|/k\mathcal{T} \ll 1$. The one trouble with Eq. (4) is that it erroneously predicts $\overline{v^2} = \infty$ when $G_v(f)$ is integrated over all f. However, you seldom have to deal directly with $\overline{v^2}$ since $v(t)$ always is subject to the filtering effects of other circuitry. That topic will be examined shortly. Meanwhile, we'll

---

[†] Equation (2a) omits a term corresponding to the zero-point energy, which is independent of temperature and plays no part in thermal noise transferred to a load.

Figure 5.3-2 Thermal resistance noise. (a) Thévenin equivalent circuit; (b) Norton equivalent circuit.

use Eq. (4) to construct the Thévenin equivalent model of a thermal resistor in Fig. 5.3-2a. Here the resistance is replaced by a fictitious *noiseless* resistance of the same value and the noise is represented by a mean square voltage generator. Similarly, Fig. 5.3-2b is the Norton equivalent with a mean square current generator having $G_i(f) = G_v(f)/R^2 = 2k\mathcal{T}/R$. Both generators are shaded to indicate their special nature.

Instead of dealing with mean square voltage or current, describing thermal noise by its *available power* cleans up the notation and speeds calculations. Available power is the *maximum* power that can be delivered to a load from a source having fixed nonzero source resistance. The familiar maximum-power-transfer theorem states that this power is delivered only when the load impedance is the complex conjugate of the source impedance. The load is then said to be *matched* to the source, a condition usually desired in a communication system.

Let the sinusoidal source in Fig. 5.3-3a have impedance $Z_S = R_S + jX_S$, and let the open-circuit voltage be v_S. If the load is matched so that $Z_L = Z_S^* = R_S - jX_S$, the the terminal voltage is $v_S/2$, and the available power is

$$P_a = \frac{\langle [v_S(t)/2]^2 \rangle}{R_S} = \frac{\langle v_S^2(t) \rangle}{4R_S}$$

Using the Thévenin model, we can extend this concept to a thermal resistor viewed as a noise source in Fig. 5.3-3b. By comparison, the *available spectral density* at the load resistance is

$$G_a(f) = \frac{G_v(f)}{4R} = \tfrac{1}{2}k\mathcal{T} \qquad \text{watts per hertz (W/Hz)} \tag{5}$$

which depends only on the temperature. A thermal resistor therefore delivers a maximum of $k\mathcal{T}/2$ W/Hz to a matched load, regardless of the value of R!

Figure 5.3-3 Available power. (a) AC source with matched load; (b) thermal resistance with matched load.

Exercise 5.3-1 Calculate from Eq. (1) the rms noise voltage across the terminals of a $1 - k\Omega$ resistance at $\mathcal{T} = 29$ K. Then use Eq. (2b) to find the percentage of the mean square voltage that comes from frequency components in the range $|f| < 1$ GHz.

White Noise and Filtered Noise

Besides thermal resistors, many other types of noise sources are gaussian and have a flat spectral density over a wide range of frequencies. Such a spectrum has all frequency components in equal proportion and is called *white* noise, by analogy to white light.

We write the spectral density of white noise in general as

$$G(f) = \frac{\eta}{2} \tag{6a}$$

where the seemingly extraneous factor of $\frac{1}{2}$ is included to indicate that half the power is associated with positive frequency and half with negative frequency (Fig. 5.3-4a). Alternatively, η is the *positive-frequency* power density. Since $G(f)$ is known, the autocorrelation function follows immediately by Fourier transformation

$$R(\tau) = \int_{-\infty}^{\infty} \frac{\eta}{2} e^{j\omega\tau} \, df = \frac{\eta}{2} \delta(\tau) \tag{6b}$$

as in Fig. 5.3-4b.

From Eq. (6b) or Fig. 5.3-4b we see that $R (\tau \neq 0) = 0$, so any two different samples of a gaussian white noise signal are *uncorrelated* and hence *statistically independent*. This observation, coupled with the constant power spectrum, leads to an interesting conclusion: if white noise is displayed on an oscilloscope, successive sweeps are always different from each other; yet the waveform always "looks" the same, no matter what sweep speed is used, since all rates of time variation (frequency components) are contained in equal proportion. Similarly, if white noise drives a loudspeaker, it always sounds the same, somewhat like a

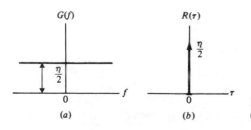

Figure 5.3-4 White noise. (*a*) Power spectrum; (*b*) autocorrelation.

waterfall. As Pierce† puts it:

> Mathematically, white gaussian noise ... is the epitome of the various and unexpected. It is the least predictable, the most original of sounds. To a human being, however, all gaussian white noise sounds alike. Its subleties are hidden from him, and he says that it is dull and monotonous.

The value of η in Eqs. (6a) and (6b) depends on two factors: the type of noise source and the type of spectral density, i.e., mean square voltage, mean square current, or available power. If the source is a *thermal resistor*, then

$$\eta_v = 4Rk\mathcal{T} \qquad \eta_i = \frac{4k\mathcal{T}}{R} \qquad \eta_a = k\mathcal{T} \tag{7}$$

where subscripts identify the type of spectral density. Moreover, any *thermal noise source* has $\eta_a = k\mathcal{T}$, by definition. Other white noise sources are *nonthermal* in the sense that the available power is unrelated to a physical temperature. Nonetheless, we can speak of the *noise temperature* $\mathcal{T}_N$ of any white noise source, thermal or nonthermal, by defining

$$\mathcal{T}_N \triangleq \frac{2G_a(f)}{k} = \frac{\eta_a}{k} \tag{8a}$$

where $\eta_a/2$ is the maximum noise power the source can deliver per unit frequency. Then, given a source's noise temperature,

$$\eta_a = k\mathcal{T}_N \tag{8b}$$

It must be emphasized that $\mathcal{T}_N$ is not necessarily a physical temperature. For instance, certain electronic noise generators have $\mathcal{T}_N \approx 10\mathcal{T}_0 \approx 3000$ K (5000°F) but the devices surely aren't that hot.

Now consider gaussian white noise $x(t)$ with spectral density $G_x(f) = \eta/2$ applied to an LTI filter having transfer function $H(f)$. The resulting output $y(t)$ will be gaussian noise described by

$$G_y(f) = \frac{\eta}{2}|H(f)|^2 \tag{9a}$$

$$R_y(\tau) = \frac{\eta}{2}\mathcal{F}^{-1}[|H(f)|^2] \tag{9b}$$

$$\overline{y^2} = \frac{\eta}{2}\int_{-\infty}^{\infty}|H(f)|^2\,df \tag{9c}$$

Pay careful attention to Eq. (9a) which shows that the spectral density of filtered white noise takes the *shape* of $|H(f)|^2$. We therefore say that filtering white noise

† J. R. Pierce, "Symbols, Signals, and Noise," Harper & Row, Publishers, Incorporated, New York, 1961.

$G_y(f)$ $R_y(\tau)$

Figure 5.3-5 White noise passed by an ideal LPF. (a) Power spectrum; (b) autocorrelation.

produces *colored* noise with frequency content primarily in the range passed by the filter.

As an illustration of filtered noise, let $H(f)$ be an ideal lowpass function with unit gain and bandwidth B. Then

$$G_y(f) = \frac{\eta}{2} \Pi\left(\frac{f}{2B}\right) \qquad R_y(\tau) = \eta B \text{ sinc } 2B\tau \qquad (10a)$$

which are plotted in Fig. 5.3-5. Besides the spectral shaping, we see that lowpass filtering causes the output noise to be correlated over time intervals of about $1/2B$. We also see that

$$\overline{y^2} = \eta B \qquad (10b)$$

so the output power is directly proportional to the filter's bandwidth.

Example 5.3-1 Thermal noise in an RC circuit To pull together several of the topics so far, consider the RC circuit in Fig. 5.3-6a where the resistor is at temperature $\mathcal{T}$. Replacing this thermal resistor with its Thévenin model leads to Fig. 5.3-6b, a white noise source with $G_x(f) = 2Rk\mathcal{T}$ V²/Hz applied to a *noiseless* RC LPF. Since $|H(f)|^2 = [1 + (f/B)^2]^{-1}$, the output spectral density is

$$G_y(f) = |H(f)|^2 G_x(f) = \frac{2Rk\mathcal{T}}{1 + (f/B)^2} \qquad B = \frac{1}{2\pi RC} \qquad (11a)$$

The inverse transform then yields

$$R_y(\tau) = 2Rk\mathcal{T}\pi Be^{-2\pi B|\tau|} = \frac{k\mathcal{T}}{C} e^{-|\tau|/RC} \qquad (11b)$$

$G_x(f) = 2Rk\mathcal{T}$ $\overline{y^2} = k\mathcal{T}/C$

Figure 5.3-6 (a) RC circuit with resistance noise; (b) noise equivalent circuit.

which shows that the interval over which the filtered noise has appreciable correlation approximately equals the circuit's time constant RC, as might have been suspected.

We can further say that $y(t)$ is a gaussian random signal with no DC component — since $x(t)$ is a zero-mean gaussian — and

$$\overline{y^2} = R_y(0) = \frac{k\mathcal{T}}{C} \tag{12}$$

Surprisingly, $\overline{y^2}$ depends on C but not on R, even though the noise source is the thermal resistor! This paradox will be explained shortly; here we conclude our example with a numerical calculation.

Suppose the resistor is at room temperature $\mathcal{T}_0$ and $C = 0.1 \ \mu F$; then

$$\overline{y^2} = \frac{4 \times 10^{-21}}{10^{-7}} = 4 \times 10^{-14} \qquad V^2$$

and the rms output voltage is $\sigma_y = 2 \times 10^{-7} = 0.2$ microvolts (μV). Such exceedingly small values are characteristic of thermal noise, which is why thermal noise goes unnoticed in ordinary situations. However, the received signal in a long-distance communication system may be of this same order of magnitude or even smaller.

Noise Equivalent Bandwidth

Filtered white noise usually has finite power. To emphasize this, we designate *average noise power* by $N = \overline{y^2}$ and write Eq. (9c) in the form

$$N = \frac{\eta}{2} \int_{-\infty}^{\infty} |H(f)|^2 \ df = \eta \int_0^{\infty} |H(f)|^2 \ df$$

Noting that the integral depends only on the filter transfer function, we can simplify discussion of noise power by defining a *noise equivalent bandwidth* B_N as

$$B_N \triangleq \frac{1}{g} \int_0^{\infty} |H(f)|^2 \ df \tag{13}$$

where $g^{1/2} = |H(f)|_{max}$ is the center-frequency amplitude ratio, i.e., the voltage gain. (This definition assumes the filter has a meaningful center frequency.) Hence the filtered noise power is

$$N = g\eta B_N \tag{14}$$

Examining Eq. (14) shows that the effect of the filter has been separated into two parts: the *relative frequency selectivity*, as described by B_N, and the *power gain* (or attentuation), represented by g. Thus, as illustrated in Fig. 5.3-7 for a bandpass filter, B_N equals the bandwidth of an ideal rectangular filter that would pass as much white noise power as the filter in question, their maximum gains being equal.

$|H(f)|^2$

$\leftarrow B_N \rightarrow$

g

Equal areas

0

f

Figure 5.3-7 Noise equivalent bandwidth of a bandpass filter.

Let's apply Eqs. (13) and (14) to the RC LPF in Example 5.3-1. The filter's center frequency is $f = 0$ so

$$g = |H(0)|^2 = 1$$

and

$$B_N = \int_0^\infty \frac{df}{1 + (f/B)^2} = \frac{\pi}{2} B = \frac{1}{4RC} \tag{15}$$

The reason why $\overline{y^2}$ in Eq. (12) is independent of R now becomes apparent if we write $\overline{y^2} = N = \eta B_N = (4Rk\mathcal{T}) \times (1/4RC)$. Thus, increasing R increases the noise density η (as it should) but decreases the noise bandwidth B_N. These two effects precisely cancel each other and $\overline{y^2} = k\mathcal{T}/C$.

By definition, the noise equivalent bandwidth of an ideal filter is its actual bandwidth. For practical filters, B_N is somewhat greater than the 3-dB bandwidth; e.g., a first-order lowpass filter has $B_N = \pi B/2$, so B_N is about 50 percent greater than B. However, as the filter becomes more selective (sharper cutoff characteristics), its noise bandwidth approaches the 3-dB bandwidth, and for most applications one is not too far off in taking them to be equal.

In summary, if $y(t)$ is filtered white noise of zero mean, then

$$\overline{y^2} = \sigma_y^2 = N = g\eta B_N$$
$$\sigma_y = \sqrt{N} = \sqrt{g\eta B_N} \tag{16}$$

This means that given a source of white noise, an average-power meter (or mean square voltage meter) will read $\overline{y^2} = N = \eta B_N$, where B_N is the noise equivalent bandwidth of the meter itself. Working backward, the source power density can be inferred via $\eta = N/B_N$, providing you're sure that the noise is white over the frequency-response range of the meter.

Exercise 5.3-2 Consider an nth-order Butterworth LPF defined by Eq. (4b), Sect. 3.4. Show that the noise equivalent bandwidth is related to the 3-dB bandwidth by

$$B_N = \frac{\pi B}{2n \sin (\pi/2n)} \tag{17}$$

and hence $B_N \rightarrow B$ as $n \rightarrow \infty$.

System Measurements Using White Noise★

Because white noise contains all frequencies in equal proportion, it's a convenient signal for system measurements and experimental design work. Consequently, white noise sources with calibrated power density have become standard laboratory instruments. A few of the measurements that can be made with these sources are outlined below.

Noise equivalent bandwidth Suppose the gain of an amplifier is known and we wish to find its noise equivalent bandwidth. To do so, we can apply white noise to the input and measure the average output power with a meter whose frequency response is essentially constant over the amplifier's passband. The noise bandwidth in question is then, from Eq. (14), $B_N = N/g\eta$.

Amplitude response To find the amplitude response of a given system, we apply white noise to the input so the output power spectrum is proportional to $|H(f)|^2$. Then we scan the output spectrum with a tunable bandpass filter whose bandwidth is constant and small compared to the variations of $|H(f)|^2$. Figure 5.3-8a diagrams the experimental setup. If the scanning filter is centered at f_c, the rms noise voltage at its output is proportional to $|H(f_c)|$. By varying f_c, a point-by-point plot of $|H(f)|$ is obtained.

Impulse response Figure 5.3-8b shows a method for measuring the impulse response $h(t)$ of a given system. The instrumentation required is a white noise source, a variable time delay, a multiplier, and an averaging device.

Denoting the input noise as $x(t)$, the system output is $h * x(t)$, and the delayed signal is $x(t - t_d)$. Thus, the output of the multiplier is

$$z(t) = x(t - t_d)[h * x(t)] = x(t - t_d) \int_{-\infty}^{\infty} h(\lambda)x(t - \lambda) \, d\lambda$$

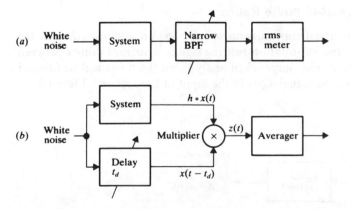

Figure 5.3-8 System measurements using white noise. (a) Amplitude response; (b) impulse response.

Now suppose that $z(t)$ is averaged over a long enough time to obtain $\langle z(t) \rangle$. If the noise source is ergodic and the system is linear and time-invariant, the averaged output approximates the ensemble average

$$E[z(t)] = \int_{-\infty}^{\infty} h(\lambda) E[x(t - t_d) x(t - \lambda)] \, d\lambda = \int_{-\infty}^{\infty} h(\lambda) R_x(\lambda - t_d) \, d\lambda$$

But, with $x(t)$ being white noise, Eq. (6b) says that $R_x(\lambda - t_d) = (n/2) \, \delta(\lambda - t_d)$. Hence

$$\langle z(t) \rangle \approx \frac{\eta}{2} \int_{-\infty}^{\infty} h(\lambda) \, \delta(\lambda - t_d) \, d\lambda = \frac{\eta}{2} \, h(t_d)$$

and $h(t)$ can be measured by varying the time delay t_d.

The measurement techniques in Fig. 5.3-8 have special value for industrial processing or control systems. A conventional sinusoidal or pulsed input could drive such a system out of its linear operating region and, possibly, cause damage to the system. Low-level white noise then provides an attractive alternative for the test input signal.

5.4 SIGNAL TRANSMISSION WITH NOISE

At last we're ready to investigate the effects of noise on electrical communication. We begin here by studying signal transmission systems with additive noise. The signal-to-noise ratio is introduced as a measure of system performance, especially in regard to analog communication. We also examine pulse detection in noise.

Throughout, we restrict our consideration to a linear system that does not include modulation. This elementary type of signal transmission will be called *baseband communication*. The results obtained for baseband systems serve as a benchmark for comparison when we get to modulation systems in later chapters.

Additive Noise and Signal-to-Noise Ratios

Contaminating noise in signal transmission usually has an *additive* effect in the sense that noise adds to the information-bearing signal at various points between the source and destination. For purposes of analysis, all the noise will be lumped into one source added to the signal $x_R(t)$ at the input of the receiver. Figure 5.4-1

Figure 5.4-1 Model of received signal with additive noise.

diagrams our model of additive noise. This model emphasizes the fact that the most vulnerable point in a transmission system is the receiver input where the signal level has its weakest value. Furthermore, noise sources at other points can be "referred" to the receiver input using techniques covered in App. A.

Since the receiver is linear, its combined input produces output *signal plus noise*. Accordingly, we write the output waveform as

$$y_D(t) = x_D(t) + n_D(t) \tag{1}$$

where $x_D(t)$ and $n_D(t)$ stand for the signal and noise waveforms at the destination. The total output power is then found by averaging

$$y_D^2(t) = x_D^2(t) + 2x_D(t)n_D(t) + n_D^2(t)$$

To calculate this average, we'll treat the signal as a sample function of an ergodic process and we'll make two reasonable assumptions about additive noise:

1. The noise comes from an ergodic source with zero mean and power spectral density $G_n(f)$.
2. The noise is physically independent of the signal and therefore uncorrelated with it.

Under these conditions the statistical average of the crossproduct $x_D(t)n_D(t)$ equals zero because $x_D(t)$ and $n_D(t)$ are *incoherent*. Thus, the statistical average of $y_D^2(t)$ yields

$$\overline{y_D^2} = \overline{x_D^2} + \overline{n_D^2} \tag{2}$$

which states that we have *superposition of signal and noise power* at the destination.

Let's underscore the distinction between desired signal power and unwanted noise power by introducing the notation

$$S_D \triangleq \overline{x_D^2} \qquad N_D \triangleq \overline{n_D^2} \tag{3a}$$

so that

$$\overline{y_D^2} = S_D + N_D \tag{3b}$$

The *signal-to-noise ratio* will now be defined as the ratio of signal power to noise power, symbolized by

$$(S/N)_D \triangleq S_D/N_D = \overline{x_D^2}/\overline{n_D^2} \tag{4}$$

This ratio provides an important and handy indication of the degree to which the signal has been contaminated with additive noise. But note that the interpretation of signal-to-noise ratio is meaningful only when Eq. (2) holds. Otherwise, $\overline{y_D^2}$ would include an additional term involving the crossproduct of signal *times* noise.

Superposition of signal and noise power is a helpful condition in experimental work because you can't turn off the noise to determine S_D alone. Instead, you must measure N_D alone (with the signal off) and measure $\overline{y_D^2} = S_D + N_D$ (with the signal on); given these measured values, you can calculate $(S/N)_D$ from $\overline{y_D^2}/N_D = (S_D + N_D)/N_D = (S/N)_D + 1$.

For analytic work, we generally take the case of *white noise* with $G_n(f) = \eta/2$. If the receiver has power gain g_R and noise equivalent bandwidth B_N, the destination noise power becomes

$$N_D = g_R \eta B_N \tag{5}$$

The noise density η may also be expressed in terms of the noise temperature $\mathscr{T}_N$ referred to the receiver input, so that

$$\eta = k\mathscr{T}_N = k\mathscr{T}_0(\mathscr{T}_N/\mathscr{T}_0)$$

$$\approx 4 \times 10^{-21}(\mathscr{T}_N/\mathscr{T}_0) \qquad \text{W/Hz} \tag{6}$$

where we've introduced the standard temperature $\mathscr{T}_0$ for numerical convenience. Typical values of $\mathscr{T}_N$ range from about $0.2\mathscr{T}_0$ (60 K) for a carefully designed low-noise system up to $10\mathscr{T}_0$ or more for a noisy system.

Unless stated to the contrary, we'll assume additive white noise described by Eqs. (5) and (6) in the remainder of this chapter. When the noise has a gaussian distribution, this case is called *additive white gaussian noise* (AWGN).

Analog Baseband Transmission

Figure 5.4-2 represents a simple analog baseband transmission system. The information source generates a *message waveform* $x(t)$ which is to be reproduced at the destination. We'll model the source as an ergodic process characterized by a *message bandwidth* W such that any sample function $x(t)$ has negligible spectral content for $|f| > W$. The channel is assumed to be distortionless over the message band so that

$$x_D(t) = Kx(t - t_d)$$

where K and t_d account for the total amplification and time delay of the system. We'll concentrate on the contaminating effects of additive white noise, as measured by the system's signal-to-noise ratio at the destination.

The average signal power generated at the source can be represented as

$$S_x \triangleq \overline{x^2}$$

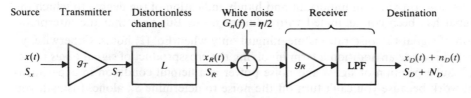

Figure 5.4-2 Analog baseband transmission system with noise.

Since the channel does not require equalization, the transmitter and receiver merely act as amplifiers over the message band with power gains g_T and g_R compensating for the transmission loss L. Thus, the transmitted signal power, the received signal power, and the destination signal power are related by

$$S_T = g_T \overline{x^2} = g_T S_x$$
$$S_R = \overline{x_R^2} = S_T/L \tag{7}$$
$$S_D = \overline{x_D^2} = g_R S_R$$

These parameters are labeled at the corresponding locations in Fig. 5.4-2.

The figure also shows a *lowpass filter* as part of the receiver. This filter has the crucial task of passing the message while reducing the noise at the destination. Obviously, the filter should reject all noise frequency components that fall outside the message band—which calls for an *ideal* LPF with bandwidth $B = W$. The resulting destination noise power will be

$$N_D = g_R \eta W$$

obtained from Eq. (5) with $B_N = B = W$.

We now see that the receiver gain g_R amplifies signal and noise equally. Therefore, g_R cancels out when we divide S_D by N_D, and

$$(S/N)_D = S_R/\eta W \tag{8}$$

This simple result gives the destination signal-to-noise ratio in terms of three fundamental system parameters: the signal power S_R and noise density η at the receiver input, and the message bandwidth W. We can also interpret the denominator ηW as the *noise power in the message band* before amplification by g_R. Consequently, a wideband signal suffers more from noise contamination than a narrowband signal.

Being a power ratio, the value of $(S/N)_D$ is commonly stated in decibels. For dB calculations we'll express the signal power in milliwatts (or dBm) and write the noise power in terms of the noise temperature $\mathcal{T}_N$. Thus,

$$(S/N)_{D_{dB}} = 10 \log_{10}(S_R/k\mathcal{T}_N W)$$

$$= S_{R_{dBm}} + 174 - 10 \log_{10}\left(\frac{\mathcal{T}_N}{\mathcal{T}_0} W\right) \tag{9}$$

where the constant 174 dB comes from Eq. (6) converted to milliwatts.

Table 5.4-1 lists typical dB values of $(S/N)_D$ along with the frequency range needed for various types of analog communication systems. The upper limit of the frequency range equals the nominal message bandwidth W. The lower limit also has design significance because many transmission systems include transformers or coupling capacitors that degrade the low-frequency response.

The destination signal-to-noise ratio doesn't depend on the receiver gain, which only serves to produce the desired signal level at the output. However,

Table 5.4-1 Typical transmission requirements for selected analog signals

Signal type	Frequency range	Signal-to-noise ratio, dB
Barely intelligible voice	500 Hz–2 kHz	5–10
Telephone-quality voice	200 Hz–3.2 kHz	25–35
AM broadcast-quality audio	100 Hz–5 kHz	40–50
High-fidelity audio	20 Hz–20 kHz	55–65
Television video	60 Hz–4.2 MHz	45–55

$(S/N)_D$ will be affected by any gains or losses that enter the picture *before* the noise has been added. Specifically, substituting $S_R = S_T/L$ in Eq. (8) yields

$$(S/N)_D = S_T/L\eta W \tag{10}$$

so $(S/N)_D$ is directly proportional to the transmitted power S_T and inversely proportional to the transmission loss L — a rather obvious but nonetheless significant conclusion.

When all the parameters in Eq. (10) are fixed and $(S/N)_D$ turns out to be too small, consideration should be given to the use of *repeaters* (as in Fig. 3.3-2) to improve system performance. In particular, suppose that the transmission path is divided into m equal sections, each having loss L_1. If a repeater amplifier with noise density η is inserted at the end of the first section, it's output signal-to-noise ratio will be

$$(S/N)_1 = \frac{S_T}{L_1 \eta W}$$

which follows immediately from Eq. (10). Repeaters are usually designed so the amplifier gain just equals the section loss. The analysis in App. A then shows that if the system consists of m identical repeater sections (including the receiver), the overall signal-to-noise ratio becomes

$$(S/N)_D \approx \frac{1}{m}(S/N)_1 = \frac{L}{mL_1}\left(\frac{S_T}{L\eta W}\right) \tag{11}$$

Compared to direct transmission, this result represents potential improvement by a factor of L/mL_1.

It should be stressed that all of our results have been based on distortionless transmission, additive white noise, and ideal filtering. Consequently, Eqs. (8)–(11) represent *upper bounds* on $(S/N)_D$. If the noise bandwidth of the lowpass filter in an actual system is appreciably greater than the message bandwidth, the signal-to-noise ratio will be reduced by the factor W/B_N. System nonlinearities that cause the output to include signal-times-noise terms also reduce $(S/N)_D$. However, nonlinear companding may yield a net improvement; see Benett (1970, chap. 3).

Example 5.4-1 Consider a cable system having $L = 140$ dB $= 10^{14}$ and $\mathcal{T}_N = 5\mathcal{T}_0$. If you want high-fidelity audio transmission with $W = 20$ kHz and $(S/N)_D \geq 60$ dB, the necessary signal power at the receiver can be found from Eq. (9) written as

$$S_{R_{dBm}} + 174 - 10 \log_{10} (5 \times 20 \times 10^3) \geq 60 \text{ dB}$$

Hence, $S_R \geq -64$ dBm $\approx 4 \times 10^{-7}$ mW and the transmitted power must be $S_T = LS_R \geq 4 \times 10^7$ mW $= 40,000$ W.

Needless to say, you wouldn't even try to put 40 kW on a signal transmission cable! Instead, you might insert a repeater at the midpoint so that $L_1 = 70$ dB $= 10^7$. (Recall from Sect. 3.3 that cable loss in dB is directly proportional to length.) The resulting improvement factor in Eq. (11) is

$$\frac{L}{mL_1} = \frac{10^{14}}{2 \times 10^7} = 5 \times 10^6$$

which reduces the transmitted-power requirement to $S_T \geq 4 \times 10^7$ mW $\div$ $5 \times 10^6 = 8$ mW—a much more realistic value. You would probably take S_T in the range of 10–20 mW to provide a margin of safety.

Exercise 5.4-1 Repeat the calculations in Example 5.4-1 for the case of video transmission with $W = 4.2$ MHz and $(S/N)_D \geq 50$ dB.

Pulse Detection and Matched Filters

Pulse detection in noise differs from analog signal transmission in two major respects. First, we're usually concerned with determining the presence or absence of a pulse and, perhaps, measuring its amplitude, rather than reproducing the waveform. Second, we usually know the pulse shape in advance, but not its amplitude or arrival time. The pulse-shape information makes it possible to design *optimum* receiving filters for detecting pulses buried in noise of known spectral density. Such optimum filters, called *matched filters*, have extensive applications in digital communication and radar systems.

Figure 5.4-3a will be our model for pulse detection. The received pulse has known shape $p(t)$ but unknown amplitude A_R and arrival time t_0, so

$$x_R(t) = A_R p(t - t_0) \tag{12a}$$

This pulse is an *energy signal* with Fourier transform

$$X_R(f) = A_R P(f)e^{-j\omega t_0} \tag{12b}$$

where $P(f) = \mathscr{F}[p(t)]$. The total received *pulse energy* is

$$E_R = \int_{-\infty}^{\infty} |X_R(f)|^2 \, df = A_R^2 \int_{-\infty}^{\infty} |P(f)|^2 \, df \tag{12c}$$

To detect the pulse in the face of additive noise, the receiving filter should compress the pulse energy into an output pulse with *peak amplitude* A at some specific instant, say $t = t_0 + t_d$, and should also minimize the rms output noise. The

Figure 5.4-3 Pulse detection in noise. (a) Model; (b) filtered output.

output waveform would thus look something like Fig. 5.4-3b. We seek the filter transfer function $H(f)$ that achieves this goal, given $p(t)$ and $G_n(f)$.

First, we write the peak pulse amplitude in terms of $H(f)$ and $P(f)$, namely

$$A = \mathscr{F}^{-1}[H(f)X_R(f)]\Big|_{t=t_0+t_d}$$

$$= A_R \int_{-\infty}^{\infty} H(f)P(f)e^{+j\omega t_d}\, df \tag{13}$$

Next, assuming the noise has zero mean, its output variance is

$$\sigma^2 = \int_{-\infty}^{\infty} |H(f)|^2 G_n(f)\, df$$

We want to maximize the ratio of peak output amplitude A to rms noise σ or, equivalently,

$$\left(\frac{A}{\sigma}\right)^2 = A_R^2 \frac{\left|\int_{-\infty}^{\infty} H(f)P(f)e^{j\omega t_d}\, df\right|^2}{\int_{-\infty}^{\infty} |H(f)|^2 G_n(f)\, df} \tag{14}$$

where $H(f)$ is the only function at our disposal. Normally, optimization problems require the methods of variational calculus. But this particular problem (and a few others) can be solved by adroit application of *Schwarz's inequality*.

For our present purposes we draw upon Eq. (17), Sect. 3.5, in the form

$$\frac{\left|\int_{-\infty}^{\infty} V(f)W^*(f)\, df\right|^2}{\int_{-\infty}^{\infty} |V(f)|^2\, df} \leq \int_{-\infty}^{\infty} |W(f)|^2\, df$$

where $V(f)$ and $W(f)$ are arbitrary functions of f. The left-hand side of this inequality is identical to Eq. (14) with

$$V(f) = H(f)\sqrt{G_n(f)}$$

$$W^*(f) = A_R H(f)P(f)e^{j\omega t_d}/V(f) = A_R P(f)e^{j\omega t_d}/\sqrt{G_n(f)}$$

and the inequality becomes an equality when $V(f)$ is proportional to $W(f)$. Therefore, if $V(f) = KW(f)/A_R$, we obtain the maximum value

$$\left(\frac{A}{\sigma}\right)^2_{\max} = A_R^2 \int_{-\infty}^{\infty} \frac{|P(f)|^2}{G_n(f)} \, df \tag{15}$$

The corresponding optimum filter must have

$$H_{\text{opt}}(f) = K \frac{P^*(f)e^{-j\omega t_d}}{G_n(f)} \tag{16}$$

where K is an arbitrary constant. Observe that $|H_{\text{opt}}(f)|$ is proportional to $|P(f)|$ and inversely proportional to $G_n(f)$. Hence, the optimum filter *emphasizes* those frequencies where the signal spectrum is large and *deemphasizes* those frequencies where the noise spectrum is large — a very sensible thing to do.

In the special but important case of *white* noise, Eq. (15) reduces to

$$\left(\frac{A}{\sigma}\right)^2_{\max} = \frac{2A_R^2}{\eta} \int_{-\infty}^{\infty} |P(f)|^2 \, df = \frac{2E_R}{\eta} \tag{17}$$

which brings out the significance of pulse energy E_R for detection. The impulse response of the optimum filter is then

$$h_{\text{opt}}(t) = \mathscr{F}^{-1}\left[\frac{2K}{\eta} P^*(f)e^{-j\omega t_d}\right] = \frac{2K}{\eta} p(t_d - t) \tag{18}$$

The name *matched* filter comes from the fact that $h_{\text{opt}}(t)$ has the shape as the pulse $p(t)$ reversed in time and shifted by t_d. The value of t_d equals the delay between the pulse arrival and the output peak, and may be chosen by the designer.

Sometimes the matched filter turns out to be physically *unrealizable* because $h_{\text{opt}}(t)$ has precursors for $t < 0$. However, a reasonable approximation may be obtained by taking t_d large enough so that the precursors are negligible. Optimization with a realizability constraint imposed on $H_{\text{opt}}(f)$ is discussed in Thomas (1969, chap. 5).

Exercise 5.4-2 Consider the rectangular pulse shape $p(t) = u(t) - u(t - T)$. Sketch $h_{\text{opt}}(t)$ from Eq. (18), and find the condition on t_d for realizability of the matched filter. Then use the convolution $h_{\text{opt}}(t) * x_R(t)$ to obtain the output pulse and its peak value A when $K = \eta/2$.

5.5 PROBLEMS

5.1-1 The r.v. X has a uniform distribution over $0 \leq x \leq 2$. Find $\overline{v(t)}$, $R_v(t_1, t_2)$, and $\overline{v^2(t)}$ for the random process $v(t) = 6e^{Xt}$.

5.1-2 Do Prob. 5.1-1 with $v(t) = 6 \cos Xt$.

5.1-3 Let X and Y be independent r.v.'s. Given that X has a uniform distribution over $-1 \leq x \leq 1$ and that $\bar{y} = 2$ and $\overline{y^2} = 6$, find $\overline{v(t)}$, $R_v(t_1, t_2)$, and $\overline{v^2(t)}$ for the random process $v(t) = (Y + 3Xt)t$.

5.1-4 Do Prob. 5.1-3 with $v(t) = Ye^{Xt}$.

5.1-5 Do Prob. 5.1-3 with $v(t) = Y \cos Xt$.

5.1-6‡ Let $v(t) = A \cos (2\pi Ft + \Phi)$ where A is a constant and F and Φ are random variables. If Φ has a uniform distribution over 2π radians and F has an arbitrary PDF $p_F(f)$, show that

$$R_v(t_1, t_2) = \frac{A^2}{2} \int_{-\infty}^{\infty} \cos 2\pi\lambda(t_1 - t_2)\, p_F(\lambda)\, d\lambda$$

Also find $\overline{v(t)}$ and $\overline{v^2(t)}$.

5.1-7 Let X and Y be independent r.v.'s, both having zero mean and variance σ^2. Find the cross-correlation function of the random processes

$$v(t) = X \cos \omega_0 t + Y \sin \omega_0 t$$

$$w(t) = Y \cos \omega_0 t - X \sin \omega_0 t$$

5.1-8 Consider the process $v(t)$ defined in Prob. 5.1-7. (a) Find $\overline{v(t)}$ and $R_v(t_1, t_2)$ to confirm that the process is wide-sense stationary. (b) Show, from $E[v^2(t)]$ and $\langle v_i^2(t) \rangle$, that the process is not ergodic.

5.1-9 Let $v(t) = A \cos (\omega_0 t + \Phi)$, where A and Φ are independent r.v.'s and Φ has a uniform distribution over 2π radians. (a) Find $\overline{v(t)}$ and $R_v(t_1, t_2)$ to confirm that the process is wide-sense stationary. (b) Show, from $E[v^2(t)]$ and $\langle v_i^2(t) \rangle$, that the process is not ergodic.

5.1-10 Let $z(t) = v(t) - v(t + T)$, where $v(t)$ is a stationary nonperiodic process and T is a constant. Find the mean and variance of $z(t)$ in terms of $R_v(\tau)$.

5.1-11 Do Prob. 5.1-10 with $z(t) = v(t) + v(t - T)$.

5.2-1 A certain random signal has $R_v(\tau) = 16e^{-(8\tau)^2} + 9$. Find and sketch the power spectrum. Also determine the signal's DC value, average power, and rms value.

5.2-2 Do Prob. 5.2-1 with $R_v(\tau) = 32 \operatorname{sinc}^2 8\tau + 4 \cos 8\tau$.

5.2-3 Consider the signal in Prob. 5.1-6. Show that

$$G_v(f) = \frac{A^2}{4} \left[p_F(f) + p_F(-f) \right]$$

Then simplify this expression for the case when $F = f_0$, where f_0 is a constant.

5.2-4 Consider the spectral estimate $\tilde{G}_v(f) = |V_T(f, s)|^2/T$. Use Eq. (10) to show that $E[\tilde{G}_v(f)] = (T \operatorname{sinc}^2 fT) * G_v(f)$. What happens in the limit as $T \to \infty$?

5.2-5‡ Let $v(t)$ be a randomly phased sinusoid. Show that Eq. (9) yields Eq. (6b).

5.2-6 Modify Eq. (13) for the case when $v(t)$ and $w(t)$ are independent stationary signals with $m_v m_w \neq 0$. Show from your results that $R_z(\pm\infty) = (m_v \pm m_m)^2$ and that $\overline{z^2} > 0$.

5.2-7 Let $v(t)$ and $w(t)$ be jointly stationary, so that $R_{vw}(t_1, t_2) = R_{vw}(t_1 - t_2)$. Show that

$$R_{wv}(\tau) = R_{vw}(-\tau)$$

What's the corresponding relationship between the cross-spectral densities?

5.2-8 Let $z(t) = v(t) - v(t + T)$, where $v(t)$ is a stationary random signal and T is a constant. Start with $R_z(t_1, t_2)$ to find $R_z(\tau)$ and $G_z(f)$ in terms of $R_v(\tau)$ and $G_v(f)$.

5.2-9 Do Prob. 5.2-8 with $z(t) = v(t) + v(t - T)$.

5.2-10 Let $z(t) = A \cos (2\pi f_1 t + \Phi_1) \cos (2\pi f_2 t + \Phi_2)$, where A, f_1, and f_2 are constants, and Φ_1 and Φ_2 are independent r.v.'s, both uniformly distributed over 2π radians. Find $G_z(f)$ and sketch it taking $f_1 \ll f_2$ and $f_1 = f_2$.

5.2-11 Confirm that $R_y(\tau) = h(-\tau) * R_{yx}(\tau)$.

5.2-12 Let $y(t) = dx(t)/dt$. Find $R_y(\tau)$ and $R_{yx}(\tau)$ in terms of $R_x(\tau)$ by taking the inverse transform of $G_y(f)$ and $G_{yx}(f)$.

5.2-13 Let $y(t) = x(t) - \alpha x(t - T)$, where α and T are constants. Obtain expressions for $G_y(f)$ and $R_y(\tau)$.

5.2-14 Use the relation in Prob. 5.2-7 to show that

$$R_{x\dot{x}}(\tau) = -\dot{R}_x(\tau).$$

5.2-15‡ The *moving average* of a random signal $x(t)$ is defined as

$$y(t) = \frac{1}{T} \int_{t-T/2}^{t+T/2} x(\lambda) \, d\lambda$$

Find $H(f)$ such that $y(t) = h(t) * x(t)$ and show that

$$R_y(\tau) = \frac{1}{T} \int_{-T}^{T} \left(1 - \frac{|\lambda|}{T}\right) R_x(\tau - \lambda) \, d\lambda$$

5.3-1 Derive Eq. (2b) by using series approximations in Eq. (2a).

5.3-2 Use Eq. (18), Sect. 5.2, to show that Eq. (9b) can be written in the alternative form

$$R_y(\tau) = \frac{\eta}{2} \int_{-\infty}^{\infty} h(t) h(t + \tau) \, dt$$

5.3-3 Find and sketch $G_y(f)$ and $R_y(\tau)$ and evaluate $\overline{y^2}$ when white noise is filtered by the zero-order hold in Example 3.1-3.

5.3-4 Do Prob. 5.3-3 with a gaussian filter having $H(f) = K e^{-(af)^2}$.

5.3-5 Do Prob. 5.3-3 with an ideal BPF having gain K and delay t_0 over $f_0 - B/2 \le |f| \le f_0 + B/2$.

5.3-6 Do Prob. 5.3-3 with an ideal HPF having gain K and delay t_0 over $|f| \ge f_0$.

5.3-7 Figure P5.3-7 represents a white-noise voltage source connected to a noiseless RL circuit. Find $G_y(f)$, $R_y(\tau)$, and $\overline{y^2}$ taking $y(t)$ as the voltage across R.

$G_v(f) = \eta_v/2$

Figure P5.3-7

5.3-8‡ Do Prob. 5.3-7 taking $y(t)$ as the voltage across L.

5.3-9 The spectral density of the current $i(t)$ in Fig. P5.3-7 is $G_i(f) = \eta_v/2|R + j\omega L|^2$. If the source represents thermal noise from the resistance, then the *equipartition theorem* of statistical mechanics requires that $\frac{1}{2} L \overline{i^2} = \frac{1}{2} k \mathcal{T}$. As a consequence, show that $\eta_v = 4Rk\mathcal{T}$, in agreement with Eq. (4).

5.3-10 Thermal noise from a 10-kΩ resistance at room temperature is applied to an ideal LPF with $B = 2.5$ MHz and unit gain. The filtered noise voltage is then applied to a full-wave rectifier, producing $z(t) = |y(t)|$. Calculate the mean and rms value of $z(t)$.

5.3-11‡ Do Prob. 5.3-10 with a half-wave rectifier, so that $z(t) = 0$ for $y(t) < 0$.

5.3-12 Thermal noise from a 10-kΩ resistance at room temperature is applied to an ideal LPF with $B = 2.5$ MHz and unit gain. The filtered noise voltage is then applied to a delay line, producing $z(t) = y(t - T)$. Use Eqs. (15) and (16), Sect. 4.4, to find the joint PDF of the random variables $Y = y(t)$ and $Z = z(t)$ when $T = 1$ μs.

5.3-13‡ Do Prob. 5.3-12 with $T = 0.1$ μs.

5.3-14 Find B_N for the gaussian LPF in Prob. 5.3-4 and compare with the 3-dB bandwidth.

5.3-15‡ *Impulse noise*, which occurs in some communication systems, can be modeled as the random signal

$$x(t) = \sum_{k=-\infty}^{\infty} A_k \, \delta(t - T_k)$$

where the A_k and T_k are independent sets of random variables. The impulse weights A_k are independent and have zero mean and variance σ^2. The T_k are governed by a Poisson process such that the expected number of impulses in time T equals μT. Use Eq. (9), Sect. 5.2, to show that impulse noise has a constant spectral density $G_x(f) = \mu\sigma^2$.

5.4-1 A transmission system with $W = 5$ kHz has $(S/N)_D = 46$ dB when $S_T = 100$ mW. If the receiver bandwidth is changed accordingly, what value of S_T is appropriate to: (a) upgrade the system for high-fidelity audio; (b) downgrade the system for telephone-quality voice?

5.4-2 Consider an AWGN transmission system with $W = 10$ kHz. Express $(S/N)_D$ in a form like Eq. (8) when the receiving filter is: (a) a first-order LPF with $B = 15$ kHz; (b) a second-order Butterworth LPF with $B = 12$ kHz.

5.4-3 Consider a transmission system with additive white noise and a distorting channel having $|H_C(f)|^2 = 1/L[1 + (f/W)^2]$. The distortion is equalized by a receiving filter with $|H_R(f)|^2 = [K/|H_C(f)|]^2 \Pi(f/2W)$. Obtain an expression for $(S/N)_D$ in the form like Eq. (10).

5.4-4 Do Prob. 5.4-3 with $|H_C(f)|^2 = 1/L[1 + (2f/W)^4]$.

5.4-5 A signal with $W = 5$ kHz is transmitted 40 km via a cable whose loss is $\alpha = 3$ dB/km. The receiver has $\mathcal{T}_N = 10\mathcal{T}_0$. (a) Find S_T needed to get $(S/N)_D = 60$ dB. (b) Repeat the calculation assuming a repeater at the midpoint.

5.4-6 A cable transmission system with $L = 240$ dB has $m = 6$ equal-length repeater sections and $(S/N)_D = 30$ dB. Find the new value of $(S/N)_D$ if: (a) m is increased to 12; (b) m is decreased to 4.

5.4-7 The cable for a 400-km repeater system has $\alpha = 0.5$ dB/km. Find the minimum number of equal-length repeater sections needed to get $(S/N)_D \geq 30$ dB if $S_T/\eta W = 80$ dB.

5.4-8 If all other parameters of a cable transmission system are fixed, show that the number of equal-length repeater sections that maximizes $(S/N)_D$ is $m = 0.23L_{dB}$.

5.4-9‡ The open-circuit signal output of an oscillator is $A \cos 2\pi f_0 t$. The oscillator has a noiseless source resistance R and internally generated thermal at temperature $\mathcal{T}_N$. A capacitance C is connected across the output terminals to improve the signal-to-noise ratio. Obtain an expression for S/N and find the value of C that maximizes S/N.

5.4-10 Let the shape of a received pulse be $p(t) = e^{-bt}u(t)$. Find the characteristics of the matched filter assuming white noise. Then find the condition on t_d so that a good approximation of this filter can be realized.

5.4-11‡ A rectangular pulse with duration τ has been contaminated by white noise. The receiving filter is a first-order LPF with bandwidth B, rather than a matched filter. (a) Obtain an expression for $(A/\sigma)^2$ in terms of E_R/η. (b) Sketch $(A/\sigma)^2$ versus $\pi B \tau$ to estimate $(A/\sigma)^2_{max}$ and the corresponding value of B.

TWO

ANALOG COMMUNICATION

CHAPTER SIX

LINEAR CW MODULATION

The several purposes of modulation were itemized in Chap. 1 along with a qualitative description of the process. To briefly recapitulate: *modulation* is the systematic alteration of one waveform, called the *carrier*, according to the characteristics of another waveform, the modulating signal or message. The fundamental goal is producing an information-bearing modulated wave whose properties are best suited to the given communication task.

We now embark on a four-chapter tour of *continuous-wave* (CW) *modulation systems*. The carrier in these systems is a sinusoidal wave modulated by an analog signal — AM and FM radio being familiar examples. The abbreviation CW also refers to on-off keying of a sinusoid, as in radio telegraphy, but that process is more accurately termed *interrupted* continuous wave (ICW).

This chapter deals specifically with *linear* CW modulation, which involves direct frequency translation of the message spectrum. Double-sideband modulation (DSB) is precisely that. Minor modifications of the translated spectrum yield conventional amplitude modulation (AM), single-sideband modulation (SSB), or vestigial-sideband modulation (VSB). Each of these variations has its own distinct advantages and significant practical applications. Each will be given due consideration, including such matters as waveforms and spectra, modulation methods, transmitters, and demodulation. The chapter begins with a general discussion of bandpass signals and systems, pertinent to all forms of CW modulation.

6.1 BANDPASS SYSTEMS AND SIGNALS

Effective communication over appreciable distance usually requires a high-frequency sinusoidal carrier. Consequently, most long-haul transmission systems have a *bandpass* frequency response, similar to a bandpass filter, and any signal transmitted on such a system must have a bandpass spectrum. Our purpose here is to present the characteristics and methods of analysis unique to bandpass systems and signals.

Bandpass Systems

The simplest bandpass system is the parallel resonant or *tuned* circuit represented by Fig. 6.1-1*a*. The voltage transfer function plotted in Fig. 6.1-1*b* can be written as

$$H(f) = \frac{1}{1 + jQ\left(\dfrac{f}{f_0} - \dfrac{f_0}{f}\right)} \tag{1a}$$

in which the resonant frequency f_0 and quality factor Q are related to the element values by

$$f_0 = \frac{1}{2\pi\sqrt{LC}} \qquad Q = R\sqrt{\frac{C}{L}}$$

The 3-dB bandwidth between the lower and upper cutoff frequencies is

$$B = f_u - f_\ell = \frac{f_0}{Q} \tag{1b}$$

Since practical tuned circuits usually have $10 < Q < 100$, the 3-dB bandwidth falls between 1 and 10% of the center-frequency value.

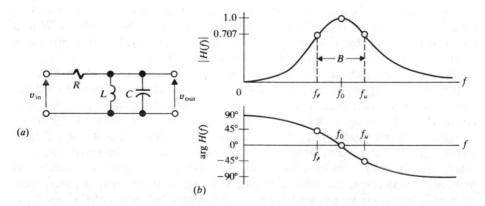

Figure 6.1-1 (*a*) Tuned circuit; (*b*) transfer function.

Table 6.1-1 Selected carrier frequencies and nominal bandwidth

Frequency band	Carrier frequency	Bandwidth
Longwave radio	100 kHz	2 kHz
Shortwave radio	5 MHz	100 kHz
VHF	100 MHz	2 MHz
Microwave	5 GHz	100 MHz
Millimeterwave	100 GHz	2 GHz
Optical	5×10^{14} Hz	10^{13} Hz

A complete bandpass system consists of the transmission channel plus tuned amplifiers and coupling devices connected at each end. Hence, the overall frequency response has a more complicated shape than that of a simple tuned circuit. Nonetheless, various physical effects result in a loose but significant connection between the system's bandwidth and the carrier frequency f_c — similar to Eq. (1b).

For instance, the antennas in a radio system produce considerable distortion unless the frequency range is small compared to f_c. Moreover, designing a reasonably distortionless bandpass amplifier turns out to be quite difficult if B is either very large or very small compared to f_c. As a rough rule of thumb, the *fractional bandwidth* B/f_c should be kept within the range

$$0.01 < \frac{B}{f_c} < 0.1 \tag{2}$$

Otherwise, the signal distortion may be beyond the scope of practical equalizers.

From Eq. (2) we see that *large bandwidths require high carrier frequencies*. This observation is underscored by Table 6.1-1, which lists selected carrier frequencies and the corresponding nominal bandwidth $B \approx 0.02 f_c$ for different frequency bands. Larger bandwidths can be achieved, of course, but at substantially greater cost. As a further consequence of Eq. (2), the terms *bandpass* and *narrowband* are virtually synonymous in signal transmission.

Example 6.1-1 Bandpass pulse transmission We found in Sect. 3.4 that transmitting a pulse of duration τ requires a lowpass bandwidth $B \geq 1/2\tau$. We also found in Example 2.3-2 that frequency translation converts a pulse to a bandpass waveform and *doubles* its spectral width. Putting these two observations together, we conclude that bandpass pulse transmission requires

$$B \geq 1/\tau$$

Since Eq. (2) imposes the additional constraint $0.1\, f_c > B$, the carrier frequency must satisfy

$$f_c > 10/\tau$$

These relations have long served as useful guidelines in radar work and related fields. To illustrate, if $\tau = 1\ \mu s$ then bandpass transmission requires $B \geq 1$ MHz and $f_c > 10$ MHz.

Bandpass Signals

The preceding example invoked knowledge of lowpass transmission to draw some conclusions about bandpass transmission. Subsequently, we'll develop a general method of analysis using lowpass-to-bandpass transformation. The first step in the method involves the representation of a bandpass signal and its equivalent *lowpass* signal.

Consider a real energy signal $v_{bp}(t)$ whose spectrum $V_{bp}(f)$ has the bandpass characteristic sketched in Fig. 6.1-2a. This spectrum exhibits hermitian symmetry, because $v_{bp}(t)$ is real, but $V_{bp}(f)$ is not necessarily symmetrical with respect to $\pm f_c$. We define a *bandpass signal* by the frequency domain property

$$V_{bp}(f) = 0 \qquad |f| < f_c - W$$
$$|f| > f_c + W \tag{3}$$

which just states that the signal has no spectral content outside a band of width $2W$ centered at f_c. The values of f_c and W may be somewhat arbitrary, as long as they satisfy Eq. (3) with $W < f_c$.

The corresponding bandpass waveform in Fig. 6.1-2b looks like a sinusoid at frequency f_c with slowly changing amplitude and phase angle. Formally we write

$$v_{bp}(t) = A(t) \cos [\omega_c t + \phi(t)] \tag{4}$$

where $A(t)$ is the *envelope* and $\phi(t)$ the *phase*, both being functions of time. The envelope, shown as a dashed line, is defined to be nonnegative, so that $A(t) \geq 0$. Negative "amplitudes," when they occur, are absorbed in the phase by adding $\pm 180°$.

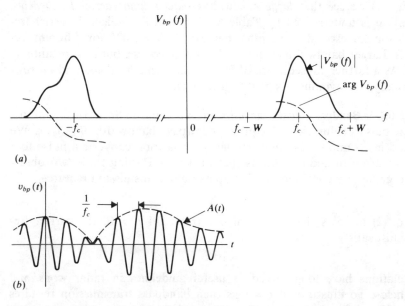

Figure 6.1-2 Bandpass signal. (*a*) Spectrum; (*b*) waveform.

Figure 6.1-3*a* depicts $v_{bp}(t)$ as a complex-plane vector whose length equals $A(t)$ and whose angle equals $\omega_c t + \phi(t)$. But the angular term $\omega_c t$ represents a steady counterclockwise rotation at f_c revolutions per second and can just as well be suppressed, leading to Fig. 6.1-3*b*. This phasor representation, used regularly hereafter, relates to Fig. 6.1-3*a* in the following manner: If you pin the origin of Fig. 6.1-3*b* and rotate the entire figure counterclockwise at the rate f_c, it becomes Fig. 6.1-3*a*.

Further inspection of Fig. 6.1-3*a* suggests another way of writing $v_{bp}(t)$. If we let

$$v_i(t) \triangleq A(t) \cos \phi(t) \qquad v_q(t) \triangleq A(t) \sin \phi(t) \tag{5}$$

then

$$v_{bp}(t) = v_i(t) \cos \omega_c t - v_q(t) \sin \omega_c t$$
$$= v_i(t) \cos \omega_c t + v_q(t) \cos (\omega_c t + 90°) \tag{6}$$

Equation (6) is called the *quadrature-carrier* description of a bandpass signal, as distinguished from the *envelope-and-phase* description in Eq. (4). The functions $v_i(t)$ and $v_q(t)$ are named the *in-phase* and *quadrature components*, respectively. The quadrature-carrier designation comes about from the fact that the two terms in Eq. (6) may be represented by phasors with the second at an angle of $+90°$ compared to the first.

While both descriptions of a bandpass signal are useful, the quadrature-carrier version has advantages for the frequency-domain interpretation. Specifically, Fourier transformation of Eq. (6) yields

$$V_{bp}(f) = \frac{1}{2}\left[V_i(f - f_c) + V_i(f + f_c) \right] + \frac{j}{2}\left[V_q(f - f_c) - V_q(f + f_c) \right] \tag{7}$$

where

$$V_i(f) = \mathscr{F}[v_i(t)] \qquad V_q(f) = \mathscr{F}[v_q(t)]$$

(a)   (b)

Figure 6.1-3 (*a*) Rotating phasor; (*b*) phasor diagram with rotation suppressed.

To obtain Eq. (7) we have used the modulation theorem from Eq. (7), Sect. 2.3, along with $e^{\pm j90°} = \pm j$. The envelope-and-phase description does not readily convert to the frequency domain since, from Eq. (5) or Fig. 6.1-3b,

$$A(t) = \sqrt{v_i^2(t) + v_q^2(t)} \qquad \phi(t) = \arctan \frac{v_q(t)}{v_i(t)} \qquad (8)$$

which are not Fourier-transformable expressions.

An immediate implication of Eq. (7) is that, in order to satisfy the bandpass condition in Eq. (3), the in-phase and quadrature functions must be *lowpass* signals with

$$V_i(f) = V_q(f) = 0 \qquad |f| > W$$

In other words, $V_{bp}(f)$ consists of two lowpass spectra that have been translated and, in the case of $V_q(f)$, quadrature phase shifted. We'll capitalize upon this property in the definition of the *lowpass equivalent spectrum*

$$V_{\ell p}(f) \triangleq \frac{1}{2}[V_i(f) + jV_q(f)] \quad (\text{NOTES } 3/4 \text{ p. 5}) \qquad (9a)$$

$$= V_{bp}(f - f_c)u(f + f_c) \quad V_{bp}(f + f_c)u\ldots \qquad (9b)$$

As shown in Fig. 6.1-4, $V_{\ell p}(f)$ simply equals the positive-frequency portion of $V_{bp}(f)$ translated down to the origin.

Going from Eq. (9) to the time domain, we obtain the *lowpass equivalent signal*

$$v_{\ell p}(t) = \mathscr{F}^{-1}[V_{\ell p}(f)] = \frac{1}{2}[v_i(t) + jv_q(t)] \qquad (10a)$$

Thus, $v_{\ell p}(t)$ is a fictitious *complex* signal whose real part equals $\frac{1}{2}v_i(t)$ and whose imaginary part equals $\frac{1}{2}v_q(t)$. Alternatively, rectangular-to-polar conversion yields

$$v_{\ell p}(t) = \frac{1}{2}A(t)e^{j\phi(t)} \qquad (10b)$$

where we've drawn on Eq. (8) to write $v_{\ell p}(t)$ in terms of the envelope and phase functions. The complex nature of the lowpass equivalent signal can be traced back to its spectrum $V_{\ell p}(f)$, which lacks the hermitian symmetry required for the transform of a real time function. Nonetheless, $v_{\ell p}(t)$ does represent a *real* bandpass signal.

The connection between $v_{\ell p}(t)$ and $v_{bp}(t)$ is derived from Eqs. (4) and (10b) as follows:

$$v_{bp}(t) = \text{Re} \{A(t)e^{j[\omega_c t + \phi(t)]}\}$$

$$= 2 \text{ Re} [\frac{1}{2}A(t)e^{j\omega_c t}e^{j\phi(t)}]$$

$$= 2 \text{ Re} [v_{\ell p}(t)e^{j\omega_c t}] \qquad (11)$$

This result expresses the *lowpass-to-bandpass transformation* in the time domain. The corresponding frequency-domain transformation is

$$V_{bp}(f) = V_{\ell p}(f - f_c) + V_{\ell p}^*(-f - f_c) \qquad (12a)$$

Figure 6.1-4 Lowpass equivalent spectrum.

whose first term constitutes the positive-frequency portion of $V_{bp}(f)$ while the second term constitutes the negative-frequency portion. Since we'll deal only with real bandpass signals, we can keep the hermitian symmetry of $V_{bp}(f)$ in mind and use the simpler expression

$$V_{bp}(f) = V_{\ell p}(f - f_c) \qquad f > 0 \tag{12b}$$

which follows from Figs. 6.1-2a and 6.1-4.

Exercise 6.1-1 Let $z(t) = v_{\ell p}(t)e^{j\omega_c t}$ and use $2\,\mathrm{Re}\,[z(t)] = z(t) + z^*(t)$ to derive Eq. (12a) from Eq. (11).

Bandpass Transmission

Now we have the tools needed to analyze bandpass transmission represented by Fig. 6.1-5a where a bandpass signal $x_{bp}(t)$ applied to a bandpass system with transfer function $H_{bp}(f)$ produces the bandpass output $y_{bp}(t)$. Obviously, you could attempt direct bandpass analysis via $Y_{bp}(f) = H_{bp}(f)X_{bp}(f)$. But it's usually easier to work with the lowpass equivalent spectra related by

$$Y_{\ell p}(f) = H_{\ell p}(f)X_{\ell p}(f) \tag{13a}$$

where

$$H_{\ell p}(f) = H_{bp}(f + f_c)u(f + f_c) \tag{13b}$$

which is the *lowpass equivalent transfer function.*

Equation (13) permits us to replace a bandpass system with the lowpass equivalent model in Fig. 6.1-5b. Besides simplifying analysis, the lowpass model provides valuable insight to bandpass phenomena by analogy with known lowpass relationships. We move back and forth between the bandpass and lowpass models with the help of our previous results for bandpass signals.

(a)   (b)

Figure 6.1-5 (a) Bandpass system; (b) lowpass model.

In particular, after finding $Y_{\ell p}(f)$ from Eq. (13), you can take its inverse Fourier transform

$$y_{\ell p}(t) = \mathscr{F}^{-1}[Y_{\ell p}(f)] = \mathscr{F}^{-1}[H_{\ell p}(f)X_{\ell p}(f)]$$

The lowpass-to-bandpass transformation in Eq. (11) then yields the output signal $y_{bp}(t)$. Or you can get the output quadrature components or envelope and phase immediately from $y_{\ell p}(t)$ as

$$y_i(t) = 2 \text{ Re } [y_{\ell p}(t)] \qquad y_q(t) = 2 \text{ Im } [y_{\ell p}(t)]$$

$$A_y(t) = 2| y_{\ell p}(t)| \qquad \phi_y(t) = \arg [y_{\ell p}(t)] \tag{14}$$

which follow from Eq. (10). The example below illustrates an important application of these techniques.

Example 6.1-2 Carrier and envelope delay Consider a bandpass system having constant amplitude ratio but *nonlinear* phase shift $\theta(f)$ over its passband. Thus,

$$H_{bp}(f) = Ke^{j\theta(f)} \qquad f_\ell < |f| < f_u$$

and

$$H_{\ell p}(f) = Ke^{j\theta(f+f_c)}u(f+f_c) \qquad f_\ell - f_c < f < f_u - f_c$$

as sketched in Fig. 6.1-6. Assuming the phase nonlinearities are relatively smooth, we can write the approximation

$$\theta(f+f_c) \approx -2\pi(t_0 f_c + t_1 f)$$

where

$$t_0 \triangleq - \frac{\theta(f_c)}{2\pi f_c} \qquad t_1 \triangleq - \frac{1}{2\pi} \frac{d\theta(f)}{df}\bigg|_{f=f_c} \tag{15}$$

This approximation comes from the first two terms of the Taylor series expansion of $\theta(f+f_c)$.

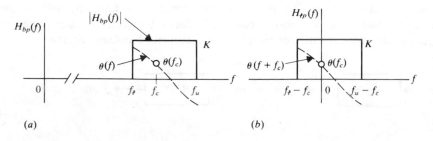

Figure 6.1-6 (a) Bandpass transfer function; (b) lowpass equivalent.

To interpret the parameters t_0 and t_1, let the input signal have zero phase so that $x_{bp}(t) = A_x(t) \cos \omega_c t$ and $x_{\ell p}(t) = \frac{1}{2} A_x(t)$. If the input spectrum $X_{bp}(f)$ falls entirely within the system's passband, then, from Eq. (13),

$$Y_{\ell p}(f) = Ke^{j\theta(f + f_c)}X_{\ell p}(f) \approx Ke^{-j2\pi(t_0 f_c + t_1 f)}X_{\ell p}(f)$$

$$\approx Ke^{-j\omega_c t_0}[X_{\ell p}(f)e^{-j2\pi f t_1}]$$

Recalling the time-delay theorem, we see that the second term corresponds to $x_{\ell p}(t)$ delayed by t_1. Hence,

$$y_{\ell p}(t) \approx Ke^{-j\omega_c t_0}x_{\ell p}(t - t_1) = Ke^{-j\omega_c t_0} \frac{1}{2} A_x(t - t_1)$$

and Eq. (11) yields the bandpass output

$$y_{bp}(t) \approx KA_x(t - t_1) \cos \omega_c(t - t_0)$$

Based on this result, we conclude that t_0 is the *carrier delay* while t_1 is the *envelope delay* of the system. And since t_1 is independent of frequency, at least to the extent of our approximation for $\theta(f + f_c)$, the envelope has not suffered delay distortion. Envelope delay is also called the *group* delay.

We'll later describe *multiplexing* systems in which several bandpass signals at different carrier frequencies are transmitted over a single channel. Plots of $d\theta/df$ versus f are used in this context to evaluate the channel's delay characteristics. If the curve is not reasonably flat over a proposed band, phase equalization may be required to prevent excessive envelope distortion.

Exercise 6.1-2 Suppose a bandpass system has zero phase shift but $|H_{bp}(f)| = K_0 + (K_1/f_c)(f - f_c)$ for $f_\ell < f < f_u$, where $K_0 > (K_1/f_c)(f_\ell - f_c)$. Sketch $H_{\ell p}(f)$ taking $f_1 < f_c$ and $f_u > f_c$. Now show that if $x_{bp}(t) = A_x(t) \cos \omega_c t$ then the quadrature components of $y_{bp}(t)$ are

$$y_i(t) = K_0 A_x(t) \qquad y_q(t) = -\frac{K_1}{2\pi f_c}\frac{dA_x(t)}{dt}$$

provided that $X_{bp}(f)$ falls entirely within the pass band of the system.

6.2 DOUBLE-SIDEBAND AMPLITUDE MODULATION

There are two types of double-sideband amplitude modulation: standard amplitude modulation (AM), and suppressed-carrier double-sideband modulation (DSB). We'll examine both types and show that the minor theoretical difference between them has major repercussions in practical applications. Before plunging into the details, let's establish some conventions regarding the modulating signal.

Analog Message Conventions

Whenever possible, our study of analog communication will be couched in terms of an arbitrary message waveform $x(t)$—which might stand for a sample function from the ensemble of possible messages produced by an information source. The one essential condition imposed on $x(t)$ is that it must have a reasonably well defined *message bandwidth* W, so there's negligible spectral content for $|f| > W$. Accordingly, Fig. 6.2-1 represents a typical message spectrum $X(f) = \mathscr{F}[x(t)]$ assuming the message is an energy signal.

For mathematical convenience, we'll also scale or normalize all messages to have a magnitude not exceeding unity, so

$$|x(t)| \leq 1 \tag{1}$$

This normalization puts an upper limit on the average message power, namely

$$S_x = \langle x^2(t) \rangle \leq 1 \tag{2}$$

when we assume $x(t)$ is a deterministic power signal. (The same limit applies to a random signal from an ergodic message source.) Both energy-signal and power-signal models will be used for $x(t)$, depending on which one best suits the circumstances at hand.

Occasionally, analysis with arbitrary $x(t)$ turns out to be difficult if not impossible. As a fall-back position we may then resort to the specific case of sinusoidal or *tone* modulation, taking

$$x(t) = A_m \cos \omega_m t \qquad A_m \leq 1 \qquad f_m < W \tag{3}$$

Tone modulation allows us to work with one-sided line spectra and simplifies power calculations. Moreover, if you can find the response of the modulation system at a particular frequency f_m, you can infer the response for all frequencies in the message band—barring any nonlinearities. To reveal potential nonlinear effects, you must use *multitone* modulation such as

$$x(t) = A_1 \cos \omega_1 t + A_2 \cos \omega_2 t + \cdots$$

with $A_1 + A_2 + \cdots \leq 1$ to satisfy Eq. (1).

Figure 6.2-1 Message spectrum with bandwidth W.

AM Signals and Spectra

The unique property of AM is that the *envelope* of the modulated carrier has the same *shape* as the message. If A_c denotes the unmodulated carrier amplitude, modulation by $x(t)$ produces the modulated envelope

$$A(t) = A_c[1 + \mu x(t)] \qquad (4)$$

where μ is a positive constant called the *modulation index*. The complete AM signal $x_c(t)$ is then

$$\begin{aligned} x_c(t) &= A_c[1 + \mu x(t)] \cos \omega_c t \\ &= A_c \cos \omega_c t + A_c \mu x(t) \cos \omega_c t \end{aligned} \qquad (5)$$

Since $x_c(t)$ has no time-varying phase, its in-phase and quadrature components are

$$x_{ci}(t) = A(t) \qquad x_{cq}(t) = 0$$

as obtained from Eqs. (4) and (5), Sect. 6.1, with $\phi(t) = 0$. Actually, we should include a constant carrier phase shift to emphasize that the carrier and message come from independent and unsynchronized sources. However, putting a constant phase in Eq. (5) increases the notational complexity without adding to the physical understanding.

Figure 6.2-2 shows part of a typical message and the resulting AM signal

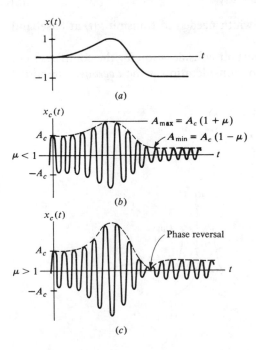

Figure 6.2-2 AM waveforms. (*a*) Message; (*b*) AM wave with $\mu < 1$; (*c*) AM wave with $\mu > 1$.

with two values of μ. The envelope clearly reproduces the shape of $x(t)$ if

$$f_c \gg W \qquad \mu \le 1 \tag{6}$$

When these conditions are satisfied, the message $x(t)$ is easily extracted from $x_c(t)$ by use of a simple *envelope detector* whose circuitry will be described in Sect. 6.5.

The condition $f_c \gg W$ ensures that the carrier oscillates rapidly compared to the time variation of $x(t)$; otherwise, an envelope could not be visualized. The condition $\mu \le 1$ ensures that $A_c[1 + \mu x(t)]$ does not go negative. With 100-percent modulation ($\mu = 1$), the envelope varies between $A_{min} = 0$ and $A_{max} = 2A_c$. *Overmodulation* ($\mu > 1$), causes *phase reversals* and *envelope distortion* illustrated by Fig. 6.2-2c.

Going to the frequency domain, Fourier transformation of Eq. (5) yields

$$X_c(f) = \frac{1}{2} A_c \, \delta(f - f_c) + \frac{\mu}{2} A_c X(f - f_c) \qquad f > 0 \tag{7}$$

where we've written out only the positive-frequency half of $X_c(f)$. The negative-frequency half will be the hermitian image of Eq. (7) since $x_c(t)$ is a real bandpass signal. Both halves of $X_c(f)$ are sketched in Fig. 6.2-3 with $X(f)$ from Fig. 6.2-1. The AM spectrum consists of carrier-frequency impulses and *symmetrical sidebands* centered at $\pm f_c$. The presence of *upper* and *lower sidebands* accounts for the name *double-sideband* amplitude modulation. It also accounts for the AM *transmission bandwidth*

$$B_T = 2W \tag{8}$$

Note that AM requires twice the bandwidth needed to transmit $x(t)$ at baseband without modulation.

Transmission bandwidth is an important consideration for the comparison of modulation systems. Another important consideration is the *average transmitted power*

$$S_T \triangleq \langle x_c^2(t) \rangle$$

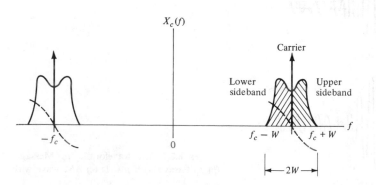

Figure 6.2-3 AM spectrum.

Upon expanding $x_c^2(t)$ from Eq. (5), we have

$$S_T = \frac{1}{2}A_c^2\langle 1 + 2\mu x(t) + \mu^2 x^2(t)\rangle + \frac{1}{2}A_c^2\langle[1 + \mu x(t)]^2 \cos 2\omega_c t\rangle$$

whose second term averages to zero under the condition $f_c \gg W$. Thus, if $\langle x(t)\rangle = 0$ and $\langle x^2(t)\rangle = S_x$ then

$$S_T = \frac{1}{2}A_c^2(1 + \mu^2 S_x) \tag{9}$$

The statistical average $E[x_c^2(t)]$ gives the same result. The assumption that the message has zero average value (or no DC component) anticipates the conclusion from Sect. 6.5 that ordinary AM is not practical for transmitting signals with significant low-frequency content.

We bring out the interpretation of Eq. (9) by putting it in the form

$$S_T = P_c + 2P_{sb}$$

where

$$P_c = \frac{1}{2}A_c^2 \qquad P_{sb} = \frac{1}{4}A_c^2 \mu^2 S_x = \frac{1}{2}\mu^2 S_x P_c \tag{10}$$

The term P_c represents the *unmodulated carrier power*, since $S_T = P_c$ when $\mu = 0$; the term P_{sb} represents the *power per sideband* since, when $\mu \neq 0$, S_T consists of the power in the carrier plus two symmetric sidebands. The modulation constraint $|\mu x(t)| \leq 1$ requires that $\mu^2 S_x \leq 1$, so $P_{sb} \leq \frac{1}{2}P_c$ and

$$P_c = S_T - 2P_{sb} \geq \frac{1}{2}S_T \qquad P_{sb} \leq \frac{1}{4}S_T \tag{11}$$

Consequently, at least 50% of the total transmitted power resides in a carrier term that's independent of $x(t)$ and thus conveys no message information.

DSB Signals and Spectra

The "wasted" carrier power in amplitude modulation can be eliminated by setting $\mu = 1$ and suppressing the unmodulated carrier-frequency component. The resulting modulated wave becomes

$$x_c(t) = A_c x(t) \cos \omega_c t \tag{12}$$

which is called *double-sideband suppressed-carrier modulation* — or DSB for short. (The abbreviations DSB–SC and DSSC are also used.) The transform of Eq. (12) is simply

$$X_c(f) = \frac{1}{2}A_c X(f - f_c) \qquad f > 0$$

and the DSB spectrum looks like an AM spectrum without the unmodulated carrier impulses. The transmission bandwidth thus remains unchanged at $B_T = 2W$.

Although DSB and AM are quite similar in the frequency domain, the time-domain picture is another story. As illustrated by Fig. 6.2-4, the DSB envelope and phase are

$$A(t) = A_c |x(t)| \qquad \phi(t) = \begin{cases} 0 & x(t) > 0 \\ \pm 180° & x(t) < 0 \end{cases} \tag{13}$$

Figure 6.2-4 DSB waveforms.

The envelope here takes the shape of $|x(t)|$, rather than $x(t)$, and the modulated wave undergoes a *phase reversal* whenever $x(t)$ crosses zero. Full recovery of the message requires knowledge of these phase reversals, and could not be accomplished by an envelope detector. Suppressed-carrier DSB thus involves more than just "amplitude" modulation and, as we'll see in Sect. 6.5, calls for a more sophisticated demodulation process.

However, carrier suppression does put all of the average transmitted power into the information-bearing sidebands. Thus

$$S_T = 2P_{sb} = \tfrac{1}{2}A_c^2 S_x \tag{14}$$

which holds even when $x(t)$ includes a DC component. From Eqs. (14) and (11) we see that DSB makes better use of the total average power available from a given transmitter. Practical transmitters also impose a limit on the *peak envelope power* A_{max}^2. We'll take account of this peak-power limitation by examining the ratio P_{sb}/A_{max}^2 under maximum modulation conditions. Using Eq. (14) with $A_{max} = A_c$ for DSB and using Eq. (10) with $A_{max} = 2A_c$ for AM, we find that

$$P_{sb}/A_{max}^2 = \begin{cases} S_x/4 & \text{DSB} \\ S_x/16 & \text{AM with } \mu = 1 \end{cases} \tag{15}$$

Hence, if A_{max}^2 is fixed and other factors are equal, a DSB transmitter produces four times the sideband power of an AM transmitter.

The foregoing considerations suggest a trade-off between power efficiency and demodulation methods. DSB conserves power but requires complicated demodulation circuitry, whereas AM requires increased power to permit simple envelope detection.

Example 6.2-1 Consider a radio transmitter rated for $S_T \le 3$ kW and $A_{max}^2 \le 8$ kW. Let the modulating signal be a tone with $A_m = 1$ so $S_x = A_m^2/2 = \tfrac{1}{2}$. If the modulation is DSB, the maximum possible power per sideband equals the lesser of the two values determined from Eqs. (14) and (15).

Thus

$$P_{sb} = \tfrac{1}{2}S_T \leq 1.5 \text{ kW} \qquad P_{sb} = \tfrac{1}{8}A_{max}^2 \leq 1.0 \text{ kW}$$

which gives the upper limit $P_{sb} = 1.0$ kW.

If the modulation is AM with $\mu = 1$, then Eq. (15) requires that $P_{sb} = A_{max}^2/32 \leq 0.25$ kW. To check on the average-power limitation, we note from Eq. (10) that $P_{sb} = P_c/4$ so $S_T = P_c + 2P_{sb} = 6P_{sb}$ and $P_{sb} = S_T/6 \leq 0.5$ kW. Hence, the peak power limit again dominates and the maximum sideband power is $P_{sb} = 0.25$ kW. Since transmission range is proportional to P_{sb}, the AM path length would be only 25% of the DSB path length with the same transmitter.

Exercise 6.2-1 Let the modulating signal be a *square wave* that switches periodically between $x(t) = +1$ and $x(t) = -1$. Sketch $x_c(t)$ when the modulation is AM with $\mu = 0.5$, AM with $\mu = 1$, and DSB. Indicate the envelopes by dashed lines.

Exercise 6.2-2 Suppose a voice signal has $|x(t)|_{max} = 1$ and $S_x = 1/5$. Calculate the values of S_T and A_{max}^2 needed to get $P_{sb} = 10$ W for DSB and for AM with $\mu = 1$.

Tone Modulation and Phasor Analysis

Setting $x(t) = A_m \cos \omega_m t$ in Eq. (12) gives the tone-modulated DSB waveform

$$x_c(t) = A_c A_m \cos \omega_m t \cos \omega_c t$$

$$= \frac{A_c A_m}{2} \cos (\omega_c - \omega_m)t + \frac{A_c A_m}{2} \cos (\omega_c + \omega_m)t \qquad (16a)$$

where we have used the trigonometric expansion for the product of cosines. Similar expansion of Eq. (5) yields the tone-modulated AM wave

$$x_c(t) = A_c \cos \omega_c t + \frac{A_c \mu A_m}{2} \cos (\omega_c - \omega_m)t + \frac{A_c \mu A_m}{2} \cos (\omega_c + \omega_m)t \qquad (16b)$$

Figure 6.2-5 shows the positive-frequency line spectra obtained from Eqs. (16a) and (16b).

(a)   (b)

Figure 6.2-5 Line spectra for tone modulation. (a) DSB; (b) AM.

It follows from Fig. 6.2-5 that tone-modulated DSB or AM can be viewed as a sum of ordinary *phasors*, one for each spectral line. This viewpoint prompts the use of phasor analysis to find the envelope-and-phase or quadrature-carrier terms. Phasor analysis is especially helpful for studying the effects of transmission distortion, interference, etc., as demonstrated in the example below.

Example 6.2-2 Let's take the case of tone-modulated AM with $\mu A_m = \frac{2}{3}$ for convenience. The phasor diagram is constructed in Fig. 6.2-6a by adding the sideband phasors to the tip of the horizontal carrier phasor. Since the carrier frequency is f_c, the sideband phasors at $f_c \pm f_m$ rotate with speeds of $\pm f_m$ relative to the carrier phasor. The resultant of the sideband phasors is seen to be colinear with the carrier, and the phasor sum equals the envelope $A_c(1 + \frac{2}{3} \cos \omega_m t)$.

But suppose a transmission channel completely removes the lower sideband, so we get the diagram in Fig. 6.2-6b. Now the envelope becomes

$$A(t) = [(A_c + \tfrac{1}{3}A_c \cos \omega_m t)^2 + (\tfrac{1}{3}A_c \sin \omega_m t)^2]^{1/2}$$
$$= A_c\sqrt{\tfrac{10}{9} + \tfrac{2}{3} \cos \omega_m t}$$

from which the *envelope distortion* can be determined. Also note that the transmission amplitude distortion has produced a *time-varying phase* $\phi(t)$.

Exercise 6.2-3 Draw the phasor diagram for tone-modulated DSB with $A_m = 1$. Then find $A(t)$ and $\phi(t)$ when the amplitude of the lower sideband is cut in half.

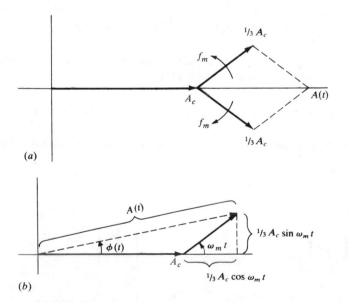

(a)

(b)

Figure 6.2-5 Phasor diagrams for Example 6.2-2.

6.3 MODULATORS AND TRANSMITTERS

The sidebands of an AM or DSB signal contain new frequencies that were not present in the carrier or message. The modulator must therefore be a *time-varying* or *nonlinear* system, because LTI systems never produce new frequency components. This section describes the operating principles of modulators and transmitters that employ product, square-law, or switching devices. Detailed circuit designs are given in the references cited in the Supplementary Reading.

Product Modulators

Figure 6.3-1a is the block diagram of a *product modulator* for AM based on the equation $x_c(t) = A_c \cos \omega_c t + \mu x(t) A_c \cos \omega_c t$. The schematic diagram in Fig. 6.3-1b implements this modulator with an analog multiplier and an op-amp summer. Of course, a DSB product modulator needs only the multiplier to produce $x_c(t) = x(t) A_c \cos \omega_c t$. In either case, the crucial operation is *multiplying* two analog signals.

Analog multiplication can be carried out electronically in a number of different ways. One popular integrated-circuit design is the *variable transconductance multiplier* illustrated by Fig. 6.3-2. Here, input voltage v_1 is applied to a differential amplifier whose gain depends on the transconductance of the transistors which, in turn, varies with the total emitter current. Input v_2 controls the emitter current by means of a voltage-to-current converter, so the differential output equals $K v_1 v_2$. Other circuits achieve multiplication directly with *Hall-effect* devices, or indirectly with *log* and *antilog amplifiers* arranged to produce antilog $(\log v_1 + \log v_2) = v_1 v_2$. However, most analog multipliers are limited to low power levels and relatively low frequencies.

(a)

(b)

Figure 6.3-1 (*a*) Product modulator for AM; (*b*) schematic diagram with analog multiplier.

Figure 6.3-2 Circuit for variable transconductance multiplier.

Square-Law and Balanced Modulators

Signal multiplication at higher frequencies can be accomplished by the *square-law modulator* diagrammed in Fig. 6.3-3a. The circuit realization in Fig. 6.3-3b uses a field-effect transistor as the nonlinear element and a parallel RLC circuit as the filter. We assume the nonlinear element approximates the square-law transfer curve

$$v_{\text{out}} = a_1 v_{\text{in}} + a_2 v_{\text{in}}^2$$

Thus, with $v_{\text{in}}(t) = x(t) + \cos \omega_c t$,

$$v_{\text{out}}(t) = a_1 x(t) + a_2 x^2(t) + a_2 \cos^2 \omega_c t + a_1 \left[1 + \frac{2a_2}{a_1} x(t) \right] \cos \omega_c t \qquad (1)$$

(a)

(b)

Figure 6.3-3 (a) Square-law modulator; (b) FET circuit realization.

Figure 6.3-4 Spectral components in Eq. (1).

The last term is the desired AM wave, with $A_c = a_1$ and $\mu = 2a_2/a_1$, provided it can be separated from the rest.

As to the feasibility of separation, Fig. 6.3-4 shows the spectrum $V_{out}(f) = \mathscr{F}[v_{out}(t)]$ taking $X(f)$ as in Fig. 6.2-1. Note that the $x^2(t)$ term in Eq. (1) becomes $X * X(f)$, which is bandlimited in $2W$. Therefore, if $f_c > 3W$, there is no spectral overlapping and the required separation can be accomplished by a bandpass filter of bandwidth $B_T = 2W$ centered at f_c. Also note that the carrier-frequency impulse disappears and we have a DSB wave if $a_1 = 0$—corresponding to the *perfect* square-law curve $v_{out} = a_2 v_{in}^2$.

Unfortunately, perfect square-law devices are rare, so high-frequency DSB is obtained in practice using *two* AM modulators arranged in a balanced configuration to cancel out the carrier. Figure 6.3-5 shows such a *balanced modulator* in block-diagram form. Assuming the AM modulators are identical, save for the reversed sign of one input, the outputs are $A_c[1 + \frac{1}{2} x(t)] \cos \omega_c t$ and $A_c[1 - \frac{1}{2} x(t)] \cos \omega_c t$. Subtracting one from the other yields $x_c(t) = x(t)A_c \cos \omega_c t$, as required. Hence, a balanced modulator is a multiplier. You should observe that if the message has a *DC term*, that component is *not* canceled out in the modulator, even though it appears at the carrier frequency in the modulated wave.

A balanced modulator using switching circuits is discussed in Chap. 10 under the heading of *bipolar choppers*. Other circuit realizations can be found in the literature.

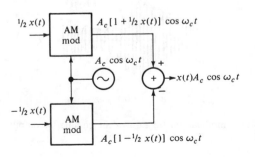

Figure 6.3-5 Balanced modulator.

Exercise 6.3-1 Suppose the AM modulators in Fig. 6.3-5 are constructed with identical nonlinear elements having $v_{\text{out}} = a_1 v_{\text{in}} + a_2 v_{\text{in}}^2 + a_3 v_{\text{in}}^3$. Take an arbitrary input $x(t)$ and show that the AM signals have second-harmonic distortion but, nonetheless, the final output is undistorted DSB.

Switching Modulators

In view of the heavy filtering required, square-law modulators are used primarily for *low-level* modulation, i.e., at power levels lower than the transmitted value. Substantial linear amplification is then necessary to bring the power up to S_T. But RF power amplifiers of the required linearity are not without problems of their own, and it often is better to employ *high-level* modulation if S_T is to be large.

Efficient high-level modulators are arranged so that undesired modulation products never fully develop and need not be filtered out. This is usually accomplished with the aid of a *switching* device, whose detailed analysis is postponed to Chap. 10. However, the basic operation of the supply-voltage modulated class C amplifier is readily understood from its idealized equivalent circuit and waveforms in Fig. 6.3-6.

The active device, a vacuum tube or transistor, serves as a switch driven at the carrier frequency, closing briefly every $1/f_c$ seconds. The RLC load, called a *tank* circuit, is tuned to resonate at f_c, so the switching action causes the tank circuit to "ring" sinusoidally. The steady state load voltage in absence of modulation is then $v(t) = V \cos \omega_c t$. Adding the message to the supply voltage, say via transformer, gives $v(t) = [V + Nx(t)] \cos \omega_c t$, where N is the transformer turns ratio. If V and N are correctly proportioned, the desired modulation has been accomplished without appreciable generation of undesired components.

A complete AM transmitter is diagramed in Fig. 6.3-7 for the case of high-level modulation. The carrier wave is generated by a crystal-controlled oscillator to ensure stability of the carrier frequency. Because high-level modulation demands husky input signals, both the carrier and message are amplified before modulation. The modulated signal is then delivered directly to the antenna.

(a)   (b)

Figure 6.3-6 Class C amplifier with supply-voltage modulation. (a) Equivalent circuit; (b) output waveform.

Figure 6.3-7 AM transmitter with high-level modulation.

6.4 SUPPRESSED-SIDEBAND AMPLITUDE MODULATION

Conventional amplitude modulation is wasteful of both transmission power and bandwidth. Suppressing the carrier reduces the transmission power. Suppressing one sideband, in whole or part, reduces transmission bandwidth and leads to single-sideband modulation (SSB) or vestigial-sideband modulation (VSB) discussed in this section.

SSB Signals and Spectra

The upper and lower sidebands of DSB are uniquely related by symmetry about the carrier frequency, so either one contains *all* the message information. Hence, transmission bandwidth can be cut in half if one sideband is suppressed along with the carrier.

Figure 6.4-1a presents a conceptual approach to single-sideband modulation. Here, the DSB signal from a balanced modulator is applied to a sideband filter that suppresses one sideband. If the filter removes the lower sideband, the output spectrum $X_c(f)$ consists of the upper sideband alone, as illustrated by Fig. 6.4-1b. We'll label this a USSB spectrum to distinguish it from the LSSB spectrum containing just the lower sideband, as illustrated by Fig. 6.4-1c. The resulting signal in either case has

$$B_T = W \qquad S_T = P_{sb} = \frac{1}{4}A_c^2 S_x \qquad (1)$$

which follow directly from our DSB results.

Although SSB is readily visualized in the frequency domain, the time-domain description is not immediately obvious — save for the special case of tone modulation. By referring back to the DSB line spectrum in Fig. 6.2-5a, we see that removing one sideband line leaves only the other line. Hence,

$$x_c(t) = \frac{1}{2}A_c A_m \cos{(\omega_c \pm \omega_m)t} \qquad (2)$$

in which the upper sign stands for USSB and the lower for LSSB, a convention employed hereafter. Note that the frequency of a tone-modulated SSB wave is offset from f_c by $\pm f_m$ and the envelope is a *constant* proportional to A_m. Obviously, envelope detection won't work for SSB.

Figure 6.4-1 Single-sideband modulation. (*a*) Modulator; (*b*) USSB spectrum; (*c*) LSSB spectrum.

To analyze SSB with an arbitrary message $x(t)$, we'll draw upon the fact that the sideband filter in Fig. 6.4-1a is a *bandpass* system with a bandpass DSB input $x_{bp}(t) = A_c x(t) \cos \omega_c t$ and a bandpass SSB output $y_{bp}(t) = x_c(t)$. Hence, we'll find $x_c(t)$ by applying the *equivalent lowpass method* from Sect. 6.1. Since $x_{bp}(t)$ has no quadrature component, the lowpass equivalent input is simply

$$x_{\ell p}(t) = \tfrac{1}{2}A_c x(t) \qquad X_{\ell p}(f) = \tfrac{1}{2}A_c X(f)$$

The bandpass filter transfer function for USSB is plotted in Fig. 6.4-2a along with the equivalent lowpass function

$$H_{\ell p}(f) = H_{bp}(f + f_c)u(f + f_c) = u(f) - u(f - W)$$

The corresponding transfer functions for LSSB are plotted in Fig. 6.4-2b, where

$$H_{\ell p}(f) = u(f + W) - u(f)$$

Both lowpass transfer functions can be represented by

$$H_{\ell p}(f) = \tfrac{1}{2}(1 \pm \operatorname{sgn} f) \qquad |f| \leq W \tag{3}$$

You should confirm for yourself that this rather strange expression does include both parts of Fig. 6.4-2.

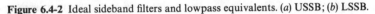

Figure 6.4-2 Ideal sideband filters and lowpass equivalents. (*a*) USSB; (*b*) LSSB.

Multiplying $H_{\ell p}(f)$ and $X_{\ell p}(f)$ yields the lowpass equivalent spectrum for either USSB or LSSB, namely

$$Y_{\ell p}(f) = \tfrac{1}{4}A_c(1 \pm \text{sgn } f)X(f) = \tfrac{1}{4}A_c[X(f) \pm (\text{sgn } f)X(f)]$$

Now recall that $(-j \text{ sgn } f)X(f) = \mathscr{F}[\hat{x}(t)]$, where $\hat{x}(t)$ is the *Hilbert transform* of $x(t)$ defined in Sect. 3.4. Therefore, $\mathscr{F}^{-1}[(\text{sgn } f)X(f)] = j\hat{x}(t)$ and

$$y_{\ell p}(t) = \tfrac{1}{4}A_c[x(t) \pm j\hat{x}(t)]$$

Finally, we perform the lowpass-to-bandpass transformation $x_c(t) = y_{bp}(t) = 2 \text{ Re } [y_{\ell p}(t)e^{j\omega_c t}]$ to obtain

$$x_c(t) = \tfrac{1}{2}A_c[x(t) \cos \omega_c t \mp \hat{x}(t) \sin \omega_c t] \tag{4}$$

This is our desired result for the SSB waveform in terms of an arbitrary message $x(t)$.

Closer examination reveals that Eq. (4) has the form of a quadrature-carrier expression. Hence, the in-phase and quadrature components are

$$x_{ci}(t) = \tfrac{1}{2}A_c x(t) \qquad x_{cq}(t) = \pm \tfrac{1}{2}A_c \hat{x}(t)$$

while the SSB envelope is

$$A(t) = \tfrac{1}{2}A_c \sqrt{x^2(t) + \hat{x}^2(t)} \tag{5}$$

The complexity of Eqs. (4) and (5) make it a difficult task to sketch SSB waveforms or to determine the peak envelope power. Instead, we must infer time-domain properties from simplified cases such as tone modulation or pulse modulation.

Example 6.4-1 SSB with pulse modulation Whenever the SSB modulating signal has abrupt transitions, the Hilbert transform $\hat{x}(t)$ contains sharp peaks. These peaks then appear in the envelope $A(t)$, giving rise to the effect known as envelope *horns*. To demonstrate this effect, let's take the rectangular pulse $x(t) = u(t) - u(t - \tau)$ so we can use $\hat{x}(t)$ found in Example 3.4-1. The resulting SSB envelope plotted in Fig. 6.4-3 exhibits infinite peaks at $t = 0$ and $t = \tau$,

Figure 6.4-3 Envelope of SSB with pulse modulation.

the instants when $x(t)$ has stepwise discontinuities. Clearly, a transmitter couldn't handle the peak envelope power needed for these infinite horns. Also note the *smears* in $A(t)$ before and after each peak.

We thus conclude that SSB is not appropriate for pulse transmission, digital data, or similar applications. We further conclude that more suitable modulating signals (such as audio waveforms) should still be lowpass filtered before modulation in order to smooth out any abrupt transitions that might cause excessive horns or smearing.

Exercise 6.4-1 Show that Eqs. (4) and (5) agree with Eq. (2) when $x(t) = A_m \cos \omega_m t$ so $\hat{x}(t) = A_m \sin \omega_m t$.

SSB Generation

Our conceptual SSB generation system (Fig. 6.4-1*a*) calls for the ideal filter functions in Fig. 6.4-2. But a perfect cutoff at $f = f_c$ cannot be synthesized, so a real sideband filter will either pass a portion of the undesired sideband or attenuate a portion of the desired sideband. (Doing both is tantamount to vestigial-sideband modulation.) Fortunately, many modulating signals of practical interest have little or no low-frequency content, their spectra having "holes" at zero frequency as shown in Fig. 6.4-4*a*. Such spectra are typical of audio signals (voice and

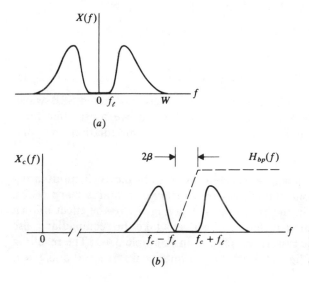

Figure 6.4-4 (*a*) Message spectrum with zero-frequency hole; (*b*) practical sideband filter.

Figure 6.4-5 Two-step SSB generation.

music), for example. After translation by the balanced modulator, the zero-frequency hole appears as a vacant space centered about the carrier frequency into which the transition region of a practical sideband filter can be fitted. Figure 6.4-4b illustrates this point.

As a rule of thumb, the width 2β of the transition region cannot be much smaller than 1% of the nominal cutoff frequency, which imposes the limit $f_{co} <$ 200β. Since 2β is constrained by the width of the spectral hole and f_{co} should equal f_c, it may not be possible to obtain a sufficiently high carrier frequency with a given message spectrum. For these cases the modulation process can be carried out in two (or more) steps using the system in Fig. 6.4-5 (see Prob. 6.4-6).

Another method for SSB generation is based on writing Eq. (4) in the form

$$x_c(t) = \frac{A_c}{2} x(t) \cos \omega_c t \mp \frac{A_c}{2} \hat{x}(t) \cos (\omega_c t - 90°) \tag{6}$$

This expression suggests that an SSB signal consists of *two* DSB waveforms with quadrature carriers and modulating signals $x(t)$ and $\hat{x}(t)$. Figure 6.4-6 diagrams a system that implements Eq. (6) and produces either USSB or LSSB, depending upon the sign at the summer. This system, known as the *phase-shift method*, bypasses the need for sideband filters. Instead, the DSB sidebands are phased such that they cancel out on one side of f_c and add on the other side to create a single-sideband output.

However, the quadrature phase shifter $H_Q(f)$ is itself an unrealizable network that can only be approximated — usually with the help of additional but identical phase networks in both branches of Fig. 6.4-6. Approximation imperfections generally cause low-frequency signal distortion, and the phase-shift system works best with message spectra of the type in Fig. 6.4-4a. A third method for SSB generation that avoids both sideband filters and quadrature phase shifters is considered in Prob. 6.4-10.

Figure 6.4-6 Phase-shift method for SSB generation.

Exercise 6.4-2 Take $x(t) = \cos \omega_m t$ in Fig. 6.4-6 and confirm the sideband cancellation by sketching line spectra at appropriate points.

VSB Signals and Spectra★

Consider a modulating signal of very large bandwidth having significant low-frequency content. Principal examples are television video, facsimile, and high-speed data signals. Bandwidth conservation argues for the use of SSB, but practical SSB systems have poor low-frequency response. On the other hand, DSB works quite well for low message frequencies but the transmission bandwidth is twice that of SSB. Clearly, a compromise modulation scheme is desired; that compromise is VSB.

VSB is derived by filtering DSB (or AM) in such a fashion that one sideband is passed almost completely while just a trace, or *vestige*, of the other sideband is included. The key to VSB is the sideband filter, a typical transfer function being that of Fig. 6.4-7a. While the exact shape of the response is not crucial, it must have odd symmetry about the carrier frequency and a relative response of $1/2$ at f_c. Therefore, taking the upper sideband case, we have

$$H(f) = u(f - f_c) - H_\beta(f - f_c) \qquad f > 0 \tag{7a}$$

where

$$H_\beta(-f) = -H_\beta(f) \qquad \text{and} \qquad H_\beta(f) = 0 \qquad |f| > \beta \tag{7b}$$

as shown in Fig. 6.4-7b.

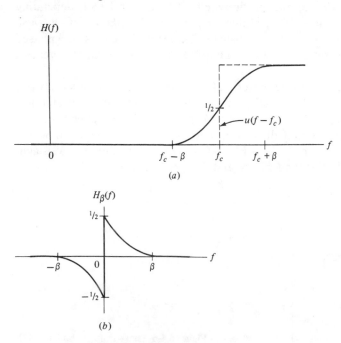

(a)

(b)

Figure 6.4-7 VSB filter characteristics.

The VSB filter is thus a *practical* sideband filter with transition width 2β. Because the width of the partial sideband is one-half the filter transition width, the transmission bandwidth is

$$B_T = W + \beta \approx W \tag{8}$$

However, in some applications the vestigial filter symmetry is achieved primarily at the receiver, so the transmission bandwidth must be slightly larger than $W + \beta$.

When $\beta \ll W$, which is usually true, the VSB spectrum looks essentially like an SSB spectrum. The similarity also holds in the time domain, and a VSB waveform can be expressed as a modification of Eq. (4). Specifically,

$$x_c(t) = \tfrac{1}{2} A_c [x(t) \cos \omega_c t - x_q(t) \sin \omega_c t] \tag{9a}$$

where $x_q(t)$ is the quadrature message component defined by

$$x_q(t) = \hat{x}(t) + x_\beta(t) \tag{9b}$$

with

$$x_\beta(t) = j2 \int_{-\beta}^{\beta} H_\beta(f) X(f) e^{j\omega t} \, df \tag{9c}$$

If $\beta \ll W$, VSB approximates SSB and $x_\beta(t) \approx 0$; conversely, for large β, VSB approximates DSB and $\hat{x}(t) + x_\beta(t) \approx 0$. The transmitted power S_T is not easy to determine exactly, but is bounded by

$$\tfrac{1}{4} A_c^2 S_x \leq S_T \leq \tfrac{1}{2} A_c^2 S_x \tag{10}$$

depending on the vestige width β.

Finally, suppose an AM wave is applied to a vestigial sideband filter. This modulation scheme, termed VSB plus carrier (VSB + C), is used for television video transmission. The unsuppressed carrier allows for *envelope detection*, as in AM, while retaining the bandwidth conservation of suppressed sideband. Distortionless envelope modulation actually requires symmetric sidebands, but VSB + C can deliver a fair approximation.

To analyze the envelope of VSB + C, we incorporate a carrier term and modulation index μ in Eq. (9) which becomes

$$x_c(t) = A_c \{[1 + \mu x(t)] \cos \omega_c t - \mu x_q(t) \sin \omega_c t\} \tag{11}$$

The in-phase and quadrature components are then

$$x_{ci}(t) = A_c [1 + \mu x(t)] \qquad x_{cq}(t) = A_c \mu x_q(t)$$

so the envelope is $A(t) = [x_{ci}^2(t) + x_{cq}^2(t)]^{1/2}$ or

$$A(t) = A_c [1 + \mu x(t)] \left\{ 1 + \left[\frac{\mu x_q(t)}{1 + \mu x(t)} \right]^2 \right\}^{1/2} \tag{12}$$

Hence, if μ is not too large and β not too small, then $|\mu x_q(t)| \ll 1$ and

$$A(t) \approx A_c[1 + \mu x(t)]$$

as desired. Empirical studies with typical signals are needed to find values for μ and β that provide a suitable compromise between the conflicting requirements of distortionless envelope modulation, power efficiency, and bandwidth conservation.

6.5 FREQUENCY CONVERSION AND DEMODULATION

Linear CW modulation—be it AM, DSB, SSB, or VSB—produces upward translation of the message spectrum. *Demodulation* therefore implies *downward frequency translation* in order to recover the message from the modulated wave. Demodulators that perform this operation fall into the two broad categories of *synchronous detectors* and *envelope detectors*.

Frequency translation, or *conversion*, is also used to shift a modulated signal to a new carrier frequency (up or down) for amplification or other processing. Thus, translation is a fundamental concept in linear modulation systems and includes modulation and detection as special cases. Before examining detectors, we'll look briefly at the general process of frequency conversion.

Frequency Conversion

Frequency conversion starts with multiplication by a sinusoid. Consider, for example, the DSB wave $x(t) \cos \omega_1 t$. Multiplying by $\cos \omega_2 t$, we get

$$x(t) \cos \omega_1 t \cos \omega_2 t = \tfrac{1}{2}x(t) \cos (\omega_1 + \omega_2)t + \tfrac{1}{2}x(t) \cos (\omega_1 - \omega_2)t \qquad (1)$$

The product consists of the *sum* and *difference frequencies*, $f_1 + f_2$ and $|f_1 - f_2|$, each modulated by $x(t)$. We write $|f_1 - f_2|$ for clarity, since $\cos (\omega_2 - \omega_1)t = \cos (\omega_1 - \omega_2)t$. Assuming $f_2 \neq f_1$, multiplication has translated the signal spectra to *two* new carrier frequencies. With appropriate filtering, the signal is up-converted or down-converted. Devices that carry out this operation are called *frequency converters* or *mixers*. The operation itself is termed *heterodyning* or *mixing*.

Figure 6.5-1 diagrams the essential components of a frequency converter. Implementation of the multiplier follows the same line as the modulator circuits discussed in Sect. 6.3. Converter applications include beat-frequency oscillators, regenerative frequency dividers, speech scramblers, and spectrum analyzers, in addition to their roles in transmitters and receivers.

Figure 6.5-1 Frequency converter.

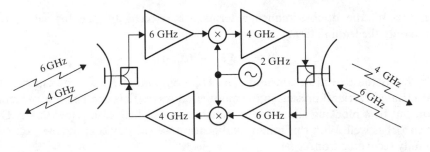

Figure 6.5-2 Satellite transponder with frequency conversion.

Example 6.5-1 Figure 6.5-2 represents a simplified *transponder* in a satellite relay that provides two-way communication between two ground stations. Different carrier frequencies, 6 GHz and 4 GHz, are used on the uplink and downlink to prevent self-oscillation due to positive feedback from the transmitting side to the receiving side. A frequency converter translates the spectrum of the amplified uplink signal to the passband of the downlink amplifier.

Exercise 6.5-1 Sketch the spectrum of Eq. (1) for $f_2 < f_1, f_2 = f_1,$ and $f_2 > f_1,$ taking $X(f)$ as in Fig. 6.2-1.

Synchronous Detection

All types of linear modulation can be detected by the *product* demodulator of Fig. 6.5-3. The incoming signal is first multiplied with a locally generated sinusoid and then lowpass-filtered, the filter bandwidth being the same as the message bandwidth W or somewhat larger. It is assumed that the local oscillator (LO) is *exactly synchronized* with the carrier, in both phase and frequency, accounting for the name *synchronous* or *coherent detection*.

For purposes of analysis, we'll write the input signal in the generalized form

$$x_c(t) = [K_c + K_\mu x(t)] \cos \omega_c t - K_\mu x_q(t) \sin \omega_c t \qquad (2)$$

which can represent any type of linear modulation with proper identification of K_c, K_μ, and $x_q(t)$—i.e., take $K_c = 0$ for suppressed carrier, $x_q(t) = 0$ for double sideband, etc. The filter input is thus the product

$$x_c(t)A_{\mathrm{LO}} \cos \omega_c t$$

$$= \frac{A_{\mathrm{LO}}}{2} \{[K_c + K_\mu x(t)] + [K_c + K_\mu x(t)] \cos 2\omega_c t - K_\mu x_q(t) \sin 2\omega_c t\}$$

Figure 6.5-3 Synchronous product detection.

Since $f_c > W$, the double-frequency terms are rejected by the lowpass filter, leaving only the leading term

$$y_D(t) = K_D[K_c + K_\mu x(t)] \tag{3}$$

where K_D is the detection constant. The DC component $K_D K_c$ corresponds to the translated carrier if present in the modulated wave. This can be removed from the output by a blocking capacitor or transformer — which also removes any DC term in $x(t)$ as well. With this minor qualification we can say that the message has been fully recovered from $x_c(t)$.

Although perfectly correct, the above manipulations fail to bring out what goes on in the demodulation of VSB. This is best seen in the frequency domain with the message spectrum taken to be constant over W (Fig. 6.5-4a) so the modulated spectrum takes the form of Fig. 6.5-4b. The downward-translated spectrum at the filter input will then be as shown in Fig. 6.5-4c. Again, high-frequency terms are eliminated by filtering, while the down-converted sidebands overlap around zero frequency. Recalling the symmetry property of the vestigial filter, we find that the portion removed from the upper sideband is exactly restored by the corresponding vestige of the lower sideband, so $X(f)$ has been reconstructed at the output and the detected signal is proportional to $x(t)$.

Theoretically, product demodulation borders on the trivial; in practice, it can be rather tricky. The crux of the problem is *synchronization* — synchronizing an oscillator to a sinusoid that is not even present in the incoming signal if carrier is suppressed. To facilitate the matter, suppressed-carrier systems may have a small amount of carrier reinserted in $x_c(t)$ at the transmitter. This *pilot carrier* is picked off at the receiver by a narrow bandpass filter, amplified, and used in place of an LO. The system, shown in Fig. 6.5-5, is called *homodyne detection*. (Actually, the amplified pilot more often serves to synchronize a separate oscillator rather than being used directly.)

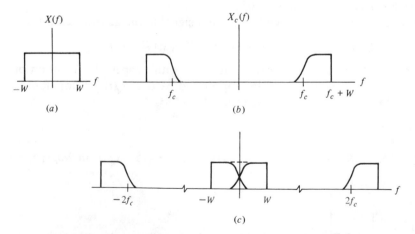

Figure 6.5-4 VSB spectra. (a) Message; (b) modulated signal; (c) frequency-translated signal before lowpass filtering.

Figure 6.5-5 Homodyne detection.

A variety of other techniques are possible for synchronization, including *phase-lock loops* (to be covered in Sect. 8.3) or the use of highly stable, crystal-controlled oscillators at transmitter and receiver. Nonetheless, some degree of asynchronism must be expected in synchronous detectors. It is therefore important to investigate the effects of phase and frequency drift in various applications. This we'll do for DSB and SSB in terms of tone modulation.

Let the local oscillator wave be $\cos(\omega_c t + \omega' t + \phi')$, where ω' and ϕ' represent slowly drifting frequency and phase errors compared to the carrier. For double sideband with tone modulation, the detected signal becomes

$$y_D(t) = K_D \cos \omega_m t \cos(\omega' t + \phi') \tag{4}$$

$$= \begin{cases} \dfrac{K_D}{2}[\cos(\omega_m + \omega')t + \cos(\omega_m - \omega')t] & \phi' = 0 \\[2ex] K_D \cos \omega_m t \cos \phi' & \omega' = 0 \end{cases}$$

Similarly, for single sideband with $x_c(t) = \cos(\omega_c \pm \omega_m)t$, we get

$$y_D(t) = K_D \cos[\omega_m t \mp (\omega' t + \phi')] \tag{5}$$

$$= \begin{cases} K_D \cos(\omega_m \mp \omega')t & \phi' = 0 \\ K_D \cos(\omega_m t \mp \phi') & \omega' = 0 \end{cases}$$

All of the foregoing expressions come from simple trigonometric expansions.

Clearly, in both DSB and SSB, a frequency drift that's not small compared to W will substantially alter the detected tone. The effect is more severe in DSB since a *pair* of tones, $f_m + f'$ and $f_m - f'$, is produced. If $f' \ll f_m$, this sounds like warbling or the beat note heard when two musical instruments play in unison but slightly out of tune. While only one tone is produced with SSB, this too can be disturbing, particularly for music transmission. To illustrate, the major triad chord consists of three notes whose frequencies are related as the integers 4, 5, and 6. Frequency error in detection shifts each note by the same absolute amount, destroying the harmonic relationship and giving the music an oriental flavor. (Note that the effect is *not* like playing recorded music at the wrong speed, which preserves the frequency ratios.) For voice transmission, subjective listener tests have shown that frequency drifts of less than ± 10 Hz are tolerable, otherwise, everyone sounds rather like Donald Duck.

As to phase drift, again DSB is more sensitive, for if $\phi' = \pm 90°$ (LO and carrier in quadrature), the detected signal vanishes entirely. With slowly varying ϕ', we get an apparent *fading* effect. Phase drift in SSB appears as *delay distortion*, the extreme case being when $\phi' = \pm 90°$ and the demodulated signal becomes

$\hat{x}(t)$. However, as was remarked before, the human ear can tolerate sizeable delay distortion, so phase drift is not so serious in voice-signal SSB systems.

To summarize, phase and frequency synchronization requirements are rather modest for voice transmission via SSB. But in data, facsimile, and video systems with suppressed carrier, careful synchronization is a necessity. Consequently, television broadcasting employs VSB + C rather than suppressed-carrier VSB.

Envelope Detection

Very little was said above about synchronous demodulation of AM for the simple reason that it's almost never used. True, synchronous detectors work for AM, but so does an *envelope detector*, which is much simpler. Because the envelope of an AM wave has the same shape as the message, independent of carrier frequency and phase, demodulation can be accomplished by extracting the envelope with no worries about synchronization.

A simplified envelope detector and its waveforms are shown in Fig. 6.5-6, where the diode is assumed to be piecewise-linear. In absence of further circuitry, the voltage v would be just the half-rectified version of the input v_{in}. But $R_1 C_1$ acts as a lowpass filter, responding only to variations in the peaks of v_{in} provided that

$$W \ll \frac{1}{R_1 C_1} \ll f_c \tag{6}$$

Thus, as noted earlier, we need $f_c \gg W$ so the envelope is clearly defined. Under these conditions, C_1 discharges only slightly between carrier peaks, and v approx-

(a)

(b)

Figure 6.5-6 Envelope detection. (*a*) Circuit; (*b*) waveforms.

Figure 6.5-7 Envelope reconstruction for suppressed-carrier modulation.

imates the envelope of v_{in}. More sophisticated filtering produces further improvement if needed. Finally, $R_2 C_2$ acts as a DC block to remove the bias of the unmodulated carrier component. Since the DC block distorts low-frequency message components, conventional envelope detectors are inadequate for signals with important low-frequency content.

The voltage v may also be filtered to remove the *envelope* variations and produce a DC voltage proportional to the carrier amplitude. This voltage in turn is fed back to earlier stages of the receiver for *automatic volume control* (AVC) to compensate for fading.

Despite the nonlinear element, Fig. 6.5-6 is termed a *linear envelope detector*; the output is linearly proportional to the input envelope. Power-law diodes can also be used, but then v will include terms of the form v_{in}^2, v_{in}^3, etc., and there may be appreciable second-harmonic distortion unless $\mu \ll 1$. In fact, the design of linear envelope detectors has several interesting problems not immediately apparent in our discussion. See, for instance, Clarke and Hess (1971, chap. 10).

Some DSB and SSB demodulators employ the method of *envelope reconstruction* diagrammed in Fig. 6.5-7. The addition of a large, locally generated carrier to the incoming signal reconstructs the envelope for recovery by an envelope detector. This method eliminates signal multiplication but does not get around the synchronization problem, for the local carrier must be as well synchronized as the LO in a product demodulator.

Exercise 6.5-2 Let the input in Fig. 6.5-7 be SSB with tone modulation, and let the LO have a phase error ϕ' but no frequency error. Use a phasor diagram to obtain an expression for the resulting envelope. Then show that $A(t) \approx A_{\text{LO}} + \frac{1}{2} A_c A_m \cos(\omega_m t \mp \phi')$ if $A_{\text{LO}} \gg A_c A_m$.

6.6 PROBLEMS

6.1-1 Derive Eq. (1b) by obtaining expressions for f_ℓ and f_u from Eq. (1a).

6.1-2 Let $f = (1 + \delta) f_0$ in Eq. (1a) and assume that $|\delta| \ll 1$. Derive the handy approximation

$$H(f) \approx 1/[1 + j2Q(f - f_0)/f_0]$$

which holds for $f > 0$ and $|f - f_0| \ll f_0$.

6.1-3 A *stagger-tuned* bandpass system centered at $f = f_c$ has $H(f) = 2H_1(f)H_2(f)$, where $H_1(f)$ is given by Eq. (1a) with $f_0 = f_c - b$ and $Q = f_0/2b$ while $H_2(f)$ is given by Eq. (1a) with $f_0 = f_c + b$ and $Q = f_0/2b$. Use the approximation in Prob. 6.1-2 to plot $|H(f)|$ for $f_c - 2b < f < f_c + 2b$ and compare it with a simple tuned circuit having $f_0 = f_c$ and $B = 2\sqrt{2b}$.

6.1-4 Use a phasor diagram to obtain expressions for $v_i(t)$, $v_q(t)$, $A(t)$, and $\phi(t)$ when $v_{bp}(t) = v_1(t) \cos \omega_c t + v_2(t) \cos (\omega_c t + \alpha)$. Then simplify $A(t)$ and $\phi(t)$ assuming $|v_2(t)| \ll |v_1(t)|$.

6.1-5 Do Prob. 6.1-4 with $v_{bp}(t) = v_1(t) \cos (\omega_c - \omega_0)t + v_2(t) \cos (\omega_c + \omega_0)t$.

6.1-6 Let $v_i(t)$ and $v_q(t)$ in Eq. (6) be lowpass signals with energy E_i and E_q, respectively, and bandwidth $W < f_c$. (a) Use Eq. (17), Sect. 2.2, to prove that

$$\int_{-\infty}^{\infty} v_{bp}(t) \, dt = 0$$

(b) Now show that the bandpass signal energy equals $(E_i + E_q)/2$.

6.1-7 Find $v_{\ell p}(t)$, $v_i(t)$, and $v_q(t)$ when $f_c = 1000$ and

$$V_{bp}(f) = \begin{cases} 1 & 900 \leq |f| \leq 1300 \\ 0 & \text{otherwise} \end{cases}$$

6.1-8 Do Prob. 6.1-7 with

$$V_{bp}(f) = \begin{cases} 1 & 1000 \leq |f| < 1100 \\ 1/2 & 950 \leq |f| < 1000 \\ 0 & \text{otherwise} \end{cases}$$

6.1-9 Let $v_{bp}(t) = 2z(t) \cos [(\omega_c \pm \omega_0)t + \alpha]$. Find $v_i(t)$ and $v_q(t)$ to obtain

$$v_{\ell p}(t) = z(t) \exp j(\pm \omega_0 t + \alpha)$$

6.1-10 Use lowpass *time-domain* analysis to find and sketch $y_{bp}(t)$ when $x_{bp}(t) = A \cos \omega_c t \, u(t)$ and $H_{bp}(f) = 1/[1 + j2(f - f_c)/B]$ for $f > 0$, which corresponds to the tuned-circuit approximation in Prob. 6.1-2.

6.1-11 Do Prob. 6.1-10 with $H_{bp}(f) = \Pi[(f - f_c)/B]e^{-j\omega t_d}$ for $f > 0$, which corresponds to an ideal BPF. *Hint:* See Eq. (9), Sect. 3.4.

6.1-12‡ The bandpass signal in Prob. 6.1-9 has $z(t) = 2u(t)$ and is applied to an ideal BPF with unit gain, zero time delay, and bandwidth B centered at f_c. Use lowpass frequency-domain analysis to obtain an approximation for the bandpass output signal when $B \ll f_0$.

6.1-13‡ Consider a BPF with bandwidth B centered at f_c, unit gain, and parabolic phase shift $\theta(f) = (f - f_c)^2/b$ for $f > 0$. Obtain a quadrature-carrier approximation for the output signal when $|b| \gg (B/2)^2$ and $x_{bp}(t) = z(t) \cos \omega_c t$, where $z(t)$ has a bandlimited lowpass spectrum with $W \leq B/2$.

6.2-1 Let $x(t) = \cos 2\pi f_m t \, u(t)$ with $f_m \ll f_c$. Sketch $x_c(t)$ and indicate the envelope when the modulation is AM with $\mu < 1$, AM with $\mu > 1$, and DSB.

6.2-2 Do Prob. 6.2-1 for $x(t) = 0.5u(t) - 1.5u(t - T)$ with $T \gg 1/f_c$.

6.2-3 From Fig. 6.2-2b, devise a formula for calculating μ using measurements of an AM envelope. What value does your formula predict if $\mu > 1$?

6.2-4 Redo the calculations in Example 6.2-1 with a radio transmitter rated for $S_T \leq 600$ W and $A_{max}^2 \leq 3200$ W.

6.2-5 Calculate the peak envelope power of an AM wave with 100% tone modulation and $S_T = 6$ kW.

6.2-6 Consider a radio transmitter rated for 4-kW peak envelope power. Find the maximum allowable value of μ for AM with tone modulation and $S_T = 1$ kW.

6.2-7 The multitone modulating signal $x(t) = 2K (\cos 10\pi t + 4 \cos 30\pi t)$ is the input to an AM transmitter with $\mu = 1$ and $f_c = 1000$. Find K so that $x(t)$ is properly normalized, draw the positive-frequency line spectrum of the modulated wave, and calculate the upper bound on $2P_{sb}/S_T$.

6.2-8 Do Prob. 6.2-7 with $x(t) = 4K(1 + \cos 10\pi t) \cos 30\pi t$.

6.2-9 Let $x(t)$ be a stationary gaussian random signal with $\bar{x} = 0$ and $\overline{x^2} = S_x$. Since the normalization in Eq. (1) does not hold, find the condition on S_x such that an AM wave with $\mu = 1$ will be overmodulated no more than 10% of the time. Then compute the resulting upper bound on $2P_{sb}/S_T$.

6.2-10 Let $x_c(t) = A_c[1 + \mu x(t)] \cos (\omega_c t + \Phi)$ where $x(t)$ is an ergodic random signal with zero mean and Φ is an independent random phase angle uniformly distributed over 2π radians. Show that $E[x_c^2(t)]$ yields Eq. (9).

6.3-1 Let $v(t) = x(t)$ in Fig. P6.3-1. Show that an appropriate choice of the gain K produces DSB modulation without output filtering.

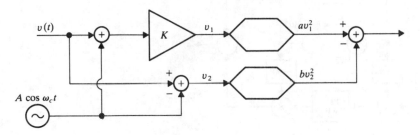

Figure P6.3-1

6.3-2 Find K and $v(t)$ so that the system in Fig. P6.3-1 produces AM without output filtering.

6.3-3 Suppose the nonlinear elements in Fig. P6.3-1 have a *half square-law characteristic*, so the square-law relationship requires $v_1 > 0$ and $v_2 > 0$. Show that an appropriate choice of K and $v(t)$ produces AM without output filtering, providing that $\mu < 1$.

6.3-4 Design in block-diagram form a DSB modulator using a nonlinear element with $v_{out} = a_1 v_{in} + a_3 v_{in}^3$. What's the condition on f_c in terms of W?

6.3-5 Design in block-diagram form an AM modulator using a nonlinear element with $v_{out} = a_1 v_{in} + a_3 v_{in}^3$ and a frequency doubler. What's the condition on f_c in terms of W?

6.3-6 Find the output signal in Fig. 6.3-5 when the AM modulators are *unbalanced*, so one nonlinear element has $v_{out} = a_1 v_{in} + a_2 v_{in}^2 + a_3 v_{in}^3$ while the other has $v_{out} = b_1 v_{in} + b_2 v_{in}^2 + b_3 v_{in}^3$.

6.4-1 Derive Eq. (4) from $y_{\ell p}(t)$.

6.4-2 Take the transform of Eq. (4) to obtain the SSB spectrum

$$X_c(f) = \tfrac{1}{4} A_c\{[1 \pm \text{sgn} (f - f_c)]X(f - f_c) + [1 \mp \text{sgn} (f + f_c)]X(f + f_c)\}.$$

6.4-3 Confirm that the expression for $X_c(f)$ in Prob. 6.4-2 agrees with Figs. 6.4-1b and 6.4-1c.

6.4-4 Find the SSB envelope when $x(t) = \cos \omega_m t - \tfrac{1}{3} \cos 3\omega_m t$, which approximates a square wave. Sketch $A(t)$ taking $A_c = 2$ and compare with $x(t)$.

6.4-5 When $x(t) = \text{sinc } 2Wt$, show that the SSB envelope is $A(t) = \tfrac{1}{2} A_c |\text{sinc } Wt|$. Then sketch $A(t)$ taking $A_c = 2$ and compare with $x(t)$. *Hint:* See Prob. 3.4-14 and use the fact that $\sin^2 \theta = \tfrac{1}{2}(1 - \cos 2\theta)$.

6.4-6 The system in Fig. 6.4-5 produces USSB with $f_c = f_1 + f_2$ when the lower cutoff frequency of the first BPF equals f_1 and the lower cutoff frequency of the second BPF equals f_2. Demonstrate the system's operation by taking $X(f)$ as in Fig. 6.4-4a and sketching spectra at appropriate points. How should the system be modified to produce LSSB?

6.4-7 Suppose the system in Fig. 6.4-5 is designed for USSB as described in Prob. 6.4-6. Let $x(t)$ be a typical voice signal, so $X(f)$ has negligible content outside $200 < |f| < 3200$ Hz. Sketch the spectra at appropriate points to find the maximum permitted value of f_c when the transition regions of the BPFs must satisfy $2\beta \geq 0.01 f_{co}$.

6.4-8 Suppose the carrier phase shift in Fig. 6.4-6 is actually $-90° + \delta$, where δ is a small angular error. Obtain approximate expressions for $x_c(t)$ and $A(t)$ at the output.

6.4-9 Obtain an approximate expression for $x_c(t)$ at the output in Fig. 6.4-6 when $x(t) = \cos \omega_m t$ and the quadrature phase shifter has $|H_Q(f_m)| = 1 - \epsilon$ and $\arg H_Q(f_m) = -90° + \delta$, where ϵ and δ are small errors. Write your answer as a sum of two sinusoids.

6.4-10 *Weaver's SSB modulator* is diagrammed in Fig. P6.4-10. Analyze its operation taking $x(t) = \cos 2\pi f_m t$ with $f_m < W$.

Figure P6.4-10

6.4-11 Obtain an expression for VSB with tone modulation taking $f_m < \beta$ so the VSB filter has $H(f_c \pm f_m) = 0.5 \pm a$. Then show that $x_c(t)$ reduces to DSB when $a = 0$ or SSB when $a = \pm 0.5$.

6.4-12‡ Apply the equivalent lowpass analysis method to derive Eq. (9a).

6.4-13 Consider VSB + C with tone modulation at $f_m > \beta$, so Eq. (11) becomes

$$x_c(t) = A_c \left[\cos \omega_c t + (\mu/2) \cos (\omega_c + \omega_m)t \right]$$

(*a*) Construct the phasor diagram and find $A(t)$. (*b*) Assuming $\mu \ll 1$, obtain the condition on P_{sb}/S_T such that the second-harmonic envelope distortion is less than 5%.

6.5-1 Given a 60-MHz bandpass amplifier, design a television transponder that receives a signal on Channel 9 (187.25 MHz) and transmits it on Channel 4 (67.25 MHz). Use only one oscillator.

6.5-2 Do Prob. 6.5-1 with the received signal on Channel 2 (55.25 MHz) and the transmitted signal on Channel 7 (175.25 MHz).

6.5-3 The system in Fig. 6.4-5 becomes a *scrambler* when the first BPF passes only the upper side-band, the second oscillator frequency is $f_2 = f_1 + W$, and the second BPF is replaced by an LPF with $B = W$. Sketch the output spectrum taking $X(f)$ as in Fig. 6.4-4a, and explain why this output would be unintelligible when $x(t)$ is a voice signal. How can the output signal be unscrambled?

6.5-4 Figure P6.5-4 is a *regenerative frequency divider* constructed using a frequency multiplier. Assume a sinusoidal input at frequency f_0 and show that an appropriate BPF yields the output frequency f_0/M, where M is an integer. Then obtain an upper bound on $f_u - f_\ell$ for successful operation at any input frequency in the range $f_\ell \le f_0 \le f_u$.

Figure P6.5-4

6.5-5 Take $x_c(t)$ as in Eq. (2) and find the output of a synchronous detector whose local oscillator produces $2 \cos (\omega_c t + \phi)$, where ϕ is a constant phase error. Then write separate answers for AM, DSB, SSB, and VSB by appropriate substitution of the modulation parameters.

6.5-6 Suppose the diode in Fig. 6.5-6 is such that $v = v_{in}^2$ in absence of filtering. Find and discuss the resulting demodulated signal.

6.5-7 Sketch a half-rectified AM wave having tone modulation with $\mu A_m = 1$ and $f_m = W$. Use your sketch to determine upper and lower limits on the time constant $R_1 C_1$ of the envelope detector in Fig. 6.5-6. From these limits find the minimum practical value of f_c/W.

6.5-8 Take $x_c(t)$ as given in Eq. (2) and find the output of the envelope reconstruction system in Fig. 6.5-7 when the LO output is $A_{LO} \cos [\omega_c t + \theta(t)]$, where $\theta(t)$ accounts for any phase or frequency error. Simplify your result when $A_{LO} \gg A_c$.

SEVEN

EXPONENTIAL CW MODULATION

Two properties of *linear* CW modulation bear repetition at the outset of this chapter: the modulated spectrum is basically the translated message spectrum, and the transmission bandwidth never exceeds twice the message bandwidth. A third property, derived in Chap. 9, is that the destination signal-to-noise ratio $(S/N)_D$ is no better than baseband transmission and can be improved only by increasing the transmitted power. *Exponential* modulation differs on all three counts.

In contrast to linear modulation, exponential modulation is a *nonlinear* process; therefore, it should come as no surprise that the modulated spectrum is not related in a simple fashion to the message spectrum. Moreover, it turns out that the transmission bandwidth is usually much greater than twice the message bandwidth. Compensating for the bandwidth liability is the fact that exponential modulation can provide increased signal-to-noise ratios without increased transmitted power. Exponential modulation thus allows you to *trade bandwidth for power* in the design of a communication system.

We begin our study of exponential modulation by defining the two basic types, *phase modulation* (PM) and *frequency modulation* (FM). We'll examine the signals and spectra, investigate the transmission bandwidth and distortion problem, and describe typical hardware for generation and detection. The analysis of interference at the end of the chapter brings out the value of FM for radio broadcasting and sets the stage for our consideration of noise in Chap. 9.

7.1 PHASE AND FREQUENCY MODULATION

This section introduces the concepts of instantaneous phase and frequency for the definition of PM and FM signals. Then, since the nonlinear nature of exponential modulation precludes spectral analysis in general terms, we must work instead with the spectra resulting from particular cases such as narrowband modulation and tone modulation.

PM and FM Signals

Consider a CW signal with constant envelope but time-varying phase, so

$$x_c(t) = A_c \cos \left[\omega_c t + \phi(t) \right] \tag{1}$$

Upon defining the *total instantaneous angle*

$$\theta_c(t) \triangleq \omega_c t + \phi(t)$$

we can express $x_c(t)$ as

$$x_c(t) = A_c \cos \theta_c(t) = A_c \operatorname{Re} \left[e^{j\theta_c(t)} \right]$$

Hence, if $\theta_c(t)$ contains the message information $x(t)$, we have a process that may be termed either *angle* modulation or *exponential* modulation. We'll use the latter name because it emphasizes the nonlinear relationship between $x_c(t)$ and $x(t)$.

As to the specific dependence of $\theta_c(t)$ on $x(t)$, *phase modulation* (PM) is defined by

$$\phi(t) \triangleq \phi_\Delta x(t) \qquad \phi_\Delta \leq 180° \tag{2}$$

so that

$$x_c(t) = A_c \cos \left[\omega_c t + \phi_\Delta x(t) \right] \tag{3}$$

These equations state that the instantaneous phase varies directly with the modulating signal. The constant ϕ_Δ represents the *maximum phase shift* produced by $x(t)$, since we're still keeping our normalization convention $|x(t)| \leq 1$. The upper bound $\phi_\Delta \leq 180°$ (or π radians) limits $\phi(t)$ to the range $\pm 180°$ and prevents phase ambiguities—after all, there's no physical distinction between angles of $+270$ and $-90°$, for instance. The bound on ϕ_Δ is analogous to the restriction $\mu \leq 1$ in AM, and ϕ_Δ can justly be called the *phase modulation index*, or the *phase deviation*.

The rotating-phasor diagram in Fig. 7.1-1 helps interpret phase modulation and leads to the definition of frequency modulation. The total angle $\theta_c(t)$ consists of the constant rotational term $\omega_c t$ plus $\phi(t)$, which corresponds to angular shifts relative to the dashed line. Consequently, the phasor's instantaneous rate of rotation in cycles per second will be

$$f(t) \triangleq \frac{1}{2\pi} \theta_c(t) = f_c + \frac{1}{2\pi} \dot{\phi}(t) \tag{4}$$

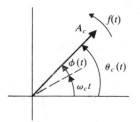

Figure 7.1-1 Rotating-phasor representation of exponential modulation.

in which the dot notation stands for the time derivative, that is, $\dot{\phi}(t) = d\phi(t)/dt$, etc. We call $f(t)$ the *instantaneous frequency* of $x_c(t)$. Although $f(t)$ is measured in *hertz*, it should *not* be equated with spectral frequency. Spectral frequency f is the independent variable of the frequency domain, whereas instantaneous frequency $f(t)$ is a time-dependent property of waveforms with exponential modulation.

In the case of *frequency modulation* (FM), the instantaneous frequency of the modulated wave is defined to be

$$f(t) \triangleq f_c + f_\Delta x(t) \qquad f_\Delta < f_c \tag{5}$$

so $f(t)$ varies in proportion with the modulating signal. The proportionality constant f_Δ, called the *frequency deviation*, represents the maximum shift of $f(t)$ relative to the carrier frequency f_c. The upper bound $f_\Delta < f_c$ simply ensures that $f(t) > 0$. However, we usually want $f_\Delta \ll f_c$ in order to preserve the bandpass nature of $x_c(t)$.

Equations (4) and (5) show that an FM wave has $\dot{\phi}(t) = 2\pi f_\Delta x(t)$, and integration yields the phase modulation

$$\phi(t) = 2\pi f_\Delta \int_{t_0}^{t} x(\lambda) \, d\lambda + \phi(t_0) \qquad t \geq t_0 \tag{6a}$$

If t_0 is taken such that $\phi(t_0) = 0$, we can drop the lower limit of integration and use the informal expression

$$\phi(t) = 2\pi f_\Delta \int^{t} x(\lambda) \, d\lambda \tag{6b}$$

The FM waveform is then written as

$$x_c(t) = A_c \cos \left[\omega_c t + 2\pi f_\Delta \int^{t} x(\lambda) \, d\lambda \right] \tag{7}$$

But it must be assumed that the message has no *DC component* so the above integrals do not diverge when $t \to \infty$. Physically, a DC term in $x(t)$ would produce a constant carrier-frequency shift equal to $f_\Delta \langle x(t) \rangle$.

A comparison of Eqs. (3) and (7) implies little difference between PM and FM, the essential distinction being the integration of the message in FM. Moreover, nomenclature notwithstanding, both FM and PM have both time-varying phase and frequency, as underscored by Table 7.1-1. These relations clearly indicate that, with the help of integrating and differentiating networks, a phase

Table 7.1-1 Comparison of PM and FM

	Instantaneous phase $\phi(t)$	Instantaneous frequency $f(t)$
PM	$\phi_\Delta x(t)$	$f_c + \dfrac{1}{2\pi}\phi_\Delta \dot{x}(t)$
FM	$2\pi f_\Delta \displaystyle\int^t x(\lambda)\,d\lambda$	$f_c + f_\Delta x(t)$

modulator can produce frequency modulation and vice versa. In fact, in the case of tone modulation it's nearly impossible visually to distinguish FM and PM waves.

On the other hand, a comparison of exponential modulation with linear modulation reveals some pronounced differences. For one thing, the amplitude of an FM or PM wave is *constant*; therefore, regardless of the message $x(t)$, the average transmitted power is

$$S_T = \tfrac{1}{2}A_c^2 \tag{8}$$

For another, the zero crossings of an exponentially modulated wave are not periodic, whereas they are always periodic in linear modulation. Indeed, because of the constant-amplitude property of FM and PM, it can be said that the message resides in the *zero crossings alone*, providing the carrier frequency is large. Finally, since exponential modulation is a nonlinear process, the modulated wave does not look at all like the message waveform.

Figure 7.1-2 illustrates some of these points by showing typical AM, FM, and PM waves. As a mental exercise you may wish to check these waveforms against

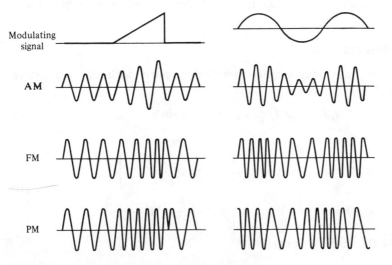

Figure 7.1-2 Illustrative AM, FM, and PM waveforms.

the corresponding modulating signals. For FM and PM this is most easily done by considering the instantaneous frequency rather than by substituting $x(t)$ in Eqs. (3) and (7).

Despite the many similarities of PM and FM, frequency modulation turns out to have superior noise-reduction properties and thus will receive most of our attention. To gain a qualitative appreciation of FM noise reduction, suppose a demodulator simply extracts the instantaneous frequency $f(t) = f_c + f_\Delta x(t)$ from $x_c(t)$. The demodulated output is then proportional to the frequency deviation f_Δ, which can be increased without increasing the transmitted power S_T. If the noise level remains constant, increased signal output is equivalent to reduced noise. However, noise reduction does require increased transmission bandwidth to accommodate large frequency deviations.

Ironically, frequency modulation was first conceived as a means of *bandwidth reduction*, the argument going somewhat as follows: If, instead of modulating the carrier amplitude, we modulate the frequency by swinging it over a range of, say, ± 50 Hz, then the transmission bandwidth will be 100 Hz regardless of the message bandwidth. As we'll soon see, this argument has a serious flaw, for it ignores the distinction between *instantaneous* and *spectral frequency*. Carson (1922) recognized the fallacy of the bandwidth-reduction notion and cleared the air on that score. Unfortunately, he and many others also felt that exponential modulation had no advantages over linear modulation with respect to noise. It took some time to overcome this belief but, thanks to Armstrong (1936), the merits of exponential modulation were finally appreciated. Before we can understand them quantitatively, we must address the problem of spectral analysis.

Exercise 7.1-1 Suppose FM had been defined in direct analogy to AM by writing $x_c(t) = A_c \cos \omega_c(t)t$ with $\omega_c(t) = \omega_c[1 + \mu x(t)]$. Demonstrate the physical impossibility of this definition by finding $f(t)$ when $x(t) = \cos \omega_m t$.

Narrowband PM and FM

Our spectral analysis of exponential modulation starts with the quadrature-carrier version of Eq. (1), namely

$$x_c(t) = x_{ci}(t) \cos \omega_c t - x_{cq}(t) \sin \omega_c t \qquad (9)$$

where

$$x_{ci}(t) = A_c \cos \phi(t) = A_c \left[1 - \frac{1}{2!} \phi^2(t) + \cdots \right]$$

$$x_{cq}(t) = A_c \sin \phi(t) = A_c \left[\phi(t) - \frac{1}{3!} \phi^3(t) + \cdots \right] \qquad (10)$$

Now we impose the simplifying condition

$$|\phi(t)| \ll 1 \text{ rad} \qquad (11a)$$

so that

$$x_{ci}(t) \approx A_c \qquad x_{cq}(t) \approx A_c \, \phi(t) \qquad (11b)$$

Then it becomes an easy task to find the spectrum $X_c(f)$ of the modulated wave in terms of an arbitrary message spectrum $X(f)$.

Specifically, the transforms of Eqs. (9) and (11b) yield

$$X_c(f) = \frac{1}{2} A_c \, \delta(f - f_c) + \frac{j}{2} A_c \, \Phi(f - f_c) \qquad f > 0 \qquad (12a)$$

in which

$$\Phi(f) = \mathscr{F}[\phi(t)] = \begin{cases} \phi_\Delta X(f) & \text{PM} \\ -jf_\Delta X(f)/f & \text{FM} \end{cases} \qquad (12b)$$

The FM expression comes from the integration theorem applied to $\phi(t)$ in Eq. (6).

Based on Eq. (12), we conclude that if $x(t)$ has message bandwidth $W \ll f_c$, then $x_c(t)$ will be a bandpass signal with bandwidth $2W$. But this conclusion holds only under the conditions of Eq. (11). For larger values of $|\phi(t)|$, the terms $\phi^2(t)$, $\phi^3(t)$, ... cannot be ignored in Eq. (10) and will *increase* the bandwidth of $x_c(t)$. Hence, Eqs. (11) and (12) describe the special case of *narrowband* phase or frequency modulation (NBPM or NBFM).

Example 7.1-1 An informative illustration of Eq. (12) is provided by taking $x(t) = \text{sinc } 2Wt$, so $X(f) = (1/2W)\Pi(f/2W)$. The resulting NBPM and NBFM spectra are depicted in Fig. 7.1-3. Both spectra have carrier-frequency impulses and bandwidth $2W$. However, the lower sideband in NBFM is 180° out of phase (represented by the negative sign), whereas both NBPM sidebands have a 90° phase shift (represented by j). Except for the phase shift, the NBPM spectrum looks just like an AM spectrum with the same modulating signal.

(a)

(b)

Figure 7.1-3 Narrowband modulated spectra with $x(t) = \text{sinc } 2Wt$. (a) PM; (b) FM.

Exercise 7.1-2 Use the second-order approximations $x_{ci}(t) \approx A_c[1 - \frac{1}{2}\phi^2(t)]$ and $x_{cq}(t) \approx A_c \phi(t)$ to find and sketch the components of the PM spectrum when $x(t) = \text{sinc } 2Wt$.

Tone Modulation

The study of FM and PM with tone modulation can be carried out jointly by the simple expedient of allowing a 90° difference in the modulating tones. For if we take

$$x(t) = \begin{cases} A_m \sin \omega_m t & \text{PM} \\ A_m \cos \omega_m t & \text{FM} \end{cases}$$

then Eqs. (2) and (6) both give

$$\phi(t) = \beta \sin \omega_m t \tag{13a}$$

where

$$\beta \triangleq \begin{cases} \phi_\Delta A_m & \text{PM} \\ (A_m/f_m)f_\Delta & \text{FM} \end{cases} \tag{13b}$$

The parameter β serves as the *modulation index* for PM or FM with tone modulation. This parameter equals the *maximum phase deviation* and is proportional to the tone amplitude A_m in both cases. Note, however, that β for FM is inversely proportional to the tone frequency f_m since the integration of $\cos \omega_m t$ yields $(\sin \omega_m t)/\omega_m$.

Narrowband tone modulation requires $\beta \ll 1$, and Eq. (9) simplifies to

$$x_c(t) \approx A_c \cos \omega_c t - A_c \beta \sin \omega_m t \sin \omega_c t$$

$$\approx A_c \cos \omega_c t - \frac{A_c \beta}{2} \cos (\omega_c - \omega_m)t + \frac{A_c \beta}{2} \cos (\omega_c + \omega_m)t \tag{14}$$

The corresponding line spectrum and phasor diagram are shown in Fig. 7.1-4. Observe how the phase reversal of the lower sideband line produces a component perpendicular or *quadrature* to the carrier phasor. This quadrature relationship is precisely what's needed to create phase or frequency modulation instead of amplitude modulation.

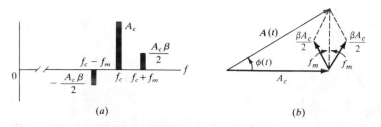

Figure 7.1-4 NBFM with tone modulation. (a) Line spectrum; (b) phasor diagram.

Now, to determine the line spectrum with an arbitrary value of the modulation index, we drop the narrowband approximation and write

$$x_c(t) = A_c[\cos \phi(t) \cos \omega_c t - \sin \phi(t) \sin \omega_c t]$$

$$= A_c[\cos (\beta \sin \omega_m t) \cos \omega_c t - \sin (\beta \sin \omega_m t) \sin \omega_c t] \qquad (15)$$

Then we use the fact that, even though $x_c(t)$ is not necessarily periodic, the terms $\cos (\beta \sin \omega_m t)$ and $\sin (\beta \sin \omega_m t)$ are periodic and each can be expanded as a trigonometric Fourier series with $f_0 = f_m$. Indeed, a well-known result from applied mathematics states that

$$\cos (\beta \sin \omega_m t) = J_0(\beta) + \sum_{\substack{n=1 \\ n \text{ even}}}^{\infty} 2J_n(\beta) \cos n\omega_m t$$

$$\sin (\beta \sin \omega_m t) = \sum_{\substack{n=1 \\ n \text{ odd}}}^{\infty} 2J_n(\beta) \sin n\omega_m t \qquad (16)$$

where n is positive and

$$J_n(\beta) \triangleq \frac{1}{2\pi} \int_{-\pi}^{\pi} e^{j(\beta \sin \lambda - n\lambda)} \, d\lambda \qquad (17)$$

The coefficients $J_n(\beta)$ are *Bessel functions* of the first kind, of order n and argument β. With the aid of Eq. (17), you should encounter little difficulty in deriving the trigonometric expansions given in Eq. (16).

Substituting Eq. (16) into Eq. (15) and expanding products of sines and cosines finally yields

$$x_c(t) = A_c J_0(\beta) \cos \omega_c t$$

$$+ \sum_{\substack{n=1 \\ n \text{ odd}}}^{\infty} A_c J_n(\beta)[\cos (\omega_c + n\omega_m)t - \cos (\omega_c - n\omega_m)t]$$

$$+ \sum_{\substack{n=1 \\ n \text{ even}}}^{\infty} A_c J_n(\beta)[\cos (\omega_c + n\omega_m)t + \cos (\omega_c - n\omega_m)t] \qquad (18a)$$

Alternatively, taking advantage of the property that $J_{-n}(\beta) = (-1)^2 J_n(\beta)$, we get [*should be n, not 2*] the more compact but less informative expression

$$x_c(t) = A_c \sum_{n=-\infty}^{\infty} J_n(\beta) \cos (\omega_c + n\omega_m)t \qquad (18b)$$

In either form, Eq. (18) is the mathematical representation for a constant-amplitude wave whose instantaneous frequency varies sinusoidally. A phasor interpretation, to be given shortly, will shed more light on the matter.

Examining Eq. (18), we see that the FM spectrum consists of a carrier-frequency line plus an *infinite* number of sideband lines at frequencies $f_c \pm nf_m$. As illustrated in the typical spectrum of Fig. 7.1-5, all lines are equally spaced by the modulating frequency and the odd-order lower sideband lines are reversed in phase or inverted relative to the unmodulated carrier. Since we're dealing with a

Figure 7.1-5 Line spectrum of FM with tone modulation.

positive-frequency line spectrum, any apparent negative frequencies ($f_c + nf_m < 0$) must be folded back to the positive values $| f_c + nf_m |$. But such components will be negligible as long as $\beta f_m \ll f_c$. In general, the relative amplitude of a line at $f_c + nf_m$ is given by $J_n(\beta)$, so before we can say more about the spectrum, we must examine the behavior of Bessel functions.

Figure 7.1-6a shows a few Bessel functions of various order plotted versus the argument β. Several important properties emerge from this plot.

1. The relative amplitude of the carrier line $J_0(\beta)$ varies with the modulation index and hence depends on the modulating signal. Thus, in contrast to linear modulation, the carrier-frequency component of an FM wave "contains" part of the message information. Nonetheless, there will be spectra in which the carrier line has zero amplitude since $J_0(\beta) = 0$ when $\beta = 2.4$, 5.5, etc.
2. The number of sideband lines having appreciable relative amplitude also depends on β. With $\beta \ll 1$ only J_0 and J_1 are significant, so the spectrum will consist of carrier and two sideband lines as in Fig. 7.1-4a. But if $\beta \gg 1$, there will be many sideband lines, giving a spectrum quite unlike linear modulation.
3. Large β implies a large bandwidth to accommodate the extensive sideband structure, agreeing with the physical interpretation of large frequency deviation.

Table 7.1-2 Selected values of $J_n(\beta)$

n	$J_n(0.1)$	$J_n(0.2)$	$J_n(0.5)$	$J_n(1.0)$	$J_n(2.0)$	$J_n(5.0)$	$J_n(10)$	n
0	1.00	0.99	0.94	0.77	0.22	−0.18	−0.25	0
1	0.05	0.10	0.24	0.44	0.58	−0.33	0.04	1
2			0.03	0.11	0.35	0.05	0.25	2
3				0.02	0.13	0.36	0.06	3
4					0.03	0.39	−0.22	4
5						0.26	−0.23	5
6						0.13	−0.01	6
7						0.05	0.22	7
8						0.02	0.32	8
9							0.29	9
10							0.21	10
11							0.12	11
12							0.06	12
13							0.03	13
14							0.01	14

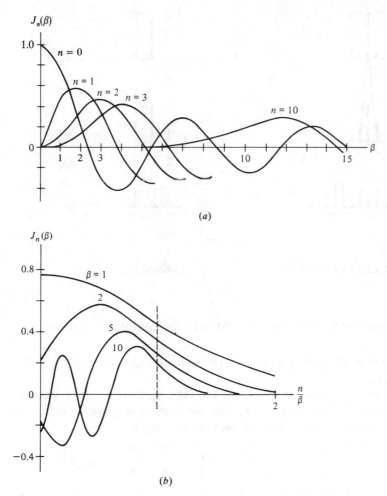

Figure 7.1-6 Plots of Bessel functions. (a) Fixed order n, variable argument β; (b) fixed argument β, variable order n.

Some of the above points are better illustrated by Fig. 7.1-6b, which gives $J_n(\beta)$ as a function of n/β for various *fixed* values of β. These curves represent the "envelope" of the sideband lines if we multiply the horizontal axis by βf_m to obtain the line position $n f_m$ relative to f_c. Observe in particular that all $J_n(\beta)$ decay monotonically for $n/\beta > 1$ and that $|J_n(\beta)| \ll 1$ if $|n/\beta| \gg 1$. Table 7.1-2 lists selected values of $J_n(\beta)$, rounded off at the second decimal place. Blanks in the table correspond to $|J_n(\beta)| < 0.01$. More extensive tabulations are given in Jahnke and Emde (1945).

Line spectra drawn from the data in Table 7.1-2 are shown in Fig. 7.1-7, omitting the sign inversions. Part a of the figure has β increasing with f_m held fixed, and applies to FM and PM. Part b applies only to FM and illustrates the effect of increasing β by *decreasing* f_m with $A_m f_\Delta$ held fixed. The dashed lines help

$\beta = 0.2$

$\beta = 1$

$\beta = 5$

$2\beta f_m$

$\beta = 10$

$2\beta f_m$

f_c

f_c

(a)   (b)

Figure 7.1-7 Tone-modulated line spectra. (a) FM or PM with f_m fixed; (b) FM with $A_m\, f_\Delta$ fixed.

bring out the concentration of significant sideband lines within the range $f_c \pm \beta f_m$ as β becomes large.

For the phasor interpretation of $x_c(t)$ in Eq. (18), we first return to the narrowband approximation and Fig. 7.1-4. The envelope and phase constructed from the carrier and first pair of sideband lines are seen to be

$$A(t) \approx \sqrt{A_c^2 + \left(2\,\frac{\beta}{2}\, A_c \sin \omega_m t\right)^2} \approx A_c\left[1 + \frac{\beta^2}{4} - \frac{\beta^2}{4}\cos 2\omega_m t\right]$$

$$\phi(t) \approx \arctan\left[\frac{2(\beta/2)A_c \sin \omega_m t}{A_c}\right] \approx \beta \sin \omega_m t$$

Thus the phase variation is approximately as desired, but there is an additional *amplitude* variation at twice the tone frequency. To cancel out the latter we should include the second-order pair of sideband lines that rotate at $\pm 2f_m$ relative to the carrier and whose resultant is collinear with the carrier. While the second-order pair virtually wipes out the undesired amplitude modulation, it also distorts $\phi(t)$. The phase distortion is then corrected by adding the third-order pair, which again introduces amplitude modulation, and so on ad infinitum.

When all spectral lines are included, the odd-order pairs have a resultant in quadrature with the carrier that provides the desired frequency modulation plus unwanted amplitude modulation. The resultant of the even-order pairs, being collinear with the carrier, corrects for the amplitude variations. The net effect is then as illustrated in Fig. 7.1-8. The tip of the resultant sweeps through a circular arc, reflecting the constant amplitude A_c.

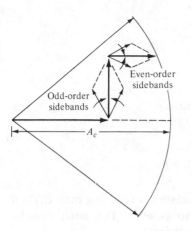

Figure 7.1-8 FM phasor diagram for arbitrary β.

Exercise 7.1-3 Consider tone-modulated FM with $A_c = 100$, $A_m f_\Delta = 8$ kHz, and $f_m = 4$ kHz. Draw the line spectrum for $f_c = 30$ kHz and for $f_c = 11$ kHz.

Multitone and Periodic Modulation★

The Fourier series technique used to arrive at Eq. (18) also can be applied to the case of FM with multitone modulation. For instance, suppose that $x(t) = A_1 \cos \omega_1 t + A_2 \cos \omega_2 t$, where f_1 and f_2 are not harmonically related. The modulated wave is first written as

$$x_c(t) = A_c[(\cos \alpha_1 \cos \alpha_2 - \sin \alpha_1 \sin \alpha_2) \cos \omega_c t$$

$$- (\sin \alpha_1 \cos \alpha_2 + \cos \alpha_1 \sin \alpha_2) \sin \omega_c t]$$

where $\alpha_1 = \beta_1 \sin \omega_1 t$, $\beta_1 = A_1 f_\Delta / f_1$, etc. Terms of the form $\cos \alpha_1$, $\sin \alpha_1$, etc., are then expanded according to Eq. (16), and after some routine manipulations we arrive at the compact result

$$x_c(t) = A_c \sum_{n=-\infty}^{\infty} \sum_{m=-\infty}^{\infty} J_n(\beta_1) J_m(\beta_2) \cos (\omega_c + n\omega_1 + m\omega_2)t \qquad (19)$$

This technique can be extended to include three or more nonharmonic tones; the procedure is straightforward but tedious.

To interpret Eq. (19) in the frequency domain, the spectral lines can be divided into four categories: (1) the carrier line of amplitude $A_c J_0(\beta_1) J_0(\beta_2)$; (2) sideband lines at $f_c \pm n f_1$ due to one tone alone; (3) sideband lines at $f_c \pm m f_2$ due to the other tone alone; and (4) sideband lines at $f_c \pm n f_1 \pm m f_2$ which appear to be beat-frequency modulation at the sum and difference frequencies of the modulating tones and their harmonics. (This last category would not occur in linear modulation where simple superposition of sideband lines is the rule.) A double-tone FM spectrum showing the various types of spectral lines is given in Fig. 7.1-9 for $f_1 \ll f_2$ and $\beta_1 > \beta_2$. Under these conditions there exists the curious property that each sideband line at $f_c \pm m f_2$ looks like another FM carrier with tone modulation of frequency f_1.

Figure 7.1-9 Double-tone FM line spectrum with $f_1 \ll f_2$ and $\beta_1 > \beta_2$.

When the tone frequencies are harmonically related—meaning that $x(t)$ is a *periodic* waveform—then $\phi(t)$ is periodic and so is $e^{j\phi(t)}$. The latter can be expanded in an exponential Fourier series with coefficients

$$c_n = \frac{1}{T_0} \int_{T_0} \exp j[\phi(t) - n\omega_0 t] \, dt \qquad (20a)$$

Therefore

$$x_c(t) = A_c \, \text{Re} \left[\sum_{n=-\infty}^{\infty} c_n e^{j(\omega_c + n\omega_0)t} \right] \qquad (20b)$$

and $A_c |c_n|$ equals the magnitude of the spectral line at $f = f_c + nf_0$.

Example 7.1-2 FM with pulse-train modulation Let $x(t)$ be a unit-amplitude rectangular pulse train with period T_0, pulse duration τ, and duty cycle $d = \tau/T_0$. After removing the DC component $\langle x(t) \rangle = d$, the instantaneous frequency of the resulting FM wave is as shown in Fig. 7.1-10a. The time origin is chosen such that $\phi(t)$ plotted in Fig. 7.1-10b has a peak value $\phi_\Delta = 2\pi f_\Delta \tau$

Figure 7.1-10 FM with pulse-train modulation. (a) Instantaneous frequency; (b) phase; (c) line spectrum for $d = 1/4$.

at $t = 0$. We've also taken the constant of integration such that $\phi(t) \geq 0$. Thus

$$\phi(t) = \begin{cases} \phi_\Delta(1 + t/\tau) & -\tau < t < 0 \\ \phi_\Delta[1 - t/(T_0 - \tau)] & 0 < t < T_0 - \tau \end{cases}$$

which defines the range of integration for Eq. (20a).

The evaluation of c_n is a nontrivial exercise involving exponential integrals and trigonometric relations. The final result can be written as

$$c_n = \left[\frac{\sin \pi(\beta - n)d}{\pi(\beta - n)} + \frac{(1 - d) \sin \pi(\beta - n)d}{\pi(\beta - n)d + \pi n} \right] e^{j\pi(\beta + n)d}$$

$$= \frac{\beta d}{(\beta - n)d + n} \text{ sinc } (\beta - n)d \; e^{j\pi(\beta + n)d}$$

where we've let

$$\beta = f_\Delta T_0 = f_\Delta/f_0$$

which plays a role similar to the modulation index for single-tone modulation.

Figure 7.1-10c plots the magnitude line spectrum for the case of $d = 1/4$, $\beta = 4$, and $A_c = 1$. Note the absence of symmetry here and the peaking around $f = f_c - \frac{1}{4}f_\Delta$ and $f = f_c + \frac{3}{4}f_\Delta$, the two values taken on by the instantaneous frequency. The fact that the spectrum contains other frequencies as well underscores the difference between spectral frequency and instantaneous frequency. The same remarks apply for the *continuous* spectrum of FM with a single modulating pulse—demonstrated by our results in Example 2.4-1.

7.2 TRANSMISSION BANDWIDTH AND DISTORTION

The spectrum of a signal with exponential modulation has infinite extent, in general. Hence, generation and transmission of *pure* FM requires *infinite bandwidth*, whether or not the message is bandlimited. But practical FM systems having finite bandwidth do exist and perform quite well. Their success depends upon the fact that, sufficiently far away from the carrier frequency, the spectral components are quite small and may be discarded. True, omitting any portion of the spectrum will cause *distortion* in the demodulated signal; but the distortion can be minimized by keeping all significant spectral components.

We'll formulate in this section estimates of transmission bandwidth requirements by drawing upon results from the previous section. Then we'll look at distortion produced by linear and nonlinear systems. Topics encountered in passing include the concept of *wideband* FM and that important piece of FM hardware known as a *limiter*. We'll concentrate primarily on FM, but minor modifications make the analyses applicable to PM.

Transmission Bandwidth Estimates

Determination of FM transmission bandwidth boils down to the question: How much of the modulated signal spectrum is *significant*? Of course, significance standards are not absolute, being contingent upon the amount of distortion that can be tolerated in a specific application. However, rule-of-thumb criteria based on studies of tone modulation have met with considerable success and lead to useful approximate relations. Our discussion of FM bandwidth requirements therefore begins with the significant sideband lines for tone modulation.

Figure 7.1-6 indicated that $J_n(\beta)$ falls off rapidly for $|n/\beta| > 1$, particularly if $\beta \gg 1$. Assuming that the modulation index β is large, we can say that $|J_n(\beta)|$ is significant only for $|n| \leq \beta = A_m f_\Delta/f_m$. Therefore, all significant lines are contained in the frequency range $f_c \pm \beta f_m = f_c \pm A_m f_\Delta$, a conclusion agreeing with intuitive reasoning. On the other hand, suppose the modulation index is small; then *all* sideband lines are small compared to the carrier, since $J_0(\beta) \gg J_{n \neq 0}(\beta)$ when $\beta \ll 1$. But we must retain at least the first-order sideband pair, else there would be no frequency modulation at all. Hence, for small β, the significant sideband lines are contained in $f_c \pm f_m$.

To put the above observations on a quantitative footing, all sideband lines having relative amplitude $|J_n(\beta)| > \epsilon$ are *defined* as being significant, where ϵ ranges from 0.01 to 0.1 according to the application. Then, if $|J_M(\beta)| > \epsilon$ and $|J_{M+1}(\beta)| < \epsilon$, there are M significant sideband *pairs* and $2M + 1$ significant lines all told. The bandwidth is thus written as

$$B = 2M(\beta)f_m \qquad M(\beta) \geq 1 \tag{1}$$

since the lines are spaced by f_m and M depends on the modulation index β. The condition $M(\beta) \geq 1$ has been included in Eq. (1) to account for the fact that B cannot be less than $2f_m$.

Figure 7.2-1 shows M as a continuous function of β for $\epsilon = 0.01$ and 0.1. Experimental studies indicate that the former is often overly conservative, while the latter may result in small but noticeable distortion. Values of M between these two bounds are acceptable for most purposes and will be used hereafter.

But the bandwidth B is not the transmission bandwidth B_T; rather it's the minimum bandwidth necessary for modulation by a tone of specified amplitude and frequency. To estimate B_T, we should calculate the *maximum* bandwidth required when the tone parameters are constrained by $A_m \leq 1$ and $f_m \leq W$. For this purpose, the dashed line in Fig. 7.2-1 depicts the approximation

$$M(\beta) \approx \beta + 2 \tag{2}$$

which falls midway between the solid lines for $\beta \geq 2$. Inserting Eq. (2) into Eq. (1) gives

$$B \approx 2(\beta + 2)f_m = 2\left(\frac{A_m f_\Delta}{f_m} + 2\right)f_m = 2(A_m f_\Delta + 2f_m)$$

Now, bearing in mind that f_Δ is a property of the modulator, what tone produces the maximum bandwidth? Clearly, it is the *maximum-amplitude maximum-frequency* tone having $A_m = 1$ and $f_m = W$. The worst-case tone-modulation

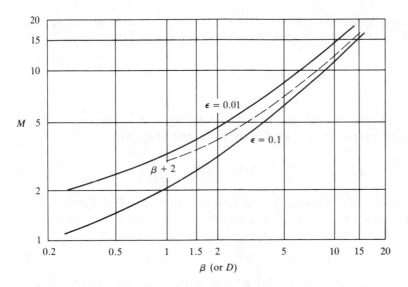

Figure 7.2-1 The number of significant sideband pairs as a function of β (or D).

bandwidth is then

$$B_T \approx 2(f_\Delta + 2W) \qquad \text{if } \beta > 2$$

Note carefully that the corresponding modulation index $\beta = f_\Delta/W$ is not the maximum value of β but rather the value which, combined with the maximum modulating frequency, yields the maximum bandwidth. Any other tone having $A_m < 1$ or $f_m < W$ will require less bandwidth even though β may be larger.

Finally, consider a reasonably smooth but otherwise *arbitrary modulating signal* having the message bandwidth W and satisfying the normalization convention $|x(t)| \leq 1$. We'll estimate B_T directly from the worst-case tone-modulation analysis, assuming that any component in $x(t)$ of smaller amplitude or frequency will require a smaller bandwidth than B_T. Admittedly, this procedure ignores the fact that superposition is not applicable to exponential modulation. However, our investigation of multitone spectra has shown that the beat-frequency sideband pairs are contained primarily within the bandwidth of the dominating tone alone, as illustrated by Fig. 7.1-9.

Therefore, extrapolating tone modulation to an arbitrary modulating signal, we define the *deviation ratio*.

$$D \triangleq \frac{f_\Delta}{W} \tag{3}$$

which equals the maximum deviation divided by the maximum modulating frequency, analogous to the modulation index of worst-case tone modulation. The transmission bandwidth required for $x(t)$ is then

$$B_T = 2M(D)W \tag{4}$$

where D is treated just like β to find $M(D)$, say from Fig. 7.2-1.

Lacking appropriate curves or tables for $M(D)$, there are several approximations to B_T that can be invoked. With extreme values of the deviation ratio we find that

$$B_T = \begin{cases} 2DW = 2f_\Delta & D \gg 1 \\ 2W & D \ll 1 \end{cases}$$

paralleling our results for tone modulation with β very large or very small. Both of these approximations are combined in the convenient relation

$$B_T \approx 2(f_\Delta + W) = 2(D + 1)W \quad \begin{matrix} D \gg 1 \\ D \ll 1 \end{matrix} \tag{5}$$

known as *Carson's rule*. Perversely, the majority of actual FM systems have $2 < D < 10$, for which Carson's rule somewhat underestimates the transmission bandwidth. A better approximation for equipment design is then

$$B_T \approx 2(f_\Delta + 2W) = 2(D + 2)W \quad D > 2 \tag{6}$$

which would be used, for example, to determine the 3-dB bandwidths of FM amplifiers.

Physically, the deviation ratio represents the *maximum phase deviation* of an FM wave under worst-case bandwidth conditions. Our FM bandwidth expressions therefore apply to *phase* modulation if we replace D with the maximum phase deviation ϕ_Δ of the PM wave. Accordingly, the transmission bandwidth for PM with arbitrary $x(t)$ is estimated to be

$$B_T = 2M(\phi_\Delta)W \quad M(\phi_\Delta) \geq 1 \tag{7a}$$

or

$$B_T \approx 2(\phi_\Delta + 1)W \tag{7b}$$

which is the approximation equivalent to Carson's rule. These expressions differ from the FM case in that ϕ_Δ is independent of W.

You should review our various approximations and their conditions of validity. In deference to most of the literature, we'll usually take B_T as given by Carson's rule in Eqs. (5) and (7b). But when the modulating signal has discontinuities — a rectangular pulse train, for instance — the bandwidth estimates become invalid and we must resort to brute-force spectral analysis.

Example 7.2-1 Commercial FM bandwidth Commercial FM broadcast stations in the United States are limited to a maximum frequency deviation of 75 kHz, and modulating frequencies typically cover 30 Hz to 15 kHz. Letting $W = 15$ kHz, the deviation ratio is $D = 75$ kHz/15 kHz $= 5$ and Eq. (6) yields $B_T \approx 2(5 + 2) \times 15$ kHz $= 210$ kHz. High-quality FM radios have bandwidths of at least 200 kHz. Carson's rule in Eq. (5) underestimates the bandwidth, giving $B_T \approx 180$ kHz.

If a single modulating tone has $A_m = 1$ and $f_m = 15$ kHz, then $\beta = 5$, $M(\beta) \approx 7$, and Eq. (1) shows that $B = 210$ kHz. A lower-frequency tone, say

3 kHz, would result in a larger modulation index ($\beta = 25$), a greater number of significant sideband pairs ($M = 27$), but a smaller bandwidth since $B = 2 \times 27 \times 3$ kHz $= 162$ kHz.

Exercise 7.2-1 Calculate B_T/W for $D = 0.3$, 3, and 30 using Eqs. (5) and (6) where applicable.

Wideband FM

Previously, we defined *narrowband* exponential modulation by the property $|\phi(t)| \ll 1$ radian. Now we can say that NBPM has $\phi_\Delta \ll 1$ and NBFM has $D \ll 1$ so $B_T \approx 2W$ in both cases. Going to the opposite extreme, *wideband* FM (WBFM) is characterized by the properties

$$D = f_\Delta/W \gg 1 \qquad B_T \approx 2f_\Delta \gg W \tag{8}$$

Hence, the bandwidth of WBFM is large compared to the message bandwidth W and depends only on f_Δ. Wideband PM does not exist in the same sense because of the phase-shift limitation $\phi_\Delta \leq \pi$ radians.

Although we can't develop a simplified expression for the WBFM waveform, as we did for NBFM, we can approximate the *power spectral density*. The procedure, called the *quasi-static approximation*, stems from the following observation: the instantaneous frequency of an FM wave with $f_\Delta \gg W$ varies so slowly compared to $1/W$ that $x_c(t)$ looks more or less like an ordinary sinusoid at frequency $f(t) = f_c + f_\Delta x(t)$. If $x(t)$ happens to be constant, say $x(t) = X$, then all the signal power S_T would be concentrated at $f_c + f_\Delta X$ in the frequency domain. When $x(t)$ changes with time, S_T must be distributed over frequency in accordance with the relative amount of time that $f(t)$ falls in a given differential band df.

To formulate these notions mathematically, let $p_X(x)$ represent the relative amount of time that $x(t)$ equals some particular value X. In the case of a *random* modulating signal, $p_X(x)$ would be the same as its *probability distribution function* (PDF). But $x(t) = [f(t) - f_c]/f_\Delta$ for either a deterministic or random signal, so the positive-frequency power distribution is proportional to $(1/f_\Delta)p_X[(f - f_c)/f_\Delta]$ if $D \gg 1$. Therefore, we write the two-sided power spectral density for WBFM as

$$G(f) = \frac{S_T}{2f_\Delta}\left[p_X\left(\frac{f - f_c}{f_\Delta}\right) + p_X\left(\frac{f + f_c}{f_\Delta}\right)\right] \tag{9}$$

Peebles (1976, chap. 6) fills in the details of the quasi-static approximation that we've omitted here.

Example 7.2-2 WBFM with tone modulation The spectrum of WBFM with unit-amplitude tone modulation can be obtained from Eq. (9) using $p_X(x) = (\pi\sqrt{1 - x^2})^{-1}$ for $|x| \leq 1$, as determined in Example 4.2-3. Figure 7.2-2 shows $G(f)$ for $f > 0$. The shape of $G(f)$ represents the limiting form of the line spectra in Fig. 7.1-7 when $\beta \to \infty$.

Figure 7.2-2 WBFM power spectrum with tone modulation.

Exercise 7.2-2 Find $G(f)$ when the modulating signal is a triangular wave with peak values $x(t) = \pm 1$ and period $T_0 \gg 1/f_\Delta$.

Linear Distortion

The analysis of distortion produced in an FM or PM wave by a linear network is an exceedingly knotty problem—so much so that several different approaches to it have been devised, none of them easy. Panter (1965) devotes three chapters to the subject and serves as a reference guide. Since we're limited here to a few pages, we can only view the "tip of the iceberg." Nonetheless, we'll gain some valuable insights regarding linear distortion of FM and PM.

Figure 7.2-3 represents an exponentially modulated bandpass signal $x_c(t)$ applied to a linear system with transfer function $H(f)$, producing the output $y_c(t)$. The constant-amplitude property of $x_c(t)$ allows us to write the lowpass equivalent input

$$x_{\ell p}(t) = \tfrac{1}{2} A_c \, e^{j\phi(t)} \tag{10}$$

where $\phi(t)$ contains the message information. In terms of $X_{\ell p}(f)$, the lowpass equivalent output spectrum is

$$Y_{\ell p}(f) = H(f + f_c)u(f + f_c)X_{\ell p}(f) \tag{11}$$

Lowpass-to-bandpass transformation finally gives the output as

$$y_c(t) = 2 \, \text{Re} \, [y_{\ell p}(t)e^{j\omega_c t}] \tag{12}$$

While this method appears simple on paper, the calculations of $X_{\ell p}(f) = \mathcal{F}[x_{\ell p}(t)]$ and $y_{\ell p}(t) = \mathcal{F}^{-1}[Y_{\ell p}(f)]$ generally prove to be major stumbling blocks. Computer-aided numerical techniques are then necessary.

One of the few cases for which Eqs. (10)–(12) yield closed-form results is the transfer function plotted in Fig. 7.2-4. The gain $|H(f)|$ equals K_0 at f_c and increases (or decreases) linearly with slope K_1/f_c; the phase-shift curve corre-

$x_c(t) \longrightarrow \boxed{H(f)} \longrightarrow y_c(t)$

Figure 7.2-3

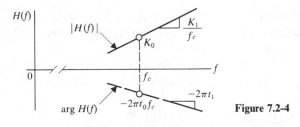

Figure 7.2-4

sponds to carrier delay t_0 and group delay t_1, as discussed in Example 6.1-2. The lowpass equivalent of $H(f)$ is

$$H(f + f_c)u(f + f_c) = \left(K_0 + \frac{K_1}{f_c}f\right)e^{-j2\pi(t_0 f_c + t_1 f)}$$

and Eq. (11) becomes

$$Y_{\ell p}(f) = K_0 e^{-j\omega_c t}[X_{\ell p}(f)e^{-j2\pi t_1 f}] + \frac{K_1}{j\omega_c}e^{-j\omega_c t}[(j2\pi f)X_{\ell p}(f)e^{-j2\pi t_1 f}]$$

Invoking the time-delay and differentiation theorems for $\mathcal{F}^{-1}[Y_{\ell p}(f)]$ we see that

$$y_{\ell p}(t) = K_0 e^{-j\omega_c t}x_{\ell p}(t - t_1) + \frac{K_1}{j\omega_c}e^{-j\omega_c t}\dot{x}_{\ell p}(t - t_1)$$

where

$$\dot{x}_{\ell p}(t - t_1) = \frac{d}{dt}\left[\frac{1}{2}A_c e^{j\phi(t - t_1)}\right] = \frac{j}{2}A_c \dot{\phi}(t - t_1)e^{j\phi(t - t_1)}$$

obtained from Eq. (10).

Inserting these expressions into Eq. (12) gives the output signal

$$y_c(t) = A(t)\cos\left[\omega_c(t - t_0) + \phi(t - t_1)\right] \tag{13a}$$

which has a time-varying amplitude

$$A(t) = A_c\left[K_0 + \frac{K_1}{\omega_c}\dot{\phi}(t - t_1)\right] \tag{13b}$$

In the case of an FM input, $\dot{\phi}(t) = 2\pi f_\Delta x(t)$ so

$$A(t) = A_c\left[K_0 + \frac{K_1 f_\Delta}{f_c}x(t - t_1)\right] \tag{14}$$

Equation (14) has the same form as the envelope of an AM wave with $\mu = K_1 f_\Delta / K_0 f_c$. We thus conclude that $|H(f)|$ in Fig. 7.2-4 produces *FM-to-AM conversion*, along with the carrier delay t_0 and group delay t_1 produced by arg $H(f)$. (By the way, a second look at Example 6.2-2 reveals that amplitude distortion of an AM wave can produce AM-to-PM conversion.)

FM-to-AM conversion does not present an insurmountable problem for FM or PM transmission, as long as $\phi(t)$ suffers no ill effects other than time delay. We therefore ignore the amplitude distortion from any reasonably smooth gain curve. But delay distortion from a nonlinear phase-shift curve can be quite severe and must be equalized in order to preserve the message information.

A simplified approach to phase-distortion effects is provided by the *quasi-static approximation*. For if the system's response to a carrier-frequency sinusoid is

$$y_c(t) = A_c \, |H(f_c)| \cos [\omega_c t + \arg H(f_c)]$$

and if $x_c(t)$ has a slowly changing instantaneous frequency $f(t)$, then

$$y_c(t) \approx A_c \, |H[f(t)]| \cos \{\omega_c t + \phi(t) + \arg H[f(t)]\} \tag{15}$$

It can be shown that this approximation requires the condition

$$|\ddot{\phi}(t)|_{\max} \left| \frac{1}{H(f)} \frac{d^2 H(f)}{df^2} \right|_{\max} \ll 8\pi^2 \tag{16}$$

in which $|\ddot{\phi}(t)| \leq 4\pi^2 f_\Delta W$ for tone-modulated FM with $f_m \leq W$. If $H(f)$ represents a single-tuned bandpass filter with 3-dB bandwidth B, then the second term in Eq. (16) equals $8/B^2$ and the condition becomes $4f_\Delta W/B^2 \ll 1$ which is satisfied by the transmission bandwidth requirement $B \geq B_T$.

Now suppose that Eq. (16) holds and the system has a nonlinear phase shift such as $\arg H(f) = \alpha f^2$, where α is a constant. Upon substituting $f(t) = f_c + \phi(t)/2\pi$ we get

$$\arg H[f(t)] = \alpha f_c^2 + \frac{\alpha f_c}{\pi} \phi(t) + \frac{\alpha}{4\pi^2} \phi^2(t)$$

Thus, the total phase in Eq. (15) will be distorted by the addition of $\phi(t)$ and $\phi^2(t)$.

Exercise 7.2-3 Let $|H(f)| = 1$ and $\arg H(f) = -2\pi t_1(f - f_c)$. Show that Eqs. (13) and (15) give the same result with $\phi(t) = \beta \sin \omega_m t$ provided that $\omega_m t_1 \ll \pi$.

Nonlinear Distortion and Limiters

Amplitude distortion of an FM wave produces FM-to-AM conversion. Here we'll show that the resulting AM can be eliminated through the use of controlled *nonlinear* distortion and filtering.

For purposes of analysis, let the input signal in Fig. 7.2-5 be

$$v_{in}(t) = A(t) \cos \theta_c(t)$$

where $\theta_c(t) = \omega_c t + \phi(t)$ and $A(t)$ is the amplitude. The nonlinear element is assumed to be *memoryless*—meaning no energy storage—so the input and

Nonlinear element

$v_{in}(t)$ $v_{out}(t) = T\,[v_{in}(t)]$

Figure 7.2-5

output are related by an instantaneous nonlinear transfer characteristic $v_{out} = T[v_{in}]$. We'll also assume for convenience that $T[0] = 0$.

Although $v_{in}(t)$ is not necessarily periodic in time, it may be viewed as a *periodic function of* θ_c with period 2π. (Try to visualize plotting v_{in} versus θ_c with time held fixed.) Likewise, the output is a periodic function of θ_c and can be expanded in the trigonometric Fourier series

$$v_{out} = \sum_{n=1}^{\infty} |2a_n| \cos (n\theta_c + \arg a_n) \tag{17a}$$

where

$$a_n = \frac{1}{2\pi} \int_{2\pi} T[v_{in}] e^{-jn\theta_c} \, d\theta_c \tag{17b}$$

The time variable t does not appear explicitly here, but v_{out} depends on t via the time-variation of θ_c. Additionally, the coefficients a_n may be functions of time when the amplitude of v_{in} has time variations.

But we'll first consider the case of an undistorted FM input, so $A(t)$ equals the constant A_c and all the a_n are constants. Hence, writing out Eqs. (17a) term by term with t explicitly included, we have

$$v_{out}(t) = |2a_1| \cos [\omega_c t + \phi(t) + \arg a_1]$$
$$+ |2a_2| \cos [2\omega_c t + 2\phi(t) + \arg a_2]$$
$$+ \cdots \tag{18}$$

This expression reveals that the nonlinear distortion produces additional FM waves at *harmonics* of the carrier frequency, the nth harmonic having constant amplitude $|2a_n|$ and phase modulation $n\phi(t)$ plus a constant phase shift $\arg a_n$.

If these waves don't overlap in the frequency domain, the *undistorted* input can be recovered by applying the distorted output to a *bandpass filter*. Thus, we say that FM enjoys considerable immunity from the effects of memoryless nonlinear distortion.

Now let's return to FM with unwanted amplitude variations $A(t)$. Those variations can be flattened out by an *ideal hard limiter* or *clipper* whose transfer characteristic is plotted in Fig. 7.2-6a. Figure 7.2-6b shows a clipper circuit

Figure 7.2-6 Hard limiter. (a) Transfer characteristic; (b) circuit realization with Zener diodes.

(a) $A(t) \cos [\omega_c t + \phi(t)]$ ⟶ [limiter] ⟶ [BPF at f_c] ⟶ $\dfrac{4V_0}{\pi} \cos [\omega_c t + \phi(t)]$

(b) $A_c \cos [\omega_c t + \phi(t)]$ ⟶ [nonlinear] ⟶ [BPF at nf_c] ⟶ $|2a_n| \cos [n\omega_c t + n\phi(t) + \arg a_n]$

Figure 7.2-7 Nonlinear processing circuits. (a) Amplitude limiter; (b) frequency multiplier.

employing back-to-back Zener diodes with breakdown voltage V_0 at the output of a high-gain amplifier.

The clipper output looks essentially like a square wave, since $T[v_{in}] = V_0 \operatorname{sgn} v_{in}$ and

$$v_{out} = \begin{cases} +V_0 & v_{in} > 0 \\ -V_0 & v_{in} < 0 \end{cases}$$

The coefficients are then found from Eq. (17b) to be

$$a_n = \begin{cases} 4V_0/\pi n & n = 1, 5, 9, \ldots \\ -4V_0/\pi n & n = 3, 7, 11, \ldots \\ 0 & n = 2, 4, 6, \ldots \end{cases}$$

which are independent of time because $A(t) \geq 0$ does not affect the sign of v_{in}. Therefore,

$$v_{out}(t) = \frac{4V_0}{\pi} \cos [\omega_c t + \phi(t)] - \frac{4V_0}{3\pi} \cos [3\omega_c t + 3\phi(t)] + \cdots \qquad (19)$$

and bandpass filtering yields a constant-amplitude FM wave if the components of $v_{out}(t)$ have no spectral overlap. Incidentally, this analysis lends support to the previous statement that message information resides entirely in the *zero-crossings* of an FM or PM wave.

Figure 7.2-7 summarizes our results. The limiter plus BPF in part *a* removes unwanted amplitude variations from an AM or PM wave, and would be used in a receiver. The nonlinear element in part *b* distorts a constant-amplitude wave, but the BPF passes only the undistorted term at the *n*th harmonic. This combination acts as a *frequency multiplier* if $n > 1$, and is used in certain types of transmitters.

7.3 GENERATION AND DETECTION OF FM AND PM

The operating principles of several methods for the generation and detection of exponential modulation are presented in this section. Other FM and PM systems that involve phase-lock loops will be mentioned in Sect. 8.4. Additional methods and information regarding specific circuit designs can be found in the radio electronics texts cited at the back of the book.

When considering equipment for exponential modulation, you should keep in mind that the instantaneous phase or frequency varies linearly with the message waveform. Devices are thus required that produce or are sensitive to phase or frequency variation in a linear fashion. Such characteristics can be approximated in a variety of ways, but it is sometimes difficult to obtain a suitably linear relationship over a wide operating range.

On the other hand, the constant-amplitude property of exponential modulation is a definite advantage from the hardware viewpoint. For one thing, the designer need not worry about excessive power dissipation or high-voltage breakdown due to extreme envelope peaks. For another, the relative immunity to nonlinear distortion allows the use of nonlinear electronic devices that would hopelessly distort a signal with linear modulation. Consequently, considerable latitude is possible in the design and selection of equipment. As a case in point, the microwave repeater links of long-distance telephone communications employ FM primarily because the wideband linear amplifiers required for amplitude modulation are unavailable at microwave frequencies.

Direct FM and VCOs

Conceptually, direct FM is straightforward and requires nothing more than a *voltage-controlled oscillator* (VCO) whose oscillation frequency has a linear dependence on applied voltage. This is readily implemented in the microwave band ($f_c \geq 1$ GHz), where devices such as the klystron tube have linear VCO characteristics over a substantial frequency range, typically several megahertz. If a lower carrier frequency is desired, the modulated signal can be down-converted by heterodyning with the output of a fixed-frequency oscillator. Some laboratory test generators use precisely this technique.

At lower carrier frequencies, it's possible to modulate a conventional tuned-circuit oscillator by introducing a *variable-reactance* element as part of the LC parallel resonant circuit. If the equivalent capacitance has a time dependence of the form

$$C(t) = C_0 - Cx(t)$$

and if $Cx(t)$ is "small enough" and "slow enough," then the oscillator produces $x_c(t) = A_c \cos \theta_c(t)$ where

$$\theta_c(t) = \frac{1}{\sqrt{LC(t)}} = \frac{1}{\sqrt{LC_0}} \left[1 - \frac{C}{C_0} x(t) \right]^{-1/2}$$

Letting $\omega_c = 1/\sqrt{LC_0}$ and assuming $|(C/C_0)x(t)| \ll 1$, the binomial series expansion gives $\theta_c(t) \approx \omega_c[1 + (C/2C_0)x(t)]$, or

$$\theta_c(t) \approx 2\pi f_c\, t + 2\pi \frac{C}{2C_0} f_c \int^t x(\lambda)\, d\lambda \tag{1}$$

Figure 7.3-1 VCO circuit with varactor diode for variable reactance.

which constitutes frequency modulation with $f_\Delta = (C/2C_0)f_c$. Since $|x(t)| \le 1$, the approximation is good to within 1 percent when $C/C_0 < 0.013$ so the attainable frequency deviation is limited by

$$f_\Delta = \frac{C}{2C_0} f_c \le 0.006 f_c \qquad (2)$$

This limitation quantifies our meaning of $Cx(t)$ being "small" and it seldom imposes a design hardship. Similarly, the usual condition $W \ll f_c$ ensures that $Cx(t)$ is "slow enough."

The variable reactance can be obtained in numerous ways. The workhorse of early direct FM transmitters was the reactance tube, a pentode biased such that its output impedance contains a capacitive term proportional to grid voltage. Other VCOs exploit Miller-effect capacitance in transistor circuits or the voltage-variable capacitance of varactor diodes. Figure 7.3-1 shows a tuned-circuit oscillator with a varactor diode biased to get $Cx(t)$. The input transformer, RF choke (RFC), and DC block serve to isolate the low-frequency, high-frequency, and DC voltages.

The principal advantage of direct FM is that large frequency deviations are possible without additional operations. The major disadvantage is that the carrier frequency tends to drift and must be stabilized by rather elaborate feedback frequency control (see Prob. 7.3-2). Because satisfactory stabilization techniques have emerged only recently, many older FM transmitters are of the indirect type.

Phase Modulators and Indirect FM

Although we seldom transmit a PM wave, we're still interested in phase modulators because: (1) the implementation is relatively easy; (2) the carrier can be supplied by a stable frequency source, such as a crystal-controlled oscillator; and (3) integrating the input signal to a phase modulator produces a *frequency-modulated* output.

Figure 7.3-2 depicts a narrowband phase modulator derived from the approximation $x_c(t) \approx A_c \cos \omega_c t - A_c \phi_\Delta x(t) \sin \omega_c t$—see Eqs. (9) and (11), Sect. 7.1. The evident simplicity of this modulator depends upon the approximation condition $|\phi_\Delta x(t)| \ll 1$ radian, and phase deviations greater than 10° result in distorted modulation.

$\phi_\Delta x(t)$ —→ (×) —————→ (+) ——→ $x_c(t)$

$-A_c \sin \omega_c t$ | +90° |

$A_c \cos \omega_c t$

Figure 7.3-2 Narrowband phase modulator.

Larger phase shifts can be achieved by the *switching-circuit* modulator in Fig. 7.3-3a. The typical waveforms shown in Fig. 7.3-3b help explain the operation. The modulating signal and a sawtooth wave at twice the carrier frequency are applied to a comparator. The comparator's output voltage goes high whenever $x(t)$ exceeds the sawtooth wave, and the flip-flop switches states at each rising edge of a comparator pulse. The flip-flop thus produces a phase-modulated square wave (like the output of a hard limiter), and bandpass filtering yields $x_c(t)$.

Now consider the indirect FM transmitter diagrammed in Fig. 7.3-4. The integrator and phase modulator constitute a *narrowband frequency modulator* that generates an initial NBFM signal with instantaneous frequency

$$f_1(t) = f_{c_1} + \frac{\phi_\Delta}{2\pi T} x(t)$$

(a)

(b)

Figure 7.3-3 Switching-circuit phase modulator. (a) Schematic diagram; (b) waveforms.

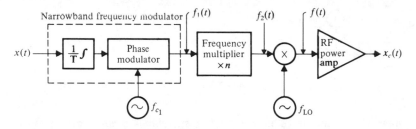

Figure 7.3-4 Indirect FM transmitter.

where T is the integrator's proportionality constant. The initial frequency deviation therefore equals $\phi_\Delta/2\pi T$ and must be increased to the desired value f_Δ by a *frequency multiplier*.

The frequency multiplier produces n-fold multiplication of *instantaneous frequency*, so

$$f_2(t) = nf_1(t) = nf_{c_1} + f_\Delta x(t) \tag{3}$$

where

$$f_\Delta = n\left(\frac{\phi_\Delta}{2\pi T}\right)$$

Typical frequency multipliers consist of a chain of doublers and triplers, each unit constructed as shown in Fig. 7.2-7b. Note that this multiplication is a subtle process, affecting the *range* of frequency variation but not the *rate*. Frequency multiplication of a tone-modulated signal, for instance, increases the carrier frequency and modulation index but not the modulation frequency, so the amplitude of the sideband lines are altered while the line spacing remains the same. (Compare the spectra in Fig. 7.1-7a with $\beta = 5$ and $\beta = 10$.)

The amount of multiplication required to get f_Δ usually results in nf_{c_1} being much higher than the desired carrier frequency. Hence, Fig. 7.3-4 includes a *frequency converter* that translates the spectrum intact down to $f_c = |nf_{c_1} \pm f_{LO}|$ and the final instantaneous frequency becomes $f(t) = f_c + f_\Delta x(t)$. (The frequency conversion may actually be performed in the middle of the multiplier chain to keep the frequencies at reasonable values.) The last system component is a *power amplifier*, since all of the previous operations must be carried out at low power levels.

Example 7.3-1 The indirect FM system originally designed by Armstrong employed a narrowband phase modulator in the form of Fig. 7.3-2 and produced a minute initial frequency deviation. As an illustration with representative numbers, suppose that $\phi_\Delta/2\pi T \approx 15$ Hz (which ensures negligible modulation distortion) and that $f_{c_1} = 200$ kHz (which falls near the lower limit of practical crystal-oscillator circuits). A broadcast FM output with $f_\Delta = 75$ kHz requires frequency multiplication by the factor $n \approx 75,000 \div$

$15 = 5000$. This could be achieved with a chain of four triplers and six doublers, so $n = 3^4 \times 2^6 = 5184$. But $nf_{c_1} \approx 5000 \times 200 \text{ kHz} = 1000 \text{ MHz}$, and a down-converter with $f_{\text{LO}} \approx 900 \text{ MHz}$ is needed to put f_c in the FM band of 88–108 MHz.

Exercise 7.3-1 Show that the phase at the output of Fig. 7.3-2 is given by

$$\phi(t) = \phi_\Delta x(t) - \tfrac{1}{3}\phi_\Delta^3 x^3(t) + \tfrac{1}{5}\phi_\Delta^5 x^5(t) + \cdots \tag{4}$$

Hence, $\phi(t)$ contains *odd-harmonic distortion* unless ϕ_Δ is quite small.

Triangular-Wave FM★

Triangular-wave FM is a modern and rather novel method for frequency modulation that overcomes the inherent problems of conventional VCOs and indirect FM systems. The method generates virtually distortionless modulation at carrier frequencies up to 30 MHz, and is particularly well suited for instrumentation applications.

We'll define triangular FM by working backwards from $x_c(t) = A_c \cos \theta_c(t)$ with

$$\theta_c(t) = \omega_c t + \phi(t) - \phi(0)$$

where the initial phase shift $-\phi(0)$ has been included so that $\theta_c(0) = 0$. This phase shift does not affect the instantaneous frequency

$$f(t) = \frac{1}{2\pi} \dot{\theta}_c(t) = f_c + f_\Delta x(t)$$

Expressed in terms of $\theta_c(t)$, a unit-amplitude triangular FM signal is

$$x_\Lambda(t) = \frac{2}{\pi} \arcsin \left[\cos \theta_c(t)\right] \tag{5a}$$

which defines a triangular waveform when $\phi(t) = 0$. Even with $\phi(t) \neq 0$, Eq. (5a) represents a *periodic triangular function of θ_c*, as plotted in Fig. 7.3-5a. Thus,

$$x_\Lambda = \begin{cases} 1 - \dfrac{2}{\pi} \theta_c & 0 < \theta_c < \pi \\[2mm] -3 + \dfrac{2}{\pi} \theta_c & \pi < \theta_c < 2\pi \end{cases} \tag{5b}$$

and so forth for $\theta_c > 2\pi$.

Figure 7.3-5b shows the block diagram of a system that produces $x_\Lambda(t)$ from the voltage

$$v(t) = \frac{2}{\pi} \dot{\theta}_c(t) = 4[f_c + f_\Delta x(t)]$$

which is readily derived from the message waveform $x(t)$. The system consists of an analog inverter, an integrator, and a Schmitt trigger controlling an electronic switch. The trigger puts the switch in the upper position whenever $x_\Lambda(t)$ increases to $+1$ and puts the switch in the lower position whenever $x_\Lambda(t)$ decreases to -1.

Suppose the system starts operating at $t = 0$ with $x_\Lambda(0) = +1$ and the switch in the upper position. Then, for $0 < t < t_1$,

$$x_\Lambda(t) = 1 - \int_0^t v(\lambda)\, d\lambda = 1 - \frac{2}{\pi}[\theta_c(t) - \theta_c(0)]$$

$$= 1 - \frac{2}{\pi}\theta_c(t) \qquad 0 < t < t_1$$

so $x_\Lambda(t)$ traces out the downward ramp in Fig. 7.3-5a until time t_1 when $x_\Lambda(t_1) = -1$, corresponding to $\theta_c(t_1) = \pi$. Now the trigger throws the switch to the lower position and

$$x_\Lambda(t) = -1 + \int_{t_1}^t v(\lambda)\, d\lambda = -1 + \frac{2}{\pi}[\theta_c(t) - \theta_c(t_1)]$$

$$= -3 + \frac{2}{\pi}\theta_c(t) \qquad t_1 < t < t_2$$

so $x_\Lambda(t)$ traces out the upward ramp in Fig. 7.3-5a. The upward ramp continues until time t_2 when $\theta_c(t_2) = 2\pi$ and $x_\Lambda(t_2) = +1$. The switch then triggers back to the upper position, and the operating cycle goes on periodically for $t > t_2$.

(a)

(b)

Figure 7.3-5 Triangular-wave FM. (a) Waveform; (b) modulation system.

A sinusoidal FM wave is obtained from $x_\Lambda(t)$ using a nonlinear waveshaper with transfer characteristics $T[x_\Lambda(t)] = A_c \sin[(\pi/2)x_\Lambda(t)]$, which performs the inverse of Eq. (5a). Or $x_\Lambda(t)$ can be applied to a hard limiter to produce *square-wave* FM. A laboratory test generator might have all three outputs available.

Frequency Detection

A *frequency detector*, often called a *discriminator*, produces an output voltage that should vary linearly with the instantaneous frequency of the input. There are perhaps as many different circuit designs for frequency detection as there are designers who have considered the problem. However, almost every circuit falls into one of the following four operational categories:

1. FM-to-AM conversion
2. Phase-shift discrimination
3. Zero-crossing detection
4. Frequency feedback

We'll look at illustrative examples from the first three categories, postponing frequency feedback to Sect. 8.4. Analog *phase* detection is not discussed here because it's seldom needed in practice and, if needed, can be accomplished by integrating the output of a frequency detector.

Any device or circuit whose output equals the *time derivative* of the input produces *FM-to-AM conversion*. To be more specific, let $x_c(t) = A_c \cos \theta_c(t)$ with $\dot\theta_c(t) = 2\pi[f_c + f_\Delta x(t)]$; then

$$\dot x_c(t) = -A_c \dot\theta_c(t) \sin \theta_c(t)$$
$$= 2\pi A_c[f_c + f_\Delta x(t)] \sin [\theta_c(t) \pm 180°] \qquad (6)$$

Hence, an *envelope detector* with input $\dot x_c(t)$ yields an output proportional to $f(t) = f_c + f_\Delta x(t)$.

Figure 7.3-6a diagrams a conceptual frequency detector based on Eq. (6). The diagram includes a limiter at the input to remove any spurious amplitude variations from $x_c(t)$ before they reach the envelope detector. It also includes a DC block to remove the constant carrier-frequency offset from the output signal. Typical waveforms are sketched in Fig. 7.3-6b taking the case of tone modulation.

For actual hardware implementation of FM-to-AM conversion, we draw upon the fact that an ideal differentiator has $|H(f)| = 2\pi f$. Slightly above or below resonance, the transfer function of an ordinary tuned circuit shown in Fig. 7.3-7a approximates the desired linear amplitude response over a small frequency range. Thus, for instance, a detuned AM receiver will roughly demodulate FM via *slope detection*.

Extended linearity is achieved by the *balanced discriminator* circuit in Fig. 7.3-7b. A balanced discriminator includes two resonant circuits, one tuned

(a)

(b)

Figure 7.3-6 (a) Frequency detector with limiter and FM-to-AM conversion; (b) waveforms.

above f_c and the other below, and the output equals the difference of the two envelopes. The resulting frequency-to-voltage characteristic takes the form of the well-known S curve in Fig. 7.3-7c. No DC block is needed, since the carrier-frequency offset cancels out, and the circuit has good performance at low modulating frequencies. The balanced configuration easily adapts to the microwave band, with resonant cavities serving as tuned circuits and crystal diodes for envelope detectors.

Phase-shift discriminators involve circuits with linear *phase* response, in contrast to the linear amplitude response of slope detection. The underlying principle comes from an approximation for time differentiation, namely

$$\dot{v}(t) \approx \frac{1}{t_1}\left[v(t) - v(t - t_1)\right] \tag{7}$$

providing that t_1 is small compared to the variation of $v(t)$. Now an FM wave has $\dot{\phi}(t) = 2\pi f_\Delta x(t)$ so

$$\phi(t) - \phi(t - t_1) \approx t_1 \dot{\phi}(t) = 2\pi f_\Delta t_1 x(t) \tag{8}$$

The term $\phi(t - t_1)$ can be obtained with the help of a delay line or, equivalently, a linear phase-shift network.

(a)

(b)

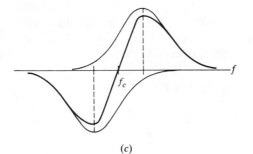

(c)

Figure 7.3-7 (a) Slope detection with a tuned circuit; (b) balanced discriminator circuit; (c) frequency-to-voltage characteristic.

Figure 7.3-8 represents a phase-shift discriminator built with a network having group delay t_1 and carrier delay t_0 such that $\omega_c t_0 = 90°$—which accounts for the name *quadrature detector*. From Eq. (13), Sect. 7.2, the phase-shifted signal is proportional to $\cos\left[\omega_c t - 90° + \phi(t - t_1)\right] = \sin\left[\omega_c t + \phi(t - t_1)\right]$. Multiplication by $\cos\left[\omega_c t + \phi(t)\right]$ followed by lowpass filtering yields an output proportional to

$$\sin\left[\phi(t) - \phi(t - t_1)\right] \approx \phi(t) - \phi(t - t_1)$$

assuming t_1 is small enough that $|\phi(t) - \phi(t - t_1)| \ll \pi$. Therefore,

$$y_D(t) \approx K_D f_\Delta x(t)$$

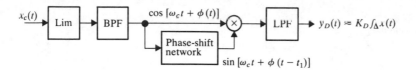

Figure 7.3-8 Phase-shift discriminator or quadrature detector.

where the detection constant K_D includes t_1. Despite these approximations, a quadrature detector provides better linearity than a balanced discriminator and is often found in high-quality receivers.

Other phase-shift circuit realizations include the *Foster-Seely discriminator* and the popular *ratio detector*. The latter is particularly ingenious and economical, for it combines the operations of limiting and demodulation into one unit. See Clarke and Hess (1971, chap. 12) or Krauss, Bostian, and Raab (1980, chap. 10) for further details.

Lastly, Fig. 7.3-9 gives the diagram and waveforms for a simplified *zero-crossing detector*. The square-wave FM signal from a hard limiter triggers a monostable pulse generator, which produces a short pulse of fixed amplitude A and duration τ at each upward (or downward) zero crossing of the FM wave. If we invoke the quasi-static viewpoint and consider a time interval T such that $W \ll 1/T \ll f_c$, the monostable output $v(t)$ looks like a rectangular pulse train with nearly constant period $1/f(t)$. Thus, there are $n_T \approx Tf(t)$ pulses in this interval, and continually integrating $v(t)$ over the past T seconds yields

$$\frac{1}{T} \int_{t-T}^{t} v(\lambda)\, d\lambda = \frac{1}{T}\, n_T A\tau \approx A\tau f(t)$$

which becomes $y_D(t) \approx K_D f_\Delta x(t)$ after the DC block.

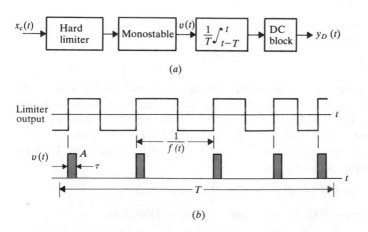

(a)

(b)

Figure 7.3-9 Zero-crossing detector. (a) Diagram; (b) waveforms.

Commercial zero-crossing detectors may have better than 0.1% linearity and operate at center frequencies from 1 Hz to 10 MHz. A divide-by-ten counter inserted after the hard limiter extends the range up to 100 MHz.

Exercise 7.3-2 Given a delay line with time delay $t_0 \ll 1/f_c$, devise a frequency detector based on Eqs. (6) and (7).

7.4 INTERFERENCE

Interference refers to the contamination of an information-bearing signal by another similar signal, usually from a human source. This occurs in radio communication when the receiving antenna picks up two or more signals in the same frequency band. Interference may also result from multipath propagation, or from electromagnetic coupling between transmission cables. Regardless of the cause, severe interference prevents successful recovery of the message information.

Our study of interference begins with the simple but nonetheless informative case of interfering sinusoids, representing unmodulated carrier waves. This simplified case helps bring out the differences between interference effects in AM, FM, and PM. Then we'll see how the technique of *deemphasis filtering* improves FM performance in the face of interference. We conclude with a brief examination of the FM *capture effect*.

Interfering sinusoids

Consider a receiver tuned to some carrier frequency f_c. Let the total received signal be

$$v(t) = A_c \cos \omega_c t + A_i \cos [(\omega_c + \omega_i)t + \phi_i]$$

The first term represents the desired signal as an unmodulated carrier, while the second term is an interfering carrier with amplitude A_i, frequency $f_c + f_i$, and relative phase angle ϕ_i.

To put $v(t)$ in the envelope-and-phase form $v(t) = A_v(t) \cos [\omega_c t + \phi_v(t)]$, we'll introduce

$$\rho \triangleq A_i/A_c \qquad \theta_i(t) \triangleq \omega_i t + \phi_i \tag{1}$$

Hence, $A_i = \rho A_c$ and the phasor construction in Fig. 7.4-1 gives

$$A_v(t) = A_c \sqrt{1 + \rho^2 + 2\rho \cos \theta_i(t)}$$

$$\phi_v(t) = \arctan \frac{\rho \sin \theta_i(t)}{1 + \rho \cos \theta_i(t)} \tag{2}$$

These expressions show that interfering sinusoids produce both *amplitude* and *phase modulation*. In fact, if $\rho \ll 1$ then

$$A_v(t) \approx A_c[1 + \rho \cos (\omega_i t + \phi_i)]$$

$$\phi_v(t) \approx \rho \sin (\omega_i t + \phi_i) \tag{3}$$

Figure 7.4-1 Phasor diagram of interfering carriers.

which looks like *tone modulation* at frequency f_i with AM modulation index $\mu = \rho$ and FM or PM modulation index $\beta = \rho$. At the other extreme, if $\rho \gg 1$ then

$$A_v(t) \approx A_i[1 + \rho^{-1} \cos(\omega_i t + \phi_i)]$$

$$\phi_v(t) \approx \omega_i t + \phi_i$$

so the envelope still has tone modulation but the phase corresponds to a shifted carrier frequency $f_c + f_i$ plus the constant ϕ_i.

Next we investigate what happens when $v(t)$ is applied to an ideal envelope, phase, or frequency demodulator with detection constant K_D. We'll take the weak interference case ($\rho \ll 1$) and use the approximation in Eq. (3) with $\phi_i = 0$. Thus, the demodulated output is

$$y_D(t) \approx \begin{cases} K_D(1 + \rho \cos \omega_i t) & \text{AM} \\ K_D \rho \sin \omega_i t & \text{PM} \\ K_D \rho f_i \cos \omega_i t & \text{FM} \end{cases} \qquad (4)$$

provided that $|f_i| \leq W$ — otherwise, the lowpass filter at the output of the demodulator would reject $|f_i| > W$. The constant term in the AM result would be removed if the demodulator includes a DC block. As written, this result also holds for *synchronous* detection in DSB and SSB systems since we've assumed $\phi_i = 0$. The multiplicative factor f_i in the FM result comes from the instantaneous frequency deviation $\dot{\phi}_v(t)/2\pi$.

Equation (4) reveals that weak interference in a linear modulation system or phase modulation system produces a spurious output tone with amplitude proportional to $\rho = A_i/A_c$, independent of f_i. But the tone amplitude is proportional to ρf_i in an FM system. Consequently, FM will be less vulnerable to interference from a *cochannel* signal having the same carrier frequency, so $f_i \approx 0$, and more vulnerable to *adjacent-channel* interference ($f_i \neq 0$). Figure 7.4-2 illustrates this difference in the form of a plot of demodulated interference amplitude versus $|f_i|$. (The crossover point would correspond to $|f_i| = 1$ Hz if all three detector constants had the same numerical value.)

Figure 7.4-2 Amplitude of demodulated interference from a carrier at frequency $f_c + f_i$.

The analysis of demodulated interference becomes a much more difficult task with arbitrary values of ρ and/or modulated carriers. We'll return to that problem after exploring the implications of Fig. 7.4-2.

Exercise 7.4-1 Let $A_i = A_c$ so $\rho = 1$ in Eq. (2). Take $\phi_i = 0$ and use trigonometric identities to show that

$$A_v(t) = 2A_c \mid \cos (\omega_i t/2) \mid \qquad \phi_v(t) = \omega_i t/2$$

Then sketch the demodulated output waveform for envelope, phase, and frequency detection assuming $f_i \ll W$.

Deemphasis and Preemphasis Filtering

The fact that detected FM interference is most severe at large values of $\mid f_i \mid$ suggests a method for improving system performance with selective postdetection filtering, called *deemphasis filtering*. Suppose the demodulator is followed by a lowpass filter having an amplitude ratio that begins to decrease gradually *below* W; this will *deemphasize* the high-frequency portion of the message band and thereby reduce the more serious interference. A sharp-cutoff (ideal) lowpass filter is still required to remove any residual components above W, so the complete demodulator consists of a frequency detector, deemphasis filter, and lowpass filter, as in Fig. 7.4-3.

Obviously deemphasis filtering also attenuates the high-frequency components of the message itself, causing distortion of the output signal unless corrective measures are taken. But it's a simple matter to compensate for deemphasis distortion by *predistorting* or *preemphasizing* the modulating signal at the transmitter before modulation. The preemphasis and deemphasis filter characteristics should be related by

$$H_{\text{pe}}(f) = \frac{1}{H_{\text{de}}(f)} \qquad |f| \le W \tag{5}$$

to yield net undistorted transmission. In essence, we preemphasize the message before modulation (where the interference is absent) so we can deemphasize the interference relative to the message after demodulation.

Preemphasis/deemphasis filtering offers potential advantages whenever undesired contaminations tend to predominate at certain portions of the message band. For instance, high frequencies are preemphasized in audio disk recording so that high-frequency surface noise (scratch) can be deemphasized during playback. The *Dolby* system for tape recording goes even further and dynamically adjusts the amount of preemphasis/deemphasis in inverse proportion to the high-frequency signal content; see Stremler (1982, app. *F*) for details. However, little is

Figure 7.4-3 Complete FM demodulator.

gained from decmphasizing phase modulation or linear modulation because the demodulated interference amplitude does not depend on the frequency.

The FM deemphasis filter is usually a simple first-order network having

$$H_{de}(f) = \left[1 + j\left(\frac{f}{B_{de}}\right)\right]^{-1} \approx \begin{cases} 1 & |f| \ll B_{de} \\ \dfrac{B_{de}}{jf} & |f| \gg B_{de} \end{cases} \tag{6}$$

where the 3-dB bandwidth B_{de} is considerably less than the message bandwidth W. Since the interference amplitude increases linearly with $|f_i|$ in the absence of filtering, the deemphasized interference response is $|H_{de}(f_i)| \times |f_i|$, as sketched in Fig. 7.4-4. Note that, like PM, this becomes constant for $|f_i| \gg B_{de}$. Therefore, FM can be superior to PM for both adjacent-channel and cochannel interference.

At the transmitting end, the corresponding preemphasis filter function should be

$$H_{pe}(f) = \left[1 + j\left(\frac{f}{B_{de}}\right)\right] \approx \begin{cases} 1 & |f| \ll B_{de} \\ \dfrac{jf}{B_{de}} & |f| \gg B_{de} \end{cases} \tag{7}$$

which has little effect on the lower message frequencies. At higher frequencies, however, the filter acts as a *differentiator*, the output spectrum being proportional to $fX(f)$ for $|f| \gg B_{de}$. But differentiating a signal before frequency modulation is equivalent to *phase modulation*! Hence, preemphasized FM is actually a combination of FM and PM, combining the advantages of both with respect to interference. As might be expected, this turns out to be equally effective for reducing *noise*.

Referring to $H_{pe}(f)$ as given above, we see that the amplitude of the maximum modulating frequency is increased by a factor of W/B_{de}, which means that the frequency deviation is increased by this same factor. Generally speaking, the increased deviation requires a greater transmission bandwidth, so the preemphasis-deemphasis improvement is not without price. Fortunately, many modulating signals of interest, particularly audio signals, have relatively little energy in the high-frequency end of the message band, and therefore the higher frequency components do not develop maximum deviation, the transmission bandwidth being dictated by lower components of larger amplitude. Adding high-frequency preemphasis tends to equalize the message spectrum so that all components require the same bandwidth. Under this condition, the transmission bandwidth need not be increased.

Figure 7.4-4 Demodulated interference amplitude with FM deemphasis filtering.

Figure 7.4-5 (a) Deemphasis filter; (b) preemphasis filter.

Example 7.4-1 Typical deemphasis and preemphasis networks for commercial FM are shown in Fig. 7.4-5 along with their Bode diagrams. The RC time constant in both circuits equals 75 μs, so $B_{de} = 1/2\pi RC \approx 2.1$ kHz. The preemphasis filter has an upper break frequency at $f_u = (R + r)/2\pi RrC$, usually chosen to be well above the audio range, say $f_u \geq 30$ kHz.

Exercise 7.4-2 Suppose an audio signal is modeled as a sum of tones with low-frequency amplitudes $A_m \leq 1$ for $f_m \leq 1$ kHz and high-frequency amplitudes $A_m \leq 1$ kHz/f_m for $f_m > 1$ kHz. Use Eqs. (1) and (2), Sect. 7.2, to estimate the bandwidth required for a single tone at $f_m = 15$ kHz whose amplitude has been preemphasized by $|H_{pe}(f)|$ given in Eq. (7) with $B_{de} = 2$ kHz. Assume $f_\Delta = 75$ kHz and compare your result with $B_T \approx 210$ kHz.

FM Capture Effect★

Capture effect is a phenomenon that takes place in FM systems when two signals have nearly equal amplitudes at the receiver. Small variations of relative amplitude then cause the stronger of the two to dominate the situation, suddenly displacing the other signal at the demodulated output. You may have heard the annoying results when listening to a distant FM station.

For a reasonably tractable analysis of capture effect, we'll consider an unmodulated carrier with modulated cochannel interference ($f_i = 0$). The resultant phase $\phi_v(t)$ is then given by Eq. (2) with $\theta_i(t) = \phi_i(t)$, where $\phi_i(t)$ denotes the phase modulation of the interfering signal. Thus, if $K_D = 2\pi$ for simplicity, the demodulated signal becomes

$$y_D(t) = \dot{\phi}_v(t) = \frac{d}{dt}\left[\arctan \frac{\rho \sin \phi_i(t)}{1 + \rho \cos \phi_i(t)}\right]$$

$$= \alpha(\rho, \phi_i)\dot{\phi}_i(t) \tag{8a}$$

where

$$\alpha(\rho, \phi_i) \triangleq \frac{\rho^2 + \rho \cos \phi_i}{1 + \rho^2 + 2\rho \cos \phi_i} \tag{8b}$$

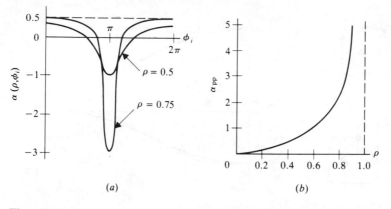

(a)   (b)

Figure 7.4-6

The presence of $\dot{\phi}_i(t)$ in Eq. (8a) indicates potentially *intelligible* interference (or *crosstalk*) to the extent that $\alpha(\rho, \phi_i)$ remains constant with time. After all, if $\rho \gg 1$ then $\alpha(\rho, \phi_i) \approx 1$ and $y_D(t) \approx \dot{\phi}_i(t)$.

But capture effect occurs when $A_i \approx A_c$, so $\rho \approx 1$ and Eq. (8b) does not immediately simplify. Instead, we note that

$$\alpha(\rho, \phi_i) = \begin{cases} \rho/(1+\rho) & \phi_i = 0, \pm 2\pi, \ldots \\ \rho^2/(1+\rho^2) & \phi_i = \pm\pi/2, \pm 3\pi/2, \ldots \\ -\rho/(1-\rho) & \phi_i = \pm\pi, \pm 3\pi, \ldots \end{cases}$$

and we resort to plots of $\alpha(\rho, \phi_i)$ versus ϕ_i as shown in Fig. 7.4-6a. Except for the negative spikes, these plots approach $\alpha(\rho, \phi_i) = 0.5$ as $\rho \to 1$, and thus $y_D(t) = 0.5 \cdot \dot{\phi}_i(t)$. For $\rho < 1$, the strength of the demodulated interference essentially depends on the peak-to-peak value

$$\alpha_{pp} = \alpha(\rho, 0) - \alpha(\rho, \pi) = 2\rho/(1 - \rho^2)$$

which is plotted versus ρ in Fig. 7.4-6b. This knee-shaped curve reveals that if transmission fading causes ρ to vary around a nominal value of about 0.7, the interference almost disappears when $\rho < 0.7$ whereas it takes over and "captures" the output when $\rho > 0.7$.

Panter (1965, chap. 11) presents a detailed analysis of FM interference, including waveforms that result when both carriers are modulated.

7.5 PROBLEMS

7.1-1 Sketch and label $\phi(t)$ and $f(t)$ for PM and FM when $x(t) = A\Lambda(t/\tau)$. Take $\phi(-\infty) = 0$ in the FM case.

7.1-2 Do Prob. 7.1-1 with $x(t) = A \sin(\pi t/\tau)\Pi(t/2\tau)$.

7.1-3 Do Prob. 7.1-1 with $x(t) = 2At/(1 + t^4)$.

7.1-4 A *frequency-sweep generator* produces a sinusoidal output whose instantaneous frequency increases linearly from f_1 at $t = 0$ to f_2 at $t = T$. Write $\theta_c(t)$ for $0 \leq t \leq T$.

7.1-5 Besides PM and FM, two other possible forms of exponential modulation are *phase-integral modulation*, with $\phi(t) = K\ dx(t)/dt$, and *phase-acceleration modulation*, with

$$f(t) = f_c + K \int^t x(\lambda)\ d\lambda$$

Add these to Table 7.1-1 and find the maximum values of $\phi(t)$ and $f(t)$ for all four types when $x(t) = \cos 2\pi f_m t$.

7.1-6 Use Eq. (16) to obtain Eq. (18a) from Eq. (15).

7.1-7‡ Derive Eq. (16) by finding the exponential Fourier series of the complex periodic function $\exp(j\beta \sin \omega_m t)$.

7.1-8 Tone modulation is applied simultaneously to a frequency modulator and a phase modulator and the two output spectra are identical. Describe how these two spectra will change when: (a) the tone amplitude is increased or decreased; (b) the tone frequency is increased or decreased; (c) the tone amplitude and frequency are increased or decreased in the same proportion.

7.1-9 Consider a tone-modulated FM or PM wave with $f_m = 20$ kHz, $\beta = 1.0$, $A_c = 100$, and $f_c = 30$ kHz. (a) Write an expression for $f(t)$. (b) Draw the line spectrum and show therefrom that $S_T < A_c^2/2$.

7.1-10 Do Prob. 7.1-9 with $f_c = 40$ kHz, in which case $S_T > A_c^2/2$.

7.1-11 Construct phasor diagrams for tone-modulated FM with $A_c = 1$ and $\beta = 0.5$ when $\omega_m t = 0$, $\pi/4$, and $\pi/2$. Calculate A and ϕ from each diagram and compare with the theoretical values.

7.1-12 Do Prob. 7.1-11 with $\beta = 1.0$.

7.1-13 A tone-modulated FM signal with $\beta = 1.0$ and $f_m = 200$ Hz is applied to an ideal BPF with $B = 500$ Hz centered at f_c. Draw the line spectrum, phasor diagram, and envelope of the output signal.

7.1-14 Do Prob. 7.1-13 with $\beta = 2.0$.

7.1-15 The carrier-frequency component of a tone-modulated FM signal is zero whenever $J_0(\beta) = 0$. Exploit this property to devise an experimental procedure for calibrating a frequency modulator. How would you check for possible nonlinear frequency deviation?

7.1-16‡ Consider FM with periodic square-wave modulation defined by $x(t) = 1$ for $0 < t < T_0/2$ and $x(t) = -1$ for $-T_0/2 < t < 0$. (a) Take $\phi(0)$ and plot $\phi(t)$ for $-T_0/2 < t < T_0/2$. Then use Eq. (20a) to obtain

$$c_n = \frac{1}{2} e^{jn\beta} \left[\operatorname{sinc}\left(\frac{n+\beta}{2}\right) e^{jn\pi/2} + \operatorname{sinc}\left(\frac{n-\beta}{2}\right) e^{-jn\pi/2} \right]$$

where $\beta = f T_0$. (b) Sketch the resulting magnitude line spectrum when β is a large integer.

7.2-1 A message has $W = 10$ kHz. Estimate the FM transmission bandwidth for $f_\Delta = 0.1, 0.5, 1, 5, 10, 50, 100$, and 500 kHz.

7.2-2 Do Prob. 7.2-1 with $W = 2.5$ kHz.

7.2-3 An FM system has $f_\Delta = 10$ kHz. Use Table 5.4-1 and Fig. 7.2-1 to estimate the bandwidth for: (a) barely intelligible voice transmission; (b) telephone-quality voice transmission; (c) high-fidelity audio transmission.

7.2-4 A television signal with $W = 5$ MHz is to be transmitted via FM with $f_\Delta = 20$ MHz. Find the minimum carrier frequency consistent with fractional-bandwidth considerations.

7.2-5 An analog data signal with $W = 5$ kHz is to be transmitted on an FM system having $B_T = 30$ kHz. Find the maximum value of f_Δ for high-fidelity reproduction.

7.2-6 Do Prob. 7.2-5 with $B_T = 60$ kHz.

7.2-7 An FM system with $f_\Delta = 30$ kHz has been designed for $W = 10$ kHz. Approximately what percentage of B_T is occupied when the modulating signal is a unit-amplitude tone at $f_m = 0.1, 1.0,$ or 5.0 kHz? Repeat your calculations for a PM system with $\phi_\Delta = 3$ rad.

7.2-8 Consider phase-integral and phase-acceleration modulation defined in Prob. 7.1-5. Investigate the bandwidth requirements for tone modulation, and obtain transmission bandwidth estimates. Discuss your results.

7.2-9 If the WBFM power spectrum in Eq. (9) does not yield a well-defined bandwidth, then the transmission bandwidth can be defined as the frequency range that includes 90% of the modulated signal power, so

$$2 \int_{f_c - B_T/2}^{f_c + B_T/2} G(f) \, df = 0.9 S_T$$

Sketch $G(f)$ for $f > 0$ and find B_T in terms of the modulating signal power S_x when $f_c \gg f_\Delta$ and $x(t)$ is a voice signal with a Laplace PDF given by $p_X(x) = (\alpha/2)e^{-\alpha|x|}$ where $\alpha = \sqrt{2/S_x}$.

7.2-10 Do Prob. 7.2-9 for a random gaussian modulating signal with zero mean and $\sigma^2 = S_x$.

7.2-11‡ The *rms bandwidth* of a bandpass signal with power spectrum $G(f)$ centered at f_c is defined by

$$B_{rms}^2 = (8/S_T) \int_0^\infty (f - f_c)^2 G(f) \, df$$

(a) Show that if Eq. (9) holds and if $f_c \gg f_\Delta$ and $\bar{x} = 0$, then $B_{rms} = 2f_\Delta \sqrt{S_x}$, where $S_x = \overline{x^2}$. (b) Calculate B_{rms} for a zero-mean gaussian modulating signal normalized such that $P[\,|x(t)| > 1\,] = 0.2$.

7.2-12 The transfer function of a single-tuned BPF is $H(f) \approx 1/[1 + j2Q(f - f_c)/f_c]$ over the positive-frequency passband. Use Eq. (12) to obtain an expression for the output signal and its instantaneous phase when the input is an NBPM signal.

7.2-13 Use Eq. (12) to obtain an expression for the output signal and its amplitude when an FM signal is distorted by a system having $H(f) = K_0 - K_2(f - f_c)^2$ over the positive-frequency passband.

7.2-14 Use Eq. (15) to obtain an expression for the output signal and its instantaneous frequency when an FM signal is distorted by a system having $|H(f)| = 1$ and $\arg H(f) = \alpha_1(f - f_c) + \alpha_2(f - f_c)^2$ over the positive-frequency passband.

7.2-15 An FM signal is applied to the BPF in Prob. 7.2-12. Let $\alpha = 2Qf_\Delta/f_c \ll 1$ and use Eq. (15) to obtain an approximate expression for the output signal and its instantaneous frequency.

7.2-16 Let the input to the system in Fig. 7.2-7a be an FM signal with $D = f_\Delta/W$ and spurious amplitude variations. Sketch the spectrum at the output of the limiter, and show that successful operation requires $f_\Delta < (f_c - W)/2$.

7.2-17 The input to the system in Fig. 7.2-7b is an FM signal with $D = f_\Delta/W$ and the BPF is centered at $2f_c$, corresponding to a frequency doubler. Sketch the spectrum at the filter's input, and obtain a condition on f_Δ in terms of f_c and W that ensures successful operation.

7.2-18 Do Prob. 7.2-17 with the BPF centered at $4f_c$, corresponding to a frequency quadrupler.

7.3-1 The equivalent tuning capacitance in Fig. 7.3-1 is $C(t) = C_1 + C_v(t)$ where $C_v(t) = C_2/\sqrt{V_B + x(t)/N}$. Show that $C(t) \approx C_0 - C_x(t)$ with 1% accuracy if $NV_B \geq 300/4$. Then show that the corresponding limitation on the frequency deviation is $f_\Delta < f_c/300$.

7.3-2 Figure P7.3-2 is a direct FM generator using a VCO whose output has $\theta_c(t) = 2\pi[f_c + K_1v(t) + \epsilon(t)]$, where $\epsilon(t)$ is a random frequency drift that changes slowly compared to $x(t)$. The stabilizing feedback loop includes a frequency detector to produce $y(t) = (K_2/2\pi)[\theta_1(t) - \omega_1]$. Show that if $K_2 K_1 \gg 1$, then the instantaneous output frequency is $f(t) \approx f_c + K_1[x(t) - \langle x(t) \rangle] + \epsilon(t)/K_2 K_1$, so feedback reduces drift and removes the carrier-frequency offset when $\langle x(t) \rangle \neq 0$.

Figure P7.3-2

7.3-3 Confirm that $x_c(t) = A_c \cos \theta_c(t)$ is a solution of the integrodifferential equation $\dot{x}_c(t) = -\dot{\theta}_c(t) \int \dot{\theta}_c(t) x_c(t) \, dt$. Then draw the block diagram of a direct FM generator based on this relationship.

7.3-4 The quasi-static approximation in Sect. 7.2 may also be invoked when a system includes a time-varying element and the input is an *unmodulated* sinusoid at frequency f_c, so we have a time-varying transfer function $H(f_c, t)$. Use this approach to show that an RC lowpass filter becomes a phase modulator if $R = R(t) = -R_0 x(t)$, providing that $\phi_\Delta \leq 10°$ for 1% accuracy.

7.3-5 Apply the method in Prob. 7.3-4 to the circuit in Fig. P7.3-5, which generates phase modulation with 1% accuracy if $\phi_\Delta \leq 20°$.

$$R(t) = -R_0 x(t)$$

Figure P7.3-5

7.3-6 The audio portion of a television transmitter is an indirect FM system having $W = 10$ kHz, $D = 2.5$, and $f_c = 4.5$ MHz. Devise a block diagram of this system with $\phi_\Delta/2\pi T < 20$ Hz and $f_c = 200$ kHz. Use the shortest possible multiplier chain consisting of frequency triplers and doublers, and locate the down-converter such that no frequency exceeds 100 MHz.

7.3-7 An indirect FM transmitter is to be designed for $W = 4$ kHz, $D = 3$, $f_c = 10$ MHz, and $f_{c_1} = 100$ kHz. The multiplier chain will consist of six triplers like Fig. 7.2-7a. All BPFs must satisfy the fractional-bandwidth constraint of Eq. (2), Sect. 6.1, but their bandwidths may be larger than the signal bandwidth at any point as long as unwanted signals are rejected. Draw a block diagram of this system, specifying the values for $\phi_\Delta/2\pi T$, f_{LO}, and all BPF bandwidths and center frequencies. Locate the down-converter such that no frequency exceeds 12 MHz.

7.3-8 Suppose the phase modulator in Fig. 7.3-4 is implemented as in Fig. 7.3-2. Take $x(t) = A_m \cos \omega_m t$ and let $\beta = (\phi_\Delta/2\pi T)(A_m/f_m)$. (a) Show that if $\beta \ll 1$, then

$$f_1(t) \approx f_c + \beta f_m[\cos \omega_m t + (\beta/2)^2 \cos 3\omega_m t]$$

(b) Obtain a condition on $\phi_\Delta/2\pi T$ so the third-harmonic distortion does not exceed 1% when $A_m \leq 1$ and 30 Hz $\leq f_m \leq 15$ kHz, as in FM broadcasting.

7.3-9 Let the input to Fig. 7.3-6a be an FM signal with $f_\Delta \ll f_c$ and let the differentiator be implemented by a tuned circuit with $H(f) = 1/[1 + j(2Q/f_0)(f - f_0)]$ for $f \approx f_0$. Use the quasi-static method to show that $y_D(t) \approx K_D f_\Delta x(t)$ when $f_0 = f_c + b$ provided that $f_\Delta \ll b \ll f_0/2Q$.

7.3-10 Let the input to Fig. 7.3-6a be an FM signal with $f_\Delta \ll f_c$ and let the differentiator be implemented by a first-order lowpass filter with $B = f_c$. Use quasi-static analysis to show that $y_D(t) \approx -K_1 f_\Delta x(t) + K_2 f_\Delta^2 x^2(t)$. Then take $x(t) = \cos \omega_m t$ and obtain a condition on f_Δ/f_c so the second-harmonic distortion is less than 1%.

7.3-11 The tuned circuits in Fig. 7.3-7b have transfer functions of the form $H(f) = 1/[1 + j(2Q/f_0)(f - f_0)]$ for $f \approx f_0$. Let the two center frequencies be $f_0 = f_c \pm b$ with $f_\Delta \leq b \ll f_c$. Use quasi-static analysis to show that if both circuits have $(2Q/f_0)b = \alpha \ll 1$, then $y_D(t) \approx K_1 x(t) - K_3 x^3(t)$ where $K_3/K_1 \ll 1$.

7.4-1 Obtain an approximate expression for the output of an amplitude demodulator when the input is an AM signal with 100% modulation plus an interfering signal $A_i[1 + x_i(t)] \cos [(\omega_c + \omega_i)t + \phi_i]$ with $\rho = A_i/A_c \ll 1$. Is the demodulated interference intelligible?

7.4-2 Obtain an approximate expression for the output of a phase demodulator when the input is an NBPM signal with 100% modulation plus an interfering signal $A_i \cos [(\omega_c + \omega_i)t + \phi_i(t)]$ with $\rho = A_i/A_c \ll 1$. Is the demodulated interference intelligible?

7.4-3 Investigate the performance of envelope detection versus synchronous detection of AM in the presence of multipath propagation, so that $v(t) = x_c(t) + \alpha x_c(t - t_d)$ with $\alpha^2 < 1$. Consider the special cases $\omega_c t_d \approx \pi/2$ and $\omega_c t_d \approx \pi$.

7.4-4 Confirm that the transfer function of the preemphasis circuit in Fig. 7.4-5b approximates Eq. (7) if $r \ll R$.

7.4-5 Justify the assertion that the preemphasis filter in Eq. (7) does not increase the FM transmission bandwidth if the energy or power spectrum of $x(t)$ satisfies the condition

$$G_x(f) \leq (B_{de}/f)^2 G_{\max} \qquad |f| > B_{de}$$

where $G_{\max}$ is the maximum of $G_x(f)$ in $|f| < B_{de}$.

7.4-6 Let $x(t)$ be a random modulating signal with $\bar{x} = 0$, $\overline{x^2} = 1$, and power spectrum $G_x(f) = 0$ for $|f| > W$. When $G_x(f)$ does not satisfy the condition in Prob. 7.4-5, the increased transmission bandwidth can be estimated using the rms preemphasized deviation ratio $D_{pe} = (\overline{x_{pe}^2})^{1/2} f_\Delta / W$, where $x_{pe}(t)$ is the preemphasized signal. (a) Taking $H_{de}(f)$ as in Eq. (6) and $H_{pe}(f) = K/H_{de}(f)$, show that

$$\overline{x_{pe}^2} = K^2 \left[1 + \int_{-\infty}^{\infty} (f/B_{de})^2 G_x(f) \, df \right]$$

(b) Evaluate the resulting bandwidth expansion when $K = 1$, $G_x(f) = (1/2W)\Pi(f/2W)$, $B_{de} = W/7$, and $f_\Delta = 5W$.

7.4-7 Equation (8) also holds for the case of unmodulated adjacent-channel interference if we let $\phi_i(t) = \omega_i t$. Sketch the resulting demodulated waveform when $\rho = 0.4, 0.8$, and 1.2.

7.4-8 When $\rho = 1$, Eq. (8b) appears to reduce to $\alpha(\rho, \phi_i) = 1/2$ for all ϕ_i. Show, however, that if $\phi_i = \pi$ and $\rho = 1 \pm \epsilon$, then $\alpha(\rho, \pi) \to \pm \infty$ as $\epsilon \to 0$.

7.4-9‡ Develop an expression for the demodulated signal when an FM signal with instantaneous phase $\phi(t)$ has interference from an unmodulated adjacent-channel carrier. Write your result in terms of $\phi(t)$, $\rho = A/A_c$, and $\theta_i(t) = \omega_i t + \phi_i$.

CHAPTER

EIGHT

CW MODULATION SYSTEMS

Communication systems that employ linear or exponential CW modulation may differ in many respects—type of modulation, carrier frequency, transmission medium, etc. But they have in common the property that a sinusoidal bandpass signal with time-varying envelope and/or phase conveys the message information. Consequently, generic hardware items such as oscillators, mixers, and bandpass filters are important building blocks for *all* CW modulation systems. Furthermore, many systems involve both linear and exponential CW modulation techniques.

This chapter therefore takes a broader look at CW modulation systems and hardware, using concepts and results from the previous two chapters. Specific topics are: CW receivers, frequency-division multiplexing, phase-lock loops, and television systems.

8.1 RECEIVERS FOR CW MODULATION

All that's really essential in a CW receiver is some tuning mechanism, demodulation, and amplification. With a sufficiently strong received signal you may even get by without amplification—witness the historic crystal radio set. However, most receivers operate on the more sophisticated *superheterodyne* principle, which we'll discuss first. Then we'll consider other types of receivers and the related scanning spectrum analyzer.

Superheterodyne Receivers

Besides demodulation, a typical broadcast receiver must perform three other operations: (1) carrier-frequency *tuning* to select the desired signal, (2) *filtering* to separate that signal from others received along with it, and (3) *amplification* to compensate for transmission loss. And at least some of the amplification should be provided before demodulation to bring the signal up to a level usable by the demodulator circuitry. In theory, all of the foregoing requirements could be met with a high-gain tunable bandpass amplifier. In practice, fractional-bandwidth and stability problems make such an amplifier expensive and difficult to build. Armstrong devised the superheterodyne or "superhet" receiver to circumvent those problems.

The superhet principle calls for two distinct amplification and filtering sections prior to demodulation, as diagrammed in Fig. 8.1-1. The incoming signal $x_c(t)$ is first selected and amplified by a *radio-frequency* (RF) section tuned to the desired carrier frequency f_c. This amplifier has a relatively broad bandwidth B_{RF} that partially passes adjacent-channel signals along with $x_c(t)$. Next, a *frequency converter* comprised of a mixer and local oscillator translates the RF output down to an *intermediate-frequency* (IF) band at $f_{IF} < f_c$. The adjustable LO frequency tracks with the RF tuning such that

$$f_{LO} = f_c + f_{IF} \quad \text{or} \quad f_{LO} = f_c - f_{IF} \tag{1}$$

and hence

$$| f_c - f_{LO} | = f_{IF} \tag{2}$$

An IF section with bandwidth $B_{IF} \approx B_T$ now removes the adjacent-channel signals. This section is a fixed bandpass amplifier, called the *IF strip*, which provides most of gain. Finally, the IF output goes to the demodulator for message recovery and baseband amplification.

The spectral drawings in Fig. 8.1-2 help clarify the action of a superhet receiver. Here we assume a modulated signal with symmetric sidebands (as distinguished from SSB or VSB), and we take $f_{LO} = f_c + f_{IF}$ so

$$f_c = f_{LO} - f_{IF}$$

The RF input spectrum in Fig. 8.1-2a includes our desired signal plus adjacent-channel signals on either side and another signal at the *image frequency*

$$f_c' = f_c + 2f_{IF} = f_{LO} + f_{IF} \tag{3}$$

Figure 8.1-1 Superheterodyne receiver.

Figure 8.1-2

The main task of the RF section is to pass $f_c \pm B_T/2$ while rejecting the image-frequency signal. For if f'_c reached the mixer, it would be down-converted to

$$f'_c - f_{LO} = (f_{LO} + f_{IF}) - f_{LO} = f_{IF}$$

and the image-frequency signal would produce an effect similar to cochannel interference. Hence, we want an RF response $|H_{RF}(f)|$ like the dashed line, with

$$B_T < B_{RF} < 2f_{IF} \qquad (4)$$

The filtered and down-converted spectrum at the IF input is shown in Fig. 8.1-2b. The indicated IF response $|H_{IF}(f)|$ with $B_{IF} \approx B_T$ completes the task of adjacent-channel rejection.

The superheterodyne structure results in several practical benefits. First, tuning takes place entirely in the "front end" so the rest of the circuitry, including the demodulator, requires no adjustment to change f_c. Second, the separation between f_c and f_{IF} eliminates potential instability due to stray feedback from the amplified output to the receiver's input. Third, most of the gain and selectivity is concentrated in the fixed-frequency IF strip. Since f_{IF} is an internal design parameter, it can be chosen to obtain a reasonable fractional bandwidth B_{IF}/f_{IF} for ease of implementation. Taken together, these benefits make it possible to build superhets with extremely high gain — 75 dB or more in the IF strip alone.

Additionally, when the receiver must cover a wide carrier-frequency range, the choice of $f_{LO} = f_c + f_{IF}$ may result in a smaller and more readily achieved LO *tuning ratio*. We should point out, however, that taking $f_{LO} > f_c$ in an SSB superhet causes *sideband reversal* in the down-converted signal, so USSB at RF becomes LSSB at IF, and vice versa.

A major disadvantage of the superhet structure is its potential for *spurious responses* at frequencies besides f_c. Image-frequency response is the most obvious problem, and satisfactory image rejection may require raising f_{IF} to increase the spacing between f_c and f'_c. Further problems come from signal feed-through and nonlinearities. For example, when a strong signal frequency near $\frac{1}{2}f_{IF}$ gets to the IF input, its second harmonic may be produced if the first stage of the IF amplifier is nonlinear. This second harmonic, being approximately f_{IF}, will then be amplified by later stages and appear at the detector input as interference.

Table 8.1-1 Parameters of AM and FM radios

	AM	FM
Carrier frequency	540–1600 kHz	88.1–107.9 MHz
Carrier spacing	10 kHz	200 kHz
Intermediate frequency	455 kHz	10.7 MHz
IF bandwidth	6–10 kHz	200–250 kHz
Audio bandwidth	3–5 kHz	15 kHz

Example 8.1-1 AM and FM broadcast radios Conventional AM and FM broadcast radios are superheterodyne receivers, like Fig. 8.1-1 with an audio-frequency (AF) amplifier after the demodulator. The frequency parameters are listed in Table 8.1-1 for reference and comparison purposes. Note that both types have $B_{IF}/f_{IF} \approx 0.02$. The selectivity made possible by mechanical IF filters was previously illustrated in Fig. 3.4-6.

An AM radio usually includes an *automatic volume control* (AVC) signal from the demodulator back to the IF, while an FM radio has *automatic frequency control* (AFC) fed back to the LO to correct small frequency drifts. With the exception of tuning coils, capacitors, and potentiometers, all of the circuitry needed for either radio is available on IC chips. Combination AM/FM radios often share a common audio amplifier and a single IF strip tuned to both 455 kHz and 10.7 MHz. (Stereo FM will be discussed in Sect. 8.2.)

Exercise 8.1-1 A superhet is to be designed for the range $f_c = 1.6$–2.6 MHz with $B_T = 6$ kHz. Unwanted strong signals in the vicinity are at 1.4–1.5 MHz and 3.0–5.6 MHz. (*a*) Take $f_c = f_{LO} - f_{IF}$ and find a value for f_{IF} that avoids image-frequency interference. Then calculate the corresponding fractional bandwidth of the IF section and the LO tuning ratio. (*b*) Repeat (*a*) with $f_c = f_{LO} + f_{IF}$.

Special-Purpose Receivers★

Other types of receivers used for special purposes include the heterodyne, tuned-RF, and double-conversion structure. A *heterodyne receiver* is a superhet without the RF section, which raises potential image-frequency problems. Such receivers can be built at microwave frequencies with a diode mixer preceded by a fixed microwave filter to reject images. The *tuned-RF* (TRF) *receiver* consists of two or more stages of tunable RF amplification followed directly by demodulation. This economical design has limited gain and tuning range, and works best with envelope detection.

The *double-conversion receiver* in Fig. 8.1-3 takes the superhet principle one step further by including two frequency converters and two IF sections. The second IF is always fixed-tuned, while the first IF and second LO may be fixed or tunable. In either case, double-conversion permits a larger value of $f_{IF\text{-}1}$ to

Figure 8.1-3 Double-conversion receiver.

improve image rejection in the RF section, and a smaller value of $f_{\text{IF-2}}$ to improve adjacent-channel rejection in the second IF. High-performance receivers for SSB and shortwave AM usually employ this design strategy.

Notice that a double-conversion SSB receiver with synchronous detection requires *three* stable oscillators plus automatic frequency control and synchronization circuitry. Fortunately, modern integrated-circuit technology has made the *frequency synthesizer* available for this application. We'll discuss frequency synthesis using phase-lock loops in Sect. 8.3.

Finally, we point out that receivers for cable transmission and other fixed-frequency systems need not be as elaborate as tunable broadcast receivers. But superhets with *up* conversion are occasionally used to get sufficient amplification when the modulated signal has a low carrier frequency.

Exercise 8.1-2 The image-frequency rejection ratio for a superheterodyne or double-conversion receiver is defined as

$$RR = |H_{\text{RF}}(f_c)/H_{\text{RF}}(f'_c)|^2 \qquad (5)$$

usually expressed in dB. Suppose the RF section is a simple tuned circuit described by Eq. (1), Sect. 6.1, with $f_0 = f_c$ and $Q = 50$. Show that achieving $RR \approx 10^6$ (60 dB) requires $f'_c/f_c \approx 20$ when $f'_c = f_c + 2f_{\text{IF}}$. This requirement could easily be satisfied by a double-conversion receiver with $f_{\text{IF-1}} \approx 9.5 f_c$.

Scanning Spectrum Analyzers★

If the LO in a superhet is replaced by a VCO, then the predection portion acts like a *voltage-tunable bandpass amplifier* with center frequency $f_0 = f_{\text{LO}} \pm f_{\text{IF}}$ and bandwidth $B = B_{\text{IF}}$. This property is at the heart of the *scanning spectrum analyzer* in Fig. 8.1-4a — a useful laboratory instrument that displays the spectral magnitude of an input signal over some selected frequency range.

The VCO is driven by a periodic ramp generator that sweeps the instantaneous frequency $f_{\text{LO}}(t)$ linearly from f_1 to f_2 in T seconds. The IF section has a narrow bandwidth B, usually adjustable, and the IF output goes to an envelope detector. Hence, the system's amplitude response at any instant t looks like Fig. 8.1-4b, where $f_0(t) = f_{\text{LO}}(t) - f_{\text{IF}}$. A fixed BPF (or LPF) at the input passes $f_1 \leq f \leq f_2$ while rejecting the image at $f_0(t) + 2f_{\text{IF}}$.

(a)

(b)

Figure 8.1-4 Scanning spectrum analyzer. (a) Block diagram; (b) amplitude response.

As $f_0(t)$ repeatedly scans past the frequency components of an input signal $v(t)$, its spectrum is displayed by connecting the envelope detector and ramp generator to the vertical and horizontal deflections of an oscilloscope. Obviously, a transient signal would not yield a fixed display, so $v(t)$ must be either a *periodic* or *quasi-periodic signal* or a *stationary random signal* over the time of observation. Correspondingly, the display represents the amplitude line spectrum or the power spectral density. (A square-law envelope detector would be used for the latter.)

Some of the operational subtleties of this system are best understood by assuming that $v(t)$ consists of two or more sinusoids. To resolve one spectral line from the others, the IF bandwidth must be smaller than the line spacing. Hence, we call B the *frequency resolution*, and the maximum number of resolvable lines equals $(f_2 - f_1)/B$. The IF output produced by a single line takes the form of a *bandpass pulse* with time duration

$$\tau = BT/(f_2 - f_1) = B/\dot{f}_0$$

where $\dot{f}_0 = (f_2 - f_1)/T$ represents the *frequency sweep rate* in hertz per second.

But a rapid sweep rate may exceed the IF pulse response. Recall that our guideline for bandpass pulses requires $B \geq 1/\tau = \dot{f}_0/B$, or

$$\dot{f}_0 = \frac{f_2 - f_1}{T} \leq B^2 \tag{6}$$

This important relation shows that accurate resolution (small B) calls for a slow rate and correspondingly long observation time. Also note that Eq. (6) involves four parameters adjustable by the user. Some scanning spectrum analyzers have built-in hardware that prevents you from violating Eq. (6); others simply have a warning light.

8.2 FREQUENCY-DIVISION MULTIPLEXING

When several communication channels are needed between the same two points, significant economies may be realized by sending all the messages on one transmission facility—a process called *multiplexing*. Applications of multiplexing range from the vital, if prosaic, telephone network to the glamour of FM stereo and space-probe telemetry systems. There are two basic multiplexing techniques: *frequency-division multiplexing* (FDM) treated here, and *time-division multiplexing* (TDM) discussed in Chap. 10.

FDM Systems

The principle of FDM is illustrated by Fig. 8.2-1a, where several input messages (three are shown) individually modulate the *subcarriers* f_{c_1}, f_{c_2}, etc., after passing

(a)

(b)

Figure 8.2-1 Typical FDM transmitter. (a) Input spectra and block diagram; (b) baseband FDM spectrum.

through LPFs to limit the message bandwidths. We show the subcarrier modulation as SSB, and it often is; but any of the CW modulation techniques could be employed, or a mixture of them. The modulated signals are then summed to produce the *baseband signal*, with spectrum $X_b(f)$, as shown in Fig. 8.2-1b. (The designation "baseband" is used here to indicate that final carrier modulation has not yet taken place.) The baseband time function $x_b(t)$ is left to your imagination.

Assuming that the subcarrier frequencies are properly chosen, the multiplexing operation has assigned a slot in the frequency domain for each of the individual messages in modulated form, hence the name *frequency-division* multiplexing. The baseband signal may then be transmitted directly or used to modulate a transmitted carrier of frequency f_c. We are not particularly concerned here with the nature of the final carrier modulation, since the baseband spectrum tells the story.

Message recovery or demodulation of FDM is accomplished in three steps portrayed by Fig. 8.2-2. First, the carrier demodulator reproduces the baseband signal $x_b(t)$. Then the modulated subcarriers are separated by a bank of bandpass filters in parallel, following which the messages are individually detected.

The major practical problem of FDM is *cross talk*, the unwanted coupling of one message into another. Intelligible cross talk (cross modulation) arises primarily because of nonlinearities in the system, which cause one message signal to appear as modulation on another subcarrier, Consequently, standard practice calls for negative feedback to minimize amplifier nonlinearity in FDM systems. (As a matter of historical fact, the FDM cross-talk problem was a primary motivation for the development of negative-feedback amplifiers.)

Unintelligible cross talk may come from nonlinear effects or from imperfect spectral separation by the filter bank. To reduce the latter, the modulated message spectra are spaced out in frequency by *guard bands* into which the filter transition regions can be fitted. For example, the guard band between $X_{c_1}(f)$ and $X_{c_2}(f)$ in Fig. 8.2-1b is of width $f_{c_2} - (f_{c_1} + W_1)$. The net *baseband bandwidth* is therefore the sum of the modulated message bandwidths plus the guard bands.

While the concept of FDM is quite simple, typical systems may be very elaborate. A case in point is the AT & T type L4 carrier system, in which 3600

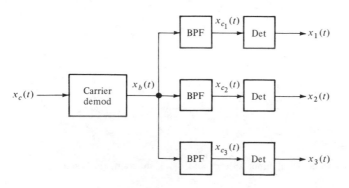

Figure 8.2-2 Typical FDM receiver.

voice channels (each with $W = 4$ kHz nominally) are multiplexed together for transmission via coaxial cable. All modulation is single sideband, both USSB and LSSB, and the final baseband spectrum runs from 0.5 to 17.5 MHz, including pilot carrier and guard bands. All subcarriers are multiples of 4 kHz synthesized from a common oscillator, and the 512-kHz pilot provides synchronization. To avoid excessive guard-band width at the upper end of the baseband spectrum, the multiplexing is done by *groups* in four stages, an arrangement that also facilitates switching and routing of the various channels.

Figure 8.2-3 depicts the first two stages of telephone multiplexing. The *group* multiplexer takes 12 input voice channels and puts them on subcarriers at 64, 68, ..., 108 kHz using LSSB modulation. The resulting group spectrum is 48-kHz wide, starting at 60 kHz. Five such groups are then LSSB modulated on subcarriers at 420, 468, ..., 612 kHz to produce a *supergroup* spectrum from 312 to 552 kHz. Note that the suppressed subcarrier for group 1 would fall in the group 3 spectrum. Also note that voice channel 1 in group 1 now appears as a USSB signal at 354 kHz. (Anyone who perchance is not fully convinced of the power of frequency-domain analysis should consider the task of designing this sytem without the aid of spectral concepts.)

Table 8.2-1 lists the AT & T FDM hierarchy for reference purposes. The indicated frequency ranges are for the L4 system, but the channel groupings and bandwidths also apply to other systems. At any point in the hierarchy, a wideband video or digital data channel of comparable bandwidth may replace the specified number of voice channels.

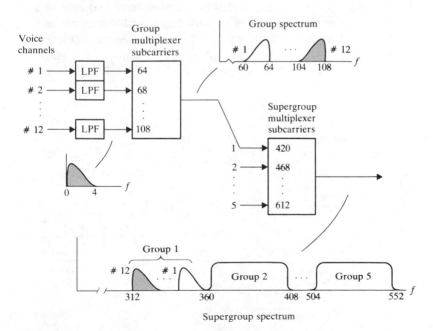

Figure 8.2-3 Group and supergroup multiplexing in voice-telephone FDM (all frequencies in kilohertz).

Table 8.2-1 AT & T FDM hierarchy

Designation	Frequency range (L4 system)	Bandwidth	Number of voice channels
Group	60–108 kHz	48 kHz	12
Supergroup	312–552 kHz	240 kHz	60
Mastergroup	564–3084 kHz	2.52 MHz	600
Jumbogroup	0.5–17.5 MHz	17 MHz	3600

Example 8.2-1 FDMA satellite systems The Intelsat global network adds a third dimension to long-distance communication. Since a particular satellite links several ground stations in different countries, various *access* methods have been devised for international telephony. One scheme, known as *frequency-division multiple access* (FDMA), assigns a fixed number of voice channels between pairs of ground stations. These channels are grouped with standard FDM hardware, and relayed through the satellite using FM carrier modulation.

For the sake of example, suppose a satellite over the Atlantic Ocean serves ground stations in the United States, Brazil, and France. Further suppose that 36 channels (3 groups) are assigned to the US-France route and 24 channels (2 groups) to the US-Brazil route. Figure 8.2-4 shows the arrangement of the US transmitter and the receivers in Brazil and France. Not shown are the French and Brazilian transmitters and the US receiver needed for two-way conversations. Additional transmitters and receivers at slightly different carrier frequencies would provide a Brazil-France route.

The FDMA scheme creates at the satellite a *composite FDM signal* assembled with the FM signals from all ground stations. The satellite equipment consists of a bank of transponders, like Fig. 6.5-2. Each trans-

Figure 8.2-4 Simplified FDMA satellite system.

ponder has 36-MHz bandwidth accommodating 336 to 900 voice channels, depending on the ground-pair assignments. More details and other access schemes can be found in Puente and Werth (1971).

Exercise 8.2-1 Suppose an FDM baseband amplifier has cube-law nonlinearity which produces a baseband component proportional to $(v_2 \cos \omega_2 t)^2 v_1 \cos \omega_1 t$, where f_1 and f_2 are two subcarrier frequencies. Show that AM subcarrier modulation with $v_1 = 1 + x_1(t)$ and $v_2 = 1 + x_2(t)$ results in both intelligible and unintelligible cross talk on subcarrier f_1. Compare with the DSB case $v_1 = x_1(t)$ and $v_2 = x_2(t)$.

FM Stereo Multiplexing★

Figure 8.2-5a diagrams the FDM system that generates the baseband signal for FM stereophonic broadcasting. The left-speaker and right-speaker signals are first matrixed and preemphasized to produce $x_L(t) + x_R(t)$ and $x_L(t) - x_R(t)$. The sum signal is heard with a monophonic receiver; matrixing is required so the monaural listener will not be subjected to sound gaps in program material having stereophonic Ping-Pong effects. The $x_L(t) + x_R(t)$ signal is then inserted directly into the baseband, while $x_L(t) - x_R(t)$ DSB modulates a 38-kHz subcarrier. Double-sideband modulation is employed to preserve fidelity at low frequencies, and a 19-kHz pilot tone is added for receiver synchronization.

The resulting baseband spectrum is sketched in Fig. 8.2-5b. Also shown is another spectral component labeled SCA, which stands for Subsidiary Communication Authorization. The SCA signal has NBFM subcarrier modulation and is

Figure 8.2-5 FM stereo multiplexing. (a) Transmitter; (b) baseband spectrum.

Figure 8.2-6 FM stereo multiplex receiver.

transmitted by some FM stations for the use of private subscribers who pay for commercial-free program material—the so-called "wallpaper music" heard in stores and offices.

For stereo broadcasting without SCA, the pilot carrier is allotted 10% of the peak frequency deviation and the seesaw relationship between $L + R$ and $L - R$ permits each to achieve nearly 90% deviation. The fact that the baseband spectrum extends to 53 kHz (or 75 kHz with SCA) does not appreciably increase the transmission bandwidth requirement because the higher frequencies produce smaller deviation ratios. High-fidelity stereo receivers typically have $B_{IF} \approx 250$ kHz.

The stereo *demultiplexing* or *decoding* system is diagrammed in Fig. 8.2-6. Notice how the pilot tone is used to actuate the stereo indicator as well as for synchronous detection. Integrated-circuit decoders employ switching circuits or phase-lock loops to carry out the functional operations.

Incidentally, discrete four-channel (quadraphonic) disk recording takes a logical extension of the FM stereo strategy to multiplex four independent signals on the two channels of a stereophonic record. Let's denote the four signals as L_F, L_R, R_F, and R_R (for left-front, left-rear, etc.). The matrixed signal $L_F + L_R$ is recorded directly on one channel along with $L_F - L_R$ multiplexed via frequency modulation of a 30-kHz subcarrier. The matrixed signals $R_F + R_R$ and $R_F - R_R$ are likewise multiplexed on the other channel. Because the resulting baseband spectrum goes up to 45 kHz, discrete quadraphonic signals cannot be transmitted in full on stereo FM. Other quadraphonic systems have only two independent channels and are compatible with FM stereo; see Jurgen (1972).

Quadrature-Carrier Multiplexing

Quadrature-carrier multiplexing, also known as *quadrature amplitude modulation* (QAM), utilizes carrier phase shifting and synchronous detection to permit two DSB signals to occupy the same frequency band. Figure 8.2-7 illustrates the

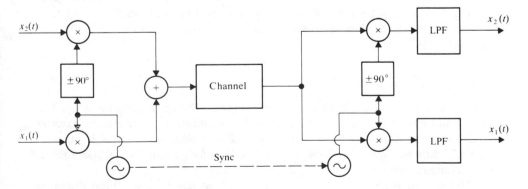

Figure 8.2-7 Quadrature-carrier multiplexing.

multiplexing and demultiplexing arrangement. The transmitted signal has the form

$$x_c(t) = A_c[x_1(t) \cos \omega_c t \pm x_2(t) \sin \omega_c t] \tag{1}$$

Since the modulated spectra overlap each other, this technique is more properly characterized as frequency-*domain* rather than frequency-*division* multiplexing.

From our prior study of synchronous detection for DSB and SSB, you should readily appreciate the fact that QAM involves more stringent synchronization than, say, an FDM system with SSB subcarrier modulation. Hence, QAM is limited to specialized applications, notably color television and digital data transmission.

A potential third application is *AM stereo broadcasting*. Two of the five methods under consideration for compatible stereo in the AM band involve QAM. In particular, Harris Corporation proposes a modulated signal in the form

$$x_c(t) = A_c[\cos \omega_c t + x_L(t) \cos (\omega_c t - \phi_0) + x_R(t) \cos (\omega_c t + \phi_0)] \tag{2}$$

where $\phi_0 = 15°$. Envelope detection in a standard AM radio yields a monaural output approximately equal to $x_L(t) + x_R(t)$, while quadrature demultiplexing yields the stereo outputs $x_L(t)$ and $x_R(t)$. Mennie (1978) describes all five proposed systems.

Exercise 8.2-2 Express Eq. (2) in the form of Eq. (1). Then use your result to diagram a system that recovers the stereo signals.

8.3 PHASE-LOCK LOOPS

The phase-lock loop (PLL) is undoubtedly the most versatile building block available for CW modulation systems. PLLs are found in modulators, demodulators, frequency synthesizers, multiplexers, and a variety of signal processors. We'll illustrate some of these applications after discussing PLL operation and lock-in conditions. Our introductory study provides a useful working knowledge of PLLs but does not go into detailed analysis of nonlinear behavior and transients. Treatments of these advanced topics are given in Blanchard (1976) and Gardner (1979).

PLL Operation and Lock-in

The basic aim of a PLL is to *lock* or *synchronize* the instantaneous angle (i.e., phase and frequency) of a VCO output to the instantaneous angle of an external bandpass signal that may have some type of CW modulation. For this purpose, the PLL must perform *phase comparison*. We therefore begin with a brief look at phase comparators.

The system in Fig. 8.3-1a is an *analog phase comparator*. It produces an output $y(t)$ that depends on the instantaneous angular difference between two

(a)

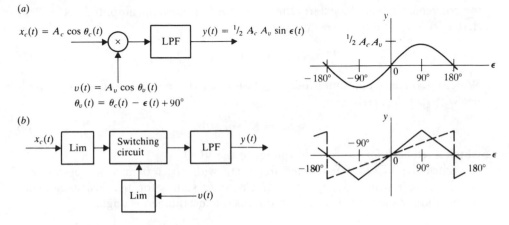

$x_c(t) = A_c \cos \theta_c(t)$

LPF

$y(t) = \frac{1}{2} A_c A_v \sin \epsilon(t)$

$v(t) = A_v \cos \theta_v(t)$
$\theta_v(t) = \theta_c(t) - \epsilon(t) + 90°$

(b)

$x_c(t)$ — Lim — Switching circuit — LPF — $y(t)$

Lim — $v(t)$

Figure 8.3-1 Phase comparators. (a) Analog; (b) digital.

bandpass input signals, $x_c(t) = A_c \cos \theta_c(t)$ and $v(t) = A_v \cos \theta_v(t)$. Specifically, if

$$\theta_v(t) = \theta_c(t) - \epsilon(t) + 90° \tag{1}$$

and if the LPF simply extracts the difference-frequency term from the product $x_c(t)v(t)$, then

$$y(t) = \frac{1}{2}A_c A_v \cos [\theta_c(t) - \theta_v(t)]$$
$$= \frac{1}{2}A_c A_v \cos [\epsilon(t) - 90°] = \frac{1}{2}A_c A_v \sin \epsilon(t)$$

We interpret $\epsilon(t)$ as the *angular error*, and the plot of y versus ϵ emphasizes that $y(t) = 0$ when $\epsilon(t) = 0$. Had we omitted the 90° shift in Eq. (1), we would get $y(t) = 0$ at $\epsilon(t) = \pm 90°$. Thus, zero output from the phase comparator corresponds to a *quadrature* phase relationship.

Also note that $y(t)$ depends on $A_c A_v$ when $\epsilon(t) \neq 0$, which could cause problems if the input signals have amplitude modulation. These problems are eliminated by the *digital phase comparator* in Fig. 8.3-1b, where hard limiters convert input sinusoids to square waves applied to a switching circuit. The resulting plot of y versus ϵ has a triangular or sawtooth shape, depending on the switching-circuit details. However, all three phase-comparison curves are essentially the same when $|\epsilon(t)| \ll 90°$ — the intended operating condition in a PLL.

Hereafter, we'll work with the analog PLL structure in Fig. 8.3-2. We assume

$x_c(t) = 2 \cos \theta_c(t)$

LPF

K_a

$y(t) = K_a \sin \epsilon(t)$

VCO

$v(t) = \cos \theta_v(t)$

Figure 8.3-2 Phase-lock loop.

for convenience that the external input signal has constant amplitude $A_c = 2$ so that $x_c(t) = 2 \cos \theta_c(t)$ where, as usual,

$$\theta_c(t) = \omega_c t + \phi(t) \qquad \omega_c = 2\pi f_c \tag{2}$$

We also assume a unit-amplitude VCO output $v(t) = \cos \theta_v(t)$ and a loop amplifier with gain K_a. Hence,

$$y(t) = K_a \sin \epsilon(t) \tag{3}$$

which is fed back for the control voltage to the VCO.

Since the VCO's *free-running frequency* with $y(t) = 0$ may not necessarily equal f_c, we'll write it as $f_v = f_c - \Delta f$ where Δf stands for the *frequency error*. Application of the control voltage produces the instantaneous angle

$$\theta_v(t) = 2\pi(f_c - \Delta f)t + \phi_v(t) + 90° \tag{4a}$$

with

$$\phi_v(t) = 2\pi K_v \int^t y(\lambda) \, d\lambda \tag{4b}$$

when K_v equals the frequency-deviation constant. The angular error is then

$$\epsilon(t) = \theta_c(t) - \theta_v(t) + 90°$$
$$= 2\pi \Delta f t + \phi(t) - \phi_v(t)$$

and differentiation with respect to t gives

$$\dot{\epsilon}(t) = 2\pi \Delta f + \dot{\phi}(t) - 2\pi K_v \, y(t)$$

Upon combining this expression with Eq. (3) we obtain the *nonlinear differential equation*

$$\dot{\epsilon}(t) + 2\pi K \sin \epsilon(t) = 2\pi \Delta f + \dot{\phi}(t) \tag{5}$$

in which we've introduced the *loop gain*

$$K \triangleq K_v K_a$$

This gain is measured in *hertz* and turns out to be a critical parameter.

Equation (5) governs the dynamic operation of a PLL, but it does not yield a closed-form solution with an arbitrary $\phi(t)$. To get a sense of PLL behavior and lock-in conditions, consider the case of a constant input phase $\phi(t) = \phi_0$ starting at $t = 0$. Then $\dot{\phi}(t) = 0$ and we rewrite Eq. (5) as

$$\frac{1}{2\pi K} \dot{\epsilon}(t) + \sin \epsilon(t) = \frac{\Delta f}{K} \qquad t \geq 0 \tag{6}$$

Lock-in with a constant phase implies that the loop attains a *steady state* with $\dot{\epsilon}(t) = 0$ and $\epsilon(t) = \epsilon_{ss}$. Hence, $\sin \epsilon_{ss} = \Delta f/K$ at lock-in, and it follows that

$$\epsilon_{ss} = \arcsin \frac{\Delta f}{K} \tag{7a}$$

$$y_{ss} = K_a \sin \epsilon_{ss} = \frac{\Delta f}{K_v} \tag{7b}$$

$$v_{ss}(t) = \cos (\omega_c t + \phi_0 - \epsilon_{ss} + 90°) \tag{7c}$$

Note that the nonzero value of y_{ss} cancels out the VCO frequency error, and $v_{ss}(t)$ is locked to the frequency of the input signal $x_c(t)$. The phase error ϵ_{ss} will be negligible if $|\Delta f/K| \ll 1$.

However, Eq. (6) has *no steady-state solution* and ϵ_{ss} in Eq. (7a) is undefined when $|\Delta f/K| > 1$. Therefore, lock-in requires the condition

$$K \geq |\Delta f| \tag{8}$$

Stated another way, a PLL will lock to any constant input frequency within the range $\pm K$ hertz of the VCO's free-running frequency f_v.

Additional information regarding PLL behavior comes from Eq. (6) when we require sufficient loop gain that $\epsilon_{ss} \approx 0$. Then, after some instant $t_0 > 0$, $\epsilon(t)$ will be small enough to justify the approximation $\sin \epsilon(t) \approx \epsilon(t)$ and

$$\frac{1}{2\pi K} \dot{\epsilon}(t) + \epsilon(t) = 0 \qquad t \geq t_0 \tag{9a}$$

This linear equation yields the well-known solution

$$\epsilon(t) = \epsilon(t_0)e^{-2\pi K(t-t_0)} \qquad t \geq t_0 \tag{9b}$$

a *transient* error that virtually disappears after five time constants have elapsed, that is, $\epsilon(t) \approx 0$ for $t > t_0 + 5/(2\pi K)$. We thus infer that if the input $x_c(t)$ has a *time-varying* phase $\phi(t)$ whose variations are slow compared to $1/(2\pi K)$, and if the instantaneous frequency $f_c + \dot{\phi}(t)/2\pi$ does not exceed the range $f_v \pm K$, then the PLL will stay in lock and track $\phi(t)$ with negligible error — provided that the LPF in the phase comparator passes the variations of $\phi(t)$ on to the VCO.

Exercise 8.3-1 The *phase-plane plot* of $\dot{\epsilon}$ versus ϵ is defined by rewriting Eq. (6) in the form

$$\dot{\epsilon} = 2\pi(\Delta f - K \sin \epsilon)$$

(a) Sketch $\dot{\epsilon}$ versus ϵ for $K = 2 \Delta f$ and show that an arbitrary initial value $\epsilon(0)$ must go to $\epsilon_{ss} = 30° \pm m\ 360°$ where m is an integer. *Hint:* $\epsilon(t)$ increases when $\dot{\epsilon}(t) > 0$ and decreases when $\dot{\epsilon}(t) < 0$. (b) Now sketch the phase-plane plot for $K < \Delta f$ to show that $|\dot{\epsilon}(t)| > 0$ for any $\epsilon(t)$ and, consequently, ϵ_{ss} does not exist.

Synchronous Detection and Frequency Synthesizers

The lock-in ability of a PLL makes it ideally suited to systems that have a pilot carrier for synchronous detection. Rather than attempting to filter the pilot out of the accompanying modulated waveform, the augmented PLL circuit in Fig. 8.3-3 can be used to generate a sinusoid synchronized with the pilot. To minimize clutter here, we've lumped the phase comparator, lowpass filter, and amplifier into a phase discriminator (PD) and we've assumed unity sinusoidal amplitudes throughout.

Initial adjustment of the tuning voltage brings the VCO frequency close to f_c and $\epsilon_{ss} \approx 0$, a condition sensed by the quadrature phase discriminator and displayed by the lock-in indicator. Thereafter, the PLL automatically tracks any phase or frequency drift in the pilot, and the phase-shifted VCO output provides the LO signal needed for the synchronous detector. Thus, the whole unit acts as a *narrowband pilot filter* with a virtually noiseless output.

Incidentally, a set-up like Fig. 8.3-3 can be used to *search* for a signal at some unknown frequency. You disconnect the VCO control voltage and apply a ramp generator to sweep the VCO frequency until the lock-in indicator shows that a signal has been found. Some radio scanners employ an automated version of this procedure.

For *synchronous detection* of DSB without a transmitted pilot, Costas invented the PLL system in Fig. 8.3-4. The modulated DSB waveform $x(t) \cos \omega_c t$ with bandwidth $2W$ is applied to a pair of phase discriminators whose outputs are proportional to $x(t) \sin \epsilon_{ss}$ and $x(t) \cos \epsilon_{ss}$. Multiplication and integration over $T \gg 1/W$ produces the VCO control voltage

$$y_{ss} \approx T\langle x^2(t)\rangle\sin \epsilon_{ss} \cos \epsilon_{ss} = \frac{T}{2} S_x \sin 2\epsilon_{ss}$$

If $\Delta f \approx 0$, the PLL locks with $\epsilon_{ss} \approx 0$ and the output of the quadrature discriminator is proportional to the demodulated message $x(t)$. Of course the loop loses lock if $x(t) = 0$ for an extended interval.

The *frequency-offset loop* in Fig. 8.3-5 translates the input frequency (and phase) by an amount equal to that of an auxiliary oscillator. The intended output

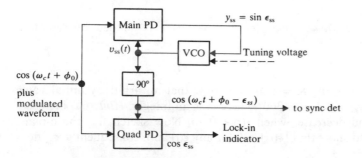

Figure 8.3-3 PLL pilot filter with two phase discriminators (PD).

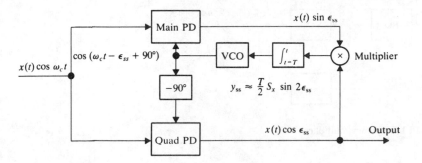

Figure 8.3-4 Costas PLL system for synchronous detection.

frequency is now $f_c + f_1$, so the free-running frequency of the VCO must be

$$f_v = (f_c + f_1) - \Delta f \approx f_c + f_1$$

The oscillator and VCO outputs are mixed and filtered to obtain the difference-frequency signal $\cos [\theta_v(t) - (\omega_1 t + \phi_1)]$ applied to the phase discriminator. Under locked conditions with $\epsilon_{ss} \approx 0$, the instantaneous angles at the input to the discriminator will differ by 90°. Hence, $\theta_v(t) - (\omega_1 t + \phi_1) = \omega_c t + \phi_0 + 90°$, and the VCO produces $\cos [(\omega_c + \omega_1)t + \phi_0 + \phi_1 + 90°]$.

By likewise equating instantaneous angles, you can confirm that Fig. 8.3-6 performs *frequency multiplication*. Like the frequency multiplier discussed in Sect. 7.2, this unit multiplies the instantaneous angle of the input by a factor of n. However, it does so with the help of a *frequency divider* which is easily implemented using a *digital counter*. Commercially available divide-by-n counters allow you to select any integer value for n from 1 to 10 or even higher. When such a counter is inserted in a PLL, you have an *adjustable* frequency multiplier.

A *frequency synthesizer* starts with the output of one crystal-controlled master oscillator; various other frequencies are synthesized therefrom by combinations of frequency division, multiplication, and translation. Thus, all resulting frequencies are stabilized by and synchronized with the master oscillator.

Figure 8.3-5 Frequency-offset loop.

Figure 8.3-6 PLL frequency multiplier.

General-purpose laboratory synthesizers incorporate additional refinements and have rather complicated diagrams. So we'll illustrate the principles of frequency synthesis by an example.

Example 8.3-1 Suppose a double-conversion SSB receiver needs fixed LO frequencies at 100 kHz (for synchronous detection) and 1.6 MHz (for the second mixer), and an adjustable LO that covers 9.90–9.99 MHz in steps of 0.01 MHz (for RF tuning). The custom-tailored synthesizer in Fig. 8.3-7 provides all the required frequencies by dividing down, multiplying up, and mixing with the output of a 10-MHz oscillator. You can quickly check out the system by putting a frequency-multiplication block in place of each PLL with a divider.

Observe here that all output frequencies are *less* than the master-oscillator frequency. This ensures that any frequency drift will be reduced rather than increased by the synthesis operations.

Figure 8.3-7 Frequency synthesizer with fixed and adjustable outputs.

Exercise 8.3-2 Draw the block diagram of a PLL system that synthesizes the output frequency nf_c/m from a master-oscillator frequency f_c. State the condition for locked operation in terms of the loop gain K and the VCO free-running frequency f_v.

Linearized PLL Models and FM Detection

Suppose that a PLL has been tuned to lock with the input frequency f_c, so $\Delta f = 0$. Suppose further that the PLL has sufficient loop gain to track the input phase $\phi(t)$ within a small error $\epsilon(t)$, so $\sin \epsilon(t) \approx \epsilon(t) = \phi(t) - \phi_v(t)$. These suppositions constitute the basis for the *linearized PLL model* in Fig. 8.3-8a, where the LPF has been represented by its impulse response $h(t)$.

Since we'll now focus on the phase variations, we view $\phi(t)$ as the input "signal" which is compared with the feedback "signal"

$$\phi_v(t) = 2\pi K_v \int^t y(\lambda)\, d\lambda$$

to produce the output $y(t)$. We emphasize that viewpoint by redrawing the linearized model as a *negative feedback system*, Fig. 8.3-8b. Note that the VCO becomes an integrator with gain $2\pi K_v$, while phase comparison becomes subtraction.

(a)

(b)

(c)

Figure 8.3-8 Linearized PLL models. (a) Time domain; (b) phase; (c) frequency domain.

Fourier transformation finally takes us to the *frequency-domain model* in Fig. 8.3-8c, where $\Phi(f) = \mathcal{F}[\phi(t)]$, $H(f) = \mathcal{F}[h(t)]$, etc. Routine analysis (or comparison with Fig. 3.1-8c) yields

$$Y(f) = \frac{K_a H(f)}{1 + K_a H(f)(K_v/jf)}\,\Phi(f) = \frac{1}{K_v}\,\frac{jfKH(f)}{jf + KH(f)}\,\Phi(f) \tag{10}$$

which expresses the frequency-domain relationship between the input phase and output voltage.

Now let $x_c(t)$ be an *FM wave* with $\phi(t) = 2\pi f_\Delta x(t)$ and, accordingly,

$$\Phi(f) = 2\pi f_\Delta X(f)/(j2\pi f) = (f_\Delta/jf)X(f)$$

Substituting for $\Phi(f)$ in Eq. (10) gives

$$Y(f) = \frac{f_\Delta}{K_v}\,H_L(f)X(f) \tag{11a}$$

where

$$H_L(f) = \frac{H(f)}{H(f) + j(f/K)} \tag{11b}$$

which we interpret as the equivalent *loop transfer function*. If $X(f)$ has message bandwidth W and if

$$H(f) = 1 \qquad |f| \le W \tag{12a}$$

then $H_L(f)$ takes the form of a *first-order lowpass filter* with 3-dB bandwidth K, namely

$$H_L(f) = \frac{1}{1 + j(f/K)} \qquad |f| \le W \tag{12b}$$

Thus, $Y(f) \approx (f_\Delta/K_v)X(f)$ when $K \ge W$ so

$$y(t) \approx \frac{f_\Delta}{K_v}\,x(t) \tag{13}$$

Under these conditions, the PLL recovers the message $x(t)$ from $x_c(t)$ and thereby serves as an *FM detector*.

A disadvantage of the first-order PLL with $H(f) = 1$ is that the loop gain K determines both the bandwidth of $H_L(f)$ and the lock-in frequency range. In order to track the instantaneous input frequency $f(t) = f_c + f_\Delta x(t)$ we must have $K \ge f_\Delta$. The large bandwidth of $H_L(f)$ may then result in excessive interference and noise at the demodulated output. For this reason, and other considerations, $H_L(f)$ is usually a more sophisticated *second-order* function in practical PLL frequency detectors.

8.4 TELEVISION

The message to be transmitted by a television system is a two-dimensional *image* with motion, and therefore a function of two spatial variables as well as time. This section introduces the theory and practice of image transmission via an electrical signal. Our initial discussion of monochrome (black-and-white) video signals and bandwidth requirements also applies to facsimile systems, which transmit only still pictures. Then we'll describe TV transmitters and receivers, in block-diagram form, and the modifications needed for color television.

There are three major types of television systems with numerous variations found in different countries. We'll concentrate primarily on the NTSC (National Television System Committee) system used in North America, South America, and Japan. Specific details regarding the electronic circuitry are given in texts such as Hansen (1969) or Herrick (1976).

Video Signals, Resolution, and Bandwidth

To start with the simplest case, consider a motion-free monochrome intensity pattern $I(h, v)$, where h and v are the horizontal and vertical coordinates. Converting $I(h, v)$ to a signal $x(t)$—and vice versa—requires a discontinuous mapping process such as the *scanning raster* diagrammed in Fig. 8.4-1. The scanning device, which produces a voltage or current proportional to intensity, starts at point A and moves with constant but unequal rates in the horizontal and vertical directions, following the path AB. Thus, if s_h and s_v are the horizontal and vertical scanning speeds, the output of the scanner is the *video signal*

$$x(t) = I(s_h t, s_v t) \tag{1}$$

since $h = s_h t$, etc. Upon reaching point B, the scanning spot quickly flies back to C (the horizontal retrace) and proceeds similarly to point D, where facsimile scanning would end.

In TV, however, image motion must be accommodated, so the spot retraces vertically to E and follows an interlaced pattern ending at F. The process is then

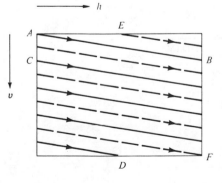

Figure 8.4-1 Scanning raster with two fields (line spacing grossly exaggerated).

Figure 8.4-2 Video waveform for one full line (NTSC standards).

repeated starting again at *A*. The two sets of lines are called the first and second *fields*; together they constitute one complete picture or *frame*. The frame rate is just rapid enough (25 to 30 per second) to create the illusion of continuous motion, while the field rate (twice the frame rate) makes the flickering imperceptible to the human eye. Hence, interlaced scanning allows the lowest possible picture repetition rate without visible flicker.

Two modifications are made to the video signal after scanning: *blanking pulses* are inserted during the retrace intervals to blank out retrace lines on the receiving picture tube; and *synchronizing pulses* are added on top of the blanking pulses to synchronize the receiver's horizontal and vertical sweep circuits. Figure 8.4-2 shows the waveform for one complete line, with amplitude levels and durations corresponding to NTSC standards. Other parameters are listed in Table 8.4-1 along with some comparable values for the European CCIR (International Radio Consultative Committee) system and a proposed *high-definition* (HDTV) system.

Analyzing the spectrum of the video signal in absence of motion is relatively easy with the aid of Fig. 8.4-3 where, instead of retraced scanning, the image has been periodically repeated in both directions so the equivalent scanning path is

Table 8.4-1 Television system parameters

	NTSC	CCIR	HDTV
Aspect ratio, horizontal/vertical	4/3	4/3	5/3
Total lines per frame	525	625	1125
Field frequency, Hz	60	50	60
Line frequency, kHz	15.75	15.625	
Line time, μs	63.5	64.0	
Video bandwidth, MHz	4.2	5.0	≈ 27
Vertical retrace, lines/field	21		
Horizontal retrace time, μs	10		

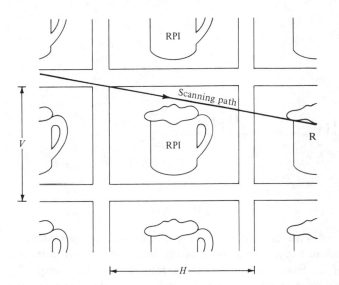

Figure 8.4-3 Periodically repeated image with unbroken scanning path.

unbroken. Now any periodic function of two variables may be expanded as a *two-dimensional Fourier series* by straightforward extension of the one-dimensional series. For the case at hand with H and V the horizontal and vertical periods (including retrace allowance), the image intensity is

$$I(h, v) = \sum_{m=-\infty}^{\infty} \sum_{n=-\infty}^{\infty} c_{mn} \exp\left[j2\pi \left(\frac{mh}{H} + \frac{nv}{V} \right) \right] \qquad (2)$$

where

$$c_{mn} = \frac{1}{HV} \int_0^H \int_0^V I(h, v) \exp\left[-j2\pi \left(\frac{mh}{H} + \frac{nv}{V} \right) \right] dh\, dv \qquad (3)$$

Therefore, letting

$$f_h = \frac{s_h}{H} \qquad f_v = \frac{s_v}{V}$$

and using Eqs. (1) and (2), we obtain

$$x(t) = \sum_{m=-\infty}^{\infty} \sum_{n=-\infty}^{\infty} c_{mn} e^{j2\pi(mf_h + nf_v)t} \qquad (4)$$

This expression represents a doubly periodic signal containing all harmonics of the *line frequency* f_h and the *field frequency* f_v, plus their sums and differences. Since $f_h \gg f_v$ and since $|c_{mn}|$ generally decreases as the product mn increases, the amplitude spectrum has the form shown in Fig. 8.4-4, where the spectral lines cluster around the harmonics of f_h and there are large gaps between clusters.

Figure 8.4-4 Video spectrum for still image.

Equation (4) and Fig. 8.4-4 are exact for a still picture, as in facsimile systems. When the image has motion, the spectral lines merge into continuous clumps around the harmonics of f_h. Even so, the spectrum remains mostly "empty" everywhere else, a property used to advantage in the subsequent development of color TV. Despite the gaps in Fig. 8.4-4, the video spectrum theoretically extends indefinitely — similar to an FM line spectrum. Determining the bandwidth required for a video signal thus involves additional considerations.

Two basic facts stand in the way of perfect image reproduction: (1) there can be only a finite number of lines in the scanning raster, which limits the image clarity or *resolution* in the vertical direction; and (2) the video signal must be transmitted with a finite bandwidth, which limits horizontal resolution. Quantitatively, we measure resolution in terms of the maximum number of discrete image lines that can be distinguished in each direction, say n_h and n_v. In other words, the most detailed image that can be resolved is taken to be a checkerboard pattern having n_h columns and n_v rows. We usually desire equal horizontal and vertical resolution in lines per unit distance, so $n_h/H = n_v/V$ and

$$\frac{n_h}{n_v} = \frac{H}{V} \tag{5}$$

which is called the *aspect ratio*.

Clearly, vertical resolution is related to the total number of raster lines N; indeed, n_v equals N if all scanning lines are active in image formation (as in facsimile but not TV) and the raster aligns perfectly with the rows of the image. Experimental studies show that arbitrary raster alignment reduces the effective resolution by a factor of about 70 percent, called the *Kerr factor*, so

$$n_v = 0.7(N - N_{vr}) \tag{6}$$

where N_{vr} is the number of raster lines lost during vertical retrace.

Horizontal resolution is determined by the baseband bandwidth B allotted to the video signal. If the video signal is a sinusoid at frequency $f_{\max} = B$, the resulting picture will be a sequence of alternating dark and light spots spaced by one-half cycle in the horizontal direction. It then follows that

$$n_h = 2B(T_{line} - T_{hr}) \tag{7}$$

where T_{line} is the total duration of one line and T_{hr} is the horizontal retrace time. Solving Eq. (7) for B and using Eqs. (5) and (6) yields

$$B = \frac{(H/V)n_v}{2(T_{line} - T_{hr})} = 0.35(H/V)\frac{N - N_{vr}}{T_{line} - T_{hr}} \tag{8}$$

Another, more versatile bandwidth expression is obtained by multiplying both sides of Eq. (8) by the frame time $T_{frame} = NT_{line}$ and explicitly showing the desired resolution. Since $N = n_v/0.7(1 - N_{vr}/N)$, this results in

$$BT_{frame} = \frac{0.714n_p}{\left(1 - \dfrac{N_{vr}}{N}\right)\left(1 - \dfrac{T_{hr}}{T_{line}}\right)} \tag{9a}$$

where

$$n_p = \frac{H}{V} n_v^2 = n_h n_v \tag{9b}$$

The parameter n_p represents the number of picture elements or *pixels*. Equation (9) brings out the fact that the bandwidth (or frame time) requirement increases in proportion to the number of pixels or as the square of the vertical resolution.

Example 8.4-1 The NTSC system has $N = 525$ and $N_{vr} = 2 \times 21 = 42$ so there are 483 active lines. The line time is $T_{line} = 1/f_h = 63.5$ μs and $T_{vr} = 10$ μs, leaving an active line time of 53.5 μs. Therefore, using Eq. (8) with $H/V = 4/3$, we get the video bandwidth

$$B = 0.35 \times \frac{4}{3} \times \frac{483}{53.5 \times 10^{-6}} \approx 4.2 \text{ MHz}$$

This bandwidth is sufficiently large to reproduce the 5-μs sync pulses with reasonably square corners.

The proposed HDTV system would have an aspect ratio of 5/3 and about one million pixels, for a picture quality rivaling that of 35-mm movie film. But the resulting 27-MHz video bandwidth implies a need for digital bandwidth compression and/or transmission by direct-broadcast satellites. See Mokhoff (1981) for further discussion.

Exercise 8.4-1 Facsimile systems require no vertical retrace and the horizontal retrace time is negligible. Calculate the time T_{frame} needed for facsimile transmission of a newspaper page, 37 by 59 cm, with a resolution of 40 lines/cm using a voice telephone channel with $B \approx 3.2$ kHz.

Monochrome Transmitters and Receivers

The large bandwidth and significant low-frequency content of the video signal, together with the desired simplicity of envelope detection, have led to the selection of VSB + C (as described in Sect. 6.4) for TV broadcasting in the United States. However, since precise vestigial sideband shaping is more easily carried out at the receiver where the power levels are small, the actual modulated-signal spectrum is as indicated in Fig. 8.4-5a. The half-power frequency of the upper sideband is about 4.2 MHz above the video carrier f_{cv} while the lower sideband has a 1-MHz bandwidth. Figure 8.4-5b shows the frequency shaping at the receiver.

The audio signal is frequency-modulated on a separate carrier $f_{ca} = f_{cv} + f_a$, with $f_a = 4.5$ MHz and frequency deviation $f_\Delta = 25$ kHz. Thus, assuming an audio bandwidth of 10 kHz, $D = 2.5$ and the modulated audio occupies about 80 kHz. TV channels are spaced by 6 MHz, leaving a 250-kHz guard band. Carrier frequencies are assigned in the VHF ranges 54–72, 76–88, and 174–216 MHz, and in the UHF range 470–806 MHz.

The essential parts of a TV transmitter are block-diagrammed in Fig. 8.4-6. The synchronizing generator controls the scanning raster and supplies blanking and sync pulses for the video signal. The DC restorer and white clipper working together ensure that the amplified video signal levels are in proportion. The video modulator is of the high-level AM type with $\mu = 0.875$, and the power amplifier removes the lower portion of the lower sideband.

The antenna has a balanced-bridge configuration such that the outputs of the audio and video transmitters are radiated by the same antenna without interfering with each other. The transmitted audio power is 50 to 70 percent of the video power.

As indicated in Fig. 8.4-7, a TV receiver is of the superheterodyne type. The main IF amplifier has f_{IF} in the 41- to 46-MHz range and provides the vestigial shaping per Fig. 8.4-5b. Note that the modulated audio signal is also passed by this amplifier, but with substantially less gain. Thus, drawing upon Eq. (11),

Figure 8.4-5 (a) Transmitted TV spectrum; (b) VSB shaping at receiver.

Figure 8.4-6 Monochrome TV transmitter.

Sect. 6.4, the total signal at the input to the envelope detector is

$$y(t) = A_{cv}[1 + \mu x(t)] \cos \omega_{cv} t - A_{cv} \mu x_q(t) \sin \omega_{cv} t$$
$$+ A_{ca} \cos [(\omega_{cv} + \omega_a)t + \phi(t)] \qquad (10)$$

where $x(t)$ is the video signal, $\phi(t)$ is the FM audio, and $\omega_a = 2\pi f_a$. Since $|\mu x_q(t)| \ll 1$ and $A_{ca} \ll A_{cv}$, the resulting envelope is approximately

$$A_y(t) = A_{cv}[1 + \mu x(t)] + A_{ca} \cos [\omega_a t + \phi(t)] \qquad (11)$$

which gives the signal at the output of the envelope detector.

The video amplifier has a lowpass filter that removes the audio component from $A_y(t)$ as well as a DC restorer that electronically clamps the blanking pulses and thereby restores the correct DC level to the video signal. The amplified and DC-restored video signal is applied to the picture tube and to a sync-pulse separator that provides synchronization for the sweep generators. The "brightness" control permits manual adjustment of the DC level while the "contrast" control adjusts the gain of the IF amplifier.

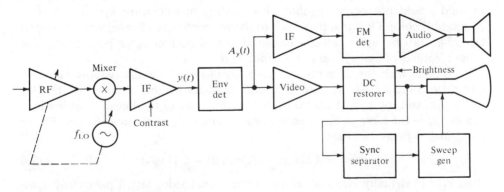

Figure 8.4-7 Monochrome TV receiver.

Equation (11) shows that the envelope detector output also includes the modulated audio. This component is picked out and amplified by another IF amplifier tuned to 4.5 MHz. FM detection and amplification then yields the audio signal.

Observe that, although the transmitted composite audio and video signal is a type of frequency-division multiplexing, separate frequency conversion is not required for the audio. This is because the video carrier acts like a local oscillator for the audio in the envelope-detection process, an arrangement called the *intercarrier-sound system* having the advantageous feature that the audio and video are always tuned in together. Successful operation depends on the fact that the video component is large compared to the audio at the envelope detector input, as made possible by the white clipper at the transmitter (which prevents the modulated video signal from becoming too small) and the relative attenuation of the audio by the receiver's IF response.

Some additional features not shown on our transmitter and receiver diagrams relate to the *vertical retrace interval*. The NTSC system allots 21 lines per field to vertical retracing, or about 1.3 ms every 1/60 sec. The first 9 lines carry control pulses, but the remaining 12 may be utilized for other purposes while the retrace goes on. Applications of these available lines include: the vertical-interval test signal (VITS) for checking transmission quality; the vertical-interval reference (VIR) for receiver servicing and/or automatic adjustments; and digital signals that generate the closed-captioning characters on special receivers for the hearing impaired.

Exercise 8.4-2 Use a phasor diagram to derive Eq. (11) from Eq. (10).

Color Television

Any color can be synthesized from a mixture of the three additive primary colors, red, green, and blue. Accordingly, a brute-force approach to color TV would involve direct transmission of three video signals, say $x_R(t)$, $x_G(t)$, and $x_B(t)$ — one for each primary. But, aside from the increased bandwidth requirement, this method would not be compatible with existing monochrome systems. A fully compatible color TV signal that fits into the monochrome channel was developed in 1954, drawing upon certain characteristics of human color perception. The salient features of that system are outlined here.

To begin with, the three primary color signals can be uniquely represented by any three other signals that are independent linear combinations of $x_R(t)$, $x_G(t)$, and $x_B(t)$. And, by proper choice of coefficients, one of the linear combinations can be made the same as the intensity or *luminance* signal of monochrome TV. In particular, it turns out that if

$$x_Y(t) = 0.30x_R(t) - 0.59x_G(t) - 0.11x_B(t) \tag{12a}$$

then $x_Y(t)$ is virtually identical to the conventional video signal previously symbolized by $x(t)$. The remaining two signals, called the *chrominance* signals, are taken as

$$x_I(t) = 0.60x_R(t) - 0.28x_G(t) - 0.32x_B(t) \tag{12b}$$

$$x_Q(t) = 0.21x_R(t) - 0.52x_G(t) - 0.31x_B(t) \tag{12c}$$

Here, the color signals are normalized such that $0 \le x_R(t) \le 1$, etc., so the lumi-nance signal is never negative while the chrominance signals are bipolar.

Understanding the chrominance signals is enhanced by introducing the *color vector*

$$x_C(t) = x_I(t) + jx_Q(t) \tag{13}$$

whose magnitude $|x_C(t)|$ is the color intensity or *saturation* and whose angle arg $x_C(t)$ is the *hue*. Figure 8.4-8 shows the vector positions of the saturated primary colors in the *IQ* plane. A partially saturated (pastel) blue-green, for instance, might have $x_R = 0$ and $x_B = x_G = 0.5$, so $x_C = -0.300 - j0.105$, $|x_C| = 0.318$, and arg $x_C = -160°$. Since the origin of the *IQ* plane represents the absence of color, the luminance signal may be viewed as a vector perpendicular to this plane.

Because $x_Y(t)$ serves as the monochrome signal, it must be allotted the entire 4.2-MHz baseband bandwidth to provide adequate horizontal resolution. Conse-quently, there would seem to be no room for the chrominance signals. Recall, however, that the spectrum of $x_Y(t)$ has periodic gaps between the harmonics of the line frequency f_h — and the same holds for the chrominance signals. More-over, subjective tests have shown that the human eye is less perceptive of chro-minance resolution than luminance resolution, so that $x_I(t)$ and $x_Q(t)$ can be restricted to about 1.5 MHz and 0.5 MHz, respectively, without significant visible degradation of the color picture. Combining these factors permits multiplexing the chrominance signals in an interleaved fashion in the baseband spectrum of the luminance signal.

The chrominance signals are multiplexed on a *color subcarrier* whose fre-quency falls exactly halfway between the 227th and 228th harmonic of f_h, namely,

$$f_{cc} = \frac{455}{2} f_h \approx 3.6 \text{ MHz} \tag{14}$$

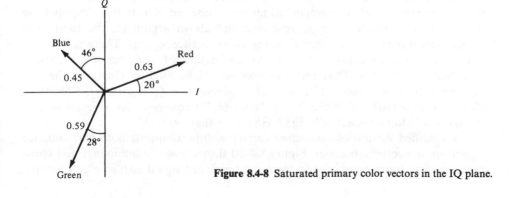

Figure 8.4-8 Saturated primary color vectors in the IQ plane.

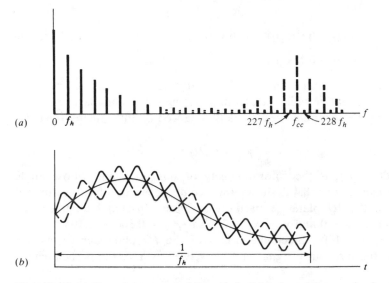

(a)

(b)

Figure 8.4-9 (a) Chrominance spectral lines (dashed) interleaved between luminance lines; (b) line-to-line phase reversal of chrominance variations on luminance.

Therefore, by extension of Fig. 8.4-3, the luminance and chrominance frequency components are interleaved as indicated in Fig. 8.4-9a, and there is 0.6 MHz between f_{cc} and the upper end of the baseband channel. The subcarrier modulation will be described shortly, after we examine frequency interleaving and compatibility.

What happens when a color signal is applied to a monochrome picture tube? Nothing, surprisingly, as far as the viewer sees. True, the color subcarrier and its sidebands produce sinusoidal variations on top of the luminance signal. But because all of these sinusoids are exactly an odd multiple of one-half the line frequency, they reverse in phase from line to line and from field to field—illustrated by Fig. 8.4-9b. This produces flickering in small areas that averages out over time and space to the correct luminance value and goes essentially unnoticed by the viewer.

By means of this averaging effect, frequency interleaving renders the color signal compatible with an unmodified monochrome receiver. It also simplifies the design of color receivers since, reversing the above argument, the luminance signal does not visibly interfere with the chrominance signals. There is a minor interference problem caused by the difference frequency $f_a - f_{cc}$ between the audio and color subcarriers. That problem was solved by slightly changing the line frequency to $f_h = f_a/286 = 15.73426$ kHz giving $f_a - f_{cc} = 4500 - 3,579.545 = 920.455$ kHz $= (107/2)f_h$ which is an "invisible" frequency. (As a result of this change, the field rate is actually 59.94 Hz rather than 60 Hz!)

A modified version of *quadrature-carrier multiplexing* puts both chrominance signals on the color subcarrier. Figure 8.4-10 shows how the luminance and chrominance signals are combined to form the baseband signal $x_b(t)$ in a color trans-

Figure 8.4-10 Color subcarrier modulation system.

mitter. Not shown is the nonlinear *gamma correction* introduced at the camera output to compensate for the brightness distortion of color picture tubes.

The gamma-corrected color signals are first matrixed to obtain $x_Y(t)$, $x_I(t)$, and $x_Q(t)$ in accordance with Eq. (12). Next, the chrominance signals are lowpass-filtered (with different bandwidths) and applied to the subcarrier modulators. Subsequent bandpass filtering produces conventional DSB modulation for the Q channel and modified VSB for the I channel—i.e. DSB for baseband frequencies of $x_I(t)$ below 0.5 MHz and LSSB for $0.5 < |f| < 1.5$ MHz. The latter keeps the modulated chrominance signals as high as possible in the baseband spectrum, thereby confining the flicker to small areas, while still allowing enough bandwidth for proper resolution of $x_I(t)$. Total sideband suppression cannot be used owing to the significant low-frequency content in $x_I(t)$ and $x_Q(t)$.

Including $x_Y(t)$, the entire baseband signal becomes

$$x_b(t) = x_Y(t) + x_Q(t) \sin \omega_{cc} t + x_I(t) \cos \omega_{cc} t + \hat{x}_{IH}(t) \sin \omega_{cc} t \qquad (15)$$

where $\hat{x}_{IH}(t)$ is the Hilbert transform of the high-frequency portion of $x_I(t)$ and accounts for the asymmetric sidebands. This baseband signal takes the place of the monochrome video signal in Fig. 8.4-6. Additionally, an 8-cycle piece of the color subcarrier known as the *color burst* is put on the trailing portion or "back porch" of the blanking pulses for purposes of synchronization.

Demultiplexing is accomplished in a color TV receiver after the envelope detector, as laid out in Fig. 8.4-11. Since the luminance signal is at baseband here, it requires no further processing save for amplification and a 3.6-MHz trap or

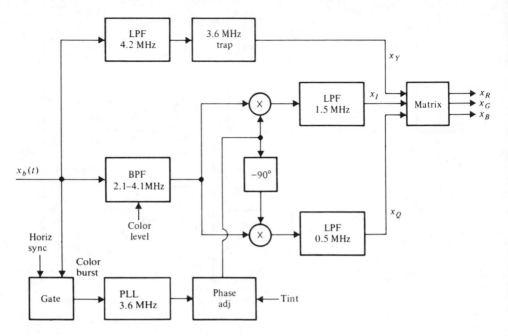

Figure 8.4-11 Color demodulation system.

rejection filter to eliminate the major flicker component; the chrominance side-bands need not be removed, thanks to frequency interleaving. The chrominance signals pass through a bandpass amplifier and are applied to a pair of synchronous detectors whose local oscillator is the VCO in a PLL synchronized by phase comparison with the received color burst. Manual controls usually labeled "color level" (i.e., saturation) and "tint" (i.e., hue) are provided to adjust the gain of the chrominance amplifier and the phase of the VCO; their effect on the picture is readily explained in terms of the color vector and Fig. 8.4-8.

Assuming good synchronization, it follows from Eq. (15) that the detected but unfiltered I- and Q-channel signals are proportional to

$$v_I(t) = x_I(t) + 2x_{YH}(t) \cos \omega_{cc} t + x_I(t) \cos 2\omega_{cc} t$$
$$+ [x_Q(t) + \hat{x}_{IH}(t)] \sin 2\omega_{cc} t \qquad (16a)$$
$$v_Q(t) = x_Q(t) + \hat{x}_{IH}(t) + 2x_{YH}(t) \sin \omega_{cc} t + x_I(t) \sin 2\omega_{cc} t$$
$$- [x_Q(t) + \hat{x}_{IH}(t)] \cos 2\omega_{cc} t \qquad (16b)$$

where $x_{YH}(t)$ represents the luminance frequency components in the ·2.1- to 4.1-MHz range. Clearly, lowpass filtering will remove the double-frequency terms, while the terms involving $x_{YH}(t)$ are "invisible" frequencies. Furthermore, $\hat{x}_{IH}(t)$ in Eq. (16b) has no components less than 0.5 MHz, so it is rejected by the LPF in the Q channel. (Imperfect filtering here results in a bothersome effect called quadrature color cross talk.) Therefore, ignoring the invisible-frequency terms, $x_I(t)$

and $x_Q(t)$ have been recovered and can then be matrixed with $x_Y(t)$ to generate the color signals for the picture tube. Specifically, by inversion of Eq. (12),

$$x_R(t) = x_Y(t) - 0.96x_I(t) + 0.62x_Q(t)$$
$$x_G(t) = x_Y(t) - 0.28x_I(t) - 0.64x_Q(t) \tag{17}$$
$$x_B(t) = x_Y(t) - 1.10x_I(t) + 1.70x_Q(t)$$

If the received signal happens to be monochrome, then the three color signals will be equal and the reproduced picture will be black-and-white. This is termed *reverse compatibility*.

The NTSC color system described here certainly ranks high as an extraordinary engineering achievement! It solved the problems of color reproduction with direct and reverse monochrome compatibility while staying within the confines of the 6-MHz channel allocation. The inevitable compromises resulted in an acceptable but not perfect color picture. The near-perfection promised by HDTV requires a separate rather than interlaced chrominance signal and much greater channel bandwidth.

8.5 PROBLEMS

8.1-1 Suppose a commercial AM superhet has been designed such that the image frequency always falls above the broadcast band. Find the minimum value of f_{IF}, the corresponding range of f_{LO}, and the bounds on B_{RF}.

8.1-2 Suppose a commercial FM superhet has been designed such that the image frequency always falls below the broadcast band. Find the minimum value of f_{IF}, the corresponding range of f_{LO}, and the bounds on B_{RF}.

8.1-3 Suppose a commercial AM superhet has $f_{IF} = 455$ kHz and $f_{LO} = 1/2\pi\sqrt{LC}$, where $L = 1$ μH and C is a variable capacitor for tuning. Find the range of C when $f_{LO} = f_c + f_{IF}$ and when $f_{LO} = f_c - f_{IF}$.

8.1-4 Suppose the RF stage of a commercial AM superhet is a tuned circuit like Fig. 6.1-1 with $L = 1$ μH and variable C for tuning. Find the range of C and the corresponding bounds on R.

8.1-5 Consider a superhet intended for USSB modulation with $W = 4$ kHz and $f_c = 3.57$–3.63 MHz. Take $f_{LO} = f_c + f_{IF}$ and choose the receiver parameters so that all bandpass stages have $B/f_0 \approx 0.02$. Then sketch $|H_{RF}(f)|$ to show that the RF stage can be fixed-tuned. Also sketch $|H_{IF}(f)|$, accounting for sideband reversal.

8.1-6 Do Prob. 8.1-5 for LSSB modulation with $W = 6$ kHz and $f_c = 7.14$–7.26 MHz.

8.1-7 Sketch the spectrum of $x_c(t) \times \cos 2\pi f_{LO} t$ to demonstrate the sideband-reversal effect in an SSB superhet when $f_{LO} = f_c + f_{IF}$.

8.1-8 For *automatic frequency control* in an FM superhet, the LO is replaced by a VCO that generates $A_{LO} \cos \theta(t)$ with $\dot\theta(t) = 2\pi[f_c - f_{IF} + Kv(t) + \epsilon(t)]$ where $\epsilon(t)$ is a slow and random frequency drift. The control voltage $v(t)$ is derived by applying the demodulated signal to an LPF with $B \ll W$. The demodulated signal is $y_D(t) = K_D \dot\phi_{IF}(t)/2\pi$ where $\phi_{IF}(t)$ is the instantaneous phase at the IF output. Analyze this AFC system by finding $y_D(t)$ in terms of $x(t)$ and $\epsilon(t)$.

8.1-9 Find suitable parameters of a double-conversion receiver having $RR = 60$ dB and intended for DSB modulation with $W = 10$ kHz and $f_c = 4$ MHz.

8.1-10 A double conversion receiver designed for $f_c = 300$ MHz has $f_{IF-1} = 30$ MHz and $f_{IF-2} = 3$ MHz, and each LO frequency is set at the higher of the two possible values. Insufficient filtering by the RF and first IF stages results in interference from *three* image frequencies. What are they?

8.1-11 Do Prob. 8.1-10 with each LO frequency set at the lower of the two possible values.

8.1-12 Specify the settings on a scanning spectrum analyzer to display the spectrum up to the 10th harmonic of a signal with a 50-ms period.

8.1-13 Specify the settings on a scanning spectrum analyzer to display the spectrum of a tone-modulated FM signal with $f_c = 100$ kHz, $f_m = 1$ kHz, and $\beta = 5$.

8.1-14‡ The magnitude spectrum of an *energy* signal $v(t)$ can be displayed by multiplying $v(t)$ with the swept-frequency wave $\cos(\omega_c t - \alpha t^2)$ and applying the product to a bandpass filter having $h_{bp}(t) = \cos(\omega_c t + \alpha t^2)$. Use equivalent lowpass time-domain analysis to show that $h_{\ell p}(t) = \frac{1}{2} e^{j\alpha t^2}$ and that the envelope of the bandpass output is proportional to $|V(f)|$ with $f = \alpha t/\pi$.

8.2-1 Four signals, each having $W = 3$ kHz, are to be multiplexed with 1-kHz guard bands between channels. The subcarrier modulation is USSB, expect for the lowest channel which is unmodulated, and the carrier modulation is AM. Sketch the spectrum of the baseband and transmitted signal, and calculate the transmission bandwidth.

8.2-2 Do Prob. 8.2-1 with AM subcarrier modulation.

8.2-3 Referring to Fig. 8.2-3 and Table 8.2-1, determine the number of different subcarrier frequencies needed to generate a mastergroup signal. How many would be needed if a mastergroup signal is generated without grouping, as in Fig. 8.2-1?

8.2-4 Let f_i be an arbitrary subcarrier in an FDM signal. Use frequency-translation sketches to show that the BPFs in Fig. 8.2-2 are not necessary if the subcarrier modulation is DSB and the detector includes an LPF. Then show that the BPFs are needed, in general, for SSB subcarrier modulation.

8.2-5 Ten signals with bandwidth W are to be multiplexed using SSB subcarrier modulation and a guard band B_g between channels. The BPFs at the receiver have $|H(f)| = \exp\{-[1.2(f - f_0)/W]^2\}$, where f_0 equals the center frequency for each subcarrier signal. Find B_g so that the adjacent-channel response satisfies $|H(f)| \leq 0.1$. Then calculate the resulting transmission bandwidth of the FDM signal.

8.2-6 Suppose the voice channels in an AT & T group signal are separated at the receiver using BPFs with $|H(f)| = \{1 + [2(f - f_0)/B]^{2n}\}^{-1/2}$. Make a careful sketch of three adjacent channels in the group spectrum, taking account of the fact that a baseband voice signal has negligible content outside $200 < |f| < 3200$ Hz. Use your sketch to determine values for B, f_0, and n so that $|H(f)| \leq 0.1$ outside the desired passband.

8.2-7 Some FDM telemetry systems employ *proportional bandwidth* FM subcarrier modulation when the signals to be multiplexed have different bandwidths. All subcarrier signals have the same deviation ratio but the ith subcarrier frequency and message bandwidth are related by $f_i = W_i/\alpha$, where α is a constant. (a) Show that the subcarrier signal bandwidth B_i is proportional to f_i, and obtain an expression for f_{i+1} in terms of f_i to provide a guard band B_g between channels. (b) Calculate the next three subcarrier frequencies when $f_1 = 2$ kHz, $B_1 = 800$ Hz, and $B_g = 400$ Hz.

8.2-8 Find the output signals of the quadrature-carrier system in Fig. 8.2-7 when the receiver local oscillator has a phase error ϕ'.

8.2-9 In one proposed system for FM *quadraphonic multiplexing*, the baseband signal in Fig. 8.2-5 is modified as follows: The unmodulated signal is $x_0(t) = L_F + L_R + R_F + R_R$ (for monophonic compatability), the 38-kHz subcarrier has quadrature-carrier multiplexing with modulating signals $x_1(t)$ and $x_2(t)$, and the SCA signal is replaced by a 76-kHz subcarrier with DSB modulation by $x_3(t) = L_F - L_R + R_F - R_R$. What should be the components of $x_1(t)$ for stereophonic compatibility? Now consider $x_0(t) \pm x_1(t) \pm x_3(t)$ to determine the components of $x_2(t)$. Draw a block diagram of the corresponding transmitter and quadraphonic receiver.

8.2-10‡ Suppose the transmission channel in Fig. 8.2-7 has linear distortion represented by the transfer function $H_C(f)$. Find the resulting spectrum at the lower output and show that the condition for no cross talk is $H_C(f - f_c) = H_C(f + f_c)$ for $|f| \leq W$. If this condition holds, what must be done to recover $x_1(t)$?

8.3-1 For one implementation of digital phase comparison, the switching circuit in Fig. 8.3-1b has a set–reset flip-flop whose output becomes $s(t) = +A$ after a positive-going zero-crossing of $x_c(t)$ and

$s(t) = -A$ after a positive-going zero-crossing of $v(t)$. (a) Take $x_c(t) = \cos \omega_c t$ and $v(t) = \cos(\omega_c t - \phi_v)$ and sketch one period of $s(t)$ for $\phi_v = 45$, 135, 180, 225, and 315°. (b) Now plot y versus $\epsilon = \phi_v - 180°$ assuming that $y(t) = \langle s(t) \rangle$. Note that this implementation requires $\pm 180°$ phase difference between the inputs for $y = 0$.

8.3-2 Do part (a) of Prob. 8.3-1 for a digital phase comparator with a switch controlled by $v(t)$ so its output is $s(t) = A$ sgn $x_c(t)$ when $v(t) > 0$ and $s(t) = 0$ when $v(t) < 0$. Now plot y versus $\epsilon = \phi_v - 90°$ assuming that $y(t) = \langle s(t) \rangle$.

8.3-3 Consider a PLL in the steady state with $\epsilon_{ss} \ll 1$ for $t < 0$. The input frequency has a step change at $t = 0$, so $\phi(t) = 2\pi f_1 t$ for $t > 0$. Solve Eq. (5) to find and sketch $\epsilon(t)$, assuming that $K \gg |\Delta f + f_1|$.

8.3-4 Explain why the Costas PLL system in Fig. 8.3-4 cannot be used for synchronous detection of SSB or VSB.

8.3-5 Consider a PLL in steady state locked conditions. If the external input is $x_c(t) = A_c \cos(\omega_c t + \phi_0)$, then the feedback signal to the phase comparator must be proportional to $\cos(\omega_c t + \phi_0 + 90° - \epsilon_{ss})$. Use this property to find the VCO output in Fig. 8.3-5 when $\epsilon_{ss} \neq 0$.

8.3-6 Use the property stated in Prob. 8.3-5 to find the VCO output in Fig. 8.3-6 when $\epsilon_{ss} \neq 0$.

8.3-7 Modify the FM stereo receiver in Fig. 8.2-6 to incorporate a PLL with $f_v \approx 38$ kHz for the subcarrier. Also include a DC stereo indicator.

8.3-8 Given a 100-kHz master oscillator and two adjustable divide-by-n counters with $n = 1$ to 10, devise a system that synthesizes any frequency from 1 kHz to 99 kHz in steps of 1 kHz. Specify the nominal free-running frequency of each VCO.

8.3-9 Referring to Table 8.1-1, devise a frequency synthesizer to generate $f_{LO} = f_c + f_{IF}$ for an FM radio. Assume you have available a master oscillator at 120.0 MHz and adjustable divide-by-n counters with $n = 1$ to 1000.

8.3-10 Referring to Table 8.1-1, devise a frequency synthesizer to generate $f_{LO} = f_c + f_{IF}$ for an AM radio. Assume you have available a master oscillator at 2105 kHz and adjustable divide-by-n counters with $n = 1$ to 1000.

8.3-11 The linearized PLL in Fig. 8.3-8 becomes a *phase* demodulator if we add an ideal integrator to get

$$z(t) = \int^t y(\lambda) \, d\lambda$$

Find $Z(f)/X(f)$ when the input is a PM signal. Compare with Eq. (11).

8.3-12 Consider the PLL model in Fig. 8.3-8c, where $E(f) = \Phi(f) - \Phi_v(f)$. (a) Find $E(f)/\Phi(f)$ and derive Eq. (10) therefrom. (b) Show that if the input is an FM signal, then $E(f) = (f_\Delta/K) H_e(f) X(f)$ with $H_e(f) = 1/[H(f) + j(f/K)]$.

8.3-13 Suppose an FM detector is a linearized first-order PLL with $H(f) = 1$. Let the input signal be modulated by $x(t) = A_m \cos 2\pi f_m t$ where $A_m \leq 1$ and $0 \leq f_m \leq W$. (a) Use the relationship in Prob. 8.3-12b to find the steady state amplitude of $\epsilon(t)$. (b) Since linear operation requires $|\epsilon(t)| \leq 0.5$ rad, so $\sin \epsilon \approx \epsilon$, show that the minimum loop gain is $K = 2f_\Delta$.

8.3-14‡ Suppose an FM detector is a *second-order* PLL with loop gain K and $H(f) = 1 + K/j2f$. Let the input signal be modulated by $x(t) = A_m \cos 2\pi f_m t$ where $A_m \leq 1$ and $0 \leq f_m \leq W$. (a) Use the relationship in Prob. 8.3-12b to show that the steady state amplitude of $\epsilon(t)$ is maximum when $f_m = K/\sqrt{2}$ if $K/\sqrt{2} \leq W$. (b) Now assume that $K/\sqrt{2} > W$ and $f_\Delta > W$. Since linear operation requires $|\epsilon(t)| < 0.5$ rad, so $\sin \epsilon \approx \epsilon$, show that the minimum loop gain is $K \approx 2\sqrt{f_\Delta W}$.

8.3-15‡ Consider the second-order PLL in Prob. 8.3-14. (a) Show that $H_L(f)$ becomes a second-order LPF with $|H_L|$ maximum at $f \approx 0.556K$ and 3-dB bandwidth $B \approx 1.14K$. (b) Use the loop-gain conditions in Probs. 8.3-13 and 8.3-14 to compare the minimum 3-dB bandwidths of a first-order and second-order PLL FM detector when $f_\Delta/W = 2, 5$, and 10.

8.4-1 Explain the following statements: (a) A TV frame should have an *odd* number of lines. (b) The waveform that drives the scanning path should be a *sawtooth*, rather than a sinusoid or triangle.

8.4-2 Consider a scanning raster with very small slope and retrace time. Sketch the video signal and its spectrum, without using Eq. (4), when the image consists of: (a) alternating black and white vertical bars of width $H/4$; (b) alternating·black and white horizontal bars of height $V/4$.

8.4-3 Consider an image that's entirely black ($I = 0$) except for a centered white rectangle ($I = 1.0$) of width αH and height βV. (a) Show that $|c_{mn}| = \alpha\beta\,|\text{sinc }\alpha m \text{ sinc }\beta n|$. (b) Sketch the resulting line spectrum when $\alpha = 1/2$, $\beta = 1/4$, and $f_v = f_h/100$.

8.4-4 Calculate the number of pixels and the video bandwidth requirement for a low-resolution TV system with a square image, 230 active lines, and 100-μs active line time.

8.4-5 Calculate the number of pixels and the video bandwidth requirement for the HDTV system in Table 8.4-1 if $N_{vr} \ll N$ and $T_{hr} = 0.2T_{\text{line}}$.

8.4-6 Calculate the number of pixels and the video bandwidth requirement for the CCIR system in Table 8.4-1 if $N_{vr} = 48$ and $T_{hr} = 10$ μs.

8.4-7 Horizontal *aperature effect* arises when the scanning process in a TV camera produces the output

$$\tilde{x}(t) = \int_{t-\tau}^{t} x(\lambda)\, d\lambda$$

where $x(t)$ is the desired video signal and $\tau \ll T_{\text{line}}$. (a) Describe the resulting TV picture. (b) Find an equalizer that will improve the picture quality.

8.4-8 Describe what happens to a color TV picture when: (a) the gain of the chrominance amplifier is too high or too low; (b) the phase adjustment of the color subcarrier is in error by ± 90 or $180°$.

8.4-9 Carry out the details leading from Eq. (15) to Eq. (16).

8.4-10‡ Obtain expressions equivalent to Eqs. (15) and (16) when all the filters in the x_Q channel (at transmitter and receiver) are the same as the x_I channel. Discuss your results.

CHAPTER NINE

NOISE IN CW MODULATION

This chapter rounds out our study of CW modulation with an examination of system performance in the presence of contaminating noise. Following the definition of system models and parameters, we consider the special properties of *bandpass* noise. The assumption of bandpass-filtered white noise from a stationary gaussian process leads to mathematical descriptions that we then employ to investigate additive noise effects in linear and exponential modulation systems.

Our work culminates in a relative comparison of the several types of CW modulation. Further comparisons will be made in Chap. 15, based upon absolute standards derived from information theory.

9.1 SYSTEM MODELS AND PARAMETERS

Although we must give individual attention to the effects of noise on specific types of modulation, all CW communication systems have the same general structure. This section summarizes the system models, parameters, and assumptions used throughout the rest of the chapter.

System Models

Figure 9.1-1 depicts our generalized model of a CW communication system. The analog message $x(t)$ comes from an ergodic process with bandwidth W, normalized such that

$$|x(t)| \leq 1 \qquad S_x = \overline{x^2} = \langle x^2(t) \rangle \leq 1$$

All signal averages will be taken in the statistical sense over the message ensemble.

Figure 9.1-1 Model of a CW communication system.

The channel has transmission loss L but otherwise provides nearly distortionless transmission and negligible time delay. (A distorting channel would require equalization. The channel equalizer, if needed, would be a bandpass device located before the detector because demodulation is a nonlinear operation — except for synchronous detection which permits baseband equalization.) To minimize notational changes, the modulated signal at the output of the channel will be represented by $x_c(t)$ with carrier amplitude A_c. The *received signal power* is then

$$S_R = \frac{S_T}{L} = \overline{x_c^2} \tag{1}$$

The corresponding transmitted waveform is $\sqrt{L}\, x_c(t)$, so our previous expressions for S_T still apply but with A_c replaced by $\sqrt{L}\, A_c$.

We model the entire predetection portion of the receiver as a *bandpass filter* $H_R(f)$ having unit gain over the transmission bandwidth B_T. In the case of a superheterodyne receiver, $H_R(f)$ is the frequency response of the IF amplifier with $f_{\text{IF}} = f_c$. We can ignore any predetection frequency translation since it has the same effect on signal and noise — which is also true for any predetection gain.

Under the foregoing conditions with the assumption of additive noise at the receiver's input, the total signal plus noise at the detector becomes

$$v(t) = x_c(t) + n(t) \tag{2}$$

where $n(t)$ represents the *predetection noise*. Eventually, we'll recast $v(t)$ in envelope-and-phase or quadrature-carrier form as

$$v(t) = A_v(t) \cos\left[\omega_c t + \phi_v(t)\right] \tag{3a}$$

$$= v_i(t) \cos \omega_c t - v_q(t) \sin \omega_c t \tag{3b}$$

which facilitates analysis of the demodulated signal $y(t)$. Specifically, the demodulation operation is represented by the following idealized mathematical models:

$$y(t) = \begin{cases} v_i(t) & \text{Synchronous detector} & (4a) \\ A_v(t) - \bar{A}_v & \text{Envelope detector} & (4b) \\ \phi_v(t) & \text{Phase detector} & (4c) \\ \dfrac{1}{2\pi}\dot{\phi}_v(t) & \text{Frequency detector} & (4d) \end{cases}$$

These models presuppose perfect synchronization, perfect amplitude limiting, etc., as appropriate. The term $\bar{A}_v = \langle A_v(t) \rangle$ in Eq. (4b) reflects the DC block normally found in an envelope detector; the other detectors may or may not have this feature. A detection constant could be included in Eq. (4) but adds nothing in the way of generality.

Lastly, the demodulated waveform goes through a *lowpass filter* $H_D(f)$ to yield the final output signal $y_D(t)$ at the destination. The *postdetection* filtering may include deemphasis or other processing operations.

Predetection S/N

The additive form of Eq. (2) together with the reasonable assumption of statistically independent signal and noise allows us to write

$$\overline{v^2} = \overline{x_c^2} + \overline{n^2} = S_R + N_R \tag{5}$$

where $N_R = \overline{n^2}$ is the *predetection noise power*. Before discussing the signal-to-noise ratio, we'll turn off the signal and examine N_R by itself.

Figure 9.1-2a isolates that part of our system diagram relevant to N_R. Here, we treat the channel noise plus any noise generated in the predetection portion of the receiver as being equivalent to *white noise* with power spectral density $G(f) = \eta/2$ at the input of $H_R(f)$. Hence, the filtered output $n(t)$ has spectral density

$$G_n(f) = \frac{\eta}{2} |H_R(f)|^2 \tag{6}$$

as sketched in Fig. 9.1-2b. This drawing is based on a predetection filter with unit gain and nearly square frequency response, so its noise bandwidth equals B_T and

$$N_R = \int_{-\infty}^{\infty} G_n(f)\, df = \eta B_T \tag{7}$$

A less selective filter would, of course, pass more noise power.

(a)

(b)

Figure 9.1-2 Bandpass filtered white noise. (a) Block diagram; (b) power spectrum.

Note that the carrier frequency f_c does not necessarily fall in the center of the passband in Fig. 9.1-2b. We've written the lower cutoff as $f_c - \alpha B_T$ to include the symmetric-sideband case ($\alpha = 1/2$) and the suppressed-sideband case ($\alpha = 0$ or 1). Needless to say, the value of B_T depends on the message bandwidth W and the type of modulation.

From Eqs. (5) and (7), we now define the *predetection signal-to-noise ratio*

$$\left(\frac{S}{N}\right)_R \triangleq \frac{S_R}{N_R} = \frac{S_R}{\eta B_T} \tag{8}$$

which looks similar to the *destination* signal-to-noise ratio $S_R/\eta W$ defined in Sect. 5.4 relative to baseband transmission. But $(S/N)_R$ should not be confused with $S_R/\eta W$. To bring out the distinction, we'll introduce the system parameter

$$\gamma \triangleq \frac{S_R}{\eta W} \tag{9}$$

such that

$$\left(\frac{S}{N}\right)_R = \frac{W}{B_T}\gamma \tag{10}$$

and hence $(S/N)_R \leq \gamma$ since $B_T \geq W$. You should keep in mind the interpretation that γ equals the *maximum* destination S/N for analog baseband transmission with identical values of S_R and η at the receiver. By the same token, Eqs. (8) and (10) are actually *upper bounds* on $(S/N)_R$. The various imperfections in a practical system inevitably degrade $(S/N)_R$ to some extent.

9.2 BANDPASS NOISE

The predetection noise in a CW modulation system has a *bandpass* or *narrowband* spectrum. Here we invoke the further assumption that the noise comes from a *stationary gaussian* random process. We'll state the statistical properties of the quadrature components, the envelope, and the phase of $n(t)$ — all being random signals. The stated properties are subsequently justified by studying the correlation functions.

Quadrature Components

Let $n(t)$ be a sample function of a stationary gaussian process, and let $G_n(f)$ be as before (Fig. 9.1-2b). Then

$$\bar{n} = 0 \qquad \overline{n^2} = \sigma_n^2 = N_R \tag{1}$$

which follows from the absence of a DC component in $G_n(f)$. These statistical averages, and others below, also equal the corresponding time averages since a stationary gaussian process is ergodic.

The shape of $G_n(f)$ suggests expressing $n(t)$ in the usual bandpass form

$$n(t) = n_i(t) \cos \omega_c t - n_q(t) \sin \omega_c t \tag{2}$$

with in-phase component $n_i(t)$ and quadrature component $n_q(t)$. These components are jointly *stationary* and *gaussian*, like $n(t)$, and have the following properties:

$$\overline{n_i} = \overline{n_q} = 0 \qquad \overline{n_i(t)n_q(t)} = 0 \tag{3a}$$

$$\overline{n_i^2} = \overline{n_q^2} = \overline{n^2} = N_R \tag{3b}$$

Equation (3a) means that the random variables $n_i(t)$ and $n_q(t)$ are independent at any instant t, so their joint probability density function is the product of identical gaussian PDFs, namely

$$p_{n_i n_q}(n_i, n_q) = p_{n_i}(n_i)p_{n_q}(n_q)$$

$$= \frac{1}{2\pi N_R} e^{-(n_i^2 + n_q^2)/2N_R} \tag{4}$$

Here, to avoid confusion with noise power denoted by N, we use the same symbols for the subscript and argument of a PDF — a minor abuse of notation.

The *power spectral densities* of the quadrature components are identical *lowpass* functions related to $G_n(f)$ by

$$G_{n_i}(f) = G_{n_q}(f) = G_n(f + f_c)u(f + f_c) + G_n(f - f_c)[1 - u(f - f_c)] \tag{5}$$

Here, the term $G_n(f + f_c)u(f + f_c)$ simply stands for the positive-frequency portion of $G_n(f)$ translated *downward*, while $G_n(f - f_c)[1 - u(f - f_c)]$ stands for the negative-frequency portion translated *upward*. These translated terms then overlap and add for $|f| \leq \alpha B_T$, as illustrated by Fig. 9.2-1a. Figure 9.2-1b, shows

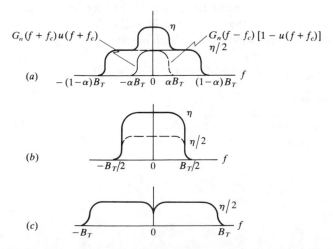

Figure 9.2-1 Lowpass power spectra of the quadrature components of bandpass noise. (a) General case; (b) symmetric-sideband case; (c) suppressed-sideband case.

the complete overlap in the symmetric-sideband case ($\alpha = 1/2$), and Fig. 9.2-1c shows the lack of overlap in the suppressed-sideband case ($\alpha = 0$ or 1). When $G_n(f)$ has *local symmetry* around $\pm f_c$, as reflected in Fig. 9.2-1b, the quadrature components are *uncorrelated* and *statistically independent* processes.

It should be emphasized that all the stated properties of $n_i(t)$ and $n_q(t)$ rest on the assumption of a stationary gaussian bandpass process. That assumption is justifiable in many CW systems.

Exercise 9.2-1 Sketch $G_n(f)$ and $G_{n_i}(f)$ when $|H_R(f)|$ has the TV VSB shaping in Fig. 8.4-5b. What's the value of $G_{n_i}(0)$?

Envelope and Phase

As an alternative to Eq. (2), we also want to express $n(t)$ in the form

$$n(t) = A_n(t) \cos [\omega_c t + \phi_n(t)] \tag{6}$$

with envelope $A_n(t)$ and phase $\phi_n(t)$. The standard phasor diagram in Fig. 9.2-2 relates our two sets of components. Clearly, at any instant of time,

$$A_n^2 = n_i^2 + n_q^2 \qquad \phi_n = \arctan \frac{n_q}{n_i} \tag{7a}$$

and conversely

$$n_i = A_n \cos \phi_n \qquad n_q = A_n \sin \phi_n \tag{7b}$$

These nonlinear relationships make spectral analysis difficult, even though we know $G_{n_i}(f)$ and $G_{n_q}(f)$. However, the lowpass spectrum of the quadrature components suggests that the time variations of $A_n(t)$ and $\phi_n(t)$ will be slow compared to f_c, in agreement with the bandpass nature of $n(t)$.

Furthermore, Eq. (7a) constitutes a rectangular-to-polar conversion of independent gaussian r.v.'s, just like the one that led to the Rayleigh distribution. We thus conclude that the PDF of the envelope is a *Rayleigh* function given by

$$p_{A_n}(A_n) = \frac{A_n}{N_R} e^{-A_n^2/2N_R} u(A_n) \tag{8}$$

with mean and second moment

$$\overline{A_n} = \sqrt{\pi N_R/2} \qquad \overline{A_n^2} = 2N_R \tag{9}$$

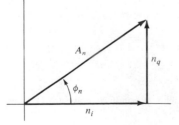

Figure 9.2-2 Phasor diagram for bandpass noise components.

The probability that A_n exceeds some specified positive value a is then

$$P(A_n > a) = e^{-a^2/2N_R} \tag{10}$$

These results follow from Eqs. (11)–(13), Sect. 4.4.

The phase ϕ_n has a *uniform* PDF over 2π radians, independent of A_n. Hence,

$$\overline{n^2} = \overline{A_n^2 \cos^2(\omega_c t + \phi_n)} = \overline{A_n^2} \times 1/2 = N_R$$

which explains the factor of 2 in Eq. (9).

Exercise 9.2-2 Suppose bandpass noise with $N_R = 1 \ \mu W$ is applied to a one-ohm resistor. Calculate the mean and rms value of the envelope voltage. Also evaluate $P(A_n > 2\overline{A_n})$.

Correlation Functions★

The properties of the quadrature components of bandpass noise were presented without proof in order to put the important results up front. Now we'll outline the derivation of those results by drawing upon various correlation functions. This analysis brings together concepts and relationships from several previous chapters to shed further light on bandpass noise.

We begin with the fictitious *lowpass equivalent* noise waveform defined by

$$n_{\ell p}(t) \triangleq \tfrac{1}{2}[n(t) + j\hat{n}(t)]e^{-j\omega_c t} \tag{11a}$$

in which $\hat{n}(t)$ is the Hilbert transform of the bandpass noise $n(t)$. The lowpass nature of $n_{\ell p}(t)$ is easily confirmed by deterministic Fourier transformation; see Prob. 9.2-9. But the quadrature components of $n(t)$ should be such that

$$\tfrac{1}{2}[n_i(t) + jn_q(t)] = n_{\ell p}(t) \tag{11b}$$

Thus, equating the real and imaginary parts of $n_{\ell p}(t)$ yields

$$n_i(t) = n(t) \cos \omega_c t + \hat{n}(t) \sin \omega_c t \tag{12a}$$

$$n_q(t) = \hat{n}(t) \cos \omega_c t - n(t) \sin \omega_c t \tag{12b}$$

which establishes explicit relationships between the quadrature components and $n(t)$. Multiplying Eq. (12a) by $\cos \omega_c t$ and adding Eq. (12b) multiplied by $-\sin \omega_c t$ then gives $n_i(t) \cos \omega_c t - n_q(t) \sin \omega_c t = n(t)$ as required.

Equation (12) contains much valuable information. First, it states the physically obvious fact that $n_i(t)$ and $n_q(t)$ depend entirely on $n(t)$—remember that $\hat{n}(t)$ represents a linear operation on $n(t)$. Second, if $n(t)$ is gaussian then $\hat{n}(t)$ is gaussian and Eq. (12) shows that the quadrature components are linear combinations of gaussian r.v.'s at any time instant, so therefore must also be gaussian. Third, Eq. (12) provides the starting point for correlation analysis. Fourth, it brings out the importance of Hilbert transforms in the study of bandpass noise.

The Hilbert transform of a random signal was previously considered in Example 5.2-3. Applying those results to the case at hand, we have

$$G_{\hat{n}}(f) = G_n(f) \qquad R_{\hat{n}}(\tau) = R_n(\tau) \tag{13a}$$

and

$$R_{\hat{n}n}(\tau) = \hat{R}_n(\tau) \qquad R_{n\hat{n}}(\tau) = -\hat{R}_n(\tau) \tag{13b}$$

Here, $\hat{R}_n(\tau)$ stands for the Hilbert transform of $R_n(\tau)$, defined by $\hat{R}_n(\tau) = h_Q(\tau) *$ $R_n(\tau)$ with $h_Q(\tau) = 1/\pi\tau$.

Having completed the necessary groundwork, we proceed to the autocorrelation function of the in-phase component $n_i(t)$. Into the basic definition

$$R_{n_i}(t, t - \tau) = E[n_i(t)n_i(t - \tau)] \tag{14}$$

we insert Eq. (12a) and perform some manipulations to get

$$
\begin{aligned}
R_{n_i}(t, t - \tau) = \tfrac{1}{2}\{ &[R_n(\tau) + R_{\hat{n}}(\tau)] \cos \omega_c \tau \\
&+ [R_{\hat{n}n}(\tau) - R_{n\hat{n}}(\tau)] \sin \omega_c \tau \\
&+ [R_n(\tau) - R_{\hat{n}}(\tau)] \cos \omega_c(2t - \tau) \\
&+ [R_{\hat{n}n}(\tau) + R_{n\hat{n}}(\tau)] \sin \omega_c(2t - \tau)\}
\end{aligned} \tag{15}
$$

This cumbersome expression then simplifies with the help of Eq. (13) to

$$R_{n_i}(t, t - \tau) = R_n(\tau) \cos \omega_c \tau + \hat{R}_n(\tau) \sin \omega_c \tau$$

which is independent of t. The same result holds for the autocorrelation of $n_q(t)$. Thus

$$R_{n_i}(\tau) = R_{n_q}(\tau) = R_n(\tau) \cos \omega_c \tau + \hat{R}_n(\tau) \sin \omega_c \tau \tag{16}$$

so the quadrature components are *stationary* and have identical autocorrelation and spectral density functions.

To obtain the power spectral density via Fourier transformation of Eq. (16), we note that

$$\mathcal{F}_\tau[R_n(\tau) \cos \omega_c \tau] = \tfrac{1}{2}[G_n(f - f_c) + G_n(f + f_c)]$$

Then, using the convolution and modulation theorems,

$$\mathcal{F}_\tau[\hat{R}_n(\tau)] = \mathcal{F}_\tau[h_Q(\tau)]\mathcal{F}_\tau[R_n(\tau)] = (-j \operatorname{sgn} f)G_n(f)$$

and

$$
\begin{aligned}
\mathcal{F}_\tau[\hat{R}_n(\tau) \sin \omega_c \tau] &= \mathcal{F}_\tau[\hat{R}_n(\tau) \cos (\omega_c \tau - \pi/2)] \\
&= -\frac{j}{2}[-j \operatorname{sgn} (f - f_c)G_n(f - f_c)] \\
&\quad + \frac{j}{2}[-j \operatorname{sgn} (f + f_c)G_n(f + f_c)]
\end{aligned}
$$

Therefore,

$$G_{n_i}(f) = G_{n_q}(f) = \tfrac{1}{2}[1 + \text{sgn}\,(f + f_c)]G_n(f + f_c)$$
$$+ \tfrac{1}{2}[1 - \text{sgn}\,(f - f_c)]G_n(f - f_c)$$

which reduces to Eq. (5) because the first term vanishes for $f < -f_c$ while the second term vanishes for $f > f_c$.

Finally, a similar analysis for the crosscorrelation of the quadrature components produces

$$R_{n_i n_q}(\tau) = R_n(\tau)\,\sin\,\omega_c\,\tau - \hat{R}_n(\tau)\,\cos\,\omega_c\,\tau \qquad (17)$$

and

$$\mathscr{F}_\tau[R_{n_i n_q}(\tau)] = j\{G_n(f + f_c)u(f + f_c) - G_n(f - f_c)[1 - u(f - f_c)]\} \qquad (18)$$

If $G_n(f)$ has local symmetry around $f = \pm f_c$, then the right-hand side of Eq. (18) equals zero for all f. This means that $R_{n_i n_q}(\tau) = 0$ for all τ, so the quadrature components are uncorrelated and statistically independent processes.

9.3 LINEAR MODULATION WITH NOISE

At last we're prepared to deal with the situation in Fig. 9.3-1. The signal $x_c(t)$ has linear modulation and is contaminated by additive gaussian white noise at the input to the receiver. Predetection bandpass filtering produces

$$v(t) = x_c(t) + n(t)$$

with $\overline{x_c^2} = S_R$ and $\overline{n^2} = N_R$ so

$$\left(\frac{S}{N}\right)_R = \frac{S_R}{N_R} = \frac{S_R}{\eta B_T} = \frac{W}{B_T}\,\gamma$$

The bandpass noise can be expressed as

$$n(t) = n_i(t)\,\cos\,\omega_c\,t - n_q(t)\,\sin\,\omega_c\,t$$

where $\overline{n_i^2} = \overline{n_q^2} = N_R = \eta B_T$.

The questions at hand are these: Given $x_c(t)$ and the type of detector, what's the final signal-plus-noise waveform $y_D(t)$ at the destination? And if the signal

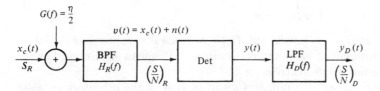

Figure 9.3-1 Model of receiver for CW modulation with noise.

and noise are additive at the output, what's the destination signal-to-noise ratio $(S/N)_D$? We'll address these questions for the two cases of synchronous and envelope detection.

The section closes with an introductory look at the intriguing properties of spread-spectrum transmission, a specialized technique that employs pseudo-noise signals to suppress interference.

Synchronous Detection

An ideal synchronous detector simply extracts the in-phase component of $v(t)$. If the modulation is DSB, then $x_c(t) = A_c x(t) \cos \omega_c t$ so

$$v(t) = [A_c x(t) + n_i(t)] \cos \omega_c t - n_q(t) \sin \omega_c t \tag{1}$$

and

$$y(t) = A_c x(t) + n_i(t)$$

Thus, if the postdetection filter is an ideal LPF with bandwidth W,

$$y_D(t) = A_c x(t) + n_i(t) \tag{2}$$

We see that the output signal and noise are, indeed, additive, and that the quadrature noise component $n_q(t)$ has been rejected by the detector.

Furthermore, if the predetection filter $H_R(f)$ has a relatively square response with bandwidth $B_T = 2W$ centered at f_c, then the output noise power spectrum will take the shape of Fig. 9.2-1b. Hence,

$$G_{n_i}(f) \approx \eta \Pi(f/2W) \tag{3}$$

which looks like lowpass-filtered white noise. Under these conditions, we really don't need any postdetection filter beyond the LPF within the synchronous detector (see Fig. 6.5-3).

Next we obtain the postdetection S/N from Eq. (2) by taking the mean square values of the signal and noise terms. Upon noting that $N_D = \overline{n_i^2}$ and $S_D = A_c^2 \overline{x^2} = A_c^2 S_x$, whereas $S_R = \overline{x_c^2} = \frac{1}{2} A_c^2 S_x$, we get

$$\left(\frac{S}{N}\right)_D = \frac{S_D}{N_D} = \frac{2S_R}{\eta B_T} = 2\left(\frac{S}{N}\right)_R \tag{4a}$$

or, since $B_T = 2W$,

$$\left(\frac{S}{N}\right)_D = \frac{S_R}{\eta W} = \gamma \qquad \text{DSB} \tag{4b}$$

Therefore, insofar as noise is concerned, DSB with ideal synchronous detection has the same performance as baseband transmission. (Problem 9.3-1 considers the effects of phase synchronization errors.)

You might have suspected a different result in view of the predetection noise power $N_R = \eta B_T = 2\eta W$. However, the signal sidebands add in a *coherent* fashion when translated to baseband, while the noise sidebands add *incoherently*. The sideband coherence in synchronous detection of DSB exactly counter-balances the double-sideband noise power passed by the predetection filter.

The preceding analysis is readily adapted to the case of an AM signal $x_c(t) = A_c[1 + x(t)] \cos \omega_c t$, in which we've taken $\mu = 1$ for simplicity. If the synchronous detector includes an ideal DC block, $y_D(t)$ will be as given in Eq. (2) so $S_D = A_c^2 S_x$ and $N_D = \overline{n_i^2}$. But when we account for the unmodulated carrier power in $S_R = \frac{1}{2} A_c^2(1 + S_x)$ we find that $S_D = 2 S_x S_R / (1 + S_x)$ and

$$\left(\frac{S}{N}\right)_D = \frac{2S_x}{1 + S_x} \left(\frac{S}{N}\right)_R = \frac{S_x}{1 + S_x} \gamma \qquad \text{AM} \qquad (5)$$

This ratio is bounded by $(S/N)_D \leq \gamma/2$ since $S_x \leq 1$.

Full-load tone modulation corresponds to $S_x = 1/2$ and $(S/N)_D = \gamma/3$, which is 5 dB below that of DSB with the same parameters. More typically, however, $S_x \approx 0.1$ and AM would be some 10 dB inferior to DSB. AM broadcasting stations combat this effect using special techniques such as volume compression and peak limiting of the modulating signal to keep the carrier fully modulated most of the time. These techniques actually distort $x(t)$ and would be unacceptable for analog data transmission.

For SSB modulation (or VSB with a small vestige) we have $x_c(t) = \frac{1}{2} A_c[x(t) \cos \omega_c t \mp \hat{x}(t) \sin \omega_c t]$ with $B_T = W$ and $S_R = \frac{1}{4} A_c^2 S_x$. Synchronous detection rejects the quadrature component of both the signal and noise, leaving

$$y_D(t) = \frac{1}{2} A_c x(t) + n_i(t) \qquad (6)$$

so $S_D = \frac{1}{4} A_c^2 S_x = S_R$. Since f_c falls at either edge of an ideal predetection filter, $G_{n_i}(f)$ has the shape of Fig. 9.2-1c. Hence,

$$G_{n_i}(f) \approx \frac{\eta}{2} \Pi(f/2W) \qquad (7)$$

and $N_D = \overline{n_i^2} = \eta W$. Therefore,

$$\left(\frac{S}{N}\right)_D = \left(\frac{S}{N}\right)_R = \gamma \qquad \text{SSB} \qquad (8)$$

which shows that SSB yields the same noise performance as baseband or DSB transmission.

Finally, consider VSB plus carrier. If the vestigial band is small compared to W, the predetection and postdetection noise will be essentially the same as SSB. But the signal will be essentially the same as AM with all the information-bearing power in one sideband. Hence,

$$\left(\frac{S}{N}\right)_D \approx \frac{S_x}{1 + S_x} \left(\frac{S}{N}\right)_R \approx \frac{S_x}{1 + S_x} \gamma \qquad \text{VSB} + \text{C} \qquad (9)$$

assuming that $B_T \approx W$.

To summarize the results in Eqs. (2)–(9), we state the following general properties of synchronously detected linear modulation with noise:

1. The message and noise are additive at the output if they are additive at the detector input.
2. If the predetection noise spectrum is reasonably flat over the transmission bandwidth, the destination noise spectrum is essentially constant over the message bandwidth.
3. Relative to $(S/N)_D$, suppressed-sideband modulation (SSB and VSB) has no particular advantage over double-sideband modulation (AM and DSB). This is because of the coherence property of double sideband, which compensates for the reduced predetection noise power of single sideband.
4. Making due allowance for the "wasted" power in unsuppressed-carrier systems, all types of linear modulation have the same performance as baseband transmission on the basis of average transmitted power and fixed noise density.

These statements presume nearly ideal systems with fixed *average* power.

Comparisons based on *peak envelope power* indicate that SSB yields a postdetection S/N about 3 dB better than DSB and 9 dB better than AM, assuming a reasonably smooth modulating signal. But SSB is inferior to DSB if the message has pronounced discontinuities causing envelope horns. Downing (1964, chap. 4) gives further details.

> **Exercise 9.3-1** Suppose the predetection filter for a USSB signal passes $f_c - 0.25W \leq |f| \leq f_c + W$. Use Fig. 9.2-1a to sketch the postdetection noise power spectrum. Then show that $(S/N)_D$ will be about 1 dB less than the value predicted by Eq. (8).

Envelope Detection and Threshold Effect

Inasmuch as AM is normally demodulated by an envelope detector, we should examine how this differs from synchronous detection when noise is present. At the detector input we have

$$v(t) = A_c[1 + x(t)] \cos \omega_c t + [n_i(t) \cos \omega_c t - n_q(t) \sin \omega_c t] \qquad (10)$$

The phasor construction of Fig. 9.3-2 shows that the resultant envelope and phase are

$$A_v(t) = \sqrt{\{A_c[1 + x(t)] + n_i(t)\}^2 + [n_q(t)]^2}$$

$$\phi_v(t) = \arctan \frac{n_q(t)}{A_c[1 + x(t)] + n_i(t)} \qquad (11)$$

Clearly, further progress calls for some simplifications, so let's assume the signal is either very large or very small.

Figure 9.3-2 Phasor diagram for AM plus noise.

Taking the signal to dominate, say $A_c^2 \gg \overline{n^2}$, then $A_c[1 + x(t)]$ will be large compared to $n_i(t)$ and $n_q(t)$, at least most of the time. The envelope can then be approximated by

$$A_v(t) \approx A_c[1 + x(t)] + n_i(t) \tag{12}$$

which shows the envelope modulation due to noise, similar to interference modulation. An ideal envelope detector reproduces the envelope minus its DC component, so

$$y_D(t) = A_v(t) - \overline{A_v} = A_c\, x(t) + n_i(t)$$

identical to that of a synchronous detector. The postdetection S/N is then as previously given in Eq. (5). Likewise, Eq. (9) will hold for envelope detection of VSB + C.

But bear in mind the condition for these results, namely, $A_c^2 \gg \overline{n^2}$. Since $A_c^2/\overline{n^2}$ is proportional to $S_R/\eta B_T$, an equivalent requirement is $(S/N)_R \gg 1$. (There is no such condition with synchronous detection.) Thus, providing the predetection signal-to-noise ratio is large, envelope demodulation in the presence of noise has the same performance quality as synchronous demodulation.

At the other extreme, $(S/N)_R \ll 1$, the situation is quite different. For if $A_c^2 \ll \overline{n^2}$, the noise dominates in a fashion similar to strong interference, and we can think of $x_c(t)$ as modulating $n(t)$ rather than the reverse. To expedite the analysis, $n(t)$ is represented in envelope-and-phase form

$$n(t) = A_n(t) \cos\left[\omega_c t + \phi_n(t)\right]$$

leading to the phasor diagram of Fig. 9.3-3. In this figure the noise phasor is the reference because we are taking $n(t)$ to be dominant. By inspection, the envelope is approximately

$$A_v(t) \approx A_n(t) + A_c[1 + x(t)]\cos \phi_n(t) \tag{13}$$

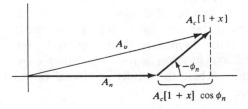

Figure 9.3-3 Phasor diagram for AM plus noise with $(S/N)_R \ll 1$.

from which

$$y(t) = A_n(t) + A_c\, x(t)\, \cos\, \phi_n(t) - \overline{A_n} \tag{14}$$

where $\overline{A_n} = \sqrt{\pi N_R/2}$.

The principal output component is obviously the noise envelope $A_n(t)$, as expected. Furthermore, there is no term in Eq. (14) strictly proportional to the message; though signal and noise were *additive* at the input, the detected message term is *multiplied* by noise in the form of $\cos\, \phi_n(t)$, which is random. The message is therefore hopelessly *mutilated*, and its information has been lost. Under these circumstances, an output signal-to-noise ratio is difficult to define, if not meaningless.

The mutilation or loss of message at low predetection signal-to-noise ratios is called *threshold effect*. The name comes about because there is some value of $(S/N)_R$ above which envelope mutilation is negligible and below which system performance rapidly deteriorates. With synchronous detection, the output signal and noise are always additive; true, the message is buried in noise if $(S/N)_R \ll 1$, but its identity is preserved.

Actually, the threshold is not a unique point unless some convention is established for its definition. Generally speaking, threshold effects are minimal if $A_c \gg A_n$ most of the time. To be more specific we define the *threshold level* as that value of $(S/N)_R$ for which $A_c \geq A_n$ with probability 0.99. Then

$$\left(\frac{S}{N}\right)_{R\text{th}} = 4\ \ln 10 \approx 10 \tag{15a}$$

or, since $(S/N)_R = \gamma/2$,

$$\gamma_{\text{th}} = 8\ \ln 10 \approx 20 \tag{15b}$$

If $(S/N)_R < (S/N)_{R\text{th}}$ (or $\gamma < \gamma_{\text{th}}$), message mutilation must be expected, along with the consequent loss of information.

Looking at the value of $(S/N)_{R\text{th}}$ and recalling that $(S/N)_D < (S/N)_R$ leads to a significant conclusion: threshold effect is not a serious limitation for AM broadcasting because *audio* transmission demands a postdetection signal-to-noise ratio of about 30 dB or more, so $(S/N)_R$ is well above the threshold level. In other words, additive noise obscures the signal long before multiplicative noise mutilates it. On the other hand, sophisticated processing techniques exist for recovering *digital* signals buried in additive noise. Hence, if AM is used for digital transmission, synchronous detection may be necessary to avoid threshold effects.

Finally, let's consider how an envelope detector can act in synchronous fashion and why this requires large $(S/N)_R$. Referring back to Fig. 6.5-6 and assuming the input noise is negligible, we see that the diode functions as a switch, closing briefly on the positive carrier peaks; therefore the switching is perfectly synchronized with the carrier. But when noise dominates, the switching is controlled primarily by the noise peaks and synchronism is lost. The latter effect never occurs in true synchronous detectors, where the locally generated carrier can always be much greater than the noise.

Exercise 9.3-2 Use Eq. (10), Sect. 9.2, to derive the threshold level in Eq. (15a). Specifically, show that if $S_x = 1$, then $P(A_c \geq A_n) = 0.99$ requires $(S/N)_R = 4 \ln 10 \approx 10$.

Spread Spectrum with Synchronous Detection★

Spread spectrum communication systems employ special techniques designed to combat strong interference and/or to prevent message recovery by unauthorized receivers. As the name suggests, these techniques spread the transmitted signal spectrum over a frequency range much greater than the message bandwidth. The spectral spreading involves an auxiliary *pseudo-noise* (PN) *process* that looks random but can be replicated by authorized receivers. In a *frequency-hopping* system, for instance, the carrier frequency hops "randomly" from one value to another under the control of a PN process, which obviously spreads the spectrum and makes the transmitted signal difficult to demodulate unless you know the hopping pattern. Dixon (1976) discusses frequency hopping along with other spread spectrum systems and applications.

We'll investigate the *direct spreading* technique diagrammed in Fig. 9.3-4, where the message $x(t)$ is multiplied by a wideband PN waveform $c(t)$ prior to DSB carrier modulation. Multiplying by $c(t)$ effectively masks the message and spreads the spectrum of the modulating signal $\tilde{x}(t) = x(t)c(t)$. Although such systems are used most often for digital communication, we'll continue to work with an analog message in order to focus on the operating principles. For the same reason, the details of PN generation are postponed to Chap. 11. Here we'll show that if the receiver includes synchronous detection and an identical PN generator, then direct spreading significantly reduces interference effects without degrading the signal-to-noise ratio.

The PN generator produces a *pseudo-random binary wave* $c(t)$, illustrated in Fig. 9.3-5a, consisting of rectangular pulses called *chips*. Each chip has duration T_c and amplitude ± 1 so that $c^2(t) = 1$—an essential condition for message recovery. To facilitate analysis, we'll assume that $c(t)$ has the same properties as the random digital wave in Example 5.1-2 when $D = T_c$, $a_k = \pm 1$, and $\sigma^2 = \overline{c^2} = 1$. Thus, from our previous studies,

$$R_c(\tau) = \Lambda(\tau/T_c) \qquad G_c(f) = T_c \operatorname{sinc}^2 \frac{f}{W_c} \tag{16}$$

Figure 9.3-4 Modulator for DSB spread spectrum.

Figure 9.3-5 Pseudo-random binary wave. (a) Waveform; (b) autocorrelation and power spectrum.

which are sketched in Fig. 9.3-5b. The parameter

$$W_c \triangleq \frac{1}{T_c}$$

serves as a measure of the PN bandwidth.

Next, consider the "chipped message" $\tilde{x}(t) = x(t)c(t)$. Treating $x(t)$ as the output of an ergodic information process independent of $c(t)$, we have

$$\overline{\tilde{x}^2} = E[x^2(t)c^2(t)] = \overline{x^2} = S_x$$

Recall further that multiplying independent random signals corresponds to multiplying their autocorrelation functions and *convolving* their power spectra. For clarity we denote the message bandwidth by W_x such that $G_x(f) \approx 0$ for $|f| > W_x$ and

$$G_{\tilde{x}}(f) = G_x(f) * G_c(f) = \int_{-W_x}^{W_x} G_x(\lambda)G_c(f - \lambda) \, d\lambda$$

But effective spectral spreading calls for $W_c \gg W_x$, in which case $G_c(f - \lambda) \approx G_c(f)$ over $|\lambda| \le W_x$. Therefore,

$$G_{\tilde{x}}(f) \approx \left[\int_{-W_x}^{W_x} G_x(\lambda) \, d\lambda\right] G_c(f) = S_x \, G_c(f) \tag{17}$$

and we see that $\tilde{x}(t)$ has a spread spectrum whose bandwidth essentially equals the PN bandwidth W_c. The bandwidth expansion ratio W_c/W_x ranges from 10 to 100 in typical systems.

DSB carrier modulation produces a transmitted signal proportional to $\tilde{x}(t) \cos \omega_c t$, requiring transmission bandwidth $B_T \gg W_c \gg W_x$. This wideband

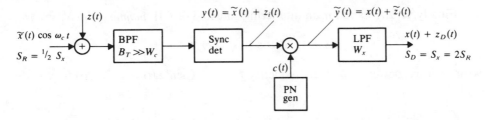

Figure 9.3-6 Receiver for DSB spread spectrum with additive noise or interference $z(t)$.

noise-like signal would be hard to distinguish from the background noise at an unauthorized receiver; accordingly, we say that spread-spectrum transmission has a *low probability of intercept*. However, an authorized receiver configured as in Fig. 9.3-6 will recover the message without increased output noise, despite the increased bandwidth. Moreover, the receiver structure suppresses stray inter-ference or hostile jamming.

We'll analyze the system's performance taking unit-amplitude carrier at the receiver, so $S_R = \frac{1}{2}\overline{\tilde{x}^2} = \frac{1}{2}S_x$, and we'll let $z(t)$ stand for additive noise or inter-ference with in-phase component $z_i(t)$. Synchronous detection after bandpass fil-tering yields $y(t) = \tilde{x}(t) + z_i(t)$ which is multiplied by the locally generated PN wave to get

$$\tilde{y}(t) = [\tilde{x}(t) + z_i(t)]c(t)$$
$$= x(t)c^2(t) + z_i(t)c(t) = x(t) + \tilde{z}_i(t) \qquad (18)$$

Notice that this multiplication spreads the spectrum of $z_i(t)$ but *de-spreads* $\tilde{x}(t)$ and recovers $x(t)$, assuming near perfect synchronization of the local PN gener-ator. Final lowpass filtering removes the out-of-band portion of $\tilde{z}_i(t)$, leaving $y_D(t) = x(t) + z_D(t)$ with output signal power $S_D = S_x = 2S_R$. Our next step is to find the contaminating power $\overline{z_D^2}$ at the output.

When $z(t)$ stands for white noise $n(t)$, the in-phase component $n_i(t)$ has the lowpass power spectrum back in Fig. 9.2-1b and

$$R_{n_i}(\tau) = \mathcal{F}[G_{n_i}(f)] \approx \eta B_T \operatorname{sinc} B_T \tau$$

The autocorrelation of the chipped lowpass noise $\tilde{n}_i(t)$ equals the product $R_{n_i}(\tau)R_c(\tau)$. Since $R_{n_i}(\tau)$ becomes quite small for $|\tau| \geq 1/B_T \ll 1/W_c$ while $R_c(\tau) \approx 1$ for $|\tau| \ll T_c = 1/W_c$, we conclude that $R_{\tilde{n}_i}(\tau) \approx R_{n_i}(\tau)$ and hence

$$G_{\tilde{n}_i}(f) \approx G_{n_i}(f) \approx \eta\Pi(f/B_T)$$

The output noise power from the lowpass filter is then $N_D = 2\eta W_x$, so

$$(S/N)_D = 2S_R/2\eta W_x = S_R/\eta W_x$$

Comparing this result with Eq. (4b) confirms that spread spectrum with synchro-nous detection has the same noise performance as a conventional DSB system.

Finally, let $z(t)$ stand for an interfering sinusoid or *CW jamming signal*, say

$$z(t) = \sqrt{2J} \cos \left[(\omega_c + \omega_z)t + \theta \right]$$

with average power $\overline{z^2} = J$ at frequency $f_c + f_z$. Then $z_i(t) = \sqrt{2J} \cos (\omega_z t + \theta)$ and

$$G_{z_i}(f) = \frac{J}{2} \left[\delta(f - f_z) + \delta(f + f_z) \right] \tag{19}$$

Multiplication by $c(t)$ spreads this spectrum so that $G_{\tilde{z}_i}(f) = G_{z_i}(f) * G_c(f)$, a routine convolution with impulses. If you sketch the result you'll see that maximum jamming power in the lowpass range $|f| \leq W_x$ occurs when $|f_z| \ll W_c$. The corresponding upper bound on the output jamming power is

$$\overline{z_D^2} = \int_{-W_x}^{W_x} G_{\tilde{z}_i}(f) \, df \leq 2W_x \frac{J}{W_c} \tag{20a}$$

and the signal-to-jamming ratio becomes

$$\left(\frac{S}{J} \right)_D \geq \frac{W_c}{W_x} \frac{S_R}{J} \tag{20b}$$

Thus, relative to the output signal power, the jamming power has been reduced by the factor $W_x/W_c \ll 1$. The bandwidth expansion ratio W_c/W_x is therefore also called the *processing gain*. Similar results apply in the case of *broadband jamming* when $z(t)$ acts like additive white noise.

Exercise 9.3-3 Sketch $G_{\tilde{z}_i}(f) = G_{z_i}(f) * G_c(f)$ taking $G_{z_i}(f)$ in Eq. (19) with $|f_z| \ll W_c$. Then derive Eq. (20a) by showing that $G_{\tilde{z}_i}(f) \leq J/W_c$.

9.4 EXPONENTIAL MODULATION WITH NOISE

This section deals with noise in PM and FM systems. As we saw in Chap. 7, the inherent nonlinear nature of exponential modulation leads to analytic difficulties—all the more so when noise must be considered. We'll therefore begin with the large-signal condition $(S/N)_R \gg 1$ to determine the postdetection noise characteristics and signal-to-noise ratios for PM and FM. Our efforts here pay off in results that quantify the valuable wideband *noise-reduction* property, a property further enhanced by postdetection FM deemphasis filtering.

But wideband noise reduction involves a *threshold effect* which, unlike the AM case, may pose a significant performance limitation. We'll qualitatively discuss operation near threshold, and take a brief look at the FM feedback receiver as one technique for threshold extension.

Postdetection Noise

The predetection portion of an exponential modulation receiver has the structure previously diagrammed in Fig. 9.3-1. The received signal is

$$x_c(t) = A_c \cos \left[\omega_c t + \phi(t) \right]$$

where $\phi(t) = \phi_\Delta x(t)$ for a PM wave or $\dot\phi(t) = 2\pi f_\Delta x(t)$ for an FM wave. In either case, the carrier amplitude remains constant so

$$S_R = \tfrac{1}{2} A_c^2 \qquad \left(\frac{S}{N} \right)_R = \frac{A_c^2}{2\eta B_T} \tag{1}$$

and $(S/N)_R$ is often called the *carrier-to-noise ratio* (CNR). The predetection BPF is assumed to have a nearly ideal response with bandwidth B_T centered at f_c.

Figure 9.4-1 portrays our model for the remaining portion of the receiver with the detector input

$$v(t) = x_c(t) + n(t) = A_v(t) \cos \left[\omega_c t + \phi_v(t) \right]$$

The limiter suppresses any amplitude variation $A_v(t)$. To find the signal and noise contained in $\phi_v(t)$, we express $n(t)$ in envelope-and-phase form and write

$$v(t) = A_c \cos \left[\omega_c t + \phi(t) \right] + A_n(t) \cos \left[\omega_c t + \phi_n(t) \right] \tag{2}$$

The phasor construction of Fig. 9.4-2 then shows that

$$\phi_v(t) = \phi(t) + \arctan \frac{A_n(t) \sin \left[\phi_n(t) - \phi(t) \right]}{A_c + A_n(t) \cos \left[\phi_n(t) - \phi(t) \right]} \tag{3}$$

The first term of $\phi_v(t)$ is the signal phase by itself, but the contaminating second term involves both noise *and* signal. Clearly, this expression is very unclear and we can't go much further without some simplifications.

A logical simplification comes from the large-signal condition $(S/N)_R \gg 1$, so $A_c \gg A_n(t)$ most of the time and we can use the small-argument approximation for the arctangent function. A less obvious simplification ignores $\phi(t)$ in the second term of Eq. (3), replacing $\phi_n(t) - \phi(t)$ with $\phi_n(t)$ alone. We justify this step for purposes of noise analysis by recalling that ϕ_n has a *uniform* distribution over

$$v(t) = x_c(t) + n(t) = A_v(t) \cos \left[\omega_c t + \phi_v(t) \right]$$

$$\left(\frac{S}{N} \right)_R = \frac{A_c^2 / 2}{\eta B_T}$$

$$y(t) = \begin{cases} \phi_v(t) & \text{PM} \\ \dfrac{1}{2\pi} \dot\phi_v(t) & \text{FM} \end{cases}$$

Figure 9.4-1 Model for detection of exponential modulation plus noise.

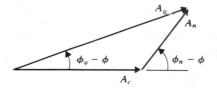

Figure 9.4-2 Phasor diagram of exponential modulation plus noise.

2π radians; hence, in the sense of ensemble averages, $\phi_n - \phi$ differs from ϕ_n only by a shift of the mean value. With these two simplifications Eq. (3) becomes

$$\phi_v(t) \approx \phi(t) + \psi(t) \tag{4}$$

where

$$\psi(t) \triangleq \frac{A_n(t)\sin\phi_n(t)}{A_c} = \frac{1}{\sqrt{2S_R}}\, n_q(t) \tag{5}$$

in which we've substituted $n_q = A_n \sin\phi_n$ and $S_R = \frac{1}{2}A_c^2$.

Equation (4) says that the signal phase $\phi(t)$ and the equivalent phase noise $\psi(t)$ are *additive* under the large-signal condition. Equation (5) brings out the fact that $\psi(t)$ depends on the *quadrature* component of $n(t)$ and decreases with increasing signal power.

Now let $\phi(t) = 0$ and consider the resulting noise $\psi(t)$ at the output of a *phase discriminator*. The PM postdetection noise power spectrum has the shape of $G_{n_q}(f)$ in Fig. 9.2-1b, but multiplied by $1/2S_R$ because $\overline{\psi^2} = \overline{n_q^2}/2S_R$. Hence,

$$G_\psi(f) \approx \frac{\eta}{2S_R}\, \Pi\!\left(\frac{f}{B_T}\right) \tag{6}$$

which is essentially flat over $|f| \leq B_T/2$, as sketched in Fig. 9.4-3.

Since $B_T/2$ exceeds the message bandwidth W, save for the special case of NBPM, the receiver should include a *postdetection filter* $H_D(f)$ to remove the out-of-band noise. If $H_D(f)$ is an ideal LPF with unit gain and bandwidth W, then the output noise power at the destination will be

$$N_D = \int_{-W}^{W} G_\psi(f)\, df = \frac{\eta W}{S_R} \qquad \text{PM} \tag{7}$$

The shaded area in Fig. 9.4-3 equals N_D.

Figure 9.4-3 PM postdetection noise spectrum.

Next consider a *frequency* discriminator with input $\phi_v(t) = \psi(t)$, so the output is the instantaneous *frequency noise*

$$\xi(t) \triangleq \frac{1}{2\pi} \dot{\psi}(t) = \frac{1}{2\pi\sqrt{2S_R}} \dot{n}_q(t) \tag{8}$$

Thus, from Eq. (22), Sect. 5.2, we get the FM postdetection noise power spectrum

$$G_\xi(f) = (2\pi f)^2 \frac{1}{8\pi^2 S_R} G_{n_q}(f) = \frac{\eta f^2}{2S_R} \Pi\left(\frac{f}{B_T}\right) \tag{9}$$

This *parabolic* function sketched in Fig. 9.4-4 has components beyond $W < B_T/2$, like PM, but increases as f^2.

If we again take the postdetection filter to be an ideal LPF that passes the shaded region in Fig. 9.4-4, the destination noise power will be

$$N_D = \int_{-W}^{W} G_\xi(f)\, df = \frac{\eta W^3}{3S_R} \qquad \text{FM} \tag{10}$$

However, if we also incorporate *deemphasis* filtering such that

$$|H_D(f)| = |H_{de}(f)|\, \Pi(f/2W)$$

with

$$|H_{de}(f)| = [1 + (f/B_{de})^2]^{-1/2}$$

then

$$N_D = \int_{-W}^{W} |H_{de}(f)|^2 G_\xi(f)\, df$$

$$= \frac{\eta B_{de}^3}{S_R}\left[\left(\frac{W}{B_{de}}\right) - \arctan\left(\frac{W}{B_{de}}\right)\right] \tag{11a}$$

In the usual case where $W/B_{de} \gg 1$, Eq. (11a) simplifies to

$$N_D \approx \eta B_{de}^2 W/S_R \qquad \text{Deemphasized FM} \tag{11b}$$

since $\arctan(W/B_{de}) \approx \pi/2 \ll W/B_{de}$.

Let's summarize and comment on our results up to this point. First, the postdetection noise spectral densities in PM and FM have out-of-band components that call for postdetection filtering. Second, the PM noise spectrum is flat whereas

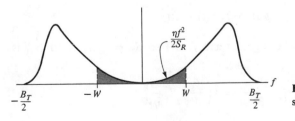

Figure 9.4-4 FM postdetection noise spectrum.

the FM noise spectrum increases parabolically. Consequently, higher baseband signal frequencies in FM suffer more noise contamination than lower frequencies. Deemphasis filtering compensates for this effect, providing that the message has been pre-emphasized at the transmitter. (Another compensating technique suited to FDM systems is described in Prob. 9.5-7.) Third, the destination noise power N_D in PM and FM decreases as S_R increases, a phenomenon known as *noise quieting*. You can hear noise quieting on an FM radio with the volume up when you tune between stations.

> **Exercise 9.4-1** Sketch the deemphasized noise spectrum $|H_{de}(f)|^2 G_\xi(f)$ *without* a lowpass cutoff for $|f| > W$. Then estimate the total area N_D assuming $B_T \gg W \gg B_{de}$. Compare your result with Eq. (11b).

Destination S/N

Here we calculate the destination signal-to-noise ratios for PM, FM, and deemphasized FM. We continue the large-signal condition $(S/N)_R \gg 1$ so our previous results for N_D still hold. The presence of a signal phase $\phi(t)$ does not void those results even though we replaced $\phi_n(t) - \phi(t)$ with $\phi_n(t)$ in Eq. (3). Had we included $\phi(t)$ in the phase noise, a more complicated analysis would have revealed additional components in the postdetection noise spectrum that fall outside the message band and are rejected by $H_D(f)$; see Downing (1964, chap. 4).

The demodulated signal plus noise in a PM system with $\phi(t) = \phi_\Delta x(t)$ is

$$y(t) = \phi_v(t) = \phi_\Delta x(t) + \psi(t)$$

The postdetection filter passes the signal term $\phi_\Delta x(t)$ so $S_D = \phi_\Delta^2 \overline{x^2} = \phi_\Delta^2 S_x$, and the output noise power N_D is given by Eq. (7). Hence,

$$\left(\frac{S}{N}\right)_D = \frac{\phi_\Delta^2 S_x}{(\eta W/S_R)} = \phi_\Delta^2 S_x \frac{S_R}{\eta W}$$
$$= \phi_\Delta^2 S_x \gamma \qquad \text{PM} \qquad (12)$$

Since γ equals the output S/N for baseband transmission (or suppressed-carrier linear modulation) with received power S_R, bandwidth W, and noise density η, we see that PM gives an improvement over baseband of exactly $\phi_\Delta^2 S_x$. But because of the ambiguity constraint $\phi_\Delta \leq \pi$, the PM improvement is no greater than $\phi_\Delta^2 S_x|_{max} = \pi^2$, or about 10 dB at best. If $\phi_\Delta^2 S_x < 1$, PM performance is inferior to baseband while the transmission bandwidth is still $B_T \geq 2W$.

The demodulated signal plus noise in an FM system with $\dot{\phi}(t) = 2\pi f_\Delta x(t)$ is

$$y(t) = \frac{1}{2\pi} \dot{\phi}_v(t) = f_\Delta x(t) + \xi(t)$$

The postdetection filter passes the signal term $f_\Delta x(t)$ so $S_D = f_\Delta^2 S_x$, and N_D is given by Eq. (10). Hence,

$$\left(\frac{S}{N}\right)_D = \frac{f_\Delta^2 S_x}{(\eta W^3/3S_R)} = 3\left(\frac{f_\Delta}{W}\right)^2 S_x \frac{S_R}{\eta W}$$

in which we spot the *deviation ratio* $D = f_\Delta/W$. We therefore write

$$\left(\frac{S}{N}\right)_D = 3D^2 S_x \gamma \qquad \text{FM} \tag{13}$$

and it now appears that $(S/N)_D$ can be made *arbitrarily large* by increasing D without increasing the signal power S_R—a conclusion that requires further qualification and will be reexamined shortly.

Meanwhile, recall that the transmission bandwidth requirement B_T increases with the deviation ratio. Therefore, Eq. (13) represents *wideband noise reduction* or, equivalently, the latitude to exchange increased bandwidth for reduced transmitted power while keeping $(S/N)_D$ fixed. To emphasize this property, take the case of wideband FM with $D \gg 1$ and $B_T = 2f_\Delta \gg W$. Then $D = B_T/2W$ and Eq. (13) becomes

$$\left(\frac{S}{N}\right)_D = \frac{3}{4}\left(\frac{B_T}{W}\right)^2 S_x \gamma \qquad \text{WBFM} \tag{14}$$

which shows that $(S/N)_D$ increases as the square of the bandwidth ratio B_T/W. With smaller deviation ratios, the break-even point compared to baseband transmission occurs when $3D^2 S_x = 1$ or $D = 1/\sqrt{3S_x} \geq 0.6$. The dividing line between NBFM and WBFM is sometimes designated to be $D \approx 0.6$ for this reason.

Finally, if the receiver includes *deemphasis filtering* and $B_{\text{de}} \ll W$, the output noise is further reduced in accordance with Eq. (11*b*). Thus,

$$\left(\frac{S}{N}\right)_D \approx \left(\frac{f_\Delta}{B_{\text{de}}}\right)^2 S_x \gamma \qquad \text{Deemphasized FM} \tag{15}$$

and we have a deemphasis improvement factor of about $\frac{1}{3}(W/B_{\text{de}})^2$. This improvement requires preemphasis filtering at the transmitter and may carry a hidden penalty. For if the message amplitude spectrum does not roll off at least as fast as $1/f$, like an audio signal, preemphasis increases the deviation ratio and the transmission bandwidth requirement.

Example 9.4-1 Just how much can be gained from wideband noise reduction is well illustrated with the broadcast FM parameters $f_\Delta = 75$ kHz, $W = 15$ kHz, and $D = 5$. Taking $S_x = 1/2$ for a representative value, Eq. (13) gives

$$(S/N)_D = (3 \times 5^2 \times \tfrac{1}{2})\gamma \approx 38\gamma$$

or about 16 dB better than baseband transmission. Deemphasis filtering with $B_{\text{de}} = 2.1$ kHz increases the S/N to $(S/N)_D \approx 640\gamma$. Thus, other factors being equal, a 1-W FM system with deemphasis could replace a 640-W baseband system. The cost of this transmitted power reduction is increased transmission bandwidth, since FM with $D = 5$ requires $B_T \approx 14W$.

Several practical factors work against full realization of increased bandwidth in exchange for reduced transmitted power S_T. And indeed the goal of

FM broadcasting is maximum $(S/N)_D$ rather than minimum S_T. However, other applications involve minimizing S_T or squeezing as much as possible from every available watt. The large-signal condition $(S/N)_R \gg 1$ poses a serious limitation for such applications, and the FM threshold effect becomes a matter of grave concern.

Exercise 9.4-2 Calculate the minimum transmitter power needed if a PM system replaces the 1-W FM system in Example 9.4-1 and the value of $(S/N)_D$ remains the same.

FM Threshold Effect

The small-signal condition $(S/N)_R \ll 1$ can be represented by a phasor diagram like Fig. 9.4-2 with signal and noise phasors interchanged. Then, since $A_n(t) \gg A_c$ most of the time, the resultant phase at the detector input is

$$\phi_v(t) \approx \phi_n(t) + \frac{A_c \sin\left[\phi(t) - \phi_n(t)\right]}{A_n(t)} \tag{16}$$

The noise now dominates and the message, contained in $\phi(t)$, has been *mutilated* beyond all hope of recovery.

Actually, significant mutilation begins to occur when $(S/N)_R \approx 1$ and $\overline{A_n} \approx A_c$. With phasors of nearly equal length, we have a situation similar to cochannel interference when $\rho = A_i/A_c \approx 1$. Small noise variations may then produce large *spikes* in the demodulated FM output. The phasor diagram in Fig. 9.4-5a illustrates this point taking $\phi(t) = 0$ and $\phi_n(t_1) \approx -\pi$ so $\phi_v(t_1) \approx -\pi$. If the variations of $A_n(t)$ and $\phi_n(t)$ follow the dashed locus from t_1 to t_2 then $\phi_v(t_2) \approx +\pi$. Correspondingly, the waveform $\phi_v(t)$ in Fig. 9.4-5b has a step of height 2π and the

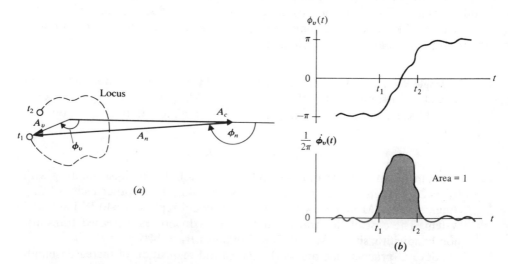

Figure 9.4-5 FM near threshold. (a) Phasor diagram; (b) instantaneous phase and frequency.

output $y(t) = \dot{\phi}_v(t)/2\pi$ has a unit-area spike. These spikes would be heard on an FM radio as a crackling or clicking sound that masks the signal.

We infer from this qualitative picture that the output noise spectrum is no longer parabolic but tends to fill in at DC, the output spikes producing appreciable low-frequency content. This conclusion has been verified through detailed analysis using the "click" approach as refined by Rice. The analysis is complicated (and placed beyond our scope) by the fact that the spike characteristics change when the carrier is modulated, the so-called modulation-suppression effect. Thus, quantitative results are obtained only for specific modulating signals. In the case of tone modulation, the total output noise becomes

$$N_D = \frac{\eta W^3}{3S_R}\left[1 + \frac{12D}{\pi}\gamma e^{-(W/B_T)\gamma}\right] \tag{17}$$

where the second term is the contribution of the spikes. See Rice (1948) and Stumpers (1948) for the original work, or Taub and Schilling (1971, chap. 10) for a tutorial treatment of click analysis.

Figure 9.4-6 shows $(S/N)_D$ in decibels plotted versus γ (also in decibels) for two values of the deviation ratio D, taking tone modulation and N_D given by Eq. (17). The rather sudden drop-off of these curves, traced to the exponential factor in Eq. (17), is the FM threshold effect. When the system is operating near the "knee" of a curve, small variations of signal power cause sizable changes in the output signal; one moment it's there, the next moment it's gone. Below threshold, noise *captures* the output just like strong cochannel interference.

Experimental studies indicate that noise mutilation is negligible in most cases of interest if $(S/N)_R \geq 10$ or thereabouts. Hence, we define the threshold point to be at

$$(S/N)_{R_{th}} = 10 \tag{18}$$

Figure 9.4-6 FM noise performance (without deemphasis).

Equivalently, since $(S/N)_R = (W/B_T)\gamma$,

$$\gamma_{th} = 10\frac{B_T}{W} = 20M(D) \tag{19a}$$

$$\approx 20(D+2) \qquad D > 2 \tag{19b}$$

where use has been made of the FM bandwidth equation $B_T = 2M(D)W \approx 2(D+2)W$. Equations (18) and (19) also apply to PM with D replaced by ϕ_Δ.

Figure 9.4-6 correctly demonstrates that FM performance above threshold is quite impressive—after all, baseband transmission at best gives $(S/N)_D = \gamma$. And the curves do not include the additional improvement afforded by deemphasis filtering. But observe what happens if we attempt to make $(S/N)_D$ arbitrarily large by increasing only the deviation ratio while holding γ fixed, say at 20 dB. With $D = 2$ ($B_T \approx 7W$) we are just above threshold, and $(S/N)_D \approx 28$ dB; but with $D = 5$ ($B_T \approx 14W$) we are below threshold, and the output signal is useless because of mutilation. We therefore cannot achieve an unlimited exchange of bandwidth for signal-to-noise ratio, and system performance may actually deteriorate with increased deviation.

Swapping bandwidth in favor of reduced power is likewise restricted. Suppose, for example, a 30-dB signal-to-noise ratio is desired with a minimum of transmitted power but the transmission bandwidth can be as large as $B_T = 14W$. Were it not for threshold, we could use FM with $D = 5$ and $\gamma = 14$ dB, a power saving compared to baseband of 16 dB. But the threshold point for $D = 5$ is $\gamma_{th} \approx 22$ dB, for which $(S/N)_D \approx 37$ dB. Thus, the design is dictated by the threshold point rather than the desired signal-to-noise ratio, and the potential power reduction cannot be fully realized.

In view of these considerations, it's useful to calculate $(S/N)_D$ at the threshold point. Thus, again omitting deemphasis, we substitute Eq. (19) into Eq. (13) to get

$$\left(\frac{S}{N}\right)_{D_{th}} = 3D^2 S_x \gamma_{th} \tag{20a}$$

$$\approx 60D^2(D+2)S_x \qquad D > 2 \tag{20b}$$

which equals the minimum value of $(S/N)_D$ as a function of D. Given a specified value for $(S/N)_D$ and no bandwidth constraint, you can solve Eq. (20) for the deviation ratio D that yields the *most efficient performance* in terms of signal power. Of course, some allowance must be made for possible signal fading since it is unadvisable to operate with no margin relative to the threshold point.

Example 9.4-2 Suppose a minimum-power FM system is to be designed such that $(S/N)_D \approx 50$ dB, given $S_x = 1/2$, $W = 10$ kHz, $\eta = 10^{-8}$ W/Hz, and no constraint on B_T. With the given values and the assumption $D > 2$, Eq. (20) becomes

$$10^5 \approx 60D^2(D+2)/2$$

and trial-and-error solution yields $D \approx 14$, so $B_T \approx 2(D+2)W = 320$ kHz.

Then, from Eq. (19),

$$S_R/\eta W \geq \gamma_{\text{th}} = 10 \times 32$$

which requires $S_R \geq 320\eta W = 32$ mW.

Exercise 9.4-3 Find the minimum useful value of $(S/N)_D$ for a deemphasized FM system with $B_T = 5W, f_\Delta = 10B_{\text{de}}$, and $S_x = 1/2$.

Threshold Extension by FM Feedback★

Because the threshold limitation is a serious constraint on the design of minimum-power FM systems, there has been considerable interest in *threshold-extension* techniques—especially for space communication. But long before the space applications arose, Chaffee (1939) proposed a means for extending the FM threshold point using a frequency-following or frequency-compressive feedback loop in the receiver, called an FM feedback (FMFB) receiver.

The FMFB receiver diagrammed in Fig. 9.4-7 embodies features of a phase-lock loop within the superheterodyne structure. Specifically, the superhet's LO has been replaced by a VCO whose free-running frequency equals $f_c - f_{\text{IF}}$. The control voltage for the VCO comes from the demodulated output $y_D(t)$. If the loop has sufficient gain K and $(S/N)_D$ is reasonably large, the VCO tracks the instantaneous phase of $x_c(t)$. This tracking action reduces the frequency deviation from f_Δ to $f_\Delta/(1 + K)$, as well as translating the signal down to the IF band. Thus, if K is such that $f_\Delta/(1 + K)W < 1$, the IF input becomes a *narrowband* FM signal and the IF bandwidth need be no larger than $B_{\text{IF}} \approx 2W$.

VCO tracking likewise reduces the noise frequency deviation by the same factor, so $(S/N)_D$ equals that of a conventional receiver when $(S/N)_R \gg 1$. But note that the IF output has a larger predetection signal-to-noise ratio, namely $(S/N)_{\text{IF}} = S_R/\eta B_{\text{IF}} \approx (B_T/2W)(S/N)_R$. Since threshold effects now depend primarily on the value of $(S/N)_{\text{IF}}$, the threshold level has been extended down to a lower value of $(S/N)_R$.

Experimental studies confirm a threshold extension of 5 to 7 dB for FMFB receivers—a significant factor for minimum-power designs. A conventional receiver with a PLL demodulator also provides threshold extension and has the advantage of simpler implementation. Further discussion of FMFB and PLL receivers is given in Taub and Schilling (1971, chap. 10) and Klapper and Frankle (1972).

Figure 9.4-7 FMFB receiver for threshold extension.

9.5 COMPARISON OF CW MODULATION SYSTEMS

At last we're in position to make a meaningful comparison of the various types of CW modulation. Table 9.5-1 summarizes the points to be compared: normalized transmission bandwidth, $b \triangleq B_T/W$; destination signal-to-noise ratio $(S/N)_D$ normalized by γ; threshold point, if any; DC or low-frequency response; and instrumentation complexity. The table also includes baseband transmission for reference purposes. As before, $\gamma \triangleq S_R/\eta W$ where W is the message bandwidth, S_R is the received signal power, and $\eta = k\mathcal{T}_N$ is the noise density referred to the receiver input. Nearly ideal systems are assumed, so the values of $(S/N)_D$ are upper bounds.

Of the several types of linear modulation, suppressed carrier methods are superior to conventional AM on several counts; signal-to-noise ratios are better, and there is no threshold effect. When *bandwidth conservation* is important, single sideband and vestigial sideband are particularly attractive. But you seldom get something for nothing in this world, and the price of efficient linear modulation is the increased complexity of instrumentation, especially at the receiver. Synchronous detection, no matter how it's accomplished, requires highly sophisticated circuitry compared to the envelope detector. For *point-to-point* communication (one transmitter, one receiver) the price may be worthwhile. But for *broadcast* systems (one transmitter, *many* receivers) economic considerations tip the balance toward the simplest possible receiver, and hence envelope detection.

From an instrumentation viewpoint AM is the least complex linear modulation, while suppressed-carrier VSB, with its special sideband filter and synchronization requirements, is the most complex. Of DSB and SSB (in their proper applications) the latter is less difficult to instrument because synchronization is not so critical. In addition, improved filter technology has made the required sideband filters more readily available. Similarly, VSB + C is classed as of "moderate" complexity, despite the vestigial filter, since envelope detection is sufficient.

Compared to baseband or linear modulation, exponential modulation can provide substantially increased values of $(S/N)_D$ — especially FM with deemphasis — with only moderately complex instrumentation. Figure 9.5-1 illustrates this in a form similar to Fig. 9.4-6 (again taking $S_x = \frac{1}{2}$) except that a representative 12-dB deemphasis improvement has been added to the FM curves and performance below threshold is omitted. All curves are labeled with the bandwidth ratio b.

Clearly, for equal values of b, FM is markedly superior to PM insofar as noise performance is concerned. (Phase modulation does have advantages over FM for *digital* signal transmission, to be discussed in Chap. 14.) And as long as the system is above threshold, the improvement can be made arbitrarily large by increasing b, whereas PM is limited to $b \leq 10$ since $\phi_\Delta \leq \pi$.

The penalty for this improvement is excessive transmission bandwidth. Therefore, wideband exponential modulation is most appropriate when extremely clean output signals are desired and bandwidth conservation is a secondary

Table 9.5-1 Comparison of CW modulation systems

Type	$b = B_T/W$	$(S/N)_D/\gamma$	γ_{th}	DC	Complexity	Comments	Typical applications
Baseband	1	1	...	No†	Minor	No modulation	Short-haul links
AM	2	$\dfrac{\mu^2 S_x}{1+\mu^2 S_x}$	20	No	Minor	Envelope detection $\mu \leq 1$	Broadcast ratio
DSB	2	1	...	Yes	Major	Synchronous detection	Analog data, multiplexing
SSB	1	1	...	No	Moderate	Synchronous detection	Point-to-point voice, multiplexing
VSB	1+	1	...	Yes	Major	Synchronous detection	Digital data
VSB + C	1+	$\dfrac{\mu^2 S_x}{1+\mu^2 S_x}$	20	Yes‡	Moderate	Envelope detection $\mu < 1$	Television video
PM§	$2M(\phi_\Delta)$	$\phi_\Delta^2 S_x$	$10b$	Yes	Moderate	Phase detection $\phi_\Delta \leq \pi$	Digital data
FM§¶	$2M(D)$	$3D^2 S_x$	$10b$	Yes	Moderate	Frequency detection	Broadcast radio, microwave relay, satellite systems

† Unless direct-coupled.
‡ With electronic DC restoration.
§ $b \geq 2$.
¶ Deemphasis not included.

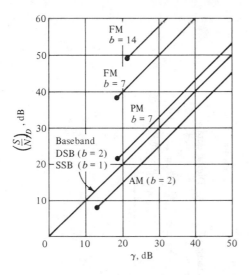

Figure 9.5-1 Performance of CW modulation systems, including 12-dB preemphasis improvement for FM.

factor. Typical applications include commercial FM and television audio. At microwave frequencies, both the noise-reduction and constant-amplitude properties are advantageous, so that most microwave systems have exponential carrier modulation.

As to power conservation, FM with moderate values of b does offer a saving over linear modulation, threshold limitations notwithstanding. In this capacity it is used for point-to-point radio, especially mobile systems, where synchronous detection would be a nuisance. Moreover, thanks to threshold-extension techniques, wideband FM is employed in certain satellite and space-probe communication systems, where transmitter power is at a premium.

Regarding transmission of modulating signals having significant low-frequency components, we have already argued the superiority of DSB and VSB; this explains their use in data transmission, both analog and digital. For facsimile and TV video, electronic DC restoration makes envelope-detected VSB possible and desirable. (AM could also be used in this way, but the bandwidth is prohibitive. Suppressed-carrier single sideband is virtually out of the question.) Also we noted in Sect. 7.3 that a balanced discriminator has excellent low-frequency response; hence the low-frequency performance of FM can equal that of DSB or VSB, and without troublesome synchronization. Consequently, small-deviation FM or PM subcarrier modulation is often used in telemetry systems for the less active input signals. For similar reasons, high quality magnetic-tape recorders are equipped with an FM mode in which the input is recorded as a frequency-modulated wave.

Not shown in the table is relative system performance in the face of time-varying transmission characteristics, frequency-selective fading, multiple-path propagation, etc. An unstable transmission medium has a *multiplicative* effect which is particularly disastrous for envelope detection. (Late-night listeners to distant AM stations are familiar with the garbled result.) Similarly, transmission instabilities often preclude wideband modulation.

To summarize briefly is impossible. There is no universal solution to all communication problems. The communication engineer must therefore approach each new task with an open mind and a careful review of all available information.

9.6 PROBLEMS

9.2-1 White noise with $\eta = 10$ is applied to a BPF having $|H_R(f)|^2$ plotted in Fig. P9.2-1. Sketch $G_{n_i}(f)$ taking $f_c = f_1$, and show therefrom that $\overline{n_i^2} = \overline{n^2}$.

Figure P9.2-1

9.2-2 Do Prob. 9.2-1 taking $f_c = f_2$.

9.2-3 White noise is applied to a tuned circuit whose transfer function is given by Eq. (1a), Sect. 6.1, with $f_0 = f_c$ and $Q = 4/3$. Evaluate $G_n(f)/\eta$ at $f/f_c = 0, \pm 0.5, \pm 1, \pm 1.5$, and ± 2. Then plot $G_{n_i}(f)$.

9.2-4 The BPF in Fig. 9.1-2 will have *local symmetry* around $\pm f_c$ if its lowpass equivalent function $H_{\ell p}(f) \triangleq H_R(f + f_c)u(f + f_c)$ has the even-symmetry property $|H_{\ell p}(-f)| = |H_{\ell p}(f)|$. (a) Let $G_{\ell p}(f) \triangleq (\eta/2)|H_{\ell p}(f)|^2$ and show that we can write

$$G_n(f) = G_{\ell p}(f - f_c) + G_{\ell p}(f + f_c) = \begin{cases} G_{\ell p}(f - f_c) & f > 0 \\ G_{\ell p}(f + f_c) & f < 0 \end{cases}$$

(b) Use this result to show that $G_{n_i}(f) = 2G_{\ell p}(f)$.

9.2-5 A tuned circuit with $Q = f_c/B_T \gg 1$ approximates the local-symmetry property in Prob. 9.2-4, and has $H_R(f) \approx 1/[1 + j2(f - f_c)/B_T]$ for $f > 0$. (a) Find $G_{\ell p}(f)$ and compare with Eq. (11a), Sect. 5.3. (b) Evaluate $\overline{n^2}$ by calculating $\overline{n_i^2}$.

9.2-6 Let $y(t) = 2n(t) \cos (\omega_c t + \theta)$, where $n(t)$ is bandpass noise centered at f_c. Show that $y(t)$ consists of a lowpass component and a bandpass component. Find the mean value and the variance of each component in terms of the properties on $n(t)$.

9.2-7 Bandpass gaussian noise with $\sigma_n = 2$ is applied to an envelope detector described by Eq. (4b), Sect. 9.1. Find and sketch the PDF of the output $y(t)$, and calculate σ_y.

9.2-8 Bandpass gaussian noise with variance σ_n^2 is applied to a square-law device, producing $y(t) = A_n^2(t)$. Find $\bar{y}, \overline{y^2}$, and the PDF of $y(t)$.

9.2-9 Let $v_{\ell p}(t) = \frac{1}{2}[v(t) + j\hat{v}(t)]e^{-j\omega_c t}$. If $v(t)$ is Fourier transformable, show that $V_{\ell p}(f) = \mathcal{F}[v_{\ell p}(t)] = V(f + f_c)u(f + f_c)$. Then sketch $V_{\ell p}(f)$ taking $V(f)$ as in Fig. P9.2-1 with $f_c = f_1 + b$.

9.2-10 Let $H_R(f)$ in Fig. 9.1-2 be an ideal BPF with unit gain and let $\alpha = 1/2$, so f_c is the center frequency. Find $R_n(\tau)$ and $\hat{R}_n(\tau)$ by taking the inverse transform of $G_n(f)$ and $(-j \operatorname{sgn} f)G_n(f)$. Then obtain $R_{n_i}(\tau)$ from Eq. (16) and confirm that it equals the inverse transform of $G_{n_i}(f)$.

9.2-11 Do Prob. 9.2-10 with $\alpha = 0$, so f_c is the lower cutoff frequency.

9.2-12 Suppose $G_n(f)$ has local symmetry around $\pm f_c$, as detailed in Prob. 9.2-4. Write the inverse transforms of $G_{n_i}(f)$ and $G_n(f)$ to show that $R_n(\tau) = R_{n_i}(\tau) \cos \omega_c \tau$. Then show from Eq. (16) that $\hat{R}_n(\tau) = R_{n_i}(\tau) \sin \omega_c \tau$.

9.2-13 Derive Eq. (16) starting from $E[n_q(t)n_q(t - \tau)]$.

9.2-14 Derive Eq. (17) starting from $E[n_i(t)n_q(t - \tau)]$.

9.2-15 Let $G_n(f)$ have local symmetry around $\pm f_c$. Prove that $R_{n_i n_q}(\tau) = 0$ for all τ using: (a) the correlation relations in Prob. 9.2-12; (b) the spectral relations in Prob. 9.2-4.

9.2-16 Let $H_R(f)$ in Fig. 9.1-2 be an ideal BPF with unit gain and let $\alpha = 0$, so f_c is the lower cutoff frequency. Use Eq. (18) to find and sketch $R_{n_i n_q}(\tau)$.

9.2-17 Do Prob. 9.2-16 with $\alpha = 1/4$, so the lower cutoff frequency is $f_c - B_T/4$.

9.3-1 A DSB signal plus noise is demodulated by a product detector (Fig. 6.5-3) with a phase error ϕ'. Take the local oscillator signal to be $2 \cos(\omega_c t + \phi')$ and show that $(S/N)_D = \gamma \cos^2 \phi'$.

9.3-2 Rewrite Eqs. (4b) and (5) in terms of $\gamma_p = S_p/\eta W$, where S_p is the peak envelope power of the DSB or AM signal.

9.3-3 Let $x_c(t)$ have quadrature-carrier multiplexing as in Eq. (1), Sect. 8.2, where $x_1(t)$ and $x_2(t)$ are independent signals and $\overline{x_1^2} = \overline{x_2^2}$. Assume an ideal receiver with AWGN and two synchronous detectors. Find the output signal plus noise for each channel, and express $(S/N)_D$ in terms of γ.

9.3-4 Explain why an SSB receiver should have a nearly rectangular BPF with bandwidth $B_T = W$, whereas predetection filtering is not critical for DSB.

9.3-5 Modify Eq. (8) for LSSB when $|H_R(f)|^2$ has the shape in Fig. P9.2-1 with $f_2 = f_c$ and $2b = W/2$.

9.3-6 Some receivers have additive "one-over-f" noise with power spectral density $G(f) = \eta f_c/2|f|$ for $|f| > 0$. Obtain the resulting expressions for $(S/N)_D$ in terms of γ and W/f_c for USSB and DSB modulation. Compare your results when $W/f_c = 1/5$ and $1/50$.

9.3-7‡ When a demodulated signal includes *multiplicative* noise or related effects, the postdetection S/N cannot be defined unambiguously. An alternative performance measure is then the *normalized mean-square error*

$$\epsilon^2 \triangleq E\{[x(t) - Ky_D(t)]^2\}/S_x$$

where K is chosen such that $Ky_D(t) = x(t)$ in absence of multiplicative effects. Find $y_D(t)$ and show that $\epsilon^2 = 2[1 - \cos \phi] + 1/\gamma$ when a USSB signal with AWGN is demodulated by a producter detector (Fig. 6.5-3) whose local-oscillator signal is $2 \cos[\omega_c t + \phi(t)]$, where $\phi(t)$ is a slowly drifting random phase. *Hint:* From Example 5.2-3 it follows that $\overline{\hat{x}^2} = \overline{x^2}$ and that $\overline{x\hat{x}} = -\hat{R}_x(0) = 0$ since $\hat{R}_x(\tau)$ is an odd function.

9.3-8 Explain why an AM receiver should have a nearly rectangular BPF with bandwidth $B_T = 2W$ for envelope detection, whereas predetection filtering is not critical for synchronous detection.

9.3-9 An AM system with envelope detection is operating at the threshold point. Find the power gain in dB needed at the transmitter to get up to $(S/N)_D = 40$ dB with full-load tone modulation.

9.3-10 An AM system with envelope detection has $(S/N)_D = 30$ dB under full-load tone-modulation conditions with $W = 8$ kHz. If all bandwidths are increased accordingly, while other parameters remain fixed, what is the largest usable value of W?

9.3-11‡ Consider an AM system with envelope detection operating below threshold. Find ϵ^2 defined in Prob. 9.3-7 assuming that $y_D(t) = y(t)$, $\overline{x^2} = 1$, and $\bar{x} = 0$. Express your answer in terms of γ.

9.4-1 Suppose an nth-order Butterworth LPF is used for the postdetection filter in an FM receiver. Obtain an upper bound on N_D and simplify for $n \gg 1$.

9.4-2 Sketch $G_e(f)$ when the predetection BPF in an FM receiver has $H_R(f)$ as given in Prob. 9.2-5. Then calculate N_D and simplify taking $B_T \gg W$.

9.4-3 Obtain an expression for $(S/N)_D$ for PM with deemphasis filtering. Simplify your result taking $B_{de} \ll W$.

9.4-4 The signal $x(t) = \cos 2\pi 200t$ is sent via FM without preemphasis. Calculate $(S/N)_D$ when $f_\Delta = 1$ kHz, $S_R = 500\eta$, and the postdetection filter is an ideal BPF passing $100 \leq f \leq 300$ Hz.

9.4-5 Obtain an expression for $(S/N)_D$ for FM with a gaussian deemphasis filter having $|H_{de}(f)|^2 = \exp[-(f/B_{de})^2]$. Calculate the resulting deemphasis improvement factor when $B_{de} = W/7$.

9.4-6 A certain PM system has $(S/N)_D = 30$ dB. Find the new value of $(S/N)_D$ when the modulation is changed to preemphasized FM with $B_{de} = W/10$, while B_T and all other parameters are held fixed.

9.4-7 Modify Eq. (20) to include deemphasis filtering, and rework the calculations in Example 9.4-2.

9.4-8 Obtain an expression like Eq. (20) for PM, and determine the upper bound on $(S/N)_D$ at threshold.

9.4-9‡ Referring to Prob. 7.4-6a, suppose K is chosen such that $\overline{x_{pe}^2} = \overline{x^2}$, so preemphasis does not significantly increase the FM transmission bandwidth. (a) Show that if $\overline{x^2} = 1$, then

$$(S/N)_D = f_\Delta^2 \gamma \left[B_{de}^2 + \int_{-\infty}^{\infty} f^2 G_x(f)\, df \right]^{-1}$$

(b) Simplify this expression when $G_x(f) = [f_0/\pi(f^2 + f_0^2)]\Pi(f/2W)$ with $f_0 = B_{de} \ll W$. Compare with Eqs. (13) and (15).

9.4-10 Consider the FMFB receiver in Fig. 9.4-7. Let $v(t)$ be a noise-free FM signal with deviation ratio D and take the VCO output to be $2\cos[(\omega_c - \omega_{IF})t + K\phi_D(t)]$ where $\dot\phi_D(t) = 2\pi y_D(t)$. Show that the deviation ratio of the IF signal is $D_{IF} = D/(1 + K)$.

9.5-1 A communication system has $\overline{x^2} = 1/2$, $W = 10$ kHz, $\eta = 10^{-15}$ W/Hz, and transmission loss $L = 100$ dB. Calculate S_T needed to get $(S/N)_D = 40$ dB when the modulation is: (a) SSB; (b) AM with $\mu = 1$ and $\mu = 0.5$; (c) PM with $\phi_\Delta = \pi$; (d) FM with $D = 1, 5$, and 10. Omit deemphasis in the FM case, but check for threshold limitations.

9.5-2 Do Prob. 9.5-1 with $\overline{x^2} = 1$ and $W = 20$ kHz.

9.5-3 A communication system has $\overline{x^2} = 1/2$, $W = 10$ kHz, $S_T = 10$ W, and $\eta = 10^{-13}$ W/Hz. Calculate the path length corresponding to $(S/N)_D = 40$ dB for a transmission cable with loss $\alpha = 1$ dB/km when the modulation is: (a) SSB; (b) AM with $\mu = 1$; (c) FM with $D = 2$ and 8.

9.5-4 Do Prob. 9.5-3 for radio transmission at $f_c = 300$ MHz with antenna gains of 26 dB at transmitter and receiver.

9.5-5 A signal with $\overline{x^2} = 1/2$ is transmitted via AM with $\mu = 1$ and $(S/N)_D = 13$ dB. If the modulation is changed to FM (without deemphasis) and the bandwidths increased while other parameters remain fixed, what's the largest usable deviation ratio and the resulting value of $(S/N)_D$?

9.5-6‡ An FDM system has USSB subcarrier modulation and FM carrier modulation without preemphasis. There are K independent input signals, each having bandwidth W_0, and the subcarrier frequencies are $f_k = (k - 1)W_0$, $k = 1, 2, \ldots, K$. The baseband signal at the FM modulator is $x_b(t) = \sum \alpha_k x_k(t)$, where the α_k are constants and $x_k(t)$ is the kth subcarrier signal, each having $\overline{x_k^2} = 1$. The discriminator at the receiver is followed by a bank of BPFs and synchronous detectors. Use the property that the predetection and postdetection S/N are equal for SSB to show that the output of the kth channel has $(S/N)_k = f_\Delta^2 \alpha_k^2/N_k$ where $N_k = (3k^2 - 3k + 1)\eta W_0^3/3S_R$.

9.5-7‡ In the system described in Prob. 9.5-6, the α_k are chosen such that $\overline{x_b^2} = 1$ and all channels have the same output S/N. Obtain expressions for α_k and $(S/N)_k$ under these conditions.

9.5-8‡ Consider the FM stereo multiplexing system in Figs. 8.2-5 and 8.2-6 where, for purposes of analysis, the preemphasis filtering may be done after matrixing and the deemphasis filtering before matrixing. Let the input signals before preemphasis be $x_1(t) = x_L(t) + x_R(t)$ and $x_2(t) = x_L(t) - x_R(t)$, and let the deemphasized outputs before matrixing be $y_1(t) = f_\Delta x_1(t) + n_1(t)$ and $y_2(t) = f_\Delta x_2(t) + n_2(t)$ so the final outputs are $y_1(t) \pm y_2(t)$. (a) Show that the power spectral density of $n_2(t)$ is $|H_{de}(f)|^2(\eta/S_R)(f^2 + f_0^2)\Pi(f/2W)$, where $f_0 = 38$ kHz, and that $\overline{n_2^2} \gg \overline{n_1^2}$ when $W = 15$ kHz and $B_{de} = 2.1$ kHz. (b) Taking $\overline{x_L x_R} = 0$ and $\overline{x_L^2} = \overline{x_R^2} \approx 1/3$—so that $\overline{x_b^2} \approx 1$ if the pilot is small—show that $(S/N)_D$ for each channel in stereo transmission is about 20 dB less than $(S/N)_D$ for monaural transmission.

TEN

SAMPLING AND PULSE MODULATION

Experimental data and mathematical functions are frequently displayed as *continuous* curves, even though a finite number of *discrete points* was used to construct the graphs. If these points or *samples* have sufficiently close spacing, a smooth curve drawn through them allows you to interpolate intermediate values to any reasonable degree of accuracy. It can therefore be said that the continuous curve is adequately described by the sample points alone.

In similar fashion, an electric *signal* satisfying certain requirements can be reproduced from an appropriate set of *instantaneous samples*. Sampling therefore makes it possible to transmit a message in the form of *pulse modulation*, rather than as a continuous signal. Usually the pulses are quite short compared to the time between them, so a pulse-modulated wave has the property of being "off" most of the time.

This property of pulse modulation offers two potential advantages over CW modulation. First, the transmitted power can be concentrated into short bursts instead of being generated continuously. The system designer then has greater latitude for equipment selection, and may choose devices such as lasers and high-power microwave tubes that operate only on a pulsed basis. Second, the time interval between pulses can be filled with sample values from other signals, a process called *time-division multiplexing* (TDM).

But pulse modulation has the disadvantage of requiring very large transmission bandwidth compared to the message bandwidth. Consequently, the methods of *analog* pulse modulation discussed in this chapter are used primarily as *message processing* for TDM and/or prior to CW modulation. *Digital* or *coded* pulse modulation has additional advantages that compensate for the increased bandwidth, as we'll see in Chap. 12.

10.1 SAMPLING THEORY AND PRACTICE

The theory of sampling presented here sets forth the conditions for signal sampling and reconstruction from sample values. We'll also examine practical implementation of the theory and some related applications.

Chopper Sampling

A simple but highly informative approach to sampling theory comes from the switching operation of Fig. 10.1-1*a*. The switch periodically shifts between two contacts at a rate of $f_s = 1/T_s$ Hz, dwelling on the input-signal contact for τ seconds and on the grounded contact for the remainder of each period. The output $x_s(t)$ then consists of short segments of the input $x(t)$, as shown in Fig. 10.1-1*b*. Figure 10.1-1*c* is an electronic version of Fig. 10.1-1*a*; the output voltage equals the input voltage except when the clock signal forward-biases the diodes and thereby clamps the output to zero. This operation variously called *single-ended* or *unipolar chopping*, is not instantaneous sampling in the strict sense. Nonetheless, $x_s(t)$ will be designated the sampled wave and f_s the sampling frequency.

We now ask: Are the sampled segments sufficient to describe the original input signal and, if so, how can $x(t)$ be retrieved from $x_s(t)$? The answer to this question lies in the frequency domain, in the spectrum of the sampled wave.

As a first step toward finding the spectrum, we introduce a *switching function* $s(t)$ such that

$$x_s(t) = x(t)s(t) \tag{1}$$

(*a*)   (*b*)

(*c*)

Figure 10.1-1 Switching sampler. (*a*) Functional diagram; (*b*) waveforms; (*c*) circuit realization with diode bridge.

Figure 10.1-2 Sampling as multiplication. (a) Functional diagram; (b) switching function.

Thus the sampling operation becomes multiplication by $s(t)$, as indicated schematically in Fig. 10.1-2a, where $s(t)$ is nothing more than the periodic pulse train of Fig. 10.1-2b. Since $s(t)$ is periodic, it can be written as a Fourier series. Using the results of Example 2.1-1 we have

$$s(t) = \sum_{n=-\infty}^{\infty} f_s \tau \text{ sinc } nf_s \tau \, e^{j2\pi nf_s t} = c_0 + \sum_{n=1}^{\infty} 2c_n \cos n\omega_s t \qquad (2)$$

where

$$c_n = f_s \tau \text{ sinc } nf_s \tau \qquad \omega_s = 2\pi f_s$$

Combining Eq. (2) with Eq. (1) yields the term-by-term expansion

$$x_s(t) = c_0 x(t) + 2c_1 x(t) \cos \omega_s t + 2c_2 x(t) \cos 2\omega_c t + \cdots \qquad (3)$$

Thus, if the input message spectrum is $X(f) = \mathscr{F}[x(t)]$, the output spectrum is

$$X_s(f) = c_0 X(f) + c_1[X(f-f_s) + X(f+f_s)]$$
$$+ c_2[X(f-2f_s) + X(f+2f_s)]$$
$$+ \cdots \qquad (4)$$

which follows directly from the modulation theorem.

While Eq. (4) appears rather messy, the spectrum of the sampled wave is readily sketched if the input signal is assumed to be *bandlimited*. Figure 10.1-3 shows a convenient $X(f)$ and the corresponding $X_s(f)$ for two cases, $f_s > 2W$ and $f_s < 2W$. This figure reveals something quite surprising: the sampling operation has left the message spectrum *intact*, merely repeating it periodically in the frequency domain with a spacing of f_s. We also note that the first term of Eq. (4) is precisely the message spectrum, attenuated by the *duty cycle* $c_0 = f_s \tau = \tau/T_s$.

If sampling preserves the message spectrum, it should be possible to recover or *reconstruct* $x(t)$ from the sampled wave $x_s(t)$. The reconstruction technique is not at all obvious from the time-domain relations in Eqs. (1) and (3). But referring again to Fig. 10.1-3, we see that $X(f)$ can be separated from $X_s(f)$ by *lowpass filtering*, provided that the spectral sidebands don't overlap. And if $X(f)$ alone is filtered from $X_s(f)$, we have recovered $x(t)$. Two conditions obviously are necessary to prevent overlapping spectral bands: the message must be bandlimited, and the

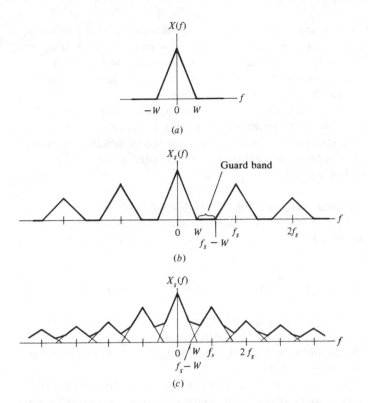

Figure 10.1-3 Spectra for switching sampling. (a) Message; (b) sampled message, $f_s > 2W$; (c) sampled message, $f_s < 2W$.

sampling frequency must be sufficiently great that $f_s - W \geq W$. Thus, we require

$$X(f) = 0 \qquad |f| > W$$

and

$$f_s \geq 2W \qquad \text{or} \qquad T_s \leq \frac{1}{2W} \tag{5}$$

The minimum sampling frequency $f_{s_{\min}} = 2W$ is called the *Nyquist rate*. When Eq. (5) is satisfied and $x_s(t)$ is filtered by an ideal LPF, the output signal will be proportional to $x(t)$; message reconstruction from the sampled signal therefore has been achieved. The exact value of the filter bandwidth B is unimportant as long as

$$W \leq B \leq f_s - W \tag{6}$$

so the filter passes $X(f)$ and rejects all higher components in Fig. 10.1-3b. Sampling at $f_s > 2W$ creates a *guard band* into which the transition region of a practical LPF can be fitted.

This analysis has shown that if a bandlimited signal is sampled at a frequency greater than the Nyquist rate, it can be *completely* reconstructed from the sampled wave. Reconstruction is accomplished by lowpass filtering. These conclusions may be difficult to believe at first exposure; they certainly test our faith in spectral analysis. Nonetheless, they are quite correct.

Finally, it should be pointed out that our results are independent of the sample-pulse duration, save as it appears in the duty cycle. If τ is made very small, $x_s(t)$ approaches a string of *instantaneous sample points*, which corresponds to *ideal sampling*. We'll pursue ideal sampling theory after a brief look at the *bipolar* chopper, which has $\tau = T_s/2$.

Example 10.1-1 Bipolar choppers Figure 10.1-4a depicts the circuit and waveforms for a bipolar chopper. The equivalent switching function is a *square wave* alternating between $s(t) = +1$ and -1. From the series expansion of $s(t)$ we get

$$x_s(t) = \frac{4}{\pi} x(t) \cos \omega_s t - \frac{4}{3\pi} x(t) \cos 3\omega_s t + \frac{4}{5\pi} x(t) \cos 5\omega_s t - \cdots \qquad (7)$$

whose spectrum is sketched in Fig. 10.1-4b for $f \geq 0$. Note that $X_s(f)$ contains no DC component and only the odd harmonics of f_s. Clearly, we can't recover $x(t)$ by lowpass filtering. Instead, the practical applications of bipolar choppers involve *bandpass* filtering.

If we apply $x_s(t)$ to a BPF centered at some odd harmonic nf_s, the output will be proportional to $x(t) \cos n\omega_s t$—a double-sideband suppressed-carrier waveform. Thus, a bipolar chopper serves as a *balanced modulator*. It also serves as a *synchronous detector* when the input is a DSB or SSB signal and

(a)

(b)

Figure 10.1-4 Bipolar chopper. (a) Circuit and waveforms; (b) spectrum.

the output is lowpass filtered. These properties are combined in the *chopper-stabilized amplifier*, which makes possible DC and low-frequency amplification using a high-gain AC amplifier. Additionally, Prob. 10.1-4 indicates how a bipolar chopper can be modified to produce the baseband mutliplexed signal for FM stereo.

Ideal Sampling and Reconstruction

By definition, ideal sampling is *instantaneous* sampling. The switching device of Fig. 10.1-1a yields instantaneous values only if $\tau \to 0$; but then $f_s \tau \to 0$, and so does $x_s(t)$. Conceptually, we overcome this difficulty by multiplying $x_s(t)$ by $1/\tau$ so that, as $\tau \to 0$ and $1/\tau \to \infty$, the sampled wave becomes a train of *impulses* whose *areas* equal the instantaneous sample values of the input signal. Formally, we write the rectangular pulse train as

$$s(t) = \sum_{k=-\infty}^{\infty} \Pi\left(\frac{t-kT_s}{\tau}\right)$$

from which we define the *ideal sampling function*

$$s_\delta(t) \triangleq \lim_{\tau\to 0} \frac{1}{\tau} s(t) = \sum_{k=-\infty}^{\infty} \delta(t-kT_s) \tag{8}$$

The *ideal sampled wave* is then

$$x_\delta(t) \triangleq x(t)s_\delta(t) \tag{9a}$$

$$= x(t) \sum_{k=-\infty}^{\infty} \delta(t-kT_s)$$

$$= \sum_{k=-\infty}^{\infty} x(kT_s)\,\delta(t-kT_s) \tag{9b}$$

since $x(t)\,\delta(t-kT_s) = x(kT_s)\,\delta(t-kT_s)$.

To obtain the corresponding spectrum $X_\delta(f) = \mathscr{F}[x_\delta(t)]$ we note that $(1/\tau)x_s(t) \to x_\delta(t)$ as $\tau \to 0$ and, likewise, $(1/\tau)X_s(f) \to X_\delta(f)$. But each coefficient in Eq. (4) has the property $c_n/\tau = f_s$ sinc $nf_s\tau = f_s$ when $\tau = 0$. Therefore,

$$X_\delta(f) = f_s X(f) + f_s[X(f-f_s) + X(f+f_s)] + \cdots$$

$$= f_s \sum_{n=-\infty}^{\infty} X(f-nf_s) \tag{10}$$

which is illustrated in Fig. 10.1-5 for the message spectrum of Fig. 10.1-3a taking $f_s > 2W$. We see that $X_\delta(f)$ is periodic in frequency with period f_s, a crucial observation in the study of sampled-data systems.

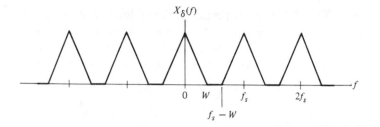

$X_\delta(f)$

0   W f_s $2f_s$ f

$f_s - W$

Figure 10.1-5 Spectrum of ideally sampled message.

Somewhat parenthetically, we can also develop an expression for $S_\delta(f) = \mathscr{F}[s_\delta(t)]$ as follows. From Eq. (9a) and the convolution theorem, $X_\delta(f) = X(f) * S_\delta(f)$ whereas Eq. (10) is equivalent to

$$X_\delta(f) = X(f) * \left[\sum_{n=-\infty}^{\infty} f_s\, \delta(f - nf_s) \right]$$

Therefore, we conclude that

$$S_\delta(f) = f_s \sum_{n=-\infty}^{\infty} \delta(f - nf_s) \tag{11}$$

so the spectrum of a periodic string of unit-weight impulses in the time domain is a periodic string of impulses in the frequency domain with spacing $f_s = 1/T_s$; in both domains we have a function that looks like a picket fence.

Returning to the main subject and Fig. 10.1-5, it's immediately apparent that if we invoke the same conditions as before—$x(t)$ bandlimited in W and $f_s \geq 2W$—then a filter of suitable bandwidth will reconstruct $x(t)$ from the ideal sampled wave. Specifically, for an ideal LPF of gain K, time delay t_d, and bandwidth B, the transfer function is

$$H(f) = K\Pi\left(\frac{f}{2B}\right) e^{-j\omega t_d}$$

so filtering $x_\delta(t)$ produces the output spectrum

$$Y(f) = H(f)X_\delta(f) = Kf_s X(f) e^{-j\omega t_d}$$

assuming B satisfies Eq. (6). The output time function is then

$$y(t) = \mathscr{F}^{-1}[Y(f)] = Kf_s x(t - t_d) \tag{12}$$

which is the original signal amplified by Kf_s and delayed by t_d.

Further confidence in the sampling process can be gained by examining reconstruction in the time domain. The impulse response of the LPF is

$$h(t) = 2BK \operatorname{sinc} 2B(t - t_d)$$

Figure 10.1-6 Ideal reconstruction.

And since the input $x_\delta(t)$ is a train of weighted impulses, the output is a train of weighted *impulse responses*, namely,

$$y(t) = h(t) * x_\delta(t) = \sum_k x(kT_s)h(t - kT_s)$$

$$= 2BK \sum_{k=-\infty}^{\infty} x(kT_s) \text{ sinc } 2B(t - t_d - kT_s) \qquad (13)$$

Now suppose for simplicity that $B = f_s/2$, $K = 1/f_s$, and $t_d = 0$, so

$$y(t) = \sum_k x(kT_s) \text{ sinc } (f_s t - k)$$

We can then carry out the reconstruction process graphically, as shown in Fig. 10.1-6. Clearly the correct values are reconstructed at the sampling instants $t = kT_s$, for all sinc functions are zero at these times save one, and that one yields $x(kT_s)$. Between sampling instants $x(t)$ is *interpolated* by summing the precursors and postcursors from *all* the sinc functions. For this reason the LPF is often called an *interpolation filter*, and its impulse is called the *interpolation function*.

The above results are well summarized by stating the important theorem of *uniform (periodic) sampling*. While there are many variations of this theorem, the following form is best suited to our purposes.

If a signal contains no frequency components for $|f| \geq W$, it is completely described by instantaneous sample values uniformly spaced in time with period $T_s \leq 1/2W$. If a signal has been sampled at the Nyquist rate or greater $(f_s \geq 2W)$ and the sample values are represented as weighted impulses, the signal can be exactly reconstructed from its samples by an ideal LPF of bandwidth B, where $W \leq B \leq f_s - W$.

Another way to express the theorem comes from Eqs. (12) and (13) with $K = T_s$ and $t_d = 0$. Then $y(t) = x(t)$ and

$$x(t) = 2BT_s \sum_{k=-\infty}^{\infty} x(kT_s) \text{ sinc } 2B(t - kT_s) \qquad (14)$$

provided $T_s \leq 1/2W$ and B satisfies Eq. (6). Therefore, just as a periodic signal is completely described by its Fourier series coefficients, a bandlimited signal is completely described by its instantaneous sample values *whether or not the signal actually is sampled.*

Several additional theorems pertaining to signal sampling have been developed. They are discussed in Black (1953, chap. 4) and Peebles (1976, chap. 7).

Exercise 10.1-1 Consider a sampling pulse train of the general form

$$s_p(t) = \sum_{k=-\infty}^{\infty} p(t - kT_s) \tag{15a}$$

whose pulse shape $p(t)$ equals zero for $|t| > T_s/2$ but is otherwise arbitrary. Use an exponential Fourier series and Eq. (21), Sect. 2.2, to show that

$$S_p(f) = f_s \sum_{n=-\infty}^{\infty} P(nf_s)\,\delta(f - nf_s) \tag{15b}$$

where $P(f) = \mathcal{F}[p(t)]$. Then let $p(t) = \delta(t)$ to obtain Eq. (11).

Practical Sampling and Aliasing

Practical sampling systems differ from ideal sampling in three obvious aspects:

1. The sampled wave consists of pulses having finite amplitude and duration, rather than impulses.
2. Practical reconstruction filters are not ideal filters.
3. The messages to be sampled are *timelimited* signals whose spectra are not and cannot be strictly bandlimited.

The first two differences may present minor problems, while the third leads to the more troublesome effect known as aliasing.

Regarding pulse-shape effects, our investigation of the unipolar chopper and the results of Exercise 10.1-1 correctly imply that almost any pulse shape $p(t)$ will do when sampling takes the form of a *multiplication* operation $x(t)s_p(t)$. Another operation produces *flat-top sampling*, described in the next section. This type of sampling may require equalization, but it does not alter our conclusion that pulse shapes are relatively inconsequential.

Regarding practical reconstruction filters, we consider the typical filter response superimposed on a sampled-wave spectrum in Fig. 10.1-7. If the filter is reasonably flat over the message band, its output will consist of $x(t)$ plus spurious frequency components at $|f| > f_s - W$, outside the message band. In audio systems these components would sound like high-frequency hissing or "noise."

Figure 10.1-7 Practical reconstruction filter.

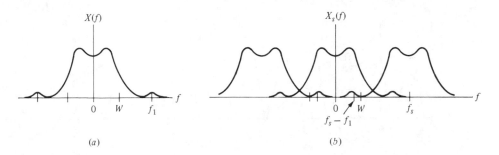

Figure 10.1-8 Aliasing effect. (*a*) Spectrum of timelimited message; (*b*) sampled message spectrum with overlaps.

However, they are considerably attenuated and their strength is proportional to $x(t)$, so they disappear when $x(t) = 0$. When $x(t) \neq 0$, the message tends to mask their presence and render them more tolerable. The combination of careful filter design and an adequate guard band created by taking $f_s > 2W$ makes practical reconstruction filtering nearly equivalent to ideal reconstruction.

Regarding the timelimited nature of real signals, a message spectrum like Fig. 10.1-8*a* may be viewed as a bandlimited spectrum if the frequency content above W is small and presumably unimportant for conveying the information. When such a message is sampled, there will be unavoidable overlapping of spectral components shown in Fig. 10.1-8*b*. In reconstruction, frequencies originally outside the nominal message band will appear at the filter output in the form of much *lower* frequencies. Thus, for example, $f_1 > W$ becomes $f_s - f_1 < W$, as indicated in the figure.

This phenomenon of downward frequency translation occurs whenever a frequency component is *undersampled*, so that $f_s < 2f_1$, and is given the descriptive name of *aliasing*. The aliasing effect is far more serious than spurious frequencies passed by nonideal reconstruction filters, for the latter fall *outside* the message band, whereas aliased components fall *within* the message band. Aliasing is combated by filtering the message as much as possible *before* sampling and, if necessary, sampling at much greater than the nominal Nyquist rate.

As illustration, the average voice spectrum (Fig. 3.2-1) extends well beyond 10 kHz, though most of the energy is concentrated in the range 100 to 600 Hz and a bandwidth of 3 kHz is sufficient for intelligibility. If a voice wave is prefiltered by a 3.3-kHz LPF and then sampled at $f_s = 8$ kHz, the aliased components are typically 30 dB below the desired signal and go unnoticed by the listener. Incidentally, these are the standard values for voice-telephone sampling.

Example 10.1-2 Sampling Oscilloscopes A practical application of aliasing occurs in the sampling oscilloscope, which exploits undersampling to display high-speed *periodic* waveforms that would otherwise be beyond the capability of the electronics. To illustrate the principle, consider the periodic waveform $x(t)$ with period $T_x + 1/f_x$ in Fig. 10.1-9*a*. If we use a sampling interval T_s

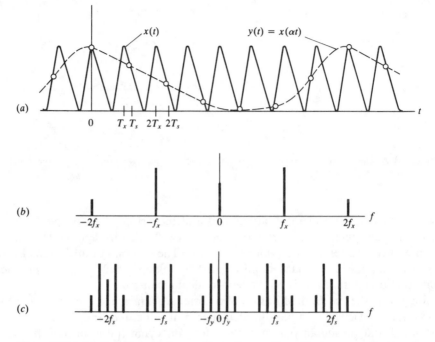

Figure 10.1-9 (a) Periodic waveform with undersampling; (b) spectrum of $x(t)$; (c) spectrum of $y(t) = x(\alpha t)$, $\alpha < 1$.

slightly greater than T_x and interpolate the sample points, we get the *expanded* waveform $y(t) = x(\alpha t)$ shown as a dashed curve. The corresponding sampling frequency is

$$f_s = (1 - \alpha)f_x \qquad 0 < \alpha < 1$$

so $f_s < f_x$ and even the fundamental frequency of $x(t)$ will be undersampled. Now let's find out if and how this system actually works by going to the frequency domain.

We assume that $x(t)$ has been prefiltered to remove any frequency components higher than the mth harmonic. Figure 10.1-9b shows a typical two-sided line spectrum of $x(t)$, taking $m = 2$ for simplicity. Since sampling translates all frequency components up and down by nf_s, the fundamental will appear in the spectrum of the sampled signal at

$$\pm f_y = \pm | f_x - f_s | = \pm \alpha f_x$$

as well as at $\pm f_x$ and at $f_x \pm nf_s = (1 + n)f_x \pm nf_y$. Similar translations applied to the DC component and second harmonic yield the spectrum in Fig. 10.1-9c, which contains a *compressed* image of the original spectrum centered at each multiple of f_s. Therefore, a lowpass filter with $B = f_s/2$ will construct $y(t) = x(\alpha t)$ from $x_s(t)$ provided that

$$\alpha < \frac{1}{2m + 1}$$

which prevents spectral overlap.

Exercise 10.1-2 Demonstrate the aliasing effect for yourself by making a careful sketch of $\cos 2\pi 10t$ and $\cos 2\pi 70t$ for $0 \le t \le \frac{1}{10}$. Put both sketches on the same set of axes and find the sample values at $t = 0, \frac{1}{80}, \frac{2}{80}, \dots, \frac{8}{80}$, which corresponds to $f_s = 80$. Also, convince yourself that no other waveform bandlimited in $10 < W < 40$ can be interpolated from the sample values of $\cos 2\pi 10t$.

10.2 ANALOG PULSE MODULATION

If a message waveform is adequately described by periodic sample values, it can be transmitted using analog pulse modulation wherein the sample values modulate some parameter of a pulse train. Pulse parameters suitable for modulation include amplitude, duration, and position. The corresponding processes are designated *pulse-amplitude modulation* (PAM), *pulse-duration modulation*† (PDM), and *pulse-position modulation* (PPM).

Figure 10.2-1 illustrates these types of pulse modulation with a representative message waveform. The modulated pulse parameter—amplitude, duration, or

† Also called pulse-*width* modulation (PWM).

Figure 10.2-1 Types of analog pulse modulation.

relative position — varies in direct proportion to the sample values of $x(t)$. For clarity, the pulses are shown as rectangular and the pulse durations have been grossly exaggerated. Actual modulated waves would also be delayed slightly compared to the message, because the pulses can't be generated before the sampling instants.

It should be evident from these waveforms that a modulated pulse train has significant DC content and that the bandwidth required to preserve the pulse shape far exceeds the message bandwidth. Consequently, you seldom encounter a single-channel communication system with PAM, PDM, or PPM. But analog pulse modulation deserves attention for its major roles in time-division multiplexing, data telemetry, and instrumentation systems.

Flat-Top Sampling and PAM

Although a PAM wave could be obtained from a chopper circuit, a more popular method employs the *sample-and-hold* (S/H) technique. This operation produces flat-top pulses, as in Fig. 10.1-2, rather than curved-top chopper pulses. We therefore begin here with the properties of *flat-top sampling*.

A rudimentary S/H circuit consists of two FET switches and a capacitor, connected as shown in Fig. 10.2-2a. A gate pulse at G1 briefly closes the sampling switch and the capacitor holds the sampled voltage until discharged by a pulse applied to G2. (Commercial integrated-circuit S/H units have further refinements, including isolating op-amps at input and output.) Periodic gating of the sample-and-hold circuit generates the sampled wave

$$x_p(t) = \sum_k x(kT_s)p(t - kT_s) \tag{1}$$

illustrated by Fig. 10.2-2b. Note that each output pulse of duration τ represents a single instantaneous sample value.

To analyze flat-top sampling, we'll draw upon the relation $p(t - kT_s) = p(t) * \delta(t - kT_s)$ and write

$$x_p(t) = p(t) * \left[\sum_k x(kT_s)\, \delta(t - kT_s) \right] = p(t) * x_\delta(t)$$

(a)   *(b)*

Figure 10.2-2 Flat-top sampling. (a) Sample-and-hold circuit; (b) waveforms.

Fourier transformation of this *convolution* operation yields

$$X_p(f) = P(f)\left[f_s \sum_n X(f - nf_s)\right] = P(f)X_\delta(f) \tag{2}$$

Figure 10.2-3 provides a graphical interpretation of Eq. (2), taking $X(f) = \Pi(f/2W)$. We see that flat-top sampling is equivalent to passing an ideal sampled wave through a network having the transfer function $P(f) = \mathscr{F}[p(t)]$.

The high-frequency rolloff characteristic of a typical $P(f)$ acts like a *lowpass* filter and attenuates the upper portion of the message spectrum. This loss of high-frequency content is called *aperture effect*. The larger the pulse duration or aperture τ, the larger the effect. Aperture effect can be corrected in reconstruction by including an *equalizer* with

$$H_{eq}(f) = Ke^{-j\omega t_d}/P(f) \tag{3}$$

However, little if any equalization is needed when $\tau/T_s \ll 1$.

Now consider a unipolar flat-top PAM signal defined by

$$x_p(t) = \sum_k A_0[1 + \mu x(kT_s)]p(t - kT_s) \tag{4}$$

The constant A_0 equals the unmodulated pulse amplitude, and the *modulation index* μ controls the amount of amplitude variation. The condition

$$1 + \mu x(t) > 0 \tag{5}$$

ensures a unipolar (single-polarity) waveform with no missing pulses. The resulting constant pulse rate f_s is particularly important for synchronization in time-division multiplexing.

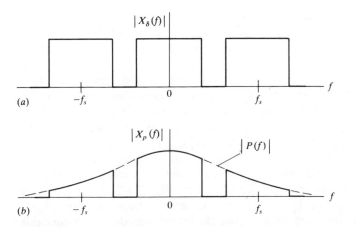

Figure 10.2-3 (a) Spectrum for ideal sampling when $X(f) = \Pi(f/2W)$; (b) aperture effect in flat-top sampling.

Comparison of Eqs. (1) and (4) shows that a PAM signal can be obtained from a sample-and-hold circuit with input $A_0[1 + \mu x(t)]$. Correspondingly, the PAM spectrum will look like Fig. 10.2-3b with $X(f)$ replaced by

$$\mathscr{F}\{A_0[1 + \mu x(t)]\} = A_0[\delta(f) + \mu X(f)],$$

which results in spectral impulses at all harmonics of f_s and at $f = 0$. Reconstruction of $x(t)$ from $x_p(t)$ therefore requires a *DC block* as well as lowpass filtering and equalization.

Clearly, PAM has many similarities to AM CW modulation—modulation index, spectral impulses, and DC blocks. (In fact, an AM wave could be derived from PAM by bandpass filtering.) But the PAM spectrum extends from DC up through several harmonics of f_s, and the estimate of required transmission bandwidth B_T must be based on time-domain considerations. For this purpose, we assume a small pulse duration τ compared to the time between pulses, so

$$\tau \ll T_s \leq \frac{1}{2W}$$

Adequate pulse resolution then requires

$$B_T \geq \frac{1}{2\tau} \gg W \tag{6}$$

Hence, practical applications of PAM are limited to those situations in which the advantages of a pulsed waveform outweigh the disadvantages of large bandwidth.

Exercise 10.2-1 Consider PAM transmission of a voice signal with $W \approx 3$ kHz. Calculate B_T if $f_s = 8$ kHz and $\tau = 0.1\ T_s$.

Pulse-Duration and Pulse-Position Modulation

We lump PDM and PPM together under one heading for two reasons. First, in both cases a *time* parameter of the pulse is being modulated, and the pulses have *constant amplitude*. Second, a close relationship exists between the modulation methods for PDM and PPM.

To demonstrate these points, Fig. 10.2-4 shows the block diagram and waveforms of a system that combines the sampling and modulation operations for either PDM or PPM. The system employs a comparator and a sawtooth-wave generator with period T_s. The output of the comparator is zero except when the message waveform $x(t)$ exceeds the sawtooth wave, in which case the output is a positive constant A. Hence, as seen in the figure, the comparator produces a PDM signal with *trailing-edge modulation* of the pulse duration. (Reversing the sawtooth results in leading-edge modulation, as in Fig. 7.3-3, while replacing the sawtooth with a triangular wave results in modulation on both edges.) Position modulation is obtained by applying the PDM signal to a monostable pulse generator that triggers on trailing edges at its input and produces short output pulses of fixed duration.

Figure 10.2-4 Generation of PDM or PPM. (*a*) Block diagram; (*b*) waveforms.

Careful examination of Fig. 10.2-4*b* reveals that the modulated duration or position depends on the message value at the time location t_k of the pulse edge, rather than the apparent sample time kT_s. Thus, the sample values are *nonuniformly* spaced. Inserting a sample-and-hold circuit at the input of the system gives uniform sampling if desired, but there's little difference between uniform and nonuniform sampling in the practical case of small amounts of time modulation such that $t_k - kT_s \ll T_s$.

If we assume nearly uniform sampling, the duration of the kth pulse in the PDM signal is

$$\tau_k = \tau_0[1 + \mu x(kT_s)] \tag{7}$$

in which the unmodulated duration τ_0 represents $x(kT_s) = 0$ and the modulation index μ controls the amount of duration modulation. Our prior condition on μ in Eq. (5) applies here to prevent missing pulses and "negative" durations when $x(kT_s) \le 0$. The PPM pulses have fixed duration and amplitude so, unlike PAM and PDM, there's no potential problem of missing pulses. The kth pulse in a PPM signal begins at time

$$t_k = kT_s + t_d + t_0 x(kT_s) \tag{8}$$

in which the unmodulated position $kT_s + t_d$ represents $x(kT_s) = 0$ and the constant t_0 controls the displacement of the modulated pulse.

The variable time parameters in Eqs. (7) and (8) make the expressions for $x_p(t)$ rather awkward. However, an informative approximation for the PDM waveform is derived by taking rectangular pulses with amplitude A centered at $t = kT_s$ and assuming that τ_k varies slowly from pulse to pulse. Series expansion then yields

$$x_p(t) \approx A f_s \tau_0 [1 + \mu x(t)] + \sum_{n=1}^{\infty} \frac{2A}{\pi n} \sin n\phi(t) \cos n\omega_s t \qquad (9)$$

where $\phi(t) = \pi f_s \tau_0 [1 + \mu x(t)]$. Without attempting to sketch the corresponding spectrum, we see from Eq. (9) that the PDM signal contains the message $x(t)$ plus a DC component and *phase-modulated* waves at the harmonics of f_s. The phase modulation has negligible overlap in the message band when $\tau_0 \ll T_s$, so $x(t)$ can be recovered by lowpass filtering with a DC block.

Another message reconstruction technique converts pulse-*time* modulation into pulse-*amplitude* modulation, and works for PDM and PPM. To illustrate this technique the middle waveform in Fig. 10.2-5 is produced by a *ramp generator* that starts at time kT_s, stops at t_k, restarts at $(k + 1)T_s$, and so forth. Both the start and stop commands can be extracted from the edges of a PDM pulse, whereas PPM reconstruction must have an auxiliary *synchronization* signal for the start command.

Regardless of the particular details, demodulation of PDM or PPM requires received pulses with short *risetime* in order to preserve accurate message information. For a specified risetime $t_r \ll T_s$, the transmission bandwidth must satisfy

$$B_T \geq \frac{1}{2t_r} \qquad (10)$$

which will be substantially greater than the PAM transmission bandwidth. In exchange for the extra bandwidth, we gain the benefit of constant-amplitude pulses that suffer no ill effects from *nonlinear distortion* in transmission since nonlinear distortion does not alter pulse duration or position.

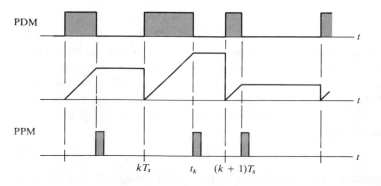

Figure 10.2-5 Conversion of PDM or PPM into PAM.

Additionally, like PM and FM CW modulation, PDM and PPM have the potential for *wideband noise reduction* — a potential more fully realized by PPM than by PDM. To appreciate why this is so, recall that the information resides in the time location of the pulse *edges*, not in the pulses themselves. Thus, somewhat like the carrier-frequency power of AM, the pulse power of pulse-time modulation is "wasted" power, and it would be more efficient to suppress the pulses and just transmit the edges! Of course we cannot transmit edges without transmitting pulses to define them. But we can send very short pulses indicating the position of the edges, a process equivalent to PPM. The reduced power required for PPM is a fundamental advantage over PDM, an advantage that becomes more apparent when we examine the signal-to-noise ratios.

Exercise 10.2-2 Derive Eq. (9) by the following procedure. First, assume constant pulse duration τ and write $x_p(t) = As(t)$ with $s(t)$ given by Eq. (2), Sect. 10.1. Then apply the *quasi-static approximation* $\tau \approx \tau_0[1 + \mu x(t)]$.

PPM Spectral Analysis★

Because PPM with nonuniform sampling is the most efficient type of analog pulse modulation for message transmission, we should take the time to analyze its spectrum. The analysis method itself is worthy of examination.

Let the kth pulse be centered at time t_k. If we ignore the constant time delay t_d in Eq. (8), nonuniform sampling extracts the sample value at t_k, rather than kT_s, so

$$t_k = kT_s + t_0 x(t_k) \tag{11}$$

By definition, the PPM wave is a summation of constant-amplitude position-modulated pulses, and can be written as

$$x_p(t) = \sum_k Ap(t - t_k) = Ap(t) * \left[\sum_k \delta(t - t_k) \right]$$

where A is the pulse amplitude and $p(t)$ the pulse shape. A simplification at this point is made possible by noting that $p(t)$ will (or should) have a very small duration compared to T_s. Hence, for our purposes, the pulse shape can be taken as impulsive, and

$$x_p(t) \approx A \sum_k \delta(t - t_k) \tag{12}$$

If desired, Eq. (12) can later be convolved with $p(t)$ to account for the non-impulsive shape.

In their present form, Eqs. (11) and (12) are unsuited to further manipulation; the trouble is the position term t_k, which cannot be solved for explicitly. Fortunately, Rowe (1965, chap. 5) has devised a technique whereby t_k can be eliminated entirely. Consider any function $g(t)$ having a single first-order zero at $t = \lambda$, such that $g(\lambda) = 0$, $g(t) \neq 0$ for $t \neq \lambda$, and $\dot{g}(t) \neq 0$ at $t = \lambda$. The distribution theory of impulses then shows that

$$\delta(t - \lambda) = |\dot{g}(t)| \, \delta[g(t)] \tag{13}$$

whose right-hand side is independent of λ. Equation (13) can therefore be used to remove t_k from $\delta(t - t_k)$ if we can find a function $g(t)$ that satisfies $g(t_k) = 0$ and the other conditions but does not contain t_k.

Suppose we take $g(t) = t - kT_s - t_0 x(t)$, which is zero at $t = kT_s + t_0 x(t)$. Now, for a given value of k, there is only one PPM pulse, and it occurs at $t_k = kT_s + t_0 x(t_k)$. Thus $g(t_k) = t_k - kT_s - t_0 x(t_k) = 0$, as desired. Inserting $\lambda = t_k$, $\dot{g}(t) = 1 - t_0 \dot{x}(t)$, etc., into Eq. (13) gives

$$\delta(t - t_k) = |1 - t_0 \dot{x}(t)| \, \delta[t - kT_s - t_0 x(t)]$$

and the PPM wave of Eq. (12) becomes

$$x_p(t) = A[1 - t_0 \dot{x}(t)] \sum_k \delta[t - t_0 x(t) - kT_s]$$

The absolute value is dropped since $|t_0 \dot{x}(t)| < 1$ for most signals of interest if $t_0 \ll T_s$. We then convert the sum of impulses to a sum of exponentials via

$$\sum_{k=-\infty}^{\infty} \delta(t - kT_s) = f_s \sum_{n=-\infty}^{\infty} e^{jn\omega_s t} \tag{14}$$

which is *Poisson's sum formula*. Thus, we finally obtain

$$x_p(t) = A f_s [1 - t_0 \dot{x}(t)] \sum_{n=-\infty}^{\infty} e^{jn\omega_s [t - t_0 x(t)]}$$

$$= A f_s [1 - t_0 \dot{x}(t)] \left\{ 1 + \sum_{n=1}^{\infty} 2 \cos [n\omega_s t - n\omega_s t_0 x(t)] \right\} \tag{15}$$

The derivation of Eq. (14) is considered in Prob. 10.2-11.

Interpreting Eq. (15), we see that PPM with nonuniform sampling is a combination of linear and exponential carrier modulation, for each harmonic of f_s is phase-modulated by the message $x(t)$ and amplitude-modulated by the derivative $\dot{x}(t)$. The spectrum therefore consists of AM and PM sidebands centered at all multiples of f_s, plus a DC impulse and the spectrum of $\dot{x}(t)$. Needless to say, sketching such a spectrum is a tedious exercise even for tone modulation. The leading term of Eq. (15) suggests that the message can be retrieved by lowpass filtering and *integrating*. However, the integration method does not take full advantage of the noise-reduction properties of PPM, so the usual procedure is conversion to PAM or PDM followed by lowpass filtering.

10.3 TIME-DIVISION MULTIPLEXING

A sampled waveform is "off" most of the time, leaving the time between samples available for other purposes. In particular, sample values from several different signals can be interlaced into a single waveform. This is the principle of *time-division multiplexing* (TDM) discussed here.

TDM Systems

The simplified system in Fig. 10.3-1 demonstrates the essential features of time-division multiplexing. Several input signals are prefiltered by the bank of input LPFs and sampled sequentially. The rotating sampling switch or *commutator* at the transmitter extracts one sample from each input per revolution. Hence, its output is a PAM waveform that contains the individual samples periodically interlaced in time. A similar rotary switch at the receiver, called a *decommutator* or *distributor*, separates the samples and distributes them to another bank of LPFs for reconstruction of the individual messages.

If all inputs have the same message bandwidth W, the commutator should rotate at the rate $f_s \geq 2W$ so that successive samples from any one input are spaced by $T_s = 1/f_s \leq 1/2W$. The time interval T_s containing one sample from each input is called a *frame*. If there are M input channels, the pulse-to-pulse spacing within a frame is $T_s/M = 1/Mf_s$. Thus, the total number of pulses per second will be

$$r = Mf_s \geq 2MW \tag{1}$$

which represents the pulse rate or *signaling rate* of the TDM signal.

(a)

(b)

Figure 10.3-1 TDM system. (a) Block diagram; (b) waveforms.

Our primitive system shows mechanical switching to generate multiplexed PAM. But almost all practical TDM systems employ electronic switching. Furthermore, other types of pulse modulation can be used instead of PAM. Therefore, a more generalized commutator might have the structure diagrammed in Fig. 10.3-2, where pulse-modulation *gates* process the individual inputs to form the TDM output. The gate control signals come from a flip-flop chain (a broken-ring counter) driven by a digital clock at frequency Mf_s. The decommutator would have a similar structure.

Regardless of the type of pulse modulation, TDM systems require careful *synchronization* between commutator and decommutator. Synchronization is a critical consideration in TDM, because each pulse must be distributed to the correct output line at the appropriate time. A popular brute-force synchronization technique devotes one time slot per frame to a distinctive *marker* pulse or nonpulse, as illustrated in Fig. 10.3-3. These markers establish the frame frequency f_s at the receiver, but the number of signal channels is reduced to $M - 1$. Other synchronization methods involve auxiliary pilot tones or the statistical properties of the TDM signal itself.

(a)

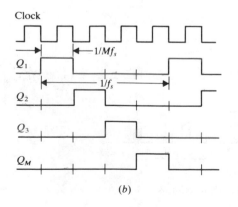

(b)

Figure 10.3-2 (a) Electronic commutator for TDM; (b) timing diagram.

Figure 10.3-3 TDM synchronization markers.

Radio-frequency transmission of TDM necessitates the additional step of CW modulation to obtain a *bandpass* waveform. For instance, a TDM signal composed of duration or position modulated pulses could be applied to an AM transmitter with 100% modulation, thereby producing a train of constant-amplitude RF pulses. The compound process would be designated PDM/AM or PPM/AM, and the required transmission bandwidth would be twice that of the baseband TDM signal. The relative simplicity of this technique suits low-speed multichannel applications such as radio control for model airplanes.

More sophisticated TDM systems may use PAM/SSB for bandwidth conservation or PAM/FM for wideband noise reduction. The complete transmitter diagram in Fig. 10.3-4a now includes a lowpass *baseband filter* with bandwidth

$$B_b = \tfrac{1}{2}r = \tfrac{1}{2}Mf_s \tag{2}$$

Baseband filtering prior to CW modulation produces a smooth modulating waveform $x_b(t)$ having the property that it passes through the individual sample values at the corresponding sample times, as portrayed in Fig. 10.3-4b. Since the

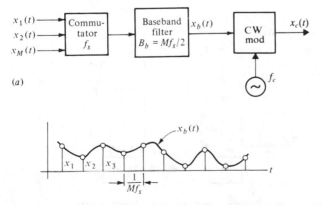

Figure 10.3-4 (*a*) TDM transmitter with baseband filtering; (*b*) baseband waveform.

interlaced sample spacing equals $1/Mf_s$, the baseband filter constructs $x_b(t)$ in the same way that an LPF with $B = f_s/2$ would reconstruct a waveform $x(t)$ from its periodic samples $x(kT_s)$ with $T_s = 1/f_s$.

If baseband filtering is employed, and if the sampling frequency is close to the Nyquist rate $f_{s_{min}} = 2W$ for the individual inputs, then the transmission bandwidth for PAM/SSB becomes

$$B_T = \tfrac{1}{2}M \times 2W = MW$$

Under these conditions, TDM approaches the theoretical *minimum bandwidth* of frequency-division multiplexing with SSB subcarrier modulation.

Although we've assumed so far that all input signals have the same bandwidth, this restriction is not essential and, moreover, would be unrealistic for the important case of analog data telemetry. The purpose of a telemetry system is to combine and transmit physical measurement data from different sources at some remote location. The sampling frequency required for a particular measurement depends on the physical process involved and can range from a fraction of one hertz up to several kilohertz. A typical telemetry system has a *main multiplexer* plus *submultiplexers* arranged to handle 100 or more data channels with various sampling rates.

Example 10.3-1 TDM Telemetry For the sake of illustration, suppose we need 5 data channels with minimum sampling rates of 3000, 700, 500, 300, and 200 Hz. If we used a 5-channel multiplexer with $f_s = 3000$ Hz for all channels, the TDM signaling rate would be $r = 5 \times 3000 = 15$ kHz—not including synchronization markers. A more efficient scheme involves an 8-channel main multiplexer with $f_s = 750$ Hz and a 2-channel submultiplexer with $f_s = 375$ Hz connected as shown in Fig. 10.3-5.

The two lowest-rate signals $x_4(t)$ and $x_5(t)$ are combined by the sub-multiplexer to create a pulse rate of $2 \times 375 = 750$ Hz for insertion into one channel of the main multiplexer. Hence, the samples of $x_4(t)$ and $x_5(t)$ will

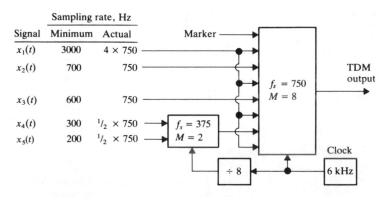

Figure 10.3-5 TDM telemetry system with main multiplexer and submultiplexer.

appear in *alternate* frames. On the other hand, the highest-rate signal $x_1(t)$ is applied to 4 inputs on the main multiplexer. Consequently, its samples appear in 4 equispaced slots within each frame, for an equivalent sampling rate of $4 \times 750 = 3000$ Hz. The total output signaling rate, including a marker, is $r = 8 \times 750$ Hz $= 6$ kHz. Baseband filtering would yield a smoothed signal whose bandwidth $B_b = 3$ kHz fits nicely into a voice telephone channel!

Exercise 10.3-1 Suppose the output in Fig. 10.3-5 is an unfiltered PAM signal with 50% duty cycle. Sketch the waveform for two successive frames, labeling each pulse with its source signal. Then calculate the required transmission bandwidth B_T from Eq. (6), Sect. 10.2.

Cross Talk and Guard Times

When a TDM system includes baseband filtering, the filter design must be done with extreme care to avoid interchannel *cross talk* from one sample value to the next in the frame. Digital signals suffer a similar problem called *intersymbol interference*, and we defer the treatment of baseband waveform shaping to Sect. 11.3.

A TDM signal without baseband filtering also has cross talk if the transmission channel results in pulses whose tails or postcursors overlap into the next time slot of the frame. Pulse overlap is controlled by establishing *guard times* between pulses, analogous to the guard bands between channels in an FDM system. Practical TDM systems have both guard times and guard bands, the former to suppress cross talk, the latter to facilitate message reconstruction with nonideal filters.

For a quantitative estimate of cross talk, let's assume that the transmission channel acts like a first-order lowpass filter with 3-dB bandwidth B. The response to a rectangular pulse then decays exponentially, as sketched in Fig. 10.3-6. The guard time T_g represents the *minimum* pulse spacing, so the pulse tail decays to a value no larger than $A_{ct} = Ae^{-2\pi BT_g}$ by the time the next pulse arrives. Accordingly, we define the *cross-talk reduction factor*

$$k_{ct} \triangleq 10 \log (A_{ct}/A)^2 \approx -54.5 \, BT_g \quad \text{dB} \tag{3}$$

Keeping the cross talk below -30 dB calls for $T_g > 1/2B$.

Guard times are especially important in TDM with pulse-duration or pulse-position modulation because the pulse edges move around within their frame

Figure 10.3-6 Cross talk in TDM.

Figure 10.3-7 TDM/PPM with guard time.

slots. Consider the PPM case in Fig. 10.3-7: here, one pulse has been position-modulated forward by an amount t_0 and the next pulse backward by the same amount. The allowance for guard time T_g requires that $T_g + 2t_0 + 2(\tau/2) \leq T_s/M$ or

$$t_0 \leq \frac{1}{2} \left(\frac{T_s}{M} - \tau - T_g \right) \tag{4}$$

A similar modulation limit applies in the case of PDM.

Exercise 10.3-2 Nine voice signals plus a marker are to be transmitted via PPM on a channel having $B = 400$ kHz. Calculate T_g such that $k_{ct} \approx -60$ dB. Then find the maximum permitted value of t_0 if $f_s = 8$ kHz and $\tau = \frac{1}{5}(T_s/M)$.

Comparison of TDM and FDM

Time-division and frequency-division multiplexing accomplish the same end by different means. Indeed, they may be classified as *dual* techniques. The individual TDM channels are assigned to distinct *time slots* but jumbled together in the frequency domain; conversely, the individual FDM channels are assigned to distinct *frequency slots* but jumbled together in the time domain. What advantages, then, does TDM offer to the communications engineer?

First and foremost, TDM involves simpler instrumentation. Recall that FDM requires an analog subcarrier modulator, bandpass filter, and demodulator for every message channel, all of which are replaced by the TDM commutator and decommutator switching circuits. And TDM synchronization is but slightly more demanding than that of suppressed-carrier FDM.

Second, TDM is invulnerable to the usual causes of cross talk in FDM, namely, imperfect bandpass filtering and nonlinear cross modulation. However, TDM cross-talk immunity does depend on the transmission bandwidth and the absence of delay distortion.

Third, the use of submultiplexers allows a TDM system to accommodate different signals whose bandwidths or pulse rates may differ by more than an order of magnitude. This flexibility has particular value for multiplexing *digital* signals, as we'll see in Sect. 12.4.

Finally, TDM may or may not be advantageous when the transmission medium is subject to *fading*. Rapid wideband fading might strike only occasional pulses in a given TDM channel, whereas all FDM channels would be affected. But slow narrowband fading wipes out all the TDM channels, whereas it might hurt only one FDM channel.

10.4 NOISE IN PULSE MODULATION

Here we complete our study of analog pulse modulation by examining system performance in the presence of noise. We'll assume throughout the case of lowpass pulse transmission with independent additive gaussian white noise at the receiver. The specific topics at hand are: pulse measurements in noise, reconstruction from noisy samples, signal-to-noise ratios, and false-pulse threshold effect.

Pulse Measurements in Noise

Consider initially the problem of measuring some parameter of a *single* pulse $p(t)$ contaminated by noise, as represented in Fig. 10.4-1a. Let the pulse be more-or-less rectangular, with amplitude A, duration τ, and energy $E_p = A^2\tau$. Let the noise be white with power spectral density $G(f) = \eta/2$ and zero mean value. A reasonably selective lowpass filter having unit gain and noise bandwidth $B_N \geq 1/2\tau$ will remove excess noise while passing $p(t)$. The output $v(t) = p(t) + n(t)$ sketched in Fig. 10.4-1b consists of noise variations superimposed on a trapezoidal pulse shape with risetime $t_r \approx 1/2B_N$.

If you want to measure the pulse amplitude, you should do so at some instant t_a near the center of the pulse. A single measurement yields the random quantity

$$v(t_a) = A + n(t_a) = A + \epsilon_A$$

where $\epsilon_A = n(t_a)$ represents the *amplitude error*. The error variance is then

$$\sigma_A^2 = \overline{n^2} = \eta B_N \tag{1}$$

which should be small compared to A^2 for an accurate measurement. Since $A^2 = E_p/\tau$ and $B_N \geq 1/2\tau$, we can write the lower bound

$$\sigma_A^2 \geq \frac{\eta}{2\tau} = \frac{\eta}{2E_p} A^2 \tag{2}$$

(a)

$G(f) = \eta/2$

(b)

Figure 10.4-1 Pulse measurement in noise. (a) Model; (b) waveform.

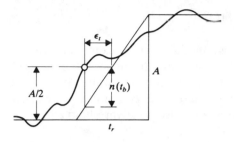

Figure 10.4-2 Time-position error caused by noise.

Any filter bandwidth less than about $1/2\tau$ would reduce the output pulse amplitude as well as the noise. Achieving the lower bound requires a *matched* filter; see Eq. (17), Sect. 5.4.

Measurements of pulse *position* or *duration* are usually carried out by detecting the instant t_b when $v(t)$ crosses some fixed level such as $A/2$. The noise perturbation $n(t_b)$ shown in the enlarged view of Fig. 10.4-2 causes a *time-position error* ϵ_t. From the similar triangles here, we see that $\epsilon_t/n(t_b) = t_r/A$ so $\epsilon_t = (t_r/A)n(t_b)$ and

$$\sigma_t^2 = (t_r/A)^2\overline{n^2} = (t_r/A)^2\eta B_N$$

Substituting $t_r \approx 1/2B_N$ and $A^2 = E_p/\tau$ yields

$$\sigma_t^2 \approx \frac{\eta}{4B_N A^2} = \frac{\eta\tau}{4B_N E_p} \tag{3}$$

which implies that we can make σ_t arbitrarily small by letting $B_N \to \infty$ so that $t_r \to 0$. But the received pulse actually has a nonzero risetime determined by the *transmission bandwidth* B_T. Hence

$$\sigma_t^2 \geq \frac{\eta}{4B_T A^2} = \frac{\eta\tau}{4B_T E_p} \tag{4}$$

and the lower bound is obtained with filter bandwidth $B_N \approx B_T$—in contrast to the lower bound on σ_A obtained with $B_N \approx 1/2\tau$.

> **Exercise 10.4-1** Suppose a 10-μs pulse is transmitted on a channel having $B_T = 800$ kHz and $\eta = E_p/50$. Calculate σ_A/A and σ_t/τ when: (a) $B_N = 1/2\tau$; (b) $B_N = B_T$.

Pulse Modulation With Noise

Regardless of the particular circuitry employed, demodulating a pulse-modulated wave boils down to message reconstruction from sample values. A generalized demodulator would therefore consist of a pulse converter that transforms the pulse-modulated wave into a train of weighted impulses from which an ideal LPF reconstructs the message. Figure 10.4-3 diagrams this demodulation model including additive noise.

Figure 10.4-3 Model for demodulation of analog pulse modulation with noise.

A noise-limiting filter at the input passes the received signal plus noise

$$v(t) = x_p(t) + n(t)$$

The converter measures the amplitude, duration, or position of each pulse in $v(t)$ and generates

$$y_\delta(t) = \sum_k [\mu_p x(kT_s) + \epsilon_k] \, \delta(t - kT_s) \tag{5}$$

where μ_p is the modulation constant relating $x(kT_s)$ to $x_p(t)$ and ϵ_k is the measurement error induced by $n(t)$. We'll assume for convenience that the reconstruction filter has bandwidth $B = f_s/2$, gain $K = T_s = 1/2B$, and zero time delay. Hence, the impulse-train input $y_\delta(t)$ produces the final output

$$y_D(t) = \sum_k [\mu_p x(kT_s) + \epsilon_k] \, \text{sinc} \, (f_s t - k)$$

$$= \mu_p x(t) + n_D(t) \tag{6a}$$

with

$$n_D(t) = \sum_k \epsilon_k \, \text{sinc} \, (f_s t - k) \tag{6b}$$

which represents the noise at the destination. The signal portion of Eq. (6a) follows from Eq. (14), Sect. 10.1.

Since the errors ϵ_k are proportional to sample values of $n(t)$ spaced by T_s, and since the noise-limiting filter has $B_N \geq 1/2\tau > 1/T_s$, the values of ϵ_k will be essentially uncorrelated and will have zero mean. We can therefore write the destination noise power as

$$N_D = \overline{n_D^2} = \overline{\epsilon_k^2} = \sigma^2$$

The signal power in terms of $S_x = \overline{x^2}$ is

$$S_D = \mu_p^2 \overline{x^2} = \mu_p^2 S_x$$

where our normalization convention $|x(t)| \leq 1$ requires $S_x \leq 1$. Hence,

$$\left(\frac{S}{N}\right)_D = \frac{S_D}{N_D} = \frac{\mu_p^2}{\sigma^2} S_x \tag{7}$$

which expresses the destination signal-to-noise ratio in terms of the error variance σ^2 caused by reconstruction from noisy samples. Our next task is to determine μ_p^2 and σ^2 for specific types of pulse modulation.

A PAM signal contains the message samples in the modulated pulse amplitude $A_0[1 + \mu x(kT_s)]$, so the modulation constant is $\mu_p = A_0 \mu \leq A_0$. The upper limit $\mu = 1$ comes from our signal normalization and Eq. (5), Sect. 10.2. The amplitude error variance is $\sigma^2 = \sigma_A^2 = \eta B_N$. Thus, under the best conditions of maximum modulation ($\mu = 1$) and minimum noise bandwidth ($B_N \approx 1/2\tau$), we have

$$\left(\frac{S}{N}\right)_D = \frac{2A_0^2 \tau}{\eta} S_x$$

Notice that $A_0^2 \tau$ equals the received signal energy in an unmodulated pulse.

The *average energy* per modulated pulse is

$$A_0^2 \overline{[1 + x(kT_s)]^2} \tau = A_0^2 (1 + S_x)\tau$$

when $\mu = 1$ and $x(t)$ has no DC component. Multiplying this average energy by the pulse rate f_s gives the *received signal power*

$$S_R = f_s A_0^2 (1 + S_x)\tau$$

We now obtain our final result in the form

$$\left(\frac{S}{N}\right)_D = \frac{S_x}{1 + S_x} \frac{2S_R}{\eta f_s} = \frac{S_x}{1 + S_x}\left(\frac{2W}{f_s}\right)\gamma \qquad \text{PAM} \qquad (8)$$

where we've introduced $\gamma = S_R/\eta W$. Equation (8) shows that $(S/N)_D \leq \gamma/2$, so PAM performance is at least 3 dB below unmodulated baseband transmission — just like AM CW modulation. The maximum value is seldom achieved in practice, nor is it sought after. The merit of PAM resides in its simplicity for multiplexing, not in its noise performance.

However, PPM and PDM do offer some improvement by virtue of wideband noise reduction. For if $B_N \approx B_T$, the time-position error variance is $\sigma^2 = \sigma_t^2 = \eta/(4B_T A^2)$. Since the pulse amplitude A is a constant, the received power can be written as $S_R = f_s A^2 \tau_0$, where τ_0 denotes the *average* pulse duration in PDM or the fixed pulse duration in PPM. Equation (7) then becomes

$$\left(\frac{S}{N}\right)_D = \frac{4\mu_p^2 B_T A^2}{\eta} S_x = 4\mu_p^2 B_T \frac{S_R}{\eta f_s \tau_0} S_x$$

$$= 4\mu_p^2 B_T \left(\frac{W}{f_s \tau_0}\right) S_x \gamma \qquad \text{PDM or PPM} \qquad (9)$$

This expression reveals that $(S/N)_D$ increases with increasing transmission bandwidth B_T. The underlying physical reason should be evident from Fig. 10.4-2 with $t_r \approx 1/2B_T$.

The PPM modulation constant is $\mu_p = t_0$, the maximum pulse displacement. The parameters t_0 and τ_0 are constrained by

$$t_0 \leq T_s/2 \qquad \tau_0 = \tau \geq 2t_r \approx 1/B_T$$

and $f_s = 1/T_s \geq 2W$. Taking all values to be optimum with respect to noise reduction, we obtain the upper bound

$$\left(\frac{S}{N}\right)_D \leq \frac{1}{8}\left(\frac{B_T}{W}\right)^2 S_x\gamma \qquad \text{PPM} \tag{10}$$

Hence, PPM performance improves as the *square* of the bandwidth ratio B_T/W. A similar optimization for PDM with $\mu_p = \mu\tau_0 \leq \tau_0$ yields the less-impressive result

$$\left(\frac{S}{N}\right)_D \leq \frac{1}{2}\frac{B_T}{W} S_x\gamma \qquad \text{PDM} \tag{11}$$

To approach the upper bound, a PDM wave must have a 50% duty cycle so that $\tau_0 \approx T_s/2$.

Practical PPM and PDM systems may fall short of the values predicted in Eqs. (10) and (11) by 10 dB or more. Consequently, the noise reduction does not measure up to that of wideband FM. But remember that the average power S_R comes from short-duration high-power pulses rather than being continuously delivered as in CW modulation. Power-supply considerations may therefore favor pulsed operation in some circumstances.

Exercise 10.4-2 Explain why a single-channel PDM system must have $\mu\tau_0 \leq 1/4W$. Then derive Eq. (11) from Eq. (9) with $\mu_p = \mu\tau_0$.

False-Pulse Threshold Effect

Suppose you try to increase the value of $(S/N)_D$ in a PDM or PPM system by making B_T very large. Since $\overline{n^2}$ increases with B_T, the noise variations in $v(t) = x_p(t) + n(t)$ will eventually dominate and be mistaken for signal pulses. If these *false pulses* occur often, the reconstructed waveform has no relationship to $x(t)$ and the message will have been completely lost. Hence, pulse-time modulation involves a false-pulse *threshold effect*, analogous to the threshold effect in wideband FM. This effect does not exist in PAM with synchronization because we always know when to measure the amplitude.

To determine the threshold level, we'll say that false pulses are sufficiently infrequent if $P(n \geq A) \leq 0.01$. For gaussian noise with $\sigma_n^2 = \overline{n^2} = \eta B_T$, the corresponding threshold condition is approximately

$$A \geq 2\sigma_n$$

so the pulse must be strong enough to "lift" the noise by at least twice its rms value. (This is the same condition as the *tangential sensitivity* in pulsed radar systems.) Using the fact that $A^2 = S_R/\tau_0 f_s$, we have

$$\frac{S_R}{\tau_0 f_s} \geq 4\eta B_T$$

or

$$\gamma_{th} = \left(\frac{S_R}{\eta W}\right)_{min} = 4\tau_0 f_s \frac{B_T}{W} \geq 8 \tag{12}$$

This threshold level is appreciably less than that of FM, so PPM could be advantageous for those situations where FM would be below its threshold point.

10.5 PROBLEMS

10.1-1 Consider the chopper-sampled waveform in Eq. (3) with $\tau = T_s/2$, $f_s = 100$ Hz, and $x(t) = 2 + 2 \cos 2\pi 30t + \cos 2\pi 80t$. Draw and label the one-sided line spectrum of $x_s(t)$ for $0 \leq f \leq 300$ Hz. Then find the output waveform when $x_s(t)$ is applied to an ideal LPF with $B = 75$ Hz.

10.1-2 Do Prob. 10.1-1 with $x(t) = 2 + 2 \cos 2\pi 30t + \cos 2\pi 140t$.

10.1-3 The usable frequency range of a certain amplifier is f_ℓ to $f_\ell + B$, with $B \gg f_\ell$. Devise a system that employs bipolar choppers and allows the amplifier to handle signals having significant DC content and bandwidth $W \ll B$.

10.1-4 The baseband signal for FM stereo (Fig. 8.2-5b) is

$$x_b(t) = [x_L(t) + x_R(t)] + [x_L(t) - x_R(t)] \cos \omega_s t + A \cos \omega_s t/2$$

with $f_s = 38$ kHz. The chopper system in Fig. P10.1-4 is intended to generate this signal. The LPF has gain K_1 for $|f| \leq 15$ kHz, gain K_2 for $23 \leq |f| \leq 53$ kHz, and rejects $|f| \geq 99$ kHz. Use a sketch to show that $x_s(t) = x_L(t)s(t) + x_R(t)[1 - s(t)]$, where $s(t)$ is a unipolar switching function with $\tau = T_s/2$. Then find the necessary values of K_1 and K_2.

Figure P10.1-4

10.1-5 A popular *stereo decoder* circuit employs transistor switches to generate $v_L(t) = x_1(t) - Kx_2(t)$ and $v_R(t) = x_2(t) - Kx_1(t)$ where K is a constant, $x_1(t) = x_b(t)s(t)$, $x_2(t) = x_b(t)[1 - s(t)]$, $x_b(t)$ is the FM stereo baseband signal in Prob. 10.1-4, and $s(t)$ is a unipolar switching function with $\tau = T_s/2$. (a) Determine K such that lowpass filtering of $v_L(t)$ and $v_R(t)$ yields the desired left- and right-channel signals. (b) What's the disadvantage of a simpler switching circuit that has $K = 0$?

10.1-6 Derive Eq. (11) using Eq. (14), Sect. 2.4.

10.1-7 Suppose $x(t)$ has the spectrum in Fig. P10.1-7 with $f_u = 25$ kHz and $W = 10$ kHz. Sketch $x_\delta(f)$ for $f_s = 60$, 45, and 25 kHz. Comment in each case on the possible reconstruction of $x(t)$ from $x_\delta(t)$.

Figure P10.1-7

10.1-8 Consider the bandpass signal spectrum in Fig. P10.1-7 whose Nyquist rate is $f_s = 2f_u$. However, the *bandpass sampling theorem* states that $x(t)$ can be reconstructed from $x_\delta(t)$ by *bandpass filtering* if $f_s = 2f_u/m$ and the integer m satisfies $(f_u/W) - 1 < m \le f_u/W$. (a) Find m and plot f_s/W versus f_u/W for $0 < f_u/W \le 5$. (b) Check the theorem by plotting $X_\delta(f)$ when $f_u = 2.5W$ and $f_s = 2.5W$. Also show that the higher rate $f_s = 4W$ would not be acceptable.

10.1-9 The signal $x(t) = \text{sinc}^2 5t$ is ideally sampled at $t = 0, \pm 0.1, \pm 0.2, \ldots$, and reconstructed by an ideal LPF with $B = 5$, unit gain, and zero time delay. Carry out the reconstruction process graphically, as in Fig. 10.1-6, for $|t| \le 0.2$.

10.1-10 A rectangular pulse with $\tau = 2$ is ideally sampled and reconstructed using an ideal LPF with $B = f_s/2$. Sketch the resulting output waveforms when $T_s = 0.8$ and 0.4, assuming one sample time is at the center of the pulse.

10.1-11 Suppose an ideally sampled wave is reconstructed using a *zero-order hold* (Example 3.1-3) with time delay $T = T_s$. (a) Find and sketch $y(t)$ to show that the reconstructed waveform is a *staircase* approximation of $x(t)$. (b) Sketch $|Y(f)|$ for $X(f) = \Pi(f/2W)$ with $W \ll f_s$. Comment on the significance of your result.

10.1-12‡ The reconstruction system in Fig. P10.1-12 is called a *first-order hold*. Each block labeled ZOH is a zero-order hold (Example 3.1-3) with time delay $T = T_s$. (a) Find $h(t)$ and sketch $y(t)$ to interpret the reconstruction operation. (b) Show that $H(f) = T_s(1 + j2\pi f T_s)(\text{sinc}^2 f T_s) \exp(-j2\pi f T_s)$. Then sketch $|Y(f)|$ for $X(f) = \Pi(f/2W)$ with $W < f_s/2$.

Figure P10.1-12

10.1-13‡ Use Parseval's theorem and Eq. (14) with $T_s = 1/2W$ and $B = W$ to show that the energy of a bandlimited signal is related to its sample values by

$$E = (1/2W) \sum_{k=-\infty}^{\infty} |x(k/2W)|^2$$

10.1-14 The *frequency-domain sampling theorem* says that if $x(t)$ is a *timelimited* signal, such that $x(t) = 0$ for $|t| \ge T$, then $X(f)$ is completely determined by its sample values $X(nf_0)$ with $f_0 \le 1/2T$. Prove this theorem by writing the Fourier series for the periodic signal $v(t) = x(t) * [\sum_k \delta(t - kT_0)]$, where $T_0 \ge 2T$, and using the fact that $x(t) = v(t)\Pi(t/2T)$.

10.1-15 A signal with period $T_x = 0.08 \, \mu s$ is to be displayed using a sampling oscilloscope whose internal high-frequency response cuts off at $B = 6$ MHz. Determine maximum values for the sampling frequency and the bandwidth of the pre-sampling LPF.

10.1-16 Explain why the sampling oscilloscope in Prob. 10.1-15 will not provide a useful display when $T_x < 1/3B$.

10.2-1 Sketch $|X_p(f)|$ and find $H_{eq}(f)$ for flat-top sampling with $\tau = T_s/2$, $f_s = 2.5W$, and $p(t) = \Pi(t/\tau)$. Is equalization essential in this case?

10.2-2 Do Prob. 10.2-1 for $p(t) = (\cos \pi t/\tau)\Pi(t/\tau)$.

10.2-3‡ Some sampling devices extract from $x(t)$ its *average value* over the sampling duration, so $x(kT_s)$ in Eq. (1) is replaced by

$$\bar{x}(kT_s) \triangleq \frac{1}{\tau} \int_{kT_s-\tau}^{kT_s} x(\lambda)\, d\lambda$$

(a) Devise a frequency-domain model of this process using an averaging filter, with input $x(t)$ and output $\bar{x}(t)$, followed by instantaneous flat-top sampling. Then obtain the inpulse response of the averaging filter and write the resulting expression for $X_p(f)$. (b) Find the equalizer needed when $p(t)$ is a rectangular pulse.

10.2-4 Consider the PAM signal in Eq. (4). (a) Show that its spectrum is

$$X_p(f) = A_0\, f_s\, P(f)\left\{ \sum_n [\delta(f - nf_s) + \mu X(f - nf_s)] \right\}$$

(b) Sketch $|X_p(f)|$ when $p(t) = \Pi(t/\tau)$ with $\tau = T_s/2$ and $\mu x(t) = \cos 2\pi f_m t$ with $f_m < f_s/2$.

10.2-5 Suppose the PAM signal in Eq. (4) is to be transmitted over a transformer-coupled channel, so the pulse shape is taken as $p(t) = \Pi[(t - \tau/2)/\tau] - \Pi[(t + \tau/2)/\tau]$ to eliminate the DC component of $x_p(t)$. (a) Use the expression in Prob. 10.2-4a to sketch $|X_p(f)|$ when $\tau = T_s/4$, $X(f) = \Pi(f/2W)$, and $f_s > 2W$. (b) Find an appropriate equalizer, assuming that $x(t)$ has negligible frequency content for $|f| < f_\ell < W$. Why is this assumption necessary?

10.2-6 Calculate the transmission bandwidth needed for voice PDM with $f_s = 8$ kHz, $|\mu x(t)| \le 0.8$, and $\tau_0 = T_s/5$ when we want $t_r \le 0.25\tau_{min}$.

10.2-7 A voice PDM signal with $f_s = 8$ kHz and $|\mu x(t)| \le 0.8$ is to be transmitted over a channel having $B_T = 500$ kHz. Obtain bounds on τ_0 such that $\tau_{max} \le T_s/3$ and $\tau_{min} \ge 3t_r$.

10.2-8 A pulse-modulated wave is generated by uniformly sampling the signal $x(t) = \cos 2\pi t/T_m$ at $t = kT_s$, where $T_s = T_m/3$. Sketch and label $x_p(t)$ when the modulation is: (a) PDM with $\mu = 0.8$, $\tau_0 = 0.4T_s$, and leading edges fixed at $t = kT_s$; (b) PPM with $t_d = 0.5T_s$ and $t_0 = \tau = 0.2T_s$.

10.2-9 Do Prob. 10.2-8 with $T_s = T_m/6$.

10.2-10 Use Eq. (15) to devise a system that employs a PPM generator and produces narrowband phase modulation with $f_c = mf_s$.

10.2-11 *Poisson's sum formula* states in general that

$$\sum_{n=-\infty}^{\infty} e^{\pm j2\pi n\lambda/L} = L \sum_{m=-\infty}^{\infty} \delta(\lambda - mL)$$

where λ is an independent variable and L is a constant. (a) Derive the time-domain version as given in Eq. (14) by taking $\mathscr{F}^{-1}[S_\delta(f)]$. (b) Derive the frequency-domain version by taking $\mathscr{F}[s_\delta(t)]$.

10.2-12‡ Let $g(t)$ be any continuous function that monotonically increases or decreases over $a \le t \le b$ and crosses zero at $t = \lambda$ within this range. Justify Eq. (13) by making the change-of-variable $v = g(t)$ in

$$\int_a^b \delta[g(t)]\, dt$$

10.3-1 Twenty-four voice signals are to be transmitted via multiplexed PAM with a marker pulse for frame synchronization. The sampling frequency is 8 kHz and the TDM signal has a 50% duty cycle. Calculate the signaling rate, pulse duration, and minimum transmission bandwidth.

10.3-2 Do Prob. 10.3-1 with a 6-kHz sampling frequency and 30% duty cycle.

10.3-3 Twenty signals, each with $W = 4$ kHz, are sampled at a rate that allows a 2-kHz guard band for reconstruction filtering. The multiplexed samples are transmitted on a CW carrier. Calculate the required transmission bandwidth when the modulation is: (a) PAM/AM with 25% duty cycle; (b) PAM/SSB with baseband filtering.

10.3-4 Ten signals, each with $W = 2$ kHz, are sampled at a rate that allows a 1-kHz guard band for reconstruction filtering. The multiplexed samples are transmitted on a CW carrier. Calculate the required transmission bandwidth when the modulation is: (a) PPM/AM with 20% duty cycle; (b) PAM/FM with baseband filtering and $f_\Delta = 75$ kHz.

10.3-5 Given a 6-channel main multiplexer with $f_s = 8$ kHz, devise a telemetry system similar to Fig. 10.3-5 (including a marker) that accommodates six input signals having the following bandwidths: 8.0, 3.5, 2.0, 1.8, 1.5, and 1.2 kHz. Make sure that successive samples of each input signal are equispaced in time. Calculate the resulting baseband bandwidth and compare with the minimum transmission bandwidth for an FDM-SSB system.

10.3-6 Do Prob. 10.3-5 for seven input signals having the following bandwidths: 12.0, 4.0, 1.0, 0.9, 0.8, 0.5, and 0.3 kHz.

10.3-7 Do Prob. 10.3-5 for eight input signals having the following bandwidths: 12.0, 3.5, 2.0, 0.5, 0.4, 0.3, 0.2, and 0.1 kHz.

10.3-8 Calculate the bandwidth required so the cross talk does not exceed -40 dB when 25 voice signals are transmitted via PPM-TDM with $f_s = 8$ kHz and $t_0 = \tau = 0.2(T_s/M)$.

10.3-9 Find the maximum number of voice signals that can be transmitted via TDM-PPM with $f_s = 8$ kHz and $t_0 = \tau = 0.25(T_s/M)$ when the channel has $B = 500$ kHz and the cross talk is to be kept below -30 dB.

10.3-10 Cross talk also occurs when a transmission system has inadequate *low-frequency response*, usually as a result of transformer coupling or blocking capacitors. Demonstrate this effect by sketching the pulse response of a high-pass filter whose step response is $g(t) = \exp(-2\pi f_\ell t) u(t)$. Consider the extreme cases $f_\ell \tau \ll 1$ and $f_\ell \tau \gg 1$.

10.4-1 A pulse transmission system has $\tau = 5$ μs, $B_T = 1$ MHz, and $\eta = 10^{-12}$. Find the minimum value of E_p so that $\sigma_A \leq A/100$, and calculate the corresponding value of σ_t/τ.

10.4-2 A pulse transmission system has $\tau = 50$ μs, $B_T = 1$ MHz, and $\eta = 10^{-12}$. Find the minimum value of E_p so that $\sigma_t \leq \tau/100$, and calculate the corresponding value of σ_A/A.

10.4-3 A pulse transmission system has $\tau = 1$ ms, $B_T = 1$ MHz, and $\eta = 10^{-12}$. Find the minimum value of E_p and the corresponding value of B_N so that $\sigma_A \leq A/100$ and $\sigma_t \leq \tau/1000$.

10.4-4 Calculate the upper bound on $(S/N)_D$ for PPM transmission with $B_T = 20W$ and compare with the actual value when $t_0 = 0.3T_s$, $\tau = 0.2T_s$, and $f_s = 2.5W$.

10.4-5 Equation (9) also holds for each output of a TDM system if we rewrite A^2 in terms of the average power in the multiplexed waveform. Use this method to find $(S/N)_D$ for the system in Exercise 10.3-2 when $S_x = 1$, $W = 3.2$ kHz, and the synchronization marker has duration 3τ. Express your result in terms of $\gamma/9$, since there are nine transmitted signals.

10.4-6 Consider an M-channel TDM-PPM system with $T_g = \tau \geq 1/B_T$ and no synchronization marker, so $S_R = Mf_s A^2 \tau$. Start with Eq. (9) to show that $(S/N)_D < \frac{1}{8}(B_T/MW)^2 S_x(\gamma/M)$.

DIGITAL COMMUNICATION

CHAPTER

ELEVEN

BASEBAND DIGITAL TRANSMISSION

This chapter launches our study of digital communication systems. We focus initially on baseband transmission in order to emphasize the generic concepts and problems of digital communication, with or without carrier modulation, and to bring out the differences between digital and analog transmission. Following an overview of digital signals and systems, we'll analyze the limitations imposed by additive noise and transmission bandwidth. We'll also look at practical design considerations such as regenerative repeaters, equalization, and synchronization.

11.1 DIGITAL SIGNALS AND SYSTEMS

Fundamentally, a digital message is nothing more than an *ordered sequence of symbols* produced by a discrete information source. The source draws from an *alphabet* of $M \geq 2$ different symbols, and produces output symbols at some average *rate* r. For instance, a typical computer terminal has an alphabet of $M \approx 90$ symbols, equal to the number of character keys multiplied by two to account for the shift key. When you operate the terminal as fast as you can, you become a discrete information source producing a digital message at a rate of perhaps $r \approx 5$ symbols per second. The computer itself works with just $M = 2$ internal symbols, represented by LOW and HIGH electrical states. We usually associate these two symbols with the *binary digits* 0 and 1, known as *bits* for short. Data transfer rates within a computer may exceed $r = 10^8$.

The task of a digital communication system is to transfer a digital message from the source to the destination. But finite transmission bandwidth sets an upper limit to the symbol rate, and noise causes errors to appear in the output message. Thus, *signaling rate* and *error probability* play roles in digital communication similar to those of bandwidth and signal-to-noise ratio in analog communication. As preparation for the analysis of signaling rate and error probability, we must first develop the description and properties of digital signals.

Digital PAM Signals

Digital message representation at baseband commonly takes the form of an *amplitude-modulated pulse train*. We express such signals by writing

$$x(t) = \sum_k a_k p(t - kD) \tag{1}$$

where the modulating amplitude a_k represents the kth symbol in the message sequence, so the amplitudes belong to a set of M discrete values. The index k ranges from $-\infty$ to $+\infty$ unless otherwise stated. Equation (1) defines a digital PAM signal, as distinguished from those rare cases when pulse-duration or pulse-position modulation is used for digital transmission.

The unmodulated pulse $p(t)$ may be rectangular or some other shape, subject to the conditions

$$p(t) = \begin{cases} 1 & t = 0 \\ 0 & t = \pm D, \pm 2D, \ldots \end{cases} \tag{2}$$

This condition ensures that we can recover the message by sampling $x(t)$ periodically at $t = KD$, $K = 0, \pm 1, \pm 2, \ldots$, since

$$x(KD) = \sum_k a_k p(KD - kD) = a_K$$

The rectangular pulse $p(t) = \Pi(t/\tau)$ satisfies Eq. (2) if $\tau \le D$, as does any time-limited pulse with $p(t) = 0$ for $|t| \ge D/2$.

Note that D does not necessarily equal the pulse duration but rather the pulse-to-pulse interval or the time allotted to one symbol. Thus, the *signaling rate* is

$$r \triangleq 1/D \tag{3a}$$

measured in symbols per second or *baud*. In the special but important case of *binary* signaling ($M = 2$), we write $D = T_b$ for the bit duration and the bit rate is†

$$r_b = 1/T_b \tag{3b}$$

measured in bits per second, abbreviated bps or b/s. The notation T_b and r_b will be used to identify results that apply only for binary signaling.

---

† The more common notation R for bit rate risks confusion with autocorrelation functions and with information rate defined in Chap. 15.

Figure 11.1-1 depicts various PAM *formats* for the binary message 10110100, taking rectangular pulses for clarity. The simple *on-off* waveform in part *a* represents each 0 by an "off" pulse ($a_k = 0$) and each 1 by an "on" pulse with amplitude $a_k = A$ and duration $T_b/2$ followed by a return to the zero level. We therefore call this a *return-to-zero* (RZ) format. A *nonreturn-to-zero* (NRZ) format has "on" pulses for full bit duration T_b, as indicated by the dashed lines. Internal computer waveforms are usually of this type. The NRZ format puts more energy into each pulse, but requires synchronization at the receiver because there's no separation between adjacent pulses.

The unipolar nature of an on-off signal results in a *DC component* that carries no information and wastes power. The *polar* signal in part *b* has opposite polarity pulses, either RZ or NRZ, so its DC component will be zero if the message contains 1s and 0s in equal proportion. This property also applies to the *bipolar*

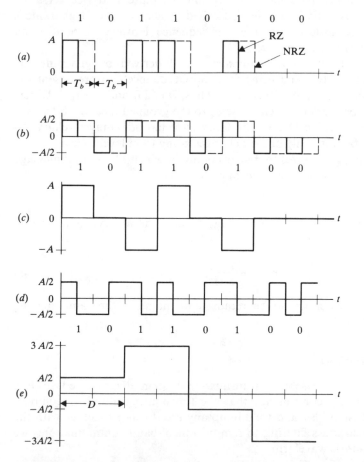

Figure 11.1-1 Binary PAM formats with rectangular pulses. (*a*) Unipolar RZ and NRZ; (*b*) polar RZ and NRZ; (*c*) bipolar NRZ; (*d*) split-phase Manchester; (*e*) polar quaternary NRZ.

Table 11.1-1

a_k	Natural code	Gray code
$3A/2$	11	10
$A/2$	10	11
$-A/2$	01	01
$-3A/2$	00	00

signal in part *c*, where successive 1s are represented by pulses with alternating polarity. The bipolar format, also known as *pseudo-trinary* or *alternate mark inversion* (AMI), eliminates ambiguities that might be caused by transmission *sign inversions* — a problem characteristic of switched telephone links.

The *split-phase Manchester* format in part *d* represents 1s with a positive half-interval pulse followed by a negative half-interval pulse, and vice versa for the representation of 0s. This format is also called *twinned binary*. It guarantees zero DC component regardless of the message sequence. However, it requires an absolute sense of polarity at the receiver.

Finally, Fig. 11.1-1*e* shows a *quaternary* signal derived by grouping the message bits in blocks of two and using four amplitude levels to prepresent the four possible combinations 00, 01, 10, and 11. Thus, $D = 2T_b$ and $r = r_b/2$. Different assignment rules or *codes* may relate the a_k to the grouped message bits. Two such codes are listed in Table 11.1-1. The Gray code has advantages relative to noise-induced errors because only one bit changes going from level to level.

Quaternary coding generalizes to *M-ary* coding in which blocks of *n* message bits are represented by an *M*-level waveform with

$$M = 2^n \tag{4a}$$

Since each pulse now corresponds to $n = \log_2 M$ bits, the *M*-ary signaling rate has been decreased to

$$r = \frac{r_b}{\log_2 M} \tag{4b}$$

But increased signal power would be required to maintain the same spacing between amplitude levels.

Transmission Limitations

Now consider the linear baseband transmission system diagrammed in Fig. 11.1-2*a*. We'll assume for convenience that the transmitting amplifier compensates for the transmission loss, and we'll lump any interference together with the additive noise. After lowpass filtering to remove out-of-band contaminations, we have the signal-plus-noise waveform

$$y(t) = \sum_k a_k \tilde{p}(t - t_d - kD) + n(t)$$

Figure 11.1-2 (a) Baseband transmission system; (b) signal-plus-noise waveform.

where t_d is the transmission delay and $\tilde{p}(t)$ stands for the pulse shape with transmission distortion. Figure 11.1-2b illustrates what $y(t)$ might look like when $x(t)$ is
the unipolar NRZ signal in Fig. 11.1-1a.

Recovering the digital message from $y(t)$ is the task of the *regenerator*. An
auxiliary synchronization signal may help the regeneration process by identifying
the optimum sampling times

$$t_K = KD + t_d$$

If $\tilde{p}(0) = 1$ then

$$y(t_K) = a_K + \sum_{k \neq K} a_k \tilde{p}(KD - kD) + n(t_K) \tag{5}$$

whose first term is the desired message information. The last term of Eq. (5) is the
noise contamination at t_K, while the middle term represents cross talk or spillover from other signal pulses — a phenomenon given the descriptive name *intersymbol interference* (ISI). The combined effects of noise and ISI may result in
errors in the regenerated message. For instance, at the sample time t_K indicated in
Fig. 11.1-2b, $y(t_K)$ is closer to 0 even though $a_K = A$.

We know that if $n(t)$ comes from a white-noise source, then its mean square
value can be reduced by reducing the bandwidth of the LPF. We also know that
lowpass filtering causes pulses to spread out, which would increase the ISI. Consequently, a fundamental limitation of digital transmission is the relationship
between ISI, bandwidth, and signaling rate.

This problem emerged in the early days of telegraphy, and Harry Nyquist
(1924, 1928) first stated the relationship as follows:

> Given an ideal lowpass channel of bandwidth B, it is possible to transmit
> independent symbols at a rate $r \leq 2B$ baud without intersymbol interference.
> It is not possible to transmit independent symbols at $r > 2B$.

The condition $r \leq 2B$ agrees with our pulse-resolution rule $B \geq 1/2\tau_{min}$ in Sect. 3.4 if we require $p(t)$ to have duration $\tau \leq D = 1/r$.

The second part of Nyquist's statement is easily proved by assuming that $r = 2(B + \epsilon) > 2B$. Now suppose that the message sequence happens to consist of two symbols alternating forever, such as 101010... The resulting waveform $x(t)$ then is *periodic* with period $2D = 2/r$ and contains only the fundamental frequency $f_0 = B + \epsilon$ and its harmonics. Since no frequency greater than B gets through the channel, the output signal will be zero—aside from a possible but useless DC component.

Signaling at the maximum rate $r = 2B$ requires a special pulse shape, the *sinc pulse*

$$p(t) = \text{sinc } rt = \text{sinc } t/D \tag{6a}$$

having the *bandlimited* spectrum

$$P(f) = \mathscr{F}[p(t)] = \frac{1}{r}\Pi\left(\frac{f}{r}\right) \tag{6b}$$

Since $P(f) = 0$ for $|f| > r/2$, this pulse suffers no distortion from an ideal lowpass frequency response with $B \geq r/2$ and we can take $r = 2B$. Although $p(t)$ is not timelimited, it does have *periodic zero crossings* at $t = \pm D, \pm 2D, \ldots$, which satisfies Eq. (2). (See Fig. 10.1-6 for an illustration of this property.) Nyquist also derived other bandlimited pulses with periodic zero crossings spaced by $D > 1/2B$ so $r < 2B$, a topic we set aside to pick up again in Sect. 11.3 after discussing noise and errors in Sect. 11.2.

Meanwhile, note that any real channel needs *equalization* to approach an ideal frequency response. Such equalizers often require experimental adjustment in the field because we don't know the channel characteristics exactly. An important experimental display is the so-called *eye pattern*, which further clarifies digital transmission limitations.

Consider the distorted but noise-free polar binary signal in Fig. 11.1-3a. When displayed on a long-persistence oscilloscope with appropriate synchronization and sweep time, we get the superposition of successive symbol intervals shown in Fig. 11.1-3b. The shape of this display accounts for the name "eye pattern." A distorted M-ary signal would result in $M - 1$ "eyes" stacked vertically.

(a)   (b)

Figure 11.1-3 (a) Distorted polar binary signal; (b) eye pattern.

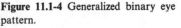

Figure 11.1-4 Generalized binary eye pattern.

Figure 11.1-4 represents a generalized binary eye pattern with labels identifying significant features. The optimum sampling time corresponds to the maximum eye opening. ISI at this time partially closes the eye and thereby reduces the *noise margin*. If synchronization is derived from the zero crossings, as it usually is, zero-crossing distortion produces *jitter* and results in nonoptimum sampling times. The slope of the eye pattern in the vicinity of the zero crossings indicates the sensitivity to timing error. Finally, any nonlinear transmission distortion would reveal itself in an asymmetric or "squinted" eye.

Exercise 11.1-1 Determine the relation between r and B when $p(t) = \mathrm{sinc}^2 at$.

Power Spectra of Digital PAM

The pulse spectrum $P(f) = \mathscr{F}[p(t)]$ provides some hints about the *power spectrum* of a digital PAM signal $x(t)$. If $p(t) = \mathrm{sinc}\ rt$, as a case in point, then $P(f)$ in Eq. (6b) implies that $G_x(f) = 0$ for $|f| > r/2$. However, detailed knowledge of $G_x(f)$ provides additional and valuable information relative to digital transmission.

A simplified random digital wave with $p(t) = \Pi(t/D)$ was tackled in Chap. 5. Under the conditions

$$E[a_k a_i] = \begin{cases} \sigma_a^2 & i = k \\ 0 & i \neq k \end{cases}$$

we found that $G_x(f) = \sigma_a^2 D\ \mathrm{sinc}^2\ fD$. Now, substituting $P(f) = D\ \mathrm{sinc}\ fD$, we write

$$G_x(f) = \frac{\sigma_a^2}{D} |P(f)|^2 \tag{7}$$

This expression holds for *any* digital PAM signal with pulse spectrum $P(f)$ when the a_k are uncorrelated and have zero mean value.

But unipolar signal formats have $\overline{a_k} \neq 0$ and, in general, we can't be sure that the message source produces uncorrelated symbols. A more realistic approach therefore models the source as a *discrete stationary random process*. Ensemble averages of the a_k are then given by the autocorrelation function

$$R_a(n) = E[a_k a_{k-n}] \tag{8}$$

analogous to writing $R_v(\tau) = E[v(t)v(t - \tau)]$ for a stationary random signal $v(t)$. The integers n and k in Eq. (8) reflect the time-discrete nature of a digital sequence.

If a digital PAM signal $x(t)$ has the pulse spectrum $P(f)$ and amplitude auto-correlation $R_a(n)$, its power spectrum is

$$G_x(f) = \frac{1}{D} |P(f)|^2 \sum_{n=-\infty}^{\infty} R_a(n) e^{-j2\pi n f D} \tag{9}$$

Despite the formidable appearance of Eq. (9), it easily reduces to Eq. (7) when $R_a(0) = \sigma_a^2$ and $R_a(n) = 0$ for $n \neq 0$. In the case of uncorrelated message symbols but $\overline{a_k} = m_a \neq 0$,

$$R_a(n) = \begin{cases} \sigma_a^2 + m_a^2 & n = 0 \\ m_a^2 & n \neq 0 \end{cases} \tag{10}$$

and

$$\sum_{n=-\infty}^{\infty} R_a(n) e^{-j2\pi n f D} = \sigma_a^2 + m_a^2 \sum_{n=-\infty}^{\infty} e^{-j2\pi n f D}$$

Then, drawing upon Poisson's sum formula,

$$\sum_{n=-\infty}^{\infty} e^{-j2\pi n f D} = \frac{1}{D} \sum_{n=-\infty}^{\infty} \delta\left(f - \frac{n}{D}\right)$$

and therefore

$$G_x(f) = \sigma_a^2 r |P(f)|^2 + (m_a r)^2 \sum_{n=-\infty}^{\infty} |P(nr)|^2 \, \delta(f - nr) \tag{11}$$

Here we have inserted $r = 1/D$ and used the sampling property of impulse multiplication.

The important result in Eq. (11) reveals that the power spectrum of $x(t)$ contains impulses at *harmonics* of the signaling rate r, unless $m_a = 0$ or $P(f) = 0$ at all $f = nr$. Hence, a synchronization signal can be obtained by applying $x(t)$ to a narrow BPF (or PLL filter) centered at one of these harmonics. We can also calculate the total *average power* $\overline{x^2}$ by integrating $G_x(f)$ over all f. Thus,

$$\overline{x^2} = \sigma_a^2 r E_p + (m_a r)^2 \sum_{n=-\infty}^{\infty} |P(nr)|^2 \tag{12}$$

where E_p equals the energy in $p(t)$. For Eq. (12), and hereafter, we presume the conditions in Eq. (10) barring information to the contrary.

The derivation of Eq. (9) starts with the definition of power spectrum from Eq. (9), Sect. 5.2. Specifically, we write

$$G_x(f) \triangleq \lim_{T \to \infty} \frac{1}{T} E[|X_T(f)|^2]$$

in which $X_T(f)$ is the Fourier transform of a truncated sample function $x_T(t) = x(t)$ for $|t| < T/2$. Next, let $T = (2K + 1)D$ so the limit $T \to \infty$ corresponds to $K \to \infty$. Then, for $K \gg 1$,

$$x_T(t) = \sum_{k=-K}^{K} a_k \, p(t - kD)$$

$$X_T(f) = \sum_{k=-K}^{K} a_k \, P(f) e^{-j\omega kD}$$

and

$$|X_T(f)|^2 = |P(f)|^2 \left(\sum_{k=-K}^{K} a_k e^{-j\omega kD} \right) \left(\sum_{i=-K}^{K} a_i e^{+j\omega iD} \right)$$

After interchanging the order of expectation and summation we have

$$E[|X_T(f)|^2] = |P(f)|^2 \rho_K(f)$$

with

$$\rho_K(f) = \sum_{k=-K}^{K} \sum_{i=-K}^{K} E[a_k a_i] e^{-j\omega(k-i)D}$$

where $E[a_k a_i] = R_a(k - i)$.

The double summation for $\rho_K(f)$ can be manipulated into the single sum

$$\rho_K(f) = (2K + 1) \sum_{n=-2K}^{2K} \left(1 - \frac{|n|}{2K + 1} \right) R_a(n) e^{-j\omega nD}$$

Substituting these expressions in the definition of $G_x(f)$ finally gives

$$G_x(f) = \lim_{K \to \infty} \frac{1}{(2K + 1)D} |P(f)|^2 \rho_K(f)$$

$$= \frac{1}{D} |P(f)|^2 \sum_{n=-\infty}^{\infty} R_a(n) e^{-j\omega nD}$$

as stated in Eq. (9).

Example 11.1-1 Consider the unipolar binary RZ signal in Fig. 11.1-1a, where $p(t) = \Pi(2r_b t)$ so

$$P(f) = \frac{1}{2r_b} \, \text{sinc} \, \frac{f}{2r_b}$$

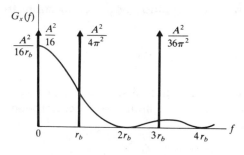

Figure 11.1-5 Power spectrum of unipolar binary RZ signal.

If the source bits are equally likely and statistically independent, then $\overline{a_k} = A/2$, $\overline{a_k^2} = A^2/2$, and Eq. (10) applies with

$$m_a^2 = \sigma_a^2 = \frac{A^2}{4}$$

Using Eq. (11) we find the power spectrum to be

$$G_x(f) = \frac{A^2}{16r_b} \text{sinc}^2 \frac{f}{2r_b} + \frac{A^2}{16} \sum_{n=-\infty}^{\infty} \left(\text{sinc}^2 \frac{n}{2} \right) \delta(f - nr_b)$$

which is sketched in Fig. 11.1-5 for $f \geq 0$. Notice the absence of impulses at the even harmonics because $P(nr_b) = 0$ when $n = \pm 2, \pm 4, \ldots$.

We could also, in principle, use Eq. (12) to calculate $\overline{x^2}$. However, it should be evident from the waveform that $\overline{x^2} = A^2/4$ when 1s and 0s are equally likely.

Exercise 11.1-2 Modify the results of Example 11.1-1 for $p(t) = \Pi(r_b t)$, corresponding to an NRZ waveform. In particular, show that the only impulse in $G_x(f)$ is at $f = 0$.

Spectral Shaping by Precoding★

Precoding refers to operations that cause the statistics of the amplitude sequence a_k to differ from the statistics of the message sequence. Usually, the purpose of precoding is to *shape* the power spectrum via $R_a(n)$, as distinguished from $P(f)$. To bring out the potential for statistical spectral shaping, we rewrite Eq. (9) in the form

$$G_x(f) = r|P(f)|^2 \left[R_a(0) + 2 \sum_{n=1}^{\infty} R_a(n) \cos(2\pi nf/r) \right] \tag{13}$$

having drawn upon the property $R_a(-n) = R_a(n)$.

Now suppose that $x(t)$ is to be transmitted over a channel having poor low-frequency response—a voice telephone channel perhaps. With appropriate precoding, we can force the bracketed term in Eq. (13) to equal zero at $f = 0$ and

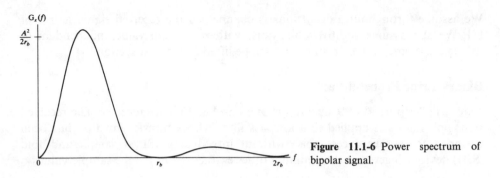

$G_x(f)$

$\frac{A^2}{2r_b}$

0 r_b $2r_b$ f

Figure 11.1-6 Power spectrum of bipolar signal.

thereby eliminate any DC component in $G_x(f)$, irrespective of the pulse spectrum $P(f)$. The *bipolar* signal format back in Fig. 11.1-1c is, in fact, a precoding technique that removes DC content.

The bipolar signal has three amplitude values, $a_k = +A$, 0, and $-A$. If 1s and 0s are equally likely in the message, then the amplitude probabilities are $P(a_k = 0) = 1/2$ and $P(a_k = +A) = P(a_k = -A) = 1/4$, so the amplitude statistics differ from the message statistics. Furthermore, the assumption of uncorrelated message bits leads to the amplitude correlation

$$R_a(n) = \begin{cases} A^2/2 & n = 0 \\ -A^2/4 & n = 1 \\ 0 & n \geq 2 \end{cases} \qquad (14a)$$

Therefore,

$$G_x(f) = r_b |P(f)|^2 \frac{A^2}{2} (1 - \cos 2\pi f/r_b)$$

$$= r_b |P(f)|^2 A^2 \sin^2 \pi f/r_b \qquad (14b)$$

which is sketched in Fig. 11.1-6 taking $p(t) = \Pi(r_b t)$.

Two other precoding techniques that remove DC content are the split-phase Manchester format (Fig. 11.1-1d) and the family of *high density bipolar codes* denoted as HDBn. The HDBn scheme is a bipolar code that also eliminates long signal "gaps" by substituting a special pulse sequence whenever the source produces n successive 0s.

Exercise 11.1-3 Sketch $G_x(f)$ for a bipolar signal with $p(t) = \text{sinc } r_b t$. Then use your sketch to show that $\overline{x^2} = A^2/2$.

11.2 NOISE AND ERRORS

Here we investigate noise, errors, and error probabilities in baseband digital transmission, starting with the binary case and generalizing to M-ary signaling.

We assume throughout a distortionless channel, so the received signal is free of ISI. We also assume additive white noise with zero mean value, independent of the signal. (Some of these restrictions will be lifted in the next section.)

Binary Error Probabilities

Figure 11.2-1 portrays the operations of a baseband binary receiver. The received signal plus noise is applied to a lowpass filter whose transfer function has been designed to remove excess noise without introducing ISI. A sample-and-hold (S/H) device triggered at the optimum times extracts from $y(t)$ the sample values

$$y(t_k) = a_k + n(t_k)$$

Comparing successive values of $y(t_k)$ with a fixed *threshold level* V completes the regeneration process. If $y(t_k) > V$, the comparator goes HIGH to indicate a 1; if $y(t_k) < V$, the comparator goes LOW to indicate 0. The regenerator thereby acts as an *analog-to-digital converter*, converting the noisy analog waveform $y(t)$ into a noiseless digital signal $x_e(t)$ with occasional *errors*.

We begin our analysis taking $x(t)$ to be a *unipolar* signal in which $a_k = A$ represents the message bit 1 and $a_k = 0$ represents the message bit 0. Intuitively, then, the threshold should be set at some intermediate level, $0 < V < A$. The regeneration process is illustrated by the waveforms in Fig. 11.2-2. Errors occur when $a_k = 0$ but a positive noise excursion results in $y(t_k) > V$, or when $a_k = A$ but a negative noise excursion results in $y(t_k) < V$.

To formulate the error probabilities, let the *random variable* Y represent $y(t_k)$ at an arbitrary sampling time and let n represent $n(t_k)$. The probability density function of Y obviously involves the noise PDF, but it also depends upon the presence or absence of the signal pulse. We therefore need to work with *conditional* probabilities. In particular, if H_0 denotes the *hypothesis* or assumption that $a_k = 0$ and $Y = n$, we can write the conditional PDF

$$p_Y(y \mid H_0) = p_N(y) \tag{1a}$$

where $p_N(n)$ is the PDF of the noise alone. Similarly, if H_1 denotes the hypothesis that $a_k = A$ and $Y = A + n$, then

$$p_Y(y \mid H_1) = p_N(y - A) \tag{1b}$$

obtained from the linear transformation of $p_N(n)$ with $n = y - A$.

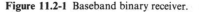

Figure 11.2-1 Baseband binary receiver.

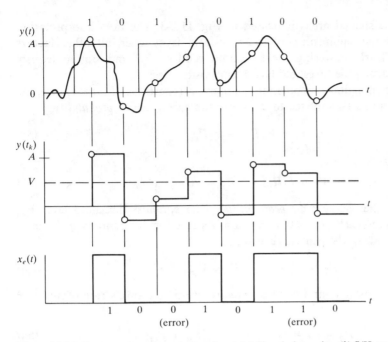

Figure 11.2-2 Regeneration of a unipolar signal. (a) Signal plus noise; (b) S/H output; (c) comparator output.

Figure 11.2-3 shows typical curves of $p_Y(y|H_0)$ and $p_Y(y|H_1)$ along with a proposed threshold V. The comparator implements the following *decision rule*:

$$\text{Choose hypothesis } H_0 \ (a_k = 0) \ \text{ if } Y < V$$

$$\text{Choose hypothesis } H_1 \ (a_k = A) \ \text{ if } Y > V$$

(We ignore the borderline case $Y = V$ whose probability of occurrence will be vanishingly small.) The corresponding regeneration error probabilities are then given by

$$P_{e_0} \triangleq P(Y > V \mid H_0) = \int_V^\infty p_Y(y \mid H_0)\, dy \tag{2a}$$

$$P_{e_1} \triangleq P(Y < V \mid H_1) = \int_{-\infty}^V p_Y(y \mid H_1)\, dy \tag{2b}$$

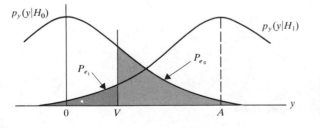

Figure 11.2-3 Conditional PDFs with decision threshold and error probabilities.

equivalent to the shaded areas indicated in Fig. 11.2-3. The area interpretation helps bring out the significance of the threshold level when all other factors remain fixed. Clearly, lowering the threshold reduces P_{e_1} and simultaneously increases P_{e_0}. Raising the threshold has the opposite effect.

But an error in digital transmission is an error, regardless of type. Hence, the threshold level should be adjusted to minimize the *average error probability*

$$P_e = P_0 P_{e_0} + P_1 P_{e_1} \tag{3a}$$

where

$$P_0 = P(H_0) \qquad P_1 = P(H_1) \tag{3b}$$

which stand for the *source digit probabilities*. The *optimum* threshold level V_{opt} must therefore satisfy $dP_e/dV = 0$, and Leibniz's rule for differentiating the integrals in Eq. (2) leads to the general relation

$$P_0 p_Y(V_{opt}|H_0) = P_1 p_Y(V_{opt}|H_1) \tag{4}$$

But we normally expect 1s and 0s to be *equally likely* in a long string of message bits, so

$$P_0 = P_1 = \tfrac{1}{2} \qquad P_e = \tfrac{1}{2}(P_{e_0} + P_{e_1}) \tag{5a}$$

and

$$p_Y(V_{opt}|H_0) = p_Y(V_{opt}|H_1) \tag{5b}$$

We'll work hereafter with the equally likely condition, unless stated otherwise.

Closer examination of Eq. (5b) reveals that V_{opt} corresponds to the point where the conditional PDF curves *intersect*. Direct confirmation of this conclusion is provided by the graphical construction in Fig. 11.2-4 labeled with four relevant areas, α_1 through α_4. The optimum threshold yields $P_{e_1} = \alpha_1 + \alpha_2$, $P_{e_0} = \alpha_3$, and $P_{e_{min}} = \tfrac{1}{2}(\alpha_1 + \alpha_2 + \alpha_3)$. A nonoptimum threshold such as $V < V_{opt}$ yields $P_{e_1} = \alpha_1$ and $P_{e_0} = \alpha_2 + \alpha_3 + \alpha_4$; thus, $P_e = \tfrac{1}{2}(\alpha_1 + \alpha_2 + \alpha_3 + \alpha_4) = P_{e_{min}} + \tfrac{1}{2}\alpha_4 > P_{e_{min}}$.

Next we make the usual assumption that the noise is *gaussian* with zero mean and variance σ^2, so

$$p_N(n) = \frac{1}{\sqrt{2\pi\sigma^2}} e^{-n^2/2\sigma^2}$$

Figure 11.2-4

Substituting this gaussian function into Eqs. (1) and (2) gives

$$P_{e0} = \int_V^\infty p_N(y)\, dy = Q\left(\frac{V}{\sigma}\right) \tag{6a}$$

$$P_{e1} = \int_{-\infty}^V p_N(y - A)\, dy = Q\left(\frac{A - V}{\sigma}\right) \tag{6b}$$

where Q is the area under the gaussian tail as previously defined in Fig. 4.4-2. Since $p_N(n)$ has even symmetry, the conditional PDFs $p_Y(y|H_0)$ and $p_Y(y|H_1)$ intersect at the *midpoint* and $V_{opt} = A/2$ when $P_0 = P_1 = 1/2$. Furthermore, with $V = V_{opt}$ in Eq. (6), $P_{e0} = P_{e1} = Q(A/2\sigma)$ so the optimum threshold yields *equal digit error probabilities* as well as minimizing the net error probability. Thus, $P_e = \frac{1}{2}(P_{e0} + P_{e1}) = P_{e0} = P_{e1}$ and

$$P_e = Q\left(\frac{A}{2\sigma}\right) \tag{7}$$

which is the minimum net error probability for binary signaling in gaussian noise when the source digits are equally likely.

Based on Eq. (7), the plot of the Q function in Table T.6 can be interpreted now as a plot of P_e versus $A/2\sigma$. This plot reveals that P_e drops off dramatically when $A/2\sigma$ increases. For instance, $P_e \approx 2 \times 10^{-2}$ at $A/2\sigma = 2.0$ whereas $P_e \approx 10^{-9}$ at $A/2\sigma = 6.0$.

Although we derived Eq. (7) for the case of a unipolar signal, it also holds in the *polar* case if $a_k = \pm A/2$ so the level spacing remains unchanged. The only difference at the receiver is that $V_{opt} = 0$, midway between the two levels. However, the transmitter needs less signal power to achieve a specified level spacing in the polar signal.

Let's bring out that advantage of polar signaling by expressing A in terms of the average received signal power S_R. If we assume equal digit probabilities and more-or-less rectangular pulses with full-interval duration T_b, then $S_R = A^2/2$ for unipolar signaling while $S_R = A^2/4$ for polar signaling. (These relations should be obvious from the NRZ waveforms in Fig. 11.1-1.) Hence,

$$A = \begin{cases} \sqrt{2S_R} & \text{Unipolar} \\ \sqrt{4S_R} & \text{Polar} \end{cases} \tag{8}$$

and the factor of $\sqrt{2}$ can make a considerable difference in the value of P_e.

Since the noise has zero mean, the variance σ^2 equals the noise power N_R at the output of the filter. Therefore, we can write $A/2\sigma$ in terms of the signal-to-noise power ratio $(S/N)_R = S_R/N_R$, namely

$$\left(\frac{A}{2\sigma}\right)^2 = \frac{A^2}{4N_R} = \begin{cases} \frac{1}{2}(S/N)_R & \text{Unipolar} \\ (S/N)_R & \text{Polar} \end{cases} \tag{9}$$

But Eq. (9) conceals the effect of the signaling rate r_b. In order to pass pulses of duration $T_b = 1/r_b$, the noise-limiting filter must have $B_N \geq r_b/2$, so

$$N_R = \eta B_N \geq \eta r_b/2 \tag{10}$$

Rapid signaling thus requires more signal power to maintain a given error probability P_e.

> **Example 11.2-1** Suppose a computer produces unipolar pulses at the rate $r_b = 10^6$ bps $= 1$ Mbps for transmission over a noisy system with $\eta = 4 \times 10^{-20}$ W/Hz. The error rate is specified to be no greater than one bit per hour, or $P_e \leq 1/3600 r_b \approx 3 \times 10^{-10}$. Table T.6 indicates that we need $A/2\sigma \geq 6.2$, and Eqs. (9) and (10) give the corresponding signal-power requirement
>
> $$S_R = 2\left(\frac{A}{2\sigma}\right)^2 N_R \geq 1.5 \times 10^{-12} = 1.5 \text{ pW}$$
>
> Clearly, any reasonable signal power ensures almost *errorless* transmission insofar as additive noise is concerned. Hardware glitches and other effects would be the limiting factors on system performance.

> **Exercise 11.2-1** Consider a unipolar system with equally likely digits and $(S/N)_R = 50$. Calculate P_{e_0}, P_{e_1}, and P_e when the threshold is set at the non-optimum value $V = 0.4A$. Compare P_e with the minimum value from Eq. (7).

Regenerative Repeaters

Long-haul transmission requires repeaters, be it for analog or digital communication. But unlike analog-message repeaters, digital repeaters can be *regenerative*. If the error probability per repeater is reasonably low and the number of hops m is large, the regeneration advantage turns out to be rather spectacular. This will be demonstrated for the case of polar binary transmission.

When analog repeaters are used and Eq. (11), Sect. 5.4, applies, the final signal-to-noise ratio is $(S/N)_R = (1/m)(S/N)_1$ and

$$P_e = Q\left[\sqrt{\frac{1}{m}\left(\frac{S}{N}\right)_1}\right] \tag{11}$$

where $(S/N)_1$ is the signal-to-noise ratio after one hop. Therefore, the transmitted power *per repeater* must be increased linearly with m just to stay even, a factor not to be sneezed at since, for example, it takes 100 or more repeaters to cross the continent. The $1/m$ term in Eq. (11) stems from the fact that the contaminating noise progressively builds up from repeater to repeater.

In contrast, a regenerative repeater station consists of a complete receiver and transmitter back to back in one package. The receiving portion converts incoming signals to message digits, making a few errors in the process; the digits are then delivered to the transmitting portion, which in turn generates a new signal for transmission to the next station. The regenerated signal is thereby completely stripped of random noise but does contain some errors.

To analyze the performance, let α be the error probability at each repeater, namely,

$$\alpha = Q\left[\sqrt{\left(\frac{S}{N}\right)_1}\right]$$

assuming identical units. As a given digit passes from station to station, it may suffer cumulative conversion errors. If the number of erroneous conversions is *even*, they cancel out, and a correct digit is delivered to the destination. (Note that this is true only for *binary* digits.) The probability of i errors in m successive conversions is given by the *binomial frequency function*, Eq. (1), Sect. 4.4,

$$P_I(i) = \binom{m}{i}\alpha^i(1 - \alpha)^{m-i}$$

Since we have a destination error only when i is *odd*,

$$P_e = \sum_{i \text{ odd}} P_I(i)$$

$$= \binom{m}{1}\alpha(1 - \alpha)^{m-1} + \binom{m}{3}\alpha^3(1 - \alpha)^{m-3} + \cdots$$

$$\approx m\alpha$$

where the approximation applies for $\alpha \ll 1$ and m not too large. Hence,

$$P_e \approx mQ\left[\sqrt{\left(\frac{S}{N}\right)_1}\right] \qquad (12)$$

so P_e increases linearly with m, which generally requires a much smaller power increase to counteract than Eq. (11).

Figure 11.2-5 illustrates the power saving provided by regeneration as a function of m, the error probability being fixed at $P_e = 10^{-5}$. Thus, for example, a 10-station nonregenerative baseband system requires about 8.5 dB more transmitted power (per repeater) than a regenerative system.

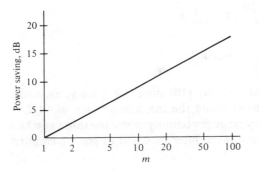

Figure 11.2-5 Power saving gained by m regenerative repeaters when $P_e = 10^{-5}$.

Matched Filtering

Every baseband digital receiver — whether at the destination or part of a regenerative repeater — should include a lowpass filter designed to remove excess noise without introducing ISI. But what's the *optimum* filter design for this purpose? For the case of timelimited pulses in white noise, the answer is a *matched filter*. We'll pursue that case here and develop the resulting minimum error probability for binary signaling in white noise.

Let the received signal be a single timelimited pulse of duration τ centered at time $t = kD$, so

$$x(t) = a_k p(t - kD)$$

where $p(0) = 1$, $p(t) = 0$ for $|t| > \tau/2$, and $\tau \le D$. Maximizing the output ratio $(a_k/\sigma)^2$ at time $t_k = kD + t_d$ will minimize the error probability. As we learned in Sect. 5.4, this maximization calls for a matched filter whose impulse response is proportional to $p(t_d - t)$. In particular, we take

$$h(t) = \frac{1}{\tau_{eq}} p(t_d - t) \tag{13a}$$

with

$$\tau_{eq} = \int_{-\infty}^{\infty} p^2(t) \, dt \qquad t_d = \tau/2 \tag{13b}$$

The delay $t_d = \tau/2$ is the minimum value that yields a causal impulse response, and the proportionality constant $1/\tau_{eq}$ has been chosen so that the peak output amplitude equals a_k. The parameter τ_{eq} can be interpreted from the property that $a_k^2 \tau_{eq}$ equals the *energy* of the pulse $x(t)$.

In absence of noise, the resulting output pulse is $y(t) = h(t) * x(t)$, with peak value $y(t_k) = a_k$ as desired. This peak occurs $\tau/2$ seconds after the peak of $x(t)$. Thus, matched filtering introduces an unavoidable *delay*. However, it does not introduce ISI at the sampling times for adjacent pulses since $y(t) = 0$ outside of $t_k \pm \tau$. Figure 11.2-6 illustrates these points taking a rectangular shape for $p(t)$, in which case $\tau_{eq} = \tau$ and $y(t)$ has a triangular shape.

When $x(t)$ is accompanied by white noise, the output noise power from the matched filter will be

$$N_R = \sigma^2 = \frac{\eta}{2} \int_{-\infty}^{\infty} |H(f)|^2 \, df$$

$$= \frac{\eta}{2} \int_{-\infty}^{\infty} |h(t)|^2 \, dt = \frac{\eta}{2\tau_{eq}} \tag{14}$$

This result agrees with the lower bound in Eq. (10) since, for binary signaling, $\tau_{eq} \le T_b = 1/r_b$. We'll use this result to evaluate the maximum value of $(A/2\sigma)^2$ and the corresponding minimum binary error probability when the noise is white and gaussian and the receiver has an optimum filter matched to the timelimited pulse shape.

Figure 11.2-6 Matched filtering with rectangular pulses. (a) Received pulse; (b) impulse response; (c) output pulse.

Consider a binary transmission system with rate r_b, average received power S_R, and noise density η. We can characterize this system in terms of two new parameters E_b and γ_b defined by

$$E_b \triangleq S_R/r_b \tag{15a}$$

$$\gamma_b \triangleq S_R/\eta r_b = E_b/\eta \tag{15b}$$

The quantity E_b corresponds to the *average energy per bit*, while γ_b represents the ratio of bit energy to noise density†. If the signal consists of timelimited pulses $p(t)$ with amplitude sequence a_k, then

$$E_b = \overline{a_k^2} \int_{-\infty}^{\infty} p^2(t)\, dt = \overline{a_k^2}\tau_{\mathrm{eq}}$$

where $\overline{a_k^2} = A^2/2$ for a unipolar signal or $\overline{a_k^2} = A^2/4$ for a polar signal. Thus, since the output noise power from a matched filter is $\sigma^2 = \eta/2\tau_{\mathrm{eq}}$, we have

$$(A/2\sigma)^2 = \begin{cases} E_b/\eta = \gamma_b & \text{Unipolar} \\ 2E_b/\eta = 2\gamma_b & \text{Polar} \end{cases}$$

† Many authors write this ratio as E_b/N_0 where $N_0 = \eta$.

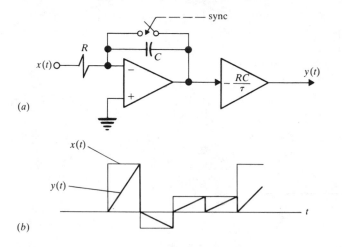

(a)

(b)

Figure 11.2-7 Integrate-and-dump filter. (a) op-amp circuit; (b) polar M-ary waveforms.

and Eq. (7) becomes

$$P_e = \begin{cases} Q(\sqrt{\gamma_b}) & \text{Unipolar} \\ Q(\sqrt{2\gamma_b}) & \text{Polar} \end{cases} \qquad (16)$$

This is the minimum possible error probability, attainable only with matched filtering.

Finally, we should give some attention to the implementation of a matched filter described by Eq. (13). The impulse response for an arbitrary $p(t)$ can be approximated with passive circuit elements, but considerable design effort must be expended to get $h(t) \approx 0$ for $t > \tau$. Otherwise, the filter may produce significant ISI. For a rectangular pulse shape, you can use an active circuit such as the one diagrammed in Fig. 11.2-7a, called an *integrate-and-dump* filter. The op-amp integrator integrates each incoming pulse so that $y(t_k) = a_k$ at the end of the pulse, after which the dump switch resets the integrator to zero — thereby ensuring no ISI at subsequent sampling times. The integrate-and-dump filter is probably the best practical implementation of matched filtering. Figure 11.2-7b illustrates its operation with a polar M-ary waveform.

> **Exercise 11.2-2** Let $x(t)$ be the unipolar RZ signal in Fig. 11.1-1a. (a) Sketch the corresponding output waveform from a matched filter and from an integrate-and-dump filter. (b) Confirm that matched filtering yields $(A/2\sigma)^2 = \gamma_b$ even though $\sigma^2 = \eta r_b$ so $N_R > \eta r_b/2$.

M-ary Error Probabilities

Binary signaling provides the greatest immunity to noise for a given S/N because it has only two amplitude levels — and you can't send information with fewer than two levels. Multilevel M-ary signaling requires more signal power but less transmission bandwidth because the signaling rate will be smaller than the bit rate of an equivalent binary signal. Consequently, M-ary signaling suits applications such as digital transmission over voice channels where the available bandwidth is limited and the signal-to-noise ratio is relatively large.

Here we calculate M-ary error probabilities in zero-mean gaussian noise. We'll take the most common case of polar signaling with an even number of equispaced levels at

$$a_k = \pm A/2, \pm 3A/2, \ldots, \pm(M-1)A/2 \tag{17}$$

We'll also assume equally likely M-ary symbols, so that

$$P_e = \frac{1}{M}(P_{e_0} + P_{e_1} + \cdots + P_{e_{M-1}}) \tag{18}$$

which is the M-ary version of Eq. (5a).

Figure 11.2-8 shows the conditional PDFs for a quaternary ($M = 4$) polar signal plus gaussian noise. The decision rule for regeneration now involves *three* threshold levels, indicated in the figure at $y = -A$, 0, and $+A$. These are the optimum thresholds for minimizing P_e, but they do not result in equal error probabilities for all symbols. For the two extreme levels at $a_k = \pm 3A/2$ we get

$$P_{e_0} = P_{e_3} = Q(A/2\sigma)$$

whereas

$$P_{e_1} = P_{e_2} = 2Q(A/2\sigma)$$

because both positive and negative noise excursions produce errors for the inner levels at $a_k = \pm A/2$. The resulting average error probability is

$$P_e = \frac{1}{4} \times 6Q\left(\frac{A}{2\sigma}\right) = \frac{3}{2}Q\left(\frac{A}{2\sigma}\right)$$

or 50% greater than binary signaling with the same level spacing.

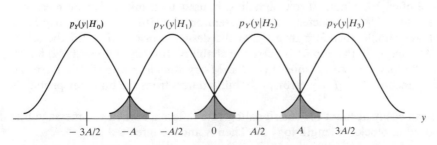

Figure 11.2-8 Conditional PDFs for a quaternary polar signal with gaussian noise.

The foregoing analysis readily generalizes to an arbitrary even value of M with $M - 1$ decision thresholds at

$$y = 0, \pm A, \pm 2A, \ldots, \pm \frac{M-2}{2} A \tag{19}$$

Then $P_{e_0} = P_{e_{M-1}} = Q(A/2\sigma)$ while the $M - 2$ inner levels have doubled error probability, yielding the average error probability

$$P_e = \frac{1}{M} \left[2 \times Q\left(\frac{A}{2\sigma}\right) + (M - 2) \times 2Q\left(\frac{A}{2\sigma}\right) \right]$$

$$= \frac{2M - 2}{M} Q\left(\frac{A}{2\sigma}\right) = 2\left(1 - \frac{1}{M}\right) Q\left(\frac{A}{2\sigma}\right) \tag{20}$$

Equation (20) clearly reduces to Eq. (7) when $M = 2$, whereas $P_e \approx 2Q(A/2\sigma)$ when $M \gg 2$.

Next, we relate $A/2\sigma$ to the signal power and noise density assuming a time-limited pulse shape $p(t)$ so the average energy per M-ary digit is $E_M = \overline{a_k^2} \tau_{eq}$ where

$$\tau_{eq} = \int_{-\infty}^{\infty} p^2(t)\, dt$$

as before. If the M amplitude levels are equally likely and given by Eq. (17), then

$$\overline{a_k^2} = 2 \times \frac{1}{M} \sum_{i=1}^{M/2} (2i - 1)^2 \left(\frac{A}{2}\right)^2 = \frac{M^2 - 1}{12} A^2 \tag{21}$$

Hence, since $S_R = rE_M$,

$$\left(\frac{A}{2\sigma}\right)^2 = \frac{3}{M^2 - 1} \frac{E_M}{\tau_{eq}\, \sigma^2}$$

$$= \frac{3}{M^2 - 1} \frac{1}{r\tau_{eq}} \frac{S_R}{N_R} \leq \frac{6}{M^2 - 1} \frac{S_R}{\eta r} \tag{22}$$

where the upper bound corresponds to $N_R = \eta/2\tau_{eq}$ obtained with matched filtering. Equations (20) and (22) constitute our final result for error probability in a polar M-ary system with gaussian white noise.

More often than not, M-ary signaling is used to transmit *binary* messages and the value of M is selected by the system engineer to best fit the available channel. We should therefore investigate the design considerations in the selection of M, especially the impact on error probability. But Eqs. (20) and (22) fail to tell the full story for two reasons: first, the M-ary signaling rate differs from the bit rate r_b; second, the M-ary error probability differs from the bit error probability.

We can easily account for the signaling-rate difference when the message bits are encoded in blocks of length $\log_2 M$. Then r_b and r are related by

$$r_b = r \log_2 M \tag{23}$$

from Eq. (4), Sect. 11.1. To relate the M-ary symbol error probability P_e to the resulting error probability per bit, we'll assume a *Gray code* and a reasonably large signal-to-noise ratio. Under these conditions a noise excursion seldom goes beyond one amplitude level in the M-ary waveform, which corresponds to just one erroneous bit in the block of $\log_2 M$ bits. Therefore,

$$P_{be} \approx P_e/\log_2 M \tag{24}$$

where P_{be} stands for the equivalent *bit error probability*, also called the bit error rate (BER).

Combining Eqs. (23) and (24) with our previous M-ary expressions, we finally have

$$P_{be} \approx 2 \frac{M-1}{M \log_2 M} Q\left(\frac{A}{2\sigma}\right) \tag{25a}$$

in which

$$\left(\frac{A}{2\sigma}\right)^2 \leq \frac{6}{M^2-1} \frac{S_R}{\eta r} = \frac{6 \log_2 M}{M^2-1} \gamma_b \tag{25b}$$

Notice that the upper bound with matched filtering has been written in terms of $\gamma_b = S_R/\eta r_b = S_R/(\eta r \log_2 M)$. This facilitates the study of M-ary signaling as a function of energy per message bit.

Example 11.2-2 Comparison of binary and M-ary signaling Suppose the channel in question has a fixed signaling rate $r = 3000$ baud $= 3$ kbaud and a fixed signal-to-noise ratio $(S/N)_R = 400 \approx 26$ dB. (These values would be typical of a voice telephone channel, for instance.) We'll assume matched filtering of NRZ rectangular pulses, so $r\tau_{eq} = 1$ and

$$\left(\frac{A}{2\sigma}\right)^2 = \frac{3}{M^2-1} (S/N)_R = \frac{6 \log_2 M}{M^2-1} \gamma_b$$

which follow from Eqs. (22) and (25b).

Binary signaling yields a vanishingly small error probability when $(S/N)_R = 400$, but at the rather slow rate $r_b = r = 3$ kbps. M-ary signaling increases the bit rate, per Eq. (23), but the error probability also increases because the spacing between amplitude levels gets smaller when you increase M with the signal power held fixed. Table 11.2-1 brings out the trade-off between bit rate and error probability for this channel.

Table 11.2-1 M-ary signaling with $r = 3$ kilobaud and $(S/N)_R = 400$

M	r_b(kbps)	$A/2\sigma$	P_{be}
2	3	20.0	3×10^{-89}
4	6	8.9	1×10^{-19}
8	9	4.4	4×10^{-6}
16	12	2.2	7×10^{-3}
32	15	1.1	6×10^{-2}

Table 11.2-2 M-ary signaling with
$r_b = 9$ kbps and $P_{be} = 4 \times 10^{-6}$

M	r (kbaud)	γ_b
2	9.00	10
4	4.50	24
8	3.00	67
16	2.25	200
32	1.80	620

Another type of trade-off is illustrated by Table 11.2-2, where the bit rate and error probability are both held fixed. Increasing M then yields a lower signaling rate r — implying a smaller transmission bandwidth requirement. However, the energy per bit must now be increased to keep the error probability unchanged. Observe that going from $M = 2$ to $M = 32$ reduces r by 1/5 but increases γ_b by more than a factor of 60. This type of trade-off will be reconsidered from the broader viewpoint of information theory in Chap. 15.

Exercise 11.2-3 Consider the three-level bipolar binary format in Fig. 11.1-1c with amplitude probabilities $P(a_k = 0) = 1/2$ and $P(a_k = +A) = P(a_k = -A) = 1/4$. Make a sketch similar to Fig. 11.2-8 and find P_e in terms of A and σ when the decision thresholds are at $y = \pm A/2$. Then calculate S_R and express P_e in a form like Eq. (16).

11.3 BANDLIMITED DIGITAL PAM SYSTEMS

This section develops design procedures for baseband digital systems when the transmission channel imposes a bandwidth limitation. By this we mean that the available transmission bandwidth is not large compared to the desired signaling rate and, consequently, rectangular signaling pulses would be severely distorted. Instead, we must use *bandlimited* pulses specially shaped to avoid ISI.

Accordingly, we begin with Nyquist's strategy for bandlimited pulse shaping. Then we consider the optimum terminal filters needed to minimize error probability. The assumption that the noise has a gaussian distribution with zero mean value will be continued, but we'll allow an arbitrary noise power spectrum. We'll also make allowance for linear transmission distortion, which leads to the subject of equalization for digital systems. The section closes with an introductory discussion of correlative coding techniques that increase the signaling rate on a bandlimited channel.

Nyquist Pulse Shaping

Our presentation of Nyquist pulse shaping will be couched in general terms of M-ary signaling with $M \geq 2$ and symbol interval $D = 1/r$. In order to focus on potential ISI problems at the receiver, we'll let $p(t)$ be the pulse shape at the output of the receiving filter. Again assuming that the transmitter gain compensates for transmission loss, the output waveform in absence of noise is

$$y(t) = \sum_k a_k p(t - t_d - kD)$$

As before, we want $p(t)$ to have the property

$$p(t) = \begin{cases} 1 & t = 0 \\ 0 & t = \pm D, \pm 2D, \dots \end{cases} \tag{1a}$$

which eliminates ISI. Now we impose the additional requirement that the pulse spectrum be *bandlimited*, such that

$$P(f) = 0 \qquad |f| \geq B \tag{1b}$$

where

$$B = \frac{r}{2} + \beta \qquad 0 \leq \beta \leq \frac{r}{2}$$

This spectral requirement permits signaling at the rate

$$r = 2(B - \beta) \qquad B \leq r \leq 2B \tag{2}$$

in which B may be interpreted as the minimum required transmission bandwidth, so that $B_T \geq B$.

Nyquist's *vestigial-symmetry theorem* states that Eq. (1) is satisfied if $p(t)$ has the form

$$p(t) = p_\beta(t) \operatorname{sinc} rt \tag{3a}$$

with

$$\mathcal{F}[p_\beta(t)] = P_\beta(f) = 0 \qquad |f| > \beta$$

$$p_\beta(0) = \int_{-\infty}^{\infty} P_\beta(f)\, df = 1 \tag{3b}$$

Clearly, $p(t)$ has the time-domain properties of Eq. (1a). It also has the frequency-domain properties of Eq. (1b) since

$$P(f) = P_\beta(f) * [(1/r)\Pi(f/r)]$$

and the convolution of two bandlimited spectra produces a new bandlimited spectrum whose bandwidth equals the sum of the bandwidths, namely, $B = \beta + r/2$. Usually we take $p_\beta(t)$ to be an even function so $P_\beta(f)$ is real and even; then $P(f)$ has vestigial symmetry around $f = \pm r/2$, like the symmetry of a vestigial sideband filter.

Infinitely many functions satisfy Nyquist's conditions, including the case when $p_\beta(t) = 1$ so $\beta = 0$ and $p(t) = \operatorname{sinc} rt$, as in Eq. (6), Sect. 11.1. We know that this pulse shape allows bandlimited signaling at the maximum rate $r = 2B$.

However, synchronization turns out to be a very touchy matter because the pulse shape falls off no faster than $1/|t|$ as $|t| \to \infty$. Consequently, a small timing error ϵ results in the sample value

$$y(t_K) = a_K \text{ sinc } r\epsilon + \sum_{k \neq K} a_k \text{ sinc } (KD - kD + r\epsilon)$$

and the ISI in the second term can be quite large.

Synchronization problems are eased by reducing the signaling rate and using pulses with a *cosine rolloff* spectrum. Specifically, if

$$P_\beta(f) = \frac{\pi}{4\beta} \cos \frac{\pi f}{2\beta} \Pi \left(\frac{f}{2\beta} \right) \tag{4a}$$

then

$$P(f) = \begin{cases} \dfrac{1}{r} & |f| < \dfrac{r}{2} - \beta \\[2ex] \dfrac{1}{r} \cos^2 \dfrac{\pi}{4\beta} \left(|f| - \dfrac{r}{2} + \beta \right) & \dfrac{r}{2} - \beta < |f| < \dfrac{r}{2} + \beta \\[2ex] 0 & |f| > \dfrac{r}{2} + \beta \end{cases} \tag{4b}$$

and the corresponding pulse shape is

$$p(t) = \frac{\cos 2\pi\beta t}{1 - (4\beta t)^2} \text{ sinc } rt \tag{5}$$

Plots of $P(f)$ and $p(t)$ are shown in Fig. 11.3-1 for two values of β along with $\beta = 0$. When $\beta > 0$, the spectrum has a smooth rolloff and the leading and trailing oscillations of $p(t)$ decay more rapidly than those of sinc rt.

Further consideration of Eqs. (4) and (5) reveals two other helpful properties of $p(t)$ in the special case when $\beta = r/2$, known as 100% rolloff. The spectrum

Figure 11.3-1 Nyquist pulse shaping. (*a*) Spectra; (*b*) waveforms.

Figure 11.3-2 Baseband waveform for 10110100 using Nyquist pulses with $\beta = r/2$.

then reduces to the *raised cosine* shape

$$P(f) = \frac{1}{r} \cos^2 \frac{\pi f}{2r} = \frac{1}{2r}\left[1 + \cos\left(\frac{\pi f}{r}\right)\right] \qquad |f| \le r \qquad (6a)$$

and

$$p(t) = \frac{\text{sinc } 2rt}{1 - (2rt)^2} \qquad (6b)$$

The half-amplitude width of this pulse exactly equals the symbol interval D, that is, $p(\pm 0.5D) = 1/2$, and there are additional zero crossings at $t = \pm 1.5D$, $\pm 2.5D$, ... A polar signal constructed with this pulse shape will therefore have zero crossings precisely halfway between the pulse centers whenever the amplitude changes polarity. Figure 11.3-2 illustrates this feature with the binary message 10110100. These zero crossings make it a simple task to extract a synchronization signal for timing purposes at the receiver. However, the penalty is a 50% reduction of signaling speed since $r = B$ rather than $2B$. Nyquist proved that the pulse shape defined by Eq. (6) is the *only* one possessing all of the aforementioned properties.

Exercise 11.3-1 Sketch $P(f)$ and find $p(t)$ for the Nyquist pulse generated by taking $P_\beta(f) = (2/r)\Lambda(2f/r)$. Compare your results with Eq. (6).

Optimum Terminal Filters

Having abandoned rectangular pulses, we must likewise abandon the conventional matched filter and reconsider the design of the optimum receiving filter that minimizes error probability. This turns out to be a relatively straightforward problem under the following reasonable conditions:

1. The signal format is polar, and the amplitudes a_k are uncorrelated and equally likely.
2. The transmission channel is linear but not necessarily distortionless.
3. The filtered output pulse $p(t)$ is to be Nyquist shaped.
4. The noise is additive and has a zero-mean gaussian distribution but may have a nonwhite power spectrum.

To allow for possible channel distortion and/or nonwhite noise, our optimization must involve filters at both the transmitter and receiver. As a bonus, the source waveform $x(t)$ may have a more-or-less arbitrary pulse shape $p_x(t)$.

Figure 11.3-3

Figure 11.3-3 lays out the system diagram, including a transmitting filter function $H_T(f)$, a channel function $H_C(f)$, and a receiving filter function $H_R(f)$. The input signal has the form

$$x(t) = \sum_k a_k p_x(t - kD) \tag{7a}$$

and its power spectrum is

$$G_x(f) = \sigma_a^2 r |P_x(f)|^2 \tag{7b}$$

where $P_x(f) = \mathscr{F}[p_x(t)]$ and

$$\sigma_a^2 = \overline{a_k^2} = \frac{M^2 - 1}{12} A^2 \tag{7c}$$

These relations follow from Eq. (12), Sect. 11.1, and Eq. (21), Sect. 11.2, with our stated conditions on a_k. Thus, the transmitted signal power will be

$$
\begin{aligned}
S_T &= \int_{-\infty}^{\infty} |H_T(f)|^2 G_x(f)\, df \\
&= \frac{M^2 - 1}{12} A^2 r \int_{-\infty}^{\infty} |H_T(f)P_x(f)|^2\, df
\end{aligned} \tag{8}
$$

a result we'll need shortly.

At the output of the receiving filter we want the input pulse $p_x(t)$ to produce a Nyquist-shaped pulse $p(t - t_d)$, where t_d represents any transmission time delay. The transfer functions in Fig. 11.3-3 must therefore obey the relationship

$$P_x(f)H_T(f)H_C(f)H_R(f) = P(f)e^{-j\omega t_d} \tag{9}$$

so both terminal filters help shape $p(t)$. But only the receiving filter controls the output noise power

$$N_R = \sigma^2 = \int_{-\infty}^{\infty} |H_R(f)|^2 G_n(f)\, df \tag{10}$$

where $G_n(f)$ is the noise power spectrum at the input to the receiver.

Equations (7)–(10) constitute the information relevant to our design problem. Specifically, since the error probability decreases as $A/2\sigma$ increases, we seek the terminal filters that maximize $(A/2\sigma)^2$ subject to two constraints: (1) the transmitted power must be held fixed at some specified value S_T, and (2) the filter transfer functions must satisfy Eq. (9).

We incorporate Eq. (9) and temporarily eliminate $H_T(f)$ by writing

$$|H_T(f)| = \frac{|P(f)|}{|P_x(f)H_C(f)H_R(f)|} \tag{11}$$

Then we use Eqs. (8) and (10) to express $(A/2\sigma)^2$ as

$$\left(\frac{A}{2\sigma}\right)^2 = \frac{3S_T}{(M^2 - 1)r} \frac{1}{I_{HR}} \tag{12a}$$

where

$$I_{HR} = \int_{-\infty}^{\infty} |H_R(f)|^2 G_n(f) \, df \int_{-\infty}^{\infty} \frac{|P(f)|^2}{|H_C(f)H_R(f)|^2} \, df \tag{12b}$$

Maximizing $(A/2\sigma)^2$ thus boils down to minimizing the product of integrals I_{HR}, in which $H_R(f)$ is the only function under our control.

Now observe that Eq. (12b) has the form of the right-hand side of Schwarz's inequality as stated in Eq. (17), Sect. 3.5. The minimum value of I_{HR} therefore occurs when the two integrands are proportional. Consequently, the optimum receiving filter has

$$|H_R(f)|^2 = \frac{g|P(f)|}{\sqrt{G_n(f)}|H_C(f)|} \tag{13a}$$

where g is an arbitrary gain constant. Equation (11) then gives the optimum transmitting filter characteristic

$$|H_T(f)|^2 = \frac{|P(f)|\sqrt{G_n(f)}}{g|P_x(f)|^2|H_C(f)|} \tag{13b}$$

These expressions specify the optimum amplitude ratios for the terminal filters. Note that the receiving filter deemphasizes those frequencies where $G_n(f)$ is large, and the transmitting filter supplies the corresponding preemphasis. The phase shifts are arbitrary, providing that they satisfy Eq. (9).

Substituting Eq. (13) into Eq. (12) yields our final result

$$\left(\frac{A}{2\sigma}\right)_{\text{max}}^2 = \frac{3S_T}{(M^2 - 1)r} \left[\int_{-\infty}^{\infty} \frac{|P(f)|\sqrt{G_n(f)}}{|H_C(f)|} \, df\right]^{-2} \tag{14}$$

from which the error probability can be calculated using Eq. (20), Sect. 11.2. As a check of Eq. (14), take the case of white noise with $G_n(f) = \eta/2$ and a distortionless channel with transmission loss L so $|H_C(f)|^2 = 1/L$; then

$$\left(\frac{A}{2\sigma}\right)_{\text{max}}^2 = \frac{6S_T/L}{(M^2 - 1)r\eta} \left[\int_{-\infty}^{\infty} |P(f)| \, df\right]^{-2}$$

But since $S_T/L = S_R$ and Nyquist-shaped pulses have

$$\int_{-\infty}^{\infty} |P(f)| \, df = 1$$

we thus obtain

$$\left(\frac{A}{2\sigma}\right)_{\text{max}}^2 = \frac{6}{M^2 - 1} \frac{S_R}{\eta r} = \frac{6 \log_2 M}{M^2 - 1} \gamma_b$$

which confirms that the optimum terminal filters yield the same upper bound as matched filtering — see Eqs. (22) and (25), Sect. 11.2.

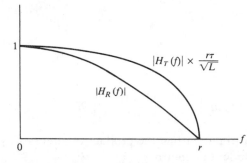

Figure 11.3-4 Amplitude ratio of optimum filters in Example 11.3-1.

Example 11.3-1 Consider a system with white noise, transmission loss L, and a distortionless channel response over $|f| \leq B_T$ where $B_T \geq r$. This transmission bandwidth allows us to use the pulse shape $p(t)$ in Eq. (6), thereby simplifying synchronization. Simplicity also suggests using the rectangular input pulse $p_x(t) = \Pi(t/\tau)$ with $\tau \leq 1/r$, so $P_x(f) = \tau \, \text{sinc} \, f\tau$. Taking the gain constant g in Eq. (13) such that $|H_R(0)| = 1$, we have

$$|H_R(f)| = \cos \frac{\pi f}{2r} \qquad |H_T(f)| = \sqrt{L} \, \frac{\cos (\pi f/2r)}{r\tau \, \text{sinc} \, f\tau} \qquad |f| \leq r$$

as plotted in Fig. 11.3-4. Notice the slight high-frequency rise in $|H_T(f)|$ compared to $|H_R(f)|$. If the input pulses have a small duration $\tau \ll 1/r$, then this rise becomes negligible and $|H_T(f)| \approx |H_R(f)|$, so one circuit design serves for both filters.

Exercise 11.3-2 Carry out the details going from Eq. (12) to Eqs. (13) and (14).

Equalization

Regardless of which particular pulse shape has been chosen, some amount of residual ISI inevitably remains in the output signal as a result of imperfect filter design, incomplete knowledge of the channel characteristics, etc. Hence, an adjustable equalizing filter is often inserted between the receiving filter and the regenerator. These "mop-up" equalizers usually have the structure of a transversal filter previously discussed in Sect. 3.2 relative to linear distortion of analog signals. However, mop-up equalization of digital signals involves different design strategies that deserve attention here.

Figure 11.3-5 shows a transversal equalizer with $2N + 1$ taps and total delay $2ND$. The distorted pulse shape $\tilde{p}(t)$ at the input to the equalizer is assumed to have its peak at $t = 0$ and ISI on both sides. The equalized output pulse will be

$$p_{eq}(t) = \sum_{n=-N}^{N} c_n \, \tilde{p}(t - nD - ND)$$

Figure 11.3-5 Transversal equalizer with $2N + 1$ taps.

and sampling at $t_k = kD + ND$ yields

$$p_{eq}(t_k) = \sum_{n=-N}^{N} c_n \tilde{p}(kD - nD) = \sum_{n=-N}^{N} c_n \tilde{p}_{k-n} \tag{15}$$

where we've introduced the shorthand notation $\tilde{p}_{k-n} = \tilde{p}[(k - n)D]$. Equation (15) thus takes the form of a *discrete convolution*.

Ideally, we would like the equalizer to eliminate all ISI, resulting in

$$p_{eq}(t_k) = \begin{cases} 1 & k = 0 \\ 0 & k \neq 0 \end{cases}$$

But this cannot be achieved, in general, because the $2N + 1$ tap gains are the only variables at our disposal. We might settle instead for choosing the tap gains such that

$$p_{eq}(t_k) = \begin{cases} 1 & k = 0 \\ 0 & k = \pm 1, \pm 2, \ldots, \pm N \end{cases} \tag{16}$$

thereby forcing N zero values on each side of the peak of $p_{eq}(t)$. The corresponding tap gains are computed from Eqs. (15) and (16) combined in the matrix equation

$$\begin{bmatrix} \tilde{p}_0 & \cdots & \tilde{p}_{-2N} \\ \vdots & & \vdots \\ \tilde{p}_{N-1} & \cdots & \tilde{p}_{-N-1} \\ \tilde{p}_N & \cdots & \tilde{p}_{-N} \\ \tilde{p}_{N+1} & \cdots & \tilde{p}_{-N+1} \\ \vdots & & \vdots \\ \tilde{p}_{2N} & \cdots & \tilde{p}_0 \end{bmatrix} \begin{bmatrix} c_{-N} \\ \vdots \\ c_{-1} \\ c_0 \\ c_1 \\ \vdots \\ c_N \end{bmatrix} = \begin{bmatrix} 0 \\ \vdots \\ 0 \\ 1 \\ 0 \\ \vdots \\ 0 \end{bmatrix} \tag{17}$$

Equation (17) describes a *zero-forcing equalizer*. This equalization strategy is optimum in the sense that it minimizes the peak intersymbol interference, and it· has the added advantage of simplicity. Other optimization criteria lead to different strategies with more complicated tap-gain relationships.

When the transmission channel is a switched telephone link or a radio link with slowly changing conditions, the values of $\tilde{p}_k$ will not be known in advance. The tap gains must then be adjusted on-line using a training sequence transmitted before the actual message sequence. An *adaptive equalizer* incorporates a microprocessor for automatic rather than manual tap-gain adjustment. More sophisticated versions adjust themselves continuously using error measures derived from the message sequence. Further details regarding adaptive equalization are given by Haykin (1983, Sect. 9.6) and Proakis (1983, Chap. 6).

Mop-up equalization, whether fixed or adaptive, does have one hidden catch in that the equalizer somewhat increases the noise power at the input to the regenerator. This increase is usually more than compensated for by the reduction of ISI.

Example 11.3-2 Suppose a three-tap zero forcing equalizer is to be designed for the distorted pulse plotted in Fig. 11.3-6a. Inserting the values of $\tilde{p}_k$ into Eq. (17) with $N = 1$, we have

$$\begin{bmatrix} 1.0 & 0.1 & 0.0 \\ -0.2 & 1.0 & 0.1 \\ 0.1 & -0.2 & 1.0 \end{bmatrix} \begin{bmatrix} c_{-1} \\ c_0 \\ c_1 \end{bmatrix} = \begin{bmatrix} 0 \\ 1 \\ 0 \end{bmatrix}$$

(a)

(b)

Figure 11.3-6 (a) Distorted pulse; (b) equalized pulse.

Therefore,

$$c_{-1} = -0.096 \qquad c_0 = 0.96 \qquad c_1 = 0.2$$

and the corresponding sample values of $p_{eq}(t)$ are plotted in Fig. 11.3-6b with an interpolated curve. As expected, there is one zero on each side of the peak. However, zero forcing has produced some small ISI at points further out where the unequalized pulse was zero.

Correlative Coding★

Correlative coding, also known as *partial-response signaling*, is a strategy for bandlimited transmission at $r = 2B$ that avoids the problems associated with $p(t) = \text{sinc } rt$. The strategy involves two key functional operations, correlative filtering and digital precoding. *Correlative filtering* purposely introduces controlled intersymbol interference, resulting in a pulse train with an increased number of amplitude levels and a correlated amplitude sequence. Nyquist's signaling-rate limitation no longer applies because the correlated symbols are not independent, and therefore higher signaling rates are possible. *Digital precoding* of the message sequence before waveform generation facilitates message recovery from the correlative-filtered pulse train.

Figure 11.3-7a shows the general model of a transmission system with correlative coding, omitting noise. The digital precoder takes the sequence of message symbols m_k and produces a coded sequence m'_k applied to an impulse generator. (In practice, the impulses would be replaced by short rectangular pulses with duration $\tau \ll D$.) The resulting input signal is an impulse train

$$x(t) = \sum_k a_k \, \delta(t - kD)$$

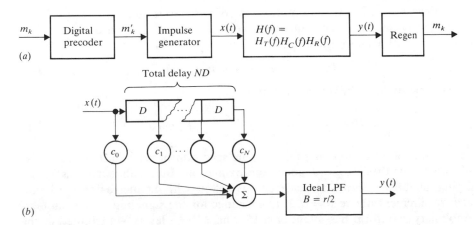

Figure 11.3-7 (a) Transmission system with correlative coding; (b) equivalent correlative filter.

whose weights a_k represent the m'_k. The terminal filters and channel have the overall transfer function $H(f)$, producing the output waveform

$$y(t) = \sum_k a_k h(t - kD) \tag{18}$$

where $h(t) = \mathcal{F}^{-1}[H(f)]$.

Although a correlative filter does not appear as a distinct block in Fig. 11.3-7a, the transfer function $H(f)$ must be equivalent to that of Fig. 11.3-7b — which consists of a transversal filter and an ideal LPF. The transversal filter has total delay ND and $N + 1$ tap gains. Since the impulse response of the LPF is sinc rt and $r = 1/D$, the cascade combination yields the overall impulse response

$$h(t) = \sum_{n=0}^{N} c_n \text{ sinc } (rt - n) \tag{19}$$

Hence, Eq. (18) becomes

$$y(t) = \sum_k a_k \left[\sum_{n=0}^{N} c_n \text{ sinc } (rt - n - k) \right]$$

$$= \sum_k a'_k \text{ sinc } (rt - k) \tag{20a}$$

where

$$a'_k \triangleq c_0 a_k + c_1 a_{k-1} + \cdots + c_N a_{k-N} = \sum_{n=0}^{N} c_n a_{k-n} \tag{20b}$$

Message regeneration must then be based on the sample values $y(t_k) = a'_k$.

Equation (20) brings out the fact that correlative filtering changes the amplitude sequence a_k into the modified sequence a'_k. We say that this sequence has a *correlation span* of N symbols, since each a'_k depends on the previous N values of a_k. Furthermore, when the a_k sequence has M levels, the a'_k sequence has $M' > M$ levels. To demonstrate that these properties of correlative filtering lead to practical bandlimited transmission at $r = 2B$, we must look at a specific case.

Duobinary signaling is the simplest type of correlative coding, having $M = 2$, $N = 1$, and $c_0 = c_1 = 1$. The equivalent correlative filter is diagrammed in Fig. 11.3-8 along with its impulse response

$$h(t) = \text{sinc } r_b t + \text{sinc } (r_b t - 1) \tag{21a}$$

and the magnitude of the transfer function

$$H(f) = \frac{2}{r_b} \cos \frac{\pi f}{r_b} e^{-j\pi f/r_b} \qquad |f| \leq r_b/2 \tag{21b}$$

The smooth rolloff of $H(f)$ is similar to the spectrum of a Nyquist-shaped pulse and can be synthesized to a good approximation. But, unlike Nyquist pulse shaping, duobinary signaling achieves this rolloff without increasing the bandwidth requirement above $B = r_b/2$. In exchange for the signaling rate advantage, a duobinary waveform has intentional ISI and $M' = 3$ levels — attributed to the property of the impulse response that $h(t) = 1$ at both $t = 0$ and $t = T_b$.

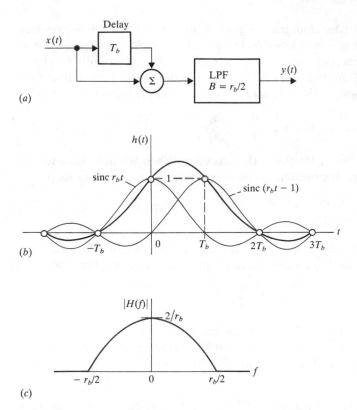

(a)

(b)

(c)

Figure 11.3-8 Duobinary signaling. (a) Equivalent correlative filter; (b) impulse response; (c) amplitude ratio.

To bring out the ISI effect, let the amplitude sequence a_k be related to the precoded binary message sequence m'_k by

$$a_k = (m'_k - 1/2)A = \begin{cases} +A/2 & m'_k = 1 \\ -A/2 & m'_k = 0 \end{cases} \tag{22}$$

which is equivalent to a polar binary format with level spacing A. Equation (20) then gives the corresponding output levels

$$y(t_k) = a'_k = a_k + a_{k-1} = (m'_k + m'_{k-1} - 1)A \tag{23a}$$

$$= \begin{cases} +A & m'_k = m'_{k-1} = 1 \\ 0 & m'_k \neq m'_{k-1} \\ -A & m'_k = m'_{k-1} = 0 \end{cases} \tag{23b}$$

In principle, you could use Eq. (23b) to recover m'_k from $y(t_k)$ if you have previously recovered m'_{k-1}. However, when noise causes an erroneous value of m'_{k-1}, all subsequent message bits will be in error until the next noise-induced error — a phenomenon called *error propagation*.

The digital precoder for duobinary signaling shown in Fig. 11.3-9a prevents error propagation and makes it possible to recover the input message sequence m_k from $y(t_k)$. The precoder consists of an exclusive-OR gate with feedback through a D-type flip-flop. Figure 11.3-9b lists the coding truth table along with the algebraic sum $m'_k + m'_{k-1}$ that appears in Eq. (23a). Substitution now yields

$$y(t_k) = \begin{cases} \pm A & m_k = 0 \\ 0 & m_k = 1 \end{cases} \tag{24}$$

which does not involve m_{k-1} thanks to the precoder. When $y(t)$ includes additive gaussian white noise, the appropriate decision rule for message regeneration is:

$$\text{Choose } m_k = 0 \quad \text{if} \quad |y| > A/2$$
$$\text{Choose } m_k = 1 \quad \text{if} \quad |y| < A/2$$

This rule is easily implemented with a rectifier and a single decision threshold set at $A/2$. Optimum terminal filter design gives the minimum error probability

$$P_e = \frac{3}{2} Q\left(\frac{\pi}{4}\sqrt{2\gamma_b}\right) \tag{25}$$

which is somewhat higher than that of a polar binary system.

When the transmission channel has poor DC response, *modified duobinary signaling* may be employed. The correlative filter has $N = 2$, $c_0 = 1$, $c_1 = 0$, and $c_2 = -1$, so that

$$H(f) = \frac{2j}{r_b} \sin \frac{2\pi f}{r_b} e^{-j2\pi f/r_b} \qquad |f| \le r_b/2 \tag{26}$$

Figure 11.3-10 shows $|H(f)|$ and the block diagram of the correlative filter. The precoder takes the form of Fig. 11.3-9 with two flip-flops in series to feed back m'_{k-2}. If the pulse generation is in accordance with Eq. (22), then

$$y(t_k) = a_k - a_{k-2} = (m'_k - m'_{k-2})A$$
$$= \begin{cases} 0 & m_k = 0 \\ \pm A & m_k = 1 \end{cases} \tag{27}$$

which can be compared with Eqs. (23) and (24).

m_k	m'_{k-1}	m'_k	$m'_k + m'_{k-1}$
0	0	0	0
0	1	1	2
1	0	1	1
1	1	0	1

(a)   (b)

Figure 11.3-9 (a) Digital precoder for duobinary signaling; (b) truth table.

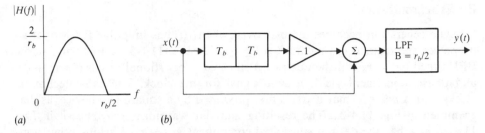

(a)   (b)

Figure 11.3-10 Correlative filter for modified duobinary signaling. (a) Amplitude ratio of correlative filter; (b) block diagram.

Exercise 11.3-3 Construct a truth table like Fig. 11.3-9b for the case of modified duobinary signaling and use it to obtain Eq. (27).

11.4 SYNCHRONIZATION TECHNIQUES

Synchronization is the art of making clocks tick together. The clocks in a digital communication system are at the transmitter and receiver, and allowance must be made for the transmission time delay between them. Besides *symbol* synchronization, most systems also require *frame* synchronization to identify the start of a message or various subdivisions within the message sequence. Additionally, carrier synchronization is essential for digital transmission with coherent carrier modulation — a topic to be discussed in Chap. 14.

Here we'll consider symbol and frame synchronization in baseband binary systems. Our attention will be focused on extracting synchronization from the received signal itself, rather than using an auxiliary sync signal. By way of an overview, Fig. 11.4-1 illustrates the position of the *bit synchronizer* relative to the clock and regenerator. Framing information is usually derived from the regenerated message and the clock, as indicated by the location of the *frame synchronizer*. We'll look at typical synchronization techniques, along with the related topics of shift-register operations for message scrambling and pseudonoise (PN) sequences for framing purposes.

Our coverage of synchronization will be primarily descriptive and illustrative. Detailed treatments of digital synchronization with additive noise are given by Stiffler (1971) and Gardner and Lindsey (1980).

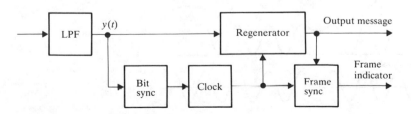

Figure 11.4-1 Synchronization in a binary receiver.

Bit Synchronization

Bit-sync generation becomes almost trivial when $y(t)$ has unipolar RZ format so its power spectrum $G_y(f)$ includes $\delta(f \pm r_b)$, like Fig. 11.1-5. A PLL or narrow BPF tuned to $f = r_b$ will then extract a sinusoid proportional to $\cos(2\pi r_b t + \phi)$, and phase adjustment yields a sync signal for the clock. The same technique works with a *polar* format if $y(t)$ is first processed by a square-law device, as diagrammed in Fig. 11.4-2a. The resulting unipolar waveform $y^2(t)$ shown in Fig. 11.4-2b now has the desired sinusoidal component at $f = r_b$. Various other nonlinear polar-to-unipolar operations on $y(t)$ achieve like results in open-loop bit synchronizers.

But a closed-loop configuration that incorporates the clock in a feedback loop provides more reliable synchronization. Figure 11.4-3 gives the diagram and explanatory waveforms for a representative closed-loop bit synchronizer. Here a zero-crossing detector generates a rectangular pulse with half-bit duration $T_b/2$ starting at each zero-crossing in $y(t)$. The pulsed waveform $z(t)$ then multiplies the square-wave clock signal $c(t)$ coming back from the *voltage-controlled clock* (VCC). The control voltage $v(t)$ is obtained by integrating and lowpass-filtering the product $z(t)c(t)$. The loop reaches steady state conditions when the edges of $c(t)$ and $z(t)$ are synchronized and offset by $T_b/4$, so the product has zero area and $v(t)$ remains constant. Practical implementations of this system usually feature digital components in place of the analog multiplier and integrator.

Both of the foregoing techniques work best when the zero-crossings of $y(t)$ are spaced by integer multiples of T_b. Otherwise, the synchronization will suffer from timing jitter. An additional problem arises if the message includes a long string of 1s or 0s, so $y(t)$ has no zero-crossings, and synchronism may be lost. Message scramblers discussed shortly help alleviate this problem.

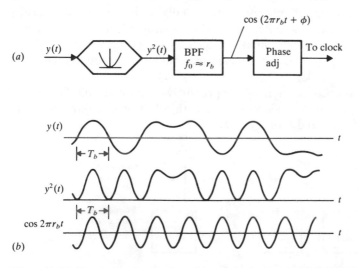

Figure 11.4-2 Bit synchronization by polar to unipolar conversion. (*a*) Block diagram; (*b*) waveforms.

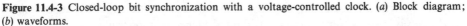

Figure 11.4-3 Closed-loop bit synchronization with a voltage-controlled clock. (*a*) Block diagram; (*b*) waveforms.

A different approach to synchronization, independent of zero-crossings, relies on the fact that a properly filtered digital signal has peaks at the optimum sampling times and is reasonably symmetric on either side. Thus, if t_k is synchronized and $\delta < T_b/2$, then

$$| y(t_k - \delta)| \approx | y(t_k + \delta)| < | y(t_k)|$$

However, a late sync signal produces the situation shown in Fig. 11.4-4*a* where $| y(t_k - \delta)| > | y(t_k + \delta)|$, while an early sync signal would result in $| y(t_k - \delta)| < |y(t_k + \delta)|$. The *early-late synchronizer* in Fig. 11.4-4*b* uses these properties to develop the control voltage for a VCC in a feedback loop. A late sync signal results in $v(t) = | y(t_k - \delta)| - | y(t_k + \delta)| > 0$, which speeds up the clock, and conversely for an early sync signal.

Scramblers and PN Sequence Generators

Scrambling is a coding operation applied to the message at the transmitter that "randomizes" the bit stream, eliminating long strings of like bits that might impair receiver synchronization. Scrambling also eliminates most periodic bit

(a)

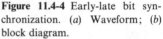

(b)

Figure 11.4-4 Early-late bit synchronization. (a) Waveform; (b) block diagram.

patterns that could produce undesirable discrete-frequency components in the power spectrum. Needless to say, the scrambled sequence must be unscrambled at the receiver so as to preserve overall *bit sequence transparency*.

Simple but effective scramblers and unscramblers are built from tapped *shift registers* having the generic form of Fig. 11.4-5, the digital counterpart of a tapped delay line. Successive bits from the binary input sequence b_k enter the register and shift from one stage to the next at each tick of the clock. The output b_k' is formed by combining the bits in the register through a set of tap gains and mod-2 adders, yielding

$$b_k' = \alpha_1 b_{k-1} \oplus \alpha_2 b_{k-2} \oplus \cdots \oplus \alpha_n b_{k-n} \tag{1}$$

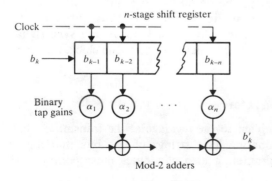

Figure 11.4-5 Tapped shift register.

The tap gains themselves are binary digits, so $\alpha_1 = 1$ simply means a direct connection while $\alpha_1 = 0$ means no connection. The symbol $\oplus$ stands for *modulo-2 addition*, defined by the properties

$$b_1 \oplus b_2 = \begin{cases} 0 & b_1 = b_2 \\ 1 & b_1 \neq b_2 \end{cases} \tag{2a}$$

and

$$b_1 \oplus b_2 \oplus b_3 = (b_1 \oplus b_2) \oplus b_3 = b_1 \oplus (b_2 \oplus b_3) \tag{2b}$$

where b_1, b_2, and b_3 are arbitrary binary digits. Mod-2 addition is implemented with exclusive-OR gates, and obeys the rules of ordinary addition except that $1 \oplus 1 = 0$.

Figure 11.4-6 shows an illustrative scrambler and unscrambler, each employing a 4-stage shift register with tap gains $\alpha_1 = \alpha_2 = 0$ and $\alpha_3 = \alpha_4 = 1$. (For clarity, we omit the clock line here and henceforth.) The binary message sequence m_k at the input to the scrambler is mod-2 added to the register output m_k'' to form the scrambled message m_k' which is also fed back to the register input. Thus, $m_k'' = m_{k-3}' \oplus m_{k-4}'$ and

$$m_k' = m_k \oplus m_k'' \tag{3a}$$

The unscrambler has essentially the reverse structure of the scrambler and reproduces the original message sequence, since

$$m_k' \oplus m_k'' = (m_k \oplus m_k'') \oplus m_k''$$
$$= m_k \oplus (m_k'' \oplus m_k'') = m_k \oplus 0 = m_k \tag{3b}$$

Equations (3a) and (3b) hold for any shift-register configuration as long as the scrambler and unscrambler have identical registers.

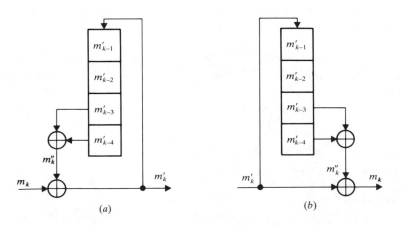

(a)   (b)

Figure 11.4-6 (a) Binary scrambler; (b) unscrambler.

Table 11.4-1

Registers contents	m'_{k-1}	0	1	0	1	0	1	1	1	1	0	0	0	1	0
	m'_{k-2}	0	0	1	0	1	0	1	1	1	1	0	0	0	1
	m'_{k-3}	0	0	0	1	0	1	0	1	1	1	1	0	0	0
	m'_{k-4}	0	0	0	0	1	0	1	0	1	1	1	1	0	0
Register output	m''_k	0	0	0	1	1	1	1	1	0	0	0	1	0	0
Input sequence	m_k	1	0	1	1	0	0	0	0	0	0	0	0	0	1
Output sequence	m'_k	1	0	1	0	1	1	1	1	0	0	0	1	0	1

The scrambling action does, of course, depend on the shift-register configuration. Table 11.4-1 portrays the scrambling produced by our illustrative scrambler when the initial state of the register is all 0s. Note that the string of nine 0s in m_k has been eliminated in m'_k. Nonetheless, there may be some specific message sequence that will result in a long string of like bits in m'_k. Of more serious concern is *error propagation* at the unscrambler, since one erroneous bit in m'_k will cause several output bit errors. Error propagation stops when the unscrambler register is full of correct bits.

Next, in preparation for the subsequent discussion of frame synchronization, we consider shift-register *sequence generation*. When a shift register has a nonzero initial state and the output is fed back to the input, the unit acts as a periodic sequence generator. Take Fig. 11.4-7, for example, where a 3-stage register with initial state 111 produces the 7-bit sequence 1110010 which repeats periodically thereafter. In general, the longest possible sequence period from a register with n stages is

$$N = 2^n - 1 \tag{4}$$

and the corresponding output is called a *maximal-length* or *pseudonoise* (PN) sequence. Figure 11.4-7 is, in fact, a PN sequence generator with $n = 3$ and $N = 7$.

The name pseudonoise comes from the *correlation* properties of PN sequences. To develop this point, let a PN sequence s_k be used to form the binary polar NRZ signal

$$s(t) = \sum_k (2s_k - 1)p(t - kT_b) \tag{5}$$

Figure 11.4-7 Shift-register sequence generator.

Figure 11.4-8 Autocorrelation of a PN sequence.

where $p(t)$ is a rectangular pulse and the amplitude of the kth pulse is

$$2s_k - 1 = \begin{cases} +1 & s_k = 1 \\ -1 & s_k = 0 \end{cases} \tag{6}$$

The signal $s(t)$ is deterministic and periodic, with period NT_b, and has a periodic autocorrelation function given by

$$R_s(\tau) = [(N + 1)\Lambda(\tau/T_b) - 1]/N \qquad |\tau| \le NT_b/2 \tag{7}$$

which is plotted in Fig. 11.4-8. If N is very large and T_b very small, then

$$R_s(\tau) \approx T_b\,\delta(\tau) - 1/N \qquad |\tau| \le NT_b/2$$

so the PN signal acts essentially like *white noise* with a small DC component. This noise-like correlation property leads to practical applications in test instruments and radar ranging, as well as spread spectrum communication described in Sect. 9.3, and digital framing, to which we now turn.

Frame Synchronization

A digital receiver needs to know when a signal is present. Otherwise, the input noise alone may produce random output bits that could be mistaken for a message. Therefore, identifying the start of a message is one aspect of frame synchronization. Another aspect is identifying subdivisions or frames within the message. To facilitate frame synchronization, binary transmission usually includes special N-bit *sync words* as represented in Fig. 11.4-9. The initial *prefix* consists of several repetitions of the sync word, which marks the begining of transmission and allows time for bit-sync acquisition. The prefix is followed by a different codeword labeling the start of the message itself. Frames are labeled by sync words inserted periodically in the bit stream.

Figure 11.4-9

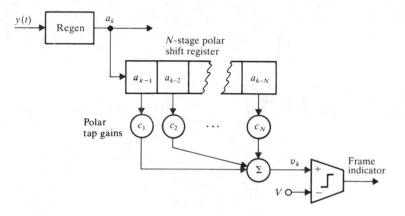

Figure 11.4-10 Frame synchronizer.

The elementary frame synchronizer in Fig. 11.4-10 is designed to detect a sync word $s_1 s_2 \cdots s_N$ whenever it appears in the regenerated sequence m_k. Output bits with the polar format

$$a_k = 2m_k - 1 = \pm 1$$

are loaded into an N-stage polar shift register having polar tap gains given by

$$c_i = 2s_{N+1-i} - 1 \tag{8}$$

This awkward-looking expression simply states that the gains equal the sync-word bits in polar form and reverse order, that is, $c_1 = 2s_N - 1$ while $c_N = 2s_1 - 1$. The tap-gain outputs are summed *algebraically* to form

$$v_k = \sum_{i=1}^{N} c_i a_{k-i} \tag{9}$$

This voltage is compared with a threshold voltage V, and the frame-sync indicator goes HIGH when $v_k > V$.

If the register word is identical to the sync word, then $a_{k-i} = c_i$ so $c_i a_{k-i} = c_i^2 = 1$ and $v_k = N$. If the register word differs from the sync word in just one bit, then $v_k = N - 2$ (why?). Setting the threshold voltage V slightly below $N - 2$ thus allows detection of error-free sync words and sync words with one bit error. Sync words with two or more errors go undetected, but that should be an unlikely event with any reasonable value of P_e. *False* frame indication occurs when N or $N - 1$ successive message bits match the sync-word bits. The probability of this event is given by the binomial model as

$$P_{ff} = \binom{N}{0}\left(\frac{1}{2}\right)^{N} + \binom{N}{1}\left(\frac{1}{2}\right)^{N-1}\left(\frac{1}{2}\right) = (1 + N)2^{-N} \tag{10}$$

assuming equally likely 1s and 0s in the bit stream.

Further examination of Eqs. (8) and (9) reveals that the frame synchronizer calculates the *cross-correlation* between the bit stream passing through the register and the sync word, represented by the tap gains. The correlation properties of a PN sequence therefore makes it an ideal choice for the sync word. In particular, suppose the prefix consists of several periods of a PN sequence. As the prefix passes through the frame-sync register, the values of v_k will trace out the shape of $R_s(\tau)$ in Fig. 11.4-8 with peaks $v_k = N$ occurring each time the initial bit s_1 reaches the end of the register. An added advantage is the ease of PN sequence generation at the transmitter, even with large values of N. For instance, getting $P_{ff} < 10^{-6}$ in Eq. (10) requires $N > 25$, which can be accomplished with a 5-stage PN generator.

11.5 PROBLEMS

11.1-1 Sketch $x(t)$ and construct the corresponding eye pattern (without transmission distortion) for binary PAM with the data sequence 1011100010 when the signal has a unipolar format and pulse shape $p(t) = \cos^2(\pi t/2T_b)\,\Pi(t/2T_b)$.

11.1-2 Do Prob. 11.1-1 with $p(t) = \cos(\pi t/2T_b)\,\Pi(t/2T_b)$.

11.1-3 Do Prob. 11.1-1 with a polar format and $p(t) = \cos(\pi t/2T_b)\,\Pi(t/2T_b)$.

11.1-4 Do Prob. 11.1-1 with a bipolar format and $p(t) = \Lambda(t/T_b)$.

11.1-5 Do Prob. 11.1-1 with a bipolar format.

11.1-6 Modify Table 11.1-1 for an *octal* signal ($M = 8$).

11.1-7 A certain computer generates binary words, each consisting of 16 bits, at the rate 20,000 words per second. (a) Find the bandwidth required to transmit the output as a binary PAM signal. (b) Find M so that the output could be transmitted as an M-ary signal on a channel having $B = 60$ kHz.

11.1-8 A certain digital tape reader produces 3000 symbols per second, and there are 128 different symbols. (a) Find the bandwidth required to transmit the output as a binary PAM signal. (b) Find M so that the output could be transmitted as an M-ary signal on a telephone link having $B = 3$ kHz.

11.1-9 Suppose a digital signal with *pulse-position modulation* (PPM) is constructed by dividing the symbol interval into $M + 1$ equal time slots, allowing for M nonoverlapping pulse positions and a one-slot guard time. Develop an expression for the minimum bandwidth B needed to transmit M-ary data at a fixed bit rate r_b. Then tabulate B/r_b for a few values of $M = 2^n$ and determine the most bandwidth-efficient choice of M.

11.1-10 Binary data is transmitted as a unipolar signal with $A = 1$ and $p(t) = u(t + T_b) - u(t)$. The transmission system's step response is $g(t) = K_0(1 - e^{-bt})u(t)$, where $b = 2/T_b$. (a) Sketch $\tilde{p}(t)$ and find K_0 such that $\tilde{p}(0) = 1$. (b) Sketch $y(t)$ for the data sequence 10110 and evaluate $y(t)$ and the ISI at the optimum sampling times.

11.1-11 Do Prob. 11.1-10 for a polar signal with $A/2 = 1$ and a transmission system having $b = 1/T_b$.

11.1-12 Consider digital transmission with a gaussian pulse shape $p(t) = \exp[-\pi(bt)^2]$, which is neither timelimited nor bandlimited and does not have periodic zero crossings. Let $p(kD) \le 0.01$ for $k \ne 0$, to limit the ISI, and let the bandwidth B be such that $P(f) \le 0.01P(0)$ for $|f| > B$. Find the resulting relationship between r and B.

11.1-13 Find and sketch the power spectrum of a binary PAM signal with polar RZ format and rectangular pulses, assuming independent and equiprobable message bits. Then show that the time-domain and frequency-domain calculations of $\overline{x^2}$ are in agreement.

11.1-14 Consider a unipolar binary PAM signal whose pulse shape is the *raised cosine* in Example 2.4-2 with total duration $2\tau = T_b$. (a) Sketch $x(t)$ for the data sequence 10110100. Then find and sketch the power spectrum, assuming independent equiprobable message bits. (b) Repeat part a with $2\tau = 2T_b$.

11.1-15 Find and sketch the power spectrum of a polar binary PAM signal with $p(t) = \text{sinc } r_b t$ when the message bits are independent but have unequal probabilities, say α and $1 - \alpha$. Use your sketch to show that $\overline{x^2}$ is independent of α.

11.1-16 Find and plot the power spectrum of a binary signal with the split-phase Manchester format in Fig. 11.1-1c, assuming independent equiprobable message bits. Compare your result with Fig. 11.1-6.

11.1-17 Verify the correlation properties of bipolar precoding given in Eq. (14a) by tabulating the possible sequences $a_{k-n} a_k$ and their probabilities for $n = 0$, $n = 1$, and $n \geq 2$. Assume independent equiprobable message bits.

11.1-18‡ Let $g(n)$ be a function of the discrete variable n. By expanding and rearranging the summation, show that

$$\sum_{k=-K}^{K} \sum_{i=-K}^{K} g(k - i) = (2K + 1) \sum_{n=-2K}^{2K} \left(1 - \frac{|n|}{2K + 1}\right) g(n)$$

Apply this result to carry out the manipulation of the sum $\rho_X(f)$ used to derive $G_x(f)$ from $E[|X_T(f)|^2]$.

11.1-19‡ Consider a binary signal constructed using *different pulse shapes* $p_1(t)$ and $p_0(t)$ to represent 1s and 0s. For purposes of analysis, we can write

$$x(t) = \sum_k [a_k p_1(t - kT_b) + b_k p_0(t - kT_b)]$$

where a_k equals the kth bit and $b_k = 1 - a_k$. (a) Assuming independent and equiprobable bits, show that $\overline{a_k} = \overline{b_k} = \overline{a_k^2} = \overline{b_k^2} = 1/2$, $\overline{a_k b_k} = 0$, and $\overline{a_k a_i} = \overline{b_k b_i} = \overline{a_k b_i} = 1/4$ for $i \neq k$. (b) Now form the truncated signal $x_T(t)$ with $T = (2K + 1)T_b \gg T_b$ and show that

$$E[|X_T(f)|^2] = \frac{2K + 1}{4} |P_1(f) - P_0(f)|^2$$

$$+ \frac{1}{4} |P_1(f) + P_0(f)|^2 \sum_{k=-K}^{K} \sum_{i=-K}^{K} e^{-j\omega(k-i)T_b}$$

where $P_1(f) = \mathscr{F}[p_1(t)]$ and $P_0(f) = \mathscr{F}[p_0(t)]$. (c) Finally, use the relationship in Prob. 11.1-18 and Poisson's sum formula to obtain

$$G_x(f) = \frac{r_b}{4} |P_1(f) - P_0(f)|^2$$

$$+ \frac{r_b^2}{4} \sum_{n=-\infty}^{\infty} |P_1(nr_b) - P_0(nr_b)|^2 \delta(f - nr_b)$$

11.2-1 Find $(S/N)_R$ such that a unipolar binary system with AWGN has $P_e = 0.001$. What would be the error probability of a polar system with the same $(S/N)_R$?

11.2-2 A binary system has AWGN with $\eta = 10^{-8}/r_b$. Find S_R for polar and unipolar signaling so that $P_e \leq 10^{-6}$.

11.2-3 Some switching circuits generate *impulse noise*, which can be modeled as filtered white noise with average power σ^2 and an exponential PDF

$$p_n(n) = \frac{1}{\sqrt{2\sigma^2}} e^{-\sqrt{2}|n|/\sigma}$$

(a) Develop an expression for P_e in terms of A and σ for polar binary signaling contaminated by impulse noise. (b) Compare the effect of impulse noise with that of gaussian noise by considering the condition $P_e \le 0.001$.

11.2-4 Consider a polar binary system with ISI and AWGN such that $y(t_k) = a_k + \epsilon_k + n(t_k)$, where the ISI ϵ_k is equally likely to be $+\alpha$ or $-\alpha$. (a) Develop an expression for P_e in terms of A, α, and σ. (b) Evaluate P_e when $A = 8\sigma$ and $\alpha = 0.1A$. Compare your result with the error probability without ISI.

11.2-5 Do Prob. 11.2-4 taking $\epsilon_k = +\alpha, 0$, and $-\alpha$ with probabilities 0.25, 0.5, and 0.25, respectively.

11.2-6 Use Eq. (4) to obtain an expression for the optimum threshold in a polar binary system with AWGN when $P_0 \ne P_1$.

11.2-7 Derive Eq. (4) using Leibniz's rule, which states that

$$\frac{d}{dz} \int_{a(z)}^{b(z)} g(z, \lambda) \, d\lambda = g[z, b(z)] \frac{db(z)}{dz}$$

$$- g[z, a(z)] \frac{da(z)}{dz} + \int_{a(z)}^{b(z)} \frac{\partial}{\partial z} [g(z, \lambda)] \, d\lambda$$

where z is an independent variable and $a(z)$, $b(z)$, and $g(z, \lambda)$ are arbitrary functions.

11.2-8 A polar binary system has 20 repeaters, each with $(S/N)_1 = 20$ dB. Find P_e when the repeaters are regenerative and nonregenerative.

11.2-9 A polar binary system with 50 repeaters is to have $P_e = 10^{-4}$. Find $(S/N)_1$ in dB when the repeaters are regenerative and nonregenerative.

11.2-10 Consider the split-phase Manchester format in Fig. 11.1-1(d), where

$$p(t) = \begin{cases} 1 & -T_b/2 < t < 0 \\ -1 & 0 < t < T_b/2 \end{cases}$$

Plot the matched filter's impulse response. Then use superposition to plot the pulse response $a_k p(t - t_0) * h(t)$ and compare with Fig. 11.2-6(c).

11.2-11 Consider a unipolar RZ binary system with $p(t) = u(t) - u(t - T_b/2)$. Instead of a matched filter, the receiver has a first-order LPF with impulse response $h(t) = K_0 e^{-bt} u(t)$ where $K_0 = b/(1 - e^{-bT_b/2})$. (a) Find and sketch $Ap(t) * h(t)$ and obtain the condition on b such that the ISI at any subsequent sampling time does not exceed $0.1A$. (b) Show that $(A/2\sigma)^2 = (4br_b/K_0^2)\gamma_b \le 0.812\gamma_b$, where the upper bound comes from the ISI condition.

11.2-12 A binary data transmission system is to have $r_b = 500$ kbps and $P_{be} \le 10^{-4}$. The noise is white gaussian with $\eta = 10^{-17}$ W/Hz. Find the minimum value of S_R when: (a) $M = 2$; (b) $M = 8$ with Gray coding.

11.2-13 Suppose the transmission bandwidth of the system in Prob. 11.2-12 is $B = 80$ kHz. Find the smallest allowed value of M and the corresponding minimum value of S_R, assuming a Gray code.

11.2-14 A binary data transmission system has AWGN, $\gamma_b = 100$, and Gray coding. What's the largest value of M that yields $P_{be} \le 10^{-5}$?

11.2-15 Derive the result stated in Eq. (21).

11.2-16 Suppose binary data is converted to $M = 4$ levels via the Gray code in Table 11.1-1. Use Fig. 11.2-8 to derive an expression for P_{be} in terms of $k = A/2\sigma$. Simplify your result by assuming $k > 1$.

11.2-17 Do Prob. 11.2-16 with the natural code in Table 11.1-1.

11.3-1 Use Eq. (3a) to find and sketch $p(t)$ and $P(f)$ when $P_\beta(f) = (1/2\beta)\Pi(f/2\beta)$ with $\beta = r/4$.

11.3-2 Use Eq. (3a) to find and sketch $p(t)$ and $P(f)$ when $P_\beta(f) = (1/2\beta)\Pi(f/2\beta)$ with $\beta = r/3$.

11.3-3 Obtain Eq. (6b) from Eq. (5) with $\beta = r/2$. Then show that $p(\pm D/2) = 1/2$.

11.3-4 Carry out the details leading to Eqs. (4b) and (5) starting from Eq. (4a).

11.3-5 A more general form of Nyquist's signaling theorem states that if $P(f) = \mathcal{F}[p(t)]$ and

$$\sum_{n=-\infty}^{\infty} P(f - nr) = 1/r \qquad -\infty < f < \infty$$

then $p(t)$ has the property in Eq. (1a) with $D = 1/r$. (a) Prove this theorem by taking the Fourier transform of both sides of

$$p(t) \sum_{k=-\infty}^{\infty} \delta(t - kD) = \sum_{k=-\infty}^{\infty} p(kD)\delta(t - kD)$$

Then use Poisson's sum formula. (b) Use a sketch to show that $P(f)$ in Eq. (4b) satisfies the foregoing condition.

11.3-6 Consider an arbitrary bandlimited pulse spectrum with $P(f) = 0$ for $|f| \geq B$. Use simple sketches to obtain the additional requirements on $P(f)$ needed to satisfy Nyquist's signaling theorem as stated in Prob. 11.3-5 when: (a) $r/2 < B < r$; (b) $B = r/2$; (c) $B < r/2$.

11.3-7 A binary data system is to be designed for $r_b = 600$ kbps and $P_{be} \leq 10^{-5}$. The waveform will have $M = 2^n$, Gray coding, and Nyquist pulse shaping. The noise is white gaussian with $\eta = 1$ pW/Hz. The transmission channel has loss $L = 50$ dB and is distortionless over the allocated bandwidth $B = 200$ kHz. Choose M to minimize the transmitted power, and find the resulting values of r, β, and S_T.

11.3-8 Do Prob. 11.3-7 with $B = 120$ kHz.

11.3-9 Do Prob. 11.3-7 with $B = 80$ kHz.

11.3-10 Consider a data transmission system with $M = 2$, $r = 20,000$, $p_x(t) = \Pi(2rt)$, $|H_C(f)| = 0.01$, $G_n(f) = 10^{-10}(1 + 3 \times 10^{-4}|f|)^2$, and $p(t)$ per Eq. (6b). (a) Find and sketch the amplitude ratio for the optimum terminal filters. (b) Calculate S_T needed to get $P_e = 10^{-6}$, assuming gaussian noise.

11.3-11 Consider a data transmission system with $M = 4$, $r = 100$, $p_x(t) = \Pi(10rt)$, $G_n(f) = 10^{-10}$, $|H_C(f)|^2 = 10^{-6}/(1 + 32 \times 10^{-4}f^2)$, and $p(t) = \mathrm{sinc}\ rt$. (a) Find and sketch the amplitude ratio for the optimum terminal filters. (b) Calculate S_T needed to get $P_e = 10^{-6}$, assuming gaussian noise.

11.3-12‡ Consider a polar system in the form of Fig. 11.3-3 with $G_n(f) = \eta/2$ and a *timelimited* input pulse shape $p_x(t)$. Let $H_T(f) = K$ and $H_R(f) = [P_x(f)H_C(f)]^* e^{-j\omega t_d}$, so the receiving filter is *matched* to the received pulse shape. Since this scheme does not shape the output pulse $p(t - t_d)$ for zero ISI, it can be used only when the duration of $p_x(t)$ is small compared to $1/r$. (a) Obtain an expression for K such that $p(0) = 1$. Then develop expressions for S_T and σ^2 to show that $(A/2\sigma)^2$ is given by Eq. (12a) with

$$I_{HR} = \frac{\eta \displaystyle\int_{-\infty}^{\infty} |P_x(f)|^2\ df}{2 \displaystyle\int_{-\infty}^{\infty} |P_x(f)H_C(f)|^2\ df}$$

(b) Show that this result is equivalent to the maximum value $(A/2\sigma)^2 = 6S_R/(M^2 - 1)\eta r$.

11.3-13‡ A certain system has been built with optimum terminal filters for Nyquist pulse shaping, assuming white noise and a distortionless channel with loss L. However, it turns out that the channel actually introduces some linear distortion, so an equalizer with $H_{eq}(f) = [\sqrt{L}H_C(f)]^{-1}$ has been added at the output. Since the terminal filters were not modified to account for the channel distortion, the system has less than optimum performance. (a) Obtain expressions for S_T and σ^2 to show that

$$\left(\frac{A}{2\sigma}\right)^2 = \frac{6S_T/L}{K(M^2 - 1)\eta r} \qquad \text{where} \qquad K = \frac{1}{L}\int_{-\infty}^{\infty} \frac{|P(f)|}{|H_C(f)|^2}\ df$$

(b) Evaluate K in dB when $P(f)$ is as given by Eq. (6a) and $H_C(f) = \{\sqrt{L}[1 + j2(f/r)]\}^{-1}$.

11.3-14 Find the tap gains for a three-tap zero-forcing equalizer when $\tilde{p}_{-1} = 0.4$, $\tilde{p}_0 = 1.0$, $\tilde{p}_1 = 0.2$, and $\tilde{p}_k = 0$ for $|k| > 1$. Then find and plot $p_{eq}(t_k)$.

11.3-15 Obtain expressions for the tap gains of a three-tap zero-forcing equalizer when $\tilde{p}_{-1} = \epsilon$, $\tilde{p}_0 = 1$, $\tilde{p}_1 = \delta$, and $\tilde{p}_k = 0$ for $|k| > 1$.

11.3-16‡ Find the tap gains for a five-tap zero-forcing equalizer for $\tilde{p}(t)$ in Fig. 11.3-6a. (You can solve the simultaneous equations by successive substitution.) Then find the resulting values of $p_{eq}(t_k)$ and compare with Fig. 11.3-6b.

11.3-17 Consider binary correlative coding with $N = 2$, $c_0 = 1$, $c_1 = 2$, and $c_2 = 1$. (a) Find and sketch $h(t)$ and $|H(f)|$. (b) Use Eqs. (20b) and (22) to develop an expression for $y(t_k)$ like Eq. (23b).

11.3-18 Do Prob. 11.3-17 with $N = 4$, $c_0 = 1$, $c_1 = 0$, $c_2 = -2$, $c_3 = 0$, and $c_4 = 1$.

11.3-19‡ Consider a duobinary system with $H(f)$ as given by Eq. (21b), $p_x(t) = \delta(t)$, $H_c(f) = 1/\sqrt{L}$, and gaussian white noise. Obtain an expression for $(A/2\sigma)^2$ and apply Schwarz's inequality to show that the optimum filters have $|H_T(f)|^2 = gL|H(f)|$ and $|H_R(f)|^2 = |H(f)|/g$. Then derive Eq. (25).

TWELVE

CODED PULSE MODULATION

Having studied the basic concepts of digital transmission in Chap. 11, we return once more to analog communication. But now we consider *digital* transmission of *analog* messages via *code pulse modulation*. Coded pulse modulation systems employ sampling, quantizing, and coding to convert analog waveforms into digital signals.

Digital coding of analog information produces a *rugged* signal with a high degree of immunity to transmission distortion, interference, and noise. Digital coding also allows the use of regenerative repeaters for long-distance analog communication. However, the quantizing process essential for digital coding results in *quantization noise* which becomes the fundamental limitation on waveform reconstruction. To keep the quantization noise small enough for suitable fidelity, a coded pulse modulation system generally requires a much larger bandwidth than a comparable analog transmission system.

We'll develop these properties first in conjunction with *pulse-code modulation* (PCM). Then we'll describe *delta modulation* (DM) and other schemes that involve *predictive coding*. The chapter closes with an introduction to *digital multiplexing*, a valuable technique that makes it possible to combine analog and digital information for transmission in the form of a multiplexed digital signal.

12.1 PULSE-CODE MODULATION

This section describes the functional operation of pulse-code modulation (PCM). We'll view PCM as a digital transmission system with an *analog-to-digital converter* (ADC) at the input and a *digital-to-analog converter* (DAC) at the output.

When the digital error probability is sufficiently small, PCM performance as an
analog communication system depends primarily on the quantization noise introduced by the ADC. Here we'll analyze analog message reconstruction with quantization noise, temporarily deferring to the next section the effects of random
noise and digital errors.

PCM Generation and Reconstruction

Figure 12.1-1*a* diagrams the functional blocks of a PCM generation system. The
analog input waveform $x(t)$ is lowpass filtered and sampled to obtain $x(kT_s)$. A
quantizer rounds off the sample values to the nearest discrete value in a set of q
quantum levels. The resulting quantized samples $x_q(kT_s)$ are discrete in time (by
virtue of sampling) and discrete in amplitude (by virtue of quantizing).

To display the relationship between $x(kT_s)$ and $x_q(kT_s)$, let the analog message
be a voltage waveform normalized such that $|x(t)| \leq 1$ V. Uniform quantization
subdivides the 2-V peak-to-peak range into q equal steps of height $2/q$ V, as
shown in Fig. 12.1-1*b*. The quantum levels are then taken to be at $\pm 1/q$,
$\pm 3/q, \ldots, \pm(q-1)/q$ in the usual case when q is an even integer. A quantized
value such as $x_q(kT_s) = 5/q$ corresponds to any sample value in the range $4/q <
x(kT_s) < 6/q$.

(*a*)

(*b*)

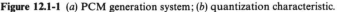

Figure 12.1-1 (*a*) PCM generation system; (*b*) quantization characteristic.

Next, an *encoder* translates the quantized samples into *digital code words*. The encoder works with M-ary digits and produces for each sample a codeword consisting of v digits in parallel. Since there are M^v possible M-ary codewords with v digits per word, unique encoding of the q different quantum levels requires that $M^v \geq q$. The parameters M, v, and q should be chosen to satisfy the equality, so that

$$q = M^v \qquad v = \log_M q \qquad (1)$$

Thus, the number of quantum levels for *binary* PCM equals some power of 2, namely $q = 2^v$.

Finally, successive codewords are read out serially to constitute the PCM waveform, an M-ary digital signal. The PCM generator thereby acts as an ADC, performing analog-to-digital conversions at the sampling rate $f_s = 1/T_s$. A timing circuit coordinates the sampling and parallel-to-serial readout.

Each encoded sample is represented by a v-digit output word, so the signaling rate becomes $r = v f_s$ with $f_s \geq 2W$. Therefore, the bandwidth needed for PCM baseband transmission is

$$B_T \geq \tfrac{1}{2} r = \tfrac{1}{2} v f_s \geq v W \qquad (2)$$

Fine-grain quantization for accurate reconstruction of the message waveform requires $q \gg 1$, which increases the transmission bandwidth by the factor $v = \log_M q$ times the message bandwidth W.

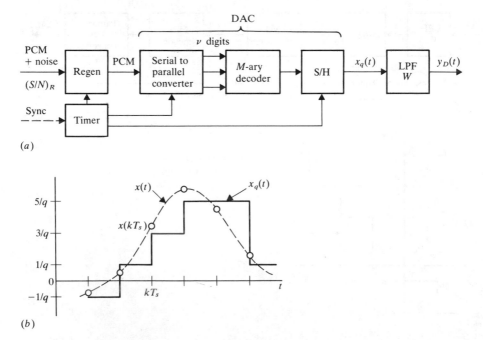

(a)

(b)

Figure 12.1-2 (a) PCM receiver; (b) reconstructed waveform.

Now consider a PCM receiver with the reconstruction system in Fig. 12.1-2a. The received signal may be contaminated by noise, but regeneration yields a clean and nearly errorless waveform if $(S/N)_R$ is sufficiently large. The DAC operations of serial-to-parallel conversion, M-ary decoding, and sample-and-hold generate the analog waveform $x_q(t)$ drawn in Fig. 12.1-2b. This waveform is a "staircase" approximation of $x(t)$, similar to flat-top sampling except that the sample values have been quantized. Lowpass filtering then produces the smoothed output signal $y_D(t)$, which differs from the message $x(t)$ to the extent that the quantized samples differ from the exact sample values $x(kT_s)$.

Perfect message reconstruction is therefore impossible in PCM, even when random noise has no effect. The ADC operation at the transmitter introduces permanent errors that appear at the receiver as quantization noise in the reconstructed signal. We'll study this quantization noise after an example of PCM hardware implementation.

Example 12.1-1 Suppose you want to build a binary PCM system with $q = 8$ so $v = \log_2 8 = 3$ bits per codeword. Figure 12.1-3a lists the 8 quantum levels and two types of binary codes. The "natural" code assigns the word 000 to

(a)

x	x_q	Natural code $b_2\ b_1\ b_0$	Sign/magnitude code $b_2\ b_1\ b_0$
	7/8	1 1 1	0 1 1
	5/8	1 1 0	0 1 0
	3/8	1 0 1	0 0 1
	1/8	1 0 0	0 0 0
	−1/8	0 1 1	1 0 0
	−3/8	0 1 0	1 0 1
	−5/8	0 0 1	1 1 0
	−7/8	0 0 0	1 1 1

(b)

(c)

Figure 12.1-3 (a) Binary PCM codes for $q = 8$; (b) direct-conversion ADC circuit for sign/magnitude code; (c) weighted-resistor decoder circuit.

the lowest level and progresses upward to 111 in the natural order of binary counting. The sign/magnitude code uses the leading bit b_2 for the algebraic sign of x_q while the remaining bits $b_1 b_0$ represent the magnitude. Other encoding algorithms are possible, and may include additional bits for error protection — the topic of Chap. 13.

A direct-conversion ADC circuit for the sign/magnitude code is shown in Fig. 12.1-3b. This circuit consists of one comparator for the sign bit and three parallel comparators plus combinational logic to generate the magnitude bits. Direct-conversion ADCs have the advantage of high operating speed and are called *flash encoders*, but they require a total of $q/2$ comparators. At lower speeds you can get by with one comparator and a feedback loop, a configuration found in the dual-slope, counter-comparison, and successive-approximation encoders.

Figure 12.1-3c shows the circuit for a *weighted-resistor decoder* that goes with a 3-bit sign/magnitude code. The sign bit operates a polarity selector switch, while the magnitude bits control the resistors to be connected to the reference voltage. The overall circuit acts as an inverting op-amp summer with output voltage $(-1)^{b_2}(4b_1 + 2b_0 + 1)/8$. Further information about DAC and ADC circuit modules is given by Mitra (1980, chap. 8).

Exercise 12.1-1 A binary channel with $r_b = 36{,}000$ bits/sec is available for PCM voice transmission. Find appropriate values of v, q, and f_s assuming $W \approx 3.2$ kHz.

Quantization Noise

Although PCM reconstruction most often takes the form of staircase filtering, as in Fig. 12.1-2, we'll find the *impulse reconstruction* model in Fig. 12.1-4 more convenient for the analysis of quantization noise. Here, a pulse converter in place of the sample-and-hold circuit generates the weighted impulse train

$$y_\delta(t) = \sum_k [x(kT_s) + \epsilon_k]\, \delta(t - kT_s) \tag{3a}$$

where ϵ_k represents the *quantization error*, namely

$$\epsilon_k = x_q(kT_s) - x(kT_s) \tag{3b}$$

Lowpass filtering with $B = f_s/2$ yields the final output

$$y_D(t) = x(t) + \sum_k \epsilon_k \operatorname{sinc}(f_s t - k) \tag{4}$$

This expression has the same form as reconstruction of analog pulse modulation with noisy samples; see Eq. (6), Sect. 10.4. Furthermore, when q is large enough

Figure 12.1-4 Impulse reconstruction model.

for reasonable signal approximation, the ϵ_k will be uncorrelated and independent of $x(t)$. Accordingly, we identify $\overline{\epsilon_k^2}$ as the mean-square *quantization noise*.

Round-off quantization with equispaced levels ensures that $|\epsilon_k| \leq 1/q$. Lacking additional information to the contrary, we assume that the quantization error has zero mean value and a uniform probability density function over $-1/q \leq \epsilon_k \leq 1/q$. Thus, the quantization noise power is

$$\sigma_q^2 = \overline{\epsilon_k^2} = \frac{1}{(2/q)} \int_{-1/q}^{1/q} \epsilon^2 \, d\epsilon = \frac{1}{3q^2} \tag{5}$$

which reflects the intuitive observation that the quantization noise decreases when the number of quantum levels increases.

Now we measure PCM performance in terms of the destination signal power $S_D = \overline{x^2} = S_x \leq 1$ and the quantization noise power σ_q^2. The destination signal-to-noise ratio then becomes

$$\left(\frac{S}{N}\right)_D = \frac{S_x}{\sigma_q^2} = 3q^2 S_x \tag{6}$$

A more informative relation for binary PCM is obtained by setting $q = 2^\nu$ and expressing $(S/N)_D$ in decibels. Thus,

$$\left(\frac{S}{N}\right)_D = 10 \log_{10}(3 \times 2^{2\nu} S_x)$$

$$\leq 4.8 + 6.0\nu \quad \text{dB} \tag{7}$$

where the upper bound holds when $S_x = 1$. Voice telephone PCM systems typically have $\nu = 8$ so $(S/N)_D \leq 52.8$ dB.

But many analog signals — especially voice and music — are characterized by a large *crest factor*, defined as the ratio of peak to rms value, $|x(t)|_{\max}/\sigma_x$. Our signal normalization establishes $|x(t)|_{\max} \leq 1$, and a large crest factor then implies that $S_x = \sigma_x^2 \ll 1$. Consequently, the actual signal-to-noise ratio will be significantly less than the theoretical upper bound. For instance, some digital audio recording systems take $\nu = 14$ to get high-fidelity quality with $(S/N)_D \approx 60$ dB, compared to $(S/N)_D \leq 88.8$ dB predicted by Eq. (7). As a bandwidth-conserving alternative to increasing ν, PCM performance may be improved through the use of *companding*, equivalent to *nonuniform quantization*.

Exercise 12.1-2 Consider binary PCM transmission of a video signal with $f_s = 10$ MHz. (*a*) Calculate the signaling rate needed to get $(S/N)_D \geq 50$ dB when $S_x = 1$. (*b*) Repeat (*a*) with $S_x = 0.1$.

Nonuniform Quantizing and Companding★

A normalized signal $x(t)$ with a large crest factor can be represented by the typical probability density function $p_X(x)$ sketched in Fig. 12.1-5. The even sym-

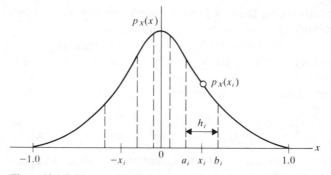

Figure 12.1-5 Message PDF with nonuniform quantization bands

metry and absence of an impulse at $x = 0$ correspond to $\bar{x} = 0$, so $\sigma_x^2 = \overline{x^2} = S_x$ and

$$S_x = \int_{-1}^{1} x^2 p_X(x)\, dx = 2 \int_{0}^{1} x^2 p_X(x)\, dx \tag{8}$$

This integration yields $S_x \ll 1$ because the PDF has a dominant peak at $x = 0$.

The shape of $p_X(x)$ also means that $|x(t)| \ll 1$ most of the time. It would therefore make good sense to use *nonuniform* quantization as indicated by the dashed lines. The quantum levels $\pm x_1, \ldots, \pm x_{q/2}$ are closely spaced near $x = 0$, but more widely spaced for the large values of $|x(t)|$ which occur infrequently. We calculate the resulting quantization noise as follows.

Consider a sample value $x = x(kT_s)$ in the band $a_i < x < b_i$ around the quantum level x_i. The quantizing error $\epsilon_i = x_i - x$ then has the mean square value

$$\overline{\epsilon_i^2} = \int_{a_i}^{b_i} (x_i - x)^2 p_X(x)\, dx \tag{9a}$$

Summing $\overline{\epsilon_i^2}$ over all q levels gives the quantization noise

$$\sigma_q^2 = 2 \sum_{i=1}^{q/2} \overline{\epsilon_i^2} \tag{9b}$$

where we've taken advantage of the even symmetry. In the usual case of $q \gg 1$, the step height $h_i = b_i - a_i$ will be small enough that $p_X(x) \approx p_X(x_i)$ over each integration band and x_i will fall roughly in the middle of the step. Under these conditions Eq. (9a) simplifies to

$$\overline{\epsilon_i^2} \approx p_X(x_i) \int_{x_i - h_i/2}^{x_i + h_i/2} (x_i - x)^2\, dx = p_X(x_i) \frac{h_i^3}{12}$$

and thus

$$\sigma_q^2 \approx \frac{1}{6} \sum_{i=1}^{q/2} p_X(x_i) h_i^3 \tag{10}$$

As a check on this expression, we note that if the signal has the uniform PDF $p_X(x) = 1/2$ and if the steps have equal height $h_i = 2/q$, then $\sigma_q^2 = (1/6)(q/2)(1/2)(2/q)^3 = 1/3q^2$ which agrees with our earlier result in Eq. (5).

Theoretically, you could *optimize* PCM performance by finding the values of x_i, a_i, and b_i that result in *minimum quantization noise*. Such optimization is a difficult procedure that requires knowledge of the signal's PDF. Additionally, the custom-tailored hardware needed for nonlinear quantizing costs far more than standard uniform quantizers. Therefore, the approach taken in practice is to use *uniform* quantizing after *nonlinear signal compression*, the compression characteristics being determined from experimental studies with representative signals.

Figure 12.1-6 plots an illustrative compressor curve $z(x)$ versus x for $0 \le x \le 1$; the complete curve must have odd symmetry such that $z(x) = -z(|x|)$ for $-1 \le x \le 0$. Uniform quantization of $z(x)$ then corresponds to nonuniform quantization of x, as shown in the figure. The nonlinear distortion introduced by the compressor is corrected after reconstruction by a complementary expander, identical to the *companding* strategy discussed in Sect. 3.2. Hence, the postdetection signal-to-noise ratio for companded PCM is $(S/N)_D = S_x/\sigma_q^2$, with σ_q^2 given by Eq. (10).

Our next task is to obtain σ_q^2 in terms of the compressor curve. For that purpose let

$$z'(x) \triangleq \frac{dz(x)}{dx}$$

so $z'(x_i)$ equals the slope of $z(x)$ at $x = x_i$. The conditions $q \gg 1$ and $h_i \ll 1$ justify the approximation $z'(x_i) \approx (2/q)/h_i$, so

$$h_i^2 \approx \left[\frac{2/q}{z'(x_i)}\right]^2 = \frac{4}{q^2[z'(x_i)]^2}$$

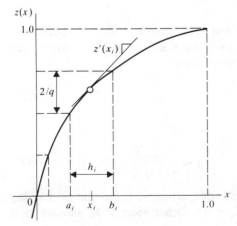

Figure 12.1-6 Compressor characteristic curve.

Equation (10) then becomes

$$\sigma_q^2 \approx \frac{4}{6q^2} \sum_{i=1}^{q/2} \frac{p_X(x_i)}{[z'(x_i)]^2} h_i \approx \frac{2}{3q^2} \int_0^1 \frac{p_X(x)}{[z'(x)]^2} dx$$

where we've passed from summation to integration via $h_i \rightarrow dx$. Therefore, our final result is

$$\left(\frac{S}{N}\right)_D = \frac{S_x}{\sigma_q^2} \approx \frac{3q^2 S_x}{K_z} \qquad (11a)$$

with

$$K_z \triangleq 2 \int_0^1 \frac{p_X(x)}{[z'(x)]^2} dx \qquad (11b)$$

which takes account of the compression. If $K_z < 1$ then $(S/N)_D > 3q^2 S_x$ and companding improves PCM performance by reducing the quantization noise.

Example 12.1-2 μ-law companding for voice PCM The popular μ-law companding for voice telephone PCM employs the compressor characteristic

$$z(x) = (\text{sgn } x) \frac{\ln (1 + \mu|x|)}{\ln (1 + \mu)} \qquad |x| \le 1$$

and hence

$$z'(x) = \frac{\mu}{\ln (1 + \mu)} \frac{1}{1 + \mu|x|}$$

The parameter μ is a large number so $z'(x) \gg 1$ for $|x| \ll 1$ whereas $z'(x) \ll 1$ for $|x| \approx 1$. Substituting $z'(x)$ into Eq. (11b) and performing the integration, we obtain

$$K_z = \frac{\ln^2 (1 + \mu)}{\mu^2} (1 + 2\mu \overline{|x|} + \mu^2 S_x) \qquad (12)$$

Now we need values for $\overline{|x|}$ and S_x to test the efficacy of μ-law companding.

Laboratory investigations have shown that the PDF of a voice signal can be modeled by a *Laplace* distribution in the form

$$p_X(x) = \frac{\alpha}{2} e^{-\alpha|x|} \qquad (13a)$$

with

$$S_x = \sigma_x^2 = \frac{2}{\alpha^2} \qquad \overline{|x|} = \frac{1}{\alpha} = \sqrt{\frac{S_x}{2}} \qquad (13b)$$

This distribution cannot be normalized for $|x(t)|_{\max} \le 1$, but the probability of $|x(t)| > 1$ will be less than 1% if $S_x < 0.1$. Other voice PDF models yield

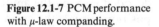

Figure 12.1-7 PCM performance with μ-law companding.

about the same relationship between $\overline{|x|}$ and S_x, which is the critical factor for evaluating K_z.

Taking the standard value $\mu = 255$ and putting $\overline{|x|} = \sqrt{S_x/2}$ in Eq. (12), we obtain

$$K_z = 4.73 \times 10^{-4}(1 + 361\sqrt{S_x} + 65{,}025S_x)$$

Numerical calculations then show that $K_z < 1$ for $S_x < 0.03$. But more significant is the fact that S_x/K_z stays nearly constant over a wide range of S_x. Consequently, μ-law companding for voice PCM provides an essentially *fixed* value of $(S/N)_D$, despite wide variations of S_x among individual talkers. Figure 12.1-7 brings out this desirable feature by plotting $(S/N)_D$ in dB versus S_x in dB, with and without companding, when $q = 2^8$. Notice the companding improvement for $S_x < -20$ dB.

12.2 PCM WITH NOISE

In this section we account for the effects of random noise in PCM transmission. The resulting *digital errors* produce *decoding noise*. After defining the error threshold level, we'll be in a position to make a meaningful comparison of PCM with analog modulation methods.

Decoding Noise

Random noise added to the PCM signal at the receiver causes regeneration errors that appear as erroneous digits in the codewords. The decoder then puts out a different quantum level than the one intended for a particular sample.

Hence, the reconstructed message waveform becomes contaminated with decoding noise as well as quantization noise.

The analysis of decoding noise is not too difficult if we restrict our attention to binary PCM with uniform quantization and a relatively small bit error probability P_e. The number of bit errors in a v-digit codeword is a random variable governed by the binomial distribution. However, when $P_e \ll 1$, the probability of one error in a given word approximately equals vP_e and the probability of two or more errors is small enough for us to ignore that event. Of course, the effect of a single error depends on where it falls in the word, because the different bit positions have different decoding interpretations.

Consider a "natural" binary codeword of the form $b_{v-1}b_{v-2} \cdots b_1 b_0$ in which the mth bit distinguishes between quantum levels spaced by 2^m times the step height $2/q$. (The sign/magnitude code illustrated back in Fig. 12.1-3 follows this pattern for the magnitude bits, but the sign bit has a variable level meaning that makes analysis more complicated.) An error in the mth bit then shifts the decoded level by the amount $\epsilon_m = \pm(2/q)2^m$, and the average of ϵ_m^2 over the v bit positions equals the mean-square decoding error for a random bit-error location. Thus,

$$\overline{\epsilon_m^2} = \frac{1}{v} \sum_{m=0}^{v-1} \left(\frac{2}{q} 2^m \right)^2 = \frac{4}{vq^2} \sum_{m=0}^{v-1} 4^m$$

$$= \frac{4}{vq^2} \frac{4^v - 1}{3} = \frac{4}{3v} \frac{q^2 - 1}{q^2} \approx \frac{4}{3v} \tag{1}$$

where we've used the formula for summing a geometric progression and substituted $4^v = 2^{2v} = q^2 \gg 1$. The *decoding noise power* is therefore

$$\sigma_d^2 = vP_e \overline{\epsilon_m^2} \approx \tfrac{4}{3}P_e \tag{2}$$

since erroneous words occur with probability vP_e.

The *total* destination noise power consists of decoding noise σ_d^2 and quantization noise $\sigma_q^2 = 1/3q^2$, which come from essentially independent processes. Therefore,

$$N_D = \sigma_q^2 + \sigma_d^2 = \frac{1 + 4q^2 P_e}{3q^2}$$

and

$$\left(\frac{S}{N} \right)_D = \frac{3q^2}{1 + 4q^2 P_e} S_x \tag{3}$$

so the effect of decoding noise depends upon the relative value of the quantity $4q^2 P_e$. Indeed, we see in Eq. (3) the two extreme conditions

$$\left(\frac{S}{N} \right)_D \approx \begin{cases} 3q^2 S_x & P_e \ll 1/4q^2 \\ \dfrac{3}{4P_e} S_x & P_e \gg 1/4q^2 \end{cases} \tag{4}$$

$$\left(\frac{S}{N}\right)_R, dB$$

Figure 12.2-1 Noise performance of PCM.

Quantization noise dominates when P_e is small, but decoding noise dominates and reduces $(S/N)_D$ when P_e is large compared to $1/4q^2$.

Now recall that the value of P_e is determined by the received signal-to-noise ratio $(S/N)_R$ at the input to the digital regenerator. Specifically, for polar binary signaling in gaussian white noise we know that $P_e = Q[\sqrt{(S/N)_R}]$. Figure 12.2-1 plots $(S/N)_D$ versus $(S/N)_R$ for this case, with $S_x = 1/2$ and two values of q. The precipitous decline of $(S/N)_D$ as $(S/N)_R$ decreases constitutes a *threshold effect* caused by increasing errors. Below the error threshold, when $P_e \gg 1/4q^2$, the errors occur so often that the reconstructed waveform bears little resemblance to the original signal and the message has been mutilated beyond recognition.

Error Threshold

The PCM error threshold level is usually defined at the point where decoding noise reduces $(S/N)_D$ by 1 dB. Unfortunately, this definition does not lend itself to analytical investigations. As a more convenient alternative, we'll say that decoding errors have negligible effect if $P_e \leq 10^{-5}$. Then we obtain the corresponding condition on $(S/N)_R$ for polar M-ary signaling using Eq. (20), Sect. 11.2, namely

$$P_e = 2\left(1 - \frac{1}{M}\right)Q\left[\sqrt{\frac{3}{M^2 - 1}\left(\frac{S}{N}\right)_R}\right] \leq 10^{-5}$$

Solving for the minimum value of $(S/N)_R$ yields the threshold level

$$\left(\frac{S}{N}\right)_{R_{th}} \approx 6(M^2 - 1) \tag{5}$$

If $(S/N)_R < 6(M^2 - 1)$, the PCM output will be hopelessly mutilated by decoding noise.

A subtle but important implication of Eq. (5) relates to the digital signaling rate and transmission bandwidth. We'll bring out that relationship with the help of the analog transmission parameter $\gamma = S_R/\eta W = (B_T/W)(S/N)_R$. The PCM transmission bandwidth is $B_T \geq r/2 \geq vW$, so

$$\gamma_{th} = (B_T/W)(S/N)_{R_{th}}$$

$$\approx 6\frac{B_T}{W}(M^2 - 1) \geq 6v(M^2 - 1) \tag{6}$$

Given v and M, this equation tells you the minimum value of γ needed for PCM operation above threshold. It also facilitates the comparison of PCM with other transmission schemes.

Example 12.2-1 Example 12.1-2 showed that a voice PCM system with $M = 2$, $v = 8$, and μ-law companding has$(S/N)_D \approx 37$ dB. Equation (6) gives the corresponding threshold level $\gamma_{th} \geq 144 \approx 22$ dB. Hence, the PCM system has a potential 15-dB advantage over direct analog baseband transmission in which $(S/N)_D = \gamma$. The full advantage would not be realized in practice where allowance must be made for $B_T > vW$ and $\gamma > \gamma_{th}$.

Exercise 12.2-1 The 1-dB definition for error threshold is equivalent to $10 \log_{10}(1 + 4q^2 P_e) = 1$. Calculate P_e from this definition when $q = 2^8$. Then find the corresponding value of $(S/N)_R$ for polar binary signaling, and compare your result with Eq. (5).

PCM versus Analog Modulation

The PCM threshold effect reminds us of analog modulation methods such as FM and PPM that have the property of *wideband noise reduction* above their threshold levels. As our initial point of comparison, let's demonstrate that PCM also provides wideband noise reduction when operated above threshold so $(S/N)_D = 3q^2 S_x$. For this purpose, we'll assume that the sampling frequency is close to the Nyquist rate and $B_T \approx vW$. Then $q = M^v \approx M^b$ where $b = B_T/W$ is the bandwidth ratio. Hence,

$$\left(\frac{S}{N}\right)_D \approx 3M^{2b}S_x \tag{7}$$

which exhibits noise reduction as an *exponential* exchange of bandwidth for signal-to-noise ratio. Granted that the noise being reduced is quantization noise, but random noise has no effect on PCM above threshold. The exponential factor in Eq. (7) is far more dramatic than that of wideband analog modulation, where $(S/N)_D$ increases proportionally to b or b^2.

For further comparison including threshold limitations, Fig. 12.2-2 illustrates the performance of several modulation types as a function of γ. All curves are calculated with $S_x = 1/2$, and the heavy dots indicate the threshold points. The PCM curves are based on Eqs. (6) and (7) with $M = 2$ and $v = b$.

Figure 12.2-2 Performance comparison of PCM and analog modulation.

Clearly, in the name of power efficiency, PCM should be operated just above threshold, since any power increase beyond $\gamma \approx \gamma_{th}$ yields no improvement of $(S/N)_D$. Near threshold, PCM does offer some advantage over FM or PPM with the same value of b and $(S/N)_D$. And even a 3-dB power advantage, being a factor of 2, may spell the difference between success and failure for some applications. But that advantage is gained at the price of more complicated and costly hardware. In fact, PCM was deemed totally impractical prior to the development of high-speed digital electronics in the late 1950s — two decades after the invention of PCM.

Two other PCM benefits don't appear in Fig. 12.2-2. One is the advantage of *regenerative repeaters* when the transmission link requires many repeater stations. Another benefit comes from the fact that PCM allows analog message transmission as a digital signal. *Digital multiplexing* then makes it possible to combine PCM and digital data signals for flexible and efficient utilization of a communication channel. Taken together, these two benefits account for PCM's preeminence in the design of new systems for long-distance telephony.

But PCM is not suited to all applications. In radio broadcasting, for instance, we want a relatively large signal-to-noise ratio, say $(S/N)_D \approx 60$ dB. Figure 12.2-2 reveals that this would require binary PCM with $b > 8$, or FM with a smaller bandwidth ratio $b = 6$ and much simpler hardware at the transmitter and receivers. Likewise, bandwidth and hardware considerations would reject PCM for most single-channel systems.

Exercise 12.2-2 Starting with Eqs. (6) and (7), show that a PCM system operated at the threshold point has

$$\left(\frac{S}{N}\right)_{D_{th}} = 3\left(1 + \frac{\gamma_{th}}{6b}\right)^b S_x \tag{8}$$

Compare this expression with that of WBFM by setting $D = b/2 \gg 1$ in Eq. (20*b*), Sect. 9.4.

12.3 DELTA MODULATION AND PREDICTIVE CODING

Sample values of analog waveforms derived from physical processes often exhibit *predictability* in the sense that the average change from sample to sample is small. Hence, you can make a reasonable guess of the next sample value based on previous values. The predicted value has some error, of course, but the range of the error should be much less than the peak-to-peak signal range. Predictive coded modulation schemes exploit this property by transmitting just the *prediction errors*. An identical prediction circuit at the destination combines the incoming errors with its own predicted values to reconstruct the waveform.

Predictive methods work especially well with audio and video signals, and much effort has been devoted to prediction strategies for efficient voice and image transmission. *Delta modulation* (DM) employs prediction to simplify hardware in exchange for increased signaling rate compared to PCM. *Differential pulse-code modulation* (DPCM) reduces signaling rate but involves more elaborate hardware. We'll discuss both DM and DPCM in this section, along with the related and fascinating topic of *speech synthesis* using prediction.

Delta Modulation

Let an analog message waveform $x(t)$ be lowpass filtered and sampled every T_s seconds. We'll find it convenient here to use *discrete-time* notation, with the integer independent variable k representing the sampling instant $t = kT_s$. We thus write $x(k)$ as a shorthand for $x(kT_s)$, etc.

When the sampling frequency is greater than the Nyquist rate, we expect that $x(k)$ roughly equals the previous sample value $x(k - 1)$. Therefore, given the quantized sample value $x_q(k - 1)$, a reasonable guess for the next value would be

$$\tilde{x}_q(k) = x_q(k - 1) \tag{1}$$

where $\tilde{x}_q(k)$ denotes our *prediction* of $x_q(k)$. A delay line with time delay T_s then serves as the prediction circuit. The difference between the predicted and actual value can be expressed as

$$x_q(k) = \tilde{x}_q(k) + \epsilon_q(k) \tag{2}$$

in which $\epsilon_q(k)$ is the *prediction error*.

If we transmit $\epsilon_q(k)$, we can use the system in Fig. 12.3-1 to generate $x_q(k)$ by delaying the current output and adding it to the input. This system implements Eqs. (1) and (2), thereby acting as an *accumulator*. The accumulation effect is

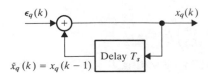

Figure 12.3-1 Accumulator for delta modulation.

brought out by writing $x_q(k) = \epsilon_q(k) + x_q(k-1)$ with $x_q(k-1) = \epsilon_q(k-1)$ $+ x_q(k-2)$, and so forth; hence

$$x_q(k) = \epsilon_q(k) + \epsilon_q(k-1) + x_q(k-2)$$
$$= \epsilon_q(k) + \epsilon_q(k-1) + \epsilon_q(k-2) + \cdots$$

An *integrator* accomplishes the same accumulation when $\epsilon_q(k)$ takes the form of brief rectangular pulses.

At the transmitting end, prediction errors are generated by the simple delta modulation system diagrammed in Fig. 12.3-2. The comparator serves as a *binary quantizer* with output values $\pm\Delta$, depending on the difference between the predicted value $\tilde{x}_q(k)$ and the unquantized sample $x(k)$. Thus, the resulting DM signal is

$$\epsilon_q(k) = [\text{sgn } \epsilon(k)] \, \Delta \tag{3a}$$

where

$$\epsilon(k) = x(k) - \tilde{x}_q(k) \tag{3b}$$

which represents the unquantized error. An accumulator (or integrator) in a feedback loop produces $\tilde{x}_q(k)$ from $\epsilon_q(k)$, similar to Fig. 12.3-1 except that the feedback signal comes from the delayed output.

Observe that this DM transmitter requires no analog-to-digital conversion other than the comparator. Also observe that an accumulator like Fig. 12.3-1 performs the digital-to-analog conversion at the receiver, reconstructing $x_q(k)$ from $\epsilon_q(k)$. A DM system thereby achieves digital transmission of analog signals with very simple hardware compared to a PCM system.

The name delta modulation reflects the fact that each input sample $x(k)$ has been encoded as a single pulse of height $+\Delta$ or $-\Delta$. But we can also view $\epsilon_q(k)$ as a *binary waveform* with signaling rate $r_b = f_s$, or *one bit per sample*. For this reason DM is sometimes called "one-bit PCM." The corresponding transmission bandwidth requirement is

$$B_T \geq r_b/2 = f_s/2 \tag{4}$$

We get by with just one bit per sample because we're transmitting prediction errors, not sample values. Nonetheless, successful operation requires rather high sampling rates, as we'll soon see.

Figure 12.3-2 DM transmitter.

Figure 12.3-3 DM waveforms.

Figure 12.3-3 depicts illustrative continuous-time waveforms $x(t)$, $\tilde{x}_q(t)$, and $\epsilon_q(t)$ involved in DM. The staircase waveform $x_q(t)$ at the receiver differs from $\tilde{x}_q(t)$ only by a time shift of T_s seconds. The transmitter starts with an arbitrary initial prediction such as $\tilde{x}_q(0) < x(0)$ so $\epsilon_q(0) = +\Delta$. Then $\epsilon_q(0)$ is fed back through the accumulator to form the updated prediction $\tilde{x}_q(T_s) = x_q(0) + \epsilon_q(0)$. Continual updating at each sampling instant causes $\tilde{x}_q(t)$ to increase by steps of Δ until the start-up interval ends when $\tilde{x}_q(kT_s) > x(kT_s)$ and $\epsilon_q(kT_s) = -\Delta$. If $x(t)$ remains constant, $\tilde{x}_q(t)$ takes on a hunting behavior. When $x(t)$ varies with time, $\tilde{x}_q(t)$ follows it in stepwise fashion as long as the rate of change does not exceed the DM tracking capability. The difference between $\tilde{x}_q(t)$ and $x(t)$ is called *granular noise*, analogous to quantization noise in PCM. The reconstructed and smoothed waveform at the receiver will be a reasonable approximation for $x(t)$ if Δ and T_s are sufficiently small.

But when $x(t)$ increases or decreases too rapidly, $\tilde{x}_q(t)$ lags behind and we have the phenomenon known as *slope overload*, a fundamental limitation of DM. Since $\tilde{x}_q(t)$ changes by $\pm\Delta$ every $T_s = 1/f_s$ seconds, the maximum DM slope is $\pm f_s \Delta$ and a sufficient condition for slope tracking is

$$f_s \Delta \geq |\dot{x}(t)|_{\max} \tag{5}$$

where $\dot{x}(t) = dx/dt$. Consider, for instance, the modulating tone $x(t) = A_m \cos 2\pi f_m t$ so $\dot{x}(t) = -2\pi f_m A_m \sin 2\pi f_m t$ and $|\dot{x}(t)|_{\max} = 2\pi f_m A_m \leq 2\pi W$, where the upper bound incorporates our message conventions $A_m \leq 1$ and $f_m \leq W$. Equation (5) therefore calls for a high sampling frequency $f_s \geq 2\pi W/\Delta \gg \pi W$, since we want $\Delta \ll 2$ to make the steps of $\tilde{x}_q(t)$ small compared to the peak-to-peak signal range $-1 \leq x(t) \leq 1$.

DM performance quality depends on the granular noise, slope-overload noise, and regeneration errors. However, only granular noise has a significant effect under normal operating conditions, which we assume hereafter. Even so, the analysis of granular noise is a difficult problem best tackled by computer simulation for accurate results or by approximations for rough results.

We'll estimate DM performance using the receiver modeled by Fig. 12.3-4a with $\epsilon_q(t - T_s) = \epsilon_q(k - 1)$ at the input to the accumulator. Equations (1) and (3b)

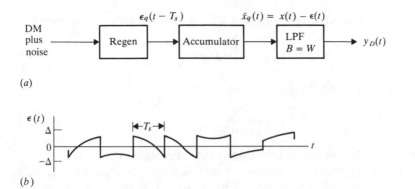

(a)

(b)

Figure 12.3-4 (a) DM receiver; (b) granular noise waveform.

then give the accumulator output as $x_q(k-1) = \tilde{x}_q(k) = x(k) - \epsilon(k)$ or

$$\tilde{x}_q(t) = x(t) - \epsilon(t) \qquad (6)$$

where $\epsilon(t)$ is the granular noise waveform sketched in Fig. 12.3-4b. The shape of $\epsilon(t)$ and the fact that $|\epsilon(t)| \leq \Delta$ suggest a uniform amplitude distribution with

$$\overline{\epsilon^2} = \Delta^2/3$$

Furthermore, experimental studies confirm that the power spectrum of $\epsilon(t)$ is essentially flat over $|f| \leq 1/T_s = f_s$. Thus,

$$G_\epsilon(f) \approx \overline{\epsilon^2}/2f_s \qquad |f| \leq f_s$$

and lowpass filtering yields

$$N_g = \int_{-W}^{W} G_\epsilon(f)\, df = \frac{W}{f_s}\overline{\epsilon^2} = \frac{W}{f_s}\frac{\Delta^2}{3} \qquad (7)$$

which is the average power of the granular noise component of $x_q(t)$.

When granular noise is the only contamination in the filtered output, we obtain the signal-to-noise ratio

$$\left(\frac{S}{N}\right)_D = \frac{S_x}{N_g} = \frac{3f_s}{\Delta^2 W}S_x \qquad (8)$$

This result is almost identical to the PCM expression $(S/N)_D = 3q^2 S_x$ if $f_s = 2W$ and $\Delta = 1/q$. But f_s and Δ must satisfy Eq. (5). We therefore need a more general relationship for the slope-tracking condition.

Recall from Eq. (22), Sect. 5.2, that if $G_x(f)$ is the power spectrum of $x(t)$, then the power spectrum of the derivative $\dot{x}(t)$ is $(2\pi f)^2 G_x(f)$. Hence, the *mean square signal slope* can be put in the form

$$\overline{|\dot{x}(t)|^2} = \int_{-\infty}^{\infty} (2\pi f)^2 G_x(f)\, df = (2\pi\sigma W_{\text{rms}})^2 \qquad (9)$$

where $\sigma = \sqrt{S_x}$ is the signal's rms value and W_{rms} is its *rms bandwidth*, defined by

$$W_{rms} \triangleq \frac{1}{\sigma} \left[\int_{-\infty}^{\infty} f^2 G_x(f) \, df \right]^{1/2} \tag{10}$$

Now we introduce the so-called *slope loading factor*

$$s \triangleq \frac{f_s \Delta}{2\pi\sigma W_{rms}} \tag{11}$$

which is the ratio of the maximum DM slope to the rms signal slope.

A reasonably large value of s ensures negligible slope overload. Hence, we'll incorporate this factor explicitly in Eq. (8) by writing Δ in terms of s from Eq. (11). Thus,

$$\left(\frac{S}{N} \right)_D = \frac{3}{4\pi^2} \frac{f_s^3}{s^2 W_{rms}^2 W}$$

$$= \frac{6}{\pi^2} \left(\frac{W}{W_{rms}} \right)^2 \frac{b^3}{s^2} \tag{12}$$

where

$$b \triangleq f_s/2W \tag{13}$$

The parameter b equals our usual bandwidth ratio B_T/W when B_T has the minimum value $f_s/2$. Equation (12) brings out the fact that DM performance falls

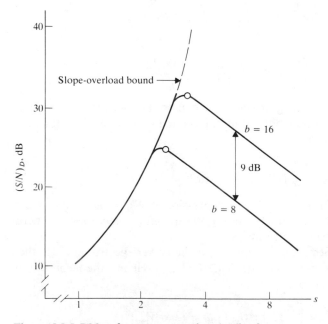

Figure 12.3-5 DM performance versus slope loading factor.

between PCM and PPM since the wideband noise reduction goes as b^3 rather than b^2 or exponentially. DM generally requires a larger transmission bandwidth than PCM to achieve the same signal-to-noise ratio, so its applications are limited to those cases where ease of implementation takes precedence over bandwidth considerations.

Computer simulations by Abate (1967) indicate that Eq. (12) holds for $\ln 2b \leq s < 8$. If $s < \ln 2b$, then *slope-overload noise* dominates and $(S/N)_D$ drops off quite rapidly. Figure 12.3-5 illustrates how $(S/N)_D$ varies with s. For a specified bandwidth ratio, DM performance is maximized by taking the empirically determined optimum slope loading factor

$$s_{\text{opt}} \approx \ln 2b \qquad (14)$$

The maximum value of $(S/N)_D$ is then given by Eq. (12) with $s = s_{\text{opt}}$.

Example 12.3-1 DM voice transmission When a typical voice signal has been prefiltered so that $W \approx 4$ kHz, its rms bandwidth will be $W_{\text{rms}} \approx 1.3$ kHz. Substituting these values and $s = s_{\text{opt}} = \ln 2b$ in Eq. (12) yields the maximum DM signal-to-noise ratio $(S/N)_D \approx 5.8b^3/(\ln 2b)^2$. If $b = 16$, then $(S/N)_D \approx 33$ dB which is comparable to binary PCM with $\nu = 7$ and μ-law companding. But the DM signal requires $r_b = f_s = 2bW = 128$ kbps and $B_T \geq 64$ kHz, whereas the PCM signal would have $r_b = \nu f_s \geq 56$ kbps and $B_T \geq 28$ kHz.

Exercise 12.3-1 Consider a signal with a uniform power spectrum $G_x(f) = (S_x/2W)\Pi(f/2W)$. Show that $W_{\text{rms}} = W/\sqrt{3}$. Then calculate the optimum value of Δ in terms of S_x when $b = 16$.

Adaptive Delta Modulation

Adaptive delta modulation (ADM) involves additional hardware designed to provide *variable step size*, thereby reducing slope-overload effects without increasing the granular noise. A reexamination of Fig. 12.3-3 reveals that slope overload appears in $\epsilon_q(t)$ as a sequence of pulses having the same polarity, whereas the polarity tends to alternate when $x_q(t)$ tracks $x(t)$. This sequence information can be utilized to adapt the step size in accordance with the signal's characteristics.

Figure 12.3-6 portrays the action of an ADM transmitter in which the step size in the feedback loop is adjusted by a variable gain $g(k)$ such that

$$\tilde{x}_q(k) = \tilde{x}_q(k-1) + g(k-1)\epsilon_q(k-1)$$

The step-size controller carries out the adjustment algorithm

$$g(k) = \begin{cases} g(k-1) \times K & \epsilon_q(k) = \epsilon_q(k-1) \\ g(k-1)/K & \epsilon_q(k) \neq \epsilon_q(k-1) \end{cases}$$

where K is a constant taken to be in the range $1 < K < 2$. Thus, the effective step size increases by successive powers of K during slope-overload conditions, as signified by $\epsilon_q(k) = \epsilon_q(k-1)$, but decreases when $\epsilon_q(k) \neq \epsilon_q(k-1)$. Another adaptive

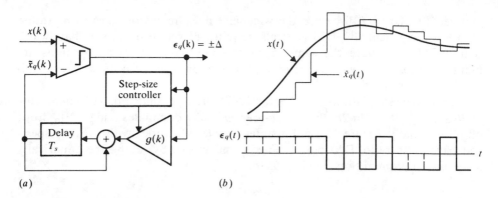

Figure 12.3-6 (a) Adaptive DM transmitter; (b) waveforms.

scheme called *continuously variable slope delta modulation* (CVSDM) provides a continuous range of step-size adjustment instead of a set of discrete values.

The signal-to-noise ratio of ADM is typically 8–14 dB better than ordinary DM. Furthermore, the variable step size yields a wider *dynamic range* for changing values of S_x, similar to the effect of μ-law companding in PCM. As a net result, ADM voice transmission gets by with a bandwidth ratio of $b = 6$–8 or $B_T = 24$–32 kHz.

Differential PCM

Differential pulse-code modulation (DPCM) combines prediction with multilevel quantizing and coding. The transmitter diagrammed in Fig. 12.3-7a has a q-level quantizer with quantum levels at $\pm\Delta, \pm 3\Delta, \ldots, \pm(q-1)\Delta$. The unquantized error $x(k) - \tilde{x}_q(k)$ is applied to the quantizer to produce the prediction error $\epsilon_q(k)$ which is then encoded as a binary word with $v = \log_2 q$ bits, just like binary PCM. DPCM transmission therefore requires

$$r_b = vf_s \qquad B_T \geq vf_s/2$$

However, since $\tilde{x}_q(k)$ changes as much as $\pm(q-1)\Delta$ from sample to sample, the slope tracking condition becomes

$$f_s(q-1)\Delta \geq |\dot{x}(t)|_{\max}$$

If $q \gg 1$ and $W_{\text{rms}} \ll W$, the sampling frequency can be nearly as low as the Nyquist rate.

Multilevel quantization of the prediction error obviously provides better information for message reconstruction at the receiver. To gain full advantage of this potential, the DPCM prediction circuit usually takes the form of a transversal filter shown in Fig. 12.3-7b, where

$$\tilde{x}_q(k) = \sum_{i=1}^{n} c_i x_q(k-i)$$

(a)

(b)

(c)

Figure 12.3-7 Differential PCM. (a) Transmitter; (b) prediction circuit; (c) receiver.

so the predictor draws upon the previous n samples. The tap gains c_i are chosen to minimize the mean square value of the error $x(k) - \tilde{x}_q(k)$. The receiver in Fig. 12.3-7c includes an identical prediction filter after the decoder.

Assuming $q \gg 1$ and no slope overload, DPCM performs essentially like PCM enhanced by a *prediction gain* G_p such that

$$\left(\frac{S}{N}\right)_D = G_p 3q^2 S_x \tag{15}$$

The gain of an optimum predictor is given by

$$G_p = \left[1 - \sum_{i=1}^{n} c_i \rho_i\right]^{-1} \tag{16a}$$

where $\rho_i = R_x(iT_s)/S_x$ is the normalized signal correlation and the tap gains satisfy the matrix relationship

$$\begin{bmatrix} \rho_0 & \rho_1 & \cdots & \rho_{n-1} \\ \rho_1 & \rho_0 & \cdots & \rho_{n-2} \\ \vdots & & & \vdots \\ \rho_{n-1} & \rho_{n-2} & \cdots & \rho_0 \end{bmatrix} \begin{bmatrix} c_1 \\ c_2 \\ \vdots \\ c_n \end{bmatrix} = \begin{bmatrix} \rho_1 \\ \rho_2 \\ \vdots \\ \rho_n \end{bmatrix} \tag{16b}$$

Jayant (1974) outlines the derivation of Eqs. (15) and (16) and presents experimental data showing that $G_p \approx 5\text{--}10$ dB for voice signals. The higher correlation of a TV video signal results in prediction gains of about 12 dB.

In contrast to delta modulation, DPCM employs more elaborate hardware than PCM for the purpose of improving performance quality or reducing the signaling rate and transmission bandwidth. *Adaptive* DPCM (ADPCM) achieves even greater improvement by adapting the quantizer or predictor or both to the signal characteristics.

Exercise 12.3-2 Suppose a DPCM predictor yields $G_p = 6$ dB. Show that the DPCM word needs one less bit than that of binary PCM, all other factors being equal. *Hint:* See Eq. (7), Sect. 12.1.

LPC Speech Synthesis

Linear predictive coding (LPC) is a new and radically different approach to digital representation of analog signals. The method uses a transversal filter (or its digital-circuit equivalent) plus some auxiliary components to *synthesize* the waveform in question. The parameters of the waveform synthesizer are then encoded for transmission, instead of the actual signal. Considerable efficiency results if the synthesizer accurately mimics the analog process. Since there already exists extensive knowledge about speech processes, LPC is particularly well suited to speech synthesis and transmission.

Figure 12.3-8 diagrams a speech synthesizer consisting of two input generators, a variable-gain amplifier, and a transversal filter in a feedback loop. The amplifier gain and filter tap gains are adjusted to model the acoustical properties of the vocal tract. Unvoiced speech (such as hissing sound) is produced by connecting the white-noise generator. Voiced speech is produced by connecting the impulse-train generator set at an appropriate pitch frequency.

If the filter has about 10 tap gains, and all parameter values are updated every 10 to 25 ms, the synthesized speech is quite intelligible although it may sound rather artificial. Some talking toys and recorded-message systems generate speech sounds by the synthesis method, drawing upon parameter values stored in a digital memory.

Figure 12.3-8 Speech synthesizer.

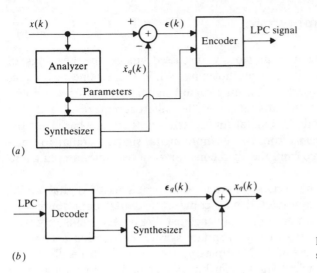

$x(k)$ + $\epsilon(k)$ LPC signal

Encoder

Analyzer $\hat{x}_q(k)$

Parameters

Synthesizer

(a)

$\epsilon_q(k)$ $x_q(k)$

LPC

Decoder

Synthesizer

(b)

Figure 12.3-9 LPC transmission system. (a) Transmitter; (b) receiver.

Now consider the LPC transmitter in Fig. 12.3-9a. Sample values of the voice input are analyzed to determine the parameters for the synthesizer, whose output is compared with the input. The resulting error is encoded along with the parameter values to form the transmitted digital signal. The receiver in Fig. 12.3-9b uses the parameter values and quantized error to reconstruct the voice waveform.

A complete LPC code word consists of about 80 bits — 1 bit for the voiced/unvoiced switch, 6 bits for the pitch frequency, 5 for the amplifier gain, 6 for each of the 10 tap gains, and a few bits for the error. Updating the parameters every 10–25 ms is equivalent to sampling at 40–100 Hz, so LPC requires a very modest bit rate in the vicinity of 3000–8000 bps. Table 12.3-1 compares LPC with other voice encoding methods. The substantial bit-rate reduction made possible by LPC has stimulated efforts to improve the quality of speech synthesis for voice communication. See Rabiner and Schafer (1978) for a general introduction to digital processing applied to speech signals.

Table 12.3-1 Comparison of voice encoding methods

Encoding method	Sampling rate, kHz	Bits per sample	Bit rate, kbps
DM	64–128	1	64–128
PCM	8	7–8	56–64
ADM	48–64	1	48–64
DPCM	8	4–6	32–48
ADPCM	8	3–4	24–32
LPC	0.04–0.1	≈ 80	3–8

12.4 DIGITAL MULTIPLEXING

Analog signal multiplexing was previously discussed under the headings of frequency-division and time-division multiplexing. While those same techniques could be applied to waveforms representing digital signals, we gain greater efficiency and flexibility by taking advantage of the inherent nature of a digital signal as a sequence of symbols. Digital multiplexing is therefore based on the principle of *interleaving symbols* from two or more digital signals, similar to time-division multiplexing but free from the rigid constraints of periodic sampling and waveform preservation.

This section presents the general concepts and problems of digital multiplexing. The signals to be multiplexed may come from digital data sources or from analog sources that have been digitally encoded. We'll deal exclusively with binary multiplexers, recognizing that appropriate symbol conversion accommodates *M*-ary signals when necessary. Multiplexing techniques will be illustrated by specific cases, including telephone system hierarchies and an introduction to computer networks.

Multiplexers and Hierarchies

A binary multiplexer (MUX) merges input bits from different sources into one signal for transmission via a digital communication system. Or, in other words, a MUX divides the capacity of the system between several pairs of input and output terminals. The multiplexed signal consists of source digits interleaved bit-by-bit or in clusters of bits (words or characters).

Successful demultiplexing at the destination requires a carefully constructed multiplexed signal with a constant bit rate. Towards this end, a MUX usually must perform four functional operations:

1. Establish a frame as the smallest time interval containing at least one bit from every input.
2. Assign to each input a number of unique bit slots within the frame.
3. Insert control bits for frame identification and synchronization.
4. Make allowance for any variations of the input bit rates.

Bit-rate variation poses the most vexing design problem in practice, and leads to three broad categories of multiplexers.

Synchronous multiplexers are used when a master clock governs all sources, thereby eliminating bit-rate variations. Synchronous multiplexing systems attain the highest throughput efficiency, but they require elaborate provision for distributing the master-clock signal.

Asynchronous multiplexers are used for digital data sources that operate in a start/stop mode, producing bursts of characters with variable spacing between bursts. Buffering and character interleaving makes it possible to

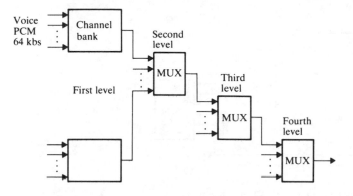

Figure 12.4-1 Multiplexing hierarchy for digital telecommunication.

merge these sources into a synchronous multiplexed bit stream, as discussed later in conjunction with computer networks.

Quasi-synchronous multiplexers are used when the input bit rates have the same nominal value but vary within specified bounds. These multiplexers, arranged in a hierarchy of increasing bit rates, constitute the building blocks of interconnected digital telecommunication systems.

Two slightly different multiplexing patterns have been adopted for digital telecommunication, the AT & T hierarchy in North America and Japan and the CCIT hierarchy in Europe. (CCIT stands for the International Telegraph and Telephone Consultive Committee of the International Telecommunications Union.) Both hierarchies are based on a 64-kbps voice PCM unit and have the same structural layout shown in Fig. 12.4-1. The third level is intended only for multiplexing purposes, whereas the other three levels are designed for point-to-point transmission as well as multiplexing. The parameters of the AT & T and CCIT hierarchies are listed in Table 12.4-1.

Observe in all cases that the output bit rate at a given level exceeds the sum of the input bit rates. This surplus allows for control bits and additional *stuff bits* needed to yield a steady output rate. Consequently, when we include the PCM bandwidth expansion, a digital network has very low *bandwidth efficiency* if devoted entirely to voice transmission. For instance, the fourth level of the

Table 12.4-1 Multiplexing hierarchies

	AT & T		CCIT	
	Number of inputs	Output rate, Mbps	Number of inputs	Output rate, Mbps
First level	24	1.544	30	2.048
Second level	4	6.312	4	8.448
Third level	7	44.736	4	34.368
Fourth level	6	274.176	4	139.264

Figure 12.4-2 Illustrative configuration of the AT & T hierarchy.

AT & T multiplexing system requires $B_T \geq r_b/2 \approx 137$ MHz to transmit $24 \times 4 \times 7 \times 6 = 4032$ voice PCM signals, so the bandwidth efficiency is $(4032 \times 4 \text{ kHz})/137$ MHz $\approx 12\%$. Going back to Table 8.2-1 we find that the Jumbogroup in the AT & T FDM hierarchy accommodates 3600 analog voice signals in $B_T = 17$ MHz, for a much higher bandwidth efficiency of $(3600 \times 4 \text{ kHz})/17$ MHz $\approx 85\%$. Thus, digital multiplexing sacrifices analog bandwidth efficiency in exchange for the advantages of digital transmission.

Previously noted advantages of digital transmission include hardware cost reduction made possible by digital integrated circuits and power cost reduction made possible by regenerative repeaters. Now we can begin to appreciate the advantage of *flexibility* made possible by digital multiplexing, since the input bit streams at any level in Fig. 12.4-1 can be any desired mix of digital data and digitally encoded analog signals.

By way of example, Fig. 12.4-2 shows an illustrative configuration of the AT & T hierarchy with voice, digital data, visual telephone (Picturephone®), and color TV signals combined for transmission on the fourth-level T4 line. The first-level T1 signals include PCM voice and multiplexed digital data. The second-level T2 signals are multiplexed T1 signals along with visual telephone signals encoded as binary differential PCM (DPCM) with $f_s \approx 2$ MHz and $v = 3$ bits per word. PCM encoding of color TV requires a 90-Mbps bit rate ($f_s \approx 10$ MHz, $v = 9$), so two third-level T3 lines are allocated to this signal. The higher-level multiplexers labeled M12, M23, and M34 belong to the quasi-synchronous class whose properties we'll examine in some detail. But let's start at the first-level synchronous multiplexer called the channel bank.

Example 12.4-1 T1 voice PCM channel bank Synchronous multiplexing of voice PCM requires that the signals be delivered in *analog* form to the channel bank. Then, as diagrammed in Fig. 12.4-3a, sequential sampling under the control of a local clock generates an analog TDM PAM signal (see Figs. 10.3-1 and 10.3-2). This signal is transformed by the encoder into TDM PCM with interleaved words. Finally, the processor appends framing and signaling information to produce the output T1 signal.

Figure 12.4-3 (*a*) T1 channel bank; (*b*) frame structure.

The T1 frame structure is represented by Fig. 12.4-3*b*. Each frame contains one 8-bit word from each of the 24 input channels plus one bit for framing, giving a total of 193 bits. The 8-kHz sampling frequency corresponds to 125-μs frame duration, so the T1 bit rate is $r_b = 193$ bits $\div$ 125 μs = 1.544 Mbps. Signaling information (dial pulses, "busy" signals, etc.) is incorporated by a method aptly known as *bit robbing*. Every sixth frame, a signaling bit replaces the least-significant bit of each channel word—denoted by the starred bit locations in the figure. Bit-robbing reduces the effective voice-PCM word length to $v = 7\frac{5}{6}$ and has inconsequential effect on reproduction quality. Yet it allows 24 signaling bits every 6 × 125 μs, or an equivalent signaling rate of 32 kbps.

T1 signals may be either combined at an M12 multiplexer or transmitted directly over short-haul links for local service up to 80 km. The T1 transmission line is a twisted-pair cable with regenerative repeaters every 2 km. A bipolar signal format eliminates the problems of absolute polarity and DC transmission.

Exercise 12.4-1 Assume that the first-level multiplexer in the CCIT hierarchy is a synchronous voice-PCM channel bank with 30 input signals, output bit rate $r_b = 2.048$ Mbps, and no bit-robbing. Find the number of framing plus signaling bits per frame.

Quasi-Synchronous Multiplexing

Quasi-synchronous multiplexing becomes necessary when the input bit streams have small variations around some nominal rate. This condition prevails at an M12 multiplexer, for instance, because the arriving T1 signals were generated by channel banks operating from stable but unsychronized local clocks. In general, the problems of quasi-synchronism arise in any digital multiplexing system that lacks overall master-clock control.

Quasi-synchronous multiplexers require three essential features: a sufficiently high output bit rate to accommodate the maximum expected input rates, special buffers called *elastic stores* for temporary storage of input bits, and *bit-stuffing* to pad the output stream. We'll introduce these features by considering the simplified rate-regulation system in Fig. 12.4-4a. Here a single input bit stream with variable rate r is to be regulated such that the output has fixed rate $r_0 \geq r_{max}$. Whenever $r < r_{max}$, the regulator stuffs extra bits into the output stream to maintain a constant rate. This system operates in the following manner.

Input bits are written sequentially into an elastic store consisting of a bank of storage cells and two switching circuits that act like commutators rotating at the input and output clock rates. Both clock signals are applied to a set-reset (SR) flip-flop whose output waveform drawn in Fig. 12.4-4b has a DC component that progressively decreases if $r_0 > r$. The control unit issues a stuff command pulse when this DC component gets too low, indicating that the read circuit is overtaking the write circuit. The stuff command inhibits the read circuit for one clock cycle and inserts a stuff bit into the output stream. Normal operation then

Figure 12.4-4 (*a*) Bit rate regulation system; (*b*) timing diagram.

Figure 12.4-5 Quasi-synchronous multiplexer.

resumes until the next time for a stuff bit. Unwanted stuff bits can be "destuffed" at the destination by another elastic store whose read circuit skips over storage cells identified by the transmitted stuff commands.

Now we put bit stuffing together with synchronous multiplexing to build the quasi-synchronous multiplexer in Fig. 12.4-5. There are N input signals whose rates vary from r_{min} to r_{max} around the nominal value $\bar{r}$. The MUX has a constant output rate $r_0 > Nr_{max}$. The control unit receives flip–flop waveforms and returns read signals to each input elastic store. This unit also supplies the control and stuff bits to the MUX.

For an analytic description of quasi-synchronous multiplexing, let T_f be the frame duration. The total number of output bits per frame is

$$T_f r_0 = D + X \tag{1}$$

where D represents the number of message and stuff bits and X represents the number of control bits. Although D is a fixed parameter, the proportion of message and stuff bits depends on the input rates. If all inputs have the same rate $\bar{r}$, then

$$D = T_f N\bar{r} + Ns \tag{2}$$

in which s stands for the average number of stuff bits per channel per frame. Thus, for instance, $s = 1/2$ means that each channel gets one stuff bit every other frame under nominal operating conditions. Using this notation, we write the output rate as

$$r_0 = \frac{D + X}{T_f} = \frac{D + X}{T_f D} T_f N\bar{r} \frac{D}{D - Ns}$$

$$= N\bar{r} \left(\frac{D + X}{D}\right) \frac{D}{D - Ns} \tag{3}$$

This expression relates r_0 and s when the other parameters have been specified.

Given r_0 and s, we determine the allowable input rate variations by worst-case analysis as follows. The maximum rate must satisfy $T_f N r_{max} \leq D$, where the upper limit corresponds to a frame with no stuff bits. At the other extreme, we usually want at most one stuff bit per channel per frame in order to simplify de-stuffing, so the minimum rate must satisfy $T_f N r_{min} \geq D - N$. Equations (1) and (2) thus yield

$$r_{max} \leq \bar{r} + \frac{r_0}{D + X} s \qquad r_{min} \geq \bar{r} - \frac{r_0}{D + X} (1 - s) \qquad (4)$$

In principle, Eq. (4) suggests taking $s = 1/2$ to permit equal rate variations around $\bar{r}$. In practice, lower average stuffing rates are employed to reduce timing jitter in the destuffed signals.

Example 12.4-2 M12 multiplexer The AT & T M12 multiplexer combines $N = 4$ T1 signals with $\bar{r} = 1.544$ Mbps. The framing pattern is divided into four subframes depicted by Fig. 12.4-6a. The letters M, C, and F denote control bits, and [48] denotes a sequence of 12 interleaved message and stuff bits from each of the four inputs. A complete frame consists of $X = 4 \times 6 = 24$ control bits and $D = 4 \times 6 \times 48 = 1152$ message and stuff bits. After canceling common factors, Eq. (3) becomes

$$r_0 = 4 \times 1.544 \text{ Mbps} \times \frac{49}{48} \times \frac{288}{288 - s}$$

The value $r_0 = 6.312$ Mbps was selected as a multiple of 8 kHz, and corresponds to $s \approx 1/3$. Each frame contains $Ns \approx 4/3$ stuff bits on the average, for a throughput efficiency of $(D - Ns)/(D + X) = N\bar{r}/r_0 \approx 98\%$.

The specific arrangement of control bits within the frame involves several considerations, some of which are made clearer with the help of Fig. 12.4-6b. The F control bits form the main framing sequence $F_0 F_1 F_0 F_1 \cdots = 0101 \cdots$ with 146 other bits between F bits. The demultiplexer synchronizes to this

(a)

Subframe I:	M_0	[48]	C_I	[48]	F_0	[48]	C_I	[48]	C_I	[48]	F_1	[48]
Subframe II:	M_1	[48]	C_{II}	[48]	F_0	[48]	C_{II}	[48]	C_{II}	[48]	F_1	[48]
Subframe III:	M_1	[48]	C_{III}	[48]	F_0	[48]	C_{III}	[48]	C_{III}	[48]	F_1	[48]
Subframe IV:	M_1	[48]	C_{IV}	[48]	F_0	[48]	C_{IV}	[48]	C_{IV}	[48]	F_1	[48]

(b)

Figure 12.4-6 M12 multiplexer. (a) Framing pattern; (b) frame synchronization sequence.

pattern and then identifies the subframes from the secondary sequence $M_0 M_1 M_1 M_1 = 0111$.

The C control bits in a given subframe serve as a stuff code. If $C_I C_I C_I = 111$, then channel I has a stuff bit in its first allocated slot after the F_1 bit; otherwise, $C_I C_I C_I = 000$ indicates no stuff bit. Stuff bits for channel II are likewise handled in the second subframe, and so forth. The triple bit pattern provides protection against regeneration errors, since the probability of two or more errors in three bits is quite small.

Exercise 12.4-2 Suppose 24 voice signals arrive at a channel bank already encoded as PCM with $\bar{r} = 64$ kbps. Further suppose that the channel bank is a quasi-synchronous multiplexer whose output frame is divided into 24 subframes, each subframe containing 3 control bits and 8 message and stuff bits. Calculate r_0 taking $s = 1/3$, and determine the throughput efficiency $N\bar{r}/r_0$.

Data Multiplexers and Computer Networks

Multiplexing for computer communication departs in two significant respects from general-purpose telecommunication service. On the one hand, complications arise from the fact that each computer and data terminal has its own independent clock and operates in an asynchronous start/stop mode. On the other hand, simplifications come from the fact that computers and data terminals don't require the nearly instantaneous response needed for two-way voice communication. Consequently, extensive buffering is essential and the associated time delay is tolerable.

A third characteristic of computer communication is relatively low bit rates compared to the megabits-per-second needed for multiplexed voice PCM. Although internal data transfers take place at very high speeds within a central processor, the limitations of electromechanical input/output devices result in much slower external transmission rates. Even high-speed printers can handle only 50 kbps, while multiplexed signals from remote data terminals are typically less than 10 kbps. We'll first look at asynchronous data multiplexers, and then describe how they are interconnected with central processors to form a computer network.

Teletypewriters and other keyboard data terminals work with alphanumeric symbols encoded as characters of 7–11 bits. These characters are transmitted asynchronously, one at a time, but the bit interval stays fixed throughout all characters. Figure 12.4-7 displays the 10-bit character format of the American

Figure 12.4-7 Format of ASCII 10-bit character code

Table 12.4-2 Selected AT & T data modems

Model	Bit rate	Modulation	Transmission channel
103A	300 bps	FSK	
202E	600 bps	FSK	Dial-up
212A	1200 bps	QAM	voice-grade line
201A	2000 bps	PSK	
203B	2400 bps	VSB	Leased
203C	3600 bps	VSB	voice-grade line
208A	7200 bps	PSK	Leased
209	9600 bps	QAM	conditioned line
303B	19.2 kbps	VSB	Half-group channel
303C	50.0 kbps	VSB	Group channel
303D	230.4 kbps	VSB	Supergroup channel

Standard Code for Information Interchange (ASCII). The start and stop bits provide character framing, the parity bit serves as error control, and the remaining 7 data bits identify the particular character. Incidentally, the signal voltage remains HIGH during idle times to keep the connection "alive."

Bit rates within the characters of low-speed asynchronous terminals range from 75 to 1200 bps. Higher-speed synchronous terminals have rates up to 7200 bps. Commercial data multiplexers are designed to combine characters from a mix of asynchronous and synchronous terminals. The multiplexed signal is then applied to a *modem* (modulator/demodulator) for transmission over a telephone channel. For purposes of illustration, Table 12.4-2 lists the maximum bit rates of selected modems available from AT & T. The carrier modulation methods will be covered in Chap. 14.

A data multiplexer operates in the same basic manner as a quasi-synchronous multiplexer, with interleaved characters and character-stuffing in place of single bits. However, the frame structure must account for the potential variety of input rates. Borrowing a strategy from TDM telemetry, frames are comprised of several characters from high-speed terminals but just one character or alternate characters from low-speed terminals. Figure 12.4-8 provides a simpli-

Figure 12.4-8 Illustrative data multiplexer.

fied example of this strategy with input rates of 110, 110, 300, and 600 bps multiplexed into a 1200-bps output, the 80-bps surplus allowing for control characters. A subframe would then contain two characters from the 600-bps input, one from the 300-bps input, and alternate characters from the 110-bps inputs. Notice the similarity to the telemetry multiplexer back in Fig. 10.3-5.

Conventional or "nonintelligent" data multiplexers have *fixed input assignments* and continue to transmit dummy characters whenever a particular terminal happens to be idle. *Demand-assignment* multiplexers achieve better efficiency by skipping over idle inputs, so the output bit rate can be reduced in proportion to the expected demand factor. Of course, the input buffers must hold several characters in order to smooth out demand variations. The more sophisticated data *concentrator* goes even further in this direction, with the help of a microprocessor and extensive memory capacity. A concentrator assembles and temporarily stores blocks of characters from several inputs, performs data-compression coding, and carries out other operations designed to minimize the output transmission bit rate.

As implied by the foregoing discussion, efficient data multiplexing involves significant processing time delays — delays that would be unacceptable in two-way voice communication. But such delays don't bother machines, and human users of data terminals have become accustomed to brief turn-around lags. Most computer networks operate in a *store-and-forward* mode that also takes advantage of time-delay tolerance.

Figure 12.4-9 diagrams a hypothetical computer network modeled after the famous ARPANET. The ARPANET was established in 1969 by the Advanced Research Projects Agency of the U.S. Department of Defense; it now interconnects more than 100 host computers via telephone lines and satellite links, mostly at 50-kbps rates. Our hypothetical network consists of five host computers, symbolized by circles, and four additional terminal interfaces, symbolized by squares. Only host computers do computational work, but every circle or square represents a communication *node* equipped with a concentrator having the ability to transmit, receive, and store data.

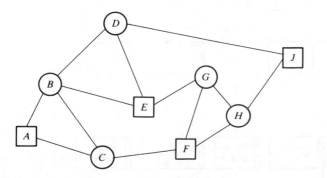

Figure 12.4-9 Computer network.

When a message is to be sent from one node to another, the source node subdivides the data into *packets* of about 1000 bits. Control bits attached to each packet include identification of the source and destination, The packets are transmitted, one by one, either directly to the destination node if possible or to the first available intermediate node. Packets arriving at an intermediate node are temporarily *stored* and then *forwarded* to the destination or to another intermediate node, depending upon the available lines. Suppose, for instance, that a two-packet message is to go from the terminal interface at node A to the host computer at node G; one packet might traverse the route A-C-F-G while the other follows the more circuitous path A-B-E-D-J-H-G.

This network illustrates the principles of store-and-forward operation with *packet-switching*. Both packet-switching and message-switching make better use of a transmission network than line-switching, which requires a complete route to be established before transmission begins. Packet-switching has the added advantage of eliminating the excessive time delay incurred when an entire message must be stored at an intermediate node.

But packet-switching has some interesting problems since, at any given time, the network contains numerous packets attempting to reach their respective destinations. Packets from one message can take different routes and may arrive at the destination in scrambled order. Furthermore, a *deadlock* situation results if too many packets have entered the network and compete for the available lines and buffer space—something like traffic gridlock in New York City on a Friday afternoon.

The packetizing technique is also used for *time-division multiple-access* (TDMA) in which a satellite or ground-based transponder links several locations. Each source location has periodic time slots assigned for packet transmission, and the transponder rebroadcasts a composite signal consisting of interleaved

Figure 12.4-10 TDMA system. (*a*) Ground station transmitter; (*b*) transponder signal.

packets. As an example, recall the FDMA satellite system in Fig. 8.2-4 with ground stations in the United States, France, and Brazil. Figure 12.4-10a shows how the U.S. transmitter might be arranged for TDMA operation. A low-speed *background processor* stores the incoming data while a high-speed *foreground processor* compresses packets into short bursts at the assigned times, adding a preamble for identification and synchronization purposes. All ground stations receive the transponder signal represented by Fig. 12.4-10b and extract the appropriate packets. Other TDMA systems have demand-assigned or random packet transmission times, rather than fixed time slots.

Analysis and design methods for packet-switching and computer networks are covered in textbooks such as Tanenbaum (1981). Spilker (1977) is devoted entirely to digital communication by satellite.

12.5 PROBLEMS

12.1-1 An analog waveform with $W = 15$ kHz is to be quantized to $q \geq 200$ levels and transmitted via an M-ary PCM signal having $M = 2^n$. Find the maximum allowed values of v and f_s and the corresponding value of n when the available transmission bandwidth is $B_T = 50$ kHz.

12.1-2 Do Prob. 12.1-1 with $B_T = 80$ kHz.

12.1-3 *Hyperquantization* is the process whereby N successive quantized sample values are represented by a single pulse with q^N possible values. Describe how PCM with hyperquantization can achieve *bandwidth compression*, so $B_T < W$.

12.1-4 Suppose the PCM quantization error ϵ_k is specified to be no greater than $\pm P\%$ of the peak-to-peak signal range. Obtain the corresponding condition on v in terms of M and P.

12.1-5 A voice signal having $W = 3$ kHz and $S_x = 1/4$ is to be transmitted via M-ary PCM. Determine values for M, v, and f_s such that $(S/N)_D \geq 40$ dB if $B_T = 16$ kHz.

12.1-6 Do Prob. 12.1-5 with $(S/N)_D \geq 36$ dB and $B_T = 20$ kHz.

12.1-7 An audio signal with $S_x = 0.3$ is to be transmitted via a PCM system whose parameters must satisfy the standards for broadcast-quality audio transmission listed in Table 5.4-1. (a) If $M = 2$, then what are the required values of v and B_T? (b) If $B_T = 4W$, then what's the minimum value of M?

12.1-8 Do Prob. 12.1-7 for high-fidelity audio transmission standards.

12.1-9‡ Consider the signal PDF in Fig. P12.1-9. Use Eq. (9) to show that uniform quantization with $q = 2^v$ levels and $v \geq 3$ yields $\sigma_q^2 = 1/3q^2$. Then calculate S_x and find $(S/N)_D$ in terms of q.

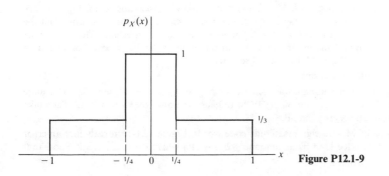

Figure P12.1-9

12.1-10‡ Consider a signal whose PDF is reasonably smooth and continuous and satisfies the condition $P[|x(t)| > 1] \ll 1$. From Eq. (9), show that uniform quantization with $q \gg 1$ yields $\sigma_q^2 \approx 1/3q^2$.

12.1-11 Let $x(t)$ have a zero-mean gaussian PDF with variance σ_x^2 such that $P[|x(t)| > 1] \le 0.01$. Use the result from Prob. 12.1-10 to obtain $(S/N)_D$ in a form like Eq. (7), assuming uniform quantization with $q \gg 1$.

12.1-12 Let $x(t)$ have a Laplace PDF, as given by Eq. (13), with variance σ_x^2 such that $P[|x(t)| > 1] \le 0.01$. Use the result from Prob. 12.1-10 to obtain $(S/N)_D$ in a form like Eq. (7), assuming uniform quantization with $q \gg 1$.

12.1-13 Consider the compressor characteristic $z(x) = (\text{sgn } x)\sqrt{|x|}$ for $|x| \le 1$. (a) Find and sketch the complementary expander characteristic $x(z)$. (b) Evaluate K_z when $x(t)$ has the PDF in Fig. P12.1-9.

12.1-14 Consider the μ-law compressor in Example 12.1-2. (a) Find the complementary expander characteristic $x(z)$. (b) Carry out the details leading to K_z as given in Eq. (12).

12.1-15 A voice signal with a Laplace distribution is applied to a compressor having $z(x) = (\text{sgn } x)(1 - e^{-3|x|})$ for $|x| \le 1$. (a) Obtain an expression for K_z in terms of α. Then show that $K_z \approx 1/9$ for $\alpha \gg 1$ and express $(S/N)_D$ in dB in terms of v and S_x for binary voice PCM with this compressor. (b) Now take $q = 2^8$ and evaluate S_x, K_z, and $(S/N)_D$ for $\alpha = 4$, 8, and 16. Use these results to make a plot like Fig. 12.1-7. Your plot should demonstrate that this compressor yields a higher $(S/N)_D$ but less dynamic range than the standard μ-law compressor.

12.1-16‡ The *A-law companding system* employs a compressor with

$$z(x) = \begin{cases} Ax/(1 + \ln A) & 0 \le x \le 1/A \\ (1 + \ln Ax)/(1 + \ln A) & 1/A < x \le 1 \end{cases}$$

and $z(x) = -z(x)$ for $-1 \le x \le 0$. (a) Assume $p_X(x)$ has even symmetry and negligible area outside $|x| \le 1$ to show that

$$K_z = (1 + \ln A)^2 \left[S_x + 2 \int_0^{1/A} \left(\frac{1}{A^2} - x^2 \right) p_X(x) \, dx \right]$$

(b) Obtain the expression for K_z in terms of A and α when $x(t)$ has a Laplace distribution, and show that $K_z \approx (1 + \ln A)^2/A^2$ when $\alpha \gg A$. (c) Let $A = 100$ and $q = 2^8$. Evaluate S_x, K_z, and $(S/N)_D$ for $\alpha = 4$, 16, 64 and $\alpha \gg 100$. Use these results to make a plot like Fig. 12.1-7. Your plot should demonstrate that A-law companding yields a lower $(S/N)_D$ but greater dynamic range than standard μ-law companding.

12.2-1 A signal with $S_x = 1/2$ and $W = 6$ kHz is to be transmitted via M-ary PCM on a channel having $\eta = 0.01$ μW/Hz and $B_T = 15$ kHz. Find the smallest value of M and the corresponding smallest value of v that yields $(S/N)_D \ge 36$ dB. The calculate the minimum value of S_R for operation above threshold, assuming the PCM signal occupies the full transmission bandwidth.

12.2-2 Do Prob. 12.2-1 with $B_T = 20$ kHz.

12.2-3 Do Prob. 12.2-1 with $B_T = 50$ kHz.

12.2-4 Consider a voice PCM system with $M = 2$, $v = 8$, and μ-law companding, so $(S/N)_D = 37$ dB. The PCM signal is transmitted via a regenerative repeater system with 20 identical sections. Find the minimum value of γ at the input to the first repeater for operation above threshold. Then determine the PCM advantage in dB by calculating the value of γ at the input to the first repeater of an analog repeater system that yields $(S/N)_D = 37$ dB after 20 sections.

12.2-5 Do Prob. 12.2-4 with 100 sections.

12.2-6 Show from Eq. (3) that decoding noise decreases $(S/N)_D$ by 1 dB when $P_e \approx 1/15q^2$. Calculate the corresponding threshold value of γ for binary PCM in gaussian noise when $v = 4$, 8, 12. Then take $S_x = 1$ and plot $(S/N)_D$ in dB versus γ_{th} in dB.

12.2-7‡ Consider binary PCM with sign/magnitude code words. Figure 12.1-1b reveals that an error in the sign bit shifts the decoded level by an amount $\pm 2(2i - 1)/q$ where $i = 1, 2, \ldots, q/2$. Show that $\overline{\epsilon_m^2} = (5q^2 - 8)/3vq^2$, so $\overline{\epsilon_m^2} \approx 5/3v$ if $q \gg 1$.

12.2-8 Suppose a PCM system has fixed values of γ and $b = B_T/W$. Taking threshold into account, obtain an expression for the maximum value of q.

12.2-9 Find and plot γ_{th} in dB versus $b = B_T/W$ for a PCM system with $q = 256$ and $v = b$.

12.3-1 The signal $x(t) = 8 \cos 2\pi Wt$ is the input to a delta modulator with $T_s = 1/24W$. Plot the sample points $x(kT_s)$ for $0 \le k \le 30$. Then take $\tilde{x}_q(0) = 0$ and plot $\tilde{x}_q(t)$ with $\Delta = 1$ and $\Delta = 3$.

12.3-2 Use the results of Example 12.3-1 to plot $(S/N)_D$ in dB versus b for DM voice transmission with $b = 4, 8,$ and 16. Compare with binary PCM transmission, taking $v = b$ and $S_x = 1/30$.

12.3-3 Use the results of Example 12.3-1 to tabulate f_s and the optimum value of Δ for DM voice transmission with $b = 4, 8,$ and 16 when $S_x = 1/9$.

12.3-4 Suppose a signal to be transmitted via DM has a dominant spectral peak at $f = f_0 < W$, so that $|\dot{x}(t)|_{max} = 2\pi f_0$. Obtain an upper bound on $(S/N)_D$ in a form like Eq. (12).

12.3-5 The power spectrum of a lowpass filtered voice signal can be approximated as $G_x(f) = [K/(f_0^2 + f^2)]\Pi(f/2W)$. Find K in terms of S_x, and obtain an expression for the rms bandwidth. Evaluate W_{rms} when $W = 4$ kHz and $f_0 = 0.8$ kHz.

12.3-6 An approximate expression for DM slope-overload noise is given by Abate (1967) as

$$N_{so} = \frac{8\pi^2}{27}\left(\frac{W_{rms}}{W}\right)^2 (3s + 1)e^{-3s}S_x$$

Write Δ^2/f_s in Eq. (7) in terms of s and b, and show that the total quantization noise $N_q + N_{so}$ is minimized by taking $s = \ln 2b$.

12.3-7 Find the tap gains and evaluate the prediction gain in dB for DPCM with a one-tap and a two-tap transversal filter when the input is a voice signal having $\rho_1 = 0.8$ and $\rho_2 = 0.6$.

12.3-8 Do Prob. 12.3-7 for a TV image signal having $\rho_1 = 0.95$ and $\rho_2 = 0.90$.

12.3-9 Consider a DPCM system with a two-tap transversal prediction filter. Assuming $q \gg 1$ so that $x_q(kT_s) \approx x(kT_s)$, find the tap gains to implement the prediction strategy

$$\tilde{x}_q(kT_s) \approx x[(k-1)T_s] + T_s \left.\frac{dx(t)}{dt}\right|_{t=(k-1)T_s}$$

12.3-10 Consider a DPCM system with a one-tap transversal prediction filter and $q \gg 1$, so $x_q(kT_s) \approx x(kT_s)$ and $\epsilon_q(kT_s) \approx x(kT_s) - \tilde{x}_q(kT_s)$. Obtain an expression for the mean square prediction error $\overline{\epsilon^2} = E[\epsilon_q^2(k)]$ in terms of S_x and the signal's autocorrelation $R_x(\tau) = E[x(t)x(t - \tau)]$. Then find the value of the tap gain that minimizes $\overline{\epsilon^2}$.

12.3-11 Use the method outlined in Prob. 12.3-10 to find the tap gains for a two-tap transversal prediction filter. Express your results in a matrix like Eq. (16b).

12.4-1 Several high-fidelity audio channels having $W = 15$ kHz are to be transmitted via binary PCM with $v = 12$. Determine how many of the PCM signals can be accommodated by the first level of the AT & T multiplexing hierarchy. Then calculate the corresponding bandwidth efficiency.

12.4-2 Do Prob. 12.4-1 for the first level of the CCIT hierarchy.

12.4-3 Consider the quasi-synchronous M12 multiplexer for the second level of the CCIT hierarchy. Assume that a frame consists of $N = 4$ subframes, each having the pattern $F[36]C[36]C[36]C[36]$ where F and C represent control bits and [36] denotes a sequence of 36 interleaved message and stuff bits. (a) Calculate the average number of stuff bits per channel per frame. (b) Find the upper limit on $(r_{max} - r_{min})/\bar{r}$, which corresponds to the maximum fractional drift of the first-level clocks.

12.4-4 Consider the quasi-synchronous M23 multiplexer for the third level of the AT & T hierarchy. Assume that a frame consists of $N = 7$ subframes, each having the pattern in Fig. 12.4-6a with $7 \times 12 = 84$ rather than 48 interleaved message and stuff bits between control bits. (a) Calculate the average number of stuff bits per channel per frame. (b) Find the upper limit on $(r_{max} - r_{min})/\bar{r}$, which corresponds to the maximum fractional drift of the second-level clocks.

12.4-5 Consider a hypothetical M13 multiplexer, intended to combine $N = 28$ T1 signals and produce an output at the third level of the AT & T hierarchy. The proposed framing pattern is a modified version of Fig. 12.4-6a, with N subframes, 6 control bits per subframe, and $6 \times N \times n$ message and stuff bits per subframe. Obtain an expression for s in terms of n, and explain why this scheme would be unsatisfactory.

12.4-6 Suppose both the input and output clock rates of the AT & T M12 multiplexer drift by a maximum factor of $1 \pm \delta$ relative to their nominal rates. Use worst-case analysis to determine the upper limit on δ.

12.4-7 A certain multiplexer combines data from several 110-bps asynchronous terminals for synchronous transmission at 4800 bps. The multiplexer inserts a synchronizing character after every 19 data characters, and it requires 2% of the output for control and stuff characters. How many terminals can be accommodated along with three 1200-bps data signals?

12.4-8 A 2400-bps synchronous signal is to be generated by multiplexing asynchronous data from one 1200-bps terminal, two 300-bps terminals, and several 110-bps terminals. The multiplexer inserts a synchronizing character after every 49 data characters, and it requires 3% of the output for control and stuff characters. Find the maximum allowed number of 110-bps terminals. Then draw a multiplexer diagram like Fig. 12.4-8 using a processor with eight input ports.

12.4-9 A 3600-bps synchronous signal is to be generated by multiplexing asynchronous data from one 1200-bps terminal, two 600-bps terminals, one 300-bps terminal, and several 150-bps terminals. The multiplexer inserts a synchronizing character after every 49 data characters, and it requires 2% of the output for control and stuff characters. Find the maximum allowed number of 150-bps terminals. Then draw a multiplexer diagram like Fig. 12.4-8 using a processor with six input ports.

THIRTEEN

ERROR-CONTROL CODING

Transmission errors in digital communication depend on the signal-to-noise ratio. If a particular system has a fixed value of S/N and the error rate is unacceptably high, then some other means of improving reliability must be sought. Error-control coding often provides the best solution.

Error-control coding involves systematic addition of extra digits to the transmitted message. These extra *check digits* convey no information by themselves, but they make it possible to detect or correct errors in the regenerated message digits. In principle, information theory holds out the promise of nearly errorless transmission, as will be discussed in Chap. 15. In practice, we seek some compromise between conflicting considerations of reliability, efficiency, and equipment complexity. A multitude of error-control codes have therefore been devised to suit various applications.

This chapter starts with an overview of error-control coding, emphasizing the distinction between *error detection* and *error correction* and systems that employ these strategies. Subsequent sections describe the two major types of code implementations, *block codes* and *convolutional codes*. We'll stick entirely to binary coding, and we'll omit formal mathematical analysis. Detailed treatments of error-control coding are provided by the references cited in the supplementary reading list.

13.1 ERROR DETECTION AND CORRECTION

Coding for error detection, without correction, is simpler than error-correction coding. When a two-way channel exists between source and destination, the receiver can request retransmission of information containing detected errors.

This error-control strategy, called *automatic repeat request* (ARQ), particularly suits data communication systems such as computer networks. However, when retransmission is impossible or impractical, error control must take the form of *forward error correction* (FEC) using an error-correcting code. Both strategies will be examined here, after an introduction to simple but illustrative coding techniques.

Repetition and Parity-Check Codes

When you try to talk to someone across a noisy room, you may need to repeat yourself to be understood. A brute-force approach to binary communication over a noisy channel likewise employs *repetition*, so each message bit is represented by a *codeword* consisting of n identical bits. Any transmission error in a received codeword alters the repetition pattern by changing a 1 to a 0 or vice versa.

If transmission errors occur randomly and independently with probability $P_e = \alpha$, then the binomial frequency function from Eq. (1), Sect. 4.4, gives the probability of i errors in an n-bit codeword as

$$P(i, n) = \binom{n}{i} \alpha^i (1 - \alpha)^{n-i}$$

$$\approx \binom{n}{i} \alpha^i \qquad \alpha \ll 1$$

(1a)

where

$$\binom{n}{i} = \frac{n!}{i!\,(n-i)!} = \frac{n(n-1)\,\cdots\,(n-i+1)}{i!}$$

(1b)

We'll proceed on the assumption that $\alpha \ll 1$—which does not necessarily imply reliable transmission since $\alpha = 0.1$ satisfies our condition but would be an unacceptable error probability for digital communication. Repetition codes improve reliability when α is sufficiently small that $P(i + 1, n) \ll P(i, n)$ and, consequently, several errors per word are much less likely than a few errors per word.

Consider, for instance, a triple-repetition code with codewords 000 and 111. All other received words, such as 001 or 101, clearly indicate the presence of errors. Depending on the decoding scheme, this code can *detect* or *correct* erroneous words. For error detection without correction, we say that any word other than 000 or 111 is a detected error. Single and double errors in a word are thereby detected, but triple errors result in an undetected *word error* with probability

$$P_{we} = P(3, 3) = \alpha^3$$

For error correction, we use majority-rule decoding based on the assumption that at least two of the three bits are correct. Thus, 001 and 101 are decoded as 000 and 111, respectively. This rule corrects words with single errors, but double or triple errors result in a decoding error with probability

$$P_{we} = P(2, 3) + P(3, 3) = 3\alpha^2 - 2\alpha^3$$

Since $P_e = \alpha$ would be the error probability without coding, we see that either decoding scheme for the triple-repetition code greatly improves reliability if, say, $\alpha \le 0.01$. However, this improvement is gained at the cost of reducing the message bit rate by a factor of $1/3$.

More efficient codes are based on the notion of *parity*. The parity of a binary word is said to be even when the word contains an even number of 1s, while odd parity means an odd number of 1s. The codewords for an *error-detecting parity-check code* are constructed with $n - 1$ message bits and one check bit chosen such that all codewords have the same parity. With $n = 3$ and even parity, the valid codewords are 000, 011, 101, and 110, the last bit in each word being the parity check. When a received word has odd parity, 001 for instance, we immediately know that it contains a transmission error—or three errors or, in general, an odd number of errors. Error correction is not possible because we don't know where the errors fall within the word. Furthermore, an even number of errors preserves valid parity and goes unnoticed.

Under the condition $\alpha \ll 1$, double errors occur far more often than four or more errors per word. Hence, the probability of an undetected error in an n-bit parity-check codeword is

$$P_{we} \approx P(2, n) \approx \frac{n(n - 1)}{2} \alpha^2 \tag{2}$$

For comparison purposes, *uncoded* transmission of words containing $n - 1$ message bits would have

$$P_{uwe} = 1 - P(0, n - 1) \approx (n - 1)\alpha$$

Thus, if $n = 10$ and $\alpha = 10^{-3}$, then $P_{uwe} \approx 10^{-2}$ whereas coding yields $P_{we} \approx 5 \times 10^{-5}$ with a rate reduction of just $9/10$. These numbers help explain the popularity of parity checking for error detection in computer systems.

As an example of parity checking for error correction, Fig. 13.1-1 illustrates an error-correcting scheme in which the codeword is formed by arranging k message bits in a square array whose rows and columns are checked by $2\sqrt{k}$ parity bits. A transmission error in one message bit causes a row and column

Figure 13.1-1 Square array for error correction by parity checking.

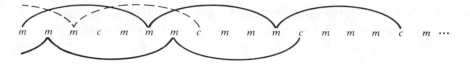

Figure 13.1-2 Interleaved check bits for error control with burst errors.

parity failure with the error at the intersection, so single errors can be corrected. This code also detects double errors.

Throughout the foregoing discussion we've assumed that transmission errors appear randomly and independently in a codeword. This assumption holds for errors caused by white noise or filtered white noise. But *impulse noise* produced by lightning and switching transients causes errors to occur in *bursts* that span several successive bits. Burst errors also appear when radio-transmission systems suffer from rapid fading. Such multiple errors wreak havoc on the performance of conventional codes and must be combated by special techniques. Parity checking controls burst errors if the check bits are *interleaved* so that the checked bits are widely spaced, as represented in Fig. 13.1-2 where a curved line connects the message bits and check bit in one parity word.

Code Vectors and Hamming Distance

Rather than continuing a piecemeal survey of particular codes, we now introduce a more general approach in terms of code *vectors*. An arbitrary n-bit codeword can be visualized in an n-dimensional space as a vector whose elements or coordinates equal the bits in the codeword. We thus write the codeword 101 in row-vector notation as $X = (1 \ 0 \ 1)$. Figure 13.1-3 portrays all possible 3-bit codewords as dots corresponding to the vector tips in a three-dimensional space. The solid dots in part (a) represent the triple-repetition code, while those in part (b) represent a parity-check code.

Notice that the triple-repetition code vectors have greater separation than the parity-check code vectors. This separation, measured in terms of the

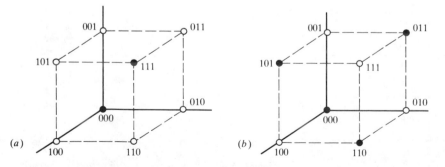

Figure 13.1-3 Vectors representing 3-bit codewords. (a) Triple-repetition code; (b) parity-check code.

Hamming distance, has direct bearing on the error-control power of a code. The Hamming distance $d(X, Y)$ between two vectors X and Y is defined to equal the number of different elements. For instance, if $X = (1\ 0\ 1)$ and $Y = (1\ 1\ 0)$ then $d(X, Y) = 2$ because the second and third elements are different.

The *minimum distance* d_{min} of a particular code is the smallest Hamming distance between valid code vectors. Consequently, error detection is always possible when the number of transmission errors in a codeword is less than d_{min} so the erroneous word is not a valid vector. Conversely, when the number of errors equals or exceeds d_{min}, the erroneous word may correspond to another valid vector and the errors cannot be detected.

Further reasoning along this line leads to the following distance requirements for various degrees of error control capability:

Detect up to ℓ errors per word $\qquad\qquad\qquad\qquad d_{min} \geq \ell + 1 \qquad (3a)$

Correct up to t errors per word $\qquad\qquad\qquad\qquad d_{min} \geq 2t + 1 \qquad (3b)$

Correct up to t errors and detect $\ell > t$ errors per word $\qquad d_{min} \geq t + \ell + 1 \quad (3c)$

By way of example, we see from Fig. 13.1-3 that the triple-repetition code has $d_{min} = 3$. Hence, this code could be used to detect $\ell \leq 3 - 1 = 2$ errors per word or to correct $t \leq (3 - 1)/2 = 1$ error per word — in agreement with our previous observations. A more powerful code with $d_{min} = 7$ could correct triple errors or it could correct double errors and detect quadruple errors.

The power of a code obviously depends on the number of bits added to each codeword for error-control purposes. In particular, suppose that the codewords consist of $k < n$ message bits and $n - k$ parity bits checking the message bits. This structure is known as an (n, k) *block code*. The minimum distance of an (n, k) block code is upper-bounded by

$$d_{min} \leq n - k + 1 \qquad (4)$$

and the code's efficiency is measured by the *code rate*

$$R_c \triangleq k/n \qquad (5)$$

Regrettably, the upper bound in Eq. (4) is realized only by repetition codes, which have $k = 1$ and very inefficient code rate $R_c = 1/n$. Considerable effort has thus been devoted to the search for powerful and reasonably efficient codes, a topic we'll return to in the next section.

FEC Systems

Now we're prepared to examine the forward error correction system diagrammed in Fig. 13.1-4. Message bits come from an information source at rate r_b. The encoder takes blocks of k message bits and constructs an (n, k) block code with

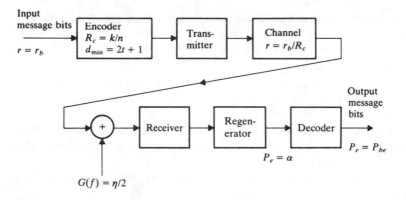

Figure 13.1-4 FEC system.

code rate $R_c = k/n < 1$. The bit rate on the channel therefore must be greater than r_b, namely

$$r = (n/k)r_b = r_b/R_c \tag{6}$$

The code has $d_{\min} = 2t + 1 \leq n - k + 1$, and the decoder operates strictly in an error-correction mode.

We'll investigate the performance of this FEC system when additive white noise causes random errors with probability $\alpha \ll 1$. The value of α depends, of course, on the signal energy and noise density at the receiver. If E_b represents the average energy per *message* bit, then the average energy per *code* bit is $R_c E_b$ and the ratio of bit energy to noise density is

$$\gamma_c \triangleq R_c E_b/\eta = R_c \gamma_b \tag{7}$$

where $\gamma_b = E_b/\eta$. Our performance criterion will be the probability of output message-bit errors, denoted by P_{be} to distinguish it from the word error probability P_{we}.

The code always corrects up to t errors per word and some patterns of more than t errors may also be correctable, depending upon the specific code vectors. Thus, the probability of a decoding word error is upper-bounded by

$$P_{we} \leq \sum_{i=t+1}^{n} P(i, n)$$

For a rough but reasonable performance estimate, we'll take the approximation

$$P_{we} \approx P(t + 1, n) \approx \binom{n}{t + 1} \alpha^{t+1} \tag{8}$$

which means that an uncorrected word typically has $t + 1$ bit errors. On the average, there will be $(k/n)(t + 1)$ message-bit errors per uncorrected word, the remaining errors being in check bits. When Nk bits are transmitted in $N \gg 1$

words, the expected total number of erroneous message bits at the output is $(k/n)(t + 1)NP_{we}$. Hence,

$$P_{be} = \frac{t + 1}{n} P_{we} \approx \binom{n - 1}{t} \alpha^{t+1} \tag{9}$$

in which we have used Eq. (1b) to combine $(t + 1)/n$ with the binomial coefficient.

If the noise has a gaussian distribution and the transmission system has been optimized (i.e., polar signaling and matched filtering), then the transmission error probability is given by Eq. (16), Sect. 11.2, as

$$\alpha = Q(\sqrt{2\gamma_c}) = Q(\sqrt{2R_c\,\gamma_b})$$
$$\approx (4\pi R_c\,\gamma_b)^{-1/2} e^{-R_c\gamma_b} \qquad R_c\,\gamma_b \geq 5 \tag{10}$$

The gaussian tail approximation invoked here follows from Eq. (10), Sect. 4.4, and is consistent with the assumption that $\alpha \ll 1$. Thus, our final result for the output error probability of the FEC system becomes

$$P_{be} = \binom{n - 1}{t} [Q(\sqrt{2R_c\,\gamma_b})]^{t+1}$$
$$\approx \binom{n - 1}{t} (4\pi R_c\,\gamma_b)^{-(t+1)/2} e^{-(t+1)R_c\gamma_b} \tag{11}$$

Uncoded transmission on the same channel would have

$$P_{ube} = Q(\sqrt{2\gamma_b}) \approx (4\pi\gamma_b)^{-1/2} e^{-\gamma_b} \tag{12}$$

since the signaling rate can be decreased from r_b/R_c to r_b.

A comparison of Eqs. (11) and (12) brings out the importance of the code parameters $t = (d_{min} - 1)/2$ and $R_c = k/n$. The added complexity of an FEC system is justified provided that t and R_c yield a value of P_{be} significantly less than P_{ube}. The exponential approximations show that this essentially requires $(t + 1)R_c > 1$. Hence, a code that only corrects single or double errors should have a relatively high code rate, while more powerful codes may succeed despite lower code rates. The channel parameter γ_b also enters into the comparison, as demonstrated by the following example.

Example 13.1-1 Suppose we have a (15, 11) block code with $d_{min} = 3$, so $t = 1$ and $R_c = 11/15$. An FEC system using this code would have $\alpha = Q[\sqrt{(22/15)\gamma_b}]$ and $P_{be} = 14\alpha^2$, whereas uncoded transmission on the same channel would yield $P_{ube} = Q(\sqrt{2\gamma_b})$. These three probabilities are plotted versus γ_b in dB in Fig. 13.1-5. If $\gamma_b > 8$ dB, we see that coding decreases the error probability by at least an order of magnitude compared to uncoded transmission. At $\gamma_b = 10$ dB, for instance, uncoded transmission yields $P_{ube} \approx 4 \times 10^{-6}$ whereas the FEC system has $P_{be} \approx 10^{-7}$ even though the higher channel bit rate increases the transmission error probability to $\alpha \approx 6 \times 10^{-5}$.

Figure 13.1-5 Curves of error
probabilities in Example
13.1-1.

If $\gamma_b < 8$ dB, however, coding does not significantly improve reliability
and actually makes matters worse when $\gamma_b < 4$ dB. Furthermore, an uncoded
system could achieve better reliability than the FEC system simply by
increasing the signal-to-noise ratio about 1.5 dB. Hence, this particular code
doesn't save much signal power, but it would be effective if γ_b has a fixed
value in the vicinity of 8–10 dB.

Exercise 13.1-1 Suppose the system in Example 13.1-1 is operated at $\gamma_b \approx$
8 dB so $\alpha = 0.001$. Evaluate $P(i, n)$ for $i = 0$, 1, 2, and 3. Do your results
support the approximation in Eq. (8)?

ARQ Systems

The automatic-repeat-request strategy for error control is based on error detection and retransmission rather than forward error correction. Consequently,
ARQ systems differ from FEC systems in three important respects. First, an (n, k)
block code designed for error detection generally requires fewer check bits and
has a higher k/n ratio than a code designed for error correction. Second, an ARQ
system needs a return transmission path and additional hardware in order to
implement repeat transmission of codewords with detected errors. Third, the
forward transmission bit rate must make allowance for repeated word transmissions. The net impact of these differences becomes clearer after we describe the
operation of the ARQ system represented by Fig. 13.1-6.

Each codeword constructed by the encoder is stored temporarily and transmitted to the destination where the decoder looks for errors. The decoder issues a
positive acknowledgement (ACK) if no errors are detected, or a *negative acknowl-*

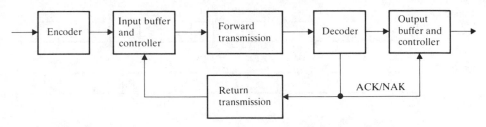

Figure 13.1-6 ARQ system.

edgement (NAK) if errors are detected. A negative acknowledgement causes the input controller to retransmit the appropriate word from those stored by the input buffer. A particular word may be transmitted just once, or it may be transmitted two or more times, depending on the occurrence of transmission errors. The function of the output controller and buffer is to assemble the output bit stream from the codewords that have been accepted by the decoder.

Compared to forward transmission, return transmission of the ACK/NAK signal involves a low bit rate and we can reasonably assume a negligible error probability on the return path. Under this condition, all codewords with detected errors are retransmitted as many times as necessary, so the only output errors appear in words with undetected errors. For an (n, k) block code with $d_{min} = \ell + 1$, the corresponding output error probabilities are

$$P_{we} = \sum_{i=\ell+1}^{n} P(i, n) \approx P(\ell + 1, n) \approx \binom{n}{\ell+1} \alpha^{\ell+1} \qquad (13)$$

$$P_{be} = \frac{\ell + 1}{n} P_{we} \approx \binom{n-1}{\ell} \alpha^{\ell+1} \qquad (14)$$

which are identical to the FEC expressions, Eqs. (8) and (9), with ℓ in place of t. Since the decoder accepts words that have either no errors or undetectable errors, the *word retransmission probability* is given by

$$p = 1 - [P(0, n) + P_{we}]$$

But a good error-detecting code should yield $P_{we} \ll P(0, n)$. Hence,

$$p \approx 1 - P(0, n) = 1 - (1 - \alpha)^n \approx n\alpha \qquad (15)$$

where we've used the approximation $(1 - \alpha)^n \approx 1 - n\alpha$ based on $n\alpha \ll 1$.

As for the retransmission process itself, there are three basic ARQ schemes illustrated by the timing diagrams in Fig. 13.1-7. The asterisk marks words received with detected errors which must be retransmitted. The *stop-and-wait* scheme in part *a* requires the transmitter to stop after every word and wait for acknowledgment from the receiver. Just one word needs to be stored by the input buffer, but the transmission time delay t_d in each direction results in an *idle time* of duration $D \geq 2t_d$ between words. Idle time is eliminated by the *go-back-N* scheme in part *b* where codewords are transmitted continuously. When the receiver sends a NAK signal, the transmitter goes back N words in the buffer and

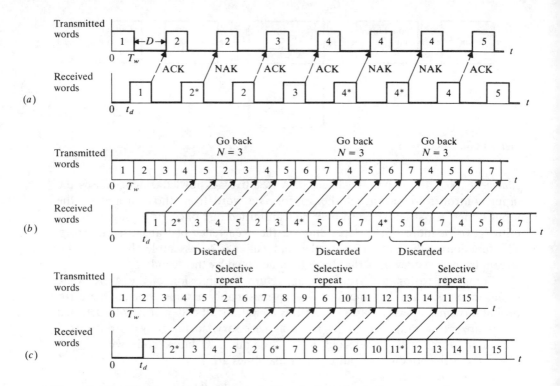

Figure 13.1-7 ARQ schemes. (a) Stop-and-wait; (b) go-back-N; (c) selective-repeat.

retransmits starting from that point. The receiver discards the $N - 1$ intervening words, correct or not, in order to preserve proper sequence. The *selective-repeat* scheme in part c puts the burden of sequencing on the output controller and buffer, so that only words with detected errors need to be retransmitted.

Clearly, a selective-repeat ARQ system has the highest *throughput efficiency*. To set this on a quantitative footing, we observe that the total number of transmissions of a given word is a discrete random variable m governed by the event probabilities $P(m = 1) = 1 - p$, $P(m = 2) = p(1 - p)$, etc. The average number of transmitted words per accepted word is then

$$\bar{m} = 1(1 - p) + 2p(1 - p) + 3p^2(1 - p) + \cdots$$

$$= (1 - p)(1 + 2p + 3p^2 + \cdots) = \frac{1}{1 - p} \tag{16}$$

since $1 + 2p + 3p^2 + \cdots = (1 - p)^{-2}$. On the average, the system must transmit $n\bar{m}$ bits for every k message bits, so the throughput efficiency is

$$R'_c = \frac{k}{n\bar{m}} = \frac{k}{n}(1 - p) \tag{17}$$

in which $p \approx n\alpha$, from Eq. (15).

We use the symbol R'_c here to reflect the fact that the forward-transmission bit rate r and the message bit rate r_b are related by

$$r = r_b/R'_c$$

comparable to the relationship $r = r_b/R_c$ in an FEC system. Thus, when the noise has a gaussian distribution, the transmission error probability α is calculated from Eq. (10) using R'_c instead of $R_c = k/n$. Furthermore, if $p \ll 1$, then $R'_c \approx k/n$. But an error-detecting code has a larger k/n ratio than an error-correcting code of equivalent error-control power. Under these conditions, the more elaborate hardware needed for selective-repeat ARQ may pay off in terms of better performance than an FEC system would yield on the same channel.

The expression for $\bar{m}$ in Eq. (16) also applies to a stop-and-wait ARQ system. However, the idle time reduces efficiency by the factor $T_w/(T_w + D)$ where $D \geq 2t_d$ is the round-trip delay and T_w is the word duration given by $T_w = n/r \leq k/r_b$. Hence,

$$R'_c = \frac{k}{n} \frac{1-p}{1+(D/T_w)} \leq \frac{k}{n} \frac{1-p}{1+(2t_d r_b/k)} \tag{18}$$

in which the upper bound comes from writing $D/T_w \geq 2t_d r_b/k$.

A go-back-N ARQ system has no idle time, but N words must be retransmitted for each word with detected errors. Consequently, we find that

$$\bar{m} = 1 + \frac{Np}{1-p} \tag{19}$$

and

$$R'_c = \frac{k}{n} \frac{1-p}{1-p+Np} \leq \frac{k}{n} \frac{1-p}{1-p+(2t_d r_b/k)p} \tag{20}$$

where the upper bound reflects the fact that $N \geq 2t_d/T_w$.

Unlike selective-repeat ARQ, the throughput efficiency of the stop-and-wait and go-back-N schemes depends on the round-trip delay. Equations (18) and (20) reveal that both of these schemes have reasonable efficiency if the delay and bit rate are such that $2t_d r_b \ll k$. However, stop-and-wait ARQ has very low efficiency when $2t_d r_b \geq k$, whereas the go-back-N scheme may still be satisfactory provided that the retransmission probability p is small enough.

Finally, we should at least describe the concept of *hybrid* ARQ systems. These systems consist of an FEC subsystem within the ARQ framework, thereby combining desirable properties of both error-control strategies. For instance, a hybrid ARC system might employ a block code with $d_{\min} = t + \ell + 1$, so the decoder can correct up to t errors per word and detect but not correct words with $\ell > t$ errors. Error correction reduces the number of words that must be retransmitted, thereby increasing the throughput without sacrificing the higher reliability of ARQ.

Example 13.1-2 Suppose a selective-repeat ARQ system uses a simple parity-check code with $k = 9$, $n = 10$, and $\ell = 1$. The transmission channel is corrupted by gaussian noise and we seek the value of γ_b needed to get $P_{be} \approx 10^{-5}$. Equation (14) yields the required transmission error probability $\alpha = (10^{-5}/9)^{1/2} \approx 1.1 \times 10^{-3}$, and the corresponding word retransmission probability in Eq. (15) is $p \approx 10\alpha \ll 1$. Hence, the throughput efficiency will be $R'_c \approx k/n = 0.9$, from Eq. (17). Since $\alpha = Q(\sqrt{2R'_c \gamma_b})$ we call upon the plot of Q in Table T.6 to obtain our final result $\gamma_b \approx 3.1^2/1.8 = 5.3$ or 7.3 dB. As a comparison, Fig. 13.1-5 shows that uncoded transmission would have $P_{ube} \approx 6 \times 10^{-4}$ if $\gamma_b = 7.3$ dB and requires $\gamma_b \approx 9.6$ dB to get $P_{ube} = 10^{-5}$. The ARQ system thus achieves a power saving of about 2.3 dB.

Exercise 13.1-2 Assume that the system in Example 13.1-2 has $r_b = 50$ kbps and $t_d = 0.2$ ms. By calculating R'_c show that the go-back-N scheme would be acceptable but not the stop-and-wait scheme when channel limitations require $r \leq 100$ kbps.

13.2 LINEAR BLOCK CODES

This section describes the structure, properties, and implementation of block codes. We start with a matrix representation of the encoding process that generates the check bits for a given block of message bits. Then we use the matrix representation to investigate decoding methods for error detection and correction. The section closes with a brief introduction to the important class of cyclic block codes.

Matrix Representation of Block Codes

An (n, k) block code consists of n-bit vectors, each vector corresponding to a unique block of $k < n$ message bits. Since there are 2^k different k-bit message blocks and 2^n possible n-bit vectors, the fundamental strategy of block coding is to choose the 2^k code vectors such that the minimum distance is as large as possible. But the code should also have some structure that facilitates the encoding and decoding processes. We'll therefore focus on the class of *systematic linear block codes*.

Let an arbitrary code vector be represented by

$$X = (x_1 \quad x_2 \quad \cdots \quad x_n)$$

where the elements $x_1, x_2, \ldots,$ are, of course, binary digits. A code is *linear* if it includes the all-zero vector and if the sum of any two code vectors produces another vector in the code. The *sum* of two vectors, say X and Z, is defined as

$$X + Z \triangleq (x_1 \oplus z_1 \quad x_2 \oplus z_2 \quad \cdots \quad x_n \oplus z_n) \tag{1}$$

in which the elements are combined according to the rules of mod-2 addition given in Eq. (2), Sect. 11.4.

As a consequence of linearity, we can determine a code's minimum distance by the following argument. Let the number of nonzero elements of a vector X be symbolized by $w(X)$, called the vector *weight*. The Hamming distance between any two code vectors X and Z is then

$$d(X, Z) = w(X + Z)$$

since $x_1 \oplus z_1 = 1$ if $x_1 \neq z_1$, etc. The distance between X and Z therefore equals the weight of another code vector $X + Z$. But if $Z = (0 \ 0 \ \cdots \ 0)$ then $X + Z = X$; hence,

$$d_{\min} - [w(X)]_{\min} \qquad X \neq (0 \ 0 \ \cdots \ 0) \tag{2}$$

In other words, the minimum distance of a linear block code equals the *smallest nonzero vector weight*.

A *systematic* block code consists of vectors whose first k elements (or last k elements) are identical to the message bits, the remaining $n - k$ elements being check bits. A code vector then takes the form

$$X = (m_1 \quad m_2 \quad \cdots \quad m_k \quad c_1 \quad c_2 \quad \cdots \quad c_q) \tag{3a}$$

where

$$q = n - k \tag{3b}$$

For convenience, we'll also write code vectors in the partitioned notation

$$X = (M \quad | \quad C)$$

in which M is a k-bit message vector and C is a q-bit check vector. Partitioned notation lends itself to the matrix representation of block codes.

Given a message vector M, the corresponding code vector X for a systematic linear (n, k) block code can be obtained by a matrix multiplication

$$X = MG \tag{4}$$

The matrix G is a $k \times n$ *generator matrix* having the general structure

$$G \triangleq [I_k \quad | \quad P] \tag{5a}$$

where I_k is the $k \times k$ *identity matrix* and P is a $k \times q$ submatrix of binary digits represented by

$$P = \begin{bmatrix} p_{11} & p_{12} & \cdots & p_{1q} \\ p_{21} & p_{22} & \cdots & p_{2q} \\ \vdots & \vdots & & \vdots \\ p_{k1} & p_{k2} & \cdots & p_{kq} \end{bmatrix} \tag{5b}$$

The identity matrix in G simply reproduces the message vector for the first k elements of X, while the submatrix P generates the check vector via

$$C = MP \tag{6a}$$

This binary matrix multiplication follows the usual rules with mod-2 addition instead of conventional addition. Hence, the jth element of C is computed using the jth column of P, and

$$c_j = m_1 p_{1j} \oplus m_2 p_{2j} \oplus \cdots \oplus m_k p_{kj} \tag{6b}$$

for $j = 1, 2, \ldots, q$. All of these matrix operations are less formidable than they appear because every element equals either 0 or 1.

The matrix representation of a block code provides a compact analytical vehicle and, moreover, leads to hardware implementations of the encoder and decoder. But it does not tell us how to pick the elements of the P submatrix to achieve specified code parameters such as $d_{\min}$ and R_c. Consequently, good codes are discovered with the help of considerable inspiration and perspiration, guided by mathematical analysis. In fact, Hamming (1950) devised the first popular block codes several years before the underlying theory was formalized by Slepian (1956).

Example 13.2-1 Hamming codes A *Hamming code* is an (n, k) linear block code with $q \geq 3$ check bits and

$$n = 2^q - 1 \qquad k = n - q \tag{7a}$$

The code rate is

$$R_c = \frac{k}{n} = 1 - \frac{q}{2^q - 1} \tag{7b}$$

and thus $R_c \approx 1$ if $q \gg 1$. Independent of q, the minimum distance is fixed at

$$d_{\min} = 3 \tag{7c}$$

so a Hamming code can be used for single-error correction or double-error detection. To construct a systematic Hamming code, you simply let the k rows of the P submatrix consist of all q-bit words with two or more 1s, arranged in any order.

For example, consider a systematic Hamming code with $q = 3$, so $n = 2^3 - 1 = 7$ and $k = 7 - 3 = 4$. According to the previously stated rule, an appropriate generator matrix is

$$G = \begin{bmatrix} 1 & 0 & 0 & 0 & 1 & 0 & 1 \\ 0 & 1 & 0 & 0 & 1 & 1 & 1 \\ 0 & 0 & 1 & 0 & 1 & 1 & 0 \\ 0 & 0 & 0 & 1 & 0 & 1 & 1 \end{bmatrix}$$

The last three columns constitute the P submatrix whose rows include all 3-bit words that have two or more 1s. Given a block of message bits $M =$

Figure 13.2-1 Encoder for (7, 4) Hamming code.

$(m_1 \; m_2 \; m_3 \; m_4)$, the check bits are determined from the set of equations

$$c_1 = m_1 \oplus m_2 \oplus m_3 \oplus 0$$

$$c_2 = 0 \oplus m_2 \oplus m_3 \oplus m_4$$

$$c_3 = m_1 \oplus m_2 \oplus 0 \oplus m_4$$

These check-bit equations are obtained by substituting the elements of P into Eq. (6).

Figure 13.2-1 depicts an encoder that carries out the check-bit calculations for this (7, 4) Hamming code. Each block of message bits going to the transmitter is also loaded into a message register. The cells of the message register are connected to exclusive-OR gates whose outputs equal the check bits. The check bits are stored in another register and shifted out to the transmitter after the message bits. An input buffer holds the next block of message bits while the check bits are shifted out. The cycle then repeats with the next block of message bits.

Table 13.2-1 lists the resulting $2^4 = 16$ codewords and their weights. The smallest nonzero weight equals 3, confirming that $d_{min} = 3$.

Table 13.2-1 Codewords for the (7, 4) Hamming code

M				C			w(X)	M				C			w(X)
0	0	0	0	0	0	0	0	1	0	0	0	1	0	1	3
0	0	0	1	0	1	1	3	1	0	0	1	1	1	0	4
0	0	1	0	1	1	0	3	1	0	1	0	0	1	1	4
0	0	1	1	1	0	1	4	1	0	1	1	0	0	0	3
0	1	0	0	1	1	1	4	1	1	0	0	0	1	0	3
0	1	0	1	1	0	0	3	1	1	0	1	0	0	1	4
0	1	1	0	0	0	1	3	1	1	1	0	1	0	0	4
0	1	1	1	0	1	0	4	1	1	1	1	1	1	1	7

Exercise 13.2-1 Consider a systematic (6, 3) block code generated by the sub-matrix

$$P = \begin{bmatrix} 1 & 1 & 0 \\ 0 & 1 & 1 \\ 1 & 0 & 1 \end{bmatrix}$$

Write the check-bit equations and tabulate the codewords and their weights to show that $d_{min} = 3$.

Syndrome Decoding

Now let Y stand for the received vector when a particular code vector X has been transmitted. Any transmission errors will result in $Y \neq X$. The decoder detects or corrects errors in Y using stored information about the code.

A direct way of performing error detection would be to compare Y with every vector in the code. This method requires storing all 2^k code vectors at the receiver and performing up to 2^k comparisons. But efficient codes generally have large values of k, which implies rather extensive and expensive decoding hardware. As an example, you need $q \geq 5$ to get $R_c \geq 0.8$ with a Hamming code; then $n \geq 31$, $k \geq 26$, and the receiver must store a total of $n \times 2^k > 10^9$ bits!

More practical decoding methods for codes with large k involve parity-check information derived from the code's P submatrix. Associated with any systematic linear (n, k) block code is a $q \times n$ matrix H called the *parity-check matrix*. This matrix is defined by

$$H^T \triangleq \begin{bmatrix} P \\ \hline I_q \end{bmatrix} \tag{8}$$

where H^T denotes the *transpose* of H and I_q is the $q \times q$ identity matrix. Relative to error detection, the parity-check matrix has the crucial property

$$XH^T = (0 \quad 0 \quad \cdots \quad 0) \tag{9}$$

provided that X belongs to the set of code vectors. However, when Y is not a code vector, the product YH^T contains at least one nonzero element.

Therefore, given H^T and a received vector Y, *error detection* can be based on

$$S = YH^T \tag{10}$$

a q-bit vector called the *syndrome*. If all elements of S equal zero, then either Y equals the transmitted vector X and there are no transmission errors, or Y equals some other code vector and the transmission errors are *undetectable*. Otherwise, errors are indicated by the presence of nonzero elements in S. Thus, a decoder for error detection simply takes the form of a syndrome calculator. A comparison of Eqs. (10) and (6) shows that the hardware needed is essentially the same as the encoding circuit.

Error correction necessarily entails more circuitry but it, too, can be based on the syndrome. We develop the decoding method by introducing an n-bit *error vector E* whose nonzero elements mark the positions of transmission errors in Y. For instance, if $X = (1\ 0\ 1\ 1\ 0)$ and $Y = (1\ 0\ 0\ 1\ 1)$ then $E = (0\ 0\ 1\ 0\ 1)$. In general,

$$Y = X + E \tag{11a}$$

and, conversely,

$$X = Y + E \tag{11b}$$

since a second error in the same bit location would cancel the original error. Substituting $Y = X + E$ into $S = YH^T$ and invoking Eq. (9), we obtain

$$S = (X + E)H^T = XH^T + EH^T = EH^T \tag{12}$$

which reveals that the syndrome depends entirely on the error pattern, not the specific transmitted vector.

However, there are only 2^q different syndromes generated by the 2^n possible n-bit error vectors, including the no-error case. Consequently, a given syndrome does not uniquely determine E. Or, putting this another way, we can correct just $2^q - 1$ patterns with one or more errors, and the remaining patterns are *uncorrectable*. We should therefore design the decoder to correct the $2^q - 1$ most likely error patterns—namely those patterns with the *fewest* errors, since single errors are more probable than double errors, and so forth. This strategy, known as *maximum-likelihood decoding*, is optimum in the sense that it minimizes the word-error probability. Maximum-likelihood decoding corresponds to choosing the code vector that has the smallest Hamming distance from the received vector.

To carry out maximum-likelihood decoding, you must first compute the syndromes generated by the $2^q - 1$ most probable error vectors. The *table-lookup decoder* diagrammed in Fig. 13.2-2 then operates as follows. The decoder calculates S from the received vector Y and looks up the assumed error vector $\hat{E}$ stored in the table. The sum $Y + \hat{E}$ generated by exclusive-OR gates finally constitutes

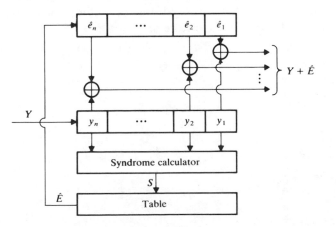

Figure 13.2-2 Table-lookup decoder.

the decoded word. If there are no errors, or if the errors are uncorrectable, then $S = (0 \ 0 \ \cdots \ 0)$ so $Y + \hat{E} = Y$. The check bits in the last q elements of $Y + \hat{E}$ may be omitted if they are of no further interest.

The relationship between syndromes and error patterns also sheds some light on the design of error-correcting codes, since each of the $2^q - 1$ nonzero syndromes must represent a specific error pattern. Now there are $\binom{n}{1} = n$ single-error patterns for an n-bit word, $\binom{n}{2}$ double-error patterns, and so forth. Hence, if a code is to correct up to t errors per word, q and n must satisfy

$$2^q - 1 \geq n + \binom{n}{2} + \cdots + \binom{n}{t} \tag{13}$$

In the particular case of a single-error-correcting code, Eq. (13) reduces to $2^q - 1 \geq n$. Furthermore, when E corresponds to a single error in the jth bit of a codeword, we find from Eq. (12) that S is identical to the jth row of H^T. Therefore, to provide a distinct syndrome for each single-error pattern and for the no-error pattern, the rows of H^T (or columns of H) must all be different and each must contain at least one nonzero element. The generator matrix of a Hamming code is designed to satisfy this requirement on H, while q and n satisfy $2^q - 1 = n$.

Example 13.2-2 Let's apply table-lookup decoding to a (7, 4) Hamming code used for single-error correction. From Eq. (8) and the P submatrix given in Example 13.2-1, we obtain the 3×7 parity-check matrix

$$H = [P^T \ | \ I_q] = \begin{bmatrix} 1 & 1 & 1 & 0 & | & 1 & 0 & 0 \\ 0 & 1 & 1 & 1 & | & 0 & 1 & 0 \\ 1 & 1 & 0 & 1 & | & 0 & 0 & 1 \end{bmatrix}$$

There are $2^3 - 1 = 7$ correctable single-error patterns, and the corresponding syndromes listed in Table 13.2-2 follow directly from the columns of H. To accommodate this table the decoder needs to store only $(q + n) \times 2^q = 80$ bits.

Table 13.2-2 Syndromes for the (7, 4) Hamming code

S			$\hat{E}$						
0	0	0	0	0	0	0	0	0	0
1	0	1	1	0	0	0	0	0	0
1	1	1	0	1	0	0	0	0	0
1	1	0	0	0	1	0	0	0	0
0	1	1	0	0	0	1	0	0	0
1	0	0	0	0	0	0	1	0	0
0	1	0	0	0	0	0	0	1	0
0	0	1	0	0	0	0	0	0	1

But suppose a received word happens to have two errors, such that $E = (1\ 0\ 0\ 0\ 0\ 1\ 0)$. The decoder calculates $S = YH^T = EH^T = (1\ 1\ 1)$ and the syndrome table gives the assumed single-error pattern $\hat{E} = (0\ 1\ 0\ 0\ 0\ 0\ 0)$. The decoded output word $Y + \hat{E}$ therefore contains *three* errors, the two transmission errors plus the erroneous correction added by the decoder.

If multiple transmission errors per word are sufficiently infrequent, we need not be concerned about the occasional extra errors committed by the decoder. If multiple errors are frequent, a more powerful code would be required. For instance, an *extended* Hamming code has an additional check bit that provides double-error detection along with single-error correction; see Prob. 13.2-12.

Exercise 13.2-2 Use Eqs. (8) and (10) to show that the jth bit of S is given by

$$s_j = y_1 p_{1j} \oplus y_2 p_{2j} \oplus \cdots \oplus y_k p_{kj} \oplus y_{k+j}$$

Then diagram the syndrome-calculation circuit for a $(7, 4)$ Hamming code, and compare it with Fig. 13.2-1.

Cyclic Codes★

The code for a forward-error-correction system must be capable of correcting $t \geq 1$ errors per word. It should also have a reasonably efficient code rate $R_c = k/n$. These two parameters are related by the inequality

$$1 - R_c \geq \frac{1}{n} \log_2 \left[\sum_{i=0}^{t} \binom{n}{i} \right] \tag{14}$$

which follows from Eq. (13) with $q = n - k = n(1 - R_c)$. This inequality under-scores the fact that if we want $R_c \approx 1$, we must use codewords with $n \gg 1$ and $k \gg 1$. However, the hardware requirements for encoding and decoding long codewords may be prohibitive unless we impose further structural conditions on the code. *Cyclic codes* are a subclass of linear block codes with a cyclic structure that leads to more practical implementation. Thus, block codes used in FEC systems are almost always cyclic codes.

To describe a cyclic code, we'll find it helpful to change our indexing scheme and express an arbitrary n-bit code vector in the form

$$X = (x_{n-1} \quad x_{n-2} \quad \cdots \quad x_1 \quad x_0) \tag{15}$$

Now suppose that X has been loaded into a *shift register* with feedback connection from the first to last stage. Shifting all bits one position to the left yields the *cyclic shift* of X, written as

$$X' \triangleq (x_{n-2} \quad x_{n-3} \quad \cdots \quad x_1 \quad x_0 \quad x_{n-1}) \tag{16}$$

A second shift produces $X'' = (x_{n-3} \cdots x_1\ x_0\ x_{n-1}\ x_{n-2})$, and so forth. A linear code is *cyclic* if every cyclic shift of a code vector X is another vector in the code.

This cyclic property can be treated mathematically by associating a code vector X with the *polynomial*

$$X(p) = x_{n-1} p^{n-1} + x_{n-2} p^{n-2} + \cdots + x_1 p + x_0 \tag{17}$$

where p is an arbitrary real variable. The powers of p denote the positions of the codeword bits represented by the corresponding coefficients of p. Formally, binary code polynomials are defined in conjunction with *Galois fields*, a branch of modern algebra that provides the theory needed for a complete treatment of cyclic codes. For our informal overview of cyclic codes we'll manipulate code polynomials using ordinary algebra modified in two respects. First, to be in agreement with our earlier definition for the sum of two code vectors, the sum of two polynomials is obtained by *mod-2 addition* of their respective coefficients. Second, since all coefficients are either 0 or 1, and since $1 \oplus 1 = 0$, the *subtraction* operation is the same as mod-2 addition. Consequently, if $X(p) + Z(p) = 0$ then $X(p) = Z(p)$.

We develop the polynomial interpretation of cyclic shifting by comparing

$$pX(p) = x_{n-1} p^n + x_{n-2} p^{n-1} + \cdots + x_1 p^2 + x_0 p$$

with the shifted polynomial

$$X'(p) = x_{n-2} p^{n-1} + \cdots + x_1 p^2 + x_0 p + x_{n-1}$$

If we sum these polynomials, noting that $(x_1 \oplus x_1)p^2 = 0$, etc., we get

$$pX(p) + X'(p) = x_{n-1} p^n + x_{n-1}$$

and hence

$$X'(p) = pX(p) + x_{n-1}(p^n + 1) \tag{18}$$

Iteration yields similar expressions for multiple shifts.

The polynomial $p^n + 1$ and its factors play major roles in cyclic codes. Specifically, an (n, k) cyclic code is defined by a *generator polynomial* of the form

$$G(p) = p^q + g_{q-1} p^{q-1} + \cdots + g_1 p + 1 \tag{19}$$

where $q = n - k$ and the coefficients are such that $G(p)$ is a factor of $p^n + 1$. Each codeword then corresponds to the polynomial product

$$X(p) = Q_M(p)G(p) \tag{20}$$

in which $Q_M(p)$ represents a block of k message bits. All such codewords satisfy the cyclic condition in Eq. (18) since $G(p)$ is a factor of both $X(p)$ and $p^n + 1$.

Any factor of $p^n + 1$ that has degree q may serve as the generator polynomial for a cyclic code, but it does not necessarily generate a good code. Table 13.2-3 lists the generator polynomials of selected cyclic codes that have been demonstrated to possess desirable parameters for FEC systems. The table includes some cyclic Hamming codes, the famous Golay code, and a few members of the important family of BCH codes discovered by Bose, Chaudhuri, and Hocquenghem. The entries under $G(p)$ denote the polynomial's coefficients; thus, for instance, 1 0 1 1 means that $G(p) = p^3 + 0 + p + 1$.

Table 13.2-3 Selected cyclic codes

Type	n	k	R_c	d_{min}				$G(p)$	
Hamming codes	7	4	0.57	3				1	011
	15	11	0.73	3				10	011
	31	26	0.84	3				100	101
BCH codes	15	7	0.46	5			111	010	001
	31	21	0.68	5		11	101	101	001
	63	45	0.71	7	1 111	000	001	011 001	111
Golay code	23	12	0.52	7		101	011	100	011

Cyclic codes may be systematic or nonsystematic, depending on the term $Q_M(p)$ in Eq. (20). For a systematic code, we define the message-bit and check-bit polynomials

$$M(p) = m_{k-1}p^{k-1} + \cdots + m_1p + m_0$$

$$C(p) = c_{q-1}p^{q-1} + \cdots + c_1p + c_0$$

and we want the codeword polynomials to be

$$X(p) = p^q M(p) + C(p) \tag{21}$$

Equations (20) and (21) therefore require $p^q M(p) + C(p) = Q_M(p)G(p)$, or

$$\frac{p^q M(p)}{G(p)} = Q_M(p) + \frac{C(p)}{G(p)} \tag{22a}$$

This expression says that $C(p)$ equals the *remainder* left over after dividing $p^q M(p)$ by $G(p)$, just as 14 divided by 3 leaves a remainder of 2 since $14/3 = 4 + 2/3$. Symbolically, we write

$$C(p) = \text{rem}\left[\frac{p^q M(p)}{G(p)}\right] \tag{22b}$$

where rem [ ] stands for the remainder of the division within the brackets.

The division operation needed to generate a systematic cyclic code is easily and efficiently performed by the shift-register encoder diagrammed in Fig. 13.2-3.

Figure 13.2-3 Shift-register encoder.

Encoding starts with the feedback switch closed, the output switch in the message-bit position, and the register initialized to the all-zero state. The k message bits are shifted into the register and simultaneously delivered to the transmitter. After k shift cycles, the register contains the q check bits. The feedback switch is now opened and the output switch is moved to deliver the check bits to the transmitter.

Syndrome calculation at the receiver is equally simple. Given a received vector Y, the syndrome is determined from

$$S(p) = \text{rem}\left[\frac{Y(p)}{G(p)}\right] \tag{23}$$

If $Y(p)$ is a valid code polynomial, then $G(p)$ will be a factor of $Y(p)$ and $Y(p)/G(p)$ has zero remainder. Otherwise we get a nonzero syndrome polynomial indicating detected errors.

Besides simplified encoding and syndrome calculation, cyclic codes have other advantages over noncyclic block codes. The foremost advantage comes from the ingenious error-correcting decoding methods that have been devised for specific cyclic codes. These methods eliminate the storage needed for table-lookup decoding and thus make it practical to use powerful and efficient codes with $n \gg 1$. Another advantage is the ability of cyclic codes to detect error bursts that span many successive bits. Detailed exposition of these properties are presented in texts such as Lin and Costello (1983).

Example 13.2-3 Consider the cyclic (7, 4) Hamming code generated by $G(p) = p^3 + 0 + p + 1$. We'll use long division to calculate the check-bit polynomial $C(p)$ when $M = (1\ 1\ 0\ 0)$. We first write the message-bit polynomial

$$M(p) = p^3 + p^2 + 0 + 0$$

so $p^q M(p) = p^3 M(p) = p^6 + p^5 + 0 + 0 + 0 + 0 + 0$. Next, we divide $G(p)$ into $p^q M(p)$, keeping in mind that subtraction is the same as addition in mod-2 arithmetic. Thus,

$$
\begin{array}{r}
Q_M(p) = p^3 + p^2 + p + 0 \\
p^3 + 0 + p + 1 \overline{\big)\ p^6 + p^5 + 0\ + 0\ + 0\ + 0 + 0} \\
\underline{p^6 + 0\ + p^4 + p^3} \\
p^5 + p^4 + p^3 + 0 \\
\underline{p^5 + 0\ + p^3 + p^2} \\
p^4 + 0\ + p^2 + 0 \\
\underline{p^4 + 0\ + p^2 + p} \\
0\ + 0\ + p + 0 \\
\underline{0\ + 0\ + 0 + 0} \\
C(p) = 0\ + p + 0
\end{array}
$$

so the complete code polynomial is

$$X(p) = p^3 M(p) + C(p) = p^6 + p^5 + 0 + 0 + 0 + p + 0$$

Input

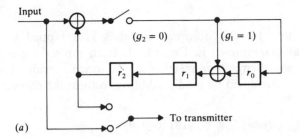

$(g_2 = 0)$ $(g_1 = 1)$

To transmitter

(a)

Input bit	Register bits before shift			Register bits after shift		
m	r_2	r_1	r_0	$r_2' = r_1$	$r_1' = r_0 \oplus r_2 \oplus m$	$r_0' = r_2 \oplus m$
1	0	0	0	0	1	1
1	0	1	1	1	0	1
0	1	0	1	0	0	1
0	0	0	1	0	1	0

(b)

Figure 13.2-4 (a) Shift-register encoder for (7, 4) Hamming code; (b) register bits when $M = (1\ 1\ 0\ 0)$.

which corresponds to the codeword

$$X = (1 \quad 1 \quad 0 \quad 0 \mid 0 \quad 1 \quad 0)$$

You'll find this codeword back in Table 13.2-1, where you'll also find the cyclic shift $X' = (1\ 0\ 0\ 0 \mid 1\ 0\ 1)$ and all multiple shifts.

Finally, Fig. 13.2-4 shows the shift-register encoder and the register bits for each cycle of the encoding process when the input is $M = (1\ 1\ 0\ 0)$. After four shift cycles, the register holds $C = (0\ 1\ 0)$—in agreement with our manual division.

Exercise 13.2-3 Let $Y(p) = X(p) + E(p)$ where $E(p)$ is the error polynomial. Use Eqs. (20) and (23) to show that the syndrome polynomial $S(p)$ depends on $E(p)$ but not on $X(p)$.

13.3 CONVOLUTIONAL CODES

Convolutional codes have a structure that effectively extends over the entire transmitted bit stream, rather than being limited to codeword blocks. The convolutional structure is especially well suited to space and satellite communication systems that require simple encoders and achieve high performance by sophisticated decoding methods. Our treatment of this important family of codes consists of selected examples that introduce the salient features of convolutional encoding and decoding.

Convolutional Encoding

The fundamental hardware unit for convolutional encoding is a tapped shift register with $L + 1$ stages, as diagrammed in Fig. 13.3-1. Each tap gain g is a binary digit representing a short-circuit connection or an open circuit. The message bits in the register are combined by mod-2 addition to form the encoded bit

$$x_j = m_{j-L} g_L \oplus \cdots \oplus m_{j-1} g_1 \oplus m_j g_0$$

$$= \sum_{i=0}^{L} m_{j-i} g_i \qquad \text{(mod-2)} \qquad (1)$$

The name *convolutional encoding* comes from the fact that Eq. (1) has the form of a *binary convolution*, analogous to the convolutional integral

$$x(t) = \int m(t - \lambda) g(\lambda) \, d\lambda$$

Notice that x_j depends on the current input m_j and on the *state* of the register defined by the previous L message bits. Also notice that a particular message bit influences a span of $L + 1$ successive encoded bits as it shifts through the register.

To provide the extra bits needed for error control, a complete convolutional encoder must generate output bits at a rate greater than the message bit rate r_b. This is achieved by connecting two or more mod-2 summers to the register and interleaving the encoded bits via a commutator switch. For example, the encoder in Fig. 13.3-2 generates $n = 2$ encoded bits

$$x'_j = m_{j-2} \oplus m_{j-1} \oplus m_j \qquad x''_j = m_{j-2} \oplus m_j \qquad (2)$$

which are interleaved by the switch to produce the output stream

$$X = x'_1 x''_1 x'_2 x''_2 x'_3 x''_3 \cdots$$

The output bit rate is therefore $2r_b$ and the code rate is $R_c = 1/2$—like an (n, k) block code with $R_c = k/n = 1/2$.

However, unlike a block code, the input bits have not been grouped into words. Instead, each message bit influences a span of $n(L + 1) = 6$ successive output bits. The quantity $n(L + 1)$ is called the *constraint length* measured in terms

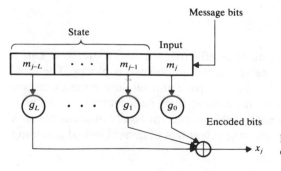

Figure 13.3-1 Tapped shift register for convolutional encoding.

Figure 13.3-2 Convolutional encoder with $n = 2$, $k = 1$, and $L = 2$.

of encoded output bits, whereas L is the encoder's *memory* measured in terms of input message bits. We say that this encoder produces an (n, k, L) convolutional code† with $n = 2$, $k = 1$, and $L = 2$.

Three different but related graphical representations have been devised for the study of convolutional encoding: the code tree, the code trellis, and the state diagram. We'll present each of these for our $(2, 1, 2)$ encoder in Fig. 13.3-2, starting with the code tree. In accordance with normal operating procedure, we presume that the register has been cleared to contain all 0s when the first message bit m_1 arrives. Hence, the initial state is $m_{-1}m_0 = 00$ and Eq. (2) gives the output $x_1'x_1'' = 00$ if $m_1 = 0$ or $x_1'x_1'' = 11$ if $m_1 = 1$.

The *code tree* drawn in Fig. 13.3-3 begins at a branch point or *node* labeled a representing the initial state. If $m_1 = 0$, you take the upper branch from node a to find the output 00 and the next state, which is also labeled a since $m_0 m_1 = 00$ in this case. If $m_1 = 1$, you take the lower branch from a to find the output 11 and the next state $m_0 m_1 = 01$ signified by the label b. The code tree progressively evolves in this fashion for each new input bit. Nodes are labeled with letters denoting the current state $m_{j-2} m_{j-1}$; you go up or down from a node, depending on the value of m_j; each branch shows the resulting encoded output $x_j'x_j''$ calculated from Eq. (2), and it terminates at another node labeled with the next state. There are 2^j possible branches for the jth message bit, but the branch pattern begins to repeat at $j = 3$ since the register length is $L + 1 = 3$.

Having observed repetition in the code tree, we can construct a more compact picture called the *code trellis* and shown in Fig. 13.3-4a. Here, the nodes on the left denote the four possible current states, while those on the right are the resulting next states. A solid line represents the state transition or branch for $m_j = 0$, and a broken line represents the branch for $m_j = 1$. Each branch is labeled with the resulting output bits $x_j'x_j''$. Going one step further, we coalesce the left and right sides of the trellis to obtain the *state diagram* in Fig. 13.3-4b. The self-loops at nodes a and d represent the state transitions a-a and d-d.

Given a sequence of message bits and the initial state, you can use either the code trellis or state diagram to find the resulting state sequence and output bits. The procedure is illustrated in Fig. 13.3-4c, starting at initial state a.

---

† Notation for convolutional codes has not been standardized and varies from author to author, as does the definition of constraint length.

$m_j = 0$

$m_j = 1$

States
$a = 00$
$b = 01$
$c = 10$
$d = 11$

Figure 13.3-3 Code tree for (2, 1, 2) encoder.

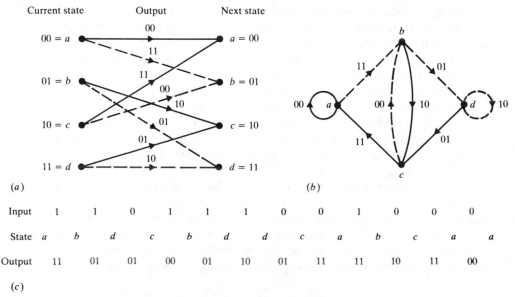

Current state   Output   Next state

(a)

(b)

Input	1	1	0	1	1	1	0	0	1	0	0	0
State	a	b	d	c	b	d	d	c	a	b	c	a	a
Output	11	01	01	00	01	10	01	11	11	10	11	00

(c)

Figure 13.3-4 (a) Code trellis; (b) state diagram for (2, 1, 2) encoder.

Numerous other convolutional codes are obtained by modifying the encoder in Fig. 13.3-2. If we just change the connections to the mod-2 summers, then the code tree, trellis, and state diagram retain the same structure since the states and branching pattern reflect only the register contents. The output bits would be different, of course, since they depend specifically on the summer connections.

If we extend the shift register to an arbitrary length $L + 1$ and connect it to $n \geq 2$ mod-2 summers, we get an (n, k, L) convolutional code with $k = 1$ and code rate $R_c = 1/n \leq 1/2$. The state of the encoder is defined by L previous input bits, so the code trellis and state diagram have 2^L different states, and the code-tree pattern repeats after $j = L + 1$ branches. Connecting one commutator terminal directly to the first stage of the register yields the encoded bit stream

$$X = m_1 x_1'' x_1''' \cdots m_2 x_2'' x_2''' \cdots m_3 x_3'' x_3''' \cdots \tag{3}$$

which defines a *systematic* convolutional code with $R_c = 1/n$.

Code rates higher than $1/n$ require $k \geq 2$ shift registers and an input distributor switch. This scheme is illustrated by the $(3, 2, 1)$ encoder in Fig. 13.3-5. The message bits are distributed alternately between $k = 2$ registers, each of length $L + 1 = 2$. We regard the pair of bits $m_{j-1} m_j$ as the current input, while the pair $m_{j-3} m_{j-2}$ constitute the state of the encoder. For each input pair, the mod-2 summers generate $n = 3$ encoded output bits given by

$$x_j' = m_{j-3} \oplus m_{j-2} \oplus m_j \qquad x_j'' = m_{j-3} \oplus m_{j-1} \oplus m_j$$

$$x_j''' = m_{j-2} \oplus m_j \tag{4}$$

Thus, the output bit rate is $3r_b/2$, corresponding to the code rate $R_c = k/n = 2/3$. The constraint length is $n(L + 1) = 6$, since a particular input bit influences a span of $n = 3$ output bits from each of its $L + 1 = 2$ register positions.

Figure 13.3-5 $(3, 2, 1)$ encoder.

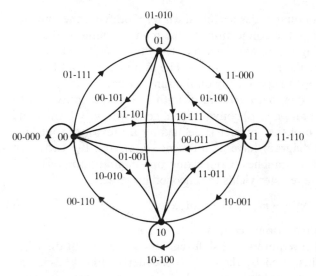

Figure 13.3-6 State diagram for (3, 2, 1) encoder.

Graphical representation becomes more cumbersome for convolutional codes with $k > 1$ because we must deal with input bits in groups of 2^k. Consequently, 2^k branches emanate and terminate at each node, and there are 2^{kL} different states. As an example, Fig. 13.3-6 shows the state diagram for the (3, 2, 1) encoder in Fig. 13.3-5. The branches are labeled with the $k = 2$ input bits followed by the resulting $n = 3$ output bits.

The convolutional codes employed for FEC systems usually have small values of n and k, while the constraint length typically falls in the range of 10 to 30. All convolutional encoders require a commutator switch at the output, as shown in Figs. 13.3-2 and 13.3-5. For codes with $k > 1$, the input distributor switch can be eliminated by using a single register of length kL and shifting the bits in groups of k. In any case, convolutional encoding hardware is simpler than the hardware for block encoding since message bits enter the register unit at a steady rate r_b and an input buffer is not needed.

Exercise 13.3-1 Consider a systematic (3, 1, 3) convolutional code. List the possible states and determine the state transitions produced by $m_j = 0$ and $m_j = 1$. Then construct and label the state diagram, taking the encoded output bits to be m_j, $m_{j-2} \oplus m_j$, and $m_{j-3} \oplus m_{j-1}$. (See Fig. P13.3-4 for a convenient eight-state pattern.)

Free Distance and Coding Gain

We previously found that the error-control power of a block code depends upon its minimum distance, determined from the weights of the codewords. A convolutional code does not subdivide into codewords, so we consider instead the weight $w(X)$ of an entire transmitted sequence X generated by some message sequence.

The *free distance* of a convolutional code is then defined to be

$$d_f \triangleq [w(X)]_{\min} \qquad X \neq 000 \cdots \tag{5}$$

The value of d_f serves as a measure of error-control power.

It would be an exceedingly dull and tiresome task to try to evaluate d_f by listing all possible transmitted sequences. Fortunately, there's a better way based on the normal operating procedure of appending a "tail" of 0s at the end of a message to clear the register unit and return the encoder to its initial state. This procedure eliminates certain branches from the code trellis for the last L transitions.

Take the code trellis in Fig. 13.3-4*a*, for example. To end up at state *a*, the next-to-last state must be either *a* or *c* so the last few branches of any transmitted sequence X must follow one of the paths shown in Fig. 13.3-7. Here the final state is denoted by *e*, and each branch has been labeled with the number of 1s in the encoded bits — which equals the *weight* associated with that branch. The total weight of a transmitted sequence X equals the sum of the branch weights along the path of X. In accordance with Eq. (5), we seek the path that has the smallest branch-weight sum, other than the trivial all-zero path.

Looking backwards $L + 1 = 3$ branches from *e*, we locate the last path that emanates from state *a* before terminating at *e*. Now suppose all earlier transitions followed the all-zero path along the top line, giving the state sequence $aa \cdots abce$. Since an *a-a* branch has weight 0, this state sequence corresponds to a *minimum-weight nontrivial path*. We therefore conclude that $d_f = 0 + 0 + \cdots + 0 + 2 + 1 + 2 = 5$. There are other minimum-weight paths, such as $aa \cdots abcae$ and $aa \cdots abcbce$, but no nontrivial path has less weight than $d_f = 5$.

Another approach to the calculation of free distance involves the *generating function* of a convolutional code. The generating function may be viewed as the transfer function of the encoder with respect to state transitions. Thus, instead of relating the input and output bits streams by *convolution*, the generating function relates the initial and final states by *multiplication*. Generating functions provide important information about code performance, including the free distance and decoding error probability.

We'll develop the generating function for our (2, 1, 2) encoder using the modified state diagram in Fig. 13.3-8*a*. This diagram has been derived from Fig.

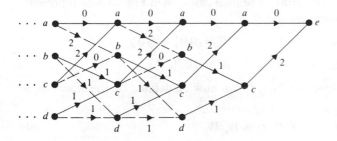

Figure 13.3-7 Termination of (2, 1, 2) code trellis.

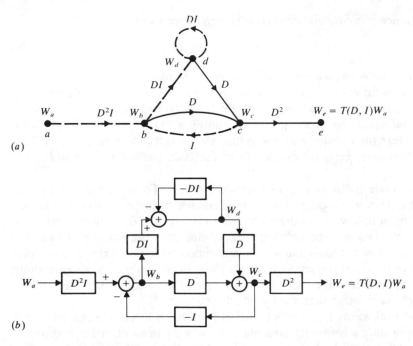

Figure 13.3-8 (a) Modified state diagram for (2, 1, 2) encoder; (b) equivalent block diagram.

13.3-4b with four modifications. First, we've eliminated the a-a loop which contributes nothing to the weight of a sequence X. Second, we've drawn the c-a branch as the final c-e transition. Third, we've assigned a *state variable* W_a at node a, and likewise at all other nodes. Fourth, we've labeled each branch with two "gain" variables D and I such that the exponent of D equals the branch weight (as in Fig. 13.3-7), while the exponent of I equals the corresponding number of nonzero message bits (as signified by the solid or dashed branch line). For instance, since the c-e branch represents $x_j' x_j'' = 11$ and $m_j = 0$, it is labeled with $D^2 I^0 = D^2$. This exponential trick allows us to perform *sums* by *multiplying* the D and I terms, which will become the independent variables of the generating function.

Our modified state diagram now looks like a *signal-flow graph* of the type sometimes used to analyze feedback systems. Specifically, if we treat the nodes as summing junctions and the DI terms as branch gains, then Fig. 13.3-8a represents the set of algebraic *state equations*

$$W_b = D^2 I W_a + I W_c \qquad W_c = D W_b + D W_d$$
$$W_d = D I W_b + D I W_d \qquad W_e = D^2 W_c \tag{6a}$$

The encoder's generating function $T(D, I)$ can now be defined by the *input-output equation*

$$T(D, I) \triangleq W_e / W_a \tag{6b}$$

These equations are also equivalent to the *block diagram* in Fig. 13.3-8b, which further emphasizes the relationships between the state variables, the branch gains, and the generating function. Note that minus signs have been introduced here so that the two feedback paths c-b and d-d correspond to negative feedback.

Next, the expression for $T(D, I)$ is obtained by algebraic solution of Eq. (6), or by block-diagram reduction of Fig. 13.3-8b using the transfer-function relations for parallel, cascade, and feedback connections in Fig. 3.1-8. (If you know *Mason's rule*, you could also apply it to Fig. 13.3-8a.) Any of these methods produces the final result

$$T(D, I) = \frac{D^5 I}{1 - 2DI} \tag{7a}$$

$$= D^5 I + 2D^6 I^2 + 4D^7 I^3 + \cdots$$

$$= \sum_{d=5}^{\infty} 2^{d-5} D^d I^{d-4} \tag{7b}$$

where we've expanded $(1 - 2DI)^{-1}$ to get the series in Eq. (7b). Keeping in mind that $T(D, I)$ represents all possible transmitted sequences that terminate with a c-e transition, Eq. (7b) has the following interpretation: for any $d \geq 5$, there are exactly 2^{d-5} valid paths with weight $w(X) = d$ that terminate with a c-e transition, and those paths are generated by messages containing $d - 4$ nonzero bits. The smallest value of $w(X)$ is the free distance, so we again conclude that $d_f = 5$.

As a generalization of Eq. (7), the generating function for an arbitrary convolutional code takes the form

$$T(D, I) = \sum_{d=d_f}^{\infty} \sum_{i=1}^{\infty} A(d, i) D^d I^i \tag{8}$$

Here, $A(d, i)$ denotes the number of different input-output paths through the modified state diagram that have weight d and are generated by messages containing i nonzero bits.

Now consider a received sequence $Y = X + E$, where E represents transmission errors. The path of Y then diverges from the path of X and may or may not be a valid path for the code in question. When Y does not correspond to a valid path, a *maximum-likelihood decoder* should seek out the valid path that has the smallest Hamming distance from Y. Before describing how such a decoder might be implemented, we'll state the relationship between generating functions, free distance, and error probability in maximum-likelihood decoding of convolutional codes.

If transmission errors occur with equal and independent probability α per bit, then the probability of a decoded message-bit error is upper-bounded by

$$P_{be} \leq \frac{1}{k} \frac{\partial T(D, I)}{\partial I} \bigg|_{D=2\sqrt{\alpha(1-\alpha)}, I=1} \tag{9}$$

The derivation of this bound is given in Lin and Costello (1983, chap. 11) or Viterbi and Omura (1979, chap. 4). When α is sufficiently small, series expansion of $T(D, I)$ yields the approximation

$$P_{be} \approx \frac{M(d_f)}{k} 2^{d_f} \alpha^{d_f/2} \qquad \sqrt{\alpha} \ll 1 \qquad (10)$$

where

$$M(d_f) = \sum_{i=1}^{\infty} i A(d_f, i)$$

The quantity $M(d_f)$ simply equals the total number of nonzero message bits over all minimum-weight input-output paths in the modified state diagram.

Equation (10) supports our earlier assertion that the error-control power of a convolutional code depends upon its free distance. For a performance comparison with uncoded transmission, we'll make the usual assumption of gaussian white noise and $(S/N)_R = 2R_c \gamma_b \geq 10$ so Eq. (10), Sect. 13.1, gives the transmission error probability

$$\alpha \approx (4\pi R_c \gamma_b)^{-1/2} e^{-R_c \gamma_b}$$

The decoded error probability then becomes

$$P_{be} \approx \frac{M(d_f) 2^{d_f}}{k(4\pi R_c \gamma_b)^{d_f/4}} e^{-(R_c d_f/2)\gamma_b} \qquad (11)$$

whereas uncoded transmission would yield

$$P_{ube} \approx \frac{1}{(4\pi\gamma_b)^{1/2}} e^{-\gamma_b} \qquad (12)$$

Since the exponential terms dominate in these expressions, we see that convolutional coding improves reliability when $R_c d_f/2 > 1$. Accordingly, the quantity $R_c d_f/2$ is known as the *coding gain*, usually expressed in dB.

Explicit design formulas for d_f do not exist, unfortunately, so good convolutional codes must be discovered by computer search and simulation. Table 13.3-1 lists the maximum free distance and coding gain of convolutional codes for selected values of n, k, and L. Observe that the free distance and coding gain increase

Table 13.3-1 Maximum free distance and coding gain of selected convolutional codes

n	k	R_c	L	d_f	$R_c d_f/2$
4	1	1/4	3	13	1.63
3	1	1/3	3	10	1.68
2	1	1/2	3	6	1.50
			6	10	2.50
			9	12	3.00
3	2	2/3	3	7	2.33
4	3	3/4	3	8	3.00

with increasing memory L when the code rate R_c is held fixed. All listed codes are nonsystematic; a systematic convolutional code has a smaller d_f than an optimum nonsystematic code with the same rate and memory.

Example 13.3-1 The (2, 1, 2) encoder back in Fig. 13.3-2 has $T(D, I) = D^5 I/(1 - 2DI)$, so $\partial T(D, I)/\partial I = D^5/(1 - 2DI)^2$. Equation (9) therefore gives

$$P_{be} \leq \frac{2^5 [\alpha(1 - \alpha)]^{5/2}}{[1 - 4\sqrt{\alpha(1 - \alpha)}]^2} \approx 2^5 \alpha^{5/2}$$

and the small-α approximation agrees with Eq. (10). Specifically, in Fig. 13.3-8a we find just one minimum-weight nontrivial path $abce$, which has $w(X) = 5 = d_f$ and is generated by a message containing one nonzero bit, so $M(d_f) = 1$.

If $\gamma_b = 10$, then $R_c \gamma_b = 5$, $\alpha \approx 8.5 \times 10^{-4}$, and maximum-likelihood decoding yields $P_{be} \approx 6.7 \times 10^{-7}$, as compared with $P_{ube} \approx 4.1 \times 10^{-6}$. This rather small reliability improvement agrees with the small coding gain $R_c d_f/2 = 5/4$.

Exercise 13.3-2 Let the connections to the mod-2 summers in Fig. 13.3-2 be changed such that $x'_j = m_j$ and $x''_j = m_{j-2} \oplus m_{j-1} \oplus m_j$.
  (a) Construct the code trellis and modified state diagram for this systematic code. Show that there are two minimum-weight paths in the state diagram, and that $d_f = 4$ and $M(d_f) = 3$. It is not necessary to find $T(D, I)$.
  (b) Now assume $\gamma_b = 10$. Calculate α, P_{be}, and P_{ube}. What do you conclude about the performance of a convolutional code when $R_c d_f/2 = 1$?

Decoding Methods

There are three generic methods for decoding convolutional codes. At one extreme, the Viterbi algorithm executes *maximum-likelihood decoding* and achieves optimum performance but requires extensive hardware for computation and storage. At thd other extreme, *feedback decoding* sacrifices performance in exchange for simplified hardware. Between these extremes, *sequential decoding* approaches optimum performance to a degree that depends upon the decoder's complexity. We'll describe how these methods work with a (2, 1, L) code. The extension to other codes is conceptually straightforward, but becomes messy to portray for $k > 1$.

Recall that a maximum-likelihood decoder must examine an entire received sequence Y and find a valid path that has the smallest Hamming distance from Y. However, there are 2^N possible paths for an arbitrary message sequence of N bits (or Nn/k bits in Y), so an exhaustive comparison of Y with all valid paths would be an absurd task in the usual case of $N \gg 1$. The *Viterbi algorithm* applies maximum-likelihood principles to limit the comparison to 2^{kL} *surviving paths*, independent of N, thereby bringing maximum-likelihood decoding into the realm of feasibility.

A Viterbi decoder assigns to each branch of each surviving path a *metric* that equals its Hamming distance from the corresponding branch of Y. (We assume here that 0s and 1s have the same transmission-error probability; if not, the branch metric must be redefined to account for the differing probabilities.) Summing the branch metrics yields the path metric, and Y is finally decoded as the surviving path with the smallest metric. To illustrate the metric calculations and explain how surviving paths are selected, we'll walk through an example of Viterbi decoding.

Suppose that our $(2, 1, 2)$ encoder is used at the transmitter, and the received sequence starts with $Y = 11\ 01\ 11$. Figure 13.3-9 shows the first three branches of the valid paths emanating from the initial node a_0 in the code trellis. The number in parentheses beneath each branch is the branch metric, obtained by counting the differences between the encoded bits and the corresponding bits in Y. The circled number at the right-hand end of each branch is the running path metric, obtained by summing branch metrics from a_0. For instance, the metric of the path $a_0\,b_1 c_2\,b_3$ is $0 + 2 + 2 = 4$.

Now observe that another path $a_0\,a_1 a_2\,b_3$ also arrives at node b_3 and has a smaller metric $2 + 1 + 0 = 3$. Regardless of what happens subsequently, this path will have a smaller Hamming distance from Y than the other path arriving at b_3 and is therefore more likely to represent the actual transmitted sequence. Hence, we discard the larger-metric path, marked by an X, and we declare the path with the smaller metric to be the survivor at this node. Likewise, we discard the larger-metric paths arriving at nodes a_3, c_3, and d_3, leaving a total of $2^{kL} = 4$ surviving paths. The fact that none of the surviving path metrics equals zero indicates the presence of *detectable errors* in Y.

Figure 13.3-9

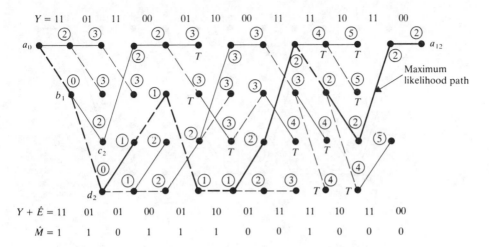

Figure 13.3-10 Illustration of the Viterbi algorithm for maximum-likelihood decoding.

Figure 13.3-10 depicts the continuation of Fig. 13.3-9 for a complete message of $N = 12$ bits, including tail 0s. All discarded branches and all labels except the running path metrics have been omitted for the sake of clarity. The letter T under a node indicates that the two arriving paths had equal running matrics, in which case we just flip a coin to choose the survivor (why?). The maximum-likelihood path follows the heavy line from a_0 to a_{12}, and the final value of the path metric signifies at least two transmission errors in Y. The decoder assumes the corresponding transmitted sequence $Y + \hat{E}$ and message sequence $\hat{M}$ written below the trellis.

A Viterbi decoder must calculate two metrics for each node and store 2^{kL} surviving paths, each consisting of N branches. Hence, decoding complexity increases exponentially with L and linearly with N. The exponential factor limits practical application of the Viterbi algorithm to codes with small values of L.

When $N \gg 1$, storage requirements can be reduced by a truncation process based on the following metric-divergence effect: if two surviving paths emanate from the same node at some point, then the running metric of the less likely path tends to increase more rapidly than the metric of the other survivor within about $5L$ branches from the common node. This effect appears several times in Fig. 13.3-10; consider, for instance, the two paths emanating from node b_1. Hence, decoding need not be delayed until the end of the transmitted sequence. Instead, the first k message bits can be decoded and the first set of branches can be deleted from memory after the first $5Ln$ received bits have been processed. Successive groups of k message bits are then decoded for each additional n bits received thereafter.

Sequential decoding, which was invented before the Viterbi algorithm, also relies on the metric-divergence effect. A simplified version of the sequential algorithm is illustrated in Fig. 13.3-11a, using the same trellis, received sequence, and

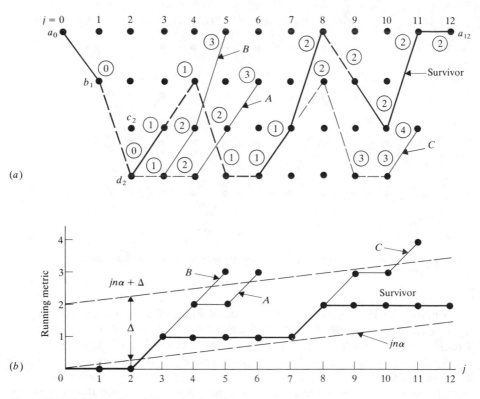

Figure 13.3-11 Illustration of sequential decoding.

metrics as in Fig. 13.3-10. Starting at a_0, the sequential decoder pursues a single path by taking the branch with the smallest branch metric at each successive node. If two or more branches from one node have the same metric, such as at node d_2, the decoder selects one at random and continues on. Whenever the current path happens to be unlikely, the running metric rapidly increases and the decoder eventually decides to go back to a lower-metric node and try another path. There are three of these abandoned paths in our example. Even so, a comparison with Fig. 13.3-10 shows that sequential decoding involves less computation than Viterbi decoding.

   The decision to backtrack and try again is based on the expected value of the running metric at a given node. Specifically, if α is the transmission error probability per bit, then the expected running metric at the jth node of the correct path equals $jn\alpha$, the expected number of bit errors in Y at that point. The sequential decoder abandons a path when its metric exceeds some specified *threshold* Δ above $jn\alpha$. If no path survives the threshold test, the value of Δ is increased and the decoder backtracks again. Figure 13.3-11b plots the running metrics versus j, along with $jn\alpha$ and the threshold line $jn\alpha + \Delta$ for $\alpha = 1/16$ and $\Delta = 2$.

Sequential decoding approaches the performance of maximum-likelihood decoding when the threshold is loose enough to permit exploration of all probable paths. However, the frequent backtracking requires more computations and results in a decoding delay significantly greater than Viterbi decoding. A tighter threshold reduces computations and decoding delay but may actually eliminate the most probable path, thereby increasing the output error probability compared to that of maximum-likelihood decoding with the same coding gain. As compensation, sequential decoding permits practical application of convolutional codes with large L and large coding gain since the decoder's complexity is essentially independent of L.

We've described sequential decoding and Viterbi decoding in terms of *algorithms* rather than block diagrams of hardware. Indeed, these methods are usually implemented as software for a computer or microprocessor that performs the metric calculations and stores the path data. When circumstances preclude algorithmic decoding, and a higher error probability is tolerable, *feedback decoding* may be the appropriate method. A feedback decoder acts in general like a "sliding block decoder" that decodes message bits one by one based on a block of L or more successive tree branches. We'll focus on the special class of feedback decoding that employs *majority logic* to achieve the simplest hardware realization of a convolutional decoder.

Consider a message sequence $M = m_1 m_2 \cdots$ and the *systematic* $(2, 1, L)$ encoded sequence

$$X = x_1' x_1'' x_2' x_2'' \cdots \tag{13a}$$

where

$$x_j' = m_j \qquad x_j'' = \sum_{i=0}^{L} m_{j-i} g_j \qquad (\text{mod-2}) \tag{13b}$$

We'll view the entire sequence X as a codeword of indefinite length. Then, borrowing from the matrix representation used for block codes, we'll define a *generator matrix G* and a *parity-check matrix H* such that

$$X = MG \qquad XH^T = 0 \quad 0 \quad \cdots$$

To represent Eq. (13), G must be a semi-infinite matrix with a diagonal structure given by

$$G = \begin{bmatrix} 1 & g_0 & 0 & g_1 & 0 & \cdots & 0 & g_L & & \\ & & 1 & g_0 & 0 & g_1 & 0 & \cdots & 0 & g_L \\ & & & \cdot & \cdot & \cdot & & & & \cdot \\ & & & & \cdot & \cdot & \cdot & & & \cdot \\ & & & & & \cdot & \cdot & \cdot & & \cdot \end{bmatrix} \tag{14a}$$

This matrix extends indefinitely to the right and down, and the triangular blank spaces denote elements that equal zero. The parity-check matrix is

$$
H = \begin{bmatrix}
g_0 & 1 & & & & & & & \\
g_1 & 0 & g_0 & 1 & & & & & \\
\cdot & \cdot & g_1 & 0 & g_0 & 1 & & & \\
\cdot & \cdot & \cdot & \cdot & & & & & \\
\cdot & \cdot & \cdot & \cdot & & & \cdot & \cdot & \\
g_L & 0 & \cdot & \cdot & & & & \cdot & \\
& & g_L & 0 & & & & & \\
& & & \cdot & \cdot & & & & \\
& & & & & \cdot & \cdot & & \\
& & & & & & & \cdot &
\end{bmatrix} \tag{14b}
$$

which also extends indefinitely to the right and down.

Next, let E be the transmission error pattern in a received sequence $Y = X + E$. We'll write these sequences as

$$
Y = y_1' y_1'' y_2' y_2'' \cdots \qquad E = e_1' e_1'' e_2' e_2'' \cdots
$$

so that $y_j' = m_j \oplus e_j'$. Hence, given the error bit e_j', the jth message bit is

$$
m_j = y_j' \oplus e_j' \tag{15}
$$

A feedback decoder estimates errors from the *syndrome* sequence

$$
S = YH^T = (X + E)H^T = EH^T
$$

Using Eq. (14b) for H, the jth bit of S is

$$
s_j = \sum_{i=0}^{L} y_{j-i}' g_i \oplus y_j'' = \sum_{i=0}^{L} e_{j-i}' g_i \oplus e_j'' \tag{16}
$$

where the sums are mod-2 and it's understood that $y_{j-i}' = e_{j-i}' = 0$ for $j - i \le 0$.

As a specific example, take a (2, 1, 6) encoder with $g_0 = g_2 = g_5 = g_6 = 1$ and $g_1 = g_3 = g_4 = 0$, so

$$
s_j = y_{j-6}' \oplus y_{j-5}' \oplus y_{j-2}' \oplus y_j' \oplus y_j'' \tag{17a}
$$

$$
= e_{j-6}' \oplus e_{j-5}' \oplus e_{j-2}' \oplus e_j' \oplus e_j'' \tag{17b}
$$

Equation (17a) leads directly to the shift-register circuit for syndrome calculation diagrammed in Fig. 13.3-12. Equation (17b) is called a *parity-check sum* and will leads us eventually to the remaining portion of the feedback decoder.

To that end, consider the parity-check table in Fig. 13.3-13a where checks indicate which error bits appear in the sums s_{j-6}, s_{j-4}, s_{j-1}, and s_j. This table brings out the fact that e_{j-6}' is checked by all four of the listed sums, while no other error bit is checked by more than one. Accordingly, this set of check sums is said to be *orthogonal* on e_{j-6}'. The tap gains of the encoder were carefully chosen to obtain orthogonal check sums.

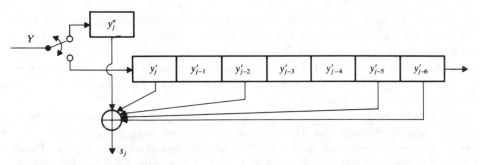

Figure 13.3-12 Shift-register circuit for syndrome calculation for a systematic (2, 1, 6) code.

	e'_{j-12}	e'_{j-11}	e'_{j-10}	e'_{j-9}	e'_{j-8}	e'_{j-7}	e'_{j-6}	e''_{j-6}	e'_{j-5}	e'_{j-4}	e''_{j-4}	e'_{j-3}	e'_{j-2}	e'_{j-1}	e''_{j-1}	e'_j	e''_j
s_{j-6}	√	√			√		√	√									
s_{j-4}			√	√			√			√	√						
s_{j-1}						√	√					√		√	√		
s_j							√		√				√			√	√

Figure 13.3-13 Parity-check table for a systematic (2, 1, 6) code.

When the transmission error probability is reasonably small, we expect to find at most one or two errors in the 17 transmitted bits represented by the parity-check table. If one of the errors corresponds to $e'_{j-6} = 1$, then the four check sums will contain three or four 1s. Otherwise, the check sums contain less than three 1s. Hence, we can apply these four check sums to a *majority-logic gate* to generate the most likely estimate of e'_{j-6}.

Figure 13.3-14 diagrams a complete majority-logic feedback decoder for our systematic (2, 1, 6) code. The syndrome calculator from Fig. 13.3-12 has two

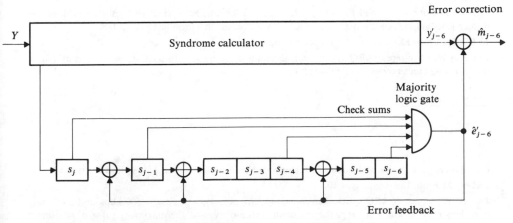

Figure 13.3-14 Majority-logic feedback decoder for a systematic (2, 1, 6) code.

outputs, y'_{j-6} and s_j. The syndrome bit goes into another shift register with taps that connect the check sums to the majority-logic gate, whose output equals the estimated error $\hat{e}'_{j-6}$. The mod-2 addition $y'_{j-6} \oplus \hat{e}'_{j-6} = \hat{m}_{j-6}$ carries out error correction based on Eq. (15). The error is also fed back to the syndrome register to improve the reliability of subsequent check sums. This feedback path accounts for the name *feedback* decoding.

Our example decoder can correct any single-error or double-error pattern in six consecutive message bits. However, more than two transmission errors produces erroneous corrections and error propagation via the feedback path. These effects result in a higher output error than that of maximum-likelihood decoding. See Lin and Costello (1983, chap. 13) for the error analysis and further treatment of majority-logic decoding.

13.4 PROBLEMS

13.1-1 Calculate the probabilities that a word has no errors, detected errors, and undetected errors when a parity-check code with $n = 4$ is used and $\alpha = 0.1$.

13.1-2 Do Prob. 13.1-1 with $n = 9$ and $\alpha = 0.05$.

13.1-3 Consider the square-array code in Fig. 13.1-1. (*a*) Confirm that if a word has two errors, then they can be detected but not corrected. (*b*) Discuss what happens when a word contains three errors.

13.1-4 An FEC system contaminated by gaussian white noise must achieve $P_{be} \leq 10^{-4}$ with minimum transmitted power. Three block codes under consideration have the following parameters:

n	k	d_{min}
31	26	3
31	21	5
31	16	7

Determine which code should be used, and calculate the power saving in dB compared to uncoded transmission.

13.1-5 Do Prob. 13.1-4 with $P_{be} \leq 10^{-6}$.

13.1-6 Calculate α, P_{be}, and P_{ube} at $\gamma_b = 2$, 5, and 10 for an FEC system with gaussian white noise using a (31, 26) block code having $d_{min} = 3$. Plot your results in a form like Fig. 13.1-5.

13.1-7 Do Prob. 13.1-6 for a (31, 21) code having $d_{min} = 5$.

13.1-8 A selective-repeat ARQ system with gaussian white noise is to have $P_{be} = 10^{-5}$ using one of the following blocks codes for error detection:

n	k	d_{min}
12	11	2
15	11	3
16	11	4

Calculate r_b/r and γ_b for each code and for uncoded transmission. Then plot γ_b in dB versus r_b/r.

13.1-9 Do Prob. 13.1-8 for $P_{be} = 10^{-6}$.

13.1-10 A go-back-N ARQ system has gaussian white noise, $\gamma_c = 6$ dB, $r = 500$ kbps, and a one-way path length of 45 km. Find P_{be}, the minimum value of N, and the maximum value of r_b using a (15, 11) block code with $d_{min} = 3$ for error detection.

13.1-11 Do Prob. 13.1-10 using a (16, 11) block code with $d_{min} = 4$.

13.1-12 A stop-and-wait ARQ system uses simple parity checking with $n = k + 1$ for error detection. The system has gaussian white noise, $r = 10$ kbps, and a one-way path length of 18 km. Find the smallest value of k such that $P_{be} \leq 10^{-6}$ and $r_b \geq 7200$ bps. Then calculate γ_b in dB.

13.1-13 Do Prob. 13.1-12 with a 60-km path length.

13.1-14‡ Derive m as given in Eq. (19) for a go-back-N ARQ system. *Hint:* If a given word has detected errors in i successive transmissions, then the total number of transmitted words equals $1 + Ni$.

13.1-15 Consider a hybrid ARQ system using a code that corrects t errors and detects $\ell > t$ errors per n-bit word. Obtain an expression for the retransmission probability p when $\alpha \ll 1$. Then take $d_{min} = 4$ and compare your result with Eq. (15).

13.1-16 Suppose a hybrid selective-repeat ARQ system uses an (n, k) block code with $d_{min} = 2t + 2$ to correct up to t errors and detect $t + 1$ errors per word. (a) Assume $\alpha \ll 1$ to obtain an approximate expression for the retransmission probability p, and show that

$$P_{be} \approx \binom{n-1}{t+1}\alpha^{t+2}$$

(b) Evaluate α and p for a (24, 12) code with $d_{min} = 8$ when $P_{be} = 10^{-5}$. Then assume gaussian white noise and find P_{be} for uncoded transmission with the same value of γ_b.

13.2-1 Let U and V be n-bit vectors. (a) By considering the number of 1's in U, V, and $U + V$, confirm that $d(U, V) \leq w(U) + w(V)$. (b) Now let $U = X + Y$ and $V = Y + Z$. Show that $U + V = X + Z$ and derive the *triangle inequality*

$$d(X, Z) \leq d(X, Y) + d(Y, Z)$$

13.2-2 Let X be a code vector, let Z be any other vector in the code, and let Y be the vector that results when X is received with i bit errors. Use the triangle inequality in Prob. 13.2-1 to show that if the code has $d_{min} \geq \ell + 1$ and if $i \leq \ell$, then the errors in Y are detectable.

13.2-3 Let X be a code vector, let Z be any other vector in the code, and let Y be the vector that results when X is received with i bit errors. Use the triangle inequality in Prob. 13.2-1 to show that if the code has $d_{min} \geq 2t + 1$ and if $i \leq t$, then the errors in Y are correctable.

13.2-4 A triple-repetition code is a systematic (3, 1) block code generated using the submatrix $P = [1 \ \ 1]$. Tabulate all possible received vectors Y and $S = YH^T$. Then determine the corresponding maximum-likelihood errors patterns and corrected vectors $Y + \hat{E}$.

13.2-5 Construct the lookup table for the (6, 3) block code in Exercise 13.2-1.

13.2-6 Consider a (5, 3) block code obtained by deleting the last column of the P submatrix in Exercise 13.2-1. Construct the lookup table, and show that this code could be used for error detection but not correction.

13.2-7 Let the P submatrix for a (15, 11) Hamming code be arranged such that the row words increasing in numerical value from top to bottom. Construct the lookup table and write the check-bit equations.

13.2-8 Suppose a block code with $t = 1$ is required to have $k = 6$ message bits per word. (a) Find the minimum value of n and the number of bits stored in the lookup table. (b) Construct an appropriate P submatrix.

13.2-9 Do Prob. 13.2-8 with $k = 8$.

13.2-10 It follows from Eq. (4) that $XH^T = MA$ with $A = GH^T$. Prove Eq. (9) by showing that any element of A has the property $a_{ij} = 0$.

13.2-11 The original (7, 4) Hamming code is a *nonsystematic* code with

$$H = \begin{bmatrix} 1 & 0 & 1 & 0 & 1 & 1 & 1 \\ 0 & 1 & 1 & 0 & 0 & 1 & 1 \\ 0 & 0 & 0 & 1 & 1 & 0 & 1 \end{bmatrix}$$

(a) Construct the lookup table and explain the rationale for this nonsystematic form. (b) Write the equations for s_1, s_2, and s_3 taking $Y = X = (x_1 \quad x_2 \quad \cdots \quad x_7)$. Then determine which are the message and check bits in X, and obtain the check-bit equations.

13.2-12 The (7, 4) Hamming code can be *extended* to form an (8, 4) code by appending a fourth check bit chosen such that all code words have even parity. (a) Apply this extension process to Table 13.2-1 to show that $d_{min} = 4$. (b) Find the corresponding equation for c_4 in terms of the message bits. (c) Explain how the decoding system in Fig. 13.2-2 should be modified to perform double-error detection as well as single-error correction.

13.2-13 Consider a systematic (7, 3) cyclic code generated by $G(p) = p^4 + p^3 + p^2 + 0 + 1$. Find $Q_M(p)$, $C(p)$, and X when $M = (1 \quad 0 \quad 1)$. Then take $Y = X'$ and confirm that $S(p) = 0$. You may carry out the divisions using binary words rather than polynomials if you wish.

13.2-14 Do Prob. 13.2-13 with $G(p) = p^4 + 0 + p^2 + p + 1$.

13.2-15 Given $G(p)$, the equivalent generator matrix of a systematic cyclic code can be found from the polynomials $R_i(p) = \text{rem} \ [p^{n-i}/G(p)]$, $1 \le i \le k$, which correspond to the rows of the P submatrix. Use this method to obtain P for a (7, 4) Hamming code generated by $G(p) = p^3 + 0 + p + 1$. (You may carry out the divisions using binary words rather than polynomials if you wish.) Compare your result with the G matrix in Example 13.2-1.

13.2-16 Do Prob. 13.2-15 with $G(p) = p^3 + p^2 + 0 + 1$.

13.2-17 Figure P13.2-17 is a shift-register circuit that divides an arbitrary mth-order polynomial $Z(p)$ by a fixed polynomial $G(p) = p^q + g_{q-1}p^{q-1} + \cdots + g_1p + 1$. If the register has been cleared before $Z(p)$ is shifted in, then the output equals the quotient and the remainder appears in the register after m shift cycles. Confirm the division operation by constructing a table similar to Fig. 13.2-4b, taking $Z(p) = p^3 M(p)$ and $G(p)$ as in Example 13.2-3.

Figure P13.2-17

13.2-18 Use the shift-register polynomial divider in Prob. 13.2-17 to diagram a circuit that produces the syndrome $S(p)$ from $Y(p)$ for the code in Example 13.2-3. Then construct a table like Fig. 13.2-4b to show that $S(p) = 0$ when $Y = (1 \quad 1 \quad 0 \quad 0 \quad 0 \quad 1 \quad 0)$.

13.3-1 Diagram the encoder for a systematic (3, 2, 3) convolutional code. Label the input and output rates and the current input and state at an arbitrary time.

13.3-2 Diagram the encoder for a systematic (4, 3 1) convolutional code. Label the input and output rates and the current input and state at an arbitrary time.

13.3-3 A (3, 1, 2) encoder achieves maximum free distance when

$$x'_j = m_{j-2} \oplus m_j \qquad x''_j = x'''_j = m_{j-1} \oplus x'_j$$

(a) Construct the code trellis and state diagram. Then find the state and output sequence produced by the input sequence 1011001111. (b) Construct the modified state diagram like Fig. 13.1-8a, identify the minimum-weight path or paths, and determine the values of d_f and $M(d_f)$.

13.3-4 Do Prob. 13.3-3 for a (2, 1, 3) encoder with

$$x'_j = m_{j-3} \oplus m_{j-1} \oplus m_j \qquad x''_j = m_{j-2} \oplus x'_j$$

which achieves maximum free distance. Use the pattern in Fig. P13.3-4 for your state diagram.

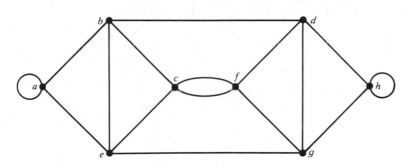

Figure P13.3-4

13.3-5 Use block-diagram reduction of Fig. 13.3-8*b* to get $T(D, I)$ as in Eq. (7*a*).

13.3-6 Derive Eq. (10) from Eqs. (8) and (9).

13.3-7 Consider a (2, 1, 1) code with $x'_j = m_j$ and $x''_j = m_{j-1} \oplus m_j$. (*a*) Construct the code trellis and modified state diagram. Then identify the minimum-weight path or paths and find the values of d_f and $M(d_f)$. (*b*) Use block-diagram reduction to obtain $T(D, I)$ in the form of Eq. (7*a*). (*c*) Find and compare P_{be} from Eqs. (9) and (10), assuming $\alpha \ll 1$.

13.3-8 Do Prob. 13.3-7 for a (3, 1, 2) code with $x'_j = m_j$, $x''_j = m_{j-2} \oplus m_j$, and $x'''_j = m_{j-2} \oplus m_{j-1} \oplus m_j$.

13.3-9 Use the Viterbi algorithm and the code trellis in Fig. 13.3-4 to find the sequences $Y + \hat{E}$ and $\hat{M}$ when $Y = 01\ 11\ 01\ 01\ 11\ 01\ 10\ 11$. If two paths arriving at a given node have equal running metrics, arbitrarily keep the upper path.

13.3-10 Construct the code trellis for a (2, 1, 2) code with $x'_j = m_{j-1} \oplus m_j$ and $x''_j = m_{j-2} \oplus m_{j-1}$. Then apply the Viterbi algorithm to find the sequences $Y + \hat{E}$ and $\hat{M}$ when $Y = 10\ 11\ 01\ 01\ 10\ 01\ 10\ 11$. If two paths arriving at a given node have equal running metrics, arbitrarily keep the upper path.

13.3-11 A systematic (2, 1, 4) code intended for feedback decoding with majority logic has $x''_j = m_{j-4} \oplus m_{j-3} \oplus m_j$. Construct a parity-check table like Fig. 13.3-13 showing all the error bits that appear in the check sums s_{j-4} through s_j. Then find an orthogonal set of three check sums, and draw a complete diagram of the decoder.

13.3-12 Do Prob. 13.3-11 with $x''_j = m_{j-4} \oplus m_{j-1} \oplus m_j$.

FOURTEEN

BANDPASS DIGITAL TRANSMISSION

Long-haul digital transmission usually requires CW modulation to generate a *bandpass* signal suited to the transmission medium — be it radio, cable, or whatever. Just as there are a multitude of modulation methods for analog signals, so too there are many ways of impressing digital information upon a carrier wave. This chapter applies concepts of baseband digital transmission and CW modulation to the study of bandpass digital transmission.

We begin with waveforms and spectral analysis of digital CW modulation for binary and *M*-ary modulating signals. Then we focus on the demodulation of binary signals in noise to bring out the distinction between *coherent* (synchronous) detection and *noncoherent* (envelope) detection. The last section deals with quadrature-carrier and *M*-ary systems, leading to a comparison of modulation methods with regard to spectral efficiency, hardware complexity, and system performance in the face of corrupting noise.

14.1 DIGITAL CW MODULATION

A digital signal can modulate the amplitude, frequency, or phase of a sinusoidal carrier wave. If the modulating waveform consists of NRZ rectangular pulses, then the modulated parameter will be switched or *keyed* from one discrete value to another. Figure 14.1-1 illustrates binary *amplitude-shift keying* (ASK), *frequency-shift keying* (FSK), and *phase-shift keying* (PSK). Also shown, for contrast, is the waveform produced by DSB modulation with Nyquist pulse shaping at baseband. Other modulation techniques combine amplitude and phase modulation, with or without baseband pulse shaping.

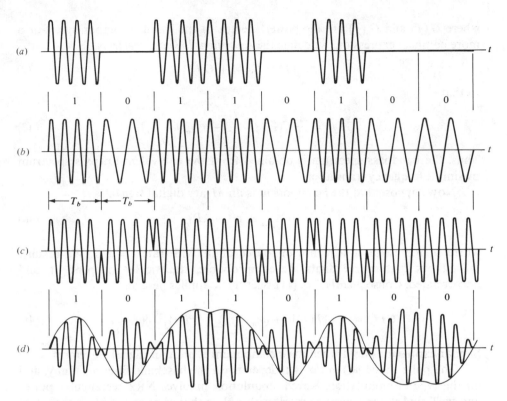

Figure 14.1-1 Binary modulated waveforms. (*a*) ASK; (*b*) FSK; (*c*) PSK; (*d*) DSB with baseband pulse shaping.

In this section we'll define specific types of digital modulation in terms of mathematical models and/or transmitter diagrams. We'll also examine their power spectra and estimate therefrom the transmission bandwidth required for a given digital signaling rate. As preparation, we first develop a technique for spectral analysis of bandpass digital signals.

Spectral Analysis of Bandpass Digital Signals

Any modulated bandpass signal may be expressed in the quadrature-carrier form

$$x_c(t) = A_c[x_i(t) \cos (\omega_c t + \theta) - x_q(t) \sin (\omega_c t + \theta)] \tag{1}$$

The carrier frequency f_c, amplitude A_c, and phase θ are constant, while the time-varying i (in-phase) and q (quadrature) components contain the message. Spectral analysis of $x_c(t)$ is relatively easy when the i and q components are statistically independent signals and at least one has zero mean. Then, from the superposition and modulation relations in Sect. 5.2, the power spectrum of $x_c(t)$ becomes

$$G_c(f) = \frac{A_c^2}{4} [G_i(f - f_c) + G_i(f + f_c) + G_q(f - f_c) + G_q(f + f_c)]$$

where $G_i(f)$ and $G_q(f)$ are the power spectra of the i and q components. For a more compact expression, we define the *equivalent lowpass spectrum*

$$G_{\ell p}(f) \triangleq G_i(f) + G_q(f) \tag{2}$$

so that

$$G_c(f) = \frac{A_c^2}{4} [G_{\ell p}(f - f_c) + G_{\ell p}(f + f_c)] \tag{3}$$

Thus, the bandpass spectrum is obtained from the equivalent lowpass spectrum by simple frequency translation.

Now suppose that the i component is an M-ary digital signal, say

$$x_i(t) = \sum_k a_k p(t - kD) \tag{4a}$$

where a_k represents a sequence of source digits with rate $r = 1/D$. We assume throughout that source digits are equiprobable, statistically independent, and uncorrelated. Consequently, Eq. (11), Sect. 11.1, applies here and

$$G_i(f) = \sigma_a^2 r |P(f)|^2 + (m_a r)^2 \sum_{n=-\infty}^{\infty} |P(nr)|^2 \, \delta(f - nr) \tag{4b}$$

Similar expressions hold when the q component is another digital waveform.

The pulse shape $p(t)$ in Eq. (4a) depends on the baseband filtering, if any, and on the type of modulation. Keyed modulation involves NRZ rectangular pulses and we'll find it convenient to work with pulses that start at $t = kD$, rather than being centered at $t = kD$ as in Chap. 11. Accordingly, let

$$p_D(t) \triangleq u(t) - u(t - D) = \begin{cases} 1 & 0 < t < D \\ 0 & \text{otherwise} \end{cases} \tag{5a}$$

whose Fourier transform yields

$$|P_D(f)|^2 = D^2 \operatorname{sinc}^2 fD = \frac{1}{r^2} \operatorname{sinc}^2 \frac{f}{r} \tag{5b}$$

If $p(t) = p_D(t)$ in Eq. (4a), then the continuous spectral term in Eq. (4b) will be proportional to $|P_D(f)|^2$. Since $\operatorname{sinc}^2 (f/r)$ is not bandlimited, we conclude from Eqs. (2) and (3) that keyed modulation requires $f_c \gg r$ in order to produce a *bandpass* signal.

Amplitude Modulation Methods

The binary ASK waveform illustrated in Fig. 14.1-1a could be generated simply by turning the carrier on and off, a process described as *on-off keying* (OOK). In general, an M-ary ASK waveform has $M - 1$ discrete "on" amplitudes as well as the "off" state. Since there are no phase reversals or other variations, we can set the q component of $x_c(t)$ equal to zero and take the i component to be a *unipolar* NRZ signal, namely

$$x_i(t) = \sum_k a_k p_D(t - kD) \qquad a_k = 0, 1, \dots, M - 1 \qquad (6a)$$

The mean and variance of the digital sequence are

$$m_a = \overline{a_k} = \frac{M-1}{2} \qquad \sigma_a^2 = \overline{a_k^2} - m_a^2 = \frac{M^2 - 1}{12} \qquad (6b)$$

Hence, the equivalent lowpass spectrum is

$$G_{\ell p}(f) = G_i(f) = \frac{M^2 - 1}{12r} \operatorname{sinc}^2 \frac{f}{r} + \frac{(M-1)^2}{4} \delta(f) \qquad (7)$$

obtained with the help of Eqs. (2), (4b), and (5b).

Figure 14.1-2 shows the resulting bandpass spectrum $G_c(f)$ for $f > 0$. Most of the signal power is contained within the range $f_c \pm r/2$, and the spectrum has a *second-order rolloff* proportional to $|f - f_c|^{-2}$ away from the carrier frequency. These considerations suggest the estimated transmission bandwidth to be $B_T \approx r$. If an M-ary ASK signal represents binary data at rate $r_b = r \log_2 M$, then $B_T \approx r_b/\log_2 M$ or

$$r_b/B_T \approx \log_2 M \qquad \text{bps/Hz} \qquad (8)$$

This ratio of bit rate to transmission bandwidth serves as our measure of modulation "speed" or spectral efficiency. Binary OOK has the poorest spectral efficiency since $r_b/B_T \approx 1$ bps/Hz when $M = 2$.

Drawing upon the principle of quadrature-carrier multiplexing, *quadrature-carrier AM (QAM)* achieves twice the modulation speed of binary ASK. Figure 14.1-3a depicts the functional blocks of a binary QAM transmitter with a *polar* binary input at rate r_b. The serial-to-parallel converter divides the input into two

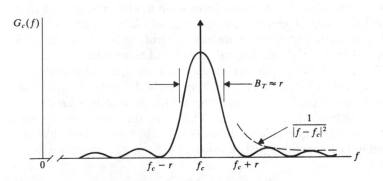

Figure 14.1-2 ASK power spectrum.

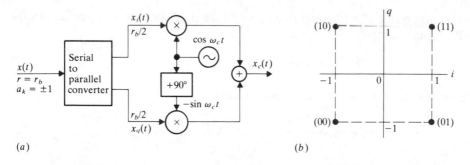

Figure 14.1-3 Binary QAM. (a) Transmitter; (b) signal constellation.

streams consisting of alternate bits at rate $r = r_b/2$. Thus, the i and q modulating signals are represented by

$$x_i(t) = \sum_k a_{2k}\, p_D(t - kD) \qquad x_q(t) = \sum_k a_{2k+1}\, p_D(t - kD)$$

where $D = 1/r = 2T_b$ and $a_k = \pm 1$. The peak modulating values are $x_i = x_q = \pm 1$ during an arbitrary interval $kD < t < (k + 1)D$. Figure 14.1-3b conveys this information as a two-dimensional *signal constellation*. The four signal points have been labeled with the corresponding pairs of source bits, known as *dibits*.

Summing the modulated carriers finally yields the QAM signal in the form of Eq. (1). The i and q components are independent but they have the same pulse shape and the same statistical values, namely, $m_a = 0$ and $\sigma_a^2 = 1$. Thus,

$$G_{\ell p}(f) = 2 \times r\,|P_D(f)|^2 = \frac{4}{r_b}\,\text{sinc}^2\,\frac{2f}{r_b} \qquad (9)$$

where we've used Eqs. (4b) and (5b) with $r = r_b/2$. Binary QAM achieves $r_b/B_T \approx$ 2 bps/Hz because the dibit rate equals one-half of the input bit rate, reducing the transmission bandwidth to $B_T \approx r_b/2$.

Keep in mind, however, that ASK and QAM spectra actually extend beyond the estimated transmission bandwidth. Such spectral "spillover" outside B_T becomes an important concern in radio transmission and frequency-division multiplexing systems when it creates interference with other signal channels. Bandpass filtering at the output of the modulator controls spillover, but heavy filtering introduces ISI in the modulated signal and should be avoided.

Spectral efficiency without spillover is achieved by the *vestigial-sideband* modulator diagrammed in Fig. 14.1-4a. This VSB method applies Nyquist pulse shaping to a polar input signal, as covered in Sect. 11.3, producing a bandlimited modulating signal with $B = (r/2) + \beta_N$. The VSB filter then removes all but a vestige of width β_V from one sideband, so $G_c(f)$ looks something like Fig. 14.1-4b — a bandlimited spectrum with $B_T = (r/2) + \beta_N + \beta_V$. Therefore, if $r = r_b/\log_2 M$, then

$$r_b/B_T \le 2 \log_2 M \qquad (10)$$

and the upper bound holds when $\beta_N \ll r$ and $\beta_V \ll r$.

Figure 14.1-4 Digital VSB. (*a*) Transmitter; (*b*) power spectrum.

Exercise 14.1-1 Binary data is to be transmitted on a 1-MHz carrier. Spill-over is not a concern, but B_T must satisfy the fractional bandwidth constraint $B_T/f_c \leq 0.1$. Estimate the maximum possible bit rate r_b when the modulation is: (*a*) OOK, (*b*) binary QAM, (*c*) VSB with $M = 8$.

Phase Modulation Methods

The binary PSK waveform back in Fig. 14.1-1*c* contains phase shifts of $\pm \pi$ radians, often described as *phase-reversal keying* (PRK). An M-ary PSK signal has phase shift ϕ_k in the time interval $kD < t < (k + 1)D$, expressed in general by

$$x_c(t) = A_c \sum_k \cos (\omega_c t + \theta + \phi_k) p_D(t - kD) \tag{11}$$

Trigonometric expansion of the cosine function yields our desired quadrature-carrier form with

$$x_i(t) = \sum_k I_k p_D(t - kD) \qquad x_q(t) = \sum_k Q_k p_D(t - kD) \tag{12a}$$

where

$$I_k = \cos \phi_k \qquad Q_k = \sin \phi_k \tag{12b}$$

To ensure the largest possible phase modulation for a given value of M, we'll take the relationship between ϕ_k and a_k to be

$$\phi_k = \pi(2a_k + N)/M \qquad a_k = 0, 1, \ldots, M - 1 \tag{13}$$

in which N is an integer, usually 0 or 1.

Examples of PSK signal constellations are shown in Fig. 14.1-5, including the corresponding binary words in Gray code. The binary words for adjacent signal points therefore differ by just one bit. The PSK signal with $M = 4$ and $N = 0$ represented in Fig. 14.1-5*a* is designated *quaternary* or *quadriphase* PSK (QPSK). Had we taken QPSK with $N = 1$, the signal points would have been identical to binary QAM (Fig. 14.1-3*b*). Indeed, you can think of binary QAM as two PRK signals on quadrature carriers. M-ary PSK differs from M-ary ASK, of course, since an ideal PSK waveform always has a *constant envelope*.

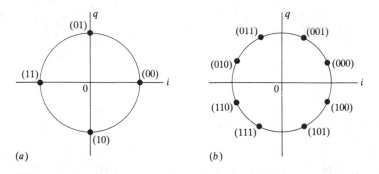

Figure 14.1-5 PSK signal constellations. (*a*) $M = 4$; (*b*) $M = 8$.

PSK spectral analysis becomes a routine task after you note from Eqs. (12*b*) and (13) that

$$\bar{I}_k = \bar{Q}_k = 0 \qquad \overline{I_k^2} = \overline{Q_k^2} = 1/2 \qquad \overline{I_k Q_j} = 0$$

Hence, the *i* and *q* components are statistically independent, and

$$G_{\ell p}(f) = 2 \times \frac{r}{2} \, |P_D(f)|^2 = \frac{1}{r} \operatorname{sinc}^2 \frac{f}{r} \tag{14}$$

Comparison with Eq. (7) reveals that $G_c(f)$ will have the same shape as an ASK spectrum (Fig. 14.1-2) without the carrier-frequency impulse. The absence of a discrete carrier component means that PSK has better power efficiency, but the spectral efficiency is the same as ASK.

Some PSK transmitters include a BPF to control spillover. However, bandpass filtering produces *envelope variations* via the FM-to-AM conversion effect discussed in Sect. 7.2. (Remember that a stepwise phase shift is equivalent to an FM impulse.) The typical nonlinear amplifier used at microwave carrier frequencies will flatten out these envelope variations and restore spillover—largely negating the function of the BPF. A special form of QPSK called *staggered* or *offset-keyed* QPSK (OQPSK) has been devised to combat this problem. The OQPSK transmitter diagrammed in Fig. 14.1-6 delays the quadrature signal such

Figure 14.1-6 Offset-keyed QPSK transmitter.

that modulated phase shifts occur every $D/2 = T_b$ seconds but they never exceed $\pm\pi/2$ radians. Cutting the maximum phase shift in half results in much smaller envelope variations after bandpass filtering.

When envelope variations are allowable, combined *amplitude-phase keying* (APK) is an attractive family of modulation methods. APK has essentially the same spectral efficiency as PSK, but it can achieve better performance with respect to noise and errors. Further discussion is postponed to Sect. 14.4.

Exercise 14.1-2 Draw the signal constellation for binary PSK with $\phi_k = \pi(2a_k - 1)/4$ and $a_k = 0, 1$. Then determine the lowpass equivalent spectrum and sketch $G_c(f)$.

Frequency Modulation Methods

There are two basic methods for digital frequency modulation. *Frequency-shift keying* (FSK) is represented conceptually by Fig. 14.1-7a, where the digital signal $x(t)$ controls a switch that selects the modulated frequency from a bank of M oscillators. The modulated signal is therefore *discontinuous* at every switching instant $t = kD$ unless the amplitude, frequency, and phase of each oscillator has been carefully adjusted. Discontinuities are avoided in *continuous-phase FSK* (CPFSK) represented by Fig. 14.1-7b, where $x(t)$ modulates the frequency of a single oscillator. Both forms of digital frequency modulation pose significant difficulties for spectral analysis, so we'll limit our consideration to some selected cases.

First, consider M-ary FSK. Let all oscillators in Fig. 14.1-7a have the same amplitude A_c and phase θ, and let their frequencies be related to a_k by

$$f_k = f_c + f_d a_k \qquad a_k = \pm 1, \pm 3, \ldots, \pm(M-1) \tag{15a}$$

which assumes that M is even, Then

$$x_c(t) = A_c \sum_k \cos(\omega_c t + \theta + \omega_d a_k t)p_D(t - kD) \tag{15b}$$

where $\omega_d = 2\pi f_d$. The parameter f_d equals the frequency shift away from f_c when $a_k = \pm 1$, and adjacent frequencies are spaced by $2f_d$. Continuity of $x_c(t)$ at $t = kD$ is assured if $2\omega_d D = 2\pi N$ where N is an integer.

$x(t)$ $x_c(t)$ $x(t)$ Frequency modulator $x_c(t)$

(a)   (b)

Figure 14.1-7 Digital frequency modulation. (a) FSK; (b) continuous-phase FSK.

We'll analyze a version of binary FSK known as *Sunde's* FSK, defined by the foregoing relations with $M = 2$, $D = T_b = 1/r_b$, and $N = 1$. Then $p_D(t) = u(t) - u(t - kT_b)$ and

$$f_d = r_b/2 \tag{16}$$

After trigonometric expansion of $x_c(t)$, we use the fact that $a_k = \pm 1$ to write

$$\cos \omega_d a_k t = \cos \omega_d t \qquad \sin \omega_d a_k t = a_k \sin \omega_d t$$

The *i* component thereby reduces to

$$x_i(t) = \cos \pi r_b t \tag{17a}$$

independent of a_k. The *q* component contains a_k in the form

$$x_q(t) = \sum_k a_k \sin (\pi r_b t)[u(t - kT_b) - u(t - kT_b - T_b)]$$

$$= \sum_k Q_k p(t - kT_b) \qquad Q_k = (-1)^k a_k \tag{17b}$$

where

$$p(t) = \sin (\pi r_b t)[u(t) - u(t - T_b)] \tag{17c}$$

The intervening manipulations are left to you as an instructive exercise.

Once again, we have independent *i* and *q* components. The *i* component, being a sinusoid, just contributes spectral impulses at $\pm r_b/2$ in the equivalent lowpass spectrum. The power spectrum of the *q* component contains no impulses since $\bar{Q}_k = 0$, whereas $\overline{Q_k^2} = \overline{a_k^2} = 1$. Thus,

$$G_{\ell p}(f) = \frac{1}{4}\left[\delta\left(f - \frac{r_b}{2} \right) + \delta\left(f + \frac{r_b}{2} \right) \right] + r_b |P(f)|^2 \tag{18a}$$

where

$$|P(f)|^2 = \frac{1}{4r_b^2}\left[\text{sinc}\, \frac{f - (r_b/2)}{r_b} + \text{sinc}\, \frac{f + (r_b/2)}{r_b} \right]^2$$

$$= \frac{4}{\pi^2 r_b^2}\left[\frac{\cos (\pi f/r_b)}{(2f/r_b)^2 - 1} \right]^2 \tag{18b}$$

The resulting bandpass spectrum is shown in Fig. 14.1-8.

Observe that the impulses correspond to the keyed frequencies $f_c \pm f_d = f_c \pm r_b/2$, and that the spectrum has a fourth-order rolloff. This rapid rolloff means that Sunde's FSK has very little spillover for $|f - f_c| > r_b$. We therefore take $B_T \approx r_b$, even though the central lobe of $G_c(f)$ is 50% wider than the central lobe of a binary ASK or PSK spectrum.

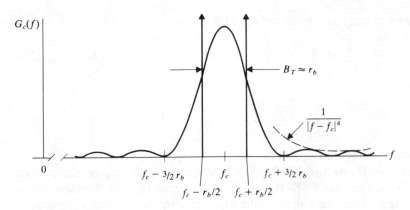

Figure 14.1-8 Power spectrum of binary FSK with $f_d = r_b/2$.

Another special case is *M-ary orthogonal FSK*, in which the M keyed frequencies are equispaced by $2f_d = 1/2D = r/2$. Without attempting the spectral analysis, we can surmise that $B_T \geq M \times 2f_d = Mr/2 = Mr_b/(2 \log_2 M)$. Therefore,

$$r_b/B_T \leq (2 \log_2 M)/M \tag{19}$$

and the modulation speed is less than M-ary ASK or PSK for $M \geq 4$. In other words, orthogonal FSK is a *wideband* modulation method.

Continuous-phase FSK may be wideband or narrowband depending on the frequency deviation. Let $x(t)$ in Fig. 14.1-7b start at $t = 0$, so

$$x(t) = \sum_{k=0}^{\infty} a_k p_D(t - kD) \qquad a_k = \pm 1, \pm 2, \ldots, \pm(M - 1)$$

and frequency modulation produces the CPFSK signal

$$x_c(t) = A_c \cos \left[\omega_c t + \theta + \omega_d \int_0^t x(\lambda) \, d\lambda \right] \qquad t \geq 0$$

To bring out the difference between CPFSK and FSK, consider the integral

$$\int_0^t x(\lambda) \, d\lambda = \sum_{k=0}^{\infty} a_k \int_0^t p_D(\lambda - kD) \, d\lambda$$

in which $p_D(\lambda - kD) = 0$ except for $kD < \lambda < (k + 1)D$ when $p_D(\lambda - kD) = 1$. Piecewise integration yields

$$\int_0^t x(\lambda) \, d\lambda = a_0 t \qquad\qquad 0 < t < D$$

$$= a_0 D + a_1(t - D) \qquad D < t < 2D$$

$$= \left(\sum_{j=0}^{k-1} a_j \right) D + a_k(t - kD) \qquad kD < t < (k + 1)D$$

Now we can express $x_c(t)$ in the summation form

$$x_c(t) = A_c \sum_{k=0}^{\infty} \cos\left[\omega_c t + \theta + \phi_k + \omega_d a_k(t - kD)\right]p_D(t - kD) \qquad (20a)$$

where $t \geq 0$ and

$$\phi_k \triangleq \omega_d D \sum_{j=0}^{k-1} a_j \qquad (20b)$$

with the understanding that $\phi_k = 0$ for $k = 0$.

Equation (20) shows that CPSK has a frequency shift $f_d a_k$ in the interval $kD < t < (k + 1)D$, just like FSK. But it also has a phase shift ϕ_k that depends on the *previous* digits. This phase shift results from the frequency-modulation process and ensures phase continuity for all t. Unfortunately, the past history embodied in ϕ_k greatly complicates CPFSK spectral analysis. Proakis (1983, chap. 3) gives further details and plots of $G_c(f)$ for various values of f_d when $M = 2$, 4, and 8. To conclude this section, we'll examine an important special case of binary CPSK called *minimum-shift keying* (MSK).

Exercise 14.1-3 Carry out the details omitted in the derivation of Eqs. (17a)–(17c). *Hint*: Show that $\sin \omega_d t = \sin\left[\omega_d(t - kT_b) + k\pi\right] = \cos(k\pi) \times \sin\left[\omega_d(t - kT_b)\right]$.

Minimum-Shift Keying★

Minimum-shift keying, also known as *fast FSK*, is binary CPFSK with

$$f_d = \frac{r_b}{4} \qquad a_k = \pm 1 \qquad \phi_k = \frac{\pi}{2}\sum_{j=0}^{k-1} a_j \qquad (21)$$

Notice that the frequency spacing $2f_d = r_b/2$ is half that of Sunde's FSK. This fact, together with the continuous-phase property, results in a more compact spectrum, free of impulses. Subsequent analysis will prove that $G_i(f) = G_q(f)$ and

$$G_{\ell p}(f) = \frac{1}{r_b}\left[\operatorname{sinc}\frac{f - (r_b/4)}{(r_b/2)} + \operatorname{sinc}\frac{f + (r_b/4)}{(r_b/2)}\right]^2$$

$$= \frac{16}{\pi^2 r_b}\left[\frac{\cos(2\pi f/r_b)}{(4f/r_b)^2 - 1}\right]^2 \qquad (22)$$

The bandpass spectrum $G_c(f)$ plotted in Fig. 14.1-9 has minuscule spillover beyond the central lobe of width $3r_b/2$. The rapid rolloff justifies taking $B_T \approx r_b/2$, so

$$r_b/B_T \approx 2 \text{ bps/Hz}$$

which is twice the modulation speed of Sunde's FSK and accounts for the name "fast" FSK.

$G_c(f)$

$B_T \approx r_b/2$

0

$f_c - {}^3/_4\, r_b \quad f_c \quad f_c + {}^3/_4\, r_b$

Figure 14.1-9 MSK power spectrum.

Our investigation of MSK starts with the usual trigonometric expansion to put $x_c(t)$ in quadrature-carrier form with

$$x_i(t) = \sum_{k=0}^{\infty} \cos \left(\phi_k + a_k c_k \right) p_{T_b}(t - kT_b)$$

$$x_q(t) = \sum_{k=0}^{\infty} \sin \left(\phi_k + a_k c_k \right) p_{T_b}(t - kT_b)$$

where

$$c_k \triangleq \frac{\pi r_b}{2} (t - kT_b) \qquad p_{T_b}(t) = u(t) - u(t - kT_b)$$

We'll also draw upon the behavior of ϕ_k versus k as displayed in the trellis pattern of Fig. 14.1-10. This pattern clearly reveals that $\phi_k = 0, \pm\pi, \pm 2\pi, \ldots$, for even values of k while $\phi_k = \pm \pi/2, \pm 3\pi/2, \ldots$, for odd values of k.

As a specific example, let the input message sequence be 100010111. The resulting phase path ϕ_k is shown in Fig. 14.1-11a, taking $a_k = +1$ for input bit 1

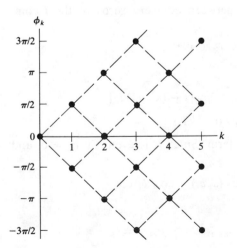

ϕ_k

$3\pi/2$

π

$\pi/2$

0

$\quad 1 \quad 2 \quad 3 \quad 4 \quad 5 \qquad k$

$-\pi/2$

$-\pi$

$-3\pi/2$

Figure 14.1-10 MSK phase trellis.

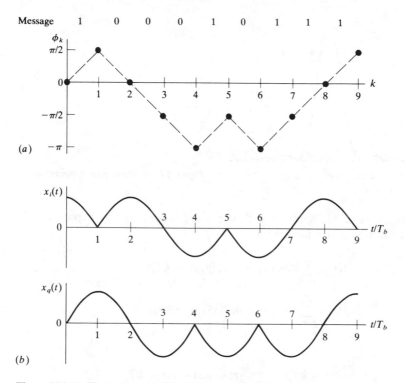

Figure 14.1-11 Illustration of MSK. (*a*) Phase path; (*b*) *i* and *q* waveforms.

and $a_k = -1$ for input bit 0. The corresponding *i* and *q* waveforms calculated from the foregoing expressions are sketched in Fig. 14.1-11*b*. We see that both waveforms have zeros spaced by $2T_b$, but staggered such that the zeros of $x_i(t)$ coincide with the peaks of $x_q(t)$, and vice versa. These observations will guide our subsequent work.

Consider an arbitrary time interval between adjacent zeros of the *i* component, i.e.,

$$(k - 1)T_b < t < (k + 1)T_b$$

with *k* being *even*. During this interval,

$$x_i(t) = \cos(\phi_{k-1} + a_{k-1}c_{k-1})p_{T_b}[t - (k - 1)T_b]$$
$$+ \cos(\phi_k + a_k c_k)p_{T_b}(t - kT_b)$$

which we seek to combine into a single term. Since *k* is even, $\sin \phi_k = 0$, and routine trigonometric manipulations yield

$$\cos(\phi_k + a_k c_k) = \cos \phi_k \cos(a_k c_k) = \cos \phi_k \cos c_k$$

Likewise, using

$$\cos \phi_{k-1} = 0 \qquad \phi_{k-1} = \phi_k - a_{k-1}\pi/2 \qquad c_{k-1} = c_k + \pi/2$$

we get

$$\cos (\phi_{k-1} + a_{k-1}c_{k-1}) = -\sin \phi_{k-1} \sin (a_{k-1}c_{k-1})$$
$$= a_{k-1}^2 \cos \phi_k \cos c_k = \cos \phi_k \cos c_k$$

Thus, for the interval in question,

$$x_i(t) = \cos \phi_k \cos c_k\{p_{T_b}[t - (k-1)T_b] + p_{T_b}(t - kT_b)\}$$
$$= \cos \phi_k \cos [(\pi r_b/2)(t - kT_b)][u(t - kT_b + T_b) - u(t - kT_b - T_b)]$$

Summing intervals to encompass all $t \geq 0$ finally yields

$$x_i(t) = \sum_{k \text{ even}} I_k p(t - kT_b) \qquad I_k = \cos \phi_k \tag{23}$$

where

$$p(t) = \cos (\pi r_b t/2)[u(t + T_b) - u(t - T_b)] \tag{24}$$

This result checks out against the waveform in Fig. 14.1-11b since $I_k = \cos \phi_k = \pm 1$ when k is even.

Now, for the q component, we consider the interval $(k - 1)T_b < t < (k + 1)T_b$ with k odd. Similar manipulations as before lead to

$$x_q(t) = \sin \phi_k \cos c_k\{p_{T_b}[t - (k-1)T_b] + p_{T_b}(t - kT_b)\}$$

Thus, for all $t \geq 0$,

$$x_q(t) = \sum_{k \text{ odd}} Q_k p(t - kT_b) \qquad Q_k = \sin \phi_k \tag{25}$$

which also agrees with Fig. 14.1-11b. Equation (22) follows from Eqs. (23)–(25) since the i and q components are independent, with $\overline{I_k} = \overline{Q_k} = 0$ and $\overline{I_k^2} = \overline{Q_k^2} = 1$.

14.2 COHERENT BINARY SYSTEMS

Coherent bandpass digital systems employ information about the carrier frequency and phase at the receiver to detect the message—like synchronous analog detection. *Noncoherent* systems don't require synchronization with the carrier phase, but they fall short of the optimum performance made possible by coherent detection.

This section examines coherent binary transmission, starting with a general treatment of optimum binary detection in the presence of additive white gaussian noise (AWGN). The results are then applied to assess the performance of specific binary modulation systems. We'll focus throughout on keyed modulation (OOK, PRK, and FSK), without baseband filtering or transmission distortion that might produce ISI in the modulated signal. The challenging problems of optimum receiver design including ISI effects are addressed in more advanced texts such as Proakis (1983, chap. 6).

Optimum Binary Detection

Any bandpass binary signal with keyed modulation can be expressed in the general quadrature-carrier form

$$x_c(t) = A_c \left\{ \left[\sum_k I_k p_i(t - kT_b) \right] \cos (\omega_c t + \theta) \right.$$

$$\left. - \left[\sum_k Q_k p_q(t - kT_b) \right] \sin (\omega_c t + \theta) \right\}$$

For practical coherent systems, the carrier wave should be synchronized with the digital modulation. Accordingly, we'll take $\theta = 0$ and impose the condition

$$f_c = N_c/T_b = N_c r_b \tag{1}$$

where N_c is an integer — usually a very large integer. Then

$$x_c(t) = A_c \sum_k [I_k p_i(t - kT_b) \cos \omega_c(t - kT_b)$$

$$- Q_k p_q(t - kT_b) \sin \omega_c(t - kT_b)]$$

and we can concentrate on a single bit interval by writing

$$x_c(t) = s_m(t - kT_b) \qquad kT_b < t < (k + 1)T_b \tag{2}$$

with

$$s_m(t) \triangleq A_c[I_k p_i(t) \cos \omega_c t - Q_k p_q(t) \sin \omega_c t]$$

Here, $s_m(t)$ stands for either of two *signaling waveforms*, $s_0(t)$ and $s_1(t)$, representing the message bits $m = 0$ and $m = 1$.

Now consider the received signal $x_c(t)$ corrupted by white gaussian noise. We showed in Sect. 11.2 that an optimum baseband receiver minimizes error probability with the help of a filter matched to the baseband pulse shape. But binary CW modulation involves two different signaling waveforms, as in Eq. (2), rather than one pulse shape with two different amplitudes. Consequently, we must redo our previous analysis in terms of $s_0(t)$ and $s_1(t)$.

Figure 14.2-1 shows the proposed receiver structure labeled with the relevant signals and noise for the interval under consideration. This bandpass receiver is just like a baseband receiver with a BPF in place of an LPF. The filtered signal

Figure 14.2-1 Bandpass binary receiver.

plus noise $y(t)$ is sampled at $t_k = (k + 1)T_b$, the end of the bit interval, and compared with a threshold level to regenerate the most likely message bit $\hat{m}$. We seek the BPF impulse response $h(t)$ and threshold level V for optimum binary detection, resulting in the smallest average regeneration error probability.

As in Sect. 11.2, let H_1 and H_0 denote the hypotheses that $m = 1$ and $m = 0$, respectively. The receiver decides between H_1 and H_0 according to the observed value of the random variable

$$Y = y(t_k) = z_m + n$$

where

$$
\begin{aligned}
z_m \triangleq z_m(t_k) &= [s_m(t - kT_b) * h(t)]\Big|_{t = t_k} \\
&= \int_{kT_b}^{(k+1)T_b} s_m(\lambda - kT_b)h(t_k - \lambda)\, d\lambda \\
&= \int_0^{T_b} s_m(\lambda)h(T_b - \lambda)\, d\lambda
\end{aligned}
\tag{3}
$$

The noise sample $n = n(t_k)$ is a gaussian r.v. with zero mean and variance σ^2, so the conditional PDFs of Y given H_1 or H_0 will be gaussian curves centered at z_1 or z_0, portrayed by Fig. 14.2-2. With the usual assumption of equally likely zeros and ones, the optimum threshold is at the intersection point, i.e.,

$$V_{\text{opt}} = \tfrac{1}{2}(z_1 + z_0)$$

Then, from the symmetry of the PDFs, $P_{e_1} = P_{e_0}$ and

$$P_e = Q(|z_1 - z_0|/2\sigma)$$

in which the absolute-value notation $|z_1 - z_0|$ includes the case of $z_1 < z_0$.

But what BPF impulse response $h(t)$ maximizes the ratio $|z_1 - z_0|/2\sigma$ or, equivalently, $|z_1 - z_0|^2/4\sigma^2$? To answer this question, we note from Eq. (3) that

$$|z_1 - z_0|^2 = \left| \int_{-\infty}^{\infty} [s_1(\lambda) - s_0(\lambda)]h(T_b - \lambda)\, d\lambda \right|^2 \tag{4a}$$

where the infinite limits are allowed since $s_m(t) = 0$ outside of $0 < t < T_b$. We also note that

$$\sigma^2 = \frac{\eta}{2}\int_{-\infty}^{\infty} |h(t)|^2\, dt = \frac{\eta}{2}\int_{-\infty}^{\infty} |h(T_b - \lambda)|^2\, d\lambda \tag{4b}$$

Figure 14.2-2 Conditional PDFs.

Application of Schwarz's inequality now yields

$$\frac{|z_1 - z_0|^2}{4\sigma^2} \leq \frac{1}{2\eta} \int_{-\infty}^{\infty} [s_1(t) - s_0(t)]^2 \, dt \tag{5}$$

and the ratio is maximum if $h(T_b - t) = K[s_1(t) - s_0(t)]$. Thus,

$$h_{opt}(t) = K[s_1(T_b - t) - s_0(T_b - t)] \tag{6}$$

with K being an arbitrary constant.

Equation (6) says that the filter for optimum binary detection should be matched to the *difference* between the two signaling waveforms. Alternatively, you could use two matched filters with $h_1(t) = Ks_1(T_b - t)$ and $h_0(t) = Ks_0(T_b - t)$ arranged in parallel per Fig. 14.2-3a; subtracting the output of the lower branch from the upper branch yields the same optimum response. In either case, any stored energy in the filters must be discharged after each sampling instant to prevent ISI in subsequent bit intervals.

Another alternative, with built-in discharge, is based on the observation that the sampled signal value from the upper branch in Fig. 14.2-3a is

$$z_{m1}(t_k) = \int_0^{T_b} s_m(\lambda) h_1(T_b - \lambda) \, d\lambda$$

$$= \int_{kT_b}^{(k+1)T_b} s_m(t - kT_b) Ks_1(t - kT_b) \, dt$$

and likewise for $z_{m0}(t_k)$. Hence, optimum filtering can be implemented by the system diagrammed in Fig. 14.2-3b, which requires two multipliers, two integrators, and stored copies of $s_0(t)$ and $s_1(t)$. This system is called a *correlation*

(a)

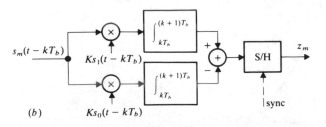

(b)

Figure 14.2-3 Optimum binary detection. (*a*) Parallel matched filters; (*b*) correlation detector.

detector because it correlates the received signal plus noise with noise-free copies of the signaling waveforms. Note that correlation detection is a generalization of the integrate-and-dump technique for matched filtering.

Regardless of the particular implementation method, the error probability with optimum binary detection depends upon the ratio maximized in Eq. (5). This ratio, in turn, depends on the signal energy per bit and on the similarity of the signaling waveforms. To pursue this point, consider the expansion

$$\int_0^{T_b} [s_1(t) - s_0(t)]^2 \, dt = E_1 + E_0 - 2E_{10}$$

where

$$E_1 \triangleq \int_0^{T_b} s_1^2(t) \, dt \qquad E_0 \triangleq \int_0^{T_b} s_0^2(t) \, dt$$

$$E_{10} \triangleq \int_0^{T_b} s_1(t)s_0(t) \, dt$$

(7)

We identify E_1 and E_0 as the respective energies of $s_1(t)$ and $s_0(t)$, while E_{10} is proportional to their correlation coefficient. Since zeros and ones are equally likely, the average signal energy per bit is

$$E_b = \frac{1}{2}(E_1 + E_0)$$

Therefore,

$$\left(\frac{z_1 - z_0}{2\sigma}\right)_{max}^2 = \frac{E_1 + E_0 - 2E_{10}}{2\eta} = \frac{E_b - E_{10}}{\eta}$$

(8a)

and

$$P_e = Q[\sqrt{(E_b - E_{10})/\eta}]$$

(8b)

Equation (8) brings out the importance of E_{10} relative to system performance when E_b and η are fixed.

Finally, substituting Eq. (6) into Eq. (3) yields $z_1 = K(E_1 - E_{10})$ and $z_0 = K(E_{10} - E_0)$, so

$$V_{opt} = \frac{1}{2}(z_1 + z_0) = \frac{K}{2}(E_1 - E_0)$$

(9)

Note that the optimum threshold does not involve E_{10}.

Exercise 14.2-1 Derive Eqs. (5) and (6) from Eqs. (4a) and (4b). Use Eq. (17), Sect. 3.5, written in the form

$$\frac{\left|\int_{-\infty}^{\infty} V(\lambda)W^*(\lambda) \, d\lambda\right|^2}{\int_{-\infty}^{\infty} |W(\lambda)|^2 \, d\lambda} \leq \int_{-\infty}^{\infty} |V(\lambda)|^2 \, d\lambda$$

and recall that the equality holds when $V(\lambda)$ and $W(\lambda)$ are proportional functions.

Coherent OOK, PRK, and FSK

Although the crude nature of ASK hardly warrants sophisticated system design, a brief look at coherent on-off keying helps clarify optimum detection concepts. The OOK signaling waveforms are just

$$s_1(t) = A_c\, p_{T_b}(t)\, \cos \omega_c t \qquad s_0(t) = 0 \tag{10}$$

Our carrier-frequency condition $f_c = N_c/T_b$ means that $s_1(t - kT_b) = A_c \cos \omega_c t$ for any bit interval while, of course, $s_0(t - kT_b) = 0$. Thus, a receiver with correlation detection simplifies to the form of Fig. 14.2-4, in which a local oscillator synchronized with the carrier provides the stored copy of $s_1(t)$. The bit sync signal actuates the sample-and-hold unit and resets the integrator. Both sync signals may be derived from a single source, thanks to the harmonic relationship between f_c and r_b.

Now we use Eqs. (7) and (10) to obtain $E_0 = E_{10} = 0$ and

$$E_1 = A_c^2 \int_0^{T_b} \cos^2 \omega_c t\; dt = \frac{A_c^2\, T_b}{2}\left[1 + \text{sinc}\, \frac{4f_c}{r_b}\right] = \frac{A_c^2\, T_b}{2}$$

so $E_b = E_1/2 = A_c^2\, T_b/4$. Setting the threshold at $V = K(E_1 - E_0)/2 = KE_b$ yields the minimum average error probability given by Eq. (8), namely

$$P_e = Q(\sqrt{E_b/\eta}) = Q(\sqrt{\gamma_b}) \tag{11}$$

Not surprisingly, the performance of coherent OOK is identical to unipolar baseband transmission.

Better performance is achieved by coherent phase-reversal keying. Let the two phase shifts be 0 and π radians, so

$$s_1(t) = A_c\, p_{T_b}(t)\, \cos \omega_c t \qquad s_0(t) = -s_1(t) \tag{12}$$

The relation $s_0(t) = -s_1(t)$ defines *antipodal* signaling, analogous to polar baseband transmission. It quickly follows that

$$E_b = E_1 = E_0 = A_c^2\, T_b/2 \qquad E_{10} = -E_b$$

so $E_b - E_{10} = 2E_b$ and

$$P_e = Q(\sqrt{2E_b/\eta}) = Q(\sqrt{2\gamma_b}) \tag{13}$$

Figure 14.2-4 Correlation receiver for OOK or PRK.

PRK therefore gets by with 3 dB less signal energy than OOK, other factors being equal.

Since $s_0(t) = -s_1(t)$, a coherent PRK receiver requires only one matched filter or correlator, just like OOK. But now $V = 0$ since $E_1 = E_0$, so the PRK threshold level need not be readjusted if the received signal undergoes fading. Furthermore, the constant envelope of PRK makes it relatively invulnerable to nonlinear distortion. PRK is therefore superior to OOK on several counts, and has the same spectral efficiency. We'll see next that PRK is also superior to binary FSK.

Consider binary FSK with frequency shift $\pm f_d$ and signaling waveforms

$$s_1(t) = A_c\, p_{T_b}(t)\, \cos 2\pi(f_c + f_d)t$$
$$s_0(t) = A_c\, p_{T_b}(t)\, \cos 2\pi(f_c - f_d)t \tag{14}$$

When $f_c \pm f_d \gg r_b$, $E_b \approx A_c^2 T_b/2$, whereas

$$E_{10} = E_b \operatorname{sinc}(4f_d/r_b) \tag{15}$$

which depends on the frequency shift. If $f_d = r_b/2$, corresponding to Sunde's FSK, then $E_{10} = 0$ and the error probability is the same as OOK.

Some improvement is possible when phase discontinuities are allowed in $x_c(t)$, but $E_b - E_{10} \le 1.22E_b$ for any choice of f_d. Hence, binary FSK does not provide any significant wideband noise reduction, and PRK has an energy advantage of at least $10 \log(2/1.22) \approx 2$ dB. Additionally, an optimum FSK receiver is more complicated than Fig. 14.2-4.

Exercise 14.2-2 Suppose the optimum receiver for Sunde's FSK is implemented in the form of Fig. 14.2-3a. Find and sketch the amplitude response of the two filters.

Timing and Synchronization

Finally, we should give some attention to the timing and synchronization problems associated with optimum coherent detection. For this purpose, consider the bandpass signaling waveform and matched filter

$$s(t) = A_c\, p_{T_b}(t)\, \cos \omega_c t \qquad f_c\, T_b = N_c \gg 1$$
$$h(t) = Ks(T_b - t) = KA_c\, p_{T_b}(t)\, \cos \omega_c t$$

When $s(t)$ is applied to its matched filter, the resulting response is

$$z(t) = s(t) * h(t) \approx KE\Lambda\left(\frac{t - T_b}{T_b}\right)\cos \omega_c t \tag{16}$$

where $E = A_c^2 T_b/2$. The sketch of $z(t)$ in Fig. 14.2-5 shows the expected maximum value $z(T_b) = KE$, and the response for $t > T_b$ would be eliminated by discharging the filter after the sampling instant.

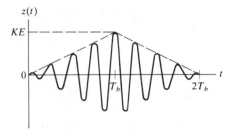

Figure 14.2-5 Response of bandpass matched filter.

But suppose there's a small timing error such that sampling actually occurs at $t_k = T_b(1 \pm \epsilon)$. Then

$$z(t_k) \approx KE \cos \theta_\epsilon \qquad \theta_\epsilon = \omega_c T_b \epsilon = 2\pi N_c \epsilon$$

so the timing error reduces the effective signal level by the factor $\cos \theta_\epsilon$. Since $|z_1 - z_0|^2$ will be reduced by $\cos^2 \theta_\epsilon$, while σ^2 remains unchanged, the error probability becomes

$$P_e = Q\left(\sqrt{\frac{E_b - E_{10}}{\eta} \cos^2 \theta_\epsilon}\right) \tag{17}$$

which follows from Eq. (8). As an example of the magnitude of this problem, take PRK with $\gamma_b = 8$, $r_b = 2$ kbps, and $f_c = 100$ kHz; perfect timing gives $P_e = Q(\sqrt{16}) \approx 3 \times 10^{-5}$, while an error of just 0.3% of the bit interval results in $\theta_\epsilon = 2\pi(100/2) \times 0.003 = 54°$ and $P_e = Q(\sqrt{16 \cos^2 54°}) \approx 10^{-2}$. These numbers illustrate why bandpass matched filter is not a practical method for coherent detection.

A correlation detector like Fig. 14.2-4 has much less sensitivity to timing error, since the integrated output does not oscillate at the carrier frequency. Correlation detection is therefore used in most coherent binary systems. However, the local oscillator must be synchronized accurately with the carrier, and a phase synchronization error θ_ϵ again reduces the effective signal level by the factor $\cos \theta_\epsilon$.

In the case of PRK, the carrier sync signal can be derived from $x_c(t)$ using techniques such as the Costas PLL system back in Fig. 8.3-4. Another approach known as *phase-comparison detection* is discussed in the next section, along with noncoherent detection of OOK and FSK.

14.3 NONCOHERENT BINARY SYSTEMS

Optimum coherent detection may not be essential if the signal is strong enough for adequate reliability with a less sophisticated receiver. A prime example of this situation is digital transmission over voice telephone channels, which have relatively large signal-to-noise ratios dictated by analog performance standards. There are also applications in which it would be very difficult and expensive to carry out coherent detection. For instance, the propagation delay on some radio channels changes too rapidly to permit accurate tracking of the carrier phase at the receiver, and unsynchronized or noncoherent detection becomes the only viable recourse.

Here we examine the suboptimum performance of noncoherent OOK and FSK systems that employ envelope detection to bypass the synchronization problems of coherent detection. We'll also look at *differentially coherent* PSK systems with phase-comparison detection. For all three cases we must first analyze the envelope of a sinusoid plus bandpass noise.

Envelope of a Sinusoid plus Bandpass Noise

Consider the sinusoid $A_c \cos (\omega_c t + \theta)$ plus gaussian bandpass noise $n(t)$ with zero mean and variance σ^2. Using the quadrature-carrier expression

$$n(t) = n_i(t) \cos (\omega_c t + \theta) - n_q(t) \sin (\omega_c t + \theta)$$

we write the sum as

$$A_c \cos (\omega_c t + \theta) + n(t) = A(t) \cos [\omega_c t + \theta + \phi(t)]$$

where, at any instant t,

$$A = \sqrt{(A_c + n_i)^2 + n_q^2} \qquad \phi = \arctan \frac{n_q}{A_c + n_i} \qquad (1)$$

We recall from Sect. 9.2 that the i and q noise components are independent r.v.'s having the same distribution as $n(t)$. Now we seek the PDF of the envelope A.

Before plunging into the analysis, let's speculate on the nature of A under extreme conditions. If $A_c = 0$, then A reduces to the noise envelope A_n, with the Rayleigh distribution

$$p_{A_n}(A_n) = \frac{A_n}{\sigma^2} e^{-A_n^2/2\sigma^2} \qquad A_n \geq 0 \qquad (2)$$

At the other extreme, if $A_c \gg \sigma$, then A_c will be large compared to the noise components most of the time, so

$$A = A_c \sqrt{1 + (2n_i/A_c) + (n_i^2 + n_q^2)/A_c^2} \approx A_c + n_i$$

which implies that A will be approximately gaussian.

For an arbitrary value of A_c, we must perform a rectangular-to-polar conversion following the procedure that led to Eq. (14), Sect. 4.4. The joint PDF of A and ϕ then becomes

$$p_{A\phi}(A, \phi) = \frac{A}{2\pi\sigma^2} \exp \left(-\frac{A^2 - 2A_c A \cos \phi + A_c^2}{2\sigma^2} \right) \qquad (3)$$

for $A \geq 0$ and $|\phi| \leq \pi$. The term $A \cos \phi$ in the exponent prevents us from factoring Eq. (3) as a product of the form $p_A(A)p_\phi(\phi)$, meaning that A and ϕ are not statistically independent. The envelope PDF must therefore be found by integrating the joint PDF over the range of ϕ, so

$$p_A(A) = \frac{A}{2\pi\sigma^2} \exp\left(-\frac{A^2 + A_c^2}{2\sigma^2}\right) \int_{-\pi}^{\pi} \exp\left(\frac{A_c A \cos \phi}{\sigma^2}\right) d\phi$$

Now we introduce the *modified Bessel function* of the first kind and order zero, defined by

$$I_0(v) \triangleq \frac{1}{2\pi} \int_{-\pi}^{\pi} \exp(v \cos \phi)\, d\phi \tag{4a}$$

with the properties

$$I_0(v) \approx \begin{cases} e^{v^2/4} & v \ll 1 \\ \dfrac{e^v}{\sqrt{2\pi v}} & v \gg 1 \end{cases} \tag{4b}$$

We then have

$$p_A(A) = \frac{A}{\sigma^2}\, e^{-(A^2 + A_c^2)/2\sigma^2} I_0\left(\frac{A_c A}{\sigma^2}\right) \qquad A \geq 0 \tag{5}$$

which is called the *Rician distribution*.

Although Eq. (5) has a formidable appearance, it easily simplifies under large-signal conditions to

$$p_A(A) \approx \sqrt{\frac{A}{2\pi A_c\, \sigma^2}}\, e^{-(A - A_c)^2/2\sigma^2} \qquad A_c \gg \sigma \tag{6}$$

obtained from the large-v approximation in Eq. (4b). Since the exponential term dominates in Eq. (6), we have confirmed that the envelope PDF is essentially a gaussian curve with variance σ^2 centered at $\bar{A} \approx A_c$. Figure 14.3-1 illustrates the transition of the envelope PDF from a Rayleigh curve to a gaussian curve as A_c becomes large compared to σ.

Figure 14.3-1 PDFs for the envelope of a sinusoid plus bandpass noise.

Noncoherent OOK

Noncoherent on-off keying is intended to be a simple system. Usually the carrier and data are unsynchronized so, for an arbitrary bit interval $kT_b < t < (k + 1)T_b$, we write

$$x_c(t) = A_c\, a_k\, p_{T_b}(t - kT_b) \cos (\omega_c t + \theta) \qquad a_k = 0, 1 \qquad (7)$$

The signaling energies are $E_0 = 0$ and

$$E_1 = \frac{A_c^2 T_b}{2}\left[1 + \frac{\sin (2\omega_c T_b + 2\theta) - \sin 2\theta}{2\omega_c T_b} \right] \approx \frac{A_c^2 T_b}{2}$$

where we've assumed that $f_c \gg r_b$. The average signal energy per bit is then $E_b = E_1/2 \approx A_c^2 T_b/4$ since we'll continue to assume that 1s and 0s are equally likely.

The OOK receiver diagrammed in Fig. 14.3-2 consists of a BPF followed by an envelope detector and regenerator. The BPF is a matched filter with

$$h(t) = K A_c\, p_{T_b}(t) \cos \omega_c t \qquad (8)$$

which ignores the carrier phase θ. The envelope detector eliminates dependence on θ by tracing out the dashed line back in Fig. 14.2-5. Thus, when $a_k = 1$, the peak signal component of the envelope $y(t)$ is $A_1 = KE_1$. Let's take $K = A_c/E_1$ for convenience, so that $A_1 = A_c$. Then

$$A_c^2/\sigma^2 = 4E_b/\eta = 4\gamma_b \qquad (9)$$

where σ^2 is the variance of the bandpass noise at the input to the envelope detector, calculated from $h(t)$ using Eq. (4b), Sect. 14.2.

Now consider the conditional PDFs of the random variable $Y = y(t_k)$. When $a_k = 0$, we have a sample value of the envelope of the noise alone; hence, $p_Y(y \mid H_0)$ is the Rayleigh function $p_{A_n}(y)$. When $a_k = 1$, we have a sample value of the envelope of a sinusoid plus noise; hence, $p_Y(y \mid H_1)$ is the Rician function $p_A(y)$. Figure 14.3-3 shows these two curves for the case of $\gamma_b \gg 1$, so the Rician PDF has a nearly gaussian shape. The intersection point defines the optimum threshold, which turns out to be

$$V_{\text{opt}} \approx \frac{A_c}{2}\sqrt{1 + \frac{2}{\gamma_b}} \approx \frac{A_c}{2} \qquad \gamma_b \gg 1$$

But we no longer have symmetry with respect to the threshold and, consequently, $P_{e1} \neq P_{e0}$ when P_e is minimum.

Figure 14.3-2 Noncoherent OOK receiver.

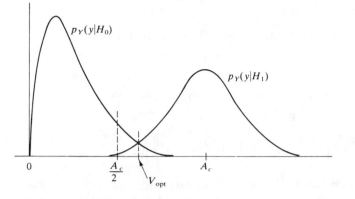

Figure 14.3-3 Conditional PDFs for noncoherent OOK.

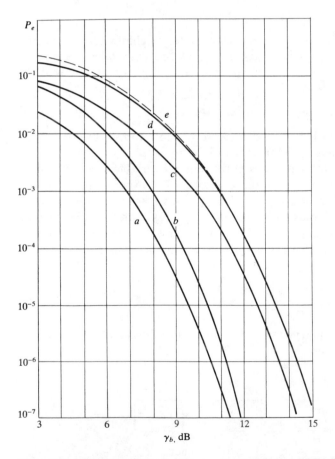

Figure 14.3-4 Binary error probability curves. (*a*) Coherent PRK; (*b*) DPSK; (*c*) coherent OOK or FSK; (*d*) noncoherent FSK; (*e*) noncoherent OOK.

Noncoherent OOK systems require $\gamma_b \gg 1$ for reasonable performance, and the threshold is normally set at $A_c/2$. The resulting error probabilities are

$$P_{e0} = \int_{A_c/2}^{\infty} p_{A_n}(y)\, dy = e^{-A_c^2/8\sigma^2} = e^{-\gamma_b/2} \tag{10a}$$

$$P_{e1} = \int_{0}^{A_c/2} p_A(y)\, dy \approx Q\left(\frac{A_c}{2\sigma}\right) = Q(\sqrt{\gamma_b})$$

$$\approx \frac{1}{\sqrt{2\pi\gamma_b}}\, e^{-\gamma_b/2} \qquad \gamma_b \gg 1 \tag{10b}$$

where we've introduced the asymptotic approximation for $Q(\sqrt{\gamma_b})$ to bring out the fact that $P_{e1} \ll P_{e0}$ when $\gamma_b \gg 1$. Finally,

$$P_e = \tfrac{1}{2}(P_{e0} + P_{e1}) = \tfrac{1}{2}[e^{-\gamma_b/2} + Q(\sqrt{\gamma_b})]$$

$$\approx \tfrac{1}{2} e^{-\gamma_b/2} \qquad \gamma_b \gg 1 \tag{11}$$

which is plotted versus γ_b in Fig. 14.3-4 along with curves for other binary systems.

Exercise 14.3-1 Consider the BPF output $z(t) = x_c(t) * h(t)$ when $x_c(t) = A_c\, p_{T_b}(t) \cos(\omega_c t + \theta)$ and $K = 2/A_c T_b$. Show that, for $0 < t < T_b$,

$$z(t) = \frac{A_c t}{T_b}\left[\cos\theta \cos\omega_c t - \left(\sin\theta - \frac{\cos\theta}{\omega_c t}\right)\sin\omega_c t\right]$$

Then find and sketch the envelope of $z(t)$ assuming $f_c \gg r_b$.

Noncoherent FSK

Although envelope detection seems an unlikely method for FSK, a reexamination of the waveform back in Fig. 14.1-1b reveals that binary FSK consists of two interleaved OOK signals with the same amplitude A_c but different carrier frequencies, $f_1 = f_c + f_d$ and $f_0 = f_c - f_d$. Accordingly, noncoherent detection can be implemented with a pair of bandpass filters and envelope detectors, arranged per Fig. 14.3-5 where

$$h_1(t) = KA_c\, p_{T_b}(t) \cos\omega_1 t \qquad h_0(t) = KA_c\, p_{T_b}(t) \cos\omega_0 t \tag{12}$$

Figure 14.3-5 Noncoherent detection of binary FSK.

We'll take $K = A_c/E_b$, noting that $E_b = E_1 = E_0 \approx A_c^2 T_b/2$. Then

$$A_c^2/\sigma^2 = 2E_b/\eta = 2\gamma_b \tag{13}$$

where σ^2 is the noise variance at the output of either filter.

We'll also take the frequency spacing $f_1 - f_0 = 2f_d$ to be an integer multiple of r_b, as in Sunde's FSK. This condition ensures that the BPFs effectively separate the two frequencies, and that the two bandpass noise waveforms are uncorrelated at the sampling instants. Thus, when $a_k = 1$, the sampled output $y_1(t_k)$ at the upper branch has the signal component $A_1 = KE_1 = A_c$ and a Rician distribution, whereas $y_0(t_k)$ at the lower branch has a Rayleigh distribution — and vice versa when $a_k = 0$.

Regeneration is based on the envelope difference $Y_1 - Y_0 = y_1(t_k) - y_0(t_k)$. Without resorting to conditional PDFs, we conclude from the symmetry of the receiver that the threshold should be set at $V = 0$, regardless of A_c. It then follows that $P_{e_1} = P(Y_1 - Y_0 < 0 | H_1)$ and $P_{e0} = P_{e1} = P_e$. Therefore,

$$P_e = P(Y_0 > Y_1 | H_1)$$

$$= \int_0^\infty p_{Y_1}(y_1 | H_1) \left[\int_{y_1}^\infty p_{Y_0}(y_0 | H_1) \, dy_0 \right] dy_1$$

where the inner integral is the probability of the event $Y_0 > Y_1$ for a *fixed* value of y_1. Inserting the PDFs $p_{Y_0}(y_0 | H_1) = p_{A_n}(y_0)$ and $p_{Y_1}(y_1 | H_1) = p_A(y_1)$ and performing the inner integration yields

$$P_e = \int_0^\infty \frac{y_1}{\sigma^2} e^{-(2y_1^2 + A_c^2)/2\sigma^2} I_0\left(\frac{A_c y_1}{\sigma^2} \right) dy_1$$

Rather amazingly, this integral can be evaluated in closed form by letting $\lambda = \sqrt{2} y_1$ and $\alpha = A_c/\sqrt{2}$ so that

$$P_e = \frac{1}{2} e^{-A_c^2/4\sigma^2} \int_0^\infty \frac{\lambda}{\sigma^2} e^{-(\lambda^2 + \alpha^2)/2\sigma^2} I_0\left(\frac{\alpha\lambda}{\sigma^2} \right) d\lambda$$

The integrand is now exactly the same function as the Rician PDF in Eq. (5), whose total area equals unity. Hence, our final result simply becomes

$$P_e = \tfrac{1}{2} e^{-A_c^2/4\sigma^2} = \tfrac{1}{2} e^{-\gamma_b/2} \tag{14}$$

having used Eq. (13).

A comparison of the performance curves for noncoherent FSK and OOK plotted in Fig. 14.3-4 reveals little difference except at small values of γ_b. However, FSK does have three advantages over OOK: constant modulated signal envelope, equal digit error probabilities, and fixed threshold level $V = 0$. These advantages usually justify the extra hardware needed for the FSK receiver.

Differentially Coherent PSK

Noncoherent detection of binary PSK would be impossible since the message information resides in the phase. Instead, the clever technique of *phase-comparison detection* gets around the phase synchronization problems associated with coherent PRK and provides much better performance than noncoherent OOK or FSK. The phase-comparison detector in Fig. 14.3-6 looks something like a correlation detector except that the local oscillator signal is replaced by the PRK signal itself after a delay of T_b. A BPF at the front end prevents excess noise from swamping the detector.

Successful operation requires f_c to be an integer multiple of r_b, as in coherent PRK. We therefore write

$$x_c(t) = A_c\, p_{T_b}(t - kT_b)\, \cos(\omega_c t + \theta + a_k \pi)$$
$$a_k = 0, 1 \qquad kT_b < t < (k+1)T_b \tag{15}$$

In the absense of noise, the phase-comparison product for the kth bit interval is

$$x_c(t) \times 2x_c(t - T_b) = 2A_c^2 \cos(\omega_c t + \theta + a_k \pi)$$
$$\times \cos[\omega_c(t - T_b) + \theta + a_{k-1}\pi]$$
$$= A_c^2\{\cos[(a_k - a_{k-1})\pi]$$
$$+ \cos[2\omega_c t + 2\theta + (a_k + a_{k-1})\pi]\}$$

where we've used the fact that $\omega_c T_b = 2\pi N_c$. Lowpass filtering then yields

$$z(t_k) = \begin{cases} +A_c^2 & a_k = a_{k-1} \\ -A_c^2 & a_k \neq a_{k-1} \end{cases} \tag{16}$$

so we have polar symmetry and the threshold should be set at $V = 0$.

Since $z(t_k)$ only tells you whether a_k differs from a_{k-1}, a PRK system with phase-comparison detection is called *differentially coherent* PSK (DPSK). Such systems generally include *differential encoding* at the transmitter, which makes it possible to regenerate the message bits directly from $z(t_k)$. Differential encoding starts with an arbitrary initial bit, say $a_0 = 1$. Subsequent bits are determined by the message sequence m_k according to the rule: $a_k = a_{k-1}$ if $m_k = 1$, $a_k \neq a_{k-1}$ if

Figure 14.3-6 Differentially coherent receiver for binary PSK.

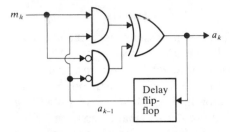

a_{k-1} Delay flip-flop

Figure 14.3-7 Logic circuit for differential encoding.

$m_k = 0$. Thus, $z(t_k) = +A_c^2$ means that $m_k = 1$ and $z(t_k) = -A_c^2$ means that $m_k = 0$. Figure 14.3-7 shows a logic circuit for differential encoding; this circuit implements the logic equation

$$a_k = a_{k-1}m_k \oplus \bar{a}_{k-1}\bar{m}_k \tag{17}$$

where the overbar stands for logical inversion. An example of differential encoding and phase-comparison detection (without noise) is given in Table 14.3-1.

To analyze the performance of DPSK with noise, we'll assume that the BPF performs most of the noise filtering, like the BPFs in an FSK receiver. Hence, the carrier amplitude and noise variance at the BPF output are related by

$$A_c^2/\sigma^2 = 2E_b/\eta = 2\gamma_b$$

We'll also exploit the symmetry and focus on the case when $a_k = a_{k-1} = 0$, so an error occurs if $y(t_k) < 0$.

Now let the delayed i and q noise components be denoted by $n_i'(t) = n_i(t - T_b)$ and $n_q'(t) = n_q(t - T_b)$. The inputs to the multiplier during the kth bit interval are $x_c(t) + n(t) = [A_c + n_i(t)] \cos(\omega_c t + \theta) - n_q(t) \sin(\omega_c t + \theta)$ and $2[x_c(t - T_b) + n(t) - T_b)] = 2[A_c + n_i'(t)] \cos(\omega_c t + \theta) - 2n_q'(t) \sin(\omega_c t + \theta)$. The LPF then removes the high-frequency terms from the product, leaving

$$Y = y(t_k) = (A_c + n_i)(A_c + n_i') + n_q n_q' \tag{18}$$

where all four noise components are independent gaussian r.v.'s with zero mean and variance σ^2.

Equation (18) has a *quadratic form* that can be simplified by a diagonalization process, resulting in

$$Y = \alpha^2 - \beta^2 \tag{19a}$$

with

$$\alpha^2 = (A_c + \alpha_i)^2 + \alpha_q^2 \qquad \beta^2 = \beta_i^2 + \beta_q^2 \tag{19b}$$

Table 14.3-1

Input message	1	0	1	1	1	0	1	0	0
Encoded message	1	1	0	0	0	1	1	0	1
Transmitted phase	π	π	0	0	0	π	π	0	π
Phase-comparison sign	+	−	+	+	+	−	+	−	−
Regenerated message	1	0	1	1	0	1	0	0	

and

$$\alpha_i \triangleq \tfrac{1}{2}(n_i + n_i') \qquad \beta_i \triangleq \tfrac{1}{2}(n_i - n_i')$$
$$\alpha_q \triangleq \tfrac{1}{2}(n_q + n_q') \qquad \beta_q \triangleq \tfrac{1}{2}(n_q - n_q') \tag{19c}$$

Note that α_i is a zero-mean gaussian r.v. with variance $\overline{\alpha_i^2} = (\overline{n_i^2} + \overline{n_i'^2})/4 = 2\sigma^2/4 = \sigma^2/2$; identical conclusions hold for the other i and q components of α and β. Therefore, α has a *Rician* PDF given by Eq. (5) with $\sigma^2/2$ in place of σ^2, while β has a *Rayleigh* PDF given by Eq. (2) with $\sigma^2/2$ in place of σ^2.

Lastly, since α and β are nonnegative, we can write the average error probability as

$$P_e = P(Y < 0 \mid a_k = a_{k-1}) = P(\alpha^2 < \beta^2) = P(\beta > \alpha)$$

and we've arrived at an expression equivalent to the one previously solved for noncoherent FSK. Substituting $\sigma^2/2$ for σ^2 in Eq. (14) now gives our DPSK result

$$P_e = \tfrac{1}{2}e^{-A_c^2/2\sigma^2} = \tfrac{1}{2}e^{-\gamma_b} \tag{20}$$

The performance curves in Fig. 14.3-4 now show that DPSK has a 3-dB energy advantage over noncoherent binary systems and a penalty of less than 1 dB compared to coherent PRK at $P_e \leq 10^{-4}$.

DPSK does not require the carrier phase synchronization essential for coherent PRK, but it does involve somewhat more hardware than noncoherent OOK or FSK—including differential encoding and carrier-frequency synchronization with r_b at the transmitter. A minor annoyance is that DPSK errors tend to occur in groups of two (why?).

Example 14.3-1 Binary data is to be sent at the rate $r_b = 100$ kbps over a channel with 60-dB transmission loss and noise density $\eta = 10^{-12}$ W/Hz at the receiver. What transmitted power S_T is needed to get $P_e = 10^{-3}$ for various types of modulation and detection?

To answer this question, we first write the received signal power as $S_R = E_b r_b = \eta \gamma_b r_b = S_T/L$ with $L = 10^6$. Thus,

$$S_T = L\eta\gamma_b r_b = 0.1\gamma_b$$

Next, using the curves in Fig. 14.3-4 or our previous formulas for P_e, we find the value of γ_b corresponding to the specified error probability and calculate S_T therefrom.

Table 14.3-2 summarizes the results. The systems have been listed here in order of increasing difficulty of implementation, bringing out the trade-off between signal power and hardware complexity.

Table 14.3-2

System	S_T, W
Noncoherent OOK or FSK	1.26
Differentially coherent PSK	0.62
Coherent PRK	0.48

Exercise 14.3-2 Suppose the system in the previous example has a limitation on the *peak envelope power*, such that $LA_c^2 \leq 2$ watts at the transmitter. Find the resulting minimum error probability for noncoherent OOK and FSK and for DPSK.

14.4 QUADRATURE-CARRIER AND *M*-ARY SYSTEMS

This section investigates the performance of *M*-ary modulation systems with coherent or phase-comparison detection, usually in a quadrature-carrier configuration. Our primary motivation here is the increased modulation speed afforded by QAM and related quadrature-carrier methods, and by *M*-ary PSK and APK modulation. These are the modulation types best suited to digital transmission on telephone lines and other bandwidth-limited channels. (Appendix C discusses wideband modulation, such as *M*-ary FSK, which is better suited to power-limited applications.)

As in previous sections, we continue to assume independent equiprobable symbols and AWGN contamination. We also assume that *M* is a power of two, consistent with binary to *M*-ary data conversion. This assumption allows a practical comparison of binary and *M*-ary systems.

Quadrature-Carrier Systems

We pointed out in Sect. 14.1 that both quadriphase PSK and keyed polar QAM are equivalent to the sum of two PRK signals impressed on quadrature carriers. Here we'll adopt that viewpoint to analyze the performance of QPSK/QAM with coherent detection. Accordingly, let the source information be grouped into *dibits* represented by $I_k Q_k$. Each dibit corresponds to one symbol from a quaternary ($M = 4$) source or two successive bits from a binary source. In the latter case, which occurs more often in practice, the dibit rate is $r = r_b/2$ and $D = 1/r = 2T_b$.

Coherent quadrature-carrier detection requires synchronized modulation, as discussed in Sect. 14.2. Thus, for the *k*th dibit interval $kD < t < (k + 1)D$, we write

$$x_c(t) = s_i(t - kD) - s_q(t - kD) \tag{1a}$$

with

$$\begin{aligned} s_i(t) &= A_c I_k p_D(t) \cos \omega_c t & I_k &= \pm 1 \\ s_q(t) &= A_c Q_k p_D(t) \sin \omega_c t & Q_k &= \pm 1 \end{aligned} \tag{1b}$$

Since f_c must be harmonically related to $r = 1/D$, the signaling energy is

$$\int_{kD}^{(k+1)D} x_c^2(t) \, dt = \tfrac{1}{2} A_c^2 (I_k^2 + Q_k^2) D = A_c^2 D$$

and we have

$$E = 2E_b \qquad E_b = A_c^2 D/2 \tag{2}$$

where E is the energy per dibit or quaternary symbol.

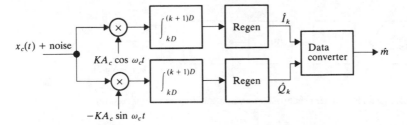

Figure 14.4-1 Quadrature-carrier receiver with correlation detectors.

From Eq. (1) and our prior study of coherent PRK, it follows that the optimum quadrature-carrier receiver can be implemented with two correlation detectors arranged as in Fig. 14.4-1. Each correlator performs coherent binary detection, independent of the other. Hence, the average error probability *per bit* is

$$P_{be} = Q(\sqrt{2E_b/\eta}) = Q(\sqrt{2\gamma_b}) \tag{3}$$

where the function $Q(\sqrt{2\gamma_b})$ denotes the area under the gaussian tail—not to be confused with Q symbolizing quadrature modulation.

We see from Eq. (3) that coherent QPSK/QAM achieves the same bit error probability as coherent PRK. But recall that the transmission bandwidth for QPSK/QAM is

$$B_T \approx r_b/2$$

whereas PRK requires $B_T \approx r_b$. This means that the additional quadrature-carrier hardware allows you to cut the transmission bandwidth in half for a given bit rate or to double the bit rate for a given transmission bandwidth. The error probability remains unchanged in either case.

Equation (3) and the bandwidth/hardware trade-off also hold for *minimum-shift keying*, whose i and q components illustrated back in Fig. 14.1-11b suggest quadrature-carrier detection. An MSK receiver has a structure like Fig. 14.4-1 modified in accordance with the pulse shaping and staggering of the i and q components. There are only two significant differences between MSK and QPSK: (1) the MSK spectrum has a broader main lobe but smaller side lobes than the spectrum of QPSK with the same bit rate; (2) MSK is inherently binary frequency modulation, whereas QPSK can be viewed as either binary or quaternary phase modulation.

When QPSK/QAM is used to transmit quaternary data, the output converter in Fig. 14.4-1 reconstructs quaternary symbols from the regenerated dibits. Since bit errors are independent, the probability of obtaining a *correct* symbol is

$$P_c = (1 - P_{be})^2$$

The average error probability *per symbol* thus becomes

$$P_e = 1 - P_c = 2Q(\sqrt{E/\eta}) - Q^2(\sqrt{E/\eta})$$
$$\approx 2Q(\sqrt{E/\eta}) \qquad E/\eta \gg 1 \tag{4}$$

where $E = 2E_b$ represents the average symbol energy.

Figure 14.4-2 PLL system for carrier synchronization in a quadrature-carrier receiver.

Various methods have been devised to generate the carrier sync signals necessary for coherent detection in quadrature-carrier receivers. Figure 14.4-2 shows a simple PLL system based on the fact that the fourth power of $x_c(t)$ contains a discrete frequency component at $4f_c$. However, since $\cos 4\omega_c t = \cos (4\omega_c t + 2\pi N)$, fourfold frequency division produces $\cos (\omega_c t + N\pi/2)$ so the output has a fixed phase error of $N\pi/2$ with N being an integer whose value depends on the lock-in transient. A known preamble may be transmitted at the start of the message to permit phase adjustment, or differential encoding may be used to nullify the phase error effects. Another carrier sync system will be described in conjunction with M-ary PSK; additional methods are covered by Lindsey (1972).

Phase-comparison detection is also possible in quadrature-carrier systems with differential encoding. From our study of DPSK in Sect. 14.3, you may correctly infer that differentially coherent QPSK (DQPSK) requires somewhat more signal energy than coherent QPSK to get a specified error probability. The difference turns out to be about 2.3 dB.

> **Example 14.4-1 DQPSK Telephone modem** Digital communication over telephone lines has been the subject of intense effort for many years. One of the earliest and most successful systems is the AT & T series 201 modem, which employs DQPSK to achieve binary transmission at $r_b = 2400$ bps within the limitations of a voice channel. (See Table 12.4-2 for a listing of other modems.) This modem is described here to illustrate how basic design concepts are augmented in practical implementation.
>
> Each modem consists of a modulator unit and a demodulator unit to allow transmission in either direction. The modulator is diagrammed in Fig. 14.4-3a, where incoming binary digits are converted to dibits that modulate the carrier at half the bit rate. The process of differential phase modulation incorporates both differential encoding and phase-shift keying, as indicated in Fig. 14.4-3b. The carrier frequency is $f_c = 1800$ Hz, derived from the 2400-Hz clock signal.
>
> Although the nominal transmission bandwidth for DQPSK is $B_T \approx r_b/2 = 1200$ Hz, the spectrum has considerable spillover that would exceed the voice channel bandwidth. Synchronized envelope modulation at 600 Hz provides pulse shaping to reduce spillover outside $f_c \pm r_b/2$. Bandpass filtering finally yields an output spectrum confined to the range 600–3000 Hz.
>
> Observe in Fig. 14.4-3b that the differential phase shift is never zero, so ϕ_k always differs from ϕ_{k-1} and there will be a phase shift in the modulated

(*a*)

Dibit	$\phi_k - \phi_{k-1}$	Q_k	I_k
0 0	$+\pi/4$	$+$	$+$
0 1	$+3\pi/4$	$+$	$-$
1 0	$-3\pi/4$	$-$	$-$
1 1	$-\pi/4$	$-$	$+$

(*b*)

(*c*)

Figure 14.4-3 DQPSK telephone modem. (*a*) Modulator; (*b*) differential encoding table; (*c*) demodulator.

wave every $D = 1/1200$ seconds. When combined with the envelope modulation, these phase shifts produce discrete frequency components at $f_c + 600 = 2400$ Hz and $f_c - 600 = 1200$ Hz used for regeneration timing.

The demodulator unit in Fig. 14.4-3c has quadrature phase-comparison detection with outputs proportional to $I_k = \sqrt{2} \cos(\phi_k - \phi_{k-1})$ and $Q_k = \sqrt{2} \sin(\phi_k - \phi_{k-1})$, from which the message bits are regenerated. The dibit sync signal is obtained by mixing the 2400- and 1200-Hz components and selecting the 1200-Hz difference frequency. This roundabout method, rather than directly filtering the 1200-Hz component, takes better account of the delay distortion induced by a typical telephone channel.

Tests have shown that the resulting error probability is $P_{be} \approx 10^{-5}$ with $\gamma_b = 15$ dB. A fully optimized DQPSK system under ideal conditions would yield the same performance with about 3 dB less energy per bit. Of course a telephone channel or any other real channel does not provide ideal conditions — a fact of life that must be anticipated by the design engineer.

Exercise 14.4-1 Consider a QPSK signal like Eq. (1) written as $x_c(t) = A_c \cos(\omega_c t + \phi_k)$ with $\phi_k = \pi/4, 3\pi/4, 5\pi/4, 7\pi/4$. Show that $x_c^4(t)$ includes an unmodulated component at $4f_c$.

M-ary PSK Systems

Now let's extend our investigation of coherent quadrature-carrier detection to encompass M-ary PSK. The carrier is again synchronized with the modulation, and f_c is harmonically related to the symbol rate r. We write the modulated signal for a given symbol interval as

$$x_c(t) = s_i(t - kD) - s_q(t - kD) \tag{5a}$$

with

$$s_i(t) = A_c \cos \phi_k \, p_D(t) \cos \omega_c t$$
$$s_q(t) = A_c \sin \phi_k \, p_D(t) \sin \omega_c t \tag{5b}$$

where

$$\phi_k = 2\pi a_k/M \qquad a_k = 0, 1, \ldots, M - 1$$

from Eq. (13), Sect. 14.1, taking $N = 0$. The signaling energy per symbol then becomes

$$E = \tfrac{1}{2} A_c^2 (\cos^2 \phi_k + \sin^2 \phi_k) D = \tfrac{1}{2} A_c^2 D \tag{6}$$

equivalent to $E_b = E/\log_2 M$ if each symbol represents $\log_2 M$ binary digits. The transmission bandwidth requirement is $B_T \approx r = r_b/\log_2 M$, from our spectral analysis in Sect. 14.1.

An optimum receiver for M-ary PSK can be modeled in the form of Fig. 14.4-4. We'll let $K = A_c/E$ so, in absence of noise, the quadrature correlators produce $z_i(t_k) = A_c \cos \phi_k$ and $z_q(t_k) = A_c \sin \phi_k$ from which $\phi_k = \arctan z_q/z_i$.

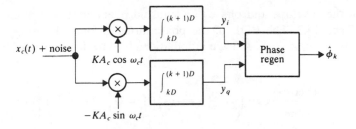

Figure 14.4-4 Coherent *M*-ary PSK receiver.

When $x_c(t)$ is contaminated by noise, message symbol regeneration is based on the noisy samples

$$y_i = A_c \cos \phi_k + n_i \qquad y_q = A_c \sin \phi_k + n_q$$

in which the i and q noise components are independent gaussian r.v.'s with zero mean and variance

$$\sigma^2 = K^2 E \eta/2 = A_c^2 \eta/2E = \eta r \tag{7}$$

The generator has M *angular* thresholds equispaced by $2\pi/M$, as illustrated in Fig. 14.4-5, and it selects the point from the signal constellation whose angle is closest to arctan y_q/y_i.

The circular symmetry of Fig. 14.4-5, together with the symmetry of the noise PDFs, means that all phase angles have the same error probability. We'll therefore focus on the case of $\phi_k = 0$, so

$$\arctan \frac{y_q}{y_i} = \arctan \frac{n_q}{A_c + n_i} = \phi$$

and we recognize ϕ as the phase of a sinusoid plus bandpass noise. Since no error results if $|\phi| < \pi/M$, the symbol error probability can be calculated using

$$P_e = P(|\phi| > \pi/M) = 1 - \int_{-\pi/M}^{\pi/M} p_\phi(\phi) \, d\phi \tag{8}$$

for which we need the PDF of the phase ϕ.

Figure 14.4-5 Decision thresholds for *M*-ary PSK.

The joint PDF for the envelope and phase of a sinusoid plus bandpass noise was given in Eq. (3), Sect. 14.3. The PDF of the phase alone is found by integrating the joint PDF over $0 \leq A < \infty$. A few manipulations lead to the awesome-looking expression

$$p_\phi(\phi) = \frac{1}{2\pi} e^{-A_c^2/2\sigma^2} + \frac{A_c \cos \phi}{\sqrt{2\pi\sigma^2}} \exp\left(-\frac{A_c^2 \sin^2 \phi}{2\sigma^2}\right)\left[1 - Q\left(\frac{A_c \cos \phi}{\sigma}\right)\right] \quad (9)$$

for $-\pi < \phi < \pi$. Under the large-signal condition $A_c \gg \sigma$, Eq. (9) simplifies to

$$p_\phi(\phi) \approx \frac{A_c \cos \phi}{\sqrt{2\pi\sigma^2}} e^{-(A_c \sin \phi)^2/2\sigma^2} \qquad |\phi| < \frac{\pi}{2} \quad (10)$$

which, for small values of ϕ, approximates a *gaussian* with $\bar{\phi} = 0$ and $\overline{\phi^2} = \sigma^2/A_c^2$. Equation (10) is invalid for $|\phi| > \pi/2$, but the probability of that event is small if $A_c \gg \sigma$. Figure 14.4-6 depicts the transition of $p_\phi(\phi)$ from a uniform distribution when $A_c = 0$ to a gaussian curve when A_c becomes large compared to σ. (See Fig. 14.3-1 for the corresponding transition of the envelope PDF.)

We'll assume that $A_c \gg \sigma$ so we can use Eq. (10) to obtain the error probability of coherent M-ary PSK with $M > 4$. (We already have the results for $M = 2$ and 4.) Inserting Eq. (10) with $A_c^2/\sigma^2 = 2E/\eta$ into Eq. (8) gives

$$P_e \approx 1 - \frac{1}{\sqrt{2\pi}} \int_{-\pi/M}^{\pi/M} \sqrt{\frac{2E}{\eta}} \cos \phi \, e^{-(2E/\eta)(\sin \phi)^2/2} \, d\phi$$

$$\approx 1 - \frac{2}{\sqrt{2\pi}} \int_0^L e^{-\lambda^2/2} \, d\lambda \quad (11)$$

Figure 14.4-6 PDFs for the phase of a sinusoid plus bandpass noise.

where we've noted the even symmetry and made the change of variable $\lambda = \sqrt{2E/\eta} \sin \phi$ so $L = \sqrt{2E/\eta} \sin (\pi/M)$. But the integrand in Eq. (11) is a gaussian function, so $P_e \approx 1 - [1 - 2Q(L)] = 2Q(L)$. Hence,

$$P_e \approx 2Q\left(\sqrt{\frac{2E}{\eta}} \sin^2 \frac{\pi}{M}\right) \tag{12}$$

which is our final result for the symbol error probability with $M > 4$. We'll discuss the equivalent bit error probability in our comparisons at the end of the chapter.

Returning to the receiver in Fig. 14.4-4, the carrier sync signals can be derived from the Mth power of $x_c(t)$ using a modified version of Fig. 14.4-2. The more sophisticated *decision-feedback* PLL system in Fig. 14.4-7 uses the estimated phase $\hat{\phi}_k$ to generate a control signal $v(t)$ that corrects any VCO phase error. The two delayors here simply account for the fact that $\hat{\phi}_k$ is obtained at the end of the kth symbol interval.

If accurate carrier synchronization proves to be impractical, then differentially coherent detection may be used instead. The noise analysis is quite complicated, but Lindsey and Simon (1973) have obtained the simple approximation

$$P_e \approx 2Q\left(\sqrt{\frac{4E}{\eta}} \sin^2 \frac{\pi}{2M}\right) \tag{13}$$

which holds for $E/\eta \gg 1$ with $M \geq 4$. We see from Eqs. (12) and (13) that M-ary DPSK achieves the same error probability as coherent PSK when the energy is increased by the factor

$$\Gamma = \frac{\sin^2 (\pi/M)}{2 \sin^2 (\pi/2M)}$$

This factor equals 2.3 dB for DQPSK ($M = 4$), as previously asserted, and it approaches 3 dB for $M \gg 1$.

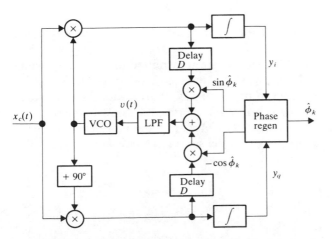

Figure 14.4-7 *M*-ary PSK receiver with decision-feedback system for carrier synchronization.

Exercise 14.4-2 Derive Eq. (7) by replacing one of the correlation detectors in Fig. 14.4-4 with an equivalent BPF, as in Fig. 14.2-3.

M-ary APK Systems

Combined amplitude-phase keying (APK) is an important new method for digital transmission on bandpass channels having limited bandwidth. Specifically, at the cost of more elaborate hardware, coherent APK provides lower error rates than other M-ary systems with keyed modulation operating at the same symbol rate. Here we'll study the class of APK systems defined by square signal constellations, after a preliminary treatment of suppressed-carrier M-ary ASK.

Consider M-ary ASK with synchronized modulation and suppressed carrier. Carrier-suppression is readily accomplished by applying a polar modulating signal. Thus, for the kth symbol interval, we write

$$x_c(t) = A_c\, I_k\, p_D(t - kD)\, \cos \omega_c t \tag{14a}$$

where

$$I_k = \pm 1,\ \pm 3,\ \ldots,\ \pm(M - 1) \tag{14b}$$

The transmission bandwidth is $B_T \approx r$, the same as M-ary PSK.

An optimum coherent receiver consists of just one correlation detector, since there's no quadrature component, and regeneration is based on the noisy samples

$$y_i = A_c I_k + n_i$$

The noise component is a zero-mean gaussian r.v. with variance $\sigma^2 = \eta r$, as in Eq. (7). Figure 14.4-8 shows the one-dimensional signal constellation and the corresponding $M - 1$ equispaced thresholds when $M = 4$. The symbol error probability for any even value of M is

$$P_e = 2\left(1 - \frac{1}{M}\right)Q\left(\frac{A_c}{\sqrt{\eta r}}\right) \tag{15}$$

obtained by the same analysis used for polar M-ary baseband transmission in Sect. 11.2.

Suppose that two of these ASK signals are transmitted on the same channel via quadrature-carrier multiplexing, which requires no more bandwidth than one signal. Let the information come from an M-ary source with $M = \mu^2$ so the message can be converted into two μ-ary digit streams, each having the same rate r. We could describe this process as μ-ary QAM, but the more common designation APK emphasizes the property that the source symbols are represented by

Thresholds

Figure 14.4-8 Decision thresholds for ASK with $M = 4$.

amplitude and phase modulation. Even so, the performance of APK fundamentally depends upon the μ-ary error rate and therefore will be superior to direct M-ary modulation with $M > \mu$.

Figure 14.4-9a diagrams the structure of our APK transmitter. The output signal for the kth symbol interval is

$$x_c(t) = s_i(t - kD) - s_q(t - kD) \tag{16a}$$

(a)

(b)

(c)

Figure 14.4-9 APK system. (a) Transmitter; (b) receiver; (c) square signal constellation and thresholds with $M = 16$.

with

$$s_i(t) = A_c I_k p_D(t) \cos \omega_c t \qquad I_k = \pm 1, \pm 3, \ldots, \pm(\mu - 1)$$
$$s_q(t) = A_c Q_k p_D(t) \sin \omega_c t \qquad Q_k = \pm 1, \pm 3, \ldots, \pm(\mu - 1)$$

$$(16b)$$

The average energy per M-ary symbol is

$$E = \tfrac{1}{2} A_c^2 (\overline{I_k^2} + \overline{Q_k^2}) D = \tfrac{1}{3} A_c^2 (\mu^2 - 1) D \qquad (17)$$

since $\overline{I_k^2} = \overline{Q_k^2} = (\mu^2 - 1)/3$.

Coherent APK detection is performed by the receiver in Fig. 14.4-9b, whose quadrature correlators produce the sample values

$$y_i = A_c I_k + n_i \qquad y_q = A_c Q_k + n_q$$

We then have a *square* signal constellation and threshold pattern, illustrated in Fig. 14.4-9c taking $M = 4^2 = 16$. Now let P denote the probability of error for I_k or Q_k, as given by Eq. (15) with M replaced by $\mu = \sqrt{M}$. The error probability per M-ary symbol is $P_e = 1 - (1 - P)^2$ and $P_e \approx 2P$ when $P \ll 1$. Therefore,

$$P_e \approx 4\left(1 - \frac{1}{\sqrt{M}}\right) Q\left[\sqrt{\frac{3E}{(M - 1)\eta}}\right] \qquad (18)$$

in which we've inserted the average symbol energy from Eq. (17).

Calculations using this result confirm the superior performance of APK. By way of example, if $M = 16$ and $E/\eta = 100$, then $P_e \approx 4 \times \tfrac{3}{4} \times Q(\sqrt{20}) = 1.2 \times 10^{-5}$, whereas an equivalent PSK system with $M = 16$ would have $P_e \approx 2Q(\sqrt{7.6}) = 6 \times 10^{-3}$.

Comparison of Digital Modulation Systems

A performance comparison of digital modulation systems should consider several factors, including: error probability, transmission bandwidth, spectral spillover, hardware requirements, and the differences between binary and M-ary signaling. To establish an equitable basis for comparison, we'll make the realistic assumption that the information comes from a binary source with bit rate r_b. This allows us to compare systems in terms of the modulation speed r_b/B_T and the energy-to-noise ratio γ_b needed to get a specified error probability per bit.

Our previous results for binary modulation systems apply directly to the comparison at hand, especially the error probability curves back in Fig. 14.3-4. Table 14.4-1 serves as a more abbreviated summary when γ_b is large enough to justify the applicable approximations. (Thus, in the case of noncoherent OOK, almost all the errors correspond to the carrier "off" state.) The table omits coherent OOK and FSK, which have little practical value, but it includes QAM and QPSK viewed as binary rather than quaternary modulation. This listing emphasizes the fact that doubled modulation speed goes hand-in-hand with coherent quadrature-carrier detection. Also recall that minimizing spectral spillover

Table 14.4-1 Summary of binary modulation systems

Modulation	Detection	r_b/B_T	P_{be}
OOK or FSK ($f_d = r_b/2$)	Envelope	1	$\frac{1}{2}e^{-\gamma_b/2}$
DPSK	Phase-comparison	1	$\frac{1}{2}e^{-\gamma_b}$
PRK	Coherent	1	$Q(\sqrt{2\gamma_b})$
MSK, QAM, or QPSK	Coherent quadrature	2	$Q(\sqrt{2\gamma_b})$

requires staggered keyed modulation (MSK or OQPSK) or additional pulse shaping.

Now consider M-ary transmission with symbol rate r and energy E per symbol. We'll take $M = 2^K$ and introduce the data-conversion factor

$$K = \log_2 M$$

which equals the number of bits per M-ary symbol. The equivalent bit rate and energy are $r_b = Kr$ and $E_b = E/K$, so

$$\gamma_b = E/K\eta$$

The modulation speed of M-ary PSK or APK is

$$r_b/B_T \approx K \tag{19}$$

since $B_T \approx r = r_b/K$. The error probability per bit is given by

$$P_{be} \approx P_e/K$$

providing that the data converter employs a Gray code, as discussed in conjunction with Eq. (24), Sect. 11.2. After incorporating these adjustments in our previous expressions, we get the comparative results listed in Table 14.4-2.

All of the quadrature-carrier and M-ary systems increase modulation speed at the expense of error probability or signal energy. Suppose, for example, that you want to keep the error probability fixed at $P_{be} \approx 10^{-4}$—a common standard for comparison purposes. The value of γ_b needed for different modulation systems with various modulation speeds then can be calculated from our tabulated

Table 14.4-2 Summary of M-ary modulation systems with $r_b/B_T = K = \log_2 M$

Modulation	Detection	P_{be}
DPSK ($M \geq 4$)	Phase-comparison quadrature	$\frac{2}{K} Q\left(\sqrt{4K\gamma_b \sin^2 \frac{\pi}{2M}}\right)$
PSK ($M \geq 8$)	Coherent quadrature	$\frac{2}{K} Q\left(\sqrt{2K\gamma_b \sin^2 \frac{\pi}{M}}\right)$
APK (K even)	Coherent quadrature	$\frac{4}{K}\left(1 - \frac{1}{\sqrt{M}}\right) Q\left(\sqrt{\frac{3K}{M-1}\gamma_b}\right)$

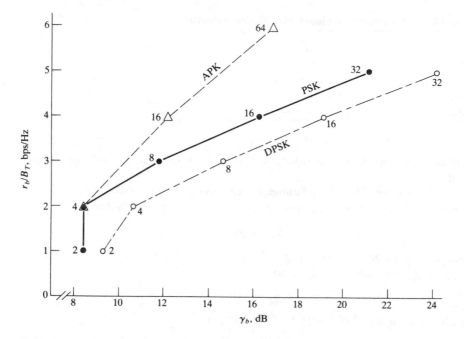

Figure 14.4-10 Performance comparison of M-ary modulation systems with $P_{be} = 10^{-4}$.

expressions. Figure 14.4-10 depicts the results as plots of r_b/B_T versus γ_b in dB, and each point is labeled with the corresponding value of M. Clearly, you would choose APK over PSK for $r_b/B_T \geq 4$ with coherent detection. M-ary DPSK eliminates the carrier-synchronization problems of coherent detection, but it requires at least 7 dB more energy than APK for $r_b/B_T \geq 4$.

As our final comparison, Table 14.4-3 combines M-ary data from Fig. 14.4-10 and calculated values for binary systems with the same error probability. The various systems are listed here in order of increasing complexity to bring out the trade-offs between modulation speed, signal energy, and hardware expense.

Table 14.4-3 Comparison of digital modulation systems with $P_{be} = 10^{-4}$

Modulation	Detection	r_b/B_T	γ_b, dB
OOK or FSK ($f_d = r_b/2$)	Envelope	1	12.3
DPSK ($M = 2$)	Phase-comparison	1	9.3
DQPSK	Phase-comparison quadrature	2	10.7
PRK	Coherent	1	8.4
MSK, QAM, or QPSK	Coherent quadrature	2	8.4
DPSK ($M = 8$)	Phase-comparison quadrature	3	14.6
PSK ($M = 8$)	Coherent quadrature	3	11.8
PSK ($M = 16$)	Coherent quadrature	4	16.2
APK ($M = 16$)	Coherent quadrature	4	12.2

You should keep two points in mind when you examine this table. First, the numerical values correspond to *ideal* systems. The modulation speed of an actual system is typically about 80% of the theoretical value, and the required energy is at least 1–2 dB higher. Second, the characteristics of specific transmission channels may impose additional considerations. In particular, rapidly changing transmission delay prohibits coherent detection, while transmission nonlinearities dictate against the envelope modulation of OOK and APK.

Other factors not covered here include the effects of interference, fading, and delay distortion. These are discussed in an excellent paper by Oetting (1979), which also contains an extensive list of references.

14.5 PROBLEMS

14.1-1 Find from Eq. (7) the average power $\overline{x_c^2}$ and the carrier-frequency power P_c of an M-ary ASK signal. Then form the ratio $P_c/\overline{x_c^2}$ and simplify for $M = 2$ and $M \gg 1$.

14.1-2 Suppose a binary ASK signal consists of RZ rectangular pulses with duration $T_b/2$, where $r_b = 1/T_b \ll f_c$. (a) Find the equivalent lowpass spectrum, and sketch and label $G_c(f)$ for $f > 0$. (b) Sketch the signal representing the sequence 010110. Then find the ratio of the carrier-frequency power P_c to the average power $\overline{x_c^2}$.

14.1-3 Consider a binary ASK signal with *raised-cosine pulse shaping* so, from Example 2.4-2

$$p(t) = \frac{1}{2}\left[1 + \cos\left(\frac{\pi t}{\tau}\right)\right]\Pi\left(\frac{t}{2\tau}\right) \qquad P(f) = \frac{\tau \operatorname{sinc} 2f\tau}{1 - (2f\tau)^2}$$

(a) Sketch the signal representing the sequence 010110 when $\tau = T_b/2$. Then find the equivalent lowpass spectrum, and sketch and label $G_c(f)$ for $f > 0$.

(b) Redo part a with $\tau = T_b$.

14.1-4 The *envelope* and *phase* variations of a QAM signal are

$$A(t) = A_c[x_i^2(t) + x_q^2(t)]^{1/2} \qquad \phi(t) = \arctan [x_q(t)/x_i(t)]$$

(a) By considering the time interval $kD < t < (k + 1)D$, obtain expressions for $A(t)$ and $\phi(t)$ with a rectangular pulse shape $p_D(t)$.

(b) Redo part a with an arbitrary pulse shape $p(t)$ whose duration does not exceed D.

14.1-5 Let a polar M-ary VSB signal have Nyquist pulse shaping per Eq. (6), Sect. 11.3. Find the equivalent lowpass spectrum before VSB filtering. Then sketch and label $G_c(f)$ for $f > 0$ when the filter has $\beta_V \ll r$.

14.1-6 Before bandpass filtering, the i and q components of the OQPSK signal generated in Fig. 14.1-6 can be written as

$$x_i(t) = \sum_k a_{2k} p(t - 2kT_b) \qquad x_q(t) = \sum_k a_{2k+1} p(t - 2kT_b - T_b)$$

where $a_k = (2A_k - 1)$ is the polar sequence corresponding to the message bit sequence A_k, and $p(t) = \Pi(t/2T_b)$ for NRZ rectangular pulse shaping.

(a) Sketch $x_i(t)$ and $x_q(t)$ for the bit sequence 10011100. Use your sketch to draw the signal constellation and to confirm that the phase $\phi(t) = \arctan [x_q(t)/x_i(t)]$ never changes by more than $\pm \pi/2$ rad.

(b) Find the equivalent lowpass spectrum.

14.1-7‡ Let $A_k B_k C_k$ denote the Gray-code binary words for the eight-phase PSK constellation in Fig. 14.1-5b. Construct a table listing $A_k B_k C_k$ and the corresponding values of I_k and Q_k expressed in terms of $\alpha = \cos \pi/8$ and $\beta = \sin \pi/8$. Then write *algebraic* expressions for I_k and Q_k as functions of

A_k, B_k, C_k, α, and β. Devise from these expressions the diagram of a quadrature-carrier transmitter to generate the PSK signal, given a serial-to-parallel converter that supplies the binary words in the inverted polar form $a_k = 1 - 2A_k$, etc.

14.1-8 Suppose a binary FSK signal with discontinuous phase is generated by switching between two oscillators with outputs $A_c \cos(2\pi f_0 t + \theta_0)$ and $A_c \cos(2\pi f_1 t + \theta_1)$. Since the oscillators are unsynchronized, the FSK signal may be viewed as the interleaved sum of two independent binary ASK signals. Use this approach to find, sketch, and label $G_c(f)$ for $f > 0$ when $f_0 = f_c - r_b/2$ and $f_1 = f_c + r_b/2$ with $f_c \gg r_b$. Estimate B_T by comparing your sketch with Figs. 14.1-2 and 14.1-8.

14.1-9 Starting with $p(t)$ in Eq. (17c), obtain both forms of $|P(f)|^2$ as given in Eq. (18b).

14.1-10‡ Consider a binary FSK signal defined by Eq. (15) with $M = 2$, $D = T_b$, and $\omega_d = N/T_b$, where N is an integer. Modify the procedure used in the text and the hint given in Exercise 14.1-3 to obtain $x_i(t)$ and $x_q(t)$. Then show that

$$G_i(f) = \frac{1}{4}\left[\delta\left(f - \frac{Nr_b}{2}\right) + \delta\left(f + \frac{Nr_b}{2}\right)\right]$$

$$G_q(f) = \frac{1}{4r_b}\left[j^{N-1}\,\text{sinc}\,\frac{f - Nr_b/2}{r_b} + (-j)^{N-1}\,\text{sinc}\,\frac{f + Nr_b/2}{r_b}\right]^2$$

which reduce to Sunde's FSK when $N = 1$.

14.1-11 Use the FSK spectral expressions given in Prob. 14.1-10 to sketch and label $G_c(f)$ for $f > 0$ when $N = 2$ and $N = 3$. Compare with Fig. 14.1-8.

14.1-12 An OQPSK signal with *cosine pulse shaping* has many similarities to MSK. In particular, let the i and q components be as given in Prob. 14.1-6 but take $p(t) = \cos(\pi r_b t/2)\Pi(t/2T_b)$.

    (a) Sketch $x_i(t)$ and $x_q(t)$ for the bit sequence 100010111. Use your sketch to draw the signal constellation and to find the phase $\phi(t) = \arctan[x_q(t)/x_i(t)]$ at $t = kT_b$, $0 \le k \le 7$. Compare these with Fig. 14.1-11.

    (b) By considering an arbitrary interval $2kT_b < t < (2k + 1)T_b$, confirm that the envelope $A(t) = A_c[x_q^2(t) + x_i^2(t)]^{1/2}$ is constant for all t.

    (c) Justify the assertion that $G_{lp}(f)$ is identical to an MSK spectrum.

14.1-13 Derive the q component of an MSK signal as given in Eq. (25).

14.2-1 Draw and label the block diagram of an optimum coherent PRK receiver with matched filtering.

14.2-2 Suppose an OOK signal has raised-cosine pulse shaping so that

$$s_1(t) = A_c \sin^2(\pi t/T_b)p_{T_b}(t) \cos \omega_c t$$

Draw and label the diagram of an optimum coherent receiver using: (a) matched filtering; (b) correlation detection.

14.2-3 Obtain an *exact* expression for E_b when a binary FSK signal has $f_c = N_c r_b$ but f_d is arbitrary. Simplify your result when $N_c - f_d T_b \gg 1$.

14.2-4 Use Table T.4 to show that taking $f_d \approx 0.35 r_b$ yields the lowest possible error probability for binary FSK with AWGN. Write the corresponding expression for P_e in terms of γ_b.

14.2-5‡ Draw the complete block diagram of an optimum receiver for Sunde's FSK. Use correlation detection and just *one* local oscillator whose frequency equals r_b. Assume that a bit-sync signal has been extracted from $x_c(t)$ and that N_c is known.

14.2-6 With perfect synchronization, a certain PRK system would have $P_e = 10^{-5}$. Use Eq. (17) to find the condition on θ_c so that $P_e < 10^{-4}$.

14.2-7 Consider a PRK receiver in the form of Fig. 14.2-4 with local-oscillator output $KA_c \cos(\omega_c t + \theta_c)$, where θ_c is a synchronization error. Show that the signal component of $y(T_b)$ is reduced in magnitude by the factor $\cos \theta_c$.

14.2-8 Find the exact expression for $z(t)$ in Eq. (16). Then take $N_c \gg 1$ to obtain the given approximation.

14.2-9 Consider a PRK signal with pilot carrier added for synchronization purposes, resulting in

$$s_1(t) = [A_c \cos \omega_c t + \alpha A_c \cos (\omega_c t + \theta)] p_{T_b}(t)$$

$$s_0(t) = [-A_c \cos \omega_c t + \alpha A_c \cos (\omega_c t + \theta)] p_{T_b}(t)$$

Take $\theta = 0$ and show that an optimum coherent receiver with AWGN yields $P_e = Q[\sqrt{2\gamma_b/(1 + \alpha^2)}]$.

14.2-10 Do Prob. 14.2-9 with $\theta = -\pi/2$.

14.2-11‡ When the noise in a coherent binary system is gaussian but has a *nonwhite* power spectrum $G_n(f)$, the noise can be "whitened" by inserting at the front end of the receiver a filter with transfer function $H_w(f)$ such that $|H_w(f)|^2 G_n(f) = \eta/2$. The rest of the receiver must then be matched to the distorted signaling waveforms $\tilde{s}_1(t)$ and $\tilde{s}_0(t)$ at the output of the whitening filter. Furthermore, the duration of the unfiltered waveforms $s_1(t)$ and $s_0(t)$ must be reduced to ensure that the whitening filter does not introduce appreciable ISI. Apply these conditions to show from Eq. (8a) that

$$\left(\frac{z_1 - z_0}{2\sigma}\right)^2_{max} = \int_{-\infty}^{\infty} \frac{|S_1(f) - S_0(f)|^2}{4G_n(f)} df$$

where $S_1(f) = \mathscr{F}[s_1(t)]$, etc. *Hint:* Recall that if $v(t)$ and $w(t)$ are real, then

$$\int_{-\infty}^{\infty} v(t)w(t)\, dt = \int_{-\infty}^{\infty} V(f)W^*(f)\, df = \int_{-\infty}^{\infty} V^*(f)W(f)\, df$$

14.3-1 A noncoherent OOK system is to have $P_e < 10^{-3}$. Obtain the corresponding bounds on γ_b and P_{e_1}.

14.3-2 Do Prob. 14.3-1 with $P_e < 10^{-5}$.

14.3-3 Obtain Eq. (9) from Eq. (8) with $K = A_c/E_1$.

14.3-4 Suppose the OOK receiver in Fig. 14.3-2 has a simple BPF with $H(f) = [1 + j2(f - f_c)/B]^{-1}$ for $f > 0$, where $2r_b \le B \ll f_c$. Assuming that $f_c \gg r_b$ and $\gamma_b \gg 1$, show that the signal energy must be increased by at least 5 dB get the same error probability as an incoherent receiver with a matched filter.

14.3-5 Consider a noncoherent system with a *trinary* ASK signal defined by Eq. (7) with $a_k = 0, 1, 2$. Let E be the average energy per symbol. Develop an expression similar to Eq. (11) for the error probability.

14.3-6 A binary transmission system with $S_T = 200$ mW, $L = 90$ dB, and $\eta = 10^{-15}$ W/Hz is to have $P_e \le 10^{-4}$. Find the maximum allowable bit rate using: (a) noncoherent FSK; (b) DPSK; (c) coherent PRK.

14.3-7 Do Prob. 14.3-6 with $P_e \le 10^{-5}$.

14.3-8 A binary transmission system with phase modulation is to have $P_e \le 10^{-4}$. Use Eq. (17), Sect. 14.2, to find the condition on the synchronization error θ_ϵ such that PRK will require less signal energy than DPSK.

14.3-9 Do Prob. 14.3-8 with $P_e \le 10^{-6}$.

14.3-10 Derive the joint PDF in Eq. (3) by rectangular-to-polar conversion, starting with $x = A_c + n_i$ and $y = n_q$.

14.4-1 Binary data is to be transmitted at the rate $r_b = 500$ kbps on a radio channel having 400-kHz bandwidth.

(a) Specify the modulation method that minimizes signal energy, and calculate γ_b in dB needed to get $P_{be} \le 10^{-6}$.

(b) Repeat part a with the additional constraint that coherent detection is not practical for the channel in question.

14.4-2 Do Prob. 14.4-1 with $r_b = 1$ Mbps.

14.4-3 Binary data is to be transmitted at the rate $r_b = 800$ kbps on a radio channel having 250-kHz bandwidth.

(a) Specify the modulation method that minimizes signal energy, and calculate γ_b in dB needed to get $P_{be} \leq 10^{-6}$.

(b) Repeat part a with the additional constraint that channel nonlinearities call for a constant-envelope signal.

14.4-4 Do Prob. 14.4-3 with $r_b = 1.2$ Mbps.

14.4-5 Let the VCO output in Fig. 14.4-7 be 2 cos $(\omega_c t + \theta_e)$. In absence of noise, show that the control voltage $v(t)$ will be proportional to sin θ_e.

14.4-6 Suppose an APK system with $M = 16$ is converted to DPSK to allow phase-comparison detection. By what factor must the symbol energy be increased to keep the error probability essentially unchanged?

14.4-7 Suppose a PSK system with $M \gg 1$ is converted to APK. By what factor can the symbol energy be reduced, while keeping the error probability essentially unchanged?

14.4-8‡ Obtain the phase PDF given in Eq. (9) from the joint PDF in Eq. (3), Sect. 14.3. *Hint:* Use the change of variable $\lambda = (A - A_c \cos \phi)/\sigma$.

FIFTEEN

INFORMATION THEORY

Throughout this book we have studied electrical communication primarily in terms of *signals*—desired information-bearing signals corrupted by noise and interference signals. Although signal theory has proved to be a valuable analysis tool, it does not come to grips with the fundamental communication process of information transfer. Recognizing the need for a broader viewpoint, Claude Shannon drew upon the earlier work of Nyquist and Hartley and the concurrent investigations of Wiener to develop and set forth in 1948 a radically new approach he called "A Mathematical Theory of Communication."

Shannon's paper isolated the central task of communication engineering in this question: Given a message-producing source, not of our choosing, how should the messages be represented for reliable transmission over a communication channel with its inherent physical limitations? To address that question, Shannon concentrated on the message information per se rather than on the signals. His approach was soon renamed *information theory*, and it has subsequently evolved into a hybrid mathematical and engineering discipline.

Information theory deals with three basic concepts: the measure of source information, the information capacity of a channel, and coding as a means of utilizing channel capacity for information transfer. The term *coding* is taken here in the broadest sense of message representation, including both discrete and continuous waveforms. The three concepts of information theory are tied together through a series of theorems that boil down to the following statement:

If the rate of information from a source does not exceed the capacity of a communication channel, then there exists a coding technique such that the information can be transmitted over the channel with an arbitrarily small frequency of errors, despite the presence of noise.

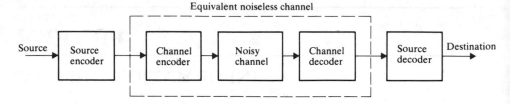

Figure 15.0 Communication system with source and channel coding.

The surprising, almost incredible aspect of this statement is its promise of *error-free* transmission on a noisy channel, a condition achieved with the help of coding.

Optimum coding matches the source and channel for maximum reliable information transfer, roughly analogous to impedance matching for maximum power transfer. The coding process generally involves two distinct encoding/decoding operations, portrayed diagrammatically by Fig. 15.0. The *channel* encoder/decoder units perform the task of error-control coding as discussed in Chap. 13 where we saw how the effects of channel noise could be reduced. Information theory goes a step further, asserting that optimum channel coding yields an equivalent *noiseless* channel with a well-defined capacity for information transmission. The *source* encoder/decoder units then match the source to the equivalent noiseless channel, provided that the source information rate falls within the channel capacity.

This chapter starts with the case of digital or *discrete* information, including information measurement, source coding, information transmission, and discrete channel capacity. Many concepts and conclusions gained from the study of discrete information carry over to the more realistic case of information transmission on a *continuous* channel, where messages take the form of time-varying signals. The *Hartley-Shannon law* defines continuous channel capacity in terms of bandwidth and signal-to-noise ratio, and thereby serves as a benchmark for comparing the performance of communication systems in the light of information theory.

15.1 INFORMATION MEASURE AND SOURCE CODING

Quite logically, we begin our study of information theory with the measure of information. Then we apply information measure to determine the information rate of discrete sources. Particular attention will be given to binary source coding for discrete memoryless sources, followed by a brief look at predictive coding for sources with memory.

Information Measure

The crux of information theory is the measure of information. Here we use *information* as a technical term not to be confused with "knowledge" or "meaning"—concepts that defy precise definition and quantitative measurement. In the context of communication, information is simply the commodity produced by the source for transfer to some user at the destination. This implies that the information was previously unavailable at the destination; otherwise, the transfer would be zero.

Suppose, for instance, that you're planning a trip to a distant city. To determine what clothes to pack, you telephone the weather bureau and hear one of the following forecasts:

The sun will rise.
There will be scattered rainstorms.
There will be a tornado.

The first message conveys virtually no information since you are quite sure in advance that the sun will rise. The forecast of rain, however, provides information not previously available to you. The third forecast gives you more information, tornados being rare and unexpected events. You might even decide to cancel the trip!

Notice that the messages have been listed in order of decreasing likelihood and increasing information. The less likely the message, the more information it conveys. We thus conclude that information measure must be related to *uncertainty*, the uncertainty of the user as to what the message will be. Alternatively, we can say that information measures the *freedom of choice* exercised by the source in selecting a message. When a source freely chooses from many different messages, the user is highly uncertain as to which message will be selected.

Whether you prefer the source or user viewpoint, it should be evident that information measure involves the *probability* of a message. If x_i denotes an arbitrary message and $P(x_i) = P_i$ is the probability of the event that x_i is selected for transmission, then the amount of information associated with x_i should be some function of P_i. Specifically, Shannon defined information measure by the logarithmic function

$$I_i \triangleq -\log_b P_i = \log_b \frac{1}{P_i} \tag{1}$$

where b is the logarithmic base. The quantity I_i is called the *self-information* of message x_i. The value of I_i depends only on P_i, irrespective of the actual message content or possible interpretations. Had you received a weather forecast that said "The sun will rain tornadoes," it would convey lots of information—being very improbable—despite the lack of substance.

Although not immediately obvious, the definition in Eq. (1) has several important and meaningful consequences, including

$$I_i \geq 0 \quad \text{for} \quad 0 \leq P_i \leq 1 \tag{2a}$$

$$I_i \to 0 \quad \text{for} \quad P_i \to 1 \tag{2b}$$

$$I_i > I_j \quad \text{for} \quad P_i < P_j \tag{2c}$$

Hence, I_i is a nonnegative quantity with the properties that $I_i = 0$ if $P_i = 1$ (no uncertainty) and $I_i > I_j$ if $P_i < P_j$ (information increases with uncertainty). Furthermore, suppose a source produces two successive and independent messages, x_i and x_j, with joint probability $P(x_i x_j) = P_i P_j$; then

$$I_{ij} = \log_b \frac{1}{P_i P_j} = \log_b \frac{1}{P_i} + \log_b \frac{1}{P_j}$$

$$= I_i + I_j \tag{3}$$

so the total information equals the sum of the individual message contributions. Shannon's information measure $\log_b (1/P_i)$ is the *only* function that satisfies all of the properties in Eqs. (2) and (3).

Specifying the logarithmic base b determines the *unit* of information. The standard convention of information theory takes $b = 2$ and the corresponding unit is the *bit*, a name coined by J. W. Tukey as the contraction for "binary digit." Equation (1) thus becomes

$$I_i = \log_2 \frac{1}{P_i} \quad \text{bits}$$

This convention normalizes information measure relative to the most elementary source, a source that selects from just two equiprobable messages. For if $P(x_1) = P(x_2) = 1/2$, then $I_1 = I_2 = \log_2 2 = 1$ bit. In other words, 1 bit is the amount of information needed to choose between two equally likely alternatives.

Binary digits enter the picture simply because any two things can be represented by the two binary digits. However, you must carefully distinguish information bits from binary digits per se — especially since a binary digit may convey more or less than one bit of information, depending upon its probability. To prevent misinterpretation, the abbreviation *binits* is used for binary digits as message or code elements.

When necessary, you can convert base-2 to natural or common logarithms via

$$\log_2 v = \frac{\ln v}{\ln 2} = \frac{\log_{10} v}{\log_{10} 2} \tag{4}$$

If $P_i = 1/10$, for instance, then $I_i = (\log_{10} 10)/0.301 = 3.32$ bits. Hereafter, all logarithms will be base-2 unless otherwise indicated.

Entropy and Information Rate

Now consider an information source that emits a sequence of symbols selected from an alphabet of M different symbols. Let X denote the entire set of symbols $x_1, x_2, \ldots, x_M$. We can treat each symbol x_i as a message that occurs with probability P_i and conveys the self-information I_i. The set of symbol probabilities, of course, must satisfy

$$\sum_{i=1}^{M} P_i = 1 \tag{5}$$

We'll assume that the source is stationary, so the probabilities remain constant over time. We'll also assume that successive symbols are statistically independent and come from the source at an average rate of r symbols per second. These properties define the model of a *discrete memoryless source*.

The amount of information produced by the source during an arbitrary symbol interval is a *discrete random variable* having the possible values $I_1, I_2, \ldots,$ I_M. The expected information per symbol is then given by the statistical average.

$$H(X) \triangleq \sum_{i=1}^{M} P_i I_i = \sum_{i=1}^{M} P_i \log \frac{1}{P_i} \qquad \text{bits/symbol} \qquad (6)$$

which is called the source *entropy*. Shannon borrowed the name and notation H from a similar expression in statistical mechanics. Subsequently, various physical and philosophical arguments have been put forth relating communication entropy to thermodynamic entropy; see Brillouin (1962) for an interesting discussion along this line.

We'll interpret Eq. (6) from the more pragmatic observation that when the source emits a sequence of $n \gg 1$ symbols, the total information to be transferred is about $nH(X)$ bits. Since the source produces r symbols per second, the time duration of this sequence is n/r. The information must therefore be transferred at the average rate $nH(X)/(n/r) = rH(X)$ bits per second. Formally, we define the source *information rate*

$$R \triangleq rH(X) \qquad \text{bits/sec} \qquad (7)$$

a critical quantity relative to transmission. Shannon asserted that information from any discrete memoryless source can be encoded as binary digits and transmitted over a noiseless channel at the signaling rate

$$r_b \geq R \qquad \text{binits/sec}$$

As preparation for the study of source coding, we need to know more about entropy.

The value of $H(X)$ for a given source depends upon the symbol probabilities P_i and the alphabet size M. Nonetheless, it always falls within the limits

$$0 \leq H(X) \leq \log M \qquad (8)$$

The lower bound corresponds to no uncertainty or freedom of choice, which occurs when one symbol has probability $P_j = 1$ while $P_i = 0$ for $i \neq j$—so the source almost always emits the same symbol. The upper bound corresponds to maximum uncertainty or freedom of choice, which occurs when $P_i = 1/M$ for all i—so all symbols are equally likely.

To illustrate the variation of $H(X)$ between these extremes, take the special but important case of a binary source ($M = 2$) with

$$P_1 = p \qquad P_2 = 1 - p$$

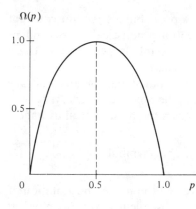

Figure 15.1-1 Binary entropy function.

Substituting these probabilities into Eq. (6) yields the binary entropy

$$H(X) = \Omega(p) \triangleq p \log \frac{1}{p} + (1 - p) \log \frac{1}{1 - p} \qquad (9)$$

in which we've introduced the "horseshoe" function $\Omega(p)$ for later use. The plot of $\Omega(p)$ in Fig. 15.1-1 displays a rather broad maximum centered at $p = 1 - p = 1/2$ where $H(X) = \log 2 = 1$ bit/symbol; $H(X)$ then decreases monotonically to zero as $p \to 1$ or $1 - p \to 1$.

Proving the lower bound in Eq. (8) with arbitrary M is easily done once you note that $v \log (1/v) \to 0$ as $v \to 0$. The proof of the upper bound $H(X) \leq \log M$ involves a few more steps, but it deserves the effort.

First, we introduce another set of probabilities $Q_1, Q_2, \ldots, Q_M$, and replace $\log (1/P_i)$ in Eq. (6) with $\log (Q_i/P_i)$. Conversion from base-2 to natural logarithms gives the quantity

$$\sum_i P_i \log \frac{Q_i}{P_i} = \frac{1}{\ln 2} \sum_i P_i \ln \frac{Q_i}{P_i}$$

where it's understood that all sums range from $i = 1$ to M. Second, we invoke the inequality

$$\ln v \leq v - 1$$

which becomes an equality only if $v = 1$, as seen in Fig. 15.1-2. Thus, letting $v = P_i/Q_i$ and using Eq. (5),

$$\sum_i P_i \ln \frac{Q_i}{P_i} \leq \sum_i P_i \left(\frac{Q_i}{P_i} - 1 \right) = \left(\sum_i Q_i \right) - 1$$

Third, we impose the condition

$$\sum_{i=1}^{M} Q_i = 1 \qquad (10a)$$

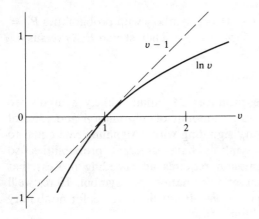

Figure 15.1-2

so it follows that

$$\sum_{i=1}^{M} P_i \log \frac{Q_i}{P_i} \le 0 \qquad (10b)$$

Finally, taking $Q_i = 1/M$ we have

$$\sum_i P_i \log \frac{1}{P_i M} = \sum_i P_i \log \frac{1}{P_i} - \sum_i P_i \log M$$

$$= H(X) - \log M \le 0$$

thereby confirming that $H(X) \le \log M$. The equality holds only in the equally likely case $P_i = 1/M$ so $v = Q_i/P_i = 1/MP_i = 1$ for all i.

Example 15.1-1 Suppose a source emits $r = 2000$ symbols/sec selected from an alphabet of size $M = 4$ with the symbol probabilities and self-information given in Table 15.1-1. Equation (6) gives the source entropy

$$H(X) = \tfrac{1}{2} \times 1 + \tfrac{1}{4} \times 2 + \tfrac{1}{8} \times 3 + \tfrac{1}{8} \times 3 = 1.75 \text{ bits/symbol}$$

which falls somewhat below the maximum value $\log M = 2$. The information rate is

$$R = 2000 \times 1.75 = 3500 \text{ bits/sec}$$

and appropriate coding should make it possible to transmit the source information at the binary signaling rate $r_b \ge 3500$ binits/sec.

Table 15.1-1

x_i	P_i	I_i
A	1/2	1
B	1/4	2
C	1/8	3
D	1/8	3

Exercise 15.1-1 Suppose a source has $M = 3$ symbols with probabilities $P_1 = p$ and $P_2 = P_3$. Show that $H(X) = \Omega(p) + 1 - p$. Then sketch $H(X)$ versus p.

Coding for a Discrete Memoryless Source

When a discrete memoryless source produces M equally likely symbols, so $R = r \log M$, all symbols convey the same amount of information and efficient transmission can take the form of M-ary signaling with a signaling rate equal to the symbol rate r. But when the symbols have different probabilities, so $R = rH(X) < r \log M$, efficient transmission requires an encoding process that takes account of the variable amount of information per symbol. Here, we'll investigate source coding with a *binary encoder*. Equivalent results for nonbinary coding just require changing the logarithmic base.

The binary encoder in Fig. 15.1-3 converts incoming source symbols to codewords consisting of binary digits produced at some fixed rate r_b. Viewed from its output, the encoder looks like a binary source with entropy $\Omega(p)$ and information rate $r_b \Omega(p) \leq r_b \log 2 = r_b$. Coding obviously does not generate additional information, nor does it destroy information providing that the code is *uniquely decipherable*. Thus, equating the encoder's input and output information rates, we have $R = rH(X) = r_b \Omega(p) \leq r_b$ or $r_b/r \geq H(X)$.

The quantity $\bar{N} = r_b/r$ is an important parameter called the *average code length*. Physically, $\bar{N}$ corresponds to the average number of binary digits per source symbol. Mathematically, we write the statistical average

$$\bar{N} = \sum_{i=1}^{M} P_i N_i \tag{11}$$

where N_i represents the length of the codeword for the ith symbol.

Shannon's *source coding theorem* states that the minimum value of $\bar{N}$ is bounded by

$$H(X) \leq \bar{N} < H(X) + \epsilon \tag{12}$$

with ϵ being a positive quantity. The value of ϵ can be made arbitrarily small, in theory, and optimum source coding achieves the lower bound $\bar{N} = H(X)$. In practice, we often settle for suboptimum coding with $\bar{N} > H(X)$ if the code has reasonably good efficiency. The ratio

$$R/r_b = H(X)/\bar{N} \leq 1$$

serves as the measure of efficiency for suboptimum codes.

The source coding theorem presumes a uniquely decipherable code to ensure that no source information is lost. This requirement imposes an additional but

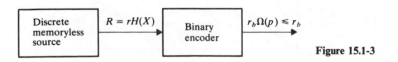

Figure 15.1-3

indirect constraint on N. Specifically, as a necessary and sufficient condition for a uniquely decipherable binary code, the word lengths N_i must be such that

$$K = \sum_{i=1}^{M} 2^{-N_i} \leq 1 \tag{13}$$

which is the *Kraft inequality*. The simplest encoding process generates a *fixed-length* code, all codewords having the same length $N_i = \bar{N}$. Equation (13) then becomes $K = M2^{-\bar{N}} \leq 1$, so decipherability requires $\bar{N} \geq \log M$ and the resulting efficiency is $H(X)/\bar{N} \leq H(X)/\log M$. When $H(X) < \log M$, higher efficiency calls for *variable-length* coding to reduce the average code length $\bar{N}$. An example should help clarify these concepts.

Table 15.1-2 lists some potential codes for the source in Example 15.1-1. Code I is a fixed-length code with $\bar{N} = \log M = 2$ binits/symbol, as compared to $H(X) = 1.75$ bits/symbol. The efficiency is $H(X)/\bar{N} = 1.75/2 \approx 88\%$ — not bad, but we can do better with a variable-length code. Application of Eqs. (11) and (13) to Code II gives

$$\bar{N} = \tfrac{1}{2} \times 1 + \tfrac{1}{4} \times 1 + \tfrac{1}{8} \times 2 + \tfrac{1}{8} \times 2 = 1.25 < H(X)$$

$$K = 2^{-1} + 2^{-1} + 2^{-2} + 2^{-2} = 1.5 > 1$$

The result $\bar{N} < H(X)$ is meaningless because $K > 1$, which tells us that this code is not uniquely decipherable. For instance, the code sequence 10011 could be decoded as BAABB or CABB or CAD, etc. Hence Code II effectively destroys source information and cannot be used. Code III, known as a *comma code*, has $K < 1$ and ensures decipherability by marking the start of each word with the digit 0. However, the extra comma digits result in $\bar{N} = 1.875 > H(X)$. Code IV is a *tree code* with the property that no codeword appears as a *prefix* in another codeword. Thus, for example, you can check that the code sequence 110010111 unambiguously represents the message CABD. This code is *optimum* for the source in question, since it has $\bar{N} = 1.75 = H(X)$ as well as $K = 1$.

Having demonstrated optimum coding by example, we turn to a general proof of the source coding theorem. For this purpose we start with Eq. (10*b*), taking

$$Q_i = 2^{-N_i}/K$$

Table 15.1-2 Illustrative source codes

x_i	P_i	Code I	Code II	Code III	Code IV
A	1/2	00	0	0	0
B	1/4	01	1	01	10
C	1/8	10	10	011	110
D	1/8	11	11	0111	111
$\bar{N}$		2.0	1.25	1.875	1.75
K		1.0	1.5	0.9375	1.0

where, to satisfy Eq. (10a), K is the same as in Eq. (13). Then

$$\sum_i P_i \log \frac{Q_i}{P_i} = \sum_i P_i \left(\log \frac{1}{P_i} - N_i - \log K \right)$$

$$= H(X) - \bar{N} - \log K \leq 0$$

or, since $\log K \leq 0$, we have

$$\bar{N} \geq H(X)$$

which establishes the lower bound in Eq. (12). The equality holds when $K = 1$ and $P_i = Q_i$. Optimum source encoding with $\bar{N} = H(X)$ therefore requires $K = 1$ and symbol probabilities of the form

$$P_i = 2^{-N_i} \qquad i = 1, 2, \ldots, M \tag{14}$$

so that $N_i = -\log P_i = I_i$. An optimum code must also have equally likely 1s and 0s in order to maximize the binary entropy at $\Omega(p) = 1$. Code IV in Table 15.1-2 exemplifies these optimal properties.

Granted, we can't expect every source to obey Eq. (14), and we certainly can't control the statistics of an information source. Even so, Eq. (14) contains a significant implication for practical source coding, to wit:

> Symbols that occur with high probability should be assigned shorter codewords than symbols that occur with low probability.

Long before Shannon, Samuel Morse applied this common-sense principle to his telegraph code for English letters. (Incidentally, Morse estimated letter probabilities by counting the distribution of type in a printer's font.) We'll invoke this principle to establish the upper bound in the source coding theorem.

Let the length of the ith codeword be the integer N_i that falls within $I_i \leq N_i < I_i + 1$ or

$$\log \frac{1}{P_i} \leq N_i < \log \frac{1}{P_i} + 1 \tag{15a}$$

You can easily confirm that this relationship satisfies the Kraft inequality. Multiplying Eq. (15a) by P_i and summing over i then yields

$$H(X) \leq \bar{N} < H(X) + 1 \tag{15b}$$

Hence, a code constructed in agreement with Eq. (15a) will be reasonably efficient if either $H(X) \gg 1$ or $N_i \approx \log (1/P_i)$ for all i. If neither condition holds, we must resort to the process called nth-*extension coding*.

For an extension code, n successive source symbols are grouped into blocks and the encoder operates on the blocks rather than on individual symbols. Since each block consists of n statistically independent symbols, the block entropy is just $nH(X)$. Thus, when the coding rule in Eq. (15a) is modified and applied to the extension code, Eq. (15b) becomes

$$nH(X) \leq n\bar{N} < nH(X) + 1$$

where $n\bar{N}$ is the average number of binary digits per block. Upon dividing by n we get

$$H(X) \leq \bar{N} < H(X) + \frac{1}{n} \qquad (16)$$

which is our final result.

Equation (16) restates the source coding theorem with $\epsilon = 1/n$. It also shows that $\bar{N} \rightarrow H(X)$ as $n \rightarrow \infty$, regardless of the source statistics. We've thereby proved that an nth-extension code comes arbitrarily close to optimum source coding What we haven't addressed is the technique of actual code construction. Systematic algorithms for efficient and practical source coding are presented in texts such as Gallager (1968, chap. 3) or Thomas (1969, chap. 8). The following example serves as an illustration.

Example 15.1-2 Shannon-Fano coding Shannon-Fano coding generates an efficient code in which the word lengths increase as the symbol probabilities decrease, but not necessarily in strict accordance with Eq. (15a). The algorithm provides a tree-code structure to ensure unique decipherability. We'll apply this algorithm to the source with $M = 8$ and $H(X) = 2.15$ whose statistics are listed in Table 15.1-3.

Table 15.1-3 Shannon-Fano coding

x_i	P_i	Coding steps						Codeword
		1	2	3	4	5	6	
A	0.50	0						0
B	0.15	1	0	0				100
C	0.15	1	0	1				101
D	0.08	1	1	0				110
E	0.08	1	1	1	0			1110
F	0.02	1	1	1	1	0		11110
G	0.01	1	1	1	1	1	0	111110
H	0.01	1	1	1	1	1	1	111111
$H(X) = 2.15$								$\bar{N} = 2.18$

The Shannon-Fano algorithm involves a succession of divide-and-conquer steps. For the first step, you draw a line that divides the symbols into two groups such that the group probabilities are as nearly equal as possible; then you assign the digit 0 to each symbol in the group above the line and the digit 1 to each symbol in the group below the line. For all subsequent steps, you subdivide each group into subgroups and again assign digits by the previous rule. Whenever a group contains just one symbol, as happens in the first and third steps in the table, no further subdivision is possible and the codeword for that symbol is complete. When all groups have been reduced to one symbol, the codewords are given by the assigned digits reading from left to right. A careful examination of Table 15.1-3 should clarify this algorithm.

The resulting Shannon-Fano code in this case has $\bar{N} = 2.18$ so the efficiency is $2.15/2.18 \approx 99\%$. Thus, if the symbol rate is $r = 1000$, then $r_b = \bar{N}r = 2180$ binits/sec — slightly greater than $R = rH(X) = 2150$ bits/sec. As comparison, a fixed-length code would require $\bar{N} = \log 8 = 3$ and $r_b = 3000$ binits/sec.

Exercise 15.1-2 Apply the Shannon-Fano algorithm to the source in Table 15.1-2. Your result should be identical to Code IV. Then confirm that this code has the optimum property that $N_i = I_i$ and that 0s and 1s are equally likely.

Predictive Coding for Sources with Memory★

Up to this point we've assumed a memoryless source whose successive symbols are statistically independent. But many information sources have *memory* in the sense that the symbol probabilities depend on one or more previous symbols. Written language, being governed by rules of spelling and grammar, provides a good illustration of a source with memory. For instance, the letter U occurs in English text with probability $P(U) = 0.02$ on the basis of relative frequency; but if the previous letter is Q, then the conditional probability becomes $P(U \mid Q) \approx 1$. Clearly, memory effect reduces uncertainty and thereby results in a lower value of entropy than would be calculated using absolute rather than conditional source statistics.

Suppose that a source has a *first-order memory*, so it "remembers" just one previous symbol. To formulate the entropy, let P_{ij} be the conditional probability that symbol x_i is chosen after symbol x_j. Substituting P_{ij} for P_i in Eq. (6), we have the *conditional* entropy

$$H(X \mid x_j) \triangleq \sum_i P_{ij} \log \frac{1}{P_{ij}} \qquad (17a)$$

which represents the average information per symbol given that the previous symbol was x_j. Averaging over all possible previous symbols then yields

$$H(X) = \sum_j P_j H(X \mid x_j) \qquad (17b)$$

An equivalent expression applies in the general case of a qth-order memory. However, the notation gets cumbersome because x_j must be replaced by the *state* of the source defined in terms of the previous q symbols, and there are M^q possible states to consider.

A source with memory is said to be *redundant* when the conditional probabilities significantly reduce $H(X)$ compared to the upper bound $\log M$. The redundancy of English text has been estimated at about 50%, meaning that roughly half of the symbols in a long passage are not essential to convey the information. By way of example, yu shd babl t read ths evntho sevrl ltrs r msng. It likewise follows that if uncertainty is reduced by the memory effect, then *predictability* is increased. Coding for efficient transmission can then be based on some prediction method, similar to predictive coding for analog signals discussed in Sect. 12.3. Here we'll analyze the scheme known as *predictive run encoding* for a discrete source with memory.

Consider the encoding system in Fig. 15.1-4a, where source symbols are first converted to a binary sequence with digit rate r. Let $x(i)$ denote the ith digit in the binary sequence, and let $\tilde{x}(i)$ be the corresponding digit generated by a predictor. Mod-2 addition of $x(i)$ and $\tilde{x}(i)$ yields a binary *error sequence* $\epsilon(i)$ such that $\epsilon(i) = 0$ when $\tilde{x}(i) = x(i)$ and $\epsilon(i) = 1$ when $\tilde{x}(i) \neq x(i)$. The error sequence is fed back to the predictor in order to update the prediction process. The decoding system in Fig. 15.1-4b employs an identical predictor to reconstruct the source symbols from the mod-2 sum $x(i) = \tilde{x}(i) \oplus \epsilon(i)$. We'll assume the source is sufficiently predictable that a predictor can be devised whose probability of a *correct* prediction is

$$p = P[\epsilon(i) = 0] > 1/2$$

We'll also assume that prediction errors are statistically independent. Hence, the error sequence entropy is $\Omega(p) < 1$ and appropriate encoding should allow information transmission at the rate $r_b < r$.

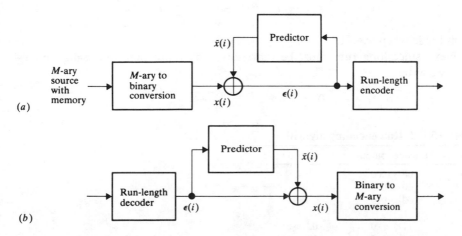

Figure 15.1-4 Predictive coding system for a source with memory. (a) Encoder; (b) decoder.

Predictive run encoding exploits the fact that the error sequence consists of more 0s than 1s—the better the prediction, the higher the percentage of 0s. We define a *run of n* to be a string of n successive 0s followed by a 1. Instead of transmitting these strings, the encoder represents the successive values of n as k-digit binary codewords. A single codeword can represent any value of n in the range $0 \le n \le m - 1$, where

$$m = 2^k - 1$$

When a run has $n \ge m$, the encoder sends a k-digit word consisting entirely of 1s, which tells the decoder to wait for the next codeword to determine n. Table 15.1-4 summarizes the encoding and decoding algorithm.

Efficient transmission results if, on the average, the codewords have fewer digits than the runs they represent. In particular, let $\bar{E}$ be the average number of error digits per run and let $\bar{N}$ be the average number of codeword digits per run, so the encoded digit rate is $r_b = (\bar{N}/\bar{E})r$. If $\bar{N}/\bar{E} < 1$ then $r_b < r$ and we say that predictive run encoding achieves *rate reduction* or *data compression*. The compression ratio $\bar{N}/\bar{E}$ is found from the following calculations.

First, note that a run consists of $E = n + 1$ error digits, including the 1 at the end, so the probability of a run of n is

$$P(n) = p^n(1 - p)$$

Hence,

$$\bar{E} = \sum_{n=0}^{\infty} (n + 1)P(n)$$

$$= (1 - p) \sum_{n=0}^{\infty} (n + 1)p^n = \frac{1}{1 - p} \tag{18}$$

where we've evaluated the summation by recalling the series expansion

$$(1 - v)^{-2} = 1 + 2v + 3v^2 + \cdots = \sum_{n=0}^{\infty} (n + 1)v^n \tag{19}$$

which holds when $v^2 < 1$.

Next, since long runs must be represented by more than one k-digit codeword, we write

$$N = Lk \qquad L = 1, 2, 3, \ldots$$

Table 15.1-4 Run encoding algorithm

n	Encoder output	Decoder output
0	$00\cdots000$	1
1	$00\cdots001$	01
2	$00\cdots010$	001
$\vdots$	$\vdots$	$\vdots$
$m - 1$	$11\cdots110$	$00\cdots01$
$\ge m$	$11\cdots111$	$00\cdots00$

Table 15.1-5 Performance of predictive run encoding

p	k_{opt}	$\bar{N}$	$\bar{E}$	$\bar{N}/\bar{E}$
0.6	1	2.50	2.5	1.00
0.8	3	3.80	5.0	0.76
0.9	4	5.04	10.0	0.50
0.95	6	6.25	20.0	0.31

where L is related to n via

$$(L - 1)m \leq n \leq Lm - 1$$

The average number of code digits per run is then given by

$$\bar{N} = k \sum_{n=0}^{m-1} P(n) + 2k \sum_{n=m}^{2m-1} P(n) + 3k \sum_{n=2m}^{3m-1} P(n) + \cdots$$

$$= k(1 - p)\left(\sum_{n=0}^{m-1} p^n\right) \sum_{i=0}^{\infty} (i + 1)(p^m)^i$$

The first summation is a standard form and the second can be evaluated using Eq. (19), with the final result that

$$\bar{N} = k(1 - p)\,\frac{p^m - 1}{p - 1}\,(1 - p^m)^{-2}$$

$$= \frac{k}{1 - p^m} = \frac{k}{1 - p^{2k-1}} \tag{20}$$

Since $\bar{N}$ depends on k and p, there is an optimum value of k that minimizes $\bar{N}$ for a given predictor.

Table 15.1-5 lists a few values of p and the corresponding optimum values of k. Also listed are the resulting values of $\bar{N}$, $\bar{E}$, and the compression ratio $\bar{N}/\bar{E} = r_b/r$. This table shows that predictive run encoding achieves substantial compression when we know enough about the source to build a predictor with $p > 0.8$.

15.2 INFORMATION TRANSMISSION ON DISCRETE CHANNELS

This section applies information theory to the study of information transmission. We'll assume that both the source and the transmission channel are discrete, so we can measure the amount of information transferred and define the *channel capacity*. Shannon asserted that, with appropriate coding, nearly errorless information transmission is possible on a noisy channel if the rate does not exceed the channel capacity. This fundamental theorem will be examined in conjunction with the *binary symmetric channel*, an important channel model for the analysis of digital communication systems.

Mutual Information

Consider the information transmission system represented by Fig. 15.2-1. A discrete source selects symbols from an alphabet X for transmission over the channel. Ideally, the transmission channel should reproduce at the destination the symbols emitted by the source. However, noise and other transmission impairments alter the source symbols, resulting in a different symbol alphabet Y at the destination. We want to measure the information transferred in this case.

Several types of symbol probabilities will be needed to deal with the two alphabets here, and we'll use the notation defined as follows:

$P(x_i)$ is the probability that the source selects symbol x_i for transmission;

$P(y_j)$ is the probability that symbol y_j is received at the destination;

$P(x_i y_j)$ is the joint probability that x_i is transmitted and y_j is received;

$P(x_i | y_j)$ is the conditional probability that x_i was transmitted given that y_j is received;

$P(y_j | x_i)$ is the conditional probability that y_j is received given that x_i was transmitted.

We'll assume that the channel is *time-invariant* and *memoryless*, so the conditional probabilities are independent of time and previous symbol transmissions. The conditional probabilities $P(y_j | x_i)$ then have special significance as the channel's *forward transition probabilities*. By way of example, Fig. 15.2-2 depicts the forward transitions for a noisy channel with two source symbols and three destination symbols. If this system is intended to deliver $y_j = y_1$ when $x_i = x_1$ and $y_j = y_2$ when $x_i = x_2$, then the symbol *error probabilities* are given by $P(y_j | x_i)$ for $j \neq i$.

Our quantitative description of information transfer on a discrete memoryless channel begins with the *mutual information*

$$I(x_i; y_j) \triangleq \log \frac{P(x_i | y_j)}{P(x_i)} \quad \text{bits} \tag{1}$$

which measures the amount of information transferred when x_i is transmitted and y_j is received. Suppose we happen to have an ideal noiseless channel such that each y_j uniquely identifies a particular x_i; then $P(x_i | y_j) = 1$ and $I(x_i; y_j) = \log [1/P(x_i)]$ — so the transferred information equals the self-information of x_i. On the other hand, suppose the channel noise has such a large effect that y_j is totally unrelated to x_i; then $P(x_i | y_j) = P(x_i)$ and $I(x_i; y_j) = \log 1 = 0$ — so no information is transferred. These extreme cases make sense intuitively and lend support to the definition of mutual information as the measure of information transfer.

Figure 15.2-1 Discrete information transmission system.

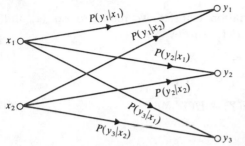

Figure 15.2-2 Forward transition probabilities for a noisy discrete channel.

Most transmission channels fall somewhere between the extremes of perfect transfer and zero transfer. To analyze the general case, we define the *average mutual information*

$$I(X; Y) \triangleq \sum_{x, y} P(x_i\, y_j) I(x_i; y_j)$$

$$= \sum_{x, y} P(x_i\, y_j) \log \frac{P(x_i|\, y_j)}{P(x_i)} \qquad \text{bits/symbol} \qquad (2)$$

where the summation subscripts indicate that the statistical average is taken over both alphabets. The quantity $I(X; Y)$ represents the average amount of source information gained per *received* symbol, as distinguished from the average information per source symbol represented by the source entropy $H(X)$.

Several different but equivalent expressions for the mutual information can be derived using the probability relationships

$$P(x_i\, y_j) = P(x_i|\, y_j)P(y_j) = P(y_j|\, x_i)P(x_i) \qquad (3a)$$

$$P(x_i) = \sum_y P(x_i\, y_j) \qquad P(y_j) = \sum_x P(x_i\, y_j) \qquad (3b)$$

obtained from Eqs. (10) and (12), Sect. 4.1. In particular, let's expand Eq. (2) as

$$I(X; Y) = \sum_{x, y} P(x_i\, y_j) \log \frac{1}{P(x_i)} - \sum_{x, y} P(x_i\, y_j) \log \frac{1}{P(x_i|\, y_j)}$$

so the first term simplifies to

$$\sum_x \left[\sum_y P(x_i\, y_j) \right] \log \frac{1}{P(x_i)} = \sum_x P(x_i) \log \frac{1}{P(x_i)} = H(X)$$

Hence,

$$I(X; Y) = H(X) - H(X|\, Y) \qquad (4)$$

where we've introduced the *equivocation*

$$H(X|\, Y) \triangleq \sum_{x, y} P(x_i\, y_j) \log \frac{1}{P(x_i|\, y_j)} \qquad (5)$$

Equation (4) says that the average information transfer per symbol equals the source entropy minus the equivocation. Correspondingly, the equivocation represents the information lost in the noisy channel.

For another perspective on information transfer, we return to Eq. (2) and note from Eq. (3b) that $P(x_i|y_j)/P(x_i) = P(y_j|x_i)/P(y_j)$ so

$$I(X; Y) = I(Y; X)$$

Therefore, upon interchanging X and Y in Eq. (4), we have

$$I(X; Y) = H(Y) - H(Y|X) \tag{6}$$

with

$$H(Y) = \sum_y P(y_j) \log \frac{1}{P(y_j)} \tag{7a}$$

$$H(Y|X) = \sum_{x,y} P(x_i y_i) \log \frac{1}{P(y_j|x_i)} \tag{7b}$$

Equation (6) says that the information transferred equals the *destination entropy* $H(Y)$ minus the *noise entropy* $H(Y|X)$ added by the channel. The interpretation of $H(Y|X)$ as noise entropy follows from our previous observation that the set of forward transition probabilities $P(y_j|x_i)$ includes the symbol error probabilities.

Example 15.2-1 The binary symmetric channel Figure 15.2-3 depicts the model of a *binary symmetric channel* (BSC). There are two source symbols with probabilities

$$P(x_1) = p \qquad P(x_2) = 1 - p$$

and two destination symbols with forward transition probabilities

$$P(y_1|x_2) = P(y_2|x_1) = \alpha$$

$$P(y_1|x_1) = P(y_2|x_2) = 1 - \alpha$$

This model could represent any binary transmission system in which errors are statistically independent and the error probabilities are the same for both symbols, so the average error probability per symbol is

$$P_e = P(x_1)P(y_2|x_1) + P(x_2)P(y_1|x_2) = p\alpha + (1-p)\alpha = \alpha$$

Since we know the forward transition probabilities, we'll use $I(X; Y) = H(Y) - H(Y|X)$ to calculate the mutual information in terms of p and α.

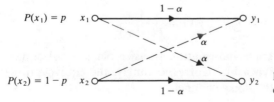

Figure 15.2-3 Binary symmetric channel.

The destination entropy $H(Y)$ is easily found by treating the output of the channel as a binary source with symbol probabilities $P(y_1)$ and $P(y_2) = 1 - P(y_1)$. We therefore write

$$H(Y) = \Omega[P(y_1)]$$

where $\Omega(\ )$ is the binary entropy function defined in Eq. (9), Sect. 15.1, and

$$P(y_1) = P(y_1|x_1)P(x_1) + P(y_1|x_2)P(x_2) = \alpha + p - 2\alpha p$$

obtained with the help of Eqs. (3a) and (3b). For the noise entropy $H(Y|X)$, we substitute $P(x_i\, y_j) = P(y_j|x_i)P(x_i)$ into Eq. (7b) to get

$$H(Y|X) = \sum_x P(x_i)\left[\sum_y P(y_j|x_i) \log \frac{1}{P(y_j|x_i)}\right]$$

which then reduces to

$$H(Y|X) = \Omega(\alpha)$$

The channel's symmetry causes this result to be independent of p.

Putting the foregoing expressions together, we finally have

$$I(X;\,Y) = \Omega(\alpha + p - 2\alpha p) - \Omega(\alpha) \tag{8}$$

so information transfer over a BSC depends on both the error probability α and the source probability p. If the noise is small, then $\alpha \ll 1$ and $I(X;\,Y) \approx \Omega(p) = H(X)$; if the noise is very large, then $\alpha = 1/2$ and $I(X;\,Y) = 0$.

Exercise 15.2-1 Confirm that $H(Y|X) = \Omega(\alpha)$ for a BSC.

Exercise 15.2-2 Consider a channel with the property that x_i and y_j are statistically independent for all i and j. Show that $H(X|Y) = H(X)$ and $I(X;\,Y) = 0$.

Discrete Channel Capacity

We've seen that discrete memoryless channels transfer a definite amount of information $I(X;\,Y)$, despite corrupting noise. A given channel usually has fixed source and destination alphabets and fixed forward transition probabilities, so the only variable quantities in $I(X;\,Y)$ are the *source* probabilities $P(x_i)$. Consequently, *maximum information transfer* requires specific source statistics — obtained, perhaps, through source encoding. Let the resulting maximum value of $I(X;\,Y)$ be denoted by

$$C_s \triangleq \max_{P(x_i)} I(X;\,Y) \qquad \text{bits/symbol} \tag{9}$$

This quantity represents the maximum amount of information transferred per channel symbol and is called the *channel capacity*. We also measure capacity in terms of information rate. Specifically, if s stands for the maximum symbol rate allowed by the channel, then the capacity per unit time is

$$C = sC_s \quad \text{bits/sec} \tag{10}$$

which represents the maximum rate of information transfer.

The significance of channel capacity becomes most evident in the light of Shannon's *fundamental theorem for a noisy channel*, stated as follows:

> If a channel has capacity C and a source has information rate $R \le C$, then there exists a coding system such that the output of the source can be transmitted over the channel with an arbitrarily small frequency of errors. Conversely, if $R > C$, then it is not possible to transmit the information without errors.

A general proof of this theorem goes well beyond our scope, but we'll attempt to make it plausible by considering two particular cases.

First, suppose we have an ideal noiseless channel with $\mu = 2^\nu$ symbols. Then $I(X; Y) = H(X)$, which is maximized if $P(x_i) = 1/\mu$ for all i. Thus

$$C_s = \max_{P(x_i)} H(X) = \log \mu = \nu \tag{11a}$$

and

$$C = s\nu \tag{11b}$$

Errorless transmission rests in this case on the fact that the channel is noiseless. However, we still need a coding system like the one diagrammed in Fig. 15.2-4 to match the source and channel. The binary source encoder generates binary digits at the rate $r_b \ge R$ for conversion to μ-ary channel symbols at the rate $s = r_b/\log \mu = r_b/\nu$. Hence,

$$R \le r_b = s\nu = C \quad \text{bits/sec}$$

and optimum source encoding achieves maximum information transfer with $R = r_b = C$. Transmission at $R > C$ would require a coding system that violates the Kraft inequality; consequently, decoding errors would occur even though the channel is noiseless.

A more realistic case, including channel noise, is the binary symmetric channel from Example 15.2-1. We previously found that $I(X; Y) = \Omega(\alpha + p - 2\alpha p) - \Omega(\alpha)$, with $\Omega(\alpha)$ being constant for a fixed symbol error probability α. But $\Omega(\alpha + p - 2\alpha p)$ varies with the source probability p and reaches a

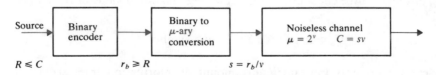

Figure 15.2-4 Encoding system for a noiseless discrete channel.

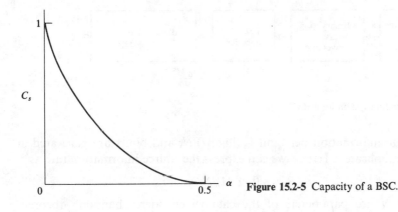

Figure 15.2-5 Capacity of a BSC.

maximum value of unity when $\alpha + p - 2\alpha p = 1/2$, which is satisfied for any α if $p = 1/2$, that is, equally likely binary input symbols. Thus, the capacity of a BSC is

$$C_s = 1 - \Omega(\alpha) \tag{12}$$

The plot of C_s versus α in Fig. 15.2-5 shows that $C_s \approx 1$ for $\alpha \ll 1$, but the capacity drops rapidly to zero as $\alpha \to 0.5$. The same curve applies for $0.5 \le \alpha \le 1$ if you replace α with $1 - \alpha$, equivalent to interchanging the two output symbols.

Reliable transmission on a BSC requires channel coding for *error control*, in addition to any source coding. Our study of error-control coding in Chap. 13 demonstrated that the error probability per binary message digit could, in fact, be made much smaller than the transmission error probability α. For instance, the (15, 11) Hamming code in Example 13.1-1 has code rate $R_c = 11/15$ and output error probability $P_{be} \approx 14\alpha^2$ per message digit; if the BSC has $\alpha = 10^{-3}$ and symbol rate s, then the Hamming code yields $P_{be} \approx 10^{-5}$ at the message digit rate $r_b = R_c s \approx 0.73s$. Shannon's theorem asserts that a better coding system would yield virtually errorless transmission at the rate $r_b = R \le C = s[1 - \Omega(\alpha)] \approx s$. We'll conclude this section with an outline of the argument Shannon used to prove the existence of such a coding system for the BSC.

Coding for the Binary Symmetric Channel★

Consider the encoding system diagrammed in Fig. 15.2-6. The BSC has symbol rate s and capacity $C_s = 1 - \Omega(\alpha)$, with $\alpha < 1/2$. The source produces information at rate R. To maximize the mutual information, an optimum binary source encoder operates on the source symbols and emits equiprobable binary digits. Channel coding for error control is then carried out in two steps. First, a converter transforms blocks of binary digits into equiprobable M-ary symbols conveying $\log M$ bits/symbol. Second, the BSC encoder represents each M-ary symbol by a *channel codeword* consisting of N binary channel symbols, hereafter called binits. (Don't confuse the channel wordlength N with the average code length $\bar{N}$ at the output of the source encoder.)

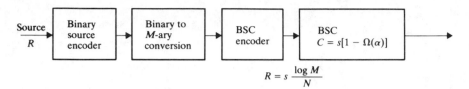

$$R = s \, \frac{\log M}{N}$$

Figure 15.2-6 Encoding system for a BSC.

The average information per binit is $(\log M)/N$, and binits are generated at the channel symbol rate s. Hence, we can express the source information rate as

$$R = s \, (\log M)/N$$

where M and N are parameters of the channel encoder. Shannon's theorem requires $R \le C = sC_s$, equivalent here to $(\log M)/N \le C_s$. The parameters M and N must therefore be related by

$$M = 2^{N(C_s - \epsilon)} \tag{13}$$

with $0 \le \epsilon < C_s$. We'll show that ϵ can be arbitrarily small, so $R \to C$, and that appropriate channel coding makes it possible to recover the M-ary symbols at the destination with arbitrarily low probability of error—providing that the wordlength N is *very large*. In fact, we'll eventually let $N \to \infty$ to ensure errorless information transfer. Ideal channel coding for the BSC therefore involves *infinite time delay*. Practical coding systems with finite time delay and finite wordlength will fall short of ideal performance.

The reasoning behind large wordlength for error control comes from the vector picture of binary codewords introduced in Sect. 13.1. Specifically, recall that all of the 2^N possible words of length N can be visualized as vectors in an N-dimensional space where distance is measured in terms of *Hamming distance*. Let the vector V in Fig. 15.2-7 represent one of the M channel codewords, and let V' be the received codeword with n erroneous binits caused by transmission errors. The Hamming distance between V' and V equals n, a random variable ranging from 0 to N. However, when N is very large, V' almost always falls within a *Hamming sphere* of radius $d < N$ around the tip of V. The decoder correctly decides that V was transmitted if V' is received within the sphere and if none of the remaining $M - 1$ code vectors falls within this sphere.

But Shannon did not propose an explicit algorithm for codeword selection. Instead, he said that *randomly* selected codewords would satisfy the requirement

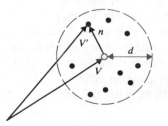

Figure 15.2-7 Vector representation of codewords.

when $N \to \infty$. Although the random coding approach met with considerable criticism at first, it has subsequently been recognized as a powerful method in coding and information theory. We'll adopt random codeword selection and we'll write the probability of a decoding error as

$$P_e = P_{ne} + P_{ce}$$

Here, P_{ne} is the probability of a noise error, so V' falls outside the Hamming sphere, and P_{ce} is the probability of a "code error" in the sense that two or more selected codewords fall within the same Hamming sphere.

A noise error corresponds to the event $n \geq d$. Since transmission errors are statistically independent and occur with probability $\alpha < 1/2$, n is governed by the *binomial distribution* with mean and variance

$$\bar{n} = N\alpha \qquad \sigma^2 = N\alpha(1 - \alpha)$$

If we take the Hamming radius as

$$d = N\beta \qquad \alpha < \beta < 1/2 \tag{14a}$$

then Chebyshev's inequality yields the upper bound

$$P_{ne} = P(n \geq d) \leq \left(\frac{\sigma}{d - \bar{n}}\right)^2 = \frac{\alpha(1 - \alpha)}{N(\beta - \alpha)^2} \tag{14b}$$

Thus, for any $\beta > \alpha$, the probability of a noise error becomes vanishingly small as $N \to \infty$.

To formulate the probability of a "code error," let the vector V in Fig. 15.2-7 represent one of the selected codewords, and let m denote the number of vectors within the Hamming sphere. The remaining $M - 1$ codewords are chosen randomly from the entire set of 2^N vectors, and the probability of selecting one of the m vectors simply equals $m/2^N$. Hence

$$P_{ce} = (M - 1)m2^{-N} < Mm2^{-N} = m2^{-N[\Omega(\alpha) + \epsilon]} \tag{15}$$

where we've inserted Eq. (13) for M and written $C_s = 1 - \Omega(\alpha)$. Now we need an upper bound on m as $N \to \infty$.

All of the m vectors within a Hamming sphere of radius d represent binary words of length N that differ in no more than d places. The number of these words that differ in i places equals the binomial coefficient $\binom{N}{i}$, so

$$m = \sum_{i=0}^{d} \binom{N}{i} = \binom{N}{0} + \binom{N}{1} + \cdots + \binom{N}{d}$$

There are $d + 1$ terms in this sum and the last term is the largest since $d = N\beta < N/2$, so we can write the upper bound

$$m \leq (d + 1)\binom{N}{d} = (d + 1)\frac{N!}{d!(N - d)!}$$

Then, for the factorial of each of the large numbers N, $d = N\beta$, and $N - d = N(1 - \beta)$, we apply *Stirling's approximation*

$$k! \approx \sqrt{2\pi k}\, k^k e^{-k} \qquad k \gg 1 \tag{16}$$

and a few manipulations lead to the expression

$$m \le \frac{N\beta + 1}{\sqrt{2\pi N\beta(1 - \beta)}}\, 2^{N\Omega(\beta)} \tag{17}$$

Combining Eqs. (15) and (17) gives

$$P_{ce} < \frac{N\beta + 1}{\sqrt{2\pi N\beta(1 - \beta)}}\, 2^{-N[\epsilon + \Omega(\alpha) - \Omega(\beta)]} \tag{18}$$

which is our final result.

Equation (18) shows that the probability of a decoding error caused by random codeword selection goes to zero as $N \to \infty$, providing that $\epsilon > \Omega(\beta) - \Omega(\alpha)$. For a given value of the channel parameter α, we can take $\beta > \alpha$ to satisfy Eq. (14) and still have an arbitrarily small value of $\epsilon > \Omega(\beta) - \Omega(\alpha)$. Under these conditions, we achieve errorless information transfer on the BSC with $R \to C$ as $N \to \infty$.

15.3 CONTINUOUS CHANNELS AND SYSTEM COMPARISONS

Having developed concepts of information transmission for the simplified discrete case, we're now ready to tackle the more realistic case of a continuous source and channel. We'll begin with the measure of information from a source that emits a continuous signal. Then we'll make some reasonable assumptions about the transmission of continuous signals to express channel capacity in terms of bandwidth and signal-to-noise ratio, a result known as the *Hartley-Shannon law*. This result leads to the definition of an ideal communication system, which serves as a standard for system comparisons and a guide for the design of improved communication systems.

Continuous Information

A continuous information source produces a time-varying *signal* $x(t)$. We'll treat the set of possible signals as an ensemble of waveforms generated by some random process, assumed to be ergodic. We further assume that the process has a finite bandwidth, meaning that $x(t)$ is completely characterized in terms of periodic *sample values*. Thus, at any sampling instant, the collection of possible sample values constitutes a *continuous random variable* X described by its probability density function $p(x)$.

The average amount of information per sample value of $x(t)$ is measured by the entropy function

$$H(X) \triangleq \int_{-\infty}^{\infty} p(x) \log \frac{1}{p(x)} \, dx \tag{1}$$

where, as before, the logarithmic base is $b = 2$. This expression has obvious similarity to the definition of entropy for a discrete source. However, Eq. (1) turns out to be a *relative* measure of information rather than an absolute measure. The *absolute* entropy of a continuous source can, in principle, be defined from the following limiting operation.

Let the continuous r.v. X be approximated by a discrete r.v. with values $x_i = i \, \Delta x$ for $i = 0, \pm 1, \pm 2, \ldots$, and probabilities $P(x_i) \approx p(x_i) \, \Delta x$. Then, based on the formula for discrete entropy, let the absolute entropy be

$$H_{abs}(X) = \lim_{\Delta x \to 0} \sum_i p(x_i) \, \Delta x \log \frac{1}{p(x_i) \, \Delta x}$$

$$= \lim_{\Delta x \to 0} \sum_i \left[p(x_i) \log \frac{1}{p(x_i)} - p(x_i) \log \Delta x \right] \Delta x$$

Passing from summation to integration yields

$$H_{abs}(X) = H(X) + H_0(X) \tag{2a}$$

where

$$H_0(X) = -\lim_{\Delta x \to 0} \log \Delta x \int_{-\infty}^{\infty} p(x) \, dx = -\lim_{\Delta x \to 0} \log \Delta x \tag{2b}$$

which is the *reference* for the relative entropy $H(X)$. Since $H_0(X) = -\log 0 = \infty$, the absolute entropy of a continuous source is always infinite—a useless but reasonable result in view of the fact that X is a continuous r.v. with an uncountable number of possible values.

Relative entropy, being finite, serves as a useful measure of information from continuous sources if you avoid misleading comparisons involving different references. In particular, consider two information signals $x(t)$ and $z(t)$. For a valid comparison of their entropy functions we write

$$H_{abs}(Z) - H_{abs}(X) = H(Z) - H(X) + H_0(Z) - H_0(X)$$

where

$$H_0(Z) - H_0(X) = -\lim_{\Delta z \to 0} \log \Delta z + \lim_{\Delta x \to 0} \log \Delta x$$

$$= -\lim_{\Delta z, \Delta x \to 0} \log \left| \frac{\Delta z}{\Delta x} \right| = -\log \left| \frac{dz}{dx} \right| \tag{3}$$

If the signals are related in some manner and $|dz/dx| = 1$, then the reference values are equal and $H(Z)$ and $H(X)$ are directly comparable.

The inherent nature of relative entropy precludes absolute bounds on $H(X)$. In fact, the value of $H(X)$ can be positive, zero, or negative, depending upon the source PDF. Nonetheless, a reasonable question to ask is: What $p(x)$ maximizes $H(X)$ for a given source? This question is significant because a given source usually has specific signal constraints, such as fixed peak value or average power, that limit the possible PDFs.

Stating the problem in more general terms, we seek the function $p = p(x)$ that maximizes an integral of the form

$$I = \int_a^b F(x, p)\, dx$$

whose integrand $F(x, p)$ is a specified function of x and p. The variable function p is subject to a set of k constraints given by

$$\int_a^b F_k(x, p)\, dx = c_k$$

with the c_k being constants. A theorem from the calculus of variations says that I is maximum (or minimum) when p satisfies the equation

$$\frac{\partial F}{\partial p} + \sum_k \lambda_k \frac{\partial F_k}{\partial p} = 0 \qquad (4)$$

where the λ_k are *Lagrange's undetermined multipliers*. The values of λ_k are found by substituting the solution of Eq. (4) into the constraint equations. For a compact treatment of the theory underlying Eq. (4), see Thomas (1969, App. F).

In the problem at hand, we wish to maximize $I = H(X)$ as defined by Eq. (1). Thus, we take

$$F(x, p) = p \log (1/p) = -p (\ln p)/\ln 2$$

Furthermore, $p(x)$ must obey the essential PDF property

$$\int_{-\infty}^{\infty} p(x)\, dx = 1 \qquad (5)$$

so we always have the constraint function and constant

$$F_1(x, p) = p \qquad c_1 = 1$$

Additional constraints come from particular source limitations, illustrated in the example that follows.

Example 15.3-1 Source entropy with fixed average power Consider the case of a source with *fixed average power* defined in terms of $p(x)$ by

$$S \triangleq \overline{x^2} = \int_{-\infty}^{\infty} x^2 p(x)\, dx \qquad (6)$$

which imposes the additional constraint function $F_2 = x^2 p$ and $c_2 = S$.

Inserting F, F_1 and F_2 into Eq. (4) yields

$$-(\ln p + 1) + \lambda_1 + \lambda_2 x^2 = 0$$

where ln 2 has been absorbed in λ_1 and λ_2. Thus

$$p(x) = e^{\lambda_1 - 1} e^{\lambda_2 x^2}$$

After using Eqs. (5) and (6) to evaluate the multipliers, we get

$$p(x) = \frac{1}{\sqrt{2\pi\sigma^2}} e^{-x^2/2\sigma^2} \qquad \sigma^2 = S \tag{7}$$

a *gaussian* function with zero mean and variance σ^2.

The corresponding maximum entropy is calculated from Eq. (1) by writing

$$\log 1/p(x) = \log \sqrt{2\pi S} + (x^2/2S) \log e$$

Therefore,

$$H(X) = \log \sqrt{2\pi S} + \tfrac{1}{2} \log e = \tfrac{1}{2} \log 2\pi e S \tag{8}$$

obtained with the help of Eqs. (5) and (6). Note that this relative entropy has a negative value when $2\pi e S < 1$. Even so, for any fixed average power S, we've established the important result that $H(X) \leq \tfrac{1}{2} \log 2\pi e S$ and that the entropy is maximum when $p(x)$ is a zero-mean gaussian PDF. Other source constraints of course lead to different results.

Exercise 15.3-1 Suppose a source has a peak-value limitation, such that $-M \leq x(t) \leq M$. (*a*) By finding the PDF that maximizes $H(X)$, show that

$$H(X) = \log 2M$$

(*b*) Now suppose the source waveform is amplified to produce $z(t) = Kx(t)$. Find $H(Z)$ and confirm that the absolute entropy remains unchanged even though $H(Z) \neq H(X)$.

Continuous Channel Capacity

Information transfer on a continuous channel takes the form of signal transmission. The source emits a signal $x(t)$ which, after corruption by transmission noise, appears at the destination as another signal $y(t)$. The *average mutual information* is defined by analogy with the discrete case to be

$$I(X; Y) \triangleq \int\int_{-\infty}^{\infty} p_{XY}(x,y) \log \frac{p_X(x|y)}{p_X(x)} \, dx \, dy \tag{9}$$

where $p_X(x)$ is the source PDF, $p_{XY}(x,y)$ is the joint PDF, etc. Averaging with respect to both X and Y removes the potential ambiguities of relative entropy.

Thus, $I(X; Y)$ measures the *absolute* information transfer per sample value of $y(t)$ at the destination. It can be shown from Eq. (9) that $I(X; Y) \geq 0$ and that $I(X; Y) = 0$ when the noise is so great that $y(t)$ is unrelated to $x(t)$.

Usually, we know the forward transition PDF $p_Y(y \mid x)$ rather than $p_X(x \mid y)$. We then calculate $I(X; Y)$ from the equivalent expression

$$I(X; Y) = H(Y) - H(Y \mid X) \tag{10}$$

in which $H(Y)$ is the destination entropy and $H(Y \mid X)$ is the noise entropy given by

$$H(Y \mid X) = \iint\limits_{-\infty}^{\infty} p_X(x) p_Y(y \mid x) \log \frac{1}{p_Y(y \mid x)} \, dx \, dy$$

If the channel has *independent additive noise* such that

$$y(t) = x(t) + n(t)$$

then

$$p_Y(y \mid x) = p_Y(x + n) = p_n(y - x)$$

where $p_n(n)$ is the noise PDF. Consequently, $H(Y \mid X)$ reduces to

$$H(Y \mid X) = \int_{-\infty}^{\infty} p_n(n) \log \frac{1}{p_n(n)} \, dn \tag{11}$$

independent of $p_X(x)$.

Now consider a channel with fixed forward transition PDF, so the maximum information transfer per sample value of $y(t)$ is

$$C_s \triangleq \max_{p_X(x)} I(X; Y) \qquad \text{bits/sample} \tag{12}$$

If the channel also has a fixed *bandwidth B*, then $y(t)$ is a bandlimited signal completely defined by sample values taken at the Nyquist rate $f_s = 2B$—identical to the maximum allowable signaling rate for a given bandwidth B. (Samples taken at greater than the Nyquist rate would not be independent and carry no additional information.) The maximum rate of information transfer then becomes

$$C \triangleq 2BC_s \qquad \text{bits/sec} \tag{13}$$

which defines the *capacity* of a bandlimited continuous channel. Shannon's fundamental theorem for a noisy channel applies here in the sense that errorless transmission is theoretically possible at any information rate $R \leq C$.

The continuous channel model of greatest interest is known as the *additive white gaussian noise* (AWGN) channel, defined by the following properties:

1. The channel provides distortion-free transmission over some bandwidth B, and any transmission loss is compensated by amplification.
2. The channel constrains the input from the source to be a bandlimited signal $x(t)$ with fixed average power $S = \overline{x^2}$.

3. The signal received at the destination is contaminated by the addition of band-limited white gaussian noise $n(t)$ with zero mean and average noise power $N = \overline{n^2} = \eta B$.

4. The signal and noise are independent so that $y(t) = x(t) + n(t)$ and

$$\overline{y^2} = \overline{x^2} + \overline{n^2} = S + N$$

These properties agree with the assumptions made in earlier chapters to obtain upper bounds on the performance of analog systems. Here, we'll determine the corresponding channel capacity.

First, we find the noise entropy $H(Y \mid X)$ via Eq. (11). Since $p_n(n)$ is a zero-mean gaussian function with variance $\sigma^2 = N$, it follows from Eqs. (7) and (8) that $H(Y \mid X) = \frac{1}{2} \log 2\pi e N$. Second, noting that $H(Y \mid X)$ does not depend on the source PDF, we use Eqs. (10) and (12) to write

$$C_s = \max_{p_X(x)} [H(Y) - H(Y \mid X)] = \left[\max_{p_X(x)} H(Y) \right] - \frac{1}{2} \log 2\pi e N$$

But the destination signal $y(t)$ has fixed average power $\overline{y^2} = S + N$, so $H(Y) \leq \frac{1}{2} \log 2\pi e(S + N)$. If $p_X(x)$ is a zero-mean gaussian function, then $y = x + n$ has a gaussian PDF and $H(Y)$ is maximized. Hence,

$$C_s = \frac{1}{2} \log 2\pi e(S + N) - \frac{1}{2} \log 2\pi e N = \frac{1}{2} \log \left(\frac{S + N}{N} \right)$$

Finally, substituting for C_s in Eq. (13) yields

$$C = B \log (1 + S/N) \tag{14}$$

where we recognize S/N as the *signal-to-noise ratio* at the destination.

Equation (14) is the famous *Hartley-Shannon law*. When coupled with the fundamental theorem, it establishes an upper limit for reliable information transmission on a bandlimited AWGN channel, namely, $R \leq B \log (1 + S/N)$ bits/sec. Additionally, since bandwidth and signal-to-noise ratio are basic transmission parameters, Eq. (14) establishes a general performance standard for the comparison of communication systems. We therefore devote the remainder of this section to implications of the Hartley-Shannon law.

Ideal Communication Systems

Virtually every real communication system is capable of transmitting continuous signals, subject to some power and bandwidth limitations. Furthermore, if we exclude radio transmission with fading, careful system design can largely eliminate all other contaminations except unavoidable thermal noise. The AWGN channel serves as a reasonable model under these conditions, and the Hartley-Shannon law gives the maximum possible rate for reliable communication. Therefore, we here define an *ideal* communication system as one that achieves nearly error-free information transmission at a rate approaching $R = B \log (1 + S/N)$.

Shannon described the operation of such an ideal system as follows. The system has the structure diagrammed in Fig. 15.3-1. Information from the source is observed for intervals of T seconds and encoded as equiprobable M-ary symbols, such that $(\log M)/T = R$. For each symbol, the signal generator emits a unique waveform $x(t)$ selected from a set of $M = 2^{RT}$ signals. These signals are generated by a bandlimited white gaussian process, and stored copies are available at the destination. The signal detector compares the received signal $y(t) = x(t) + n(t)$ with the stored copies and chooses the most probable one for decoding. The operations of encoding at the source and signal detection at the destination result in a total delay of $2T$ seconds, called the *coding delay*.

The number of different signals required depends on the desired error probability P_e, and Shannon showed that

$$\lim_{P_e \to 0} \lim_{T \to \infty} \frac{\log M}{T} = B \log \left(1 + \frac{S}{N} \right)$$

Hence, $R \to C$ in the limit as $T \to \infty$ — which means that the number of signals and the coding delay become infinitely large, so the ideal system is physically *inrealizable*. However, real systems having large but finite M and T can be designed to come as close as we wish to ideal performance. (One such system is described in App. C.) With this thought in mind, let's further consider the properties of a hypothetical ideal system.

The capacity relation $C = B \log (1 + S/N)$ underscores the fundamental role of bandwidth and signal-to-noise ratio in communication. It also shows that we can exchange increased bandwidth for decreased signal power, a feature previously observed in wideband noise-reduction systems such as FM and PCM. The Hartley-Shannon law specifies the *optimum* bandwidth-power exchange and, moreover, suggests the possibility of *bandwidth compression*.

Keeping in mind that noise power varies with bandwidth as $N = \eta B$, we explore the trade-off between bandwidth and signal power by writing

$$C = B \log \left(1 + \frac{S}{\eta B} \right) \tag{15}$$

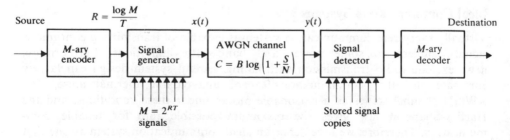

Figure 15.3-1 Ideal communication system with AWGN channel.

Thus, if η and R have fixed values, information transmission at the rate $R \leq C$ requires

$$\frac{S}{\eta R} \geq \frac{B}{R} (2^{R/B} - 1) \tag{16}$$

which becomes an equality when $R = C$. Figure 15.3-2 shows the resulting plot of $S/\eta R$ in dB versus B/R. The shaded region corresponds to $R > C$, a forbidden condition for reliable communication. This plot reveals that bandwidth compression ($B/R < 1$) demands a dramatic increase of signal power, while bandwidth expansion ($B/R > 1$) reduces $S/\eta R$ asymptotically toward a distinct limiting value of about -1.6 dB as $B/R \to \infty$.

In fact, an ideal system with *infinite bandwidth* has finite channel capacity given by

$$C_\infty \triangleq \lim_{B \to \infty} B \log \left(1 + \frac{S}{\eta B} \right)$$

$$= \frac{S}{\eta \ln 2} \approx 1.44 \frac{S}{\eta} \tag{17}$$

Equation (17) is derived from Eq. (15) written in the form

$$C = \frac{S}{\eta \lambda} \frac{\ln (1 + \lambda)}{\ln 2} \qquad \lambda = \frac{S}{\eta B}$$

The series expansion $\ln (1 + \lambda) = \lambda - \frac{1}{2}\lambda^2 + \cdots$ then shows that $[\log (1 + \lambda)]/\lambda \to 1$ as $\lambda \to 0$, corresponding to $B \to \infty$. Note that C_∞ is the maximum capacity for fixed S and η, so

$$S/\eta R \geq S/\eta C_\infty = \ln 2 \approx -1.6 \text{ dB}$$

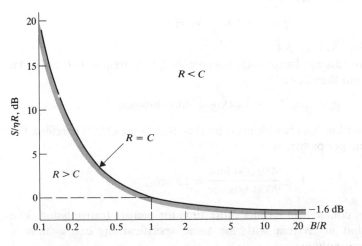

Figure 15.3-2 Trade-off between bandwidth and signal power with an ideal system.

From the shape of the curve in Fig. 15.3-2, we conclude that $C \approx C_\infty$ for $B/R > 10$.

It must be stressed that these results pertain to an ideal but unrealizable system. Consequently, they may be used to estimate upper limits on the performance of any real system whose transmission channel approximates the AWGN model. An example should help illustrate the value of such estimates.

Example 15.3-2 TV pictures from Mars A roving vehicle equipped with a monochrome TV camera has been proposed to explore the surface of Mars. The TV pictures will be digitized for transmission back to earth. We want to estimate the time required to transmit one picture, given the following system specifications.

A digitized picture consists of $n_p = 400 \times 300$ pixels (picture elements), each pixel having one of 16 possible brightness levels. Hence, the information rate R and picture transmission time T are related by

$$RT = n_p \log 16 = 480{,}000 \text{ bits}$$

assuming equally likely brightness levels.

The Mars-to-Earth link is a microwave radio system with carrier frequency $f_c = 2$ GHz and path length $\ell = 3 \times 10^8$ km. The transmitter on the vehicle delivers $S_T = 20$ W signal power to a dish antenna one meter in diameter. The earth station has a 30-m dish antenna and a low-noise receiver with noise temperature $\mathcal{T}_N = 58$ K. The signal power S at the receiver is calculated using Eqs. (10)–(12), Sect. 3.3, which give $L = 268$ dB, $g_T = 26$ dB, $g_R = 56$ dB, and

$$S = \frac{g_T g_R}{L} S_T = 5 \times 10^{-18} \text{ W}$$

The noise density at the receiver is

$$\eta = 8 \times 10^{-22} \text{ W/Hz}$$

obtained from Eq. (6), Sect. 5.4.

Since no transmission bandwidth was specified, let's assume $B/R > 10$. An ideal system would then have

$$R \leq C \approx C_\infty = 1.44 S/\eta \approx 9000 \text{ bits/sec}$$

and the corresponding bandwidth must be $B > 10R \geq 90$ kHz. Therefore, the transmission time per picture is

$$T \geq \frac{480{,}000 \text{ bits}}{9000 \text{ bits/sec}} \approx 53 \text{ sec}$$

A real system, of course, will require more time for picture transmission. The point here is that no system with the same specifications can achieve a smaller transmission time.

Exercise 15.3-2 A teletype machine with 64 different symbols is connected to a telephone channel having $B = 3$ kHz and $S/\eta B = 30$ dB. (*a*) Calculate the maximum possible symbol rate for errorless transmission. (*b*) Assuming B can be changed while other parameters are fixed, find the symbol rate with $B = 1$ kHz and $B \to \infty$.

System Comparisons

Having determined the properties of an ideal communication system, we're ready to reexamine various systems from previous chapters and compare their performance in the light of information theory. Such comparisons provide important guidelines for the development of new or improved communication systems, both digital and analog. For all comparisons, we'll assume an AWGN channel with transmission bandwidth B_T and signal-to-noise ratio $(S/N)_R = S_R/\eta B_T$ at the receiver.

Consider first the case of binary baseband transmission. In Chap. 11 we found the maximum signaling rate $2B_T$ and minimum transmission error probability $Q[\sqrt{(S/N)_R}]$, which requires the use of polar sinc-pulse signals and matched filtering. Now we'll view this system as a *binary symmetric channel* with symbol rate $s = 2B_T$, and transition error probability $\alpha = Q[\sqrt{(S/N)_R}]$. Hence, the BSC system capacity is

$$C = 2B_T[1 - \Omega(\alpha)] \qquad \alpha = Q[\sqrt{(S/N)_R}] \tag{18}$$

where $\Omega(\alpha)$ is the binary entropy function from Eq. (9), Sect. 15.1. With sufficiently elaborate error-control coding, we can obtain nearly error-free information transfer at $R \le C$.

Figure 15.3-3 shows C/B_T versus $(S/N)_R$ for a BSC and for an ideal continuous system with $C/B_T = \log[1 + (S/N)_R]$. The BSC curve flattens off at $C/B_T = 2$ as $(S/N)_R \to \infty$ because the maximum entropy of a noise-free binary signal is one bit per symbol. Hence, we must use M-ary signaling to get closer to

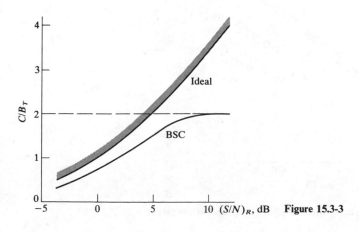

Figure 15.3-3

ideal performance when $(S/N)_R \gg 1$. At lower signal-to-noise ratios, where $C/B_T < 2$, the gap between the BSC and ideal curves suggests that there should be a way of extracting more information from a noisy binary signal. This observation has led to the sophisticated technique known as *soft-decision decoding* in which binary codewords are decoded from sample values of the continuous signal-plus-noise waveform, rather than from a regenerated (hard-decision) binary waveform; see Proakis (1983, chap. 5) for further details.

Now let's examine the performance of digital communication systems without error-control coding. We'll assume a binary source that emits equiprobable symbols at rate r_b. If the error probability is small, say $P_{be} \approx 10^{-4}$, then the information rate is $R \approx r_b$ since almost all of the binary digits correctly convey information to the destination. For transmission purposes, let the binary data be converted to a Gray-code M-ary signal with $M = 2^K$. An M-ary baseband system with polar sinc pulses and matched filtering then has

$$\frac{r_b}{B_T} = 2K \qquad P_{be} \approx \frac{2}{K}\left(1 - \frac{1}{M}\right)Q\left(\sqrt{\frac{6K}{M^2 - 1}}\,\gamma_b\right) \tag{19}$$

where $\gamma_b = E_b/\eta$ and $E_b = S_R/r_b$, which is the average energy per binary digit introduced in Chap. 11.

Upon solving Eq. (19) for the value of γ_b that gives $P_{be} \approx 10^{-4}$. we obtain the plot of r_b/B_T versus γ_b in Fig. 15.3-4. This same curve also applies for bandpass transmission via APK modulation (with M replaced by $\sqrt{M}$), while other modulation methods discussed in Sect. 14.4 have poorer performance; see Fig. 14.4-10. The comparison curve for an ideal system is calculated from Eq. (16) taking $R = r_b$ and $S/\eta R = S_R/\eta r_b = \gamma_b$. We thus see that real digital systems with $r_b/B_T \geq 2$

Figure 15.3-4 Performance comparison of ideal system and M-ary system with $P_{be} = 10^{-4}$.

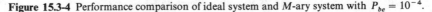

and small but nonzero error probability require at least 6–7 dB more signal energy than an ideal system. Error-control coding would reduce the energy needed for a specified error probability, thereby shifting the system performance curve closer to the ideal.

Finally, we come to analog communication in which bandwidth and signal-to-noise ratio are the crucial performance factors, rather than information rate and error probability. Consider, therefore, the analog modulation system represented by Fig. 15.3-5, where the analog signal at the destination has bandwidth W and signal-to-noise ratio $(S/N)_D$. The maximum output information rate is

$$R = W \log [1 + (S/N)_D]$$

and it cannot exceed the transmission channel capacity

$$C = B_T \log [1 + (S_R/\eta B_T)]$$

Setting $R \leq C$ and solving for $(S/N)_D$ yields

$$\left(\frac{S}{N}\right)_D \leq \left(1 + \frac{S_R}{\eta B_T}\right)^{B_T/W} - 1 = \left(1 + \frac{\gamma}{b}\right)^b - 1 \tag{20}$$

where the normalized parameters b and γ are defined by

$$b = B_T/W \qquad \gamma = S_R/\eta W$$

Equation (20) becomes an equality for an ideal system operating at full capacity, so we have a standard for the comparison of all other analog communication systems. In particular, we'll compare the analog performance curves given by $(S/N)_D$ versus γ with fixed bandwidth ratio b, and the power-bandwidth exchange given by γ versus b with fixed $(S/N)_D$.

Figure 15.3-6 repeats some of the analog performance curves from Fig. 12.2-2, together with the curve for an ideal system having $b = 6$. The heavy dots mark the threshold points, and we see that wideband noise-reduction systems (FM, PCM, and PPM) fall short of ideal performance primarily as a result of threshold limitations. Threshold extension techniques may improve performance by a few dB, but threshold effect must *always* occur in wideband analog modulation; otherwise, the curves would cross into the forbidden shaded region.

Figure 15.3-7 depicts the power-bandwidth exchange of analog modulation systems, taking $(S/N)_D = 50$ dB and $S_x = \overline{x^2} = 1/2$. SSB and DSB appear here as single points since b is fixed, and SSB is equivalent to an ideal system with $b = 1$—meaning no wideband improvement. The FM curve is calculated from Eq. (14), Sect. 9.4, and the PPM curve from Eq. (10), Sect. 10.4. The PCM curve

Figure 15.3-5 Analog modulation system.

Figure 15.3-6 Performance comparison of analog modulation systems.

corresponds to operation just above threshold, as in Eq. (8), Sect. 12.2. Observe that PCM comes closest to the ideal when we want $b > 1$ for wideband improvement.

Based on these two comparison figures, we conclude that *digital* transmission via PCM makes better use of the capacity of a continuous channel than conventional analog modulation. This difference between digital and analog transmission of analog information stems from the fundamental fact that transmission noise prevents *exact* reproduction of an analog waveform. The PCM quantization process takes advantage of this fact and provides a better match of signal to channel than does analog modulation.

Figure 15.3-7 Power-bandwidth exchange of analog modulation systems with $(S/N)_D = 50$ dB.

Shannon went further in this direction by proposing that the information rate of a continuous source be measured in terms of some specified *fidelity* or *distortion* criterion. Appropriate digital source coding then allows efficient information transmission over a channel whose capacity equals the rate in question. Shannon's proposal has subsequently been developed into the field known as *rate distortion theory*. An introduction to this theory in the context of voice and image transmission is given by Jayant and Noll (1984, App. D).

15.4 PROBLEMS

15.1-1 Suppose that equal numbers of the letter grades A, B, C, D, and F are given in a certain course. How much information in bits have you received when the instructor tells you that your grade is not F? How much more information do you need to determine your grade?

15.1-2 A card is drawn at random from an ordinary deck of 52 playing cards.

(a) Find the information in bits that you receive when you're told that the card is: a heart; a face card; a heart face card.

(b) If you're told that the card is a red face card, then how much more information do you need to identify the specific card?

15.1-3 Calculate the amount of information needed to open a lock whose combination consists of three integers, each ranging from 00 to 99.

15.1-4 Calculate $H(X)$ for a discrete memoryless source having six symbols with probabilities

$$P_A = 1/2 \qquad P_B = 1/4 \qquad P_C = 1/8 \qquad P_D = P_E = 1/20 \qquad P_F = 1/40$$

The find the amount of information contained in the messages *ABABBA* and *FDDFDF* and compare with the expected amount of information in a six-symbol message.

15.1-5 Do Prob. 15.1-4 with

$$P_A = 0.4 \qquad P_B = 0.2 \qquad P_C = 0.12 \qquad P_D = P_E = 0.1 \qquad P_F = 0.08$$

15.1-6 A certain source has eight symbols and emits data in blocks of three symbols at the rate of 1000 blocks per second. The first symbol in each block is always the same, for synchronization purposes; the remaining two places are filled by any of the eight symbols with equal probability. Find the source information rate.

15.1-7 A certain data source has 16 equiprobable symbols, each 1 ms long. The symbols are produced in blocks of 15, separated by 5-ms spaces. Find the source information rate.

15.1-8 Calculate the information rate of a telegraph source having two symbols, dot and dash. The dot duration is 0.2 sec. The dash is twice as long as the dot and half as probable.

15.1-9 Consider a source with $M = 3$. Find and sketch $H(X)$ as a function of p when $P_1 = 1/3$ and $P_2 = p$.

15.1-10 Consider a source with $M > 2$. One symbol has probability $\alpha \ll 1/M$, and the remaining symbols are equally likely. Show that $H(X) \approx \log (M - 1) + \alpha \log (1/\alpha)$.

15.1-11 Obtain the bounds on K in Eq. (13) when N_i satisfies Eq. (15a).

15.1-12 Obtain the Shannon-Fano code for the source in Prob. 15.1-4, and calculate the efficiency.

15.1-13 Show that there are two possible Shannon-Fano codes for the source whose symbol probabilities are given in Prob. 15.1-5. Then calculate the efficiency of the better code.

15.1-14 Consider a source with $M = 3$ and symbol probabilities 0.5, 0.4, and 0.1.

(a) Obtain the Shannon-Fano code and calculate its efficiency.

(b) Repeat part *a* for the second-extension code, grouping the symbols in blocks of two.

15.1-15 A binary source has symbol probabilities 0.8 and 0.2.

(a) Group the symbols in blocks of two, obtain the corresponding second-extension Shannon-Fano code, and calculate its efficiency.

(b) Repeat part a with blocks of three for the third-extension code.

15.1-16 A binary source has $P_0 = P_1 = 1/2$ and first-order memory such that $P(0|1) = P(1|0) = 3/4$. Calculate the resulting conditional entropy.

15.2-1 The *joint entropy* of a discrete system is defined as

$$H(X, Y) = \sum_{x, y} P(x_i y_j) \log \frac{1}{P(x_i y_j)}$$

Show that $H(X, Y) = H(Y) + H(X|Y)$.

15.2-2 Expand Eq. (2) to obtain $I(X; Y) = H(X) + H(Y) - H(X, Y)$, where $H(X, Y)$ is the joint entropy defined in Prob. 15.2-1.

15.2-3 Consider an ideal noiseless system with μ source symbols, μ destination symbols, and forward transition probabilities

$$P(y_j|x_i) = \begin{cases} 1 & j = i \\ 0 & j \neq i \end{cases}$$

Show that $H(Y) = H(X)$, $H(Y|X) = 0$, and therefore $I(X; Y) = H(X)$. *Hint:* Write the sums with indices $0 \leq i \leq \mu$ and $0 \leq j \leq \mu$.

15.2-4 Figure P15.2-4 represents a *nonsymmetric* binary channel. Follow the method used in Example 15.2-1 to obtain

$$I(X; Y) = \Omega[\beta + (1 - \alpha - \beta)p] - p\Omega(\alpha) - (1 - p)\Omega(\beta)$$

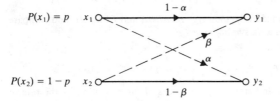

15.2-5 The channel in Prob. 15.2-4 is said to be *useless* if $\beta = 1 - \alpha$. Justify this name from intuitive and analytical arguments.

15.2-6 When a and b are constants, show that

$$\frac{d}{dp} \Omega(a + bp) = b \log \frac{1 - a - bp}{a + bp}$$

Then apply this relation to confirm that $I(X; Y)$ in Eq. (8) is maximum when $p = 1/2$.

15.2-7 The channel in Prob. 15.2-4 is called a Z channel when $\beta = 0$. Use the relation in Prob. 15.2-6 to show that if $\alpha = 1/2$, then $I(X; Y)$ is maximum when $p = 2/5$. Then evaluate C_s.

15.2-8 Let the channel in Prob. 15.2-4 have $\alpha = 1/4$ and $\beta = 0$. Use the relation in Prob. 15.2-6 to find C_s.

15.2-9 The *binary erasure channel* has two source symbols, 0 and 1, and three destination symbols, 0, 1, and E, where E denotes a detected but uncorrectable error. The forward transition probabilities are

$$P(0|0) = 1 - \alpha \qquad P(E|0) = \alpha \qquad P(1|0) = 0$$

$$P(0|1) = 0 \qquad P(E|1) = \alpha \qquad P(1|1) = 1 - \alpha$$

It follows from symmetry that $I(X; Y)$ is maximum when the source symbols are equiprobable. Find C_s in terms of α.

15.2-10 Derive the bound on $P(n \geq d)$ given in Eq. (14b).

15.3-1 Find $H(X)$ in terms of $S = \overline{x^2}$ and confirm that $H(X) < \frac{1}{2} \log 2\pi eS$ when $p(x)$ is a uniform PDF over $|x| \leq a$.

15.3-2 Find $H(X)$ in terms of $S = \overline{x^2}$ and confirm that $H(X) < \frac{1}{2} \log 2\pi eS$ when $p(x)$ is the Laplace PDF in Eq. (13), Sect. 12.1.

15.3-3 Find $H(X)$ in terms of $S = \overline{x^2}$ and confirm that $H(X) < \frac{1}{2} \log 2\pi eS$ when $p(x) = ae^{-ax}u(x)$.

15.3-4 Express $H(Z)$ in terms of $H(X)$ when $Z = aX + b$.

15.3-5 Find $p(x)$ that maximizes $H(X)$ and determine the resulting $H(X)$, given the signal constraints $x(t) \geq 0$ and $\overline{x(t)} = m$.

15.3-6 Find $p(x)$ that maximizes $H(X)$ and determine the resulting $H(X)$, given the signal constraints $x(t) \geq 0$ and $\overline{x^2(t)} = S$.

15.3-7 Write $p_X(x\,|\,y)/p_X(x) = [p_X(x)p_Y(y)/p_{XY}(x, y)]^{-1}$ and use $\ln v \leq v - 1$ to show that $I(X; Y) \geq 0$.

15.3-8 Consider an AWGN channel with $S/\eta = 10^4$. Find the maximum rate for reliable information transmission when $B = 1$ kHz, 10 kHz, and 100 kHz.

15.3-9 Consider an AWGN channel with $B = 3$ kHz. Find the minimum value of S/N in dB for reliable information transmission at $R = 2400, 4800$, and 9600 bits per second.

15.3-10 Consider an AWGN channel with $B = 1$ kHz and $\eta = 1\ \mu$W/Hz. Find the minimum value of S in milliwatts for reliable information transmission at $R = 100, 1000$, and 10,000 bits per second.

15.3-11 An engineer claims to have designed an analog communication system that has $(S/N)_D = 60$ dB when $S_R/\eta B_T = 5$ dB and $B_T = 10$ W. Do you believe this claim?

15.3-12 An ideal system has $(S/N)_D = 40$ dB when $B_T = 4$ W. Find $(S/N)_D$ when B_T is tripled while all other parameters are held fixed.

15.3-13 Do Prob. 15.3-12 with B_T reduced to $W/2$ instead of being tripled.

15.3-14 A communication system has $B_T = 12$ kHz, $L = 60$ dB, and $\eta = 100$ pW/Hz. Find the minimum value of S_T needed to get $(S/N)_D = 30$ dB when $W = 3$ kHz.

15.3-15 Do Prob. 15.3-14 with $W = 6$ kHz.

CIRCUIT AND SYSTEM NOISE

Transmission loss usually results in a very feeble signal at the input of a communication receiver. Consequently, to obtain an adequate signal level for further processing, the "front end" of a typical receiver includes several stages of amplification. High-gain amplifiers amplify any noise that accompanies the received signal, and they also add their own internally generated noise. An accurate assessment of system performance must therefore take account of the amplifier noise.

This appendix starts with circuit and device noise, as preparation for the description of amplifiers and other noisy two-port networks. Then we analyze noisy two-ports connected in cascade, a configuration that pertains to repeater systems as well as to receivers. Our analysis brings out the critical factors in receiver noise and the design principles for low-noise systems.

Circuit and Device Noise

Noise generated within an electrical circuit may come from numerous sources involving several different physical phenomena. Here we'll describe major types of device noise found in communication systems, and we'll develop appropriate circuit models and analysis methods.

The assumption of stationarity is reasonable for most electrical noise processes, thereby allowing us to represent noise sources in the frequency domain. For convenience, we'll adopt the common practice of working entirely with *positive frequency* and *one-sided frequency functions* — as distinguished from the two-sided functions used in Sect. 5.3. To clarify this distinction, let $G(f)$ be the two-sided available power spectrum of some noise source. Since $G(f)$ has even symmetry, the corresponding one-sided *available power density* will be defined by

$$\eta(f) \triangleq 2G(f) \qquad f \geq 0 \tag{1a}$$

and the *total available noise power* is

$$N = \int_0^\infty \eta(f) \, df \tag{1b}$$

Likewise, the one-sided mean square voltage and current densities equal twice the two-sided frequency functions for $f \geq 0$.

Figure A-1a shows the frequency-domain Thévenin circuit model of a stationary but otherwise arbitrary noise source with noiseless internal impedance $Z(f)$. The function $v_n^2(f)$ represents the *open-circuit mean square voltage density*, defined such that

$$\eta(f) = \frac{v_n^2(f)}{4 \operatorname{Re}\,[Z(f)]} \tag{2}$$

Alternatively, an ideal voltmeter with bandwidth df centered at some frequency f would measure the rms value $\sqrt{v_n^2(f) \, df}$ across the open terminals of this source. Converting Fig. A-1a to a Norton equivalent circuit gives the source model in Fig. A-1b, with

$$i_n^2(f) = \frac{v_n^2(f)}{|Z(f)|^2} \tag{3}$$

which represents the *short-circuit mean square current density*. Equation (3) also expresses Ohm's law in the form needed for circuit noise analysis.

Suppose the source in question happens to be a *thermal resistance R* at temperature $\mathscr{T}$, so $G(f) = \frac{1}{2}k\mathscr{T}$ from Eq. (5), Sect. 5.3. Then

$$\eta(f) = k\mathscr{T} \tag{4a}$$

and setting $Z(f) = R$ yields

$$v_n^2(f) = 4Rk\mathscr{T} \qquad i_n^2(f) = 4k\mathscr{T}/R \tag{4b}$$

These constant densities correspond to *white* noise, at least up to infrared frequencies. For computational purposes, the quantity $k\mathscr{T}$ can be rewritten as

$$k\mathscr{T} = k\mathscr{T}_0(\mathscr{T}/\mathscr{T}_0) \approx 4 \times 10^{-21}(\mathscr{T}/\mathscr{T}_0)$$

where we've inserted the standard room temperature $\mathscr{T}_0 = 290$ K.

(a)

(b)

Figure A-1 Frequency-domain models of a noise source. (a) Thévenin circuit; (b) Norton circuit.

Resistance noise occurs in almost all circuits, but reactive elements may alter the frequency density. In particular, let Fig. A-2a be a one-port (two-terminal) network containing only resistance, capacitance, and inductance, and having the equivalent impedance $Z(f) = R(f) + jX(f)$. When the resistances are in thermal equilibrium at temperature $\mathcal{T}$, *Nyquist's formula* states that

$$v_n^2(f) = 4R(f)k\mathcal{T} \tag{5}$$

Hence, the mean square voltage density takes the shape of the equivalent resistance $R(f)$. Equation (5) includes the special case of an all-resistive network whose equivalent resistance will be independent of f.

We prove Nyquist's formula by considering the complete circuit in Fig. A-2b, where the source has been connected to a load resistance R_L at the same temperature $\mathcal{T}$. This load resistance, of course, generates thermal noise with $v_L^2(f) = 4R_L k\mathcal{T}$. Since the two noise sources are physically independent, superposition applies and we can calculate the mean square current densities in each direction, namely,

$$i_L^2(f) = \frac{v_n^2(f)}{|Z(f) + R_L|^2} \qquad i_Z^2(f) = \frac{4R_L k\mathcal{T}}{|Z(f) + R_L|^2}$$

The average power delivered from the source to the load in any frequency band df is $R_L i_L^2(f)\, df$, while the load delivers $R(f)i_Z^2(f)\, df$ back to the source. Now the net power transfer must be zero at every frequency for the circuit to be in thermal equilibrium. Therefore,

$$\frac{R_L v_n^2(f)\, df}{|Z(f) + R_L|^2} = \frac{R(f)4R_L k\, df}{|Z(f) + R_L|^2}$$

and Eq. (5) follows after cancellation.

Nyquist's formula does not hold when the resistances are at different temperatures or the network contains nonthermal sources. However, such cases are easily analyzed provided that the sources are independent. You simply use superposition and sum mean square values to find the resulting frequency density.

(a)

(b)

Figure A-2

The most common type of nonthermal noise in electrical circuits is *shot noise*. This phenomenon occurs whenever charged particles cross a potential barrier — as in semiconductor junctions or vacuum tubes. Small variations of kinetic energy among the individual particles cause random fluctuations of the total current, roughly analogous to the sound of a stream of buckshot falling on a drum. Schottky first studied this effect in a *vacuum-tube diode* operated under *temperature-limited* conditions. He found that shot noise could be represented by a current source with

$$i_n^2(f) = 2qI \tag{6}$$

in which q is the electronic charge (1.6×10^{-19} coulombs) and I is the DC current. Since Eq. (6) is independent of both frequency and temperature, shot noise is *nonthermal white noise*. Subsequent investigations have shown that $i_n^2(f)$ actually decreases at frequencies above $f \approx 1/\tau$, where τ denotes the average particle transit time.

Schottky's result also holds for the *semiconductor junction diode* in Fig. A-3a. The net DC current consists of two components, given by the diode equation

$$I = I_s e^{qV/k\mathscr{T}} - I_s \tag{7a}$$

where V is the junction voltage and I_s is the reverse saturation current. The two current components produce statistically independent shot noise, so the total mean square noise current density becomes

$$i_n^2(f) = 2qI_s e^{qV/k\mathscr{T}} + 2qI_s = 2q(I + 2I_s) \tag{7b}$$

Figure A-3b shows the complete noise source model, including the diode's *dynamic resistance*

$$r = \frac{1}{dI/dV} = \frac{k\mathscr{T}}{q(I + I_s)} \tag{7c}$$

Unlike ohmic resistance, dynamic resistance is *noiseless* since it does not correspond to any power dissipation.

Now consider the *junction field-effect transistor* (JFET) in Fig. A-4a, which has a reverse-biased junction between the gate terminal G and the semiconductor channel from the source terminal S to the drain terminal D. Figure A-4b shows a simplified noise equivalent circuit with current sources representing the gate shot noise and the channel thermal noise, given by

$$i_g^2(f) = 2qI_g \qquad i_d^2(f) = 4k\mathscr{T} \times \tfrac{2}{3}g_m$$

(a)   (b)

Figure A-3 (a) Semiconductor junction diode; (b) noise model.

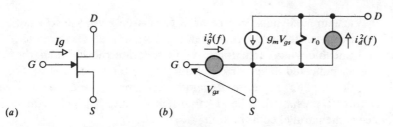

Figure A-4 (a) Junction field-effect transistor; (b) noise model.

where g_m is the transconductance. When external circuitry connects the gate and source terminals, the shot noise is amplified via the controlled current generator $g_m V_{gs}$ and adds to the output noise at the drain terminal.

An FET with an insulated gate structure avoids junction shot noise. On the other hand, a *bipolar junction transistor* (BJT) has two semiconductor junctions and two sources of shot noise in addition to thermal noise from internal ohmic resistance. Besides thermal and shot noise, all transistors generate certain types of *nonwhite* noise that may or may not be significant, depending upon the particular device and the application.

Transistors, vacuum tubes, resistors, and other devices exhibit a low-frequency phenomenon known as *flicker noise* — often called " one-over-f " noise because the mean square density is proportional to $1/f^v$ with $v \approx 1$. Some semiconductor devices also produce *burst* or "*popcorn*" *noise*, whose waveform resembles the random telegraph wave in Fig. 5.2-3a. Hence, the mean square density eventually falls off as $1/f^2$. Figure A-5 illustrates the frequency variation that results when $i_n^2(f)$ consists of flicker, burst, and white noise. Flicker and burst noise pose serious problems for low-frequency applications, but they usually can be ignored at frequencies above a few kilohertz. At much higher frequencies, capacitive coupling and various other effects tend to increase noise in electronic devices. See Robinson (1974) or Ambrózy (1982) for further discussion.

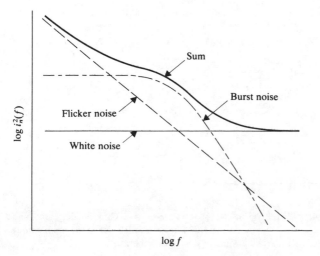

Figure A-5 Frequency variation of semiconductor noise with flicker and burst noise.

Example A-1 Vacuum diode noise generator Figure A-6a gives the schematic diagram of a noise generator employing a temperature-limited vacuum diode. The choke coil and blocking capacitor serve to separate the DC and shot-noise currents, as indicated. The corresponding noise circuit model in Fig. A-6b includes the diode's noiseless dynamic resistance r and the noisy thermal resistance R, presumed to be at room temperature $\mathcal{T}_0$. The equivalent impedance is the parallel combination

$$Z(f) = \frac{R(r + 1/j\omega C)}{R + (r + 1/j\omega C)}$$

but the nonthermal shot noise precludes the use of Nyquist's formula. Instead, we convert the shot-noise current and dynamic resistance into Thévenin form shown in Fig. A-6c. Now, applying superposition, the total short-circuit mean square current density is clearly seen to be

$$i_n^2(f) = \frac{r^2 2qI}{|r + 1/j\omega C|^2} + \frac{4k\mathcal{T}_0}{R}$$

and therefore

$$v_n^2(f) = |Z(f)|^2 i_n^2(f)$$
$$= \frac{R^2(r^2 2qI)}{|R + r + 1/j\omega C|^2} + \frac{R^2|r + 1/j\omega C|^2}{|R + r + 1/j\omega C|^2}\frac{4k\mathcal{T}_0}{R}$$

(a)

(b)

(c)

Figure A-6 Circuit diagram and noise models of a vacuum diode noise generator.

Figure A-7

Note that the first term is the open-circuit mean square voltage density produced by $v_1^2(f)$ when $i_2^2(f) = 0$, while the second term is produced by $i_2^2(f)$ when $v_1^2(f) = 0$.

In normal operation, we're not concerned with the noise at low frequencies (which would probably be dominated by flicker noise), and the dynamic resistance r is much greater than R. Under these conditions, our results simplify to

$$Z(f) \approx R \qquad v_n^2(f) \approx 2R^2qI + 4Rk\mathcal{T}_0$$

over the frequency range

$$1/2\pi rC \ll f < 1/\tau$$

Tubes designed for this application have very short transit times, typically $\tau \leq 1$ ns, so the shot noise remains constant up to about 1 GHz. Finally, from Eq. (2), the available noise power density is

$$\eta(f) \approx \tfrac{1}{2}qIR + k\mathcal{T}_0 = k(\mathcal{T}_x + \mathcal{T}_0)$$

where

$$\mathcal{T}_x = qIR/2k\mathcal{T}_0$$

which we interpret as the equivalent *noise temperature* of the shot noise. The voltage applied to the filament of the tube controls I and thus controls $\mathcal{T}_x$, whose value can be determined from an ammeter reading.

Exercise A-1 Consider the one-port network in Fig. A-7, where R_1 and R_2 are thermal resistances at different temperatures. Let $R_2 = R_1 = R$ and $L = R/2\pi$. Obtain expressions for $v_n^2(f)$ and $\eta(f)$. Then simplify your results when $\mathcal{T}_1 = \mathcal{T}_2 = \mathcal{T}$.

Amplifier Noise

A detailed circuit model showing all the individual noise sources within an amplifier would be very complicated and of little practical value. Consequently, alternative methods have been devised for the analysis of noise in amplifiers. Two particularly useful measures of amplifier noise are the *effective noise temperature* and the *noise figure*. Both of these measures involve the concept of *available power gain*, which we introduce first.

Let Fig. A-8 be the circuit model of a *noiseless* amplifier inserted between a source and a load. (For simplicity, we've omitted any reactances that might be

Figure A-8 Circuit model of a noiseless amplifier.

associated with the source, amplifier, or load impedances.) The amplifier is characterized by an input resistance r_i, output resistance r_o, and voltage transfer function $H(f)$. The source generates a mean-square voltage density $v_s^2(f)$ — representing noise or an information signal or both — and the available power density from the source is $\eta_s(f) = v_s^2(f)/4R_s$. The available power density at the output of the amplifier is

$$\eta_o(f) = \frac{v_o^2(f)}{4r_o} = \frac{|H(f)|^2 v_i^2(f)}{4r_o} = \frac{|H(f)|^2}{4r_o}\left(\frac{r_i}{R_s + r_i}\right)^2 v_s^2(f)$$

We define the amplifier's *available power gain* $g_a(f)$ as the ratio of these available power densities, i.e.,

$$g_a(f) \triangleq \frac{\eta_o(f)}{\eta_s(f)} = \frac{v_o^2(f)R_s}{v_s^2(f)r_o} = \left(\frac{|H(f)|r_i}{R_s + r_i}\right)^2 \frac{R_s}{r_o} \tag{8}$$

The actual power gain of an amplifier equals the available power gain when impedances are *matched* to obtain maximum power transfer at input and output.

We'll assume hereinafter that the source generates white noise, thermal or nonthermal, with noise temperature $\mathcal{T}_s$. Then $\eta_s(f) = k\mathcal{T}_s$ and the available noise power density at the output of a noiseless amplifier will be

$$\eta_o(f) = g_a(f)\eta_s(f) = g_a(f)k\mathcal{T}_s$$

But a *noisy* amplifier contributes additional internally generated noise. Since the internal noise is independent of the source noise, we write

$$\eta_o(f) = g_a(f)k\mathcal{T}_s + \eta_{\text{int}}(f) \tag{9}$$

where $\eta_{\text{int}}(f)$ stands for the available power density of the internal noise seen at the output. Figure A-9a depicts Eq. (9) in the form of a block diagram. Integration then yields the total available output noise power

$$N_o = \int_0^\infty \eta_o(f)\,df = k\mathcal{T}_s \int_0^\infty g_a(f)\,df + \int_0^\infty \eta_{\text{int}}(f)\,df$$

an expression that calls for some simplifications.

Most amplifiers in a communication system have a frequency-selective response, with *maximum power gain* g and *noise equivalent bandwidth* B_N. These parameters are related to $g_a(f)$ by

$$gB_N = \int_0^\infty g_a(f)\,df \tag{10}$$

(a)

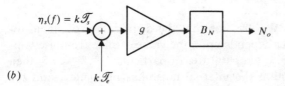

(b) $k\mathcal{T}_e$

Figure A-9 Block-diagram represent-ation of a noisy amplifier.

so the first term of N_o reduces to $k\mathcal{T}_s gB_N$. Next, to simplify the second term, we define the *effective noise temperature* of the amplifier to be

$$\mathcal{T}_e \triangleq \frac{1}{gkB_N} \int_0^\infty \eta_{\text{int}}(f) \, df \tag{11}$$

Hence, the total output noise power becomes

$$N_o = k\mathcal{T}_s gB_N + gkB_N \mathcal{T}_e = gk(\mathcal{T}_s + \mathcal{T}_e)B_N \tag{12}$$

diagrammatically portrayed by Fig. A-9b. This diagram brings out the fact that $\mathcal{T}_e$ represents the internal noise *referred to the input* and thereby expedites calculations of signal-to-noise ratios.

Now let Fig. A-10 represent a noisy amplifier with signal plus white noise at the input. The available signal power from the source is S_s and the signal spectrum falls within the passband of the amplifier so the available signal power at the output will be $S_o = gS_s$. Thus, using Eq. (12), the output signal-to-noise ratio is

$$\left(\frac{S}{N}\right)_o = \frac{gS_s}{N_o} = \frac{S_s}{k(\mathcal{T}_s + \mathcal{T}_e)B_N} \tag{13}$$

where the gain g has canceled out in numerator and denominator. Although the source noise does not necessarily have a defined bandwidth, the source signal-to-noise ratio is taken by convention to be

$$\left(\frac{S}{N}\right)_s \triangleq \frac{S_s}{k\mathcal{T}_s B_N}$$

Figure A-10 Noisy amplifier with input signal and noise.

which just corresponds to the signal-to-noise ratio produced by an ideal noiseless filter with unit gain and bandwidth B_N. However, from Eq. (13), the noisy amplifier produces

$$\left(\frac{S}{N}\right)_o = \frac{S_s}{(1 + \mathcal{T}_e/\mathcal{T}_s)k\mathcal{T}_s B_N} = \frac{1}{1 + \mathcal{T}_e/\mathcal{T}_s}\left(\frac{S}{N}\right)_s \tag{14}$$

and we see that $(S/N)_o < (S/N)_s$, We also see that the degradation of signal-to-noise ratio due to a noisy amplifier depends upon the value of effective noise temperature *relative* to the source noise temperature. In particular, if $\mathcal{T}_e \ll \mathcal{T}_s$ then $(S/N)_o \approx (S/N)_s$; under this condition the internal noise has little effect and the amplifier appears to be noiseless.

When impedances are not matched at the input or output all signal and noise powers will be less than the available powers. Nonetheless, Eqs. (13) and (14) are still valid because they express power *ratios* measured at specific points, so the impedance mismatch factor cancels out along with the gain. The effective noise temperature is therefore a significant parameter, irrespective of impedance matching.

Another measure of amplifier noise is the *noise figure†* F, defined such that

$$\left(\frac{S}{N}\right)_o = \frac{1}{F}\left(\frac{S}{N}\right)_s \qquad \text{when} \qquad \mathcal{T}_s = \mathcal{T}_0 \tag{15}$$

Since $S_o = gS_s$ and $(S/N)_s = S_s/k\mathcal{T}_0 B_N$ when $\mathcal{T}_s = \mathcal{T}_0$,

$$F = \frac{N_o}{gk\mathcal{T}_0 B_N} = 1 + \frac{\mathcal{T}_e}{\mathcal{T}_0} \tag{16}$$

and conversely

$$\mathcal{T}_e = (F - 1)\mathcal{T}_0 \tag{17}$$

A very noisy amplifier has $\mathcal{T}_e \gg \mathcal{T}_0$ and $F \gg 1$ — in which case we generally express the value of F in decibels. Then, if the source noise is at room temperature, Eq. (15) says that $(S/N)_o$ in dB equals $(S/N)_s$ in dB minus F in dB. A low-noise amplifier has $\mathcal{T}_e < \mathcal{T}_0$ and $1 < F < 2$ — in which case we usually work with the effective noise temperature and calculate $(S/N)_o$ from Eq. (13).

Table A-1 lists typical values of effective noise temperature, noise figure, and maximum power gain for various types of high-frequency amplifiers. Many low-noise amplifiers have cryogenic cooling systems to reduce the physical temperature and thus reduce internal thermal noise. Other amplifiers operate at room temperature, but the internal noise comes from nonthermal sources that may result in $\mathcal{T}_e > \mathcal{T}_0$.

---

† Also called the *noise factor* and often symbolized by n_F or NF. Some authors reserve the term noise figure for the value of the noise factor expressed in dB.

Table A-1 Noise parameters of typical amplifiers

Type	Frequency	$\mathcal{T}_e$, K	F, dB	g, dB
Maser	9 GHz	4	0.06	20–30
	6 GHz	11	0.16	
Parametric amplifier				
Room temperature	9 GHz	130	1.61	10–20
Cooled with liquid N_2	6 GHz	50	0.69	
Cooled with liquid He	4 GHz	9	0.13	
FET amplifier				
GaAs	9 GHz	330	3.3	6
	6 GHz	170	2.0	10
	1 GHz	110	1.4	12
Silicon	400 MHz	420	3.9	13
	100 MHz	226	2.5	26
Integrated-circuit amplifier	10.0 MHz	1160	7.0	50
	4.5 MHz	1860	8.7	75

Equation (16) defines the *average* or *integrated* noise figure in the sense that N_o involves the integral of $\eta_o(f)$ over all frequency. But sometimes we need to know how the internal noise varies with frequency. The *spot noise figure* $F(f)$ contains this information in the form

$$F(f) \triangleq \frac{\eta_o(f)}{k\mathcal{T}_s g_a(f)} \qquad \text{when} \qquad \mathcal{T}_s = \mathcal{T}_0 \qquad (18)$$

The value of $F(f)$ at a particular frequency serves as an estimate of the noise figure when a wideband amplifier is to be used for narrowband amplification with the help of additional filtering.

Finally, to relate system and circuit models of amplifier noise, we should mention the *two-generator model* diagrammed in Fig. A-11. This circuit model represents the internal noise in terms of fictitious voltage and current sources at the input terminals of an equivalent noiseless amplifier. The total mean square voltage density at the open-circuited output is

$$v_o^2(f) = \left(\frac{|H(f)|r_i}{R_s + r_i}\right)^2 [4R_s k\mathcal{T}_s + v_n^2(f) + R_s^2 i_n^2(f)]$$

Figure A-11 Two-generator circuit model of amplifier noise.

provided that $v_n^2(f)$ and $i_n^2(f)$ are uncorrelated. Using Eqs. (8) and (18) with $\eta_o(f) = v_o^2(f)/4r_o$, we find that

$$F(f) = 1 + \frac{v_n^2(f) + R_s^2 i_n^2(f)}{4R_s k \mathcal{T}_0} \tag{19}$$

which shows that the spot noise figure depends in part upon the external source resistance. Hence, minimizing $F(f)$ at a particular frequency often requires a transformer to obtain the optimum source resistance.

> **Example A-2 Amplifier noise measurement** Measuring *absolute* noise power is a difficult chore, so clever techniques have been developed for amplifier noise measurement with a *relative* power meter connected at the output. One technique utilizes a calibrated source of white noise, such as diode noise generator, impedance-matched to the input of the amplifier. The procedure goes as follows.
>
> First, set the source noise temperature at $\mathcal{T}_s = \mathcal{T}_0$ and record the meter reading N_1. This value corresponds to
>
> $$N_1 = CN_o = Cgk(\mathcal{T}_0 + \mathcal{T}_e)B_N$$
>
> where the proportionality constant C includes any impedance mismatch factor at the output. Second, increase the source temperature to $\mathcal{T}_s = \mathcal{T}_x + \mathcal{T}_0$ such that the meter reading has *doubled* and
>
> $$N_2 = Cgk[(\mathcal{T}_x + \mathcal{T}_0) + \mathcal{T}_e]B_N = 2N_1$$
>
> Then $N_2/N_1 = (\mathcal{T}_x + \mathcal{T}_0 + \mathcal{T}_e)/(\mathcal{T}_0 + \mathcal{T}_e) = 2$, so
>
> $$\mathcal{T}_e = \mathcal{T}_x - \mathcal{T}_0 \qquad F = \mathcal{T}_x/\mathcal{T}_0$$
>
> Note that we don't need to know g, B_N, or the constant C.

> **Exercise A-2** An amplifier with $g = 60$ dB and $B_N = 2$ MHz has $N_o = 40$ nW when the source noise is at room temperature. (*a*) Find the effective noise temperature and noise figure. (*b*) Calculate the source temperature needed for the second step of the measurement procedure in Example A-2.

System Noise Calculations

Here we take up the analysis of cascade-connected systems that include amplifiers and other noisy two-port networks. Our objective is to develop expressions for the *overall* performance of the system in terms of the parameters of the individual stages.

First, we must give consideration to *lossy* two-port networks such as transmission lines and connecting cables. Power loss implies dissipation by internal resistance. Consequently, the internal noise is *thermal noise* at the ambient temperature $\mathcal{T}_{amb}$, and $\eta_{int}(f) = k\mathcal{T}_{amb}$. However, we cannot use the model in Fig. A-9a because lossy two-ports are *bilateral*, meaning that a portion of the internal noise flows back to the input. When impedances are matched, a bilateral lossy two-port has constant gain $g < 1$ in both directions, so $g\eta_{\text{int}}(f)$ flows back to the input while $(1 - g)\eta_{\text{int}}(f)$ goes to the output. The total available noise power in bandwidth B_N at the output thus becomes

$$N_o = gk\mathscr{T}_s B_N + (1 - g)k\mathscr{T}_{\text{amb}} B_N = gk[\mathscr{T}_s + (L - 1)\mathscr{T}_{\text{amb}}]B_N$$

where

$$L = 1/g$$

which represents the transmission loss or attenuation. Comparing our expression for N_o with Eq. (12), we obtain the effective noise temperature

$$\mathscr{T}_e = (L - 1)\mathscr{T}_{\text{amb}} \tag{20a}$$

and Eq. (16) gives the noise figure

$$F = 1 + (L - 1)(\mathscr{T}_{\text{amb}}/\mathscr{T}_0) \tag{20b}$$

If a lossy two-port is at room temperature, then $\mathscr{T}_{\text{amb}} = \mathscr{T}_0$ and Eq. (20) reduces to $F = L$.

Next, consider the cascade of two noisy two-ports in Fig. A-12, where subscripts identify the maximum power gain, noise bandwidth, and effective noise temperature of each stage. We assume that both stages are linear and time invariant (LTI). We further assume that the passband of the second stage falls within the passband of the first stage, so $B_2 \le B_1$ and the overall noise bandwidth is

$$B_N \approx B_2$$

This condition reflects the sensible strategy of designing the last stage to mop up any remaining noise that falls outside the signal band. The overall power gain then equals the product

$$g = g_1 g_2$$

since the first stage amplifies everything passed by the second stage.

The total output noise power consists of three terms: source noise amplified by both stages; internal noise from the first stage, amplified by the second stage; and internal noise from the second stage. Thus,

$$N_o = gk\mathscr{T}_s B_N + g_2(g_1 k\mathscr{T}_1 B_N) + g_2 k\mathscr{T}_2 B_N$$

$$= gk\left(\mathscr{T}_s + \mathscr{T}_1 + \frac{\mathscr{T}_2}{g_1}\right)B_N$$

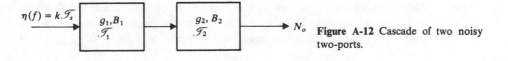

Figure A-12 Cascade of two noisy two-ports.

and the overall effective noise temperature is

$$\mathscr{T}_e = \mathscr{T}_1 + \frac{\mathscr{T}_2}{g_1}$$

The overall noise figure is

$$F = 1 + \frac{\mathscr{T}_1}{\mathscr{T}_0} + \frac{\mathscr{T}_2}{g_1 \mathscr{T}_0} = F_1 + \frac{F_2 - 1}{g_1}$$

which follows from the general relationship $F = 1 + \mathscr{T}_e/\mathscr{T}_0$.

The foregoing analysis readily generalizes to the case of three or more cascaded LTI two-ports. The overall effective noise temperature and noise figure are given by

$$\mathscr{T}_e = \mathscr{T}_1 + \frac{\mathscr{T}_2}{g_1} + \frac{\mathscr{T}_3}{g_1 g_2} + \cdots \tag{21}$$

$$F = F_1 + \frac{F_2 - 1}{g_1} + \frac{F_3 - 1}{g_1 g_2} + \cdots \tag{22}$$

Equation (21) is known as *Friis' formula*. Both expressions bring out the fact that the *first stage* plays a critical role and must be given careful attention in system design.

Suppose the first stage is a *preamplifier* with sufficiently large gain g_1 that Eq. (21) reduces to $\mathscr{T}_e \approx \mathscr{T}_1$. The system noise is then determined primarily by the preamplifier. The remaining stages provide additional amplification and filtering, amplifying the signal and noise without appreciably changing the signal-to-noise ratio. The design of low-noise receivers is usually based on this preamplification principle.

But suppose the first stage happens to be a connecting cable or any other *lossy* two-port. From the noise viewpoint, the attenuation effect is twice cursed since $g_1 = 1/L_1 < 1$ and $\mathscr{T}_1 = (L_1 - 1)\mathscr{T}_{amb}$. Equation (21) thus becomes

$$\mathscr{T}_e = (L_1 - 1)\mathscr{T}_{amb} + L_1\mathscr{T}_2 + L_1\mathscr{T}_3/g_2 + \cdots$$

which shows that L_1 multiplies the noise temperatures of all subsequent stages.

Now consider a complete communications receiver as drawn in Fig. A-13a. The receiver has been divided into two major parts, a predetection until followed by a detector. The detector processes the amplified signal plus noise and carries out a *nonlinear* operation, i.e., analog demodulation or digital regeneration. These operations are analyzed in Chaps. 9. 11, and 14 making the reasonable assumption that the detector introduces negligible noise compared to the amplified noise coming from the predetection unit. We're concerned here with the *predetection signal-to-noise ratio* denoted by $(S/N)_R$.

The predetection portion of a receiver is a cascade of noisy amplifiers and other functional blocks that act as LTI two-ports under the usual small-signal conditions. Hence, as indicated in Fig. A-13a, the entire predetection unit can be characterized by its overall effective noise temperature calculated from Eq. (21).

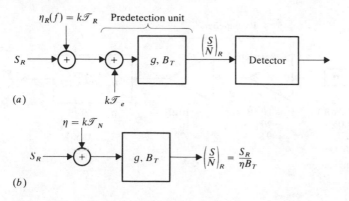

Figure A-13 (a) Communications receiver; (b) noise model of predetection unit.

(When the predetection unit includes a frequency converter, as in a superheterodyne receiver, its *conversion gain* takes the place of available power gain.) For a well-designed receiver, the predetection noise bandwidth essentially equals the transmission bandwidth B_T required for the signal. If the available signal power at the receiver input is S_R and the accompanying noise has temperature $\mathscr{T}_R$, then Eq. (13) becomes

$$(S/N)_R = S_R/\eta B_T \qquad (23a)$$

with

$$\eta \triangleq k(\mathscr{T}_R + \mathscr{T}_e) = k\mathscr{T}_N \qquad (23b)$$

The sum $\mathscr{T}_N = \mathscr{T}_R + \mathscr{T}_e$ is called the *system noise temperature*, and η represents the total noise power density referred to the input of an equivalent noiseless receiver — corresponding to the diagram in Fig. A-13b.

> **Example A-3 Satellite ground station** The signal received at a satellite ground station is extremely weak. Fortunately, the accompanying noise comes primarily from "cold" atmospheric phenomena and has a very low temperature. Minimizing the receiver noise is therefore both essential and justifiable. (In contrast, a receiving antenna pointed at or below the horizon picks up blackbody radiation from the "hot" earth; then $\mathscr{T}_R \approx \mathscr{T}_0$ and receiver noise will be relatively less important.)
>
> Figure A-14 depicts an illustrative low-noise microwave receiver for a satellite signal with frequency modulation. The waveguide is part of the antenna feed structure and introduces a small loss; the corresponding effective noise temperature is $\mathscr{T}_1 = (1.05 - 1)290 = 14.5$ K, from Eq. (20a). Two preamplifiers are employed to mitigate the noise of the high-gain FM receiver. Inserting numerical values into Eq. (21), we get the overall effective noise temperature
>
> $$\mathscr{T}_e = 14.5 + 1.05 \times 9 + \frac{1.05 \times 170}{100} + \frac{1.05 \times 1860}{100 \times 10}$$
>
> $$= 14.5 + 9.5 + 1.8 + 2.0 = 27.8 \text{ K}$$

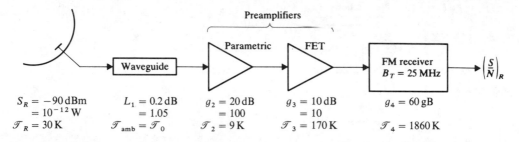

		Preamplifiers		

		Parametric	FET	FM receiver $B_T = 25$ MHz	$\left(\dfrac{S}{N}\right)_R$

| $S_R = -90$ dBm
$= 10^{-12}$ W
$\mathscr{T}_R = 30$ K | $L_1 = 0.2$ dB
$= 1.05$
$\mathscr{T}_{amb} = \mathscr{T}_0$ | $g_2 = 20$ dB
$= 100$
$\mathscr{T}_2 = 9$ K | $g_3 = 10$ dB
$= 10$
$\mathscr{T}_3 = 170$ K | $g_4 = 60$ gB
$\mathscr{T}_4 = 1860$ K |

Figure A-14 Satellite ground station.

Notice that the waveguide loss accounts for half of $\mathscr{T}_e$, while the noise from the FM receiver has been virtually washed out by the preamplification gain.

The system noise temperature is $\mathscr{T}_N = \mathscr{T}_R + \mathscr{T}_e = 57.8$ K $\approx 0.2\mathscr{T}_0$. Therefore, using Eq. (23),

$$\left(\frac{S}{N}\right)_R = \frac{10^{-12}}{4 \times 10^{-21} \times 0.2 \times 25 \times 10^6} = 50 = 17 \text{ dB}$$

This small a signal-to-noise ratio would be insufficient for analog communication, were it not for the further improvement afforded by FM demodulation.

Exercise A-3 Suppose the parametric amplifier in Fig. A-14 could be mounted directly on the antenna, ahead of the waveguide. Find $\mathscr{T}_N$ with and without the FET preamplifier.

Cable Repeater Systems

The adverse noise effect of lossy two-ports obviously cannot be avoided in cable transmission systems. However, we stated in Sect. 5.4 that inserting *repeater amplifiers* improves performance compared to cable transmission without repeaters. Now we have the tools needed to analyze such repeater systems.

A repeater system normally consists of m identical cable/repeater sections like Fig. A-15. The cable has loss L_c and ambient temperature $\mathscr{T}_{amb} \approx \mathscr{T}_0$, so $F_c = L_c$. The repeater amplifier has noise figure F_r and gain $g_r = L_c$ to compensate for the cable loss. We'll treat each section as a single unit with power gain

$$g_{cr} = (1/L_c)g_r = 1$$

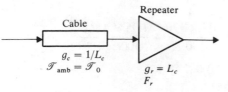

Cable
$g_c = 1/L_c$
$\mathscr{T}_{amb} = \mathscr{T}_0$

Repeater
$g_r = L_c$
F_r

Figure A-15 One section of a cable repeater system.

and noise figure

$$F_{cr} = F_c + \frac{F_r - 1}{(1/L_c)} = L_c + L_c(F_r - 1) = L_c F_r$$

calculated from Eq. (22). The overall noise figure for m cascaded sections then becomes

$$F = mF_{cr} - (m - 1) \tag{24a}$$

$$\approx mF_{cr} \tag{24b}$$

The approximation in Eq. (24b) assumes $F_{cr} \gg 1$, consistent with $L_c \gg 1$. This result explains the rule-of-thumb saying that "doubling the number of repeaters increases the noise figure by 3 dB."

Most repeater systems do have $F_{cr} \gg 1$, so the effective noise temperature is

$$\mathscr{T}_e = (F - 1)\mathscr{T}_0 \approx mF_{cr}\mathscr{T}_0$$

Furthermore, we can reasonably presume that the noise temperature at the transmitter will be small compared to $\mathscr{T}_e$. Under these conditions, the transmitted signal power S_T yields at the destination the predetection signal-to-noise ratio

$$\left(\frac{S}{N}\right)_R \approx \frac{S_T}{k\mathscr{T}_e B_T} \approx \frac{S_T}{m(kF_{cr}\mathscr{T}_0 B_T)} = \frac{1}{m}\left(\frac{S}{N}\right)_1 \tag{25}$$

where $(S/N)_1 = S_T/(kF_{cr}\mathscr{T}_0 B_T)$, which corresponds to the signal-to-noise ratio at the output of the first repeater.

Problems

A-1 Obtain expressions for $v_n^2(f)$ and $i_n^2(f)$ when resistance R_1 at temperature $\mathscr{T}_1$ is connected in series with R_2 at temperature $\mathscr{T}_2$. Check your results by taking $\mathscr{T}_1 = \mathscr{T}_2 = \mathscr{T}$.

A-2 Do Prob. A-1 for a parallel connection.

A-3 Find and sketch $v_n^2(f)$ when the circuit in Fig. A-7 has $\mathscr{T}_1 = \mathscr{T}_2 = \mathscr{T}$, $R_1 = 1$, $R_2 = 9$, and $L = 1/2\pi$.

A-4 Let the inductor in Fig. A-7 be replaced by capacitance C. Find and sketch $v_n^2(f)$ when $\mathscr{T}_1 = \mathscr{T}_2 = \mathscr{T}$ and $R_1 = R_2 = R$.

A-5 Let the voltage in Fig. A-3 be $V \gg k\mathscr{T}/q$. Write $i_n^2(f)$ in terms of r and explain why junction shot noise is sometimes called "half-thermal noise."

A-6 An amplifier with $g = 50$ dB and $B_N = 20$ kHz is found to have $N_o = 80$ pW when $\mathscr{T}_s = \mathscr{T}_0$. Find $\mathscr{T}_e$ and F, and calculate N_o when $\mathscr{T}_s = 2\mathscr{T}_0$.

A-7 When the noise temperature at the input to a certain amplifier changes from $\mathscr{T}_0$ to $2\mathscr{T}_0$, the output noise power increases by one-third. Find $\mathscr{T}_e$ and F.

A-8 A sinusoidal oscillator may be used instead of a noise source for the measurement process in Example A-2. The oscillator is connected but turned off for the first measurement, so its internal resistance provides the source noise. The oscillator is then turned on and its signal power S is adjusted to double the meter reading. Obtain an expression for F and discuss the disadvantages of this method.

A-9 Impedance matching between a 300-Ω antenna and a 50-Ω receiver is sometimes accomplished by putting a 300-Ω resistor in series with the antenna and a 50-Ω resistor across the receiver's input terminals. Find the noise figure of this resistive two-port network by calculating its power gain with a 300-Ω source resistance and a 50-Ω load resistance.

A-10 Two cascaded amplifiers have the following properties: $\mathcal{T}_1 = 3\mathcal{T}_0$, $g_1 = 10$ dB, $F_2 = 13.2$ dB, $g_2 = 50$ dB, $B_1 > B_2 = 100$ kHz. What input signal power is required to get $(S/N)_o = 30$ dB when $\mathcal{T}_s = 10\mathcal{T}_0$?

A-11 A system consists of a cable whose loss is 2 dB/km followed by an amplifier with $F = 7$ dB. If $\mathcal{T}_s = \mathcal{T}_0$, then what's the maximum possible path length for $(S/N)_o \geq 0.05(S/N)_s$?

A-12 A receiver system consists of a preamplifier with $F = 3$ dB and $g = 20$ dB, a cable with $L = 6$ dB and $\mathcal{T}_{amb} = \mathcal{T}_0$, and a receiver with $F = 13$ dB. (a) Calculate the system noise figure in dB. (b) Repeat part a with the preamplifier between the cable and the receiver.

A-13 A receiver system consists of a waveguide with $L_1 = 1.5$ dB, a preamplifier with $g_2 = 20$ dB and $\mathcal{T}_2 = 50$ K, and a receiver with $F_3 = 10$ dB. To what temperature must the waveguide be cooled so that the system has $\mathcal{T}_e \leq 150$ K?

A-14 The *Haus-Adler noise measure* for an amplifier is defined as $M = (F - 1)/(1 - 1/g)$. Show that in a cascade of two different amplifiers, the first amplifier should have the lower value of M. *Hint:* Write $F_{12} - F_{21}$ for the two possible configurations.

DISCRETE FOURIER TRANSFORMS
AND THE FFT

Fourier transformation, convolution, and other important methods of signal analysis involve integration. When the signal in question has a complicated mathematical model, or is known only in terms of experimentally observed values, it may be necessary to carry out integrations numerically using a digital computer. This appendix introduces numerical signal analysis based on the *discrete Fourier transform* (DFT).

We'll start with the definition and properties of the DFT as an operation performed on a discrete-time signal. Then we'll draw upon sampling theory to relate DFTs to the analysis of continuous-time signals. We conclude with a brief description of the *fast Fourier transform* (FFT), a family of efficient algorithms for computing DFTs.

Other applications of the DFT will be found in books devoted to digital signal processing, such as Ahmed and Natarajan (1983) or Rabiner and Gold (1975).

Discrete Fourier Transforms

A *discrete-time signal* $x(k)$ is an ordered sequence of numbers, possibly complex. Since numerical calculations are limited to a finite number of points, let $x(k)$ be given for $k = 0, 1, \ldots, N - 1$ — a total of N points. Thus, any transformation of this sequence yields another sequence consisting of no more than N independent points. In particular, the *discrete Fourier transform* of $x(k)$ is an N-point sequence defined by

$$X(n) = \text{DFT } [x(k)] \triangleq \sum_{k=0}^{N-1} x(k)W_N^{kn} \tag{1}$$

where $n = 0, 1, \ldots, N - 1$ and

$$W_N \triangleq e^{-j2\pi/N} \tag{2}$$

We'll subsequently show that $x(k)$ and $X(n)$ constitute a transform pair with many similarities to a continuous Fourier transform pair. We'll also show that if $x(k)$ represents sample values of a continuous-time signal, then $X(n)$ is related to the corresponding continuous Fourier transform. As an essential preliminary step, we must explore the properties implied by the definition of $X(n)$.

Consider first the weighting factor W_N in Eq. (2), which can be viewed as a complex-plane vector with $|W_N| = 1$ and arg $W_N = -2\pi/N$. Clearly, for any N,

$$W_N^N = e^{-j2\pi} = 1 \qquad W_N^{N/2} = e^{-j\pi} = -1 \tag{3a}$$

Furthermore, the powers of W_N have circular *periodicity* with period N, in the sense that

$$W_N^{(k \pm mN)n} = W_N^{kn} \qquad m = 0, 1, 2, \ldots \tag{3b}$$

for any integers k and n. Various powers of W_N are plotted in Fig. B-1 taking $N = 8$. Incidentally, N should be a power of 2 for computational purposes using the FFT. Hence, we assume hereafter that $N/2$ is an integer.

Now let λ be an arbitrary real number and consider the summation function

$$S_N(\lambda) \triangleq \sum_{n=0}^{N-1} W_N^{\lambda n} \tag{4a}$$

If $\lambda = \pm mN$ then $W_N^{\lambda n} = 1^n = 1$ and $S_N(\lambda) = N$. For other values of λ, we invoke the formula for the sum of a geometric series and insert Eq. (2) to obtain

$$S_N(\lambda) = \frac{W_N^{\lambda N} - 1}{W_N^{\lambda} - 1} = \frac{\sin \pi\lambda}{\sin (\pi\lambda/N)} e^{-j\pi\lambda(N-1)/N} \tag{4b}$$

Figure B-1 Powers of W_N when $N = 8$.

Figure B-2

Figure B-2 shows a sketch of $|S_N(\lambda)|$ versus λ. When λ is restricted to integer values, say $\lambda = i$, we see that

$$S_N(i) = \begin{cases} N & i = 0, \pm N, \pm 2N, \ldots \\ 0 & \text{otherwise} \end{cases} \tag{5}$$

These properties of W_N and $S_N(\lambda)$ will be drawn upon extensively hereafter.

Given a discrete transform $X(n)$, the discrete signal $x(k)$ is recovered via the *inverse discrete Fourier transform*

$$x(k) = \text{IDFT}[X(n)] \triangleq \frac{1}{N} \sum_{n=0}^{N-1} X(n) W_N^{-kn} \tag{6}$$

To prove that $\text{IDFT}[X(n)] = x(k)$, substitute Eq. (1) with a dummy index k' into Eq. (6); then

$$\text{IDFT}[X(n)] = \frac{1}{N} \sum_{n=0}^{N-1} \left[\sum_{k'=0}^{N-1} x(k') W_N^{k'n} \right] W_N^{-kn}$$

$$= \frac{1}{N} \sum_{k'=0}^{N-1} x(k') \left[\sum_{n=0}^{N-1} W_N^{(k'-k)n} \right]$$

$$= \frac{1}{N} \sum_{k'=0}^{N-1} x(k') S_N(k'-k)$$

Since k' and k are integers in the range 0 to $N - 1$, it follows from Eq. (5) that $S_N(k' - k) = 0$ except when $k' = k$ and $S_N(0) = N$. Therefore,

$$\text{IDFT}[X(n)] = x(k)$$

which is an *exact* result, with no questions of convergence or continuity.

Although we've said nothing about $x(k)$ for k outside the range of 0 to $N - 1$, the circular periodicity of the powers on W_N means that an IDFT is inherently periodic. Specifically, replacing k in Eq. (6) with $k \pm mN$ yields

$$x(k \pm mN) = x(k) \tag{7a}$$

for any integer m. Likewise, replacing n in Eq. (1) with $n \pm mN$ yields

$$X(n \pm mN) = X(n) \tag{7b}$$

Hence, both $x(k)$ and $X(n)$ are *periodic sequences* with period N. In view of this periodicity, the DFT and IDFT summations in Eqs. (1) and (6) may be carried out over *any* N successive terms—say from L to $L + N - 1$ with L being an arbitrary integer.

The discrete Fourier transform possesses several properties analogous to those of the continuous Fourier transform. Some properties, such as *duality* and *superposition*, differ only in notation from the continuous versions. Other properties are modified to reflect the periodicity. For example, when $x(k)$ is *real*, its DFT has the hermitian symmetry $X(-n) = X^*(n)$; moreover, periodicity requires that

$$X\left(\frac{N}{2} - n\right) = X^*\left(\frac{N}{2} + n\right) \tag{8}$$

so $X(n)$ has hermitian symmetry about $n = N/2$ as well as about $n = 0$.

Now suppose we *shift* a sequence $x(k)$ by m units to the right, obtaining $x(k - m)$. This operation is the discrete counterpart of *time delay*. However, the inherent periodicity of $x(k)$ produces a *wrap-around effect* in $x(k - m)$, as illustrated by Fig. B-3. The DFT of the shifted sequence is

$$\text{DFT}[x(k - m)] = \sum_{k=0}^{N-1} x(k - m)W_N^{kn} = \sum_{i=-m}^{-m+N-1} x(i)W_N^{(i+m)n}$$

$$= \left[\sum_{i=-m}^{-m+N-1} x(i)W_N^{in}\right] W_N^{mn}$$

But the sum in brackets equals $X(n)$, so

$$\text{DFT}[x(k - m)] = X(n)W_N^{mn} \tag{9}$$

Shifting a discrete signal therefore alters its transform by a *linear phase shift* $W_N^{mn} = e^{-j2\pi mn/N}$.

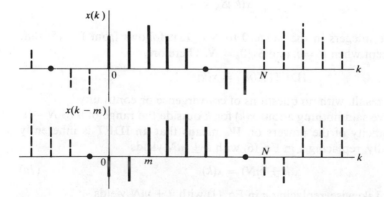

Figure B-3 A discrete-time signal $x(k)$ with periodic continuation, and the shifted signal $x(k - m)$ showing wrap-around effect.

Finally, we consider the operations of *convolution* and *multiplication*. Convolution of two N-point periodic sequences, say $h(k)$ and $x(k)$, yields another N-point periodic sequence defined by

$$y(k) \triangleq \sum_{m=0}^{N-1} h(k-m)x(m) \tag{10a}$$

Upon taking the DFT of $y(k)$ we find that

$$Y(n) = H(n)X(n) \tag{10b}$$

where $H(n) = \text{DFT}[h(k)]$, etc. Thus, discrete periodic convolution transforms to multiplication. Conversely, if

$$y(k) = h(k)x(k) \tag{11a}$$

then

$$Y(n) = \frac{1}{N} \sum_{m=0}^{N-1} H(n-m)X(m) \tag{11b}$$

which, except for the factor $1/N$, is the dual of Eq. (10a).

Example B-1 Let $x(k)$ be the 16-point sequence shown in Fig. B-4a where

$$x(k) = \begin{cases} A & k = 0, 1, 2, 14, 15 \\ 0 & 3 \le k \le 13 \end{cases}$$

Figure B-4 (a) Discrete-time signal $x(k)$; (b) $X(n) = \text{DFT}[x(k)]$.

The periodic continuation suggests that $x(k)$ could be sample values of a rectangular pulse train. The DFT is easily calculated by writing

$$X(n) = \sum_{k=-2}^{13} x(k)W_{16}^{kn} = \sum_{k=-2}^{2} Ae^{-j2\pi kn/16}$$

$$= A[1 + 2\cos(n\pi/8) + 2\cos(n\pi/4)]$$

which is plotted in Fig. B-4b. Since $x(k)$ is real and even, $X(n)$ is real and even. Thus, the periodic hermitian symmetry of $X(n)$ becomes even symmetry about $n = 0$ and $n = N/2 = 8$.

Exercise B-1 Use Eq. (9) to derive Eq. (10b).

Spectral Analysis with DFTs

When a continuous-time signal has been sampled at a uniform rate, the resulting sample values may be treated as a discrete-time signal and processed using the DFT. The computational speed of the FFT makes DFT signal processing an attractive method for numerical spectral analysis. Here we'll develop the relationship between the DFT and the Fourier transform of a continuous-time signal. We'll also look at potential problems in DFT spectral analysis and techniques for dealing with them.

Suppose a continuous-time signal $\tilde{x}(t)$ has been sampled every T_s seconds over a finite time interval $0 \le t < T_0$, as depicted by Fig. B-5. The observation interval and sample spacing are chosen such that $T_0 = NT_s$, or

$$N = T_0/T_s = f_s/f_0 \tag{12a}$$

where

$$f_s = 1/T_s \qquad f_0 = 1/T_0 \tag{12b}$$

The parameters f_s and f_0 have direct bearing upon DFT spectral analysis, which is necessarily restricted to a finite number of samples and a finite observation interval. At one extreme, $f_s/2$ represents the *highest observable frequency*, since sampling theory tells us that any higher frequency would exceed the Nyquist rate.

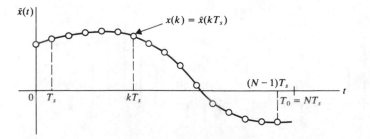

Figure B-5 A continuous-time signal sampled over a finite time interval.

At the other extreme, f_0 represents the *lowest observable frequency* (aside from a DC component), since any lower frequency would not go through a complete period in T_0 seconds.

Lacking data for $\tilde{x}(t)$ outside the interval $0 \leq t \leq T_0$, we'll presume that

$$\tilde{x}(t) = 0 \qquad t < 0, t \geq T_0 \tag{13a}$$

so the spectrum in question is

$$\tilde{X}(f) = \mathscr{F}[\tilde{x}(t)] = \int_0^{T_0} \tilde{x}(t)e^{-j2\pi ft}\, dt \tag{13b}$$

The tilde notation $\tilde{x}(t)$ and $\tilde{X}(f)$ here distinguishes continuous functions from the discrete functions introduced next. To obtain information about $\tilde{X}(f)$ from the sample values of $\tilde{x}(t)$, we let

$$x(k) = \tilde{x}(kT_s) \qquad k = 0, 1, \ldots, N-1 \tag{14a}$$

and we compute the DFT

$$X(n) = \text{DFT}[x(k)] \qquad n = 0, 1, \ldots, N-1 \tag{14b}$$

It will be shown that

$$X(n) = f_s \sum_{m=-\infty}^{\infty} \tilde{X}(nf_0 - mf_s) \tag{15}$$

which states the general relationship between a DFT and the continuous Fourier transform of a *finite-duration* signal. A similar expression relates the DFT to the Fourier series coefficients of a *periodic* signal; see Exercise B-2.

The derivation of Eq. (15) involves the ideally sampled waveform

$$\tilde{x}_\delta(t) \triangleq \sum_{k=-\infty}^{\infty} \tilde{x}(kT_s)\, \delta(t - kT_s) = \sum_{k=0}^{N-1} x(k)\, \delta(t - kT_s)$$

where we've used Eqs. (13a) and (14a) to write $\tilde{x}_\delta(t)$ as a finite sum in terms of $x(k)$. The Fourier transform of $\tilde{x}_\delta(t)$ is

$$\tilde{X}_\delta(f) = \int_{-\infty}^{\infty} \tilde{x}_\delta(t)e^{-j2\pi ft}\, dt = \sum_{k=0}^{N-1} x(k)e^{-j2\pi fkT_s}$$

Setting $f = nf_0 = n/NT_s$ and invoking Eqs. (1) and (2) yields

$$\tilde{X}_\delta(nf_0) = \sum_{k=0}^{N-1} x(k)e^{-j2\pi nk/N} = \sum_{k=0}^{N-1} x(k)W_N^{kn} = X(n)$$

But from Eq. (10), Sect. 10.1, with $f = nf_0$ we have

$$\tilde{X}_\delta(nf_0) = f_s \sum_{m=-\infty}^{\infty} \tilde{X}(nf_0 - mf_s)$$

Equation (15) then follows by equating our two expressions for $\tilde{X}_\delta(nf_0)$.

To interpret the relationship between $X(n)$ and $\tilde{X}(f)$ in Eq. (15), consider the continuous spectrum in Fig. B-6a with the property $\tilde{X}(f) \approx 0$ for $|f| \geq B$. We

Figure B-6 (a) Spectrum of $\tilde{x}(t)$; (b) relationship between $X(n) = \text{DFT}[\tilde{x}(kT_s)]$ and $\tilde{X}(f) = \mathscr{F}[\tilde{x}(t)]$.

can't assume that $\tilde{X}(f)$ is strictly bandlimited, because $x(t)$ has finite duration. Furthermore, we don't assume hermitian symmetry for $\tilde{X}(f)$, because a signal model could be a complex time function. If $f_s/2 \geq B$, then Eq. (15) leads to $X(n)$ as constructed in Fig. B-6b with the help of horizontal scales for both f and $n = f/f_0$. In the range $0 \leq n < N/2$, we see that

$$X(n) \approx f_s \tilde{X}(f)\Big|_{f=nf_0} = f_s \tilde{X}(nf_0)$$

which must be written as an approximation to account for the small but nonzero overlapping components from $\tilde{X}(f - f_s)$ and other frequency-translated terms not shown. Similarly, in the range $N/2 < n < N - 1$,

$$X(n) \approx f_s \tilde{X}(f - f_s)\Big|_{f=nf_0} = f_s \tilde{X}[(n - N)f_0]$$

Note that this range coincides with the *negative-frequency* portion of $\tilde{X}(f)$.

Our important conclusions from Fig. B-6 can be put in the following, more useful form: If $\tilde{X}(f)$ reasonably satisfies the condition

$$\tilde{X}(f) \approx 0 \qquad |f| \geq f_s/2 = Nf_0/2 \tag{16}$$

then DFT spectral analysis yields

$$\tilde{X}(nf_0) \approx T_s X(n) \qquad 0 \leq n \leq N/2 - 1 \tag{17a}$$

$$\tilde{X}(-nf_0) \approx T_s X(N - n) \qquad 1 \leq n \leq N/2 - 1 \tag{17b}$$

When $\tilde{x}(t)$ is *real*, Eq. (17b) may be replaced with

$$\tilde{X}(-nf_0) \approx T_s X^*(n) \qquad 1 \leq n \leq N/2 - 1$$

since both $\tilde{X}(f)$ and $X(n)$ will have hermitian symmetry.

Under the condition of Eq. (16), Eq. (17) states that the DFT approximates *sample values* of $\tilde{X}(f)$ at all harmonics of f_0 within the range $|nf_0| < f_s/2 = Nf_0/2$. Thus, f_0 equals the *frequency resolution* or minimum spacing between observable frequency components. Accurate DFT spectral analysis therefore requires a large value of N to cover a wide frequency range with a narrow frequency resolution.

If Eq. (16) does not hold, then the simple results in Eq. (17) become invalid due to significant overlapping components from frequency-translated terms. Such spectral overlap is, of course, the familiar *aliasing effect*. Whenever possible, the techniques discussed in Sect. 10.1 should be used to combat aliasing in DFT spectral analysis.

Now suppose the signal under consideration is a *pulse* of duration τ and more-or-less smooth shape, as sketched in Fig. B-7a where the time origin is arbitrary. We know that pulse spectra generally fall off for $|f| > 1/\tau$, so taking $f_s/2 \approx 10/\tau$ should be sufficient to avoid appreciable aliasing. (For a more irregular pulse shape, you must take f_s such that T_s is smaller than the smallest peak-to-valley spacing in the waveform.) Regardless of f_s, the choice of $T_0 = \tau$ would be unsatisfactory because the DFT then provides no spectral information in the important range $0 < |f| < f_0 = 1/\tau$.

A simple cure for inadequate frequency resolution of pulses is to construct the extended sequence $x(k)$ shown in Fig. B-7b, which consists of N_s sample

(a)

(b)

N_z augmenting zeros

N_s sample points

(c)

Augmenting zeros

Figure B-7 Augmenting zeros used with sample points of a pulse.

points followed by N_z *augmenting zeros*. The number of augmenting zeros is chosen to yield the desired resolution

$$f_0 = f_s/(N_s + N_z)$$

equivalent to $T_0 = (N_s + N_z)/f_s > \tau$. If desired, you could split the sample points and insert the augmenting zeros as shown in Fig. B-7c. This technique allows for a specified location of the time origin by exploiting the inherent periodicity of $x(k)$.

Augmenting zeros should not be used when the waveform $\tilde{x}(t)$ observed over $0 \le t < T_0$ represents just a portion of a longer signal having unknown or indefinite duration — a common situation in experimental work. The inherent periodicity of $x(k)$ then gives rise to a potential problem called *leakage effect*. This effect is clearly demonstrated by considering the single-frequency phasor signal

$$\tilde{x}(t) = Ae^{j2\pi f_c t} \qquad 0 \le t < T_0 \tag{18a}$$

Then $x(k) = \tilde{x}(kT_s) = A \exp(j2\pi f_c kT_s)$ and we find from Eqs. (1), (2), and (4) that

$$X(n) = AS_N(n - f_c T_0) \tag{18b}$$

The plot of $|X(n)|$ in Fig. B-8a reveals that the DFT appropriately consists of just one nonzero component when f_c happens to equal a harmonic of f_0. Otherwise, as seen in Fig. B-8b, the signal frequency appears to "leak out" into spurious lines at all harmonics of f_0. We've assumed here that $|f_c| < f_s/2$ so the leakage cannot be blamed on aliasing.

(a)

(b)

Figure B-8 (a) DFT of a phasor signal when f_c is a harmonic of f_0; (b) leakage effect when f_c is not a harmonic of f_0.

Figure B-9 (*a*) Periodic continuation of $\tilde{x}(t)$ with end-point discontinuities; (*b*) data window $\tilde{w}(t)$; (*c*) windowed product $\tilde{x}'(t) = \tilde{w}(t)\tilde{x}(t)$ without end-point discontinuities.

Leakage effect occurs whenever $\tilde{x}(T_0 - \epsilon) \neq \tilde{x}(0)$. Consequently, the periodic continuation of $\tilde{x}(t)$ illustrated in Fig. B-9a exhibits *discontinuities* at $t = 0$, $\pm T_0$, etc. We combat leakage by forming the *windowed sequence*

$$x'(k) = \tilde{w}(kT_s)\tilde{x}(kT_s) = w(k)x(k) \qquad (19a)$$

where $\tilde{w}(t)$ is a *data window* of the type sketched in Fig. B-9b. Figure B-9c shows that the periodic continuation of the product $\tilde{x}'(t) = \tilde{w}(t)\tilde{x}(t)$ is free of end-point discontinuities, which reduces spurious leakage in $X'(n) = \text{DFT}[x'(k)]$. However, windowing *distorts* the spectrum, since

$$X'(n) = \text{DFT}[w(k)x(k)] = \frac{1}{N}\sum_{m=0}^{N-1} W(n - m)X(m) \qquad (19b)$$

where we've drawn upon Eq. (11) with $h(k) = w(k)$ and $\text{DFT}[w(k)] = W(n)$. Considerable effort has therefore been devoted to the design of window functions that reduce leakage without introducing excessive spectral distortion.

Exercise B-2 Let $\tilde{x}(t)$ in Fig. B-5 be one period of a *periodic* signal whose exponential Fourier series coefficients are denoted by $c(nf_0)$. Relate $c(nf_0)$ to $\tilde{X}(f)$ and show from Eq. (15) that

$$X(n) = N\sum_{m=-\infty}^{\infty} c[(n - mN)f_0]$$

Then write expressions in terms of $c(nf_0)$ corresponding to Eqs. (16) and (17).

Convolution with DFTs

Here we consider numerical evaluation of the continuous-time convolution $\tilde{y}(t) = \tilde{h}(t) * \tilde{x}(t)$, in which $\tilde{h}(t)$ might represent either a signal or the impulse response of a filter. Although the convolution integral could be carried out numerically using standard methods, the speed of DFT computations via the FFT leads to another technique known as *fast convolution*. We'll ultimately demonstrate that fast convolution is analogous to the continuous-time relationship $\tilde{y}(t) = \mathcal{F}^{-1}[\tilde{H}(f)\tilde{X}(f)]$. Since correlation functions may be written as convolutions, a similar technique called *fast correlation* has become popular for numerical analysis of randon signals.

Like all numerical signal-processing methods, convolution with DFTs requires that $\tilde{h}(t)$ and $\tilde{x}(t)$ have *finite duration*. Specifically, let

$$\tilde{h}(t) = 0 \quad\quad t < 0, t > \tau_1$$
$$\tilde{x}(t) = 0 \quad\quad t < 0, t > \tau_2$$
(20)

Then $\tilde{y}(t) = 0$ for $t < 0$ and $t > \tau_1 + \tau_2$, and

$$\tilde{y}(t) = \int_0^t \tilde{h}(t - \lambda)\tilde{x}(\lambda)\, d\lambda \quad\quad 0 \le t \le \tau_1 + \tau_2$$

Taking $t = kT_s$, $\lambda = mT_s$, and $d\lambda = T_s$ gives the numerical approximation

$$\tilde{y}(kT_s) \approx T_s \sum_{m=0}^{k} \tilde{h}[(m - k)T_s]\tilde{x}(mT_s) \quad\quad 0 \le k \le N_1 + N_2$$

where

$$N_1 = \tau_1/T_s \quad\quad N_2 = \tau_2/T_s$$

which are assumed to be integers for simplicity of further discussion.

Our next step relates Eq. (20) to the discrete-time convolution

$$y(k) = \sum_{m=0}^{N-1} h(k - m)x(m) \quad\quad 0 \le k \le N - 1$$

For this purpose, we append augmenting zeros to the sample points of $\tilde{h}(t)$ and $\tilde{x}(t)$ such that

$$h(k) = \begin{cases} \tilde{h}(kT_s) & 0 \le k \le N_1 \\ 0 & N_1 < k \le N - 1 \end{cases}$$
(21a)

$$x(k) = \begin{cases} \tilde{x}(kT_s) & 0 \le k \le N_2 \\ 0 & N_2 < k \le N - 1 \end{cases}$$
(21b)

as portrayed in Fig. B-10a, including the periodic continuation. If N satisfies the condition

$$N > N_1 + N_2 = (\tau_1 + \tau_2)/T_s$$
(22)

Figure B-10 (a) Sequences extended with augmenting zeros; (b) result of discrete-time convolution when $N > N_1 + N_2$.

then the graphical interpretation of convolution gives $y(k)$ shown in Fig. B-10b, where

$$y(k) = \sum_{m=0}^{k} \tilde{h}[(m-k)T_s]\tilde{x}(mT_s) \qquad 0 \le k \le N_1 + N_2$$

and $y(k) = 0$ for $N_1 + N_2 < k \le N - 1$. Hence,

$$\tilde{y}(kT_s) \approx T_s y(k) \qquad 0 \le k \le N - 1 \tag{23}$$

which is the desired result. The condition on N in Eq. (22) is needed to prevent unwanted overlap in $y(k)$ due to the wrap-around effect.

Finally, recall from Eq. (10) that $y(k) = \text{IDFT}[Y(n)] = \text{IDFT}[H(n)X(n)]$. The procedure for convolution with DFTs therefore goes as follows:

1. Form the extended sequences $h(k)$ and $x(k)$ according to Eqs. (21) and (22).
2. Calculate $H(n) = \text{DFT}[h(k)]$ and $X(n) = \text{DFT}[x(k)]$.
3. Calculate the product $Y(n) = H(n)X(n)$.
4. Calculate $y(k) = \text{IDFT}[Y(n)]$ and insert in Eq. (23).

The computational advantage of this seemingly roundabout procedure will become apparent in our subsequent discussion of the FFT.

Exercise B-3 Redraw Fig. B-10 with $N < N_1 + N_2$ to show the wrap-round effect in $y(k)$.

Fast Fourier Transforms

DFT spectral analysis, convolution, and a host of other potential applications were brought into the realm of feasibility in 1965 when J. W. Cooley and J. W. Tukey invented the first *fast Fourier transform* algorithm. An FFT makes possible rapid computation of

$$X(n) = \sum_{k=0}^{N-1} x(k)W_N^{kn} \qquad 0 \le n \le N - 1$$

where $x(k)$ may be a complex sequence. Direct evaluation of this expression involves N complex multiplications and $N - 1$ complex additions, so the number of complex arithmetic operations required for an N-point DFT would be

$$N_{op} = N(2N - 1) \approx 2N^2 \qquad N \gg 1$$

But an FFT reduces this number to

$$N_{op} \approx N \log_2 N \qquad (24)$$

providing that N equals a power of 2.

As an example of the FFTs impact, suppose that a computer takes 5 μs to perform one complex arithmetic operation and you want a DFT with $N = 1024$ points. Direct evaluation then consumes about 10 seconds of CPU time, whereas an FFT completes the job in about 0.05 seconds! Equivalent computational reduction holds for the inverse DFT, since we can rewrite Eq. (6) as

$$x(k) = \frac{1}{N}\left[\sum_{k=0}^{N-1} X^*(n)W_N^{kn}\right]^*$$

Thus, an FFT program is easily modified to carry out both DFT and IDFT calculations.

The FFT family has grown over the years and now includes a great variety of algorithms — some of which do not need N to be a power of 2. Common to all these algorithms is an underlying strategy of "divide and conquer" exemplified by the following argument. Suppose that N equals a power of 2 and we divide the N-point DFT into two $N/2$-point DFTs. Direct evaluation then requires $N_{op} \approx 2 \times 2(N/2)^2 = N^2$, so we've cut the number of operations in half. Dividing the two $N/2$-point DFTs into four $N/4$-point DFTs likewise yields a further reduction. Continuing in this fashion, we eventually arrive at the trivial case of N one-point DFTs, and all the computational effort has gone into the subdivision process. To illustrate the divide-and-conquer strategy, we'll develop the popular FFT algorithm called *decimation in frequency*.

We begin by expanding and regrouping the summation for $X(n)$ in the form

$$X(n) = \sum_{k=0}^{N/2-1} x(k)W_N^{kn} + \sum_{k=N/2}^{N-1} x(k)W_N^{kn}$$

$$= \sum_{k=0}^{N/2-1} x(k)W_N^{kn} + W_N^{Nn/2} \sum_{k=0}^{N/2-1} x(k + N/2)W_N^{kn}$$

$$= \sum_{k=0}^{N/2-1} [x(k) + (-1)^n x(k + N/2)]W_N^{kn}$$

where we've noted from Eq. (3a) that $W_N^{Nn/2} = (-1)^n$. Giving separate attention to even and odd values of n yields

$$X(2n) = \sum_{k=0}^{N/2-1} [x(k) + x(k + N/2)]W_N^{2kn}$$

$$X(2n + 1) = \sum_{k=0}^{N/2-1} [x(k) - x(k + N/2)]W_N^k W_N^{2kn}$$

for $0 \le n \le N/2 - 1$. Now define two new sequences related to $x(k)$ by

$$\begin{aligned} x_0(k) &\triangleq x(k) + x(k + N/2) & 0 \le k \le N/2 - 1 \\ x_1(k) &\triangleq [x(k) - x(k + N/2)]W_N^k & 0 \le k \le N/2 - 1 \end{aligned} \quad (25)$$

Then, since $W_N^{2kn} = W_{N/2}^{kn}$, we have

$$X(2n) = \sum_{k=0}^{N/2-1} x_0(k)W_{N/2}^{kn} \qquad 0 \le n \le N/2 - 1$$

$$X(2n + 1) = \sum_{k=0}^{N/2-1} x_1(k)W_{N/2}^{kn} \qquad 0 \le n \le N/2 - 1 \quad (26)$$

which subdivides $X(n)$ into two $N/2$-point DFTs as desired.

The *butterfly diagram* in Fig. B-11 compactly depicts the operations defined by Eq. (25). The solid dots represent data storage locations, the open circle stands for addition and subtraction, and the arrowhead stands for multiplication by W_N^k —often called a *twiddle*. Figure B-12 shows how four butterflies convert an 8-point DFT into two 4-point DFTs in accordance with Eq. (26). Observe that the upper output block produces the even-numbered frequency points, $X(0)$, $X(2)$, ..., while the lower block produces the odd-numbered points. The output order becomes more scrambled or *decimated* as we go on, thereby explaining the algorithm's name.

Next, we expand and regroup the summations in Eq. (26) to obtain four $N/4$-point DFTs, namely,

$$X(4n) = \sum_{k=0}^{N/4-1} x_{00}(k)W_{N/4}^{kn} \qquad X(4n + 1) = \sum_{k=0}^{N/4-1} x_{01}(k)W_{N/4}^{kn}$$

$$X(4n + 2) = \sum_{k=0}^{N/4-1} x_{10}(k)W_{N/4}^{kn} \qquad X(4n + 3) = \sum_{k=0}^{N/4-1} x_{11}(k)W_{N/4}^{kn}$$

Figure B-11 Butterfly diagram for decimation-in-frequency FFT.

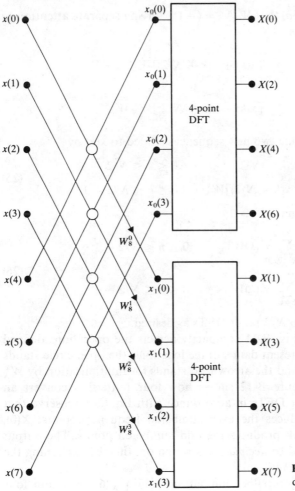

Figure B-12 Eight-point DFT subdivided into two 4-point DFTs.

for $0 \leq n \leq N/4 - 1$. The new data sequences here are defined by

$$x_{00}(k) \triangleq x_0(k) + x_0(k + N/4) \qquad x_{01}(k) \triangleq x_1(k) + x_1(k + N/4)$$

$$x_{10}(k) \triangleq [x_0(k) - x_0(k + N/4)]W_N^{2k} \qquad x_{11}(k) \triangleq [x_1(k) - x_1(k + N/4)]W_N^{2k}$$

for $0 \leq k \leq N/4 - 1$. Further interations of the subdivision process follow the same pattern.

Clearly, we arrive at the final one-point DFTs after a total of $\log_2 N$ subdivisions. Each subdivision requires $N/2$ butterflies, and each butterfly represents three operations (one addition, one subtraction, and one multiplication), so the algorithm appears to have

$$N_{\text{op}} = \tfrac{3}{2}N \log_2 N$$

However, further consideration reveals that all twiddles with $k = 0$ can be omitted because $W_N^0 = 1$. Taking $N = 8$, for instance, the complete algorithm diagrammed in Fig. B-13 has only five nontrivial multiplications and a total of 29 arithmetic operations.

A careful examination of Fig. B-13 brings out two other significant properties of the algorithm. First, we need just N data storage locations since each new sequence can be placed in the storage locations occupied by the previous sequence. Second, the decimated output order can be decoded by a simple process of *bit-reversal indexing*. To explain this process, let the storage locations be indexed by $k = 0, 1, ..., N - 1$, from top to bottom as shown on the left-hand side of Fig. B-13. If you write each value of k as a binary number consisting of $\log_2 N$ bits, then reversing the order of the bits gives a binary number equal to the index of the output contained in the kth storage location. With $N = 8$, for example, the binary version of $k = 3$ is 011 so bit reversal gives 110 corresponding to $n = 6$. The subscripts of the final x sequence in Fig. B-13 also match the binary number for n.

Another popular FFT algorithm called *decimation in time* produces unscrambled output but requires that $x(k)$ be stored initially in the scrambled order of bit-reversed indexing. This and other FFT algorithms are discussed by Rabiner and Gold (1975, chap. 6). A FORTRAN subroutine for DFT and IDFT computation is listed in Ahmed and Natarajan (1983, app. 4.1).

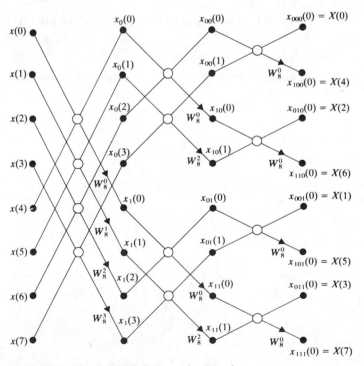

Figure B-13 Complete FFT diagram when $N = 8$.

Problems

B-1 Calculate $X(n)$ for the four-point sequences $x(k) = 2, 0, 1, 0$ and $x(k) = 2, 1, 0, 0$.

B-2 Consider the eight-point sequence $x(k) = 1, 1, 1, 0, 0, 0, 0, 0$. Use Fig. B-1 to evaluate $X(n)$ for $0 \leq n \leq 4$. Then apply Eq. (8) to find $X(n)$ for $n = 5, 6$, and 7.

B-3 Do Prob. B-2 with $x(k) = 3, 2, 1, 0, 0, 0, 0, 0$.

B-4 Prove the symmetry property in Eq. (8) when $x(k)$ is real.

B-5 Derive Eq. (11b) by taking the DFT of $y(k)$ in Eq. (11a).

B-6 Let $X(n) = \text{DFT}[x(k)]$. Obtain the DFT duality relation by expressing $\text{DFT}[X(k)]$ in terms of $x(k)$.

B-7 Modify the method used for the proof of Parseval's power theorem in Sect. 2.1 to show that

$$\sum_{k=0}^{N-1} |x(k)|^2 = \frac{1}{N} \sum_{n=0}^{N-1} |X(n)|^2$$

B-8 The discrete-time version of differentiation is the *difference sequence*. (*a*) Find the DFT of the backward difference $\nabla x(k) = x(k) - x(k-1)$. Simplify your results for $|n| \ll N$. (*b*) Do part *a* for the forward difference $\Delta x(k) = x(k+1) - x(k)$.

B-9 A voice signal has been filtered by an LPF with $B = 3$ kHz. Find N and f_s for DFT spectral analysis with a frequency resolution of 25 Hz. Take N to be a power of 2 to allow use of the FFT.

B-10 Do Prob. B-9 with $B = 5$ kHz.

B-11 DFT spectral analysis with augmenting zeros is to be applied to a pulse having duration $\tau = 2$ ms. Find N, f_s, N_s, and N_z for a frequency resolution of 50 Hz. Take N to be a power of 2 to allow use of the FFT.

B-12 Do Prob. B-11 with $\tau = 0.4$ ms.

B-13 The *Hann window function* is $\tilde{w}(t) = \frac{1}{2}[1 - \cos(\pi t/T_0)]$. Take the DFT of $x'(k) = w(k)x(k)$ to show that

$$X'(n) = \frac{1}{2}X(n) - \frac{1}{4}[X(n-1) + X(n+1)]$$

which represents a weighted moving average of three adjacent spectral components of $X(n)$. *Hint:* Express the cosine function in exponential form.

OPTIMUM DIGITAL DETECTION

This appendix is an introduction to the principles of optimum detection applied to digital communication. We'll start with the important technique of representing signals as *vectors*, a technique originated by Kotel'nikov (1959) in his theory of optimum noise immunity. Then we'll use vector concepts to design optimum receivers for digital signals with AWGN contamination. Vector concepts also facilitate the calculation of error probabilities and the selection of signals for efficient digital communication.

Our study of digital detection constitutes just one aspect of optimization theory in communication engineering. A more comprehensive treatment is given in Wozencraft and Jacobs (1965), which also includes vector analysis of coded systems, channel models, and analog communication.

Signals as Vectors

A *signal space* is a multidimensional space wherein vectors represent a specified set of signals, By describing signals as vectors, the familiar relations and insights of geometry can be brought to bear on numerous problems of signal analysis and communication system design. Here we introduce those aspects of signal-space theory relevant to digital detection, restricting our attention to the case of *real energy signals*.

Consider any two energy signals $v(t)$ and $w(t)$. If α and β are finite real constants, then the linear combination $z(t) = \alpha v(t) + \beta w(t)$ is also an energy signal.

Formally, we say that the set of all real energy signals is *closed under linear combination*, which is the essential condition for establishing a *linear space*. We thus define the signal space Γ to be a linear space wherein vectors such as v and w represent energy signals $v(t)$ and $w(t)$, and any linear combination $z = \alpha v + \beta w$ is another vector in Γ.

Vector addition and scalar multiplication in Γ obey the usual rules, and there exists a zero element such that $v + 0 = v$ and $v - v = 0$. However, to complete the geometric structure of Γ, we need appropriate measures of vector length and angle expressed in terms of signal properties. Since we're dealing with energy signals, we'll define the length or *norm* of a signal vector v to be the square root of the signal energy, written as

$$\|v\| \triangleq \sqrt{E_v} = \left[\int_{-\infty}^{\infty} v^2(t)\, dt \right]^{1/2} \tag{1}$$

This vector norm is a nonnegative quantity, and $\|v\| = 0$ only if $v(t) = 0$. We use the square root of signal energy in Eq. (1) so that the norm of the sum $z(t) = v(t) + w(t)$ satisfies the *triangle inequality*

$$\|v + w\| \le \|v\| + \|w\| \tag{2}$$

Figure C-1 shows the corresponding picture as a triangle composed of the vectors v, w, and $z = v + w$.

To measure the angle of v relative to some other vector w, we first recall from Sect. 3.5 that the crosscorrelation function measures the similarity between the shifted signal $v(t + \tau)$ and $w(t)$. Now, setting $\tau = 0$, we define the *scalar product*

$$(v,w) \triangleq R_{vw}(0) = \int_{-\infty}^{\infty} v(t)w(t)\, dt \tag{3}$$

which has the properties

$$(v,v) = \|v\|^2 \tag{4}$$

$$|(v,w)| \le \|v\|\|w\| \tag{5}$$

Equation (5) simply restates *Schwarz's inequality*, and the equality holds when $v(t)$ and $w(t)$ are proportional signals.

Justification of the scalar product for angular measurement comes from the expansion

$$\|v + w\|^2 = (v + w, v + w)$$

$$= (v,v) + (v,w) + (w,v) + (w,w)$$

$$= \|v\|^2 + 2(v,w) + \|w\|^2 \tag{6}$$

Figure C-1

If $v(t)$ and $w(t)$ are fully correlated in the sense that

$$v(t) = \pm \alpha w(t)$$

then $(v,w) = \pm \|v\| \|w\|$ and

$$\|v + w\| = \|v\| \pm \|w\|$$

Figure C-1 thus reduces to *colinear* vectors, with v parallel to w. On the other hand, if $v(t)$ and $w(t)$ are uncorrelated such that

$$(v,w) = 0$$

then

$$\|v + w\|^2 = \|v\|^2 + \|w\|^2$$

and Fig. C-1 becomes a right triangle, with *perpendicular* vectors v and w. Accordingly, we say that $v(t)$ and $w(t)$ are *orthogonal* signals when $(v,w) = 0$. Orthogonality in signal space implies *superposition of energy*, since $\|v\|^2 + \|w\|^2 = E_v + E_w$.

For the intermediate case of $0 < |(v,w)| < \|v\| \|w\|$ we can decompose v into the two orthogonal components shown in Fig. C-2. The *projection* of v on w is defined by the colinear and orthogonal conditions

$$v_w = \alpha w \qquad (v - v_w, w) = 0$$

and solving for α yields

$$v_w = \frac{(v,w)}{\|w\|^2} w \tag{7}$$

You can confirm from Schwarz's inequality that $\|v_w\| \le \|v\|$.

Next, as preparation for the study of optimum detection, suppose we have a set of M energy signals denoted by $\{s_i(t); i = 1, 2, \ldots, M\}$. Each of these signals can be represented by a vector written in the form

$$s_i = \alpha_{i1}\varphi_1 + \alpha_{i2}\varphi_2 + \cdots + \alpha_{iK}\varphi_K$$

$$= \sum_{k=1}^{K} \alpha_{ik} \varphi_k \qquad K \le M \tag{8a}$$

where the φ_k are vectors and the α_{ik} are scalar coefficients calculated from

$$\alpha_{ik} = (s_i, \varphi_k) \tag{8b}$$

Two important aspects of this representation should be stressed. First, Eq. (8a) is an exact expression, rather than a series approximation. Second, each signal is

Figure C-2 Projection of v on w.

completely specified by the scalars in Eq. (8b), so we can work with a set of K coefficients α_{ik} instead of an entire waveform $s_i(t)$. These points will become clearer from the signal-space interpretation.

The vectors φ_k in Eq. (8) represent a set of *orthonormal basis functions* $\varphi_k(t)$ with the properties

$$\|\varphi_k\|^2 = (\varphi_k,\varphi_k) = 1 \quad k = 1, 2, \ldots, K$$
$$(\varphi_k,\varphi_j) = 0 \quad j \neq k$$

(9)

These basis functions are mutually orthogonal and each one has unit norm — analogous to the *unit vectors* of ordinary two- and three-dimensional spaces. By extrapolation, we say that the φ_k span a *K-dimensional subspace* Γ_K of the signal space Γ. If Γ_K contains the signal vector s_i, its projection on a basis vector φ_k is

$$(s_i,\varphi_k)\varphi_k/\|\varphi_k\|^2 = (s_i,\varphi_k)\varphi_k = \alpha_{ik}\varphi_k$$

where we've used Eqs. (7), (9), and (8b). Therefore, the coefficients α_{ik} are the *coordinates* of the vector s_i with respect to the φ_k basis.

The subspace Γ_K and basis φ_k depend upon the specific set of signals we wish to represent. In fact, we usually generate the basis functions directly from the set of signals $\{s_i(t)\}$, thereby ensuring that Γ_K contains all of the signal vectors in question. Although a given set of signals defines a unique subspace, we have considerable flexibility in the choice of basis functions as long as they span Γ_K and satisfy Eq. (9).

A sequential method for constructing an orthonormal basis is the *Gram-Schmidt procedure*, as follows:

1. Select from the signal set any $s_1(t)$ having nonzero norm, and take $\varphi_1 = s_1/\|s_1\|$ so $\|\varphi_1\| = 1$.
2. Select another signal $s_2(t)$, compute the coordinate $\alpha_{21} = (s_2,\varphi_1)$, and define the auxiliary vector $g_2 = s_2 - \alpha_{21}\varphi_1$ in Fig. C-3a. Since g_2 is orthogonal to φ_1, take $\varphi_2 = g_2/\|g_2\|$.
3. Select another signal $s_3(t)$, compute the coordinates α_{31} and α_{32} to define $g_3 = s_3 - (\alpha_{31}\varphi_1 + \alpha_{32}\varphi_2)$ in Fig. C-3b. Then take $\varphi_3 = g_3/\|g_3\|$ so φ_3 has unit norm and is orthogonal to φ_1 and φ_2.
4. Continue in this manner until all M signals have been considered.

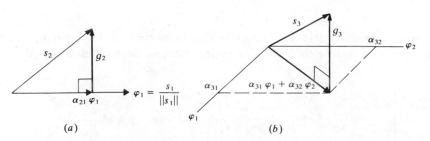

(a) (b)

Figure C-3 Vector constructions for the Gram-Schmidt procedure.

The basis resulting from this procedure can be expressed compactly by

$$\varphi_k = g_k / \| g_k \| \tag{10a}$$

with $g_1 = s_1$ and

$$g_k = s_k - \sum_{j=1}^{k-1} \alpha_{kj} \varphi_j \qquad 2 \le k \le K \tag{10b}$$

Equation (10b) yields $g_k = 0$ whenever s_k has zero norm or equals a linear combination of previous signal vectors; such cases are omitted from the basis, and the dimensionality will be $K < M$.

Having generated an orthonormal basis for a set of signals, we can compute the vector coordinates and express the norms and scalar products in terms of the coordinates. The scalar product of any two signals is found using Eqs. (8) and (9), with the result that

$$(s_i, s_m) = \left(\sum_k \alpha_{ik} \varphi_k, \sum_j \alpha_{mj} \varphi_j \right) = \sum_{k=1}^{K} \alpha_{ik} \alpha_{mk} \tag{11a}$$

Hence, letting $m = i$ we have

$$\| s_i \|^2 = (s_i, s_i) = \sum_{k=1}^{K} \alpha_{ik}^2 \tag{11b}$$

which is a K-dimensional statement of the pythagorean theorem.

Example C-1 Let's apply the Gram-Schmidt procedure to obtain an orthonormal basis for the three signals in Fig. C-4a. First, noting that $\| s_1 \|^2 = E_1 = 36$, we take $\varphi_1(t) = s_1(t)/6$ as plotted in Fig. C-4b. Then

$$\alpha_{21} = \int_{-\infty}^{\infty} s_2(t) \varphi_1(t) \, dt = 3$$

so $g_2(t) = s_2(t) - 3\varphi_1(t)$, and a sketch of $g_2(t)$ quickly yields $\| g_2 \|^2 = 9$ and $\varphi_2(t) = g_2(t) / \| g_2 \|$ in Fig. C-4c. A third basis function is not needed since we can write the linear combination $s_3(t) = s_1(t) - s_2(t)$.

By calculating the remaining coordinates, or by inspection of the waveforms, we obtain the vector representations

$$s_1 = 6\varphi_1 \qquad s_2 = 3\varphi_1 + 3\varphi_2 \qquad s_3 = 3\varphi_1 - 3\varphi_2$$

which are diagrammed in Fig. C-4d. This diagram beings out the fact that s_2 and s_3 are orthogonal and their sum equals s_1.

Exercise C-1 Let another set of signals be defined from Fig. C-4a by taking $s_1'(t) = \sqrt{2}\, s_2(t)$, $s_2'(t) = \sqrt{2}\, s_1(t)$ and $s_3'(t) = -\sqrt{2}\, s_3(t)$. (a) Apply the Gram-Schmidt procedure and draw the resulting vector diagram. (b) Calculate $\| s_2' \|^2$ using Eq. (11b) and compare with the result from Eq. (1).

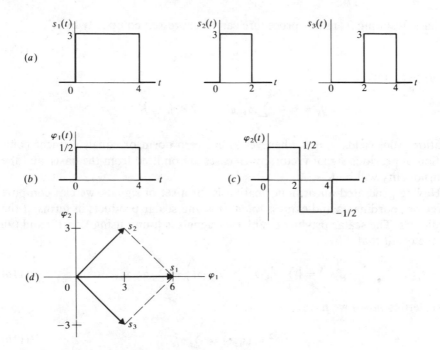

Figure C-4 (a) A signal set; (b) basis functions; (c) vector diagram.

Optimum Detection and MAP Receivers

Consider the M-ary digital communication system in Fig. C-5. The source selects a symbol m from the set $\{m_i; i = 1, 2, ..., M\}$. The transmitter generates a signal that appears at the destination as $s(t)$. This signal belongs to the set $\{s_i(t); i = 1, 2, ..., M\}$, in one-to-one correspondence with the symbol set. The received signal is corrupted by independent additive white gaussian noise $n_w(t)$ with zero mean and power spectral density $G(f) = \eta/2$. The receiver produces the estimated symbol $\hat{m}$ by comparing $y(t) = s(t) + n_w(t)$ with stored copies of the uncorrupted signals. We seek an optimum detection strategy that minimizes the average error probability $P_e = P(\hat{m} \neq m)$.

To formulate the optimization criterion more precisely, let $P(m|y)$ be the conditional probability that m was selected by the source given that $y(t)$ has been observed at the destination. Clearly, if

$$P(m_j| y) > P(m_i| y) \qquad \text{all} \qquad i \neq j \tag{12}$$

then the receiver should choose $\hat{m} = m_j$ to minimize $P(\hat{m} \neq m)$. The quantities in Eq. (12) are known as *a posteriori* probabilities, meaning "after observation," and Eq. (12) defines a *maximum a posteriori* (MAP) receiver. In the unlikely event of a tie, the receiver chooses any one of the m_i that maximizes $P(m_i| y)$. We'll assume equiprobable symbols, in which case the MAP criterion reduces to *maximum-likelihood* detection.

Figure C-5 M-ary digital communication system.

Our statement of the optimum detection problem contains only two critical assumptions: independent additive gaussian noise, and stored signal copies at the receiver. The assumption of equiprobable symbols is not critical since this condition can be ensured by *source coding*, as described in Sect. 15.1. Nor is the assumption of a white noise spectrum critical since a nonwhite noise spectrum can be equalized by a *prewhitening filter* at the front end of the receiver. The assumption that stored signal copies are available at the receiver essentially means that we're working with a *coherent* system, but we have no preconceptions about the structure of the receiver itself. Signal-space analysis eventually determines specific receiver implementations.

All of the M possible signal waveforms will be represented by vectors in a subspace Γ_K of dimensionality $K \leq M$ and spanned by some orthonormal basis φ_k. However, the received vector $y = s + n_w$ does not fall entirely within Γ_K. We therefore decompose n_w into two components

$$n_w = n + n'$$

as shown in Fig. C-6 where, for convenience, Γ_K is drawn as a plane. The component n', called the *irrelevant noise*, is orthogonal to Γ_K in the sense that

$$(n', \varphi_k) = 0 \qquad k = 1, 2, \ldots, K$$

The remaining noise vector n is contained in Γ_K, so we can write

$$n = \sum_{k=1}^{K} \beta_k \varphi_k \tag{13a}$$

where the noise coordinates β_k are given by

$$\beta_k = (n, \varphi_k) = (n_w - n', \varphi_k) = (n_w, \varphi_k) \tag{13b}$$

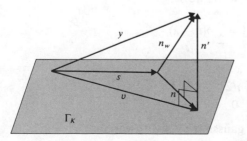

Figure C-6 Projection of $y = s + n_w$ on Γ_K.

Finally, we define the *relevant data vector*

$$v \triangleq y - n' = s + n \tag{14}$$

which is the *projection* of y on Γ_K since $y = v + n'$ and n' is orthogonal to every vector in Γ_K.

Figure C-7 illustrates the vectors in Γ_K for the case of $K = 2$ and $M = 3$. The three dots mark the tips of the uncorrupted signal vectors, and v is a typical relevant data vector when $s = s_2$. Intuitively, this diagram suggests that the receiver should choose the signal vector whose tip is closest to the tip of v. Subsequent analysis indeed confirms the *maximum-likelihood detection rule*:

Given an observed vector v, choose $\hat{m} = m_j$ with j such that $\|v - s_j\| < \|v - s_i\|$ for all $i \neq j$.

This rule defines M *decision regions* in Γ_K, one region for each signal point. In Fig. C-7, for instance, the three decision regions are bounded by the dashed lines formed by constructing perpendicular bisectors of lines connecting the signal points. These decision-region boundaries correspond to the decision thresholds introduced in Chap. 11.

The foregoing detection rule involves vector norms in Γ_K. We know the signal norms $\|s_i\| = \sqrt{E_i}$, but the noise n is a random vector whose properties require further study. From Eqs. (11b) and (13a) we have

$$\|n\|^2 = \sum_{k=1}^{K} \beta_k^2$$

where $\beta_k = (n_w, \varphi_k)$, which is a linear operation on a gaussian r.v. Hence, the β_k are also gaussian with mean value $\overline{\beta_k} = (\overline{n_w}, \varphi_k) = 0$. We turn then to the variance and the question of statistical dependence by considering the expectation

$$E[\beta_k \beta_i] = E\left[\int_{-\infty}^{\infty} n_w(t) \varphi_k(t) \, dt \int_{-\infty}^{\infty} n_w(\lambda) \varphi_i(\lambda) \, d\lambda \right]$$

$$= \iint_{-\infty}^{\infty} E[n_w(t) n_w(\lambda)] \varphi_k(t) \varphi_i(\lambda) \, dt \, d\lambda$$

But, from the autocorrelation function of white noise,

$$E[n_w(t) n_w(\lambda)] = \frac{\eta}{2} \delta(t - \lambda)$$

so we get

$$E[\beta_k \beta_i] = \begin{cases} \eta/2 & i = k \\ 0 & i \neq k \end{cases}$$

which means that the β_k are *independent* gaussian r.v.'s with variance $\sigma^2 = \eta/2$.

Figure C-7 Signal vectors and decisions regions in Γ_K with $K = 2$ and $M = 3$.

Therefore, the PDF of any noise coordinate $\beta_k = \beta$ is the zero-mean gaussian function

$$p_\beta(\beta) = \frac{1}{\sqrt{2\pi\sigma^2}} e^{-\beta^2/2\sigma^2} \qquad \sigma^2 = \frac{\eta}{2} \tag{15a}$$

The joint PDF of all K noise coordinates is then $p_n(\beta_1, \beta_2, \ldots, \beta_K) = p_\beta(\beta_1)p_\beta(\beta_2)$ $\cdots p_\beta(\beta_K)$ since the coordinates are mutually independent. Accordingly, we express the vector PDF as

$$p_n(n) = p_n(\beta_1, \beta_2, \ldots, \beta_K) = (\pi\eta)^{-K/2}e^{-\|n\|^2/\eta} \tag{15b}$$

which has *spherical symmetry* in Γ_K.

Now we're prepared to analyze optimum detection from the criterion in Eq. (12), noting that $P(m|y) = P(m|v)$. Since $P(m|v)p_v(v)\,dv = P(m)p_v(v|m)\,dv$, we can write

$$P(m|v) = P(m)p_v(v|m)/p_v(v)$$

where the unconditional PDF $p_v(v)$ does not depend on m. Thus, $P(m_j|y) > P(m_i|y)$ is equivalent to the vector PDF condition

$$P(m_j)p_v(v|m_j) > P(m_i)p_v(v|m_i) \tag{16}$$

But when $m = m_i$, $s = s_i$ so $v = s_i + n$ and

$$p_v(v|m_i) = p_n(v - s_i) = (\pi\eta)^{-K/2}e^{-\|v-s_i\|^2/\eta}$$

Substituting this expression into Eq. (16) yields, after simplification, the MAP detection rule

$$\|v - s_j\|^2 - \ln P(m_j) < \|v - s_i\|^2 - \ln P(m_i) \tag{17a}$$

When the symbols are equiprobable, $P(m_i) = 1/M$ for all i and Eq. (17a) simplifies to the maximum-likelihood detection rule

$$\|v - s_j\|^2 < \|v - s_i\|^2 \tag{17b}$$

which is the same as the decision-region condition $\|v - s_j\| < \|v - s_i\|$.

Our next task is to devise a receiver structure that implements the detection rule. Recognizing that the receiver must operate on the waveform $y(t)$ instead of the vector v, we first use Eq. (6) to write

$$\|v - s_i\|^2 = \|v\|^2 - 2(v,s_i) + \|s_i\|^2$$

where $\|s_i\|^2 = E_i$ and

$$(v,s_i) = (y - n', s_i) = (y,s_i)$$

since n' is orthogonal to any vector in Γ_K. The term $\|v\|^2$ will appear on both sides of Eq. (17) and cancel out, so we define the observable *decision function*

$$z_i \triangleq (y,s_i) - c_i \tag{18a}$$

in which

$$(y,s_i) = \int_{-\infty}^{\infty} y(t)s_i(t)\, dt \tag{18b}$$

$$c_i = \begin{cases} \frac{1}{2}E_i & \text{equiprobable } m_i \\ \frac{1}{2}[E_i - \eta \ln P(m_i)] & \text{otherwise} \end{cases} \tag{18c}$$

The detection rule thus becomes: Choose $\hat{m} = m_j$ with j such that $z_j > z_i$ for all $i \neq j$.

The scalar products (y,s_i) needed for optimum detection can be calculated at the receiver using the stored signal copies and a bank of *correlation detectors*, each correlator having the form of Fig. C-8a. If the signals are timelimited so that $s_i(t) = 0$ outside of $0 \leq t \leq D$, then correlation detection reduces to *matched filtering* and sampling as diagrammed in Fig. C-8b. In either case, the operation (y,s_i) projects y along the direction of s_i and strips off the irrelevant noise n'.

Figure C-9 shows a complete MAP receiver with a bank of correlators or matched filters that operate in parallel on $y(t)$ to produce (y,s_i), $i = 1, 2, \ldots, M$. Subtracting the bias terms c_i yields the set of decision functions $\{z_i\}$. The receiver compares the decision functions to find the largest one, say z_j, and chooses the optimum estimate $\hat{m} = m_j$.

Various simplifications of this structure are possible, depending upon the specific set of signals. When the symbols are equiprobable and all signals have the same energy, the bias terms are independent of i and may be omitted so that $z_i =$

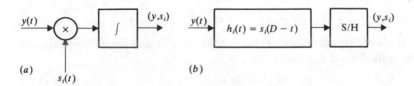

(a)

(b)

Figure C-8 Scalar product calculation. (a) Correlation detector; (b) matched filter.

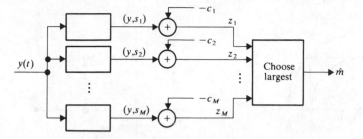

Figure C-9 MAP receiver with bank of correlators or matched filters.

(y,s_i). When the signals are such that the dimensionality of Γ_K is $K < M$, we can use Eq. (8a) to write

$$(y,s_i) = \sum_{k=1}^{K} \alpha_{ik}(y, \varphi_k) \tag{19}$$

and the scalar products can be calculated from stored copies of the K basis functions $\varphi_k(t)$ rather than the M signals. Only for this implementation do we need to know the basis functions explicitly. Modifications such as these account for the simplified receivers in Figs. 11.2-1, 14.2-4, and 14.4-4.

> **Example C-2** Suppose a set of $M = 8$ signals are generated by combined amplitude and phase modulation (APK) such that
>
> $$s_1(t) = -s_5(t) = A_c \cos \omega_c t$$
> $$s_2(t) = -s_6(t) = \sqrt{2} A_c \cos (\omega_c t + \pi/4)$$
> $$= A_c (\cos \omega_c t - \sin \omega_c t)$$
> $$s_3(t) = -s_7(t) = A_c \cos (\omega_c t + \pi/2)$$
> $$= -A_c \sin \omega_c t$$
> $$s_4(t) = -s_8(t) = \sqrt{2} A_c \cos (\omega_c t + 3\pi/4)$$
> $$= -A_c (\cos \omega_c t + \sin \omega_c t)$$
>
> All signals have duration D, and the carrier frequency f_c is an integer multiple of $1/D$.
>
> Observing that each signal can be expressed as a linear combination of $\cos \omega_c t$ and $\sin \omega_c t$, we take the basis functions to be
>
> $$\varphi_1(t) = \sqrt{2/D} \cos \omega_c t \qquad \varphi_2(t) = -\sqrt{2/D} \sin \omega_c t$$
>
> We thus obtain the two-dimensional signal set in Fig. C-10a, where $E_1 = A_c^2 D/2$ and $E_2 = 2E_1$. Figure C-10b shows the corresponding decision boundaries for maximum-likelihood detection. (These diagrams differ from the class of APK signals discussed in Sect. 14.4 which have $M = \mu^2$ points in a square array.)

Figure C-10 (a) A two-dimensional signal set; (b) decision regions; (c) optimum receiver.

Since $K < M$ and the basis functions are known, Eq. (19) leads to the simplified optimum receiver portrayed in Fig. C-10c. The computational unit uses the inputs (y,φ_1) and (y,φ_2) to calculate the decision functions

$$z_i = [\alpha_{i1}(y,\varphi_1) + \alpha_{i2}(y,\varphi_2)] - \tfrac{1}{2} E_i \qquad i = 1, 2, \ldots, 8$$

in accordance with Eqs. (18) and (19). When the source produces a sequence of symbols, the integration and decision process is repeated over each successive symbol interval $kD \le t \le (k + 1)D$.

Error Probabilities

Here we formulate general and specific expressions for the average error probability in an M-ary system with equiprobable symbols and optimum detection. As a starting point, take an arbitrary symbol m_j and let $P(e\,|\,m_j)$ stand for the probability of erroneous detection of the corresponding signal $s_j(t)$. Averaging over the M equiprobable symbols gives

$$P_e = \sum_{j=1}^{M} P(m_j)P(e\,|\,m_j) = \frac{1}{M} \sum_{j=1}^{M} P(e\,|\,m_j) \tag{20a}$$

Alternatively, we can write

$$P_e = 1 - P_c \qquad P_c = \frac{1}{M} \sum_{j=1}^{M} P(c|m_j) \tag{20b}$$

where $P(c|m_j)$ stands for the probability of correct detection. We'll use geometric vector arguments to determine $P(e|m_j)$ or $P(c|m_j)$.

Figure C-11a represents the situation in Γ_K when $s_j(t)$ has been corrupted by noise $n_w(t)$, resulting in the vector $v = s_j + n$. Also shown is another signal vector s_i whose tip-to-tip "distance" from s_j is

$$d_{ij} \triangleq \|s_i - s_j\|$$

A detection error occurs if the projection of n along the direction of $s_i - s_j$ exceeds $d_{ij}/2$, so v falls on the wrong side of the decision boundary halfway between s_i and s_j. This error event will be denoted by e_{ij}. To find the probability $P(e_{ij})$, let the vectors be translated and/or rotated until $s_i - s_j$ becomes colinear with an arbitrary basis vector φ_k as shown in Fig. C-11b. Any translation or rotation of signal points and decision boundaries does not affect error-probability calculations because $p_n(n)$ has spherical symmetry in Γ_K. But now we more readily see that $P(e_{ij})$ equals the probability that $\beta_k > d_{ij}/2$. Hence,

$$P(e_{ij}) = \int_{d_{ij}/2}^{\infty} p_\beta(\beta_k) \, d\beta_k = Q\left(\frac{d_{ij}/2}{\sqrt{\eta/2}}\right)$$

$$= Q\left(\frac{d_{ij}}{\sqrt{2\eta}}\right) = Q\left(\frac{\|s_i - s_j\|}{\sqrt{2\eta}}\right) \tag{21}$$

which follows from Eq. (15a).

Equation (21) immediately applies to the *binary* case. There are only two signal vectors, which we label s_0 and s_1 to be consistent with previous notation. Clearly, $P(e|m_0) = P(e_{10})$ and $P(e|m_1) = P(e_{01}) = P(e_{10})$, so maximum-likelihood detection of a binary signal yields the average error probability

$$P_e = Q\left(\frac{\|s_1 - s_0\|}{\sqrt{2\eta}}\right) \tag{22}$$

(a)   (b)

Figure C-11 (a) Signal-plus-noise vector; (b) noise vector projection obtained by translation and/or rotation of signal points.

A more familiar expression emerges when we note that $\|s_1 - s_0\|^2 = E_1 - 2(s_1,s_0)$ $+ E_0 = 2[E_b - (s_1,s_0)]$, where $E_b = \frac{1}{2}(E_1 + E_0)$ is the average energy per bit. If $s_0(t) = -s_1(t)$, then $(s_1,s_0) = -E_b$ and $(\|s_1 - s_0\|/\sqrt{2\eta})^2 = 4E_b/2\eta = 2\gamma_b$. Equation (22) is therefore identical to the result we obtained in Eq. (16), Sect. 11.2, for polar baseband transmission with matched filtering.

In the general M-ary case, Eq. (21) accounts for just one of the boundaries between s_j and the other $M - 1$ signal points. A complete expression for $P(e|m_j)$ or $P(c|m_j)$ necessarily involves the specific geometry of the signal set, and may or may not be easy to obtain. If the signal points form a *rectangular* array, then translation and rotation yields rectangular decision regions like Fig. C-12 where we see that correct detection of m_j requires $-a_1 < \beta_1 < b_1$ and $-a_2 < \beta_2 < b_2$. Since these orthogonal noise coordinates are statistically independent, we have

$$P(c|m_j) = \int_{-a_1}^{b_1} p_\beta(\beta_1)\, d\beta_1 \int_{-a_2}^{b_2} p_\beta(\beta_2)\, d\beta_2 \tag{23a}$$

Each integral in Eq. (23a) can be expanded in the form

$$\int_{-a}^{b} p_\beta(\beta)\, d\beta = 1 - Q\left(\frac{a}{\sqrt{\eta/2}}\right) - Q\left(\frac{b}{\sqrt{\eta/2}}\right) \tag{23b}$$

obtained from the graphical interpretation of the Q function for a gaussian PDF with zero mean and variance $\eta/2$.

When the geometry of the signal set makes exact analysis difficult, we may settle instead for an upper bound on the average error probability. To develop an upper bound, first consider the case of $M = 3$; $P(e|m_1)$ then equals the probability of the union of the error events e_{21} and e_{31}, so we write $P(e|m_1) = P(e_{21} + e_{31}) = P(e_{21}) + P(e_{31}) - P(e_{21},e_{31}) \leq P(e_{21}) + P(e_{31})$. Extrapolating to arbitrary M and m_j gives the *union bound*

$$P(e|m_j) \leq \sum_{i=1}^{M} P(e_{ij}) \qquad i \neq j$$

with $P(e_{ij})$ as in Eq. (21). There are $M - 1$ terms in this sum, and a simpler but looser bound is

$$P(e|m_j) \leq (M - 1)Q(d_j/\sqrt{2\eta})$$

where d_j stands for the "distance" between s_j and its nearest neighbor, i.e.,

$$d_j \triangleq \min_{i \neq j} \|s_i - s_j\| \tag{24a}$$

Thus, using Eq. (20a), we have

$$P_e \leq \frac{M - 1}{M} \sum_{j=1}^{M} Q\left(\frac{d_j}{\sqrt{2\eta}}\right) \tag{24b}$$

which is our final result.

Figure C-12 Noise vector coordinates relative to rectangular decision regions.

Example C-3 Consider the signal set in Fig. C-10b. The decision regions have a rather complicated pattern, but the nearest-neighbor distance is $d_j = 2a = \sqrt{E_1}$ for all eight points. We also observe that $\|s_i\|^2 = E_1 = (2a)^2$ for even i, while $\|s_i\|^2 = E_2 = 2E_1$ for odd i. The average energy per symbol is then

$$E = \tfrac{1}{8}[4 \times (2a)^2 + 4 \times 2(2a)^2] = 6a^2$$

Hence, from Eq. (24),

$$P_e \leq \frac{7}{8} 8Q\left(\frac{2a}{\sqrt{2\eta}}\right) = 7Q\left(\sqrt{\frac{E}{3\eta}}\right)$$

which is an upper bound on the error probability in terms of the average signal energy.

Closer examination of Fig. C-10b reveals that the corner points have rectangular boundaries, equivalent to Fig. C-12 with $a_1 = a_2 = a$ and $b_1 = b_2 = \infty$. Therefore, for even values of j, Eq. (23) gives the exact result

$$P(c \mid m_j) = (1 - q)^2 \qquad q = Q(a/\sqrt{\eta/2}) = Q(\sqrt{E/3\eta})$$

Furthermore, for odd values of j, we can omit the triangular area and write

$$P(c \mid m_j) > \int_{-a}^{\infty} p_\beta(\beta_1)\, d\beta_1 \int_{-a}^{a} p_\beta(\beta_2)\, d\beta_2 = (1 - q)(1 - 2q)$$

Thus, substitution in Eq. (20b) yields

$$P_c > \tfrac{1}{8}[4(1 - q)^2 + 4(1 - q)(1 - 2q)]$$

$$P_e = 1 - P_c < \tfrac{5}{2}q - \tfrac{3}{2}q^2$$

which is a more acurate result than the union bound.

Exercise C-2 Let a set of $M = 6$ signals be defined by $s_1 = -s_2 = a\varphi_1$, $s_3 = -s_4 = a\varphi_1 + 2a\varphi_2$, and $s_5 = -s_6 = a\varphi_1 - 2a\varphi_2$. Construct the decision boundaries and use Eq. (23) to show that $P_e = (7q - 4q^2)/3$, where $q = Q(\sqrt{6E/11\eta})$ with E being the average signal energy.

Signal Selection and Orthogonal Signaling ★

Having learned how to implement optimum detection given a set of signals, we come at last to the important design task of *signal selection* for digital communication. We continue to assume equiprobable symbols and gaussian noise, but we now add a constraint on the average signal energy E. In this context, we say that:

> An *optimum signal set* achieves the lowest error probability for a specified value of E or, equivalently, achieves a specified error probability with the lowest value of E.

Our vector interpretation suggests that the corresponding signal points should be arranged spherically around the origin (to minimize vector length and signal energy), with the largest possible spacing between points (to maximize separation and minimize error probability).

These optimal properties define the so-called *simplex set* in a subspace with $K = M - 1$ dimensions. The simplex signal points form the vertices of a K-dimensional pyramid whose center of gravity coincides with the orign. The pyramid reduces to an equilateral triangle when $M = 3$, and to antipodal points $(s_2 = -s_1)$ when $M = 2$. When M is large, there's little difference between the optimal simplex set and a set of *equal-energy orthogonal signals*. Since orthogonal signals are easier to generate and analyze than simplex signals, we'll focus attention on the nearly optimum case of digital communication via M-ary orthogonal signaling.

Specifically, let an M-dimensional subspace be spanned by a set of orthonormal basis functions, and let

$$s_i(t) = \sqrt{E}\,\varphi_i(t) \qquad i = 1, 2, \ldots, M \tag{25a}$$

so that

$$(s_i, s_j) = \begin{cases} E & j = i \\ 0 & j \neq i \end{cases} \tag{25b}$$

$$d_{ij}^2 = \|s_i - s_j\|^2 = 2E \qquad j \neq i \tag{25c}$$

These relations define a set of mutually orthogonal signals with average signal energy E, as illustrated in Fig. C-13 for $M = 3$. We'll also impose the timelimited condition that $s_i(t) = 0$ for $t < 0$ and $t > D$. The values of D and M are related by

$$r_b = \frac{\log_2 M}{D}$$

where r_b is the equivalent bit rate of the information source.

There are many possible sets of timelimited orthogonal signals. At baseband, for instance, the signals could be nonoverlapping pulses of duration $\tau \leq D/M$ — i.e., digital pulse-position modulation (PPM). Or, for bandpass transmission, orthogonal signaling may take the form of frequency-shift keying (FSK) with fre-

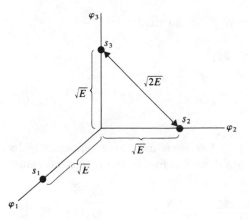

Figure C-13 Orthogonal signal set with $M = 3$.

quency spacing $1/2D$. Regardless of the particular implementation, M-ary orthogonal signaling is a *wideband* method requiring transmission bandwidth

$$B_T \geq \frac{M}{2D} = \frac{Mr_b}{2 \log_2 M} \geq r_b \tag{26}$$

We'll see that this method trades increased bandwidth for decreased error probability, similar to wideband noise-reduction in analog communication.

First note from symmetry that $P(c|m_j)$ is the same for all M signals, so

$$P_c = P(c|m_j) = P(z_j > z_i, \text{ all } i \neq j)$$

where z_i is the decision function in Eq. (18). Since the decision bias term c_i is independent of i, we can drop it and write $z_i = (y, s_i) = (v, s_i)$. When $m = m_j$, $v = s_j + n = \sqrt{E}\,\varphi_j + n$ and

$$\begin{aligned} z_i &= (\sqrt{E}\,\varphi_j + n,\ \sqrt{E}\,\varphi_i) \\ &= E(\varphi_j, \varphi_i) + \sqrt{E}(n, \varphi_i) \\ &= \begin{cases} E + \sqrt{E}\,\beta_j & i = j \\ \sqrt{E}\,\beta_i & i \neq j \end{cases} \end{aligned}$$

in which we've introduced the noise coordinates $\beta_i = (n, \varphi_i)$.

Next, consider some particular value of β_j so, for any $i \neq j$, the probability of the event $z_j > z_i$ given β_j is

$$\begin{aligned} P(z_j > z_i | \beta_j) &= P(E + \sqrt{E}\,\beta_j > \sqrt{E}\,\beta_i) \\ &= P(\beta_i < \sqrt{E} + \beta_j) = \int_{-\infty}^{\sqrt{E}+\beta_j} p_\beta(\beta_i)\, d\beta_i \end{aligned}$$

Then, since the noise coordinates are independent r.v.'s and there are $M - 1$ coordinates with $i \neq j$, the probability of correct detection given β_j is

$$P(c|\beta_j) = \left[\int_{-\infty}^{\sqrt{E}+\beta_j} p_\beta(\beta_i)\, d\beta_i\right]^{M-1}$$

Averaging $P(c \mid \beta_j)$ over all possible values of β_j finally yields

$$P_c = \int_{-\infty}^{\infty} P(c \mid \beta_j) p_\beta(\beta_j) \, d\beta_j$$

$$= \pi^{-M/2} \left[\int_{-\infty}^{\sqrt{E/\eta} + \lambda} e^{-\mu^2} \, d\mu \right]^{M-1} e^{-\lambda^2} \, d\lambda \tag{27}$$

where $\mu = \beta_i / \sqrt{\eta}$ and $\lambda = \beta_j / \sqrt{\eta}$.

The formidable expression in Eq. (27) has been evaluated numerically by Viterbi (1966), and plots of $P_e = 1 - P_c$ are shown in Fig. C-14 for selected values of M. These curves are plotted versus

$$\frac{E}{\eta \log_2 M} = \frac{SD}{\eta \log_2 M} = \frac{S}{\eta r_b} \tag{28}$$

where $S = E/T$ is the average signal power. We see that when $S/\eta r_b$ has a fixed value, the error probability can be made as small as desired by increasing M. In

$M = 2$

4

16

256

2^{20}

P_e

0.1   ln 2   1   10 $\dfrac{S}{\eta r_b}$

Figure C-14 Error probability for M-ary orthogonal signaling.

fact, Viterbi proves analytically that

$$\lim_{M \to \infty} P_e = \begin{cases} 0 & S/\eta r_b < \ln 2 \\ 1 & S/\eta r_b > \ln 2 \end{cases}$$

represented by the dashed line in Fig. C-14. Hence, if $M \to \infty$, then orthogonal signaling with optimum detection approaches *errorless* transmission at any bit rate

$$r_b < \frac{S}{\eta \ln 2} = C_\infty$$

The quantity C_∞ is the maximum capacity of an AWGN channel as predicted from information theory; see Eq. (17), Sect. 15.3.

But keep in mind that increasing M means increasing both the bandwidth and the receiver complexity, which is proportional to M. Furthermore, Fig. C-14 reveals that orthogonal signaling with fixed $M \gg 1$ has a *threshold effect* in the sense that P_e rises abruptly if $S/\eta r_b$ should happen to decrease somewhat. These observations point to the important conclusion that reliable, efficient, and practical digital communication usually involves some form of *error-control coding* as well as careful signal selection and optimum receiver design.

Problems

C-1 Consider the signals $v(t)$, $w(t)$, and $z(t) = v(t) + w(t)$ with energy $E_v = E_w = 4$ and $E_z = 8$. Construct a vector diagram based on this information, and use it to find E_x when $x(t) = 3v(t) - 0.5w(t)$.

C-2 Do Prob. C-1 with $E_v = 4$, $E_w = 12$, and $E_z = 16$.

C-3 Prove the triangle inequality in Eq. (2) from Eqs. (5) and (6).

C-4 Derive Eq. (7) and show that $\|v_w\| \le \|v\|$.

C-5 Apply the Gram-Schmidt procedure to obtain an orthonormal basis for the signals

$$s_1(t) = 1 \qquad s_2(t) = t \qquad s_3(t) = t^2$$

defined over $-1 \le t \le 1$. Then write the signal vectors in terms of the basis vectors.

C-6 Apply the Gram-Schmidt procedure to obtain an orthonormal basis for the signals

$$s_1(t) = 1 \qquad s_2(t) = \cos \pi t/2 \qquad s_3(t) = \sin \pi t$$

defined over $-1 \le t \le 1$. Then write the signal vectors in terms of the basis vectors.

C-7 Consider the following three sets of binary signals defined in terms of orthonormal basis functions φ_1 and φ_2:

    (a) $s_1 = a\varphi_1$ $s_2 = a\varphi_2$
    (b) $s_1 = a\varphi_1$ $s_2 = -a\varphi_1$
    (c) $s_1 = a\varphi_1$ $s_2 = 0$

Construct the maximum-likelihood decision regions and write the decision functions for each signal set. Then simplify and diagram optimum receivers using correlation detection.

C-8 Consider the following two sets of quaternary ($M = 4$) signals defined in terms of orthogonal basis functions φ_1 and φ_2:

    (a) $s_1 = a\varphi_1$ $s_2 = a\varphi_2$ $s_3 = -a\phi_1$ $s_4 = -a\varphi_2$
    (b) $s_1 = a\varphi_1$ $s_2 = a\varphi_2$ $s_3 = a(\varphi_1 + \varphi_2)$ $s_4 = 0$

Construct the maximum-likelihood decision regions and write the decision functions for each signal set. The simplify and diagram optimum receivers using correlation detection.

C-9 Consider the following set of quaternary ($M = 4$) signals defined in terms of orthonormal basis functions φ_1 and φ_2:

$$s_1 = -s_3 = a(\varphi_1 + \varphi_2) \qquad s_2 = -s_4 = a(\varphi_1 - \varphi_2)$$

Construct the maximum-likelihood decision regions and find the error probability with AWGN, expressing your result in terms of the average symbol energy E.

C-10 Suppose the signal $s_5 = 0$ is added to the set in Prob. C-9. Construct the maximum-likelihood decision regions and obtain an upper bound on the error probability with AWGN, expressing your result in terms of the average symbol energy E.

C-11 Suppose the signal $s_9 = 0$ is added to the set in Fig. C-10a. Construct the maximum-likelihood decision regions and find the error probability with AWGN, expressing your result in terms of the average symbol energy E. Compare with the upper bound obtained from Eq. (24).

FOURIER TRANSFORMS

Definitions

Transform

$$V(f) = \mathscr{F}[v(t)] = \int_{-\infty}^{\infty} v(t)e^{-j2\pi ft}\, dt$$

Inverse transform

$$v(t) = \mathscr{F}^{-1}[V(f)] = \int_{-\infty}^{\infty} V(f)e^{j2\pi ft}\, df$$

Integral theorem

$$\int_{-\infty}^{\infty} v(t)w^*(t)\, dt = \int_{-\infty}^{\infty} V(f)W^*(f)\, df$$

Theorems

Operation	Function	Transform
Superposition	$a_1 v_1(t) + a_2 v_2(t)$	$a_1 V_1(f) + a_2 V_2(f)$
Time delay	$v(t - t_d)$	$V(f)e^{-j\omega t_d}$
Scale change	$v(\alpha t)$	$\dfrac{1}{|\alpha|}\, V\!\left(\dfrac{f}{\alpha}\right)$
Conjugation	$v^*(t)$	$V^*(-f)$
Duality	$V(t)$	$v(-f)$
Frequency translation	$v(t)e^{j\omega_c t}$	$V(f - f_c)$
Modulation	$v(t)\cos(\omega_c t + \phi)$	$\tfrac{1}{2}[V(f - f_c)e^{j\phi} + V(f + f_c)e^{-j\phi}]$
Differentiation	$\dfrac{d^n v(t)}{dt^n}$	$(j2\pi f)^n V(f)$
Integration	$\displaystyle\int_{-\infty}^{t} v(\lambda)\, d\lambda$	$\dfrac{1}{j2\pi f}\, V(f) + \tfrac{1}{2}V(0)\, \delta(f)$
Convolution	$v * w(t)$	$V(f)W(f)$
Multiplication	$v(t)w(t)$	$V * W(f)$
Multiplication by t^n	$t^n v(t)$	$(-j2\pi)^{-n}\dfrac{d^n V(f)}{df^n}$

Transforms

Function	$v(t)$	$V(f)$
Rectangular	$\Pi\left(\dfrac{t}{\tau}\right)$	$\tau \operatorname{sinc} f\tau$
Triangular p. 48	$\Lambda\left(\dfrac{t}{\tau}\right)$	$\tau \operatorname{sinc}^2 f\tau$
Gaussian	$e^{-\pi(bt)^2}$	$(1/b)e^{-\pi(f/b)^2}$
Causal exponential	$e^{-bt}u(t)$	$\dfrac{1}{b + j2\pi f}$
Symmetric exponential	$e^{-b|t|}$	$\dfrac{2b}{b^2 + (2\pi f)^2}$
Sinc	$\operatorname{sinc} 2Wt$	$\dfrac{1}{2W}\Pi\left(\dfrac{f}{2W}\right)$
Sinc squared	$\operatorname{sinc}^2 2Wt$	$\dfrac{1}{2W}\Lambda\left(\dfrac{f}{2W}\right)$
Constant	1	$\delta(f)$
Phasor	$e^{j(\omega_c t + \phi)}$	$e^{j\phi}\,\delta(f - f_c)$
Sinusoid	$\cos(\omega_c t + \phi)$	$\frac{1}{2}[e^{j\phi}\,\delta(f - f_c) + e^{-j\phi}\,\delta(f + f_c)]$
Impulse	$\delta(t - t_d)$	$e^{-j\omega t_d}$
Sampling	$\displaystyle\sum_{k=-\infty}^{\infty} \delta(t - kT_s)$	$\displaystyle f_s \sum_{n=-\infty}^{\infty} \delta(f - nf_s)$
Signum	$\operatorname{sgn} t$	$1/j\pi f$
Step	$u(t)$	$\dfrac{1}{j2\pi f} + \frac{1}{2}\,\delta(f)$

Definitions

If $v(t)$ is periodic with fundamental period

$$T_0 = 1/f_o = 2\pi/\omega_0$$

then it can be written as the exponential Fourier series

$$v(t) = \sum_{n=-\infty}^{\infty} c_n e^{jn\omega_0 t}$$

The series coefficients are

$$c_n = \frac{1}{T_0} \int_{t_1}^{t_1 + T_0} v(t) e^{-jn\omega_0 t} \, dt$$

where t_1 is arbitrary.

When $v(t)$ is real, its Fourier series may be expressed in the trigonometric forms

$$v(t) = c_o + \sum_{n=1}^{\infty} |2c_n| \cos (n\omega_0 t + \arg c_n)$$

$$= c_o + \sum_{n=1}^{\infty} (a_n \cos n\omega_0 t + b_n \sin n\omega_0 t)$$

where

$$a_n = 2 \operatorname{Re} [c_n] \qquad b_n = -2 \operatorname{Im} [c_n]$$

Coefficient calculations

If a single period of $v(t)$ has the known Fourier transform

$$Z(f) = \mathscr{F}\left[v(t)\Pi\left(\frac{t - t_1}{T_0}\right) \right]$$

then

$$c_n = \frac{1}{T_0} Z(nf_0)$$

The following relations may be used to obtain the exponential series coefficients of a waveform that can be expressed in terms of another waveform $v(t)$ whose coefficients are known and denoted by $c_v(n)$.

Waveform	Coefficients	
$Av(t) + B$	$Ac_v(0) + B$	$n = 0$
	$Ac_v(n)$	$n \neq 0$
$v(t - t_d)$	$c_v(n)e^{-jn\omega_0 t_d}$	
$dv(t)/dt$	$(j2\pi nf_0)c_v(n)$	
$v(t) \cos (m\omega_0 t + \phi)$	$\frac{1}{2}c_v(n - m)e^{j\phi} + \frac{1}{2}c_v(n + m)e^{-j\phi}$	

Series coefficients for selected waveforms

The waveforms in the following list are periodic and are defined for one period, either $0 < t < T_0$ or $|t| < T_0/2$, as indicated. This listing may be used in conjunction with the foregoing relationships to obtain the exponential series coefficients for other waveforms.

Waveform	Coefficients	
Impulse train		
$\quad \delta(t) \qquad |t| < T_0/2$	$1/T_0$	
Rectangular pulse train		
$\quad \Pi(t/\tau) \qquad |t| < T_0/2$	$(\tau/T_0) \operatorname{sinc} nf_0 \tau$	
Square wave (odd symmetry)		
$\quad +1 \qquad 0 < t < T_0/2$	0	n even
$\quad -1 \qquad -T_0/2 < t < 0$	$-j2/\pi n$	n odd
Triangular wave (even symmetry)		
$\quad 1 - \dfrac{4|t|}{T_0} \qquad |t| < T_0/2$	0	n even
	$(2/\pi n)^2$	n odd
Sawtooth wave		
$\quad t/T_0 \qquad 0 < t < T_0$	$1/2$	$n = 0$
	$j/2\pi n$	$n \neq 0$
Half-rectified sine wave		
$\quad \sin \omega_0 t \qquad 0 < t < T_0/2$	$\dfrac{1}{\pi(1 - n^2)}$	n even
$\quad 0 \qquad\quad T_0/2 < t < T_0$	$-jn/4$	$n = \pm 1$
	0	otherwise

MATHEMATICAL RELATIONS

Certain of the mathematical relations encountered in this text are listed below for convenient reference. However, this table is not intended as a substitute for more comprehensive handbooks.

Trigonometric identities

$$e^{\pm j\theta} = \cos \theta \pm j \sin \theta$$
$$e^{j2\alpha} + e^{j2\beta} = 2 \cos (\alpha - \beta)e^{j(\alpha + \beta)}$$
$$e^{j2\alpha} - e^{j2\beta} = j2 \sin (\alpha - \beta)e^{j(\alpha + \beta)}$$

$$\cos \theta = \frac{1}{2} (e^{j\theta} + e^{-j\theta}) = \sin (\theta + 90°)$$

$$\sin \theta = \frac{1}{2j} (e^{j\theta} - e^{-j\theta}) = \cos (\theta - 90°)$$

$$\sin^2 \theta + \cos^2 \theta = 1$$
$$\cos^2 \theta - \sin^2 \theta = \cos 2\theta$$

$$\cos^2 \theta = \tfrac{1}{2}(1 + \cos 2\theta)$$
$$\cos^3 \theta = \tfrac{1}{4}(3 \cos \theta + \cos 3\theta)$$

$$\sin^2 \theta = \tfrac{1}{2}(1 - \cos 2\theta)$$
$$\sin^3 \theta = \tfrac{1}{4}(3 \sin \theta - \sin 3\theta)$$

$$\sin (\alpha \pm \beta) = \sin \alpha \cos \beta \pm \cos \alpha \sin \beta$$
$$\cos (\alpha \pm \beta) = \cos \alpha \cos \beta \mp \sin \alpha \sin \beta$$

$$\tan (\alpha \pm \beta) = \frac{\tan \alpha \pm \tan \beta}{1 \mp \tan \alpha \tan \beta}$$

$$\sin \alpha \sin \beta = \tfrac{1}{2} \cos (\alpha - \beta) - \tfrac{1}{2} \cos (\alpha + \beta)$$
$$\cos \alpha \cos \beta = \tfrac{1}{2} \cos (\alpha - \beta) + \tfrac{1}{2} \cos (\alpha + \beta)$$
$$\sin \alpha \cos \beta = \tfrac{1}{2} \sin (\alpha - \beta) + \tfrac{1}{2} \sin (\alpha + \beta)$$

$$A \cos (\theta + \alpha) + B \cos (\theta + \beta) = C \cos \theta - S \sin \theta = R \cos (\theta + \phi)$$

where

$$C = A \cos \alpha + B \cos \beta$$

$$S = A \sin \alpha + B \sin \beta$$

$$R = \sqrt{C^2 + S^2} = \sqrt{A^2 + B^2 + 2AB \cos (\alpha - \beta)}$$

$$\phi = \arctan \frac{S}{C} = \arctan \frac{A \sin \alpha + B \sin \beta}{A \cos \alpha + B \cos \beta}$$

Series expansions and approximations

$$(1 + x)^n = 1 + nx + \frac{n(n-1)}{2!} x^2 + \cdots \qquad |nx| < 1$$

$$e^x = 1 + x + \frac{1}{2!} x^2 + \cdots$$

$$a^x = 1 + x \ln a + \frac{1}{2!} (x \ln a)^2 + \cdots$$

$$\ln (1 + x) = x - \tfrac{1}{2}x^2 + \tfrac{1}{3}x^3 + \cdots$$

$$\sin x = x - \frac{1}{3!} x^3 + \frac{1}{5!} x^5 - \cdots$$

$$\cos x = 1 - \frac{1}{2!} x^2 + \frac{1}{4!} x^4 - \cdots$$

$$\tan x = x + \tfrac{1}{3}x^3 + \tfrac{2}{15}x^5 + \cdots$$

$$\arcsin x = x + \tfrac{1}{6}x^3 + \tfrac{3}{40}x^5 + \cdots$$

$$\arctan x = \begin{cases} x - \tfrac{1}{3}x^3 + \tfrac{1}{5}x^5 - \cdots & |x| < 1 \\ \dfrac{\pi}{2} - \dfrac{1}{x} + \dfrac{1}{3x^3} - \cdots & x > 1 \end{cases}$$

$$\operatorname{sinc} x = 1 - \frac{1}{3!} (\pi x)^2 + \frac{1}{5!} (\pi x)^4 - \cdots$$

$$J_n(x) = \frac{1}{n!} \left(\frac{x}{2}\right)^n - \frac{1}{(n+1)!} \left(\frac{x}{2}\right)^{n+2} + \frac{1}{2!(n+2)!} \left(\frac{x}{2}\right)^{n+4} - \cdots$$

$$J_n(x) \approx \sqrt{\frac{2}{\pi x}} \cos\left(x - \frac{\pi}{4} - \frac{n\pi}{2}\right) \qquad x \gg 1$$

$$I_0(x) \approx \begin{cases} e^{x^2/4} & x^2 \ll 1 \\ e^x / \sqrt{2\pi x} & x \gg 1 \end{cases}$$

Summations

$$\sum_{m=1}^{M} m = \frac{M(M+1)}{2}$$

$$\sum_{m=1}^{M} m^2 = \frac{M(M+1)(2M+1)}{6}$$

$$\sum_{m=1}^{M} m^3 = \frac{M^2(M+1)^2}{4}$$

$$\sum_{m=0}^{M} x^m = \frac{x^{M+1} - 1}{x - 1}$$

Definite integrals

$$\int_0^\infty \frac{x^{m-1}}{1+x^n}\,dx = \frac{\pi/n}{\sin(m\pi/n)} \qquad n > m > 0$$

$$\int_0^\infty \sin x^2\,dx = \int_0^\infty \cos x^2\,dx = \frac{1}{2}\sqrt{\frac{\pi}{2}}$$

$$\int_0^\infty \frac{\sin x}{x}\,dx = \int_0^\infty \frac{\tan x}{x}\,dx = \frac{\pi}{2}$$

$$\int_0^\infty x^n e^{-ax}\,dx = \frac{n!}{a^{n+1}} \qquad n \ge 1,\ a > 0$$

$$\int_0^\infty \frac{\sin x \cos ax}{x}\,dx = \begin{cases} \dfrac{\pi}{2} & a^2 < 1 \\[2mm] \dfrac{\pi}{4} & a^2 = 1 \\[2mm] 0 & a^2 > 1 \end{cases}$$

$$\int_0^\infty e^{-a^2 x^2}\,dx = \frac{1}{2a}\sqrt{\pi} \qquad a > 0$$

$$\int_0^\infty x^2 e^{-x^2}\,dx = \tfrac{1}{4}\sqrt{\pi}$$

$$\int_0^\infty \frac{\sin^2 x}{x^2}\,dx = \frac{\pi}{2}$$

$$\int_0^\infty e^{-ax}\cos x\,dx = \frac{a}{1+a^2} \qquad a > 0$$

$$\int_0^\infty \frac{\cos nx}{1+x^2}\,dx = \frac{\pi}{2}e^{-|n|}$$

$$\int_0^\infty e^{-ax}\sin x\,dx = \frac{1}{1+a^2} \qquad a > 0$$

$$\int_0^\infty \mathrm{sinc}\,x\,dx = \int_0^\infty \mathrm{sinc}^2\,x\,dx = \tfrac{1}{2}$$

$$\int_0^\infty e^{-a^2 x^2}\cos bx\,dx = \frac{1}{2a}\sqrt{\pi}\,e^{-(b/2a)^2}$$

Schwarz's inequality

$$\left| \int_a^b v(\lambda)w(\lambda)\,d\lambda \right|^2 \le \int_a^b |v(\lambda)|^2\,d\lambda \int_a^b |w(\lambda)|^2\,d\lambda$$

The equality holds if $v(\lambda) = Kw(\lambda)$ where K is a constant.

Poisson's sum formula

$$\sum_{n=-\infty}^\infty e^{\pm j2\pi n\lambda/L} = L \sum_{m=-\infty}^\infty \delta(\lambda - mL)$$

TABLE T.4

THE SINC FUNCTION

Numerical values of sinc $x = (\sin \pi x)/\pi x$ and its square are tabulated below for x from 0 to 3.9 in increments of 0.1.

x	sinc x	sinc^2 x	x	sinc x	sinc^2 x
0.0	1.000	1.000	2.0	0.000	0.000
0.1	0.984	0.968	2.1	0.047	0.002
0.2	0.935	0.875	2.2	0.085	0.007
0.3	0.858	0.737	2.3	0.112	0.013
0.4	0.757	0.573	2.4	0.126	0.016
0.5	0.637	0.405	2.5	0.127	0.016
0.6	0.505	0.255	2.6	0.116	0.014
0.7	0.368	0.135	2.7	0.095	0.009
0.8	0.234	0.055	2.8	0.067	0.004
0.9	0.109	0.012	2.9	0.034	0.001
1.0	0.000	0.000	3.0	0.000	0.000
1.1	−0.089	0.008	3.1	−0.032	0.001
1.2	−0.156	0.024	3.2	−0.058	0.003
1.3	−0.198	0.039	3.3	−0.078	0.006
1.4	−0.216	0.047	3.4	−0.089	0.008
1.5	−0.212	0.045	3.5	−0.091	0.008
1.6	−0.189	0.036	3.6	−0.084	0.007
1.7	−0.151	0.023	3.7	−0.070	0.005
1.8	−0.104	0.011	3.8	−0.049	0.002
1.9	−0.052	0.003	3.9	−0.025	0.001

PROBABILITY FUNCTIONS

Binomial distribution

Let the discrete r.v. I be the number of times an event A occurs in n independent trials. If $P(A) = \alpha$, then

$$P_I(i) = \binom{n}{i} \alpha^i (1 - \alpha)^{n-i} \qquad i = 0, 1, \ldots, n$$

$$\bar{i} = n\alpha \qquad \sigma_i^2 = n\alpha(1 - \alpha)$$

If $n \gg 1$, $\alpha \ll 1$, and $m = n\alpha$ remains finite, then

$$P_I(i) \approx e^{-m} m^i / i!$$

Poisson distribution

Let the discrete r.v. I be the number of times an event A occurs in time T. If $P(A) = \mu \, \Delta T \ll 1$ in a small interval ΔT, and if multiple occurrences are statistically independent, then

$$P_I(i) = e^{-\mu T} (\mu T)^i / i! \qquad \bar{i} = \mu T$$

Uniform distribution

If the continuous r.v. X is equally likely to be observed anywhere in a finite range, and nowhere else, then

$$p_X(x) = \frac{1}{b - a} \qquad a \leq x \leq b$$

$$\bar{x} = \tfrac{1}{2}(a + b) \qquad \sigma_x^2 = \tfrac{1}{12}(b - a)^2$$

Sinusoidal distribution

If X has a uniform distribution with $b - a = 2\pi$ and $Z = A \cos(X + \theta)$, where A and θ are constants, then

$$p_Z(z) = \frac{1}{\pi \sqrt{A^2 - z^2}} \qquad |z| \leq A$$

$$\bar{z} = 0 \qquad \sigma_z^2 = \tfrac{1}{2}A^2$$

Gaussian distribution

If X represents the sum of a large number of independent random components, and if each component makes only a small contribution to the sum, then

$$p_X(x) = \frac{1}{\sqrt{2\pi\sigma^2}}\, e^{-(x-m)^2/2\sigma^2}$$

$$\bar{x} = m \qquad \sigma_x^2 = \sigma^2$$

(See Table T.6 for gaussian probabilities.)

Rayleigh distribution

If $R^2 = X^2 + Y^2$, where X and Y are independent gaussian r.v.'s with zero mean and variance σ^2, then

$$p_R(r) = \frac{r}{\sigma^2}\, e^{-r^2/2\sigma^2} \qquad r \geq 0$$

$$\bar{R} = \sqrt{\pi/2}\,\sigma \qquad \overline{R^2} = 2\sigma^2$$

GAUSSIAN PROBABILITIES

The probability that a gaussian random variable with mean m and variance σ^2 will have an observed value greater than $m + k\sigma$ is given by the function

$$Q(k) \triangleq \frac{1}{\sqrt{2\pi}} \int_k^\infty e^{-\lambda^2/2} \, d\lambda$$

called the area under the gaussian tail. Thus

$$P(X > m + k\sigma) = P(X \le m - k\sigma) = Q(k)$$

$$P(|X - m| > k\sigma) = 2Q(k)$$

$$P(m < X \le m + k\sigma) = P(m - k\sigma < X \le m) = \tfrac{1}{2} - Q(k)$$

$$P(|X - m| \le k\sigma) = 1 - 2Q(k)$$

$$P(m - k_1\sigma < X \le m + k_2\sigma) = 1 - Q(k_1) - Q(k_2)$$

Other functions related to $Q(k)$ are as follows:

$$\operatorname{erf} k \triangleq \frac{2}{\sqrt{\pi}} \int_0^k e^{-\lambda^2} \, d\lambda = 1 - 2Q(\sqrt{2}\,k)$$

$$\operatorname{erfc} k \triangleq \frac{2}{\sqrt{\pi}} \int_k^\infty e^{-\lambda^2} \, d\lambda = 1 - \operatorname{erf} k = 2Q(\sqrt{2}\,k)$$

$$\Phi(k) \triangleq \frac{1}{\sqrt{2\pi}} \int_0^k e^{-\lambda^2/2} \, d\lambda = \tfrac{1}{2} - Q(k)$$

All of the foregoing relations are for $k \ge 0$. If $k < 0$, then

$$Q(-|k|) = 1 - Q(|k|)$$

Numerical values of $Q(k)$ are plotted below for $0 \le k \le 7.0$. For larger values of k, $Q(k)$ may be approximated by

$$Q(k) \approx \frac{1}{\sqrt{2\pi}\,k}\, e^{-k^2/2}$$

which is quite accurate for $k > 3$.

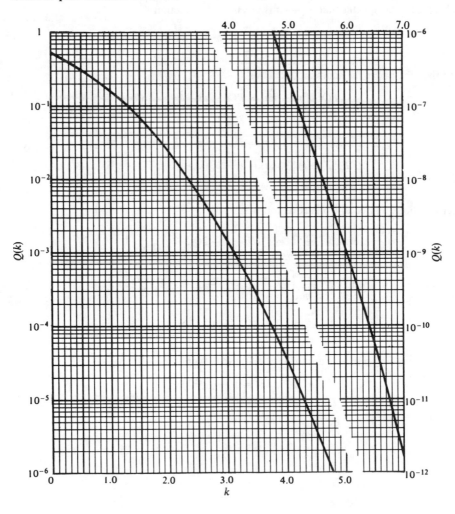

GLOSSARY OF NOTATION

Operations

z^*	Complex conjugate
Re $[z]$, Im $[z]$	Real and imaginary parts
$\|z\|$	Magnitude or absolute value
$\arg z = \arctan \dfrac{\text{Im } [z]}{\text{Re } [z]}$	Angle of a complex quantity
$\langle v(t) \rangle = \lim\limits_{T \to \infty} \dfrac{1}{T} \displaystyle\int_{-T/2}^{T/2} v(t)\, dt$	Time average
$\mathscr{F}[v(t)] = \displaystyle\int_{-\infty}^{\infty} v(t) e^{-j2\pi ft}\, dt$	Fourier transform
$\mathscr{F}^{-1}[V(f)] = \displaystyle\int_{-\infty}^{\infty} V(f) e^{j2\pi ft}\, df$	Inverse Fourier transform
$v * w(t) = \displaystyle\int_{-\infty}^{\infty} v(\lambda) w(t - \lambda)\, d\lambda$	Convolution
$\hat{v}(t) = \dfrac{1}{\pi} \displaystyle\int_{-\infty}^{\infty} \dfrac{v(\lambda)}{t - \lambda}\, d\lambda$	Hilbert transform
$R_{vw}(\tau)$	Crosscorrelation
$R_v(\tau) = R_{vv}(\tau)$	Autocorrelation
$G_v(f) = \mathscr{F}_\tau[R_v(\tau)]$	Spectral density or power spectrum
$\bar{x} = E[X]$	Mean or expected value
$E[v(t)]$	Ensemble average

Functions

$$Q(k) = \frac{1}{\sqrt{2\pi}} \int_k^\infty e^{-\lambda^2/2} \, d\lambda \qquad \text{Gaussian probability}$$

$$\exp t = e^t \qquad \qquad \text{Exponential}$$

$$\text{sinc } t = \frac{\sin \pi t}{\pi t} \qquad \qquad \text{Sinc}$$

$$\text{sgn } t = \begin{cases} 1 & t > 0 \\ -1 & t < 0 \end{cases} \qquad \text{Sign}$$

$$u(t) = \begin{cases} 1 & t > 0 \\ 0 & t < 0 \end{cases} \qquad \text{Step}$$

$$\Pi\left(\frac{t}{\tau}\right) = \begin{cases} 1 & |t| < \dfrac{\tau}{2} \\[2mm] 0 & |t| > \dfrac{\tau}{2} \end{cases} \qquad \text{Rectangle}$$

$$\Lambda\left(\frac{t}{\tau}\right) = \begin{cases} 1 - \dfrac{|t|}{\tau} & |t| < \tau \\[2mm] 0 & |t| > \tau \end{cases} \qquad \text{Triangle}$$

Miscellaneous symbols

$\triangleq$	"Equals by definition"
$\approx$	"Approximately equals"
$\displaystyle\int_T$	$\displaystyle\int_{t_1}^{t_1+T}$ where t_1 is arbitrary
$\leftrightarrow$	Denoting a Fourier transform pair
$\dbinom{n}{i} = \dfrac{n!}{i!\,(n-i)!}$	Binomial coefficient
★	Text material that may be omitted

ANSWERS TO EXERCISES

2.1-1 $v(t) = 3 \cos (2\pi 0 t \pm 180°) + 4 \cos (2\pi 15 t + 90°)$

2.1-3 $P = 107$

2.2-1 $V(f) = 2Ab/[b^2 + (2\pi f)^2]$

2.2-2 $E = A^2/2b$

2.2-3 $Z(f) = (B/2)e^{-|f|}$

2.3-3 $(A \operatorname{sinc} 2Wt)^2 \leftrightarrow (A^2/2W)\Lambda(f/2W)$

2.4-1 (a) 49; (b) $(t + 1)^2$; (c) $49\delta(t + 4)$; (d) $4(t - 3)^2$

2.4-2 $\dfrac{A}{j4\pi(f - f_c)} + \dfrac{A}{4} \delta(f - f_c) + \dfrac{A}{j4\pi(f + f_c)} + \dfrac{A}{4} \delta(f + f_c)$

2.4-3 $V(f) = (jA/\pi f)(\cos \pi f\tau - \operatorname{sinc} f\tau)$

3.1-1 $g(t) = e^{-t/RC}u(t),\ h(t) = \delta(t) - (1/RC)e^{-t/RC}u(t)$

3.1-2 $H(f) = jf/(f_\ell + jf)$

3.1-3 $y(t) \approx \begin{cases} A\tau h(t) & \tau \ll T \\ AT\Lambda[(t - T/2)/T] & \tau = T \\ Tx(t) & \tau \gg T \end{cases}$

3.2-1 $t_d = \begin{cases} 1/120 \text{ ms} & |f| \le 30 \text{ kHz} \\ 1/4f & |f| > 30 \text{ kHz} \end{cases}$

3.3-2 (a) -27 dBm $\approx 2\ \mu$W; (b) -3 dBm ≈ 0.5 mW

3.4-2 $n = 4 \ge 20/6$

3.4-3 $B \ge 100$ kHz, $t_r \approx 5\ \mu$s

3.5-2 $R_{vw}(\tau) = A^2 D\Lambda[(\tau + t_d)/D]$

4.1-1 $P(A) = 4/9,\ P(B) = 1/3,\ P(AB) = 2/9,\ P(A + B) = 5/9$

4.1-2 $P(B \mid A_1) = 1/2,\ P(B \mid A_2) = 1/5$

4.2-1 $P(-1.0 < X \leq 1.0) = 7/9$

4.2-2 $P(\pi < X \leq 3\pi/2) = P(X > 3\pi/2) = 1/4$,
$P(\pi < Z \leq 3\pi/2) = 1/4, P(\pi \leq Z \leq 3\pi/2) = 3/4$

4.2-3 $p_Z(z) = \begin{cases} 1/\pi & -\pi \leq z < 0 \\ 0 & \text{otherwise} \end{cases}$

4.3-1 $m_x = \pi, \overline{x^2} = 4\pi^2/3, \sigma_x = \pi/\sqrt{3}, P(|X - \pi| \leq 2\pi/\sqrt{3}) = 1$

4.4-1 $m = 0.5, F_I(2) \approx 0.986$

5.1-1 $\overline{v(t)} = 3t, R_v(t_1, t_2) = 5 + 9t_1 t_2, \overline{v^2(t)} = 5 + 9t^2$

5.2-3 $G_y(f) \approx \sigma^2 D/[1 + (f/B)^2], R_y(\tau) \approx \sigma^2 D\pi B e^{-\pi B|\tau|}$

5.3-1 1.26 mV, 0.1%

5.4-1 $S_T \geq 840$ kW without repeater, 168 mW with repeater

5.4-2 $t_d \geq \tau, A = A_R \tau$

6.2-2 DSB: $S_T = 20$ W, $A_{\max}^2 = 200$ W
AM: $S_T = 120$ W, $A_{\max}^2 = 800$ W

6.2-3 $A(t) = (A_c/4)\sqrt{5 + 4 \cos 2\omega_m t}$,
$\phi(t) = \arctan [(\tan \omega_m t)/3]$

6.5-2 $A^2(t) = A_{LO}^2 + a^2 + 2A_{LO} a \cos (\omega_m t \mp \phi')$

7.1-1 $f(t) = f_c(1 + \mu \cos \omega_m t - \mu\omega_m t \sin \omega_m t)$

7.1-2 $X_c(f) = \frac{1}{2}A_c\{\delta(f - f_c) + (j\phi_\Delta/2W)\Pi[(f - f_c)/2W] - (\phi_\Delta/4W)\Lambda[(f - f_c)/2W]\}$, $f \geq 0$

7.2-1

D	$2M(D)$	Approximation
0.3	3.0	$2(D + 1) = 2.6$
3.0	10	$2(D + 2) = 10$
30	...	$2(D + 1) = 62$

7.2-2 $G(f) = S_T/4f_\Delta$ for $f_c - f_\Delta \leq |f| \leq f_c + f_\Delta$

7.2-3 $\phi(t - t_1) = \beta(\cos \omega_m t_1 \sin \omega_m t - \sin \omega_m t_1 \cos \omega_m t)$
$\phi(t) + \arg H[f(t)] = \beta(\sin \omega_m t - \omega_m t_1 \cos \omega_m t)$

7.3-2 Use envelope detector with input $x_c(t) - x_c(t - t_0) \approx t_0 \dot{x}_c(t)$

7.4-2 $A_m \sqrt{1 + (f/B_{de})^2} \leq 1/2, \beta \leq 2.5, B = 2M(\beta)f_m \leq 135$ kHz $< B_T$

8.1-1 (a) $f_{IF} = 2$ MHz, $B_{IF}/f_{IF} = 0.003, f_{LO-max}/f_{LO-min} \approx 1.3$
(b) $f_{IF} = 0.6$ MHz, $B_{IF}/f_{IF} = 0.01, f_{LO-max}/f_{LO-min} = 2$

8.2-2 $x_1(t) = 1 + [x_R(t) + x_L(t)] \cos \phi_0$
$x_2(t) = [x_R(t) - x_L(t)] \sin \phi_0$

8.3-2 $K \geq |f_v - nf_c/m|$

8.4-1 $T_{\text{frame}} = 781$ sec

9.2-1 $G_{n_i}(0) = \eta/4$

9.2-2 $\bar{A}_n = 1.3$ mV, $\sigma_{A_n} = 0.655$ mV, $P(A_n > 2\bar{A}_n) = e^{-\pi} = 0.043$

9.3-1 $N_D = 1.25\eta W, (S/N)_D = 0.8$

9.3-3 $G_{z_i}(f) = \dfrac{JT_c}{2} \left| \text{sinc}^2 \dfrac{f - f_z}{W_c} + \text{sinc}^2 \dfrac{f + f_z}{W_c} \right|$

9.4-1 $N_D \approx (\eta B_{de}^2/2S_R)B_T$

9.4-2 $S_T \geq 130$ W

9.4-3 $(S/N)_D \geq 2500$

10.2-1 $B_T \geq 40$ kHz

10.3-1 $B_T \geq 6$ kHz

10.3-2 $T_g \approx 2.8$ μs, $t_0 \leq 3.6$ μs

10.4-1 (a) $\sigma_A/A = 0.1$, $\sigma_t/\tau = 0.1$

(b) $\sigma_A/A = 0.4$, $\sigma_t/\tau = 0.025$

10.4-2 $\tau_{max} = \tau_0(1 + \mu) \leq T_s$, $\tau_{min} = \tau_0(1 - \mu) \geq 0$

so $\tau_{max} - \tau_{min} = 2\mu\tau_0 \leq T_s$ and $\mu\tau_0 \leq T_s/2 \leq 1/4W$

11.1-1 Take $r = a$, so $r \leq B$

11.1-2 $G_x(f) = (A^2/4r_b)$ sinc^2 $(f/r_b) + (A^2/4)\delta(f)$

11.1-3 $G_x(f) = (A^2/r_b) \sin^2 (\pi f/r_b)\Pi(f/r_b)$

11.2-1 $P_{e_0} = Q(4)$, $P_{e_1} = Q(6)$, $P_e = 1.7 \times 10^{-5}$

11.2-3 $P_e = {}^3\!/_2 Q(A/2\sigma) = {}^3\!/_2 Q(\sqrt{\gamma_b})$

11.3-1 $p(t) =$ sinc^2 $(rt/2)$ sinc rt

11.3-3

m_k	m'_{k-2}	m'_k	$m'_k - m'_{k-2}$
0	0	0	0
0	1	1	0
1	0	1	1
1	1	0	-1

12.1-1 $q = 2^5$, $f_s = 7.2$ kHz

12.1-2 (a) 80 Mbps, (b) 110 Mbps

12.2-1 $P_e \approx 10^{-6}$, $(S/N)_R = 13.6$ dB

12.3-1 $\Delta = 0.393\sqrt{S_x}$

12.4-1 16

12.4-2 2.204 Mbps, 70%

13.1-1 0.985, 0.0148, 1.04×10^{-4}, 4.50×10^{-7}

13.1-2 $R'_c \leq 0.879$ go-back-N, $R'_c \leq 0.278$ stop-and-wait

13.2-1 $c_1 = m_1 \oplus m_3$, $c_2 = m_1 \oplus m_2$, $c_3 = m_2 \oplus m_3$

13.2-2 $s_1 = y_1 \oplus y_2 \oplus y_3 \oplus y_5$

$s_2 = y_2 \oplus y_3 \oplus y_4 \oplus y_6$

$s_3 = y_1 \oplus y_2 \oplus y_4 \oplus y_7$

13.2-3 $S(p) =$ rem $[E(p)/G(p)]$

13.3-1 $a = 000$, $b = 001$, $c = 010$, $d = 011$,

$e = 100$, $f = 101$, $g = 110$, $h = 111$

13.3-2 $\alpha = 8.5 \times 10^{-4}$, $P_{be} \approx 3.5 \times 10^{-5}$, $P_{ube} \approx 4.1 \times 10^{-6}$

14.1-1 (a) 100 kbps; (b) 200 kbps; (c) 600 kbps

14.1-2 $G_{\ell p}(f) = {}^1\!/_2\delta(f) + (1/2r_b)$ sinc^2 f/r_b

14.2-2 $|H(f)| = \dfrac{A_c T_b}{2} \left| \text{sinc} \dfrac{f - (f_c \pm r_b/2)}{r_b} + \text{sinc} \dfrac{f + (f_c \pm r_b/2)}{r_b} \right|$

14.3-2 OOK 5×10^{-2}, FSK 3×10^{-3}, DPSK 2×10^{-5}

15.3-1 (a) $p(x) = 1/2M$, $|x| \leq M$; (b) $H(Z) = \log 2KM$

15.3-2 (a) 5,000; (b) 720,000

A-1 $\eta(f) = k[\mathscr{T}_1 + (1 + f^2)\mathscr{T}_2]/(2 + f^2)$

A-2 (a) $\mathscr{T}_e = 4\mathscr{T}_0$; (b) $\mathscr{T}_i = 6\mathscr{T}_0$

A-3 42.9 K with FET, 58.7 without FET

B-2 If $c(nf_0) \approx 0$ for $|f| \geq Nf_0/2$ then

$$c(nf_0) \approx (1/N)X(n) \qquad 0 \leq n \leq N/2 - 1$$
$$c(-nf_0) \approx (1/N)X(N - n) \qquad 1 \leq n \leq N/2 - 1$$

C-1 (a) $\varphi_1 = 1/\sqrt{2}$ for $0 \leq t \leq 2$, $\varphi_2 = 1/\sqrt{2}$ for $2 \leq t \leq 4$;

(b) $\|s_2'\|^2 = 72$

C-2 $E = 11a^2/3$

$$P(c \mid m_i) = \begin{cases} (1 - 2q)(1 - q) & i = 1, 2 \\ (1 - q)^2 & i = 3, 4, 5, 6 \end{cases}$$

SUPPLEMENTARY READING

Listed below are books and papers that provide expanded coverage or alternative treatments of particular topics. Complete citations are given in the References.

Communication systems

The following texts present about the same general scope of communication systems as this book: Gregg (1977), Shanmugam (1979), Schwartz (1980), Stremler (1982), Couch (1983), and Haykin (1983).

Bennett (1970) emphasizes topics pertinent to telephone systems. Sunde (1969) addresses the problems of imperfect transmission channels. Laser communication systems are discussed by Gowar (1984). Spread-spectrum systems are included in Cooper and McGillem (1986).

Fourier signal analysis

Expanded presentations of signal analysis and Fourier methods are contained in Lathi (1965), Stuart (1966), and Frederick and Carlson (1971). Two graduate texts dealing entirely with Fourier transforms and applications are Bracewell (1965), which features a pictorial dictionary of transform pairs, and Papoulis (1962), which strikes a nice balance between rigor and lucidity.

Advanced theoretical treatments will be found in Lighthill (1958) and Franks (1969). The article by Slepian (1976) expounds on the concept of bandwidth.

Probability and random signals

Beckmann (1967), Lathi (1968), and Thomas (1969) contain several chapters on probability and random-signal theory. More comprehensive texts, in order of increasing sophistication, are Drake (1967), Peebles (1980), Cooper and McGillem (1971), Davenport and Root (1958), and Papoulis (1984).

The classic reference papers on noise analysis are Rice (1944) and Rice (1948). Bennett (1956) is an excellent tutorial article.

CW modulation and phase-lock loops

Goldman (1948), one of the earliest books on CW modulation, has numerous examples of spectral analysis. More recent texts that include chapters on this subject are Peebles (1976), Gregg (1977), Shanmugam (1979), Stark and Tuteur (1979), Stremler (1982), and Haykin (1983). Also see the books listed under Detection Theory.

Detailed analysis of FM transmission is found in Panter (1965), a valuable reference work. Taub and Schilling (1971) gives clear discussions of FM noise and threshold extension. The original papers on FM by Carson (1922) and Armstrong (1936) remain informative reading.

The theory and applications of phase-lock loops are examined in depth in Klapper and Frankle (1972), Blanchard (1976), and Gardner (1979). Digital PLLs are surveyed in the article by Lindsey and Chie (1981).

Sampling and coded pulse modulation

Although somewhat dated, Black (1953) contains a good treatment of sampling theory and applications. Further discussions are in the papers by Linden (1959) and Jerri (1977).

Oliver, Pierce, and Shannon (1948) is the landmark article on the philosophy of PCM, while Reeves (1965) recounts the history of his invention. The book by Cattermole (1969) is entirely devoted to PCM. Jayant and Noll (1984) covers the full range of digital encoding methods for analog signals. Implementation of analog/digital conversion is discussed in Schmidt (1970).

Digital transmission

Roden (1982) presents an introductory survey of digital communication systems. Undergraduate texts containing one or more chapters on digital transmission are Shanmugam (1979), Stremler (1982), Couch (1983), and Haykin (1983). Graduate-level treatments are given in Lucky, Salz, and Weldon (1968), Lindsey and Simon (1973), and Proakis (1983). Also see the books listed under Detection Theory.

The following books deal with specific aspects of digital transmission: Feher (1981), on microwave radio; Spilker (1977), on satellite systems; Dixon (1976) on spread spectrum; Schwartz (1977) and Tanenbaum (1981), on computer networks.

Of the many papers that could be mentioned here, the following have special merit: Arthurs and Dym (1962), on optimum detection; Lender (1963), on duobinary signaling; Lucky (1965), on adaptive equalization; Davey (1972), on modems; Gronemeyer and McBride (1976), on MSK and OQPSK; Oetting (1979), on digital radio.

Coding and information theory

Abramson (1963) and Hamming (1980) provide very readable introductions to both coding and information theory. Mathematically advanced treatments are given by Gallager (1968) and McEliece (1977). Viterbi and Omura (1979) includes topics at the state of the art.

Texts devoted to error-control coding and applications are Wiggert (1978) and Lin (1970) at the undergraduate level, and Berlekamp (1968) and Lin and Costello (1983) at the graduate level.

The classic papers on information theory are Nyquist (1924, 1928), Hartley (1928), and Shannon (1948, 1949). Especially recommended is Shannon (1949), which contained the first exposition of sampling theory applied to communication. A fascinating nontechnical book on information theory by Pierce (1961) discusses implications to art, literature, music, and psychology.

Detection theory

The concise monograph by Selin (1965) outlines the concepts and principles of detection theory. Applications to optimum receivers for analog and digital communication are developed in Sakrison (1968), Van Trees (1968), and Wozencraft and Jacobs (1965). The latter includes a clear and definitive presentation of vector models. Viterbi (1966) emphasizes phase-coherent detection. Tutorial introductions to matched filters are given in the papers by Turin (1960, 1976).

Electrical noise

Texts on the modeling, analysis, and measurement of electrical noise are Bennett (1960), Mumford and Scheibe (1968), Van der Ziel (1970) and Motchenbacher and Fitchen (1973). Noise in microwave systems is described by Siegman (1961) using the informative transmission-line approach. Electronic device noise is covered in Robinson (1974) and Ambrózy (1982).

Communication circuits and electronics

The design and implementation of filter circuits are covered in texts such as Hilburn and Johnson (1973) and Van Valkenburg (1982). Introductory treatments of communication electronics are found in DeFrance (1972), Alley and Atwood (1973), and Shrader (1980); more advanced details are given by Clarke and Hess (1971), Krauss, Bostian, and Raab (1980), and Smith (1986). Hansen (1969) and Herrick (1976) are devoted to television electronics.

REFERENCES

Abate, J. E. (1967), "Linear and Adaptive Delta Modulation," *Proc. IEEE*, vol. 55, pp. 298–308.

Abramson, N. (1963), *Information Theory and Coding*, McGraw-Hill, New York.

Ahmed, N., and T. Natarajan (1983), *Discrete-Time Signals and Systems*, Reston Publishing, Reston, Va.

Alley, C. L., and K. W. Atwood (1973), *Electronic Engineering*, 3d ed., Wiley, New York.

Ambrózy, A. (1982), *Electronic Noise*, McGraw-Hill, New York.

Armstrong, E. H. (1936), "A Method of Reducing Disturbances in Radio Signaling by a System of Frequency Modulation," *Proc. IRE*, vol. 24, pp. 689–740.

Arthurs, E., and H. Dym (1962), "On the Optimum Detection of Digital Signals in the Presence of White Gaussian Noise — A Geometric Interpretation and a Study of Three Basic Data Transmission Systems," *IRE Trans. Commun. Systems*, vol. CS-10, pp. 336–372.

Beckmann, P. (1967), *Probability in Communication Engineering*, Harcourt, Brace, & World, New York.

Bennett, W. R. (1956), "Methods of Solving Noise Problems," *Proc. IRE*, vol. 44, pp. 609–638.

Bennett, W. R. (1960), *Electrical Noise*, McGraw-Hill, New York.

Bennett, W. R. (1970), *Introduction to Signal Transmission*, McGraw-Hill, New York.

Berlekamp, E. R. (1968), *Algebraic Coding Theory*, McGraw-Hill, New York.

Black, H. S. (1953), *Modulation Theory*, Van Nostrand, Princeton, N.J.

Blackman, R. B., and J. W. Tukey (1958), *The Measurement of Power Spectra*, Dover, New York.

Blanchard, A. (1976), *Phase-Locked Loops*, Wiley, New York.

Bracewell, R. (1965), *The Fourier Transform and Its Applications*, McGraw-Hill, New York.

Brillouin, L. (1962), *Science and Information Theory*, 2d ed., Academic, New York.

Carson, J. R. (1922), "Notes on the Theory of Modulation," reprinted in *Proc. IEEE*, vol. 51, pp. 893–896.

Cattermole, K. W. (1969), *Principles of Pulse Code Modulation*, American Elsevier, New York.

Chaffee, J. G. (1939), "The Application of Negative Feedback to Frequency-modulation Systems," *Proc. IRE*, vol. 27, pp. 317–331.

Clarke, K. K., and D. T. Hess (1971), *Communication Circuits: Analysis and Design*, Addison-Wesley, Reading, Mass.

Cooper, G. R., and C. D. McGillem (1971), *Probabilistic Methods of Signal and System Analysis*, Holt, New York.

Cooper, G. R., and C. D. McGillem (1986), *Modern Communications and Spread Spectrum*, McGraw-Hill, New York.

Couch, L. W. (1983), *Digital and Analog Communication Systems*, Macmillan, New York.

Davenport, W. B., Jr., and W. L. Root (1958), *Introduction to Random Signals and Noise*, McGraw-Hill, New York.

Davey, J. R. (1972), "Modems," *Proc. IEEE*, vol. 60, pp. 1284–1292.

DeFrance, J. J. (1972), *Communications Electronics Circuits*, 2d ed., Rinehart, San Francisco.

Dixon, R. C. (1976), *Spread Spectrum Systems*, Wiley, New York.

Downing, J. J. (1964), *Modulation Systems and Noise*, Prentice-Hall, Englewood Cliffs, N.J.

Drake, A. W. (1967), *Fundamentals of Applied Probability Theory*, McGraw-Hill, New York.

Feher, K. (1981), *Digital Communications: Microwave Applications*, Prentice-Hall, Englewood Cliffs, N.J.

Franks, L. E. (1969), *Signal Theory*, Prentice-Hall, Englewood Cliffs, N.J.

Frederick, D. K., and A. B. Carlson (1971), *Linear Systems in Communication and Control*. Wiley, New York.

Gardner, F. M. (1979), *Phaselock Techniques*, 2d ed., Wiley, New York.

Gardner, F. M., and W. C. Lindsey (eds.) (1980), Special Issue on Synchronization, *IEEE Trans. Commun.*, vol. COM-28, no. 8.

Gallager, R. G. (1968), *Information Theory and Reliable Communication*, Wiley, New York.

Goldman, S. (1948), *Frequency Analysis, Modulation, and Noise*, McGraw-Hill, New York.

Gowar, J. (1984), *Optical Communication Systems*, Prentice-Hall, Englewood Cliffs, N.J.

Gregg, W. D. (1977), *Analog and Digital Communications*, Wiley, New York.

Gronemeyer, S. A., and A. L. McBride (1976), "MSK and Offset QPSK Modulation," *IEEE Trans. Commun.*, vol. COM-24, pp. 809–820.

Gruenberg, E. L. (ed.) (1967), *Handbook of Telemetry and Remote Control*, McGraw-Hill, New York.

Hamming, R. W. (1950), "Error Detecting and Error Correcting Codes," *Bell System Tech. J.*, vol. 29, pp. 147–160.

Hamming, R. W. (1980), *Coding and Information Theory*, Prentice-Hall, Englewood Cliffs, N.J.

Hansen, G. L. (1969), *Introduction to Solid-State Television Systems*, Prentice-Hall, Englewood Cliffs, N.J.

Hartley, R. V. (1928), "Transmission of Information," *Bell System Tech. J.*, vol. 7, pp. 535–563.

Haykin, S. (1983), *Communication Systems*, 2d ed., Wiley, New York.

Herrick, C. N. (1976), *Color Television: Theory and Servicing*, Reston Publishing, Reston, Va.

Hilburn, J. L., and D. E. Johnson (1973), *Manual of Active Filter Design*, McGraw-Hill, New York.

Jahnke, E., and F. Emde (1945), *Tables of Functions*, Dover, New York.

Jayant, N. S. (1974), "Digital Coding of Speech Waveforms: PCM, DPCM, and DM Quantizers," *Proc. IEEE*, vol. 62, pp. 611–632.

Jayant, N. S., and P. Noll (1984), *Digital Coding of Waveforms: Principles and Applications to Speech and Video*, Prentice-Hall, Englewood Cliffs, N.J.

Jerri, A. J. (1977), "The Shannon Sampling Theorem—Its Various Extensions and Applications: A Tutorial Review," *Proc. IEEE*, vol. 65, pp. 1565–1596.

Johnson, J. B. (1928), "Thermal Agitation of Electricity in Conductors," *Phys. Rev.*, vol. 32, pp. 97–109.

Jurgen, R. K. (1972), "Untangling the 'Quad' Confusion," *IEEE Spectrum*, vol. 9, pp. 55–62.

Klapper J., and J. T. Frankle (1972), *Phase-Locked and Frequency-Feedback Systems*, Academic Press, New York.

Kotel'nikov, V. A. (1959), *The Theory of Optimum Noise Immunity*, McGraw-Hill, New York.

Krauss, H. L., C. W. Bostian, and F. H. Raab (1980), *Solid State Radio Engineering*, Wiley, New York.

Lathi, B. P. (1965), *Signals, Systems and Communication*, Wiley, New York.

Lathi, B. P. (1968), *An Introduction to Random Signals and Communication Theory*, International Textbook, Scranton, Pa.

Lender, A. (1963), "The Duobinary Technique for High Speed Data Transmission," *IEEE Trans. Commun. and Electron.*, vol. 82, pp. 214–218.

Lighthill, M. J. (1958), *An Introduction to Fourier Analysis and Generalized Functions*, Cambridge, New York.

Lin, S. (1970), *An Introduction to Error-Correcting Codes*, Prentice-Hall, Englewood Cliffs, N.J.

Lin, S., and D. J. Costello, Jr. (1983), *Error Control Coding: Fundamentals and Applications*, Prentice-Hall, Englewood Cliffs, N.J.

Linden, D. A. (1959), "A Discussion of Sampling Theorems," *Proc. IRE*, vol. 47, pp. 1219–1226.

Lindsey, W. C. (1972), *Synchronization Systems in Communication and Control*, Prentice-Hall, Englewood Cliffs, N.J.

Lindsey, W. C., and C. M. Chie (1981), "A Survey of Digital Phase-Locked Loops," *Proc. IEEE*, vol. 69, pp. 410–431.

Lindsey, W. C., and M. K. Simon (1973), *Telecommunication Systems Engineering*, Prentice-Hall, Englewood Cliffs, N.J.

Lucky, R. W. (1965), "Automatic Equalization for Digital Communication," *Bell System Tech. J.*, vol. 44, pp. 547–588.

Lucky, R. W., J. Salz, and E. J. Weldon, Jr. (1968), *Principles of Data Communication*, McGraw-Hill, New York.

McEliece, R. J. (1977), *The Theory of Information and Coding*, Addison-Wesley, Reading, Mass.

Mennie, D. (1978), "AM Stereo: Five Competing Options," *IEEE Spectrum*, vol. 15, pp. 24–31.

Mitra, S. K. (1980), *An Introduction to Digital and Analog Integrated Circuits and Applications*, Harper & Row, New York.

Mokhoff, N. (1981), "A Step Toward 'Perfect' Resolution," *IEEE Spectrum*, vol. 18, 56–58.

McGillem, C. D., and G. R. Cooper (1974), *Continuous and Discrete Signal and System Analysis*, Holt, Rinehart, and Winston, New York.

Motchenbacher, C. D., and F. C. Fitchen (1973), *Low-Noise Electronic Design*, Wiley, New York.

Mumford, W. W., and E. H. Scheibe (1968), *Noise Performance Factors in Communication Systems*, Horizon House-Microwave, Dedham, Mass.

Nyquist, H. (1924), "Certain Factors Affecting Telegraph Speed," *Bell System Tech. J.*, vol. 3, pp. 324–346.

Nyquist, H. (1928), "Certain Topics in Telegraph Transmission Theory," *Trans. AIEE*, vol. 47, pp. 617–644.

Nyquist, H. (1928), "Thermal Agitation of Electric Charge in Conductors," *Phys. Rev.*, vol. 32, pp. 110–113.

Oetting, J. D. (1979), "A Comparison of Modulation Techniques for Digital Radio," *IEEE Trans. Commun.*, vol. COM-27, pp. 1752–1762.

Oliver, B. M., J. R. Pierce, and C. E. Shannon (1948), "The Philosophy of PCM," *Proc. IRE*, vol. 36, pp. 1324–1332.

Panter, P. F. (1965), *Modulation, Noise, and Spectral Analysis*, McGraw-Hill, New York.

Papoulis, A. (1962), *The Fourier Integral and Its Applications*, McGraw-Hill, New York.

Papoulis, A. (1984), *Probability, Random Variables, and Stochastic Processes*, 2d ed., McGraw-Hill, New York.

Peebles, P. Z. (1976), *Communication System Principles*, Addison-Wesley, Reading, Mass.

Peebles, P. Z. (1980), *Probability, Random Variables, and Random Signal Principles*, McGraw-Hill, New York.

Pierce, J. R. (1961), *Symbols, Signals, and Noise*, Harper & Row, New York.

Proakis, J. G. (1983), *Digital Communications*, McGraw-Hill, New York.

Puente, J. G., and A. M. Werth (1971), "Demand-assigned Service for the Intelsat Global Network," *IEEE Spectrum*, vol. 8, pp. 59–64.

Rabiner, L. R., and B. Gold (1975), *Theory and Application of Digital Signal Processing*, Prentice-Hall, Englewood Cliffs, N.J.

Rabiner, L. R., and R. W. Schafer (1978), *Digital Processing of Speech Signals*, Prentice-Hall, Englewood Cliffs, N.J.

Reeves, A. H. (1965), "The Past, Present, and Future of PCM," *IEEE Spectrum*, vol. 2, pp. 58–63.

Rice, S. O. (1944), "Mathematical Analysis of Random Noise, *Bell System Tech. J.*, vol. 23, pp. 282–332, and vol. 24, pp. 46–156.

Rice, S. O. (1948), "Statistical Properties of a Sine-wave plus Random Noise," *Bell System Tech. J.*, vol. 27, pp. 109–157.

Robinson, F. N. (1974), *Noise and Fluctuations in Electronic Devices and Circuits*, Clarendon Press, Oxford.

Roden, M. S. (1982), *Digital and Data Communication Systems*, Prentice-Hall, Englewood Cliffs, N.J.

Rowe, H. E. (1965), *Signals and Noise in Communication Systems*, Van Nostrand, Princeton, N.J.

Sakrison, D. J. (1968), *Communication Theory: Transmission of Waveforms and Digital Information*, Wiley, New York.

Schmidt, H. (1970), *Analog/Digital Conversion*, Van Nostrand Reinhold, Princeton, N.J.

Schwartz, M. (1977), *Computer Communication Network Design and Analysis*, Prentice-Hall, Englewood Cliffs, N.J.

Schwartz, M. (1980), *Information Transmission, Modulation, and Noise*, 3d ed., McGraw-Hill, New York.

Schwartz, M., and L. Shaw (1975), *Signal Processing*, McGraw-Hill, New York.

Selin, L. (1965), *Detection Theory*, Princeton University Press, Princeton, N.J.

Shanmugam, K. S. (1979), *Digital and Analog Communication Systems*, Wiley, New York.

Shannon, C. E. (1948), "A Mathematical Theory of Communication," *Bell System Tech. J.*, vol. 27, pp. 379–423 and 623–656.

Shannon, C. E. (1949), "Communication in the Presence of Noise," *Proc. IRE*, vol. 37, pp. 10–21.

Shrader, R. L. (1980), *Electronic Communication*, 4th ed., McGraw-Hill, New York.

Siegman, A. E. (1961), "Thermal Noise in Microwave Systems," *Microwave J.*, vol. 4, pp. 66–73 and 93–104.

Slepian, D. (1956), "A Class of Binary Signaling Alphabets," *Bell System Tech. J.*, vol. 35, pp. 203–234.

Slepian, D. (1976), "On Bandwidth," *Proc. IEEE*, vol. 64, pp. 292–300.

Smith, J. R. (1986), *Modern Communication Circuits*, McGraw-Hill, New York.

Spilker, J. J., Jr. (1977), *Digital Communications by Satellite*, Prentice-Hall, Englewood Cliffs, N.J.

Stark, H. S., and F. B. Tuteur (1979), *Modern Electrical Communications*, Prentice-Hall, Englewood Cliffs, N.J.

Stiffler, J. J. (1971), *Theory of Synchronous Communications*, Prentice-Hall, Englewood Cliffs, N.J.

Stremler, F. G. (1982), *Introduction to Communication Systems*, 2d ed., Addison-Wesley, Reading, Mass.

Stuart, R. D. (1966), *An Introduction to Fourier Analysis*, Methuen, London.

Stumpers, F. L. (1948), "Theory of Frequency-modulation Noise," *Proc. IRE*, vol. 36, pp. 1081–1092.

Sunde, E. D. (1969), *Communication Systems Engineering Theory*, Wiley, New York.

Tanenbaum, A. S. (1981), *Computer Networks*, Prentice-Hall, Englewood Cliffs, N.J.

Taub, H., and D. L. Schilling (1971), *Principles of Communication Systems*, McGraw-Hill, New York.

Thomas, J. B. (1969), *An Introduction to Statistical Communication Theory*, Wiley, New York.

Turin, G. (1960), "An Introduction to Matched Filters," *IRE Trans.* Inform. Theory, vol. IT-6, pp. 311–329.

Turin, G. (1976), "An Introduction to Digital Matched Filters," *Proc. IEEE*, vol. 64, pp. 1092–1112.

Van der Ziel, A. (1970), *Noise: Sources, Characterization, Measurement*, Prentice-Hall, Englewood Cliffs, N.J.

Van Trees, H. L. (1968), *Detection, Estimation, and Modulation Theory*, part I, Wiley, New York.

Van Valkenburg, M. E. (1982), *Analog Filter Design*, Holt, Rinehart, and Winston, New York.

Viterbi, A. J. (1966), *Principles of Coherent Communication*, McGraw-Hill, New York.

Viterbi, A. J., and J. K. Omura (1979), *Principles of Digital Communication and Coding*, McGraw-Hill, New York.

Wiggert, D. (1978), *Error-Control Coding and Applications*, Artech House, Dedham, Mass.

Wozencraft, J. M., and I. M. Jacobs (1965), *Principles of Communication Engineering*, Wiley, New York.

INDEX